Evolutionary Genetics of Fishes

MONOGRAPHS IN EVOLUTIONARY BIOLOGY

Series Editors:

MACROMOLECULAR SEQUENCES IN SYSTEMATIC AND EVOLUTIONARY BIOLOGY
Edited by Morris Goodman

EVOLUTIONARY GENETICS OF FISHES
Edited by Bruce J. Turner

Evolutionary Genetics of Fishes

Edited by

Bruce J. Turner

Virginia Polytechnic Institute and State University
Blacksburg, Virginia

Plenum Press • New York and London

Library of Congress Cataloging in Publication Data

Main entry under title:

Evolutionary genetics of fishes.

(Monographs in evolutionary biology)
Includes bibliographical references and index.
1. Fishes—Evolution. 2. Fishes—Genetics. I. Turner, Bruce J. II. Series.
QL618.2.E96 1984 597′.038 84-1941
ISBN 978-1-4684-4654-8 ISBN 978-1-4684-4652-4 (eBook)
DOI 10.1007/978-1-4684-4652-4

Softcover reprint of the hardcover 1st edition 1984
A Division of Plenum Publishing Corporation
233 Spring Street, New York, N.Y. 10013

Contributors

Fred W. Allendorf Department of Zoology, University of Montana, Missoula, Montana 59812

Joseph S. Balsano Biomedical Research Institute, University of Wisconsin-Parkside, Kenosha, Wisconsin 53141

Michael A. Bell Department of Ecology and Evolution, State University of New York at Stony Brook, Stony Brook, New York 11794

Richard Borowsky Department of Biology, New York University, New York, New York 10003

Donald G. Buth Department of Biology, University of California at Los Angeles, Los Angeles, California 90024

Stephen D. Ferris Department of Biochemistry, University of California, Berkeley, California 94720

Klaus D. Kallman Genetics Laboratory, Osborn Laboratories of Marine Sciences, New York Aquarium, New York Zoological Society, Brooklyn, New York 11224

Richard K. Koehn Department of Ecology and Evolution, State University of New York, Stony Brook, New York 11794

Irv Kornfield Department of Zoology and Migratory Fish Research Institute, University of Maine, Orono, Maine 04469

Paul J. Monaco Department of Biophysics, Quillen-Dishner College of Medicine, East Tennessee State University, Johnson City, Tennessee 37614

William S. Moore Department of Biological Sciences, Wayne State University, Detroit, Michigan 48202

Donald C. Morizot University of Texas Science Park, Research Division, Smithville, Texas 78957

Ellen M. Rasch Department of Biophysics, Quillen-Dishner College of Medicine, East Tennessee State University, Johnson City, Tennessee 37614

Michael J. Siciliano Department of Genetics, University of Texas System Cancer Center, M.D. Anderson Hospital and Tumor Institute, Houston, Texas 77030

Gary H. Thorgaard Program in Genetics and Cell Biology, Washington State University, Pullman, Washington 99164

Robert C. Vrijenhoek Department of Biological Sciences and Bureau of Biological Research, New Brunswick, New Jersey 08903

George C. Williams Department of Ecology and Evolution, State University of New York, Stony Brook, New York 11794.

Preface

It is my hope that this collection of reviews can be profitably read by all who are interested in evolutionary biology. However, I would like to specifically target it for two disparate groups of biologists seldom mentioned in the same sentence, classical ichthyologists and molecular biologists.

Since classical times, and perhaps even before, ichthyologists have stood in awe at the tremendous diversity of fishes. The bulk of effort in the field has always been directed toward understanding this diversity, i.e., extracting from it a coherent picture of evolutionary processes and lineages. This effort has, in turn, always been overwhelmingly based upon morphological comparisons. The practical advantages of such comparisons, especially the ease with which morphological data can be had from preserved museum specimens, are manifold. But considered objectively (outside its context of "tradition"), morphological analysis alone is a poor tool for probing evolutionary processes or elucidating relationships. The concepts of "relationship" and of "evolution" are inherently *genetic* ones, and the genetic bases of morphological traits are seldom known in detail and frequently unknown entirely. Earlier in this century, several workers, notably Gordon, Kosswig, Schmidt, and, in his salad years, Carl Hubbs, pioneered the application of genetic techniques and modes of reasoning to ichthyology. While certain that most contemporary ichthyologists are familiar with this body of work, I am almost equally certain that few of them regard it as pertinent to their own efforts. For a while, in the 1970s, the "allozyme revolution" in evolutionary biology seemed to be having an impact on ichthyology as well, and a number of younger ichthyologists began to think in genetic terms. It is my impression that this impact is now largely spent: The recent rise of "phylogenetic systematics" with its emphasis on "derived character states" (and with, at least initially, little regard for their genetic bases) has, inadvertently and unfortunately, revitalized the phenotypic typology that has always been present in ichthyology. This collection of reviews is thus designed, in part, to demonstrate the power of integrated genetic approaches to ichthyological problems. No morphological comparisons alone could have detected the role of ancient tetraploidy in the evolution of the salmonid or catostomid fishes, detected the existence of parthenoforms (much less probed their clonal population structure), or elucidated the role of natural selection in modulating plate numbers in stickleback populations. Moreover, morphological comparisons have already failed, and failed rather badly, to resolve the relationships of the Atlantic eels or to make much sense of the vast, and biologically important, radiations that are the cyprinid and cichlid fishes.

Most contemporary eukaryote molecular biologists concentrate their efforts on but a few species, none of them fishes. However, fishes offer a literal plethora of exciting "experimental systems" that could be exploited, and it seems worthwhile to point some of them out. For example, a process called *rediploidization* seems to be a common feature of the evolution of those groups of fishes (e.g., the salmonids) whose ancestries involved tetraploidization events. Yet how is the rediploidization achieved, and is it done the same way in different lineages? Are the extra genes deleted, mutated to functional silence (by what sort of regulatory mutations?), or are they somehow "coopted" by "ignorant" or "selfish" sequences? What fraction of duplicated genes becomes functionally specialized, as opposed to those that seem to disappear, and are there rules (other than chance) that govern this? Does the rediploidization occur at the same rate and in the same way in the highly repetitive, moderately repetitive, and "unique" portions of the genome? Gene duplication is clearly a fundamental (perhaps *the* fundamental) process in evolutionary genetics. The tetraploid fish lineages, quite literally, have duplicated genes "in spades" and might teach us much about how these are handled in a long-term evolutionary context. Equally tantalizing problems exist in other fish systems. The unisexual fishes, for example, offer opportunities to study the evolution of genomes that for the most part are not subject to the selective constraints of functional meiosis, and, moreover, are of hybrid origin. They may prove to be particularly useful systems in which to test current ideas of the evolution of gene sequences, expression, and genome organization. Similarly, it is now apparent that at least some aspects of sex determination in fishes may differ from those in mammals. What will these differences tell us about the ultimate molecular basis of the sex phenotypes? There are so many exciting questions amenable to contemporary molecular technology that the entry of molecular biology into the evolutionary genetics of fishes is likely to be of extraordinary significance. If this book hastens that entry to even a slight extent it will have served its purpose.

Bruce J. Turner

Contents

Chapter 2

Tetraploidy and the Evolution of the Catostomid Fishes

Stephen D. Ferris

Chapter 3

A New Look at Sex Determination in Poeciliid Fishes

Klaus D. Kallman

Chapter 4

Gene Mapping in Fishes and Other Vertebrates

Donald C. Morizot and Michael J. Siciliano

Chapter 7
Evolutionary Ecology of Unisexual Fishes
William S. Moore

Chapter 8
The Evolution of Clonal Diversity in *Poeciliopsis*
Robert C. Vrijenhoek

Chapter 9

Evolutionary Phenetics and Genetics: The Threespine Stickback, *Gasteroteus aculeatus*, and Related Species

Michael A. Bell

Chapter 10

Population Genetics of North Atlantic Catadromous Eels (*Anguilla*)

George C. Williams and Richard K. Koehn

Chapter 11

Allozymes of the Cyprinid Fishes: Variation and Application

Donald G. Buth

Chapter 12

Descriptive Genetics of Cichlid Fishes

Irv Kornfield

Evolutionary Genetics of Fishes

CHAPTER 1

Tetraploidy and the Evolution of Salmonid Fishes

FRED W. ALLENDORF and GARY H. THORGAARD

1. Introduction: Polyploidy in Vertebrate Evolution

Polyploidy has long been recognized to have played an important role in the evolution of plant species (Lewis, 1980). However, polyploidy has only recently been recognized as a potentially important process in the evolution of vertebrates (Ohno, 1974; Fisher *et al.*, 1980; Bogart, 1980; Schultz, 1980). Mammals and birds generally possess more DNA per cell than do fish and other chordates. This observation and the existence of many duplicated gene loci have led Ohno and co-workers (Ohno, 1967, 1970*a*, 1974; Ohno *et al.*, 1968) to propose that genome doubling has taken place at least once in the evolution of vertebrates. One tetraploid event apparently took place about 500 million years (Myr) ago in a common ancestor of all vertebrates. Fisher *et al.* (1980) describe isozyme studies that are consistent with this idea. Other tetraploid events may have taken place in major lineages of vertebrate evolution, possibly including a genome doubling in a reptilian ancestor of mammals (Ohno, 1967; Comings, 1972).

Results have also shown that recent polyploid events are much more common than was previously suspected. Polyploid amphibians and reptiles are surprisingly common (Bogart, 1980). It is curious that all of the polyploid frogs and toads that have been described are bisexual, while all of the polyploid salamanders and lizards are parthenogenetic or gyno-

FRED W. ALLENDORF • Department of Zoology, University of Montana, Missoula, Montana 59812. *GARY H. THORGAARD* • Program in Genetics and Cell Biology, Washington State University, Pullman, Washington 99164.

genetic triploids. In spite of the many examples of polyploidy in amphibians and reptiles, polyploidy has apparently not been an important process in progressive evolution of these groups; all of the polyploid species have closely related diploid species and no higher polyploid taxa have been found (Bogart, 1980).

A number of cases of polyploidy in fish have also been reported (Schultz, 1980). Some species of primitive fish, including the paddlefish, *Polyodon spathula* (Dingerkus and Howell, 1976), the shovelnose sturgeon, *Scaphirhynchus platorhynchus* (Ohno *et al.*, 1969*a*), the lungfish, *Protopterus dolloi* (Vervoort, 1980), and the spotted gar, *Lepisosteus productus* (Ohno *et al.*, 1969*a*), appear to be tetraploid. Closely related diploid and tetraploid forms are found in the spinous loach, *Cobitis biwae* (Kobayasi, 1976; Sezaki and Kobayasi, 1978). There also appears to be a diploid–tetraploid relationship among species of armored catfish of the genus *Corydoras* (Dunham *et al.*, 1980) and among some weatherfish of the genus *Misgurnus* (Raicu and Taisescu, 1972).

Three species of cyprinids, the barb (*Barbus barbus*), the carp (*Cyprinus carpio*), and the goldfish (*Carassius auratus*), appear to be tetrapolid (Ohno *et al.*, 1967; Muramoto *et al.*, 1968; Wolf *et al.*, 1969; Ohno, 1974). In addition, some unisexual goldfish strains are triploid or tetraploid in relation to the normal bisexual goldfish (Cherfas, 1966; Liu *et al.*, 1978).

Recent polyploids have been found in two genera of livebearing fishes (*Poecilia* and *Poeciliopis*) in Mexico (Schultz, 1980). The triploid unisexual forms all reproduce by gynogenesis and have hybrid origins. These fish and related diploids have been intensively studied and have provided useful model systems to examine the origin and adaptive value of polyploidy (Schultz, 1980). In addition, Echelle and Mosier (1981) have recently described a unisexual species of *Menidia* (Atherinidae) that is apparently triploid.

There are only two known cases of polyploidy in fish giving rise to an entire family. The suckers of the family Catostomidae are a large evolutionary group (12 genera, 58 species; Nelson, 1976) that apparently share a common tetraploid origin. Uyeno and Smith (1972) have suggested that the catostomids "evolved by tetraploidy from a cyprinid-like ancestor" over 50 Myr ago on the basis of an apparent doubling of chromosome numbers and DNA contents. They consider the catostomids to be one of the most evolutionarily successful groups of fishes in North America. The suckers have been an especially valuable group for study of evolutionary changes following a tetraploid event (Ferris, this volume, Chapter 2).

The Salmonidae are the second family of fish that apparently share a common tetraploid origin. There are three major salmonid taxa that are classified as subfamilies by most taxonomists: Coregoninae (whitefish and cisco), Salmoninae (trout, salmon, and char), and Thymallinae (grayling) (Norden, 1961; Nelson, 1976). These fish have been further classified into nine genera and some 68 species (Table I) (Nelson, 1976). The scientific and common names of all species discussed in this chapter are given in Table II.

These two tetraploid-derived families differ in that the catostomids apparently quickly returned to a diploid state of chromosome pairing, while the salmonids are still going through the "diploidization" process of restoring disomic inheritance. The evolutionary success of these two families is an exception to the common view that polyploidy is not an important factor in progressive evolution beyond the species level (Stebbins, 1977; Dobzhansky *et al.*, 1977; White, 1978). The objective of this chapter is to describe the evolution and genetics of salmonids with the principal purpose of increasing the understanding of the significance of polyploidy as an evolutionary process.

2. The Salmonid Tetraploid Event

2.1. Evidence for Ancestral Polyploidy

The idea that salmonids have a polyploid origin was first proposed by Svardson (1945). He observed that chromosome numbers in salmonid species seemed to fall into multiples of 10; Atlantic salmon have about 60 chromosomes; the brown trout, Arctic char, brook trout, and the common whitefish have about 80 chromosomes; the grayling has about 100 chromosomes. These observations, plus variations in the numbers of biarmed chromosomes among species and the observation of multivalents

Table I
Subfamilies and Genera of the Family Salmonidae[a]

Subfamily	Genus	Number of species
Coregoninae	*Coregonus*	25
	Prosopium	6
	Stenodus	1
Salmoninae	*Brachymystax*	1
	Hucho	3–5
	Oncorhynchus	7
	Salmo	13
	Salvelinus	6
Thymallinae	*Thymallus*	4

[a]Nelson (1976).

Table II
Chromosome and Arm Numbers of Salmonid Fish Species[a]

Species	Common name	Reported chromosome number	Chromosome arm number
Coregonus albula	Cisco	80	96
Coregonus artedii	Lake herring	80	96
Coregonus clupeaformis	Lake whitefish	80	100
Coregonus hoyi	Bloater	80	90
Coregonus lavaretus	Common whitefish	80,96	92–128
Coregonus nasus	Broad whitefish	80,96	92,96
Coregonus oxyrhynchus	Houting	96	96
Coregonus peled	Peled	80	92,98
Coregonus pidschian	Arctic whitefish	80,96	92–98
Coregonus reighardi	Shortnose cisco	80	92
Coregonus ussuriensis	Amur whitefish	80,82	100
Coregonus zenithicus	Shortjaw cisco	80	98
Prosopium abyssicola	Bear Lake whitefish	72	100
Prosopium coulteri	Pygmy whitefish	82	100
Prosopium cylindraceum	Round whitefish	78	100
Prosopium gemmiferum	Bonneville cisco	64	100
Prosopium spilonotus	Bonneville whitefish	74	100
Prosopium williamsoni	Mountain whitefish	78	100
Stenodus leucicnthys	Inconnu	74	108
Brachymystax lenok	Lenok	90	116
Oncorhynchus gorbuscha	Pink salmon	52	100,104
Oncorhynchus keta	Chum salmon	74	100
Oncorhynchus kisutch	Coho salmon	60	106,112
Oncorhynchus masou	Masu salmon	66	100
Oncorhynchus nerka	Sockeye salmon	56–58	102,104
Oncorhynchus rhodurus	—	66	100
Oncorhynchus tschawytscha	Chinook salmon	68	100,104
Salmo aguabonita	Golden trout	58	104
Salmo apache	Apache trout	56	106
Salmo carpio	—	80	98
Salmo clarki bouvieri	Yellowstone cutthroat trout	64	104
Salmo clarki clarki	Coastal cutthroat trout	68,70	104,106
Salmo clarki henshawi	Lahontan cutthroat trout	64	104
Salmo clarki lewisi	Westslope cutthroat trout	64–66	104
Salmo gairdneri	Rainbow trout	58–65	104
Salmo gilae	Gila trout	56	105
Salmo obtusirostris	Adriatic salmon	82	94
Salmo salar	Atlantic salmon	54–60	72–74
Salmo trutta	Brown trout	77–82	96–102
Salmo sp.	Redband trout	58	104

Table II (*continued*)

Species	Common name	Reported chromosome number	Chromosome arm number
Salvelinus alpinus	Arctic char	78,80	96,100
Salvelinus fontinalis	Brook trout	84	100
Salvelinus leucomaenis	Siberian char	84	100
Salvelinus malma	Dolly Varden	80,82	98
Salvelinus namaycush	Lake trout	84	100
Thymallus thymallus	European grayling	102	170

[a]Gold *et al.* (1980), Sola *et al.* (1981). Short arms on subtelocentric chromosomes are not counted in calculating arm numbers. Common names are from Robins *et al.* (1980) for North American species, Maitland (1977) for European species, and Ricker (1962) for Asian species.

in meiotic preparations from several species, led Svardson to propose that the basic chromosome number in salmonids is $n = 10$, and that the variations in chromosome number were the result of polyploid events. This proposal was shown to be incorrect when Rees (1964) demonstrated that cellular DNA contents and total chromosome lengths were similar in the Atlantic salmon and brown trout. Further evidence against Svardson's proposal accumulated (Booke, 1968). Svardson's proposal that salmonid fish are polyploid has subsequently been presented as a disproved example of polyploidy in animals (White, 1973, 1978).

Ohno and coworkers later proposed a new model of polyploidy in the ancestry of salmonids (Ohno *et al.*, 1968, 1969*b*; Ohno, 1970*a*,*b*, 1974). They proposed that salmonids as a group were tetraploid in comparison to related salmonoid and clupeoid fishes such as the smelt, herring, and anchovy.

There are four major lines of support for an ancestral tetraploid event in salmonid evolution (Ohno, 1970*b*). (1) Salmonid fish, with about 80% as much DNA per cell as mammals, have approximately twice the amount of DNA per cell as closely related fish. (2) Salmonids typically have about 100 chromosome arms, twice as many as closely related fish (Table II). (3) Multivalents have commonly been observed in meiotic preparations from several salmonid species. (4) Salmonids show a high incidence of duplicated enzyme loci. Subsequent studies reviewed in this chapter support such a polyploid event in an ancestral salmonid.

2.2. Time of the Event

It is difficult to determine when the tetraploid event occurred. Bailey and his co-workers (Lim *et al.*, 1975; Lim and Bailey, 1977) have attempted to do this by estimating the amount of divergence that has accumulated between duplicate loci, using quantitative immunological methods. Using two different pairs of duplicated LDH loci produced by the tetraploid event, they have found amounts of divergence that are similar to that found between two homologous lactate dehydrogenase (LDH) genes in species that have been separated an estimated 100 Myr.

This estimate is fraught with potential problems. First, it is based on the principle of the molecular clock, i.e., homologous proteins tend to evolve at similar rates in different lineages. This concept has been widely used to estimate the timing of evolutionary events, but its validity has been seriously questioned [see review by Selander (1982)].

Even if we accept the validity of the molecular clock in general, there are reasons why it may not provide an accurate estimate in this instance. This estimation of the time of the tetraploid event assumes that genes in separate species will evolve at the same rate as duplicate genes in a single species. We would expect duplicate genes to diverge more rapidly because deleterious mutations that would normally be removed by natural selection can become established at one of the loci as long as the other locus continues to perform the normal function. This effect would cause an overestimation of the time since the tetraploid event.

A third potential problem is that the divergence of the two loci resulting from the duplication of a single locus cannot begin until disomic inheritance has been established for these loci. This is demonstrated by the many pairs of duplicated loci in salmonids for which there is no evidence of divergence (Bailey *et al.*, 1970; Allendorf and Utter, 1976). The absence of detectable divergence within these pairs indicates that the disomic inheritance of these loci has been established in the comparatively recent past. In fact, there have been recent reports that some pairs of duplicate loci have still not evolved complete disomic inheritance (Wright *et al.*, 1980; May *et al.*, 1982). The length of time that the LDH loci have been diverging is therefore a minimum estimate of the time since the tetraploid event.

It should also be possible to estimate the time of the tetraploid event using the fossil record. The event had to occur after the salmonids diverged from the nearest ancestor and before the three subfamilies diverged. Unfortunately, the fossil record of the salmonids is extremely scanty for the critical time period (Norden, 1961; Cavender, 1970). The divergence of the three subfamilies apparently occurred somewhere between 25 and 100 Myr ago (R. J. Behnke, personal communication; Obruchev, 1967, cited in Schmidtke *et al.*, 1979; Norden, 1961). The estimate from the biochemical evidence is thus at least not contradicted by the fossil record. Thus, we are left with this estimate of 25–100 Myr, although we must be aware of its weaknesses.

Schmidtke *et al.* (1979) and Schmidtke and Kandt (1981) have recently seriously questioned this view of a single ancient tetraploid event in salmonid evolution, on the basis of DNA reassociation kinetics. They propose that polyploidy within the salmonids is the result of at least four separate polyploid events. They suggest that the grayling is an ancient polyploid lineage and that *Salmo, Salvelinus,* and *Coregonus* are all the products of much more recent (3 Myr ago) and separate tetraploid events.

We believe that their proposed history is untenable in view of the evidence from isozyme studies. *Salmo, Salvelinus,* and *Thymallus* species have nearly identical genetic systems of control for many different enzymes (Massaro, 1973; Section 4). Similar patterns of genetic control and tissue expression are also present in *Coregonus* and *Prosopium* species (Clayton and Franzin, 1970; Massaro, 1972; Allendorf, unpublished data). It is extremely unlikely that these same patterns of genetic control and tissue expression would have evolved independently in separate lineages. Thus, we are left with the conclusion that these similarities result from all of these species sharing a common ancestor after the polyploid event.

It is also impossible to reconcile the amount of loss and divergence of duplicate gene expression that has occurred in *Salvelinus* and *Salmo* with an ancestral polyploid event of only 3 Myr ago. Schmidtke *et al.* (1979) attempt to avoid this difficulty by suggesting that the loci reported to no longer be duplicated have not yet diverged, so that the duplication is not detectable. Both studies (Allendorf *et al.*, 1975; May, 1980) reporting an approximate 50% loss of duplicate gene expression in these genera were aware of this difficulty and therefore only included loci that were polymorphic, so that this possibility could be excluded. In addition, the amount of structural divergence found between duplicate LDH loci is incompatible with recent tetraploidy (Lim *et al.*, 1975; Lim and Bailey, 1977). May (1980) has presented evidence that approximately 90% of the loss of retention of duplicate gene expression is shared by salmonid genera, including *Thymallus*.

In addition, it is extremely unlikely that an event as rare as polyploidy would have occurred independently in all salmonid genera. We therefore believe that the available evidence overwhelmingly supports a single ancient tetraploid event in a common ancestor of all salmonid fishes.

2.3. Nature of the Tetraploid Event

There apparently was substantial homology between the contributing genomes at the time of the tetraploid event. This is supported by the presence in current salmonids of multivalents at meiosis (Table III), by the existence of some duplicate loci pairs with no evidence of any diver-

Table III
Reports of Multivalents in Meiosis of Salmonid Fishes

Species	Sex	Meiotic stage	Multivalents observed	References
Pink salmon	M	Diakinesis	No	Simon (1964)
Coho salmon	M	Metaphase I	Yes	Ohno (1970*b*)
Golden trout	M	Metaphase I	Yes	Gold and Gall (1975)
Rainbow trout	F	Metaphase I	Yes	Ohno *et al.* (1965)
	F	Pachynema	No	Thorgaard and Gall (1979)
	M	Diakinesis	Yes	Simon (1964)
	M	Metaphase I	Yes	Ohno *et al.* (1965,1968)
Atlantic salmon	M	Metaphase I	Yes	Svardson (1945) Nygren *et al.* (1972)
Brown trout	M	Metaphase I	Yes	Svardson (1945)
	M	Metaphase I	No	Nygren *et al.* (1971*a*)
Arctic char	M	Metaphase I	Yes	Svardson (1945)
	M	Metaphase I	No	Nygren *et al.* (1971*a*)
Brook trout	F	Pachynema	No	Davisson *et al.* (1973)
			Yes	Lee and Wright (1981)
	M	Metaphase I		Davisson *et al.* (1973) Lee and Wright (1981)
Lake trout	F	Pachynema	No	Davisson *et al.* (1973)
	M	Metaphase I	Yes	Davisson *et al.* (1973)

gence, and by apparent examples of tetrasomic inheritance of some of these loci (May *et al.*, 1982; Section 5.4.2). Although the possibility of segmental allopolyploidy cannot be excluded, it is likely that the salmonid genome was doubled through autopolyploidy. This is in contrast to the catostomids, which apparently had an allotetraploid origin (Ferris and Whitt, 1980).

Many possible "barriers" to tetraploidy in animal species have been proposed. One reason why polyploidy may be rare is that it might require the independently arisen tetraploids of opposite sexes to mate and start the tetraploid strain. One way of avoiding this problem is a possible two-step process as proposed by Schultz (1969) and Astaurov (1969): (1) the origin of a unisexual triploid strain, followed by (2) occasional fertilization of the unisexual triploid by a normal diploid to produce fertile tetraploids. Astaurov (1969) has successfully carried out these steps in the laboratory to produce a sexually reproducing allotetraploid strain of silkworms. As mentioned earlier, unisexual triploid strains of several species of fish are known.

For a unisexual triploid strain of fish to reproduce, it must produce a high frequency of unreduced eggs and be capable of initiating development parthenogenetically or, if using sperm from a donor male, gynogenetically. Extant salmonids are apparently capable of producing high frequencies of unreduced gametes in some cases (Thorgaard and Gall, 1979). In addition, Melander and Monten (1950) described a possible case of parthenogenetic reproduction in the common whitefish. Hybridization between species sometimes might act to suppress the second meiotic division of the egg (Uyeno, 1972) and induce polyploidy in salmonids (Capanna *et al.*, 1974) and other fishes (Vasilev *et al.*, 1975; Marian and Krasznai, 1978; Beck *et al.*, 1980). Thus, hybridization and unisexuality may both play an important role in polyploidy of fish species (Schultz, 1969, 1980).

The objection that polyploidy would upset the chromosomal sex-determining mechanism (Muller, 1925) may not apply to salmonids. The rainbow trout shows evidence of having a "dominant Y" sex-determining mechanism (Thorgaard and Gall, 1979). Such a mechanism, if present is an ancestral salmonid, would not have been subject to Muller's objection because XXXY individuals would be expected to be males and not sterile intersexes.

Another objection to polyploidy in animals has been that they might not tolerate the changes in cell size because of their tissue complextity (White, 1973). Although polyploid mammals are inviable (Niebuhr, 1974), there are many examples of viable spontaneous and induced polyploids in salmonids and other fishes (Section 5.3). It appears that none of the traditional barriers to polyploids in animals necessarily apply to salmonids.

3. Evolution of Chromosomes

3.1. Ancestral and Extant Karyotypes

The diploid ancestor of salmonids probably had 48 acrocentric chromosomes. This is the most common karyotype among fish species, and it is found in a variety of distantly related fish taxa (Ohno, 1974), including many relatives of the salmonids (Sola *et al.*, 1981). If the ancestral karyotype was not 48 acrocentric chromosomes, it is likely to have been very similar to that; the majority of fish species have chromosome numbers in the range of $2n = 44$–52 (Gold *et al.*, 1980), with predominantly acrocentric or subtelocentric chromosomes.

The ancestral tetraploid salmonid thus probably had 96 acrocentric chromosomes. Karbe (1964, in Sola *et al.*, 1981) reported such a karyotype in three species of the genus *Coregonus*, but two of these same species were later reported to have $2n = 80$ by Nygren *et al.* (1971*b*). Most salmonids have about 100 major chromosome arms and diploid chromosome numbers between 56 and 84 (Table II). The general trend since the tetraploid event has been a reduction in chromosome number by centric fusion, while conserving the chromosome arm number at about 100 (Fig. 1). Two major exceptions are the Atlantic salmon, with 74 chromosome arms, and the grayling, with 170 chromosome arms. Many pericentric inversions or unequal translocations have apparently become established in these two species.

Salmonids appear to have undergone much more rapid chromosomal divergence than other groups of fish (Fig. 1). For example, the closely related species of the genus *Oncorhynchus* (Utter *et al.*, 1973) have chromosome numbers ranging between 52 and 74. In contrast to the large variation in chromosome numbers among salmonid species, the vast majority of species of the family Cyprinidae, for example, have 48–52 chromosomes, and all species of the family Centrachidae have 46 or 48 chro-

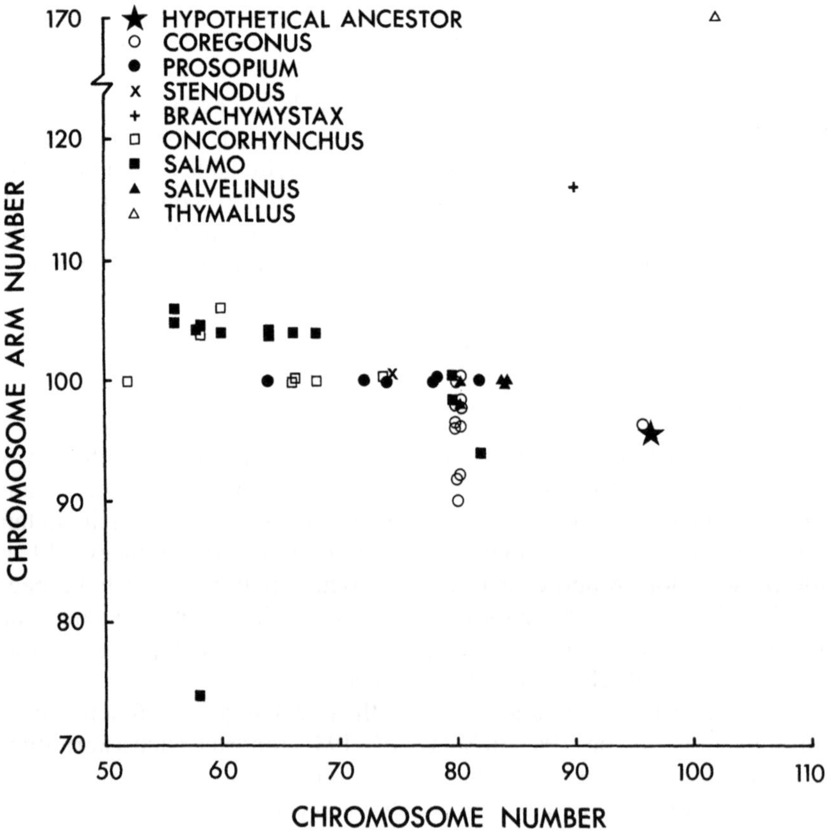

Figure 1. Chromosome arm number and chromosome number of 46 salmonid species.

mosomes (Sola *et al.*, 1981). All species of the family Catostomidae apparently have 96–100 chromosomes (Beamish and Tsuyuki, 1971; Uyeno and Smith, 1972); thus, a tetraploid ancestry alone does not ensure that rapid chromosomal divergence will take place. Perhaps the historical isolation of salmonid populations by glaciation and other geological events in combination with their reproductive homing behavior (Behnke, 1972) have promoted rapid chromosomal evolution within the salmonids. The difference in rates of chromosomal evolution in the salmonids and catostomids is probably also due, to some extent, to the difference between an auto- and allopolyploid ancestry.

3.2. Evolution of Disomy

In a new autotetraploid, or segmental allotetraploid, some multivalent formation and tetrasomic inheritance is expected. Because there is an increased frequency of non-disjunction when chromosomes associate in multivalents, this should lead to aneuploid gametes and reduced survival of zygotes. Selection for decreased infertility should cause a reduction of multivalent pairing and the restoration of disomy; what were four homologs begin to pair as two pairs of chromosomes (Sybenga, 1972). In addition to the advantage of decreased infertility, disomy might also be favored because it allows the structural and regulatory divergence of the newly duplicated gene loci.

Experiments with induced autotetraploid maize are representative of what might occur after a new tetraploid event; the average frequency of quadrivalents declined from 8.47 to 7.46 after only 10 generations (Gilles and Randolph, 1951). Similar results were found after selecting for fertility in *Brassica campestris* (Swaminathan and Sulbha, 1959). However, some naturally occurring tetraploid frog species function well with a high frequency of multivalent pairing (Bogart, 1980).

The primary mechanism for the restoration of disomic inheritance after tetraploidy is structural divergence of the four homologs into two pairs (Sybenga, 1972). A few studies have shown the effect of chromosome rearrangements on meiotic pairing in polyploids; Grell (1961), in *Drosophila,* and Shaver (1963), in maize, have shown that an inversion in a homolog can reduce the likelihood that it will pair with a normal chromosome. The effectiveness with which a rearrangement reduces pairing between homologs depends upon the meiotic mechanism of a species. In tetraploid *Rhoeo discolor,* for example, there is no preferential pairing between isosequential chromosomes, probably because the initiation of pairing in that species is restricted to small regions (Walters and Gerstel, 1948). In contrast, the well-known *Ph* locus acts to "amplify" the structural differences between the three component genomes in hexaploid wheat and allow disomic inheritance (Sears, 1976). It apparently brings this about by promoting a premeiotic somatic pairing of homologs.

The presence of multivalents at meiosis (Table III) and evidence for tetrasomic inheritance of some loci suggest that the process of return to disomic inheritance ("diploidization") is not yet complete in the salmonids. Some of the multivalents might also reflect translocation heterozygosity. As previously discussed, changes in chromosome number without a change in chromosome arm number ("Robertsonian" changes) are common among salmonid species. A number of examples of intraspecific Robertsonian polymorphisms are also known (Ohno *et al.*, 1965; Roberts, 1968, 1970; Gold and Gall, 1975; Thorgaard, 1976), which could be contributing to the observed multivalents.

A curious aspect of the meiotic multivalents and tetrasomic inheritance in salmonids is that these are both apparently restricted to males. This is difficult to explain, but might be caused by differences in the initiation of pairing at meiosis. For example, one might suggest, by analogy with the situation in *Rhoeo discolor* (Walters and Gerstel, 1948), that males may initiate pairing in small regions and not recognize the differences between homeologs that females do detect. Differences in the genetic control of meiosis between sexes in *Drosophila* (Baker *et al.*, 1976) and other animals (White, 1973) are well documented.

Many centric fusions appear to have taken place since the ancestral tetraploid event. Ohno and coworkers (1969*b*) proposed that the fusion of ancestral homologs may have been an important type of chromosomal rearrangement in the diploidization process in salmonids. If this were true, we would expect to observe loose linkage of duplicated gene loci. The present evidence, however, suggests that duplicated loci do not show classical linkage (May *et al.*, 1979*b*, 1980, 1982; Wright *et al.*, 1983). Thus the linkage data do not support Ohno's model of diploidization by fusion of homeologs.

Other types of centric fusions and structural rearrangements have presumably helped to differentiate homeologous chromosomes. The effects of specific rearrangements on meiotic pairing and the diploidization process in salmonids are not known. Similarly, it is unknown what, if any, changes to promote disomy (analogous to the *Ph* locus of wheat) have taken place in the meiotic system of salmonids. The Atlantic salmon, with only 74 chromosome arms, shows striking divergence from the typical, and probably ancestral, salmonid chromosome arm number of about 100; nevertheless, this species still shows multivalent pairing at meiosis (Table III). If these reflect homeologous pairing, it means that the many rearrangements in the lineage of the Atlantic salmon still were not sufficient to bring about disomic pairing for all chromosomes.

It will be difficult to assess what sorts of small chromosomal rearrangements have taken place and what ancestral similarities between homeologs are still present in salmonid species until improved chromosome banding techniques are developed. Zenzes and Voiculescu (1975) reported that they could arrange C-banded brown trout chromosomes into groups of four, but the apparent homologies are not striking. One case in which a rearrangement has taken place in a specific chromosome since the tetraploid event is the chromosome pair bearing satellites. In those cases reporting satellited chromosomes in salmonids, they have been present on only one pair and not on two as expected in a "fresh" tetraploid (Cuellar and Uyeno, 1972; Gold and Gall, 1975; Thorgaard, 1976, 1978; Loudenslager and Thorgaard, 1979).

4. Evolution of Proteins

A newly arisen autotetraploid is endowed with four doses of every gene at a single tetrasomically inherited locus. The four homologs are eventually transformed into two independently inherited pairs of chromosomes as discussed in the previous section. In this way, all of the protein loci are duplicated.

4.1. Possible Fate of Protein Loci

The evolution of protein loci following autotetraploidy can be conceived as occurring in three different periods (Li, 1980). The first period occurs between tetraploidization and the reestablishment of disomic segregation. During this period, the chromosome, and not the individual locus, is the unit of importance.

The second period begins with the reestablishment of disomic inheritance. During this period, the original locus is functionally duplicated. That is, there are now two genetically independent loci that are equivalent to the ancestral unduplicated locus. It is during this period that divergence of the two duplicate loci can begin.

The third period begins when structural or regulatory divergence of the duplicate loci has proceeded to a substantial extent. With regard to structural divergence, this period starts when different alleles have become "fixed," or nearly so, at the two duplicates. This period starts with regard to regulatory divergence when different ontogenetic or tissue-specific patterns of expression have been established at the two loci. It is impossible, however, to detect such regulatory divergence without the presence of some structural divergence.

The term "duplicated locus" has been applied to all three of these periods. This is a potential source of confusion. Strictly speaking, "duplicated'' loci are present only during the second period. During the third period, the loci have diverged so that they may no longer perform duplicate functions. We think it is important to differentiate among these three periods when referring to a pair of loci that share a common ancestral locus.

We therefore recommend the following nomenclature. During the first period, a locus should be referred to as being ''tetrasomic.'' A pair of genetically distinct loci that still share identical alleles, so that variant alleles cannot be unambiguously assigned to one locus or the other, will be referred to as ''isoloci.'' [Gall *et al.* (1976) have referred to these loci as being ''isoqualitic.''] After the completion of the second period, we will refer to such pairs of loci as being ''paralogous'' (Fitch, 1976). We will use the term ''duplicated'' as a general term, without reference to any of these three specific time periods.

These three time periods are not necessarily discrete. Both the reestablishment of disomy and the divergence of isoloci are continuous processes. Nevertheless, the recognition of these periods and the associated nomenclature should make it easier to understand and discuss the post-autotetraploidy evolution of protein loci.

4.1.1. Loss of Duplicate Gene Expression

Haldane (1933) first suggested that one of the duplicate loci may become nonfunctional through the accumulation and eventual fixation of deleterious (i.e., null) mutations at one locus while the other locus continued to perform the original function. The theoretically expected rates of such loss of duplicate gene expression have been recently explored by a number of authors [reviewed by Li (1980)].

The recent discovery of widespread pseudogenes in the genomes of vertebrates provides evidence that such loss of gene expression may indeed be a common evolutionary event. Pseudogenes are DNA segments that show homology to functional genes but have nucleotide changes so that they are not expressed (Proudfoot, 1980). Pseudogenes are apparently duplications that have become nonfunctional by the accumulation of mutations.

The loss of detectable function has been a common ''fate'' for enzyme loci in the salmonids (Allendorf *et al.*, 1975; May *et al.*, 1980; May, 1980; Kijima and Fujio, 1980). Approximately 50% of the additional loci created by tetraploidy are no longer detectable by their protein products. These estimates are similar to those from other tetraploid-derived fish species, which show a 25–75% loss in duplicate gene expression (Li, 1980). The existence of three nonduplicated loci in a single linkage group [Odh–Mpi–Gpi-

3 (Wright *et al.*, 1983)] suggests that part of the diploidization process may have involved whole sections of DNA.

4.1.2. Retention of Duplicate Gene Expression

Those cases in which both duplicated loci are retained fall into different categories. The first is tetrasomic loci or isoloci in which disomic inheritance has only recently evolved or is still not complete. Loss of duplicate gene expression may eventually occur in these cases. A second category is loci that have diverged structurally but not with regard to their regulation. These systems are also potential candidates for future loss of duplicate gene expression. As long as both loci are always equally expressed, a null allele at one locus may be sheltered from selection by the presence of the normal allele at the other locus.

Another category is loci that have diverged in their developmental or tissue-specific expression. Such regulatory divergence can occur in two different ways: unidirectional and bidirectional (Ferris and Whitt, 1979). In unidirectional divergence, the enzyme products of one locus predominate in all tissues in which the two loci are not equally expressed. In bidirectional divergence, there is no consistent predominance of one locus or the other in tissues showing unequal expression.

When a locus is uniquely expressed in a particular tissue or developmental stage it is unlikely to become fixed for a null allele because the other locus can no longer shield it from natural selection. Therefore, either loss of duplicate gene expression or evolution of unique expression is the probable eventual fate of all duplicated loci.

Two possible modes of selection could also "protect" against the loss of duplicate gene expression. A benefit in having large amounts of enzyme product could select against the establishment of null alleles at either locus. This is unlikely to occur for enzyme loci in eukaryotes since the quantity of enzyme present is unlikely to be rate-limiting (Kacser and Burns, 1981). Also, benefit in having two different alleles present (overdominance) would cause selection against null alleles (Spofford, 1969; Allendorf, 1978).

4.2. Salmonid Enzymes

The genetic control of a large number of enzymes in salmonids has been described. We do not intend to review all of these results; rather, we will first describe the genetic systems of control of the most studied enzymes: lactate dehydrogenase (LDH) and malate dehydrogenase (MDH). These two enzymes present extreme differences in the amount of divergence between duplicated loci. Finally, we describe the present genetic control of 29 ancestral loci in the species that has been most intensively studies: the rainbow trout. Table IV lists all of the enzymes and their abbreviations. We have used the system of nomenclature proposed by Allendorf and Utter (1979) with the additions suggested by May *et al.* (1979*b*).

4.2.1. Lactate Dehydorgenase

Lactate dehydrogenase was the first duplicated enzyme system in salmonids to be described (Goldberg, 1965; Markert and Faulhaber, 1965; Morrison and Wright, 1966). The diploid ancestor of the salmonids apparently possessed three LDH loci: a muscle-specific locus, for LDH-A; an eye-specific locus, for LDH-C; and a third locus found in all tissues, for LDH-B (Markert *et al.*, 1975). As many as eight LDH loci have been suggested to be present in extant salmonids (Massaro and Markert, 1968). This estimate, however, is based on zymogram patterns alone, without supporting inheritance studies, and is apparently inflated.

Wright *et al.* (1975) summarized their studies of the genetic control of LDH in *Salmo* and *Salvelinus*. Five loci code for LDH activity in species

Table IV
Enzymes and Proteins Examined in Rainbow Trout for Loss of Duplicate Gene Expression

Enzyme	EC number	Abbreviation
Aspartate aminotransferase	2.6.1.1	AAT
Adenosine deaminase	3.5.4.4	ADA
Alcohol dehydrogenase	1.1.1.1	ADH
Creatine kinase	2.7.3.2	CK
Diaphorase	1.6.4.3	DIA
Fructose diphosphatase	3.1.3.11	FDP
Glyceraldehyde-phosphate dehydrogenase	1.2.1.12	GAP
Glucosephosphate isomerase	5.3.1.9	GPI
Isocitrate dehydrogenase	1.1.1.42	IDH
Lactate dehydrogenase	1.1.1.27	LDH
Malate dehydrogenase	1.1.1.37	MDH
Malic enzyme	1.1.1.40	ME
Mannosephosphate isomerase	5.3.1.8	MPI
para-Albumin	—	PAL
Phosphoglucomutase	2.7.5.1	PGM
Phosphogluconate dehydrogenase	1.1.1.44	PGD
Phosphoglycerate kinase	2.7.2.3	PGK
Sorbitol dehydrogenase	1.1.1.14	SDH
Transferrin	—	TFN

of these genera. The two loci (*Ldh-1* and *-2*) descendant from the ancestral LDH-A locus have different common alleles, resulting in a common five-banded pattern for muscle-specific LDH. These two loci show no evidence of regulatory divergence as determined by patterns of tissue-specific expression.

The two loci (*Ldh-3* and *-4*) descendant from the ancestral LDH-B locus also have different common alleles. These two loci have also evolved different tissue-specific patterns of expression (Fig. 2). Only products of *Ldh-4* are found in liver tissue, while *Ldh-3* locus products predominate in heart tissue. These loci thus show bidirectional regulatory divergence. There is only one locus coding for the eye-specific form of LDH. These results have been confirmed by several other investigations (e.g., Bailey *et al.*, 1976).

A similar system of genetic control and tissue-specific expression of LDH loci is found in the grayling (Massaro, 1973; Fig. 2). However, there is only one isozyme of the muscle-specific form of LDH. This could be caused by either the comigration of the products of the *Ldh-1* and *-2* loci or by the loss of expression of one of these loci. A null allele at *Ldh-1* has been found to be common in brown trout populations and is even

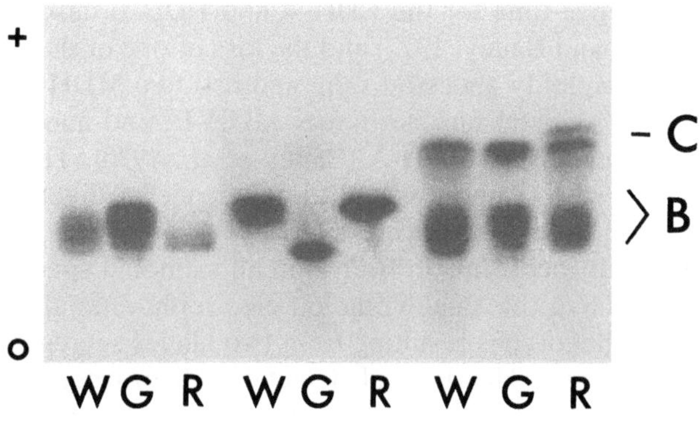

Figure 2. Tissue distribution of LDH-B and LDH-C loci in species from the three salmonid subfamilies: mountain whitefish (W), Arctic grayling (G), and rainbow trout (R). The LDH-C isozymes are present only in the eye tissue of all three species. The rainbow trout and grayling show similar patterns of expression for the two LDH-B loci: equal expression in the eye, the expression of only one locus in the liver, and the predominant expression in the heart of the locus that was not expressed in the liver. The whitefish shows equal expression of both LDH-B loci in the eye and heart and the unequal expression of both loci in the liver.

fixed in one Swedish population (Ryman *et al.*, 1979). Null alleles at the *Ldh-1* locus have also been found in rainbow trout and sockeye salmon (unpublished results). A null allele at *Ldh-2* has also been described in Snake Valley cutthroat trout (Klar and Stalnaker, 1979).

Whitefish LDH is also encoded by five loci. However, the two descendant LDH-B loci do not show the tissue-specific pattern of expression that is found in the trout, char, and grayling (Fig. 2). Rather, these two loci are nearly equally expressed in all tissues (Clayton and Franzin, 1970; Massaro, 1972). The similar expression of both loci is the primitive condition, while the highly specific pattern of expression seen in the tissues of trout, char, and grayling is a derived characteristic (synapomorphy) that is likely to have evolved only once. Thus, this result indicates that the trout, char, and grayling apparently shared a common ancestor longer after the tetraploid event than did the whitefish. The pattern of expression of other enzyme loci supports these relationships.

4.2.2. Malate Dehydrognease

The loci encoding the cytosolic form of MDH show a different picture of post-tetraploidy evolution than the LDH loci. The LDH loci have all been "diploidized" for a long time. This is supported by the estimates of 100 Myr divergence time for the LDH-A and LDH-B descendants (Lim *et al.*, 1975; Lim and Bailey, 1977) and the loss of one of the LDH-C loci.

The pre-tetraploidy ancestral salmonid had two MDH loci; one that predominated in skeletal muscle tissue, MDH-B, and another that predominated in liver tissue, MDH-A (Bailey *et al.*, 1970). The present descendants of these loci show very little evidence of divergence, either structurally or in their tissue expression.

Both MDH-B loci (*Mdh-3, Mdh-4*) in all salmonid species that have been described have the same common electrophoretic allele. Figure 3 shows the five phenotypes resulting from two alleles segregating at these loci in rainbow trout. This lack of detectable divergence indicates that the diploidization of these loci has occurred recently. Inheritance experiments indicate that this process is still apparently not complete in the rainbow trout (May *et al.*, 1982; Section 5.4.2).

A similar pattern is seen for the MDH-A loci. In most species studied, both loci have the same electrophoretic allele. In some species, however, the two MDH-A loci have diverged in structure and in tissue-specific patterns of expression. For example, the closely related brown trout and Atlantic salmon share the same three-banded pattern for MDH-A, indicating that the two descendant loci have different common alleles. In the

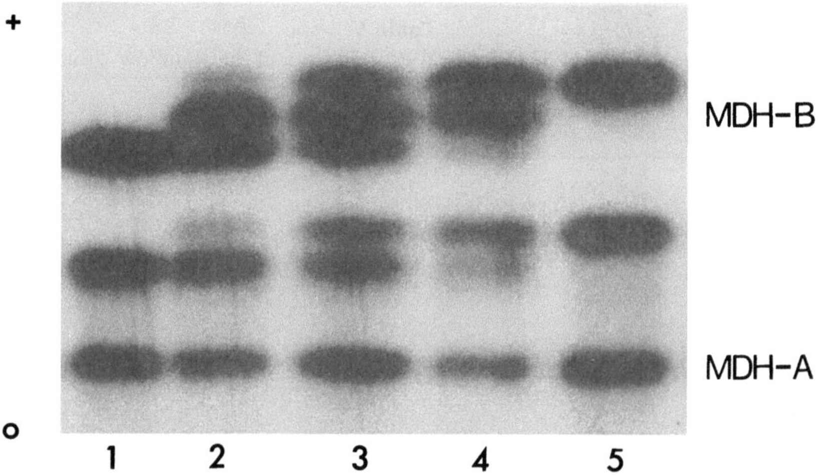

Figure 3. Electrophoretic variation for MDH-B in rainbow trout. All five possible phenotypes produced by two alleles at isoloci are shown: (1) *BBBB*, (2) *BBBb*, (3) *BBbb*, (4) *Bbbb*, and (5) *bbbb*. The most cathodal band is the homodimer of MDH-A. The next two more anodal bands are heterodimers between MDH-A and MDH-B. The three most anodal bands are the homodimers and heterodimer for the two MDH-B alleles: B [*Mdh-3, Mdh-4* (100)] and b [*Mdh-3, Mdh-4* (83)].

brown trout, however, only one of these two loci is expressed in skeletal muscle, while both of these loci are expressed in the skeletal muscle of Atlantic salmon.

4.2.3. Rainbow Trout Isozymes

Table V summarizes the status of 33 ancestral protein loci in the rainbow trout. Other species in the same subfamily differ very little from these results (May *et al.*, 1982; May, 1980). We have not included a similar table for species in the other two subfamilies because there is not sufficient information available. Many enzymes and a large number of individuals must be examined to acquire the required infromation. For example, if a single isozyme is seen for a particular enzyme (e.g., ADH), the number of loci controlling this isozyme (or electromorph) cannot be determined until electophoretic variation is detected. Many of the enzymes represented by a single electromorph in salmonids are also characterized by a low amount of genetic variation (e.g., ADH, CK-C, and GPD). Thus there is a bias against detecting loci at which duplicate gene expression has not

Table V
Present Status of 33 Ancestral Protein Loci in the Rainbow Trout

	Duplicated		
Locus	Isoloci	Diverged	Single locus
Aat-A	—	Yes	—
Aat-B	Yes	—	—
Aat-m	—	Yes	—
Adh	—	—	Yes
Ada	—	—	Yes
Ck-A	—	Yes	—
Ck-B	—	—	Yes
Ck-C	—	Yes	—
Dia	—	—	Yes
Fdp	—	—	Yes
Gap-A	—	Yes	—
Gap-B	—	Yes	—
Gpi-A	—	Yes	—
Gpi-B	—	—	Yes
Idh-A	Yes	—	—
Idh-m	—	Yes	—
Ldh-A	—	Yes	—
Ldh-B	—	Yes	—
Ldh-C	—	—	Yes
Mdh-A	Yes	—	—
Mdh-B	Yes	—	—
Mdh-m	—	Yes	—
Me-A	Yes	—	—
Me-m	Yes	—	—
Mpi	—	—	Yes
Pal	Yes	—	—
Pgd	—	—	Yes
Pgk	—	—	Yes
Pgm-A	—	Yes	—
Pgm-B	Yes	—	—
Sdh	—	Yes	—
Sod	—	—	Yes
Tfn	—	—	Yes
Total	8	13	12

been retained. Many individuals must therefore be examined in order to accurately estimate the proportion of gene duplication lost.

Thirty percent (10 out of 33) of the ancestral loci in rainbow trout do not show any evidence of duplicate gene expression. Twenty-four percent (eight out of 33) of the loci are isoloci showing no evidence of divergence.

The remaining 45% of the loci are duplicated and show evidence of structural or regulatory divergence.

4.3. Regulation of Enzyme Loci

The examination of genetic variation at enzyme-coding loci using gel electrophoresis in the last 15 years has revealed a large amount of genetic variation in plants and animals. Demonstrating the evolutionary significance of this variation has proven to be a much more difficult problem. Several authors have suggested that changes in the regulation of enzyme-producing loci may be of more evolutionary significance than changes in the enzymes themselves (Wallace, 1963; Wilson, 1976). The evidence for this view remains largely indirect. Differences between the rate of change at structural loci encoding enzymes and the rate of change in phenotypes having adaptive importance suggest that significant evolutionary changes may be due to changes in gene regulation (Wilson, 1976). Further indirect evidence of the evolutionary importance of gene regulation comes from paleontology (Vanentine and Campbell, 1975; Gould, 1980) and developmental genetics (Flickinger, 1975; Whitt, 1981).

The eventual acceptance of these views awaits direct evidence of the amount and adaptive significance of variation at regulatory genes. It is necessary to find intraspecific genetic variation for the control of enzyme-producing loci to explore the adaptive significance of gene regulation. Paralogous isozyme loci in the salmonids show considerable divergence in tissue-specific expression. This relatively rapid divergence is in contrast to the commonly observed evolutionary conservation of patterns of tissue-specific expression of enzyme loci (Markert *et al.*, 1975; Shaklee and Whitt, 1981). We would therefore expect salmonids to have a greater amount of intraspecific genetic variation for the regulation of enzyme loci because of the differential rates of divergence among pairs of duplicated loci. That is, we expect duplicate loci in the early stages of divergence to be polymorphic for the types of differences in tissue-specific patterns of expression that have been established between paralogous loci.

Salmonids are therefore potentially valuable for studying the evolutionary significance of changes in the regulation of enzyme loci. Wright *et al.* (1975) have described intraspecific variation for the ontogenetic schedule of expression of an LDH locus in the brook trout. A mutant allele resulting in a greater than 100-fold increase in the liver-specific expression of a PGM locus has been reported in the rainbow trout (Allendorf, 1980; Allendorf *et al.*, 1982). Results of inheritance experiments are consistent with a single regulatory gene having additive inheritance being responsible for the differences in the tissue-specific expression of this locus (Allendorf *et al.*, 1982). The presence or absence of this isozyme in the liver gives rise to important differences in several phenotypic characteristics of adaptive significance (developmental rate, developmental stability, body size, and age at first reproduction). We have also detected several other putative allelic variants in the tissue-specific expression of enzyme loci in other salmonid species (Allendorf, unpublished data).

5. The Current Salmonid Genetic System

All aspects of genetics and evolution in salmonids are affected by their tetraploid ancestry. Because of this we feel it is important to consider the general features of the genetic system of present-day salmonids in light of their tetraploid origin.

5.1. Sex Chromosomes and Sex Determination

Little was known about the mechanism of sex determination in salmonids until recently. The most widely studied species has been the rainbow trout; this species has an XY male, XX female sex chromosome system. This conclusion is based on chromosomal studies (Thorgaard, 1977; Thorgaard and Gall, 1979) and breeding studies with sex-reversed fish (Okada *et al.*, 1979; Johnstone *et al.*, 1979). Chromosomal evidence also supports male heterogamety in the sockeye salmon, (Thorgaard, 1978). The coho salmon also appears to be male heterogametic based on breeding studies with hormonally sex-reversed fish (Hunter *et al.*, 1982) and the production of all-female lots after diploid gynogenesis using irradiated sperm and cold shock (Refstie *et al.*, 1982).

The fact that XXY triploid rainbow trout are males (Thorgaard and Gall, 1979) suggests that a "dominant Y" sex-determining mechanism is operative. Such a system would avoid one of Muller's (1925) objections to polyploidy in animals.

5.2. Genetic Recombination

Building a linkage map for salmonids is a challenging task because of their many chromosomes and relatively long generation times. Nevertheless, several examples of classic linkage have recently been reported in salmonids [reviewed by May *et al.* (1982)]. Comparison of linkage relationships in different species should provide insight into the process of post-tetraploidy chromosomal evolution in salmonids. Unfortunately, all of the available linkage data are restricted to a single subfamily: Salmoninae. Those genera that have been studied (*Salmo, Oncorhynchus,* and *Salvelinus*) apparently show conservation of linkage groups (May *et al.,* 1982).

Males show a much lower frequency of recombination than females (Table VI). This difference between sexes is in agreement with the general observation that when there is a difference in recombination rates between the sexes, it is the heterogametic sex that shows reduced frequencies (Swanson *et al.,* 1981). The differences in recombination frequencies and in the occurrence of meiotic multivalents between sexes suggest that there are substantial differences in meiosis between male and female salmonids.

5.3 Aneuploidy and Polyploidy

Polyploid plant species tolerate aneuploidy much better than diploid species. Salmonids might also be expected to be relatively tolerant to aneuploidy because of their tetraploid origin. There are several reports of aneuploids in salmonids, supporting this contention. Davisson *et al.* (1972) found a male brook trout that was trisomic for a chromosome carrying a LDH-B enzyme locus. This fish was fertile and produced euploid and trisomic offspring in equal frequencies; the trisomics were similar in size, appearance, and viability to the normal fish. Ohno (1970*a*) reported finding "a few" trisomic and monosomic rainbow trout during a survey of nearly 500 hatchery fish; these aneuploids also had no obvious differences from normal fish. Aneuploid salmonids thus seem to be common and viable. Nevertheless, more information about the frequency and effects of aneuploidy in salmonids and other fishes is needed before conclusions about the effect of the tetraploid ancestry of salmonids on aneuploids can be made.

Polyploid salmonids also occur naturally and have been induced artifically. Spontaneously occurring triploid rainbow trout have been found

Table VI
Comparative Recombination Rates in Male and Female Brook Trout–Lake Trout Hybrids[a]

			Mean recombination rate	
Species	Loci		Females	Males
Splake trout	*Cpk-1*	*Gus*	0.199 (2)	0.088 (1)
	Gpi-3	*Mpi*	0.299 (1)	0.000 (1)
	Ada-1	*Agp-2*	0.154 (1)	0.031 (4)
	Idh-3	*Me-2*	0.086 (4)	0.048 (7)

[a]From May *et al.* (1980).

(Cuellar and Uyeno, 1972; Grammeltvedt, 1974; Thorgaard and Gall, 1979); these fish are normal in appearance but sterile. Naturally occurring polyploid brook trout have also been found; Allen and Stanley (1978) concluded on the basis of red blood cell nuclear volumes that these fish were mosaics composed mainly of triploid cells but also contained cells of other ploidy levels. Triploidy has also been induced in salmonids using a variety of treatments (Allen and Stanley, 1979; Smith and Lemoine, 1979; Chourrout, 1980; Thorgaard *et al.*, 1981). These findings suggest that triploid salmonids may not have significantly reduced viability.

Although no spontaneously occurring tetraploid salmonids have been reported, Refstie (1981) described tetraploid rainbow trout produced by treating fertilized eggs with cytochalasin B. There was an increased frequency of abnormalities among these fish, but some males were able to produce milt.

The fact that triploid, and possibly tetraploid, salmonids are reasonably viable is consistent with results with other fish species (Swarup, 1959; Purdom, 1972; Valenti, 1975; Gervai *et al.*, 1980; Schultz, 1980; Wolters *et al.*, 1981) and with the idea that an ancestral salmonid could tolerate a polyploid step.

5.4. Patterns of Genic Inheritance

The autotetraploid ancestor of present-day salmonids had four copies of each chromosome. Because the homologs were initially structurally similar, these chromosomes would pair randomly and the genes on these chromosomes would be inherited tetrasomically. The chromosomal diploidization process would eventually transform the four homologs into two pairs of two, resulting in disomic inheritance. The status and inheritance of ancestral gene loci (Table V) can be used to determine how far the diploidization process has proceeded. Duplicated loci that show no evidence of divergence are possible candidates for tetrasomic inheritance.

Most of the isozyme loci in rainbow trout have completed the diploidization process; only eight out of 33 ancestral loci remain isoloci (Table V). Thus, approximately three-quarters of the salmonid genome has become diploidized. This assumes that isozyme loci are an accurate reflection of the entire genome and that rainbow trout are typical of other salmonids. We believe that both of these assumptions are good ones. The diploidization process is a chromosomal rather than genic process. Therefore, the 33 ancestral isozyme loci spread throughout the genome should simply act as chromosomal markers. The pattern seen in rainbow trout is similar to that seen in other salmonids; we have chosen to use rainbow trout because it has been intensively studied with regard to patterns of inheritance [see Stoneking *et al.* (1979), May *et al.* (1979*b*, 1980), Wright *et al.* (1980) for analogous results with *Salvelinus* sp.].

5.4.1. Inheritance of Duplicated Loci

The many multivalents observed in meiosis (Table III) suggest that some of the duplicated loci in salmonids may be inherited tetrasomically. Several loci have been reported to be inherited tetrasomically in salmonids without any supporting inheritance results: *Idh* in rainbow trout (Wolf *et al.*, 1970); *Sdh* in rainbow trout and common whitefish (Engel *et al.*, 1970); and *H6pdh* in brook trout (Stegeman and Goldberg, 1972). It has been shown, however, that it is impossible to demonstrate tetrasomic inheritance in the absence of inheritance results (Allendorf *et al.*, 1975). These papers have continued to be cited as evidence of tetrasomic inheritance in salmonids even though they are apparently erroneous.

There are a variety of possible modes of inheritance that must be considered when distinguishing between isoloci and a single tetrasomic locus (Table VII). In a two-allele system, the critical genotype for examining inheritance is the duplex phenotype (Brunham, 1962) having two copies of both alleles, i.e., *AAaa*. The expected disomic inheritance ratios will be different depending on whether the parent is a double homozygote (*AA;aa*) or double heterozygote (*Aa;Aa*). The segregation ratios produced by the double heterozygote can be affected by linkage. Thus, the double homozygote parental genotype must be observed to establish disomic inheritance.

Table VII
Expected Segregation Ratios for Duplicated Loci with Disomic and Tetrasomic Inheritance[a]

Parental genotype	Frequency of gametes[b] *AA*	*Aa*	*aa*
AA;Aa	0.50	0.50	0
AAAa	$0.50(1+\alpha/2)$	$0.50(1-\alpha)$	0.25α
AA;aa	0	1.00	0
Aa;Aa (*trans*)	$0.50r$	$1\text{-}r$	$0.50r$
Aa;Aa (*cis*)	$0.50(1\text{-}r)$	r	$0.50(1\text{-}r$
AAaa	$0.17(1+2\alpha)$	$0.67(1-\alpha)$	$0.17(1+2\alpha)$

[a]A semicolon separates disomic loci. For example, the genotype *AA;aa* is homozygous at two disomic loci.

[b]Alpha represents the frequency of double reduction divisons and r is the frequency of recombination.

The segregation ratios produced by tetrasomic inheritance are affected by the frequency of quadrivalent formation and the distance of the locus from the centromere (Burnham, 1962). Crossovers between the locus and the centromere in a quadrivalent during the first meiotic division can produce so-called double reduction gametes that carry two copies of identical chromatids from a single chromosome. The maximum frequency of such double reduction divisions is one-sixth (Burnham, 1962). Thus, the proportion of *Aa* gametes produced by tetrasomic inheritance from the *AAaa* parental genotype varies between 67% and 55% (Table VII).

The study of isoloci presents special problems. Variant alleles cannot be assigned to a particular locus on the basis of electrophoretic phenotypes. Therefore, these loci cannot be treated individually when estimating allelic frequencies in population samples. First-generation inheritance results do not solve this problem. An enormous number of multiple-generation inheritance experiments are necessary to assign variation observed in a single population sample to one locus or the other. In addition, inheritance studies with duplicated loci have presented special problems. Beginning with Morrison and Wright (1966), inheritance ratios have been reported that do not agree with any simple genetic model of either disomic or tetrasomic inheritance.

5.4.2. Unusual Inheritance Results at Duplicated Loci

Morrison and Wright (1966) were the first to report inheritance results of duplicated loci in salmonids. They reported linkage of the two loci resulting from the duplication of the ancestral LDH-B locus in hybrids between brook and lake trout. Additional results demonstrated that these inheritance ratios could not be explained by classic linkage because the nonparental types were found to be in excess, rather than the parental types (Morrison, 1970). This phenomenon was first referred to as "pseudolinkage" by Davisson *et al.* (1973). A series of papers from the laboratory of James E. Wright has continued to further the understanding of the mechanisms producing these unusual segregation ratios (Lee and Wright, 1981; May *et al.*, 1979*a*,*b*, 1980, 1982; Wright *et al.*, 1975, 1980). In this chapter, we have chosen not to present an historical review of this important work. Rather, we believe the essential findings can be presented much more clearly by beginning with the simplest possible genetic models and then adding further complexity only when it becomes necessary.

Bailey *et al.* (1970) were the first to describe the inheritance of isoloci. They concluded that MDH-B was controlled by two disomically inherited loci in chinook salmon. As pointed out in the previous section, disomic and tetrasomic inheritance can only be distinguished in a two-allele system by examining the segregation of individuals with two doses of each allele

(*AAaa*). This was not done by Bailey and his co-workers. Therefore, their data are also compatible with tetrasomic inheritance.

One of us (Allendorf, 1975) performed a series of experimental matings with this same system in rainbow trout to distinguish between disomic and tetrasomic inheritance (Table VIII). These matings were done using Chambers Creek anadromous rainbow trout (steelhead) from the Washington State Department of Game. These results are incompatible with tetrasomic inheritance; only *Bb* gametes were produced by three male fish having the *BBbb* genotype (families C25, C37, and C39). Four additional *BBbb* types produced gametes compatible with the 1:2:1 segregation of *BB:Bb:bb* types expected with disomic inheritance. We thus concluded that MDH-B in rainbow trout is controlled by two disomically inherited loci (*Mdh-3,Mdh-4*) that have apparently only recently evolved disomic inheritance.

A more recent series of experiments with a different strain of rainbow trout (Jocko River State Trout Hatchery, Arlee, Montana) has provided some surprising results (Table IX). All four females examined produced gametes in agreement with the previous results of disomic inheritance found in the Chambers Creek fish. Five of the six males, however, produced families that can only be explained by a mixture of disomic and tetrasomic inheritance (G1, H6, H7, H18, and H19).

Families H6 and H18 demonstrate the clearest examples of tetrasomic inheritance. In these families, males having two doses of both the common

Table VIII
Observed and Expected Disomic Segregation Ratios at the MDH-B Locus in Rainbow Trout from Chambers Creek

	Parental genotype		Progeny phenotype					
Family	Female	Male	*BBBB*	*BBBb*	*BBbb*	*Bbbb*	*bbbb*	χ^2
C21	*BB;BB*	*Bb;Bb*	114	238	109	0	0	
			(115)	(230)	(115)			0.60
C22	*BB;BB*	*Bb;Bb*	21	50	20	0	0	
			(23)	(46)	(23)			1.11
C25	*BB;BB*	*BB;bb*	0	40	0	0	0	
				(40)				—
C31	*BB;BB*	*Bb;Bb*	27	51	23	0	0	
			(25)	(50)	(25)			0.33
C32	*BB;Bb*	*Bb;Bb*	12	34	38	15	0	
			(12)	(37)	(37)	(15)		0.84
C37	*BB;Bb*	*BB;bb*	0	41	39	0	0	
				(40)	(40)			0.005
C39	*BB;BB*	*BB;bb*	0	40	0	0	0	
				(40)				—

Table IX
Observed and Expected Disomic and Tetrasomic Segregation Ratios at the MDH-B Locus in Rainbow Trout from the Jocko River Hatchery[a]

	Parental genotype		Progeny phenotype					χ^2
Family	Female	Male	*BBBB*	*BBBb*	*BBbb*	*Bbbb*	*bbbb*	(1 df)
G1	*BB;Bb*	*BB;bb*	2	50	45	5	0	
			—	(56)	(56)	—	—	—
			(13)	(38)	(38)	(13)	—	17.90
			(8)	(42)	(42)	(8)	—	7.06
G5	*Bb;Bb*	*Bb;bb*	0	20	30	30	6	
			—	—	(43)	(43)	—	—
			—	(11)	(32)	(32)	(11)	1.26
			—	(7)	(36)	(36)	(7)	11.40
G6	*Bb;Bb*	*BB;BB*	32	57	19	0	0	
			—	(108)	—	—	—	—
			(27)	(54)	(27)	—	—	0.33
			(18)	(72)	(18)	—	—	9.38
H5	*BB;BB*	*Bb;Bb*	52	92	58	0	0	
			—	(202)	—	—	—	—
			(50)	(101)	(50)	—	—	1.60
			(34)	(135)	(34)	—	—	40.55
H6	*BB;BB*	*BB;bb*	14	266	9	0	0	
			—	(289)	—	—	—	—
			(72)	(144)	(72)	—	—	20.43
			(48)	(193)	(48)	—	—	83.70
H7	*BB;BB*	*Bb;Bb*	49	127	46	0	0	
			—	(222)	—	—	—	—
			(55)	(111)	(55)	—	—	4.61
			(37)	(148)	(37)	—	—	8.94
H13	*BB;bb*	*BB;BB*	0	44	0	0	0	
H17	*BB;bb*	*BB;BB*	0	249	0	0	0	
H18	*BB;BB*	*BB;bb*	12	184	5	0	0	
			—	(201)	—	—	—	—
			(50)	(100)	(50)	—	—	138.75
			(34)	(134)	(34)	—	—	55.97
H19	*BB;BB*	*Bb;Bb*	38	107	37	0	0	
			—	(182)	—	—	—	—
			(46)	(91)	(46)	—	—	5.63
			(30)	(121)	(30)	—	—	5.08

[a]The first expected rations are for disomic inheritance assuming double homozygosity (*BB;bb*), the second for double heterozygosity (*Bb;Bb*), and the third for tetrasomic inheritance with random chromosome inheritance.

and variant alleles (*BBbb*) were crossed with females having only the common allele (*BBBB*). Two different segregation ratios are possible with disomic inheritance. If the male is a double homozygote (*BB;bb*), then all progeny should be (*BB;Bb*); if the male is a double heterozygote (*Bb;Bb*), then the progeny should segregate 1:2:1 for the *BBBB:BBBb:BBbb* types. Neither of these two ratios is obtained in these families. It appears that these males were double homozygotes but that some homeologous pairing has produced infrequent *BB* and *bb* gametes.

The male-producing family G1 shows a pattern of gamete formation that is similar to the male parents of H6 and H18. This male, however, was crossed with a female segregating 1:1 for *BB* and *Bb* gamets. The male parents of H7 and H19 were apparently double heterozygotes (*Bb;Bb*); homeolog pairing in these males produced gamete frequencies intermediate between the 1:2:1 expected with disomic inheritance and the 1:4:1 expected with tetrasomic inheritance.

In this strain of rainbow trout, segregation ratios in males for MDH-B are intermediate between disomic and tetrasomic inheritance. May *et al.* (1982) have reported similar results at this locus for two males (they did not use any *BBbb* females) in another strain of rainbow trout and for an AAT locus in brook trout (Wright *et al.*, 1980). They have referred to this phenomenon as residual tetrasomic inheritance. We can thus conclude that the "diploidization" of all chromosomes in salmonids is apparently not complete.

These observations are in agreement with reports of multivalent formation in males but not females (Table III). We can apparently explain these results by the pairing of homeologs in males only, resulting from differences in meiosis between males and females. There is a problem with this simple explanation, however. The union of sperm carrying two homologs with an egg carrying two homeologs will produce progeny having three copies of one homolog and only one copy of the other homolog. Such individuals should show tetrasomic gamete formation because they do not possess the two copies of each homolog necessary for disomic segregation. This should be true regardless of the sex of the individual. Frequent homeologous pairing in one sex should thus quickly restore tetrasomic inheritance in the whole population.

We must explain the stable differences between the sexes in multivalent formation and pairing of homeologs as detected by the inheritance of isozyme variants. This may be explained by a two-stage pattern of pairing in males in which homologous chromosomes pair first followed by homeologous pairing. Disjunction so that paired chromosomes pass to opposite poles would ensure that each gamete received one copy of each homeolog. Exchanges between homeologs would produce segregation ratios approaching tetrasomic expectations (Fig. 4).

This model would produce a mixture of disomic and tetrasomic inheritance, depending upon the map distance between a locus and the centromere. Loci near the centromere would show disomic inheritance and distal loci would show ratios near those expected with tetrasomic inheritance. Exchanges between homeologs would keep the distal part of the homeologous chromosomes from diverging. Homeologs would maintain their integrity because of divergent sequences near the centromeres (see Fig. 4).

A simple extension of this model can also explain the observations of preferential production of nonparental gamete types at the paralogous *Ldh-3* and *Ldh-4* loci in brook–lake trout hybrids (Morrison and Wright, 1966; Morrison, 1970). These loci show only disomic inheritance; we would therefore assume that they lie near the centromere in the diverged chromosomal segment. The excess of nonparental gametes by males would result if there is preferential secondary pairing between homeologs from the same species followed by disjunction so that paired chromosomes pass to opposite poles (Fig. 5). The preferential pairing of homeologs from the same species is expected because they should have greater pairing affinity because of secondary tetrasomic segregation within each species, so that there would be more differences between homeologs from different species.

Many of the results upon which this model is based come from recent work by Wright and his colleagues. They have proposed a model involving diverged and undiverged chromosomal segments to explain unusual inheritance ratios seen at two AAT loci in brook trout (Wright *et al.*, 1980). Their model, however, is more specific and assumes major chromosomal

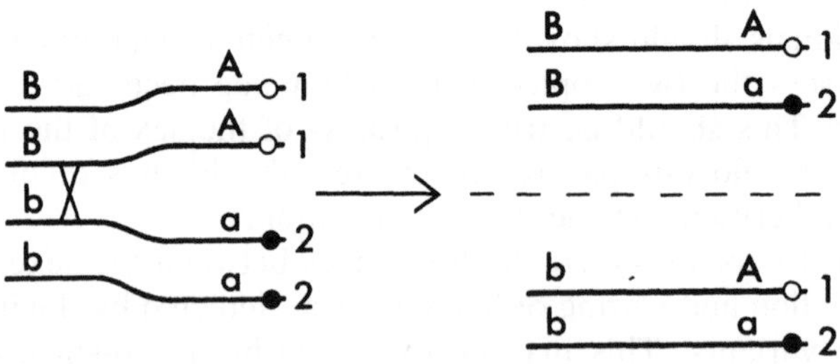

Figure 4. Diagram of proposed model of residual tetrasomic inheritance at isoloci. Loci A near the centromere show disomic inheritance while distal loci B show intermediate disomic–tetrasomic segregation because of secondary pairing during meiosis of homeologs and crossovers.

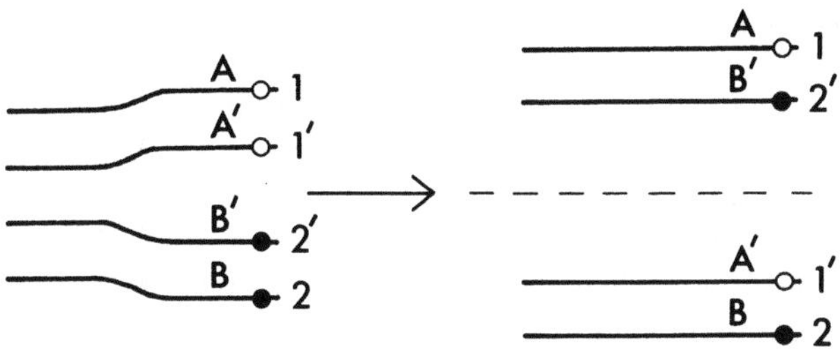

Figure 5. Diagram of proposed model of pseudolinkage resulting in an excess of nonparental gametes produced by hybrid individuals. The A and B chromosomes are from population (or species) 1 and the A′ and B′ chromosomes are from population 2. Preferential secondary homogenetic pairing of homeologs from the same population followed by separation in the first meiotic division of paired chromosomes produces an excess of nonparental gamete types.

rearrangement producing homeologous pairing between acrocentric and metacentric chromosomes. Our model is more general in that the diverged and undiverged chromosomal segments result from differential rates of divergence between homeologs because of differences in the frequency of crossovers between a locus and the centromere.

They have referred to both the partial tetrasomic ratios at isoloci and the excess of nonparental types at paralogous loci as "pseudolinkage" (Wright *et al.*, 1980; May *et al.*, 1982). We believe that the use of the term pseudolinkage should be restricted to the excess of nonparental types at paralogous loci and the term residual tetrasomic inheritance be used to refer to the intermediate disomic–tetrasomic segregation ratios seen at isoloci.

We would like to summarize the evidence for our proposed model of segregation and to suggest possible ways that it could be tested. First, the continued disomic inheritance in females implies that homologs must separate at the first meiotic division in males so that each gamete receives one copy of each homolog. Therefore, pairing of homologs must always occur in males. However, males also produce gamete types expected only with tetrasomic inheritance. Therefore, there must be some secondary pairing of homeologs in males. Preferential secondary pairing of homeologs from the same lineage followed by separation in the first meiotic division of paired chromosomes would produce an excess of nonparental types (i.e., pseudolinkage).

There is additional evidence of this pattern of segregation. Double reduction divisions (the production of gametes containing two copies of an allele carried by sister chromatids) are expected to result if three things occur (Burnham, 1962): (1) formation of multivalents, (2) crossovers between the locus and the centromere, and (3) passage of the chromatid products of such crossovers to the same pole in anaphase I followed by random separation of the chromatids in anaphase II. We have never seen any of the exceptional progeny expected from double reduction in the examination of thousands of progeny at *Mdh-3,* and *Mdh-4* in rainbow trout (Allendorf, unpublished data). However, we know that requirements 1 and 2 do occur in males. Therefore, the homeologous chromosomes that cross over in males must pass to opposite poles.

This model predicts that loci near the centromere should become diploidized more quickly than distal loci. Therefore, the same loci in different lineages should show the same relative rates of divergence. This has been found to be the case. For example, the MDH-B loci are isoloci in all salmonid species that have been studied in all three major salmonid lineages: trout, salmon, and char (as described in this chapter); whitefish (Imhoff *et al.*, 1980); and grayling (Massaro, 1972; Allendorf unpublished data). The muscle AAT loci are also isoloci in most salmonid species and aberrant inheritance ratios have been reported in both char (Wright *et al.*, 1980) and trout (Allendorf and Utter, 1976).

May *et al.* (1982) discuss this similarity among salmonid species because it is not predicted by their model involving specific chromosome rearrangements (since there is no reason to expect the same chromosomes to be involved in such rearrangements in different lineages). Such similarities are expected to result with our model simply as a function of the distance between a locus and its centromere. We do not suggest that our model can or should replace their model. Rather, ours is a simpler (and therefore perhaps more general) model sufficient to explain the available segregation data. Their model, however, is in agreement with recent cytological evidence from studies of meiosis (Wright *et al.*, 1983) and we believe that their model (or a similar one) is required to explain meiotic behavior of some salmonid chromosomes.

Our model thus predicts that loci that are commonly isoloci should be more distal than loci that should have long-established diploidization. We are presently testing this prediction by production of diploid gynogenetic offspring produced by fertilization with irradiated sperm and suppression of the second meiotic division by temperature shock (Chourrout, 1980; Thorgaard *et al.*, 1981). The amount of recombination between a locus and its centromere can be estimated by the proportion of heterozygous progeny from heterozygous females.

Segregation ratios intermediate between disomy and tetrasomy have been reported previously in tetraploid plants (Little, 1945; Gerstel and Phillips, 1958). These ratios have been explained by occasional homeologous pairing. There is no reason that the model we have described should be limited to salmonids. We would expect any tetraploid species in the process of establishing disomic inheritance to show these same effects, i.e., secondary tetrasomic segregation of distal loci and directed tetrasomic segregation in hybrids between different lineages descendent from the same polyploid event. We are not aware, however, of any segregation studies with polyploid plants showing similar results. This is at least partially due to the fact that most inheritance studies with polyploid plants have been done with recessive morphological markers in which only two phenotypes can be distinguished, in comparison to the five phenotypes that can be distinguished with isozyme markers.

Segregation ratios intermediate between disomic and tetrasomic expectations have also been reported in the autopolyploid frog *Hyla versicolor* at two isozyme loci (Danzmann and Bogart, 1982). Surprisingly enough, these loci code for the same enzymes (MDH and AAT) for which tetrasomic ratios have been discovered in salmonids. This implies that perhaps the loci encoding these enzymes are distant from the centromere in both of these groups.

5.5. Implications of Nondisomic Inheritance

Loci in the salmonid genome are inherited in a variety of different patterns, ranging from nonduplicated to at least partially tetrasomic. The possible effects of this mosaic genetic system must be taken into account when considering the inheritance of phenotypic variation in salmonids. In the next sections, we consider the possible effects on the inheritance of monogenic and quantitative traits.

5.5.1. Monogenic Traits

We are aware of only a few descriptions of the genetic control of morphological traits in salmonids. We should consider both the expected frequency and the inheritance of such traits if controlled by a tetrasomic locus or isoloci as compared to a single nonduplicated locus. Most of these traits are rare (e.g., albinism) and are apparently maintained in populations by a balance between mutation and natural selection. The simplest model to consider is one in which the mutant allele is recessive and effectively lethal when homozygous.

At equilibrium, the frequency of such a trait in a population is equal to the mutation rate u and is independent of the mode of inheritance. However, the allelic frequencies and the inheritance pattern of the trait will differ depending upon the mode of inheritance.

At a disomic locus, equilibrium will occur when the frequency of the recessive phenotype q (where q is the frequency of the recessive allele) is equal to the mutation rate. Thus, the allelic frequency at equilibrium will be u. The mode of inheritance is usually determined by crossing the variant phenotype *aa* with the common normal phenotype *AA* and then backcrossing the resulting offspring *Aa* with the variant phenotype. The expected ratios from these matings are presented in Table X.

At a tetrasomic locus, the allelic frequency at equilibrium will be equal to the fourth root of u (assuming no double reduction divisions). Thus, the allelic frequency at a tetrasomic locus will be much higher than at a disomic locus. For example, if u is 10^{-6} for albinism, then the allelic frequency at a tetrasomic locus will be 0.032, as compared to 0.001 at a disomic locus. The inheritance at a tetrasomic locus will be more complex. The expected frequencies of genotypes in a population at equilibrium are as follows:

Genotype	Frequency	
AAAA	p^4	0.878
AAAa	$4p^3q$	0.116
AAaa	$6p^2q^2$	0.006
Aaaa	$4pq^3$	0.000
aaaa	q^4	0.000

Table X

Expected Inheritance Patterns for a Recessive Trait at a Single Locus and Duplicated Loci for Matings between Affected and Normal Parents

Inheritance pattern	Parental genotype		Frequency of progeny phenotype	
	Normal	Affected	Normal	Affected
Nonduplicated	*AA*	*aa*	1.00	0
	Aa	*aa*	0.50	0.50
Isoloci[a]	*AA/AA*	*aa/aa*	1.00	0
	Aa/AA	*aa/aa*	1.00	0
	Aa/Aa	*aa/aa*	0.75	0.25
	Aa/aa	*aa/aa*	0.50	0.50
Tetrasomic[b]	*AAAA*	*aaaa*	1.00	0
	AAAa	*aaaa*	1.00	0
	AAaa	*aaaa*	0.83	0.17
	Aaaa	*aaaa*	0.50	0.50

[a]Assuming no linkage.
[b]Assuming random chromosome inheritance.

If we cross an albino with a normal fish having the genotype *AAAA*, then we expect a normal–albino ratio in backcross progeny of 5:1 and in F progeny of 35:1, as compared to the 1:1 and 3:1 ratios expected at a disomic locus. If, however, we cross an albino with an *AAAa* fish, then we will get a 1:1 mixture of *AAaa* and *Aaaa* fish in the first generation. The first of these genotypes will produce backcross and F progeny ratios identical to those if we initially used a *AAAA* normal parent. The second genotype will produce segregation ratios identical to those expected at disomic locus if the normal parent was *AA*. Thus, tetrasomic inheritance for morphological traits should produce inheritance ratios in which the recessive phenotype is much less frequent than expected with disomic inheritance (Table X).

The situation becomes more complex in the case of isoloci controlling such morphological variation. This system has been considered in some detail by Christiansen and Frydenberg (1977). There is not a single equilibrium solution for allelic frequencies. Rather, at equilibrium, the allelic frequencies will satisfy the equation

$$u = Q_a^2 \, Q_b^2$$

This equation defines a hyperbola in the plane of all possible allelic frequencies. Once allelic frequencies reach equilibrium on this line, they will be free to move along through the effects of genetic drift.

The expected inheritance results again depend on the genotype of the fish used in the original cross. This fish may be of any of the following genotypes: *AABB, AaBB, AABb, AaBb, Aabb,* or *aaBb*. The expected segregation ratios are presented in Table X. As with tetrasomic inheritance, isoloci coding for a recessive trait will result in inheritance ratios in which the recessive phenotype is observed less frequently than expected with disomic inheritance (Table X).

All reported inheritance studies of morphological variants have found nonduplicated disomic inheritance. Both albinism and golden coloration in rainbow trout have been found to be controlled by single Mendelian loci (Bridges and Von Limbach, 1972; Wright, 1972). Kincaid reported an iridescent blue color variant in rainbow trout that appeared to be inherited as a single locus recessive with incomplete penetrance (Kincaid, 1975). We would expect only those loci that have lost duplicate gene expression not to show nondisomic inheritance ratios. Assuming that isozyme loci are representative of the entire genome, we would only expect 30% of all morphological loci to show disomic inheritance.

Why have such nondisomic ratios not been reported for morphological loci? There are several possible explanations. The simplest explanation is that simply by chance a morphological locus retaining duplicate gene expression has not been studied. However, perhaps loci controlling color variation are less likely to retain duplicate gene expression than are isozyme loci. A third possibility is that there may have been some reluctance of investigators to report such unexpected results. There is very little extensive inheritance data available in the literature for any salmonid morphological locus. No conclusions can be drawn until more results are available.

5.5.2. Quantitative Traits

The two principal genetic characteristics of quantitative traits are the amount of genetic variation for a particular trait and the expected response of that trait to either natural or artificial selection. These two characteristics are usually closely related for a particular trait; the presence of more genetic variation is associated with more rapid response to selection. A tetrasomic locus is in some ways a contradiction to this rule. We expect a tetrasomic locus to possess more allelic variation but respond more slowly to selection than a disomic locus, because of the presence of the additional gene copies. This same principle also applies to a trait controlled by isoloci. We examine this effect by considering the relative rates of change expected under natural selection for traits controlled by duplicated and nonduplicated loci.

5.5.2.1. Selection for a Recessive Trait. Assume that selection changes so as to favor a recessive allele that had been previously deleterious. The present fitness of the recessive phenotype is twice that of the other genotypes. Previously, its fitness was zero, so that it was maintained in the population by selection–mutation balance.

Figure 6 shows the response to this situation at a tetrasomic and a disomic locus, using the recurrence equation of Li (1975). It will take an expected 10,000 generations at a tetrasomic locus for this trait to be incorporated into the population. This is almost exactly 10 times longer than expected at a disomic locus. This is true even though the recessive allele was initially much more common at the tetrasomic locus; 12% of the individuals in the initial tetrasomic population carried the recessive allele, as compared to only 2%, in the disomic population. Thus, the tetrasomic possessed more initial variation but responded much more slowly to selection.

This same general situation will also prevail at an isolocus, although the precise dynamics will depend upon the initial frequencies of the re-

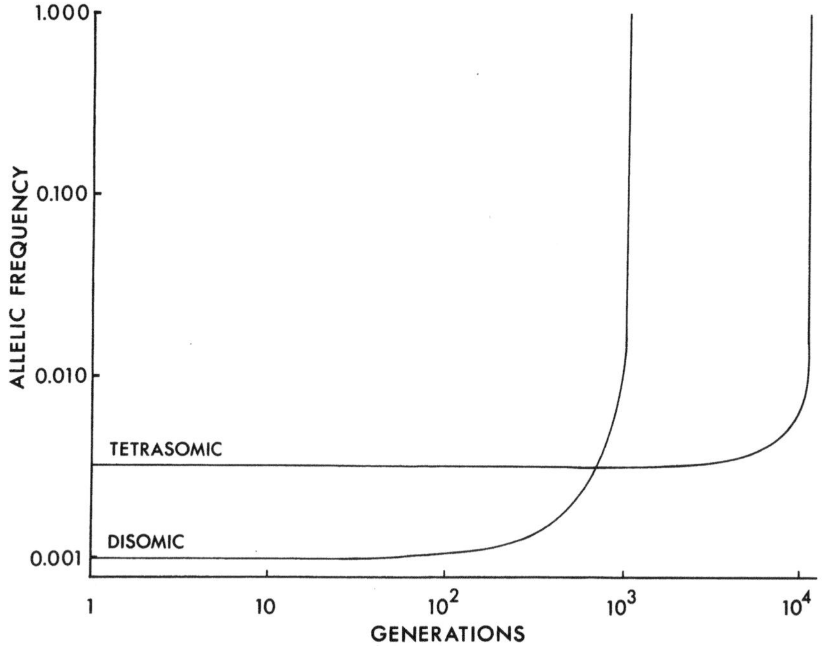

Figure 6. Effects on allelic frequencies of selection for a recessive trait at a disomic and a tetrasomic locus. The recessive genotype has a fitness twice as great as the other genotypes.

cessive alleles at the two loci. However, if the initial frequencies are equal, then the response is similar to a tetrasomic locus. Perhaps more importantly, the same mutation must occur at both loci to be expressed at isoloci. Therefore, in general, a duplicated locus will respond much more slowly to selection for a recessive trait than will a nonduplicated locus.

5.5.2.2. Selection for a Dominant Trait. Assume that a dominant advantageous mutation having a 5% advantage in fitness occurs in a population containing 100 breeding adults. Both tetrasomic and disomic loci respond very quickly to selection for a dominant allele under these conditions (Fig. 7). It is interesting to note, however, that the dominant allele is quickly nearly fixed at a disomic locus but at a tetrasomic locus there is a plateau at a frequency of about 0.8. Thus in this case, the tetrasomic locus will respond to selection nearly as rapidly as a disomic locus and will also retain a greater amount of genetic variation. In addition, a dominant advantageous mutation is more likely to occur at a tetrasomic locus because of the extra copies of the gene.

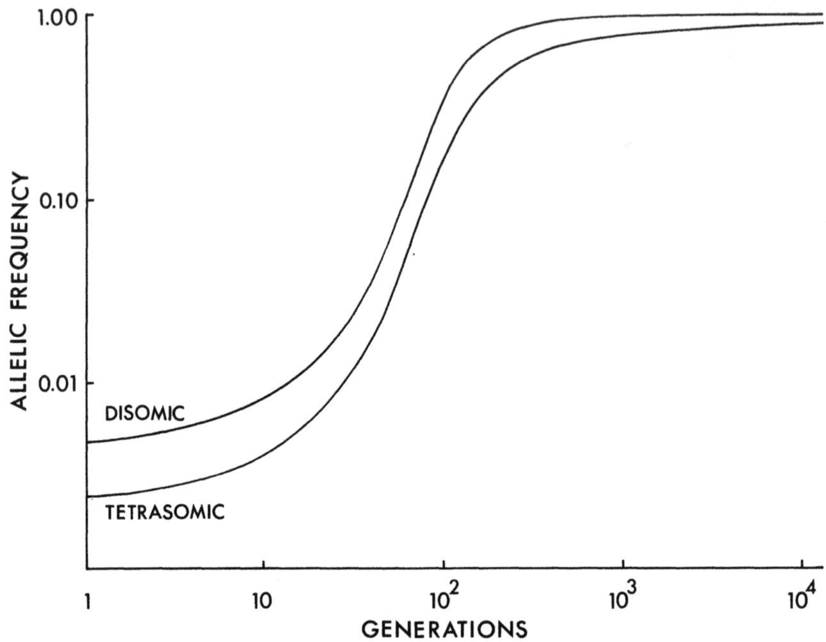

Figure 7. Effects on allelic frequencies of selection for a dominant trait at a disomic and a tetrasomic locus. The dominant genotypes have a fitness of 1.00 in comparison to a fitness of 0.95 for the recessive genotype.

The response at isoloci will be identical to a disomic locus since the mutation is likely to occur only at one locus. Thus, isoloci will phenotypically respond as quickly as a disomic locus but will maintain the original allele at the other locus.

5.5.2.3. Selection for an Overdominant Trait. Both a disomic and tetrasomic locus will initially respond similarly to selection for a mutant allele resulting in heterozygous superiority. This will be true because the initial dynamics will be identical to selection for a dominant trait. However, the equilibrium conditions will be very different. If we assume that both homozygotes have equal fitness, then the equilibrium allele frequencies will be 0.5. In the disomic case, the homozygous phenotypes with reduced fitness will be present at a frequency of $p^2 + q^2 = 0.5$; this is the so-called segregation load. At a tetrasomic locus the homozygous phenotypes will be present at a frequency of $p^4 + q^4 = 0.125$. Thus, the segregation load is much reduced at a tetrasomic locus.

At isoloci, selection will cause the alternative alleles to be fixed at alternative loci so that all individuals in the population will be fixed "heterozygotes'' (Spofford, 1969). The segregation load in the case of overdominant selection at isoloci will be zero.

5.5.3. Inbreeding Depression

Inbreeding depression is caused by two different effects: the fixation of deleterious alleles and the loss of allelic variation at loci where there is hetrozygous superiority (Falconer, 1981). Both of these effects are the result of genetic drift causing random changes in allele frequencies.

Tetrasomic loci are ''buffered'' against the effects of genetic drift by virtue of having twice as many gene copies at a locus. The relative rate of loss of genetic variation at disomic and tetrasomic loci depends upon the population size (Li, 1975, p. 342). With selfing it will take a tetrasomic locus 3.80 times as long to reach the same amount of homozygosity as a disomic locus, in comparison to 2.67 times as long with mating between full-sibs (S. Wright, 1951). Figure 8 shows the relative rate of loss of heterozygosity at a disomic and tetrasomic locus with continuous full-sib mating.

Isoloci are even more buffered against the effects of inbreeding depression. Deleterious recessive alleles would have to become frequent at two loci before their phenotype would be expressed. In addition, because there is no allelic variation present at equilibrium for overdominant selection, such loci will be completely unaffected by inbreeding.

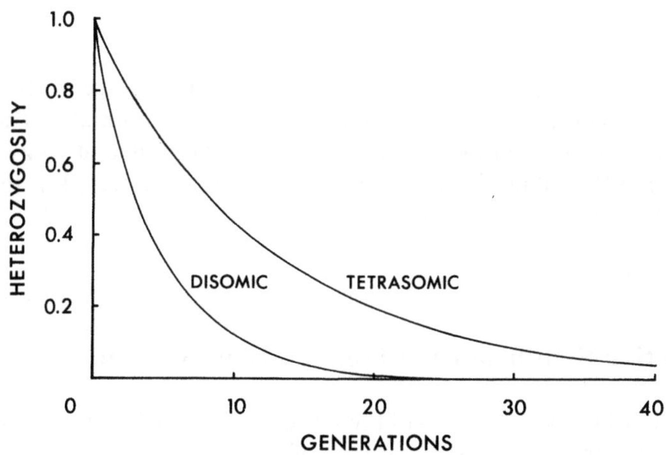

Figure 8. Loss of heterozygosity at a disomic and a tetrasomic locus with continued full-sib mating (assuming random chromosome inheritance at the tetrasomic locus).

5.5.4. Additional Considerations

Our consideration of the potential implications of the unusual genetic system of salmonids suggests a variety of experiments that are needed. First, more inheritance data are needed for morphological traits apparently controlled by single loci. Also, quantitative genetic analyses of salmonids must consider the possible effects of partial tetrasomic inheritance of polygenic traits. The assumption of disomic inheritance when estimating certain parameters (e.g., heritability) may lead to erroneous estimates. For example, the dominance covariance between half-sibs at a duplicated locus will not be zero as it is for a disomic locus. Experiments could be designed to estimate what proportion of the genotypic variation controlling polygenic traits is acting in a simple disomic manner.

The incomplete disomic inheritance of some loci may also have important practical implications. Results with autotetraploid ferns have shown that chromosomes that are inherited disomically may return to partial tetrasomic inheritance in hybrids between different lines (Hickok, 1978). This will have the effect of "releasing" genetic variation that was not present in either of the original lines. For example, consider a recessive trait for which both lines are fixed for normal and mutant alleles (*A1A1a2a2* and *A1A1a2a2*); homeologous pairing in the hybrids will produce gametes with two doses of the recessive allele (*aa*), resulting in recessive phenotypes appearing in the F_2's. Such recessive phenotypes will be at a much higher frequency in the F_2's if the two lines have become fixed for the mutant allele at different loci. The crossing of different lines (or species) of salmonids may therefore produce new strains that have more genetic variation than expected by simply adding the genetic variation of the two original lines.

Perhaps the most surprising thing about the inheritance of phenotypic variation in salmonids is that it appears to be so "normal." The chromosomal evidence and observed patterns of inheritance at isozyme loci indicate that the salmonids have an unusual genetic system. It is tempting to speculate that the evolutionary and domestic success of these fish is at least partially due to their tetraploid ancestry; we will consider this possibility in the next section.

6. Adaptive Significance of Polyploidy in Salmonids

The evolutionary success of the tetraploid salmonids and catostomids stands in contrast to the general lack of long-term evolutionary success of polyploids among reptiles, amphibians, and plants (Bogart, 1980; Stebbins, 1977). The explanations for this success currently fall into the realm of conjecture. Nevertheless, we believe a consideration of the possible adaptive significance of polyploidy in the salmonids may be valuable. Uyeno and Smith (1972) have suggested that the increase in heterozygosity of catostomids is at least partially responsible for the evolutionary success of these fish.

6.1. Short-Term Success

A newly created polyploid is faced with direct competition from its diploid progenitor. To survive, the polyploid must either avoid such competition by being adapted to a different niche or by displacing its diploid ancestral species. This competition should be more of a problem for an autopolyploid than an allopolyploid because of the absence of any genetic distinction between the diploid and polyploid forms. Schultz (1980) has reviewed the evidence of the relative success of polyploid fish and their diploid ancestors.

Perhaps the most important immediate difference between ploidy types is cell size and associated characteristics. Large increases in nuclear DNA amounts are accompanied by increased cell size, decreased metabolic rates, and slower development (Bachmann *et al.*, 1972). Cavalier-Smith (1978) has argued that differences in cell size and development rates are of fundamental importance and that autopolyploidy in plants and animals "may commonly result from selection for increased cell size."

The characteristics accompanying increased nuclear DNA amounts would be favored in *K*-selected species. The relatively few large eggs and the slow development rates of salmonids in comparison to other fish species certainly are in agreement with the hypothesis of Cavalier-Smith. In addition, salmonids have apparently maintained increased cell sizes, unlike tetraploid cyprinid species (Schmidtke *et al.*, 1975). Thus, the ancestral polyploid salmonid may have displaced its diploid progenitor because of differences associated with increased cell size.

Another possible explanation of the short-term success of the polyploid may be the possible asexual intermediate stage. A parthenogenetic form that produces all female offspring is expected to replace the competing sexual form because of the "cost" of producing 50% males (Maynard Smith, 1978). Given that all else is equal, the frequency of parthenogenetic females will continue to increase each generation until the sexual form is eliminated.

6.2. Long-Term Success

There are two conflicting views of the importance of polyploidy for major evolutionary changes. One view holds that the duplicate genes created will increase "evolutionary inertia," thereby reducing the chances of evolving new genetic types (Stebbins, 1977; White, 1978). This is correct in that to replace an allele shared by two disomic loci or a single tetrasomic locus is a much longer process than is replacing an allele at a single disomic locus. Thus, advantageous recessive mutations are much less likely to be successfully incorporated at a tetrasomic locus.

In contrast to this view, however, Ohno has argued that gene duplications are absolutely essential for progressive evolution (Ohno, 1970*a,b*). Ohno sees evolutionary changes as being caused by the addition of genes with new functions. This can only occur following gene duplication so that one copy of the gene can be conserved to perform the initial function of that gene.

Metabolic complexity has evolved by increasing the number and specificity of enzymes controlling biochemical reactions. For example, most of the secreted proteins of vertebrates, although currently widely divergent in function, have apparently originated from a few digestive enzymes secreted by an ancestral species (Hartley, 1974). Similarly, many of the dehydrogenases share homologies that apparently result from their divergence from a single ancestral gene. On a smaller scale, the at least eight loci coding for hemoglobins in humans (Harris, 1980) have all evolved from a single globin gene existing at the time of the origin of vertebrates approximately 500 Myr ago.

How has the additional tetraploid event of 25–100 Myr ago contributed to the evolutionary success of salmonids? The additional gene duplication has resulted in the further specialization of particular enzyme loci. For example, the vertebrate LDH-B gene (Markert *et al.,* 1975) is now represented in salmonids by two separate loci that are distinct in both structure and function. Other nonenzymatic classes of loci may show analogous specializations in place or time of expression.

6.2.1. Specialization of Duplicate Genes

It is tempting to suggest that the success of many salmonid species in living in different environments (e.g., freshwater and saltwater) is related to tetraploidy. Perhaps the salmonids' unparalleled anadromous success results from their having different genes expressed during the freshwater and marine parts of the life cycle.

Such a suggestion is certainly speculative, but it could be tested. One way would be to determine those genes likely to be of importance in surviving in either a freshwater or marine environment. This hypothesis predicts that salmonids would be more likely than other anadromous fishes to have different gene products expressed during the marine and freshwater phases of the life cycle. This could be tested by examining the patterns of tissue-specific expression of duplicated enzyme loci during different phases of the life cycle. We have preliminary evidence of differences between individual Atlantic salmon smolts in the expression of enzyme loci in the kidney (Leary and Allendorf, unpublished data) that may fit this pattern. Another potential way of testing this hypothesis would be to examine the mRNA produced during the marine and freshwater phases of salmonids and other anadromous fishes in critical tissues. This hypothesis predicts that salmonids would be more likely to have greater differences between the mRNA's produced during these two phases.

6.2.2. Evolutionary "Inertia" of a Polyploid Genome

It has been found that the more primitive teleost species tend to have higher cellular DNA content than the more specialized forms (Hinegardner, 1976). The salmonids with their doubled DNA content and primitive morphology are no exception to these observations. One proposed explanation for this relationship is that the presence of multiple copies of genes would have a "buffering" effect on the phenotypic effects of allelic substitutions and thus have a conservative effect on evolutionary change (Pederson, 1971).

We have already seen by looking at the effects of natural selection that such effects do result from the presence of gene duplication. These effects are also the reason that some authors feel that polyploidy in plants has not played an important role in progressive evolution beyond the species level. However, such inertia does not hold for dominant mutations; we would expect advantageous dominant mutations to be incorporated more often in a tetraploid species than a diploid species. Evolutionary conservatism, however, is not necessarily all bad. More specialized forms may be more successful in the short term but also tend to have higher exinction rates. Thus, the "inertia" inherent in the salmonid genome would resist highly specialized morphological adaptations but at the same time may have increased the probability of long-term survival of salmonid lineages.

6.2.3. Population Structure

The genetic population structure of many salmonid species is characterized by many small subpopulations or demes (Behnke, 1972; Ryman *et al.*, 1979). The loss of genetic variation within such isolated demes is a potential problem. The tetraploid genome of the salmonids will have a buffering effect against the deleterious effects of losing genetic variation.

Such isolated demes may therefore be more successful in salmonids than in a comparable diploid species.

Lande (1979) has shown that that rate of fixation of chromosomal rearrangements is inversely related to local deme size. We would therefore expect salmonid lineages to show a high rate of chromosomal divergence. This effect would be intensified by the chromosomal instability following tetraploidy. Thus the high rate of chromosomal divergence in salmonids is perhaps related to their polyploid ancestry and population structure.

Such chromosomal rearrangements usually lower the fitness of heterozygotes and are thought to be an important mechanism promoting speciation. We would therefore expect salmonid lineages to have a high potential for speciation. This notion is supported by the complex patterns of relationship among taxa within salmonid genera (Behnke, 1972).

The long-term evolutionary success of a group of organisms is dependent upon avoiding extinction of species and the creation of new species. We believe that the autopolyploid nature of the salmonid genome may have been important in both of these aspects, resulting in the evolutionary success of the salmonid fishes.

7. Summary

All the fish of the family Salmonidae are apparently descended from a single tetraploid event that occurred 25–100 Myr ago. They differ from the only other known entire family of fish with a tetraploid ancestry, the Catostomidae, in that the salmonids are still in the diploidization process of restoring disomic inheritance. Multivalents have been described at meiosis in males of several salmonid species. The salmonids have undergone rapid chromosomal divergence; chromosome numbers range from 52 in the pink salmon to 102 in the European grayling. This divergence has involved many Robertsonian changes as well as other types of structural rearrangements.

Protein loci in salmonids reflect their polyploid ancestry; only 10 of 33 ancestral loci in the rainbow trout do not show evidence of duplicate gene expression. Fifteen of 33 loci have retained duplicate gene expression but have apparently been completely "diploidized" in that the two remaining loci show structural or regulatory divergence. Eight of the 33 loci have retained duplicate gene expression and show no evidence of structural or regulatory divergence between the remaining duplicates. These pairs of loci sharing structural alleles have been termed "isoloci."

These isoloci have not diverged, because they have not yet been fully diploidized. The MDH-B isoloci in rainbow trout (*Mdh-3, Mdh-4*) show normal disomic segregation in females. However, secondary pairing of homeologs in males coupled with homeologous exchanges between the loci and the centromere produce segregation ratios approaching those expected with tetrasomic segregation.

This meiotic model can also explain the aberrant segregation ratios reported at duplicate loci showing excess of nonparental types (pseudolinkage). Preferential secondary pairing of parental homeologs combined with directed disjunction of paired chromosomes will produce an excess of nonparental types.

The autopolyploid ancestry of the salmonids provides the genetic architecture upon which the evolutionary forces of mutation, natural selection, genetic drift, and migration have acted during the history of these fishes. The presence of diploidized duplicate copies for many genes allows specialization of the two loci for different metabolic functions. This is demonstrated by the differences in tissue-specific expression of many duplicate pairs. We have suggested that the salmonids' unparalleled anadromous success may partially result from the expression of different duplicates during the freshwater and marine stages.

The unusual patterns of inheritance in salmonids have important implications for the evolutionary potential of these fishes. The duplicate copies of many gene loci will allow the accumulation of more genetic variation than in a diploid because of the greater number of mutations and relaxed selection against deleterious mutations. The duplicate copies will also act as a buffer against the harmful effects of inbreeding depression. However, the duplicate copies of many genes may also have a buffering effect on the rate of progressive evolutionary change by natural selection.

ACKNOWLEDGMENTS. One of the authors (FWA) was supported by National Science Foundation grants DEB-8004681 and ISP-8011449 and the other (GHT) by NSF grant PCM-8108787 while this manuscript was being written. We thank Bernie May for his many helpful comments on this manuscript; K. L. Knudsen for drawing the figures; and G. L. Stebbins for suggesting references. S. R. Phelps, R. F. Leary, K. L. Knudsen, and R. Vore have all cooperated in the research included in this chapter.

References

Allen, S. K., and Stanley, J. G., 1978, Reproductive sterility in polyploid brook trout *Salvelinus fontinalis*, *Trans. Am. Fish. Soc.* **107**:474–478.

Allen, S. K., Jr., and Stanley, J. G., 1979, Polyploid mosaics induced by cytochalsin B in landlocked Atlantic salmon, *Trans. Am. Fish. Soc.* **108**:462–466.

Allendorf, F. W., 1975, Genetic variability in a species possessing extensive gene duplication: Genetic interpretation of duplicate loci and examination of genetic variation in populations of rainbow trout, Ph. D. dissertation, University of Washington, Seattle.

Allendorf, F. W., 1978, Protein polymophism and the rate of loss of duplicate gene expression, *Nature* **272**:76–79.

Allendorf, F. W., 1980, Genetic control of tissue specific expression of a phosphoglucomutase locus in rainbow trout, *Genetics* **94**:s51.

Allendorf, F. W., Knudsen, K. L., and Phelps, S. R., 1982, Identification of a gene regulating the tissue expression of a phosphoglucomutase locus in rainbow trout, **Genetics** **102**:259–268.

Allendorf, F. W., and Utter, F. M., 1976, Gene duplication in the family Salmonidae III. Linkage between two duplicated loci coding for aspartate aminotransferase in the cutthroat trout (*Salmo clarki*), *Hereditas* **82**:19–24.

Allendorf, F. W., and Utter, F. M., 1979, Population genetics, in: *Fish Physiology*, Vol. 8 (W. S. Hoar, D. J. Randall, and J. R. Brett, eds.), Academic Press, New York, pp. 407–454.

Allendorf, F. W., Utter, F. M., and May, B. P., 1975, Gene duplication within the family Salmonidae: Detection and determination of the genetic control of duplicate loci through inheritance studies and the examination of populations, in: *Isozymes IV. Genetics and Evolution* (C. L. Market, ed.), Academic Press, New York, pp. 415–431.

Astaurov, B. L., 1969, Experimental polyploidy in animals, *Annu. Rev. Genet.* **3**:99–126.

Bachmann, K., Goin, O. B., and Goin, C. J., 1972, Nuclear DNA amounts in vertebrates, in: *Evolution of Genetic Systems* (H. H. Smith, ed.), Gordon and Breach, New York, pp. 419–447.

Bailey, G. S., Wilson, A. C., Halver, J. E., and Johnson, C. L., 1970, Multiple forms of supernatant malate dehydrogenase in salmonid fishes, *J. Biol. Chem.* **245**:5927–5940.

Bailey, G. S., Tsuyuki, H., and Wilson, A. C., 1976, The number of genes for lactate dehydrogenase in salmonid fishes, *J. Fish. Res. Board Can.* **33**:760–767.

Baker, B. S., Carpenter, A. T. C., Esposito, M. S., Esposito, R. E., and Sandler, L., 1976, The genetic control of meiosis, *Annu. Rev. Genet.* **10**:53–134.

Beamish, R. J., and Tsuyuki, H., 1971, A biochemical and cytological study of the longnose sucker (*Catostomus catostomus*) and large and dwarf forms of the white sucker (*Catostomus commersoni*), *J. Fish. Res. Board Can.* **28**:1745–1748.

Beck, M. L., Biggers, C. J., and Dupree, H. K., 1980, Karyological analysis of *Ctenopharyngodon idella, Aristichthys nobilis,* and their F_1 hybrid, *Trans. Am. Fish. Soc.* **109**:433–438.

Behnke, R. J., 1972, The systematics of salmonid fishes of recently glaciated lakes, *J. Fish. Res. Board Can.* **29**:639–671.

Bogart, J. P., 1980, Polyploidy in evolution of amphibians and reptiles, in: *Polyploidy: Biological Relevance* (H. L. Lewis, ed.), Plenum Press, New York, pp. 341–369.

Booke, H. E., 1968, Cytotaxomonic studies of the coregonine fishes of the Great Lakes, USA: DNA and karyotype analysis, *J. Fish. Res. Board Can.* **25**:1667–1687.

Bridges, W. R., and Von Limbach, B., 1972, Inheritance of albinism in rainbow trout, *J. Hered.* **63**:152–153.

Burnham, C. R., 1962, *Discussions in Cytogenetics,* Burgess Publishing Co., Minneapolis, Minnesota.

Capanna, E., Cataudella, S., and Volpe, R., 1974, An intergeneric hybrid between the rainbow trout and the freshwater char (*Salmo gairdneri* × *Salvelinus fontinalis*), *Boll. Pesca Pisci. Idrobiol.* **29**:101–106.

Cavalier-Smith, T., 1978, Nuclear volume control by nucleoskeletal DNA, selection for cell volume and cell growth rate, and the solution of the DNA *C*-value paradox, *J. Cell Sci.* **34**:247–278.

Cavender, T. M., 1970, A comparison of Coregonines and other Salmonids with the earliest known teleostean fishes, in: *Biology of Coregonid Fishes* (C. C. Lindsey and C. S. Woods, eds.), University of Manitoba Press, Winnipeg, Manitoba, pp. 1–30.

Cherfas, N. B., 1966, Natural triploidy in females of the unisexual form of silver carp (goldfish) (*Carassius auratus gibelio* Block), *Genetika* **2**:16–24.

Chourrout, D., 1980, Thermal induction of diploid gynogenesis and triploidy in the eggs of the rainbow trout (*Salmo gairdneri* Richardson), *Reprod. Nutr. Dev.* **20**:727–733.

Christiansen, F. B., and Frydenberg, O., 1977, Selection–mutation balance for two nonallelic recessives producing an inferior double homozygote, *Am. J. Hum. Genet.* **29**:195–207.

Clayton, J. W., and Franzin, W. G., 1970, Genetics of multiple lactate dehydrogenase isozymes in muscle tissue of lake whitefish (*Coregonus clupeaformis*), *J. Fish. Res. Board Can.* **27**:1115–1121.

Comings, D. E., 1972, Evidence for ancient tetraploidy and conservation of linkage groups in mammalian chromosomes, *Nature* **238**:455–457.

Cuellar, O., and Uyeno, T., 1972, Triploidy in rainbow trout, *Cytogenetics* **11**:508–515.

Danzmann, R. G., and Bogart, J. P. 1982, Evidence for a polymorphism in gametic segregation using a malate dehydrogenase locus in the tetraploid treefrog *Hyla versicolor, Genetics* **100**:287–306.

Davisson, M. T., Wright, J. E., and Atherton, L. M., 1972, Centric fusion and trisomy for the LDH-B locus in brook trout, *Salvelinus fontinalis, Science* **178**:992–994.

Davisson, M. T., Wright, J. E., and Atherton, L. M., 1973, Cytogenetic analysis of pseudolinkage of LDH loci in the teleost genus *Salvelinus, Genetics* **73**:645–658.

Dingerkus, G., and Howell, W. M., 1976, Karyotypic analysis and evidence of tetraploidy in the North American paddlefish *Polyodon spathula, Science* **194**:842–844.

Dobzhansky, T., Ayala, F. J., Stebbins, G. L., and Valentine, J. W., 1977, *Evolution*, W. H. Freeman, San Francisco.

Dunham, R. A., Philipp, D. P., and Whitt, G. S., 1980, Levels of duplicate gene expression in armoured catfishes, *J. Hered.* **71**:248–252.

Echelle, A. A., and Mosier, D. T., 1981, All-female fish: A cryptic species of *Menidia* (Atherinidae), *Science* **212**:1411–1413.

Engle, W., Op't Hof, J., and Wolf, U., 1970, Genduplikation durch polyploide evolution: Die isoenzyme der Sorbitdehydrogenase bei herings- und lachsartigen Fischen (Isospondyli), *Humangenetik* **9**:157–163.

Falconer, D. S., 1981, *Introduction to Quantitative Genetics*, Longman, New York.

Ferris, S. D., and Whitt G. S., 1979, Evolution of the differential regulation of duplicate genes after polyploidization, *J. Mol. Evol.* **12**:267–317.

Ferris, S. D., and Whitt, G. S., 1980, Genetic variability in species with extensive gene duplication: The tetraploid catostomid fishes, *Am. Nat.* **115**:650–666.

Fisher, S. E., Shaklee J. B., Ferris, S. D., and Whitt, G. S., 1980, Evolution of five multilocus isozyme systems in the chordates, *Genetica* **52**:73–85.

Fitch, W. M., 1976, Molecular evolutionary clocks, in: *Molecular Evolution* (F. J. Ayala, ed.), Sinauer, Sunderland, Massachusetts, pp. 160–178.

Flickinger, R., 1975, Relation of an evolutionary mechanism to differentiation, *Differentiation* **3**:155–159.

Gall, G. A. E., Busack, C. A., Smith, R. C., Gold, J. R., and Kornblatt, B. J., 1976, Biochemical genetic variation in populations of golden trout, *Salmo aguabonita, J. Hered.* **67**:330–335.

Gervai, J., Peter, S., Nagy, A., Horvath, L., and Csanyi, V., 1980, Induced triploidy in carp, *Cyprinus carpio* L., *J. Fish Biol.* **17**:667–671.

Gerstel, D. U., and Phillips, L. L., 1958, Segregation of synthetic amphiploids in *Gossypium* and *Nicotiana, Cold Spring Harbor Symp. Quant. Biol.* **23:**225–236.

Gilles, A., and Randolph, L. F., 1951, Reduction of quadrivalent frequency in autotetraploid maize during a period of ten years, *Am. J. Bot.* **38:**12–17.

Gold, J. R., and Gall, G. A. E., 1975, Chromosome cytology and polymorphism in the California High Sierra golden trout (*Salmo aquabonita*), *Can. J. Genet. Cytol.* **17:**41–53.

Gold, J. R., Karel, W. J., and Strand, M. R., 1980, Chromosome formulae of North American fishes, *Prog. Fish Cult.* **42:**10–23.

Goldberg, E., 1965, Lactate dehydrogenases in trout: Evidence for a third subunit, *Science* **148:**391–392.

Gould, S. J., 1980, Is a new and general theory of evolution emerging?, *Paleobilogy* **6:**119–130.

Grammeltvedt, A. F., 1974, A method of obtaining chromosome preparations from rainbow trout (*Salmo gairdneri*) by leukocyte culture, *Norw. J. Zool.* **22:**129–134.

Grell, E. H., 1961, Variations in preferential segregation of chromosome two in triploid females of *Drosophila melanogaster, Genetics* **46:**1267–1271.

Haldane, J. B. S., 1933, The part played by recurrent mutation in evolution, *Am. Nat.* **67:**5–19.

Hartley, B. S., 1974, Enzyme families, in: *Evolution in the Microbial World* (M. J. Carlile and J. J. Skehel, eds.), Cambridge University Press, pp. 151–182.

Harris, H., 1980, *The Principles of Human Biochemical Genetics,* 3rd ed., Elsevier, New York.

Hinegardner, R., 1976, Evolution of genome size, in: *Molecular Evolution* (F. J. Ayala, ed.), Sinaver, Sunderland, Masschusetts, pp. 179–199.

Hickok, L. G., 1978, Homoeologous chromosome pairing and restricted segregation in the fern *Ceratopteris, Am. J. Bot.* **5:**516–521.

Hunter, G. A., Donaldson, E. M., Goetz, F. W., and Edgell, P. R., 1982, Production of all female and sterile groups of coho salmon (*Oncorhynchus kisutch*) and experimental evidence for male heterogamety, *Trans. Am. Fish. Soc.* **111:**367–372.

Imhof, M., Leary, R., and Booke, H. E., 1980. Population of stock structure of lake whitefish, *Coregonus clupeaformis,* in northern Lake Michigan as assessed by isozyme electrophoresis, *Can. J. Fish, Aquat. Sci.* **37:**783–793.

Johnstone, R., Simpson, T. H., Youngson, A. F., and Whitehead, C., 1979, Sex reversal in salmonid culture. Part III, The progeny of sex-reversed rainbow trout, *Aquaculture* **18:**13–19.

Kacser, H., and Burns, J. A., 1981, The molecular basis of dominance, *Genetics* **97:**639–666.

Karbe, L., 1964, Die chromosomenverhaltinisse bei den coregen des bodensees und einiger weiterer voralpiner seen, ein beitrag zum problem der speziation in der gatung, *Coregonus, Z. Zool. Syst. Evolutionforschung* **2:**18–40.

Kijima, A., and Fujio, Y., 1980, Duplicated isozyme loci in chum salmon, *Tohoku J. Agric. Res.* **31:**159–172.

Kincaid, H. L., 1975, Iridescent metallic blue color variant in rainbow trout, *J. Hered.* **66:**100–101.

Klar, G. T., and Stalnaker, C. B., 1979, Electrophoretic variation in muscle lactate dehydrogenase in Snake Valley cutthroat trout, *Salmo clarki* subsp., *Comp. Biochem. Physiol.* **64B:**391–394.

Kobayasi, H., 1976, Comparative study of karyotypes in the small and large races of spinous loaches (*Cobitis biwae*), *Zool. Mag.* **85:**84–87.

Lande, R., 1979, Effective deme sizes during long-term evolution estimated from rates of chromosomal rearrangement, *Evolution* **33:**234–251.

Lee, G. M., and Wright, J. E., Jr., 1981, Mitotic and meiotic analyses of brook trout (*Salvelinus fontinalis*), *J. Hered.* **72:**321–327.

Lewis, W. H. (ed.), 1980, *Polyploidy: Biological Relevance,* Plenum Press, New York.

Li, C. C., 1975, *First Course in Population Genetics,* Boxwood Press, Pacific Grove, California.

Li, W. H., 1980, Rate of gene silencing at duplicate loci: A theoretical study and interpretation of data from tetraploid fishes, *Genetics* **95:**237–258.

Lim, S. T., and Bailey, G. S., 1977, Gene duplication in salmonid fish: Evidence for duplicated but catalytically equivalent A(4) lactate dehydrogenases, *Biochem. Genet.* **15:**707–721.

Lim, S. T., Kay, R. M., and Bailey, G. S., 1975, Lactate dehydrogenase isoenzymes of salmonid fish, Evidence for unique and rapid functional divergence of duplicated H4 lactate dehydrogenases, *J. Biol. Chem.* **250:**1790–1800.

Little, T. M., 1945, Gene segregation in autotetraploids, in: *The Botanical Review,* Vol. XI (E. H. Fulling, ed.), The New York Botanical Garden, Bronx, New York, pp. 60–82.

Liu, S., Sezaki, K., Hashimoto, K., Kobayasi, H., and Nakamura M., 1978, Simplified techniques for determination of polyploidy in ginbuna, *Carassius auratus langsdorf, Bull. Jpn. Soc. Sci. Fish.* **44:**601–606.

Loudenslager, E. J., and Thorgaard, G. H., 1979, Karyotypic and evolutionary relationships of the Yellowstone (*Salmo clarki bouvieri*) and West-Slope (*S. c. lewisi*) cutthroat trout, *J. Fish. Res. Board Can.* **36:**630–635.

Maitland, P. S., 1977, *The Hamlyn Guide to Freshwater Fishes of Britain and Europe,* Hamlyn, London.

Marian, T., and Krasznai, Z., 1978, Karyological investigation on *Ctenopharyngodon idella* and *Hypophthalmichthys nobilis* and their crossbreeding, *Aquacultura Hungarica* **1:**44–50.

Market, C. L., and Faulhaber, I., 1965, Lactate dehydrogenase isozyme patterns of fish, *J. Exp. Zool.* **159:**319–332.

Market, C. L., Shaklee, J. B., and Whitt, G. S., 1975, Evolution of a gene, *Science* **189:**102–114.

Massaro, E. J., 1972, Isozyme patterns of coregonus fishes: Evidence for multiple cistrons for lactate and malate dehydrogenases and achromatic bands in the tissue of *Prosopium cyclindraceum* (Pallas) and *P. coulteri, J. Exp. Zool.* **179:**247–262.

Massaro, E. J., 1973, Tissue distribution and properties of the lactate and supernatant malate dehydrogenase isozymes of the grayling, *Thymallus arcticus* (Pallas), *J. Exp. Zool.* **186:**151–158.

Massaro, E. J., and Markert, C. L., 1968, Isozyme patterns of fishes: Evidence for multiple cistrons for lactate dehydrogenase polypeptides, *J. Exp. Zool.* **168:**223–238.

May, B., 1980, The salmonid genome: Evolutionary restructuring following a tetraploid event, Ph. D. dissertation, Pennsylvania State University, University Park, Pennsylvania.

May, B., Stoneking, M., and Wright, J. E., 1979*a*, Joint segregation of malate dehydrogenase and diaphorase loci in brown trout (*Salmo trutta*), *Trans. Am. Fish. Soc.* **108:**373–377.

May, B., Wright, J. E., and Stoneking, M., 1979*b*, Joint segration of biochemical loci in Salmonidae: Results from experiments with *Salvelinus* and review of the literature on other species, *J. Fish. Res. Board Can.* **36:**1114–1128.

May, B., Stoneking, M., and Wright, J. E., 1980, Joint segregation of biochemical loci in Salmonidae: II. Linkage associations from a hybridized Salvelinus genome (*S. namaycush* × *S. fontinalis*), *Genetics* **95:**707–726.

May, B., Wright, J. E., and Johnson, K. R., 1982, Joint segregation of biochemical loci in Salmonidae. III. Linkage associations in Salmonidae including data from rainbow trout (*Salmo gairdneri*), *Biochem. Gene.* **20:**29–39.

Maynard Smith, J., 1978, *The Evolution of Sex,* Cambridge University Press, London.

Melander, Y., and Monten, E., 1950, Probable parthenogenesis in *Coregonus, Hereditas* **36:**105–106.

Morrison, W. J., 1970, Nonrandom segregation of two lactate dehydrogenase subunit loci in trout, *Trans. Am. Fish. Soc.* **1:**193–206.

Morrison, W. J., and Wright, J. E., 1966, Genetic analysis of three lactate dehydrogenase isozyme systems in trout: Evidence for linkage of genes coding subunits A and B, *J. Exp. Zool.* **163:259–270.**

Muller, H. J., 1925, Why polyploidy is rarer in animals than in plants, *Am. Nat.* **59:**345–353.

Muramoto, J. E., Ohno, S., and Atkin, N. B., 1968, On the diploid state of the fish order Ostariophysi, *Chromosoma* **24:**59–66.

Nelson, J. S., 1976, *Fishes of the World,* Wiley, New York.

Niebuhr, E., 1974, Triploidy in man: Cytogenetical and clinical aspects, *Humangenetik* **21:**103–125.

Norden, C. R., 1961, Comparative osteology of representative salmonid fishes with particular reference to the grayling (*Thymallus arcticus*) and its phylogeny, *J. Fish. Res. Board Can.* **18:**679–753.

Nygren, A., Nilsson, B., and Jahnke, M., 1971*a*, Cytological studies in *Salmo trutta* and *Salmo alpinus, Hereditas* **67:**259–268.

Nygren, A., Nilsson, B., and Jahnke, M., 1971*b*, Cytological studies in *Coregonus* from Sweden, *K. Vet. Samh. Upps. Arsb.* **15:**5–20.

Nygren, A., Nilsson, B., and Jahnke, M., 1972, Cytological studies in Atlantic salmon from Canada, in hybrids between Atlantic salmon from Canada and Sweden and in hybrids between Atlantic salmon and sea trout, *Hereditas* **70:**295–306.

Obruchev, D. V., 1967, *Fundamentals of Paleontology,* Vol. XI, *Agnatha, Pisces,* Israel Program Scientific Translations, Jerusalem.

Ohno, S., 1967, *Sex Chromosomes and Sex-Linked Genes,* Springer, Heidelberg.

Ohno, S., 1970*a*, *Evolution by Gene Duplication,* Springer Verlag, New Y ork.

Ohno, S., 1970*b*, The enormous diversity in genome sizes of fish as a reflection of nature's extensive experiments with gene duplication, *Trans. Am. Fish. Soc.* **99:**120–130.

Ohno, S., 1974, Protochordata, Cyclostomata, and Pisces, in: *Animal Cytogenetics,* Vol. 4, *Chordata 1* (B. John, ed.), Gebrüder-Bornträger, Berlin.

Ohno, S., Stenius, C., Faisst, E., and Zenzes, M. T., 1965, Post-zygotic chromosomal rearrangements in rainbow trout (*Salmo irideus* Gibbons), *Cytogenetics* **4:**117–129.

Ohno, S., Muramoto, J., Christian, L., and Atkin, N. B., 1967, Diploid–tetraploid relationship among old world members of the fish family Cyprinidae, *Chromosoma* **23:**1–9.

Ohno, S., Wolf, U., and Atkin, N. B., 1968, Evolution from fish to mammals by gene duplication, *Hereditas* **59:**169–187.

Ohno, S., Muramoto, J., Steinus, C., Christian, L., Kittrell, W. A., and Atkin, N. B., 1969*a*, Microchromosomes in holocephalian, chondrostean and holostean fishes, *Chromosoma* **26:**35–40.

Ohno, S., Muramoto, J., Klein, J., and Atkin, N. B., 1969*b*, Diploid–tetraploid relationship in clupeoid and salmonoid fish, in: *Chromosomes Today,* Vol. 21. (C. D. Darlington and K. R. Lewis, eds.), Oliver and Boyd, Edinburgh, pp. 139–147.

Okada, H., Matumoto, H., and Yamazaki, F., 1979, Functional masculinization of genetic females in rainbow trout, *Bull. Jpn. Soc. Sci. Fish.* **45:**413–419.

Pederson, R. A., 1971, DNA content, ribosomal gene multiplicity, and cell size in fish, *J. Exp. Zool.* **177:**65–78.

Proudfoot, N., 1980, Pseudogenes, *Nature* **286**:840–841.

Purdom, C. E., 1972, Induced polyploidy in plaice (*Pleuronectes platessa*) and its hybrid with the flounder (*Platichthys flesus*), *Heredity* **29**:11–24.

Raicu, P., and Taisescu, E., 1972, *Misgurnus fossilis*, a tetraploid fish species, *J. Hered.* **63**:92–94.

Rees, H., 1964, The question of polyploidy in the Salmonidae, *Chromosoma (Berl.)* **15**:275–279.

Refstie, T., 1981, Tetraploid rainbow trout produced by cytochalasin B, *Aquaculture* **25**:51–58.

Refstie, T., Stoss, J., and Donaldson, E., 1982, Production of all female coho salmon (*Oncorhynchus kisutch*) by diploid gynogenesis using irradiated sperm and cold shock, *Aquaculture* **29**:67–82.

Ricker, W. E., 1962, Russian–English glossary of names of aquatic organisms and other biological and related terms, Fisheries Research Board of Canada Circular No. 65.

Roberts, F. L., 1968, Chromosomal polymorphism in North American landlocked *Salmo salar, Can. J. Genet. Cytol.* **10**:865–875.

Roberts, F. L., 1970, Atlantic salmon (*Salmo salar* chromosomes and speciation, *Trans. Am. Fish. Soc.* **99**:105–111.

Robins, R. C., Bailey, R. M., Bond, C. E., Brooker, J. R., Lachner, E. A., Lea, R. N., and Scott, W. B., 1980, A list of common and scientific names of fishes from the United States and Canada, 4th ed., American Fisheries Society Special Publication No. 12.

Ryman, N., Allendorf, F. W., and Stahl, G., 1979, Reproductive isolation with little genetic divergence in sympatric populations of brown trout (*Salmo trutta*), *Genetics* **92**:247–262.

Schmidtke, J., and Kandt, I., 1981, Single-copy DNA relationships between diploid and tetraploid teleostean fish species, *Chromosoma* **83**:191–197.

Schmidtke, J., Atkin, N. B., and Engel, W., 1975, Gene action in fish of tetraploid origin. II. Cellular and biochemical parameters in Clupeoid and Salmonoid fish, *Biochem. Gent.* **13**:301–308.

Schmidtke, J., Schmitt, E., Matzke, E., and Engel, W., 1979, Non-repetitive DNA sequence divergence in phylogenetically diploid and tetraploid teleostan species of the family Cyprinidae and the order Isospondyli, *Chromosoma* **75**:185–198.

Schultz, R. J., 1969, Hybridization, unisexuality, and polyploidy in the teleost *Poeciliopsis* (Poeciliidae) and other vertebrates, *Am. Nat.* **108**:605–619.

Schultz, R. J., 1980, Role of polyploidy in the evolution of fishes, in: *Polyploidy: Biological Relevance* (W. H. Lewis, ed.), Plenum Press, New York, pp. 313–340.

Sears, E. R., 1976, Genetic control of chromosome pairing in wheat, *Annu. Rev. Genet.* **10**:31–51.

Selander, R. K., 1982, Phylogeny, in: *Perspectives on Evolution* (Roger Milkman, ed.), Sinauer, Sunderland, Massachusetts, pp. 32–59.

Sezaki, K., and Kobayasi, H., 1978, Comparison of erythrocyte size between diploid and tetraploid in spinous loach, *Cobitis biwae, Bull. Jpn. Soc. Sci. Fish,* **44**:851–854.

Shaklee, J. B., and Whitt, G. S., 1981, Lactate dehydrogenase isozymes of Gadiform fishes: Divergent patterns of gene expression indicate a heterogeneous taxon, *Copeia* **1981**:563–578.

Shaver, D. L., 1963, The effect of structural heterozygosity on the degree of preferential pairing in allotetraploids of *Zea, Genetics* **48**:515–524.

Simon, R. C., 1964, Cytogenetics, relationships and evolution in Salmonidae, Ph. D. Thesis, University of Washington, Seattle.

Smith, L. T., and Lemoine, H. L., 1979, Colchicine-induced polyploidy in brook trout, *Prog. Fish Cult.* **41**:86–88.

Sola, L., Cataudella, S., and Capanna, E., 1981, New developments in vertebrate cytotaxonomy III. Karyology of bony fishes: A review, *Genetica* **54**:285–328.

Spofford, J. B., 1969, Heterosis and the evolution of duplications, *Am. Nat.* **103**:407–430.

Stebbins, G. L., 1977, *Process of Organic Evolution,* 3rd ed., Prentice-Hall, Englewood Cliffs, New Jersey.

Stegeman, J. J., and Goldberg, E., 1972, Inheritance of hexose-6-phosphate dehydrogenase polymorphism in brook trout, *Biochem. Genet.* **7**:279–288.

Stoneking, M., May, B., and Wright, J. E., 1979, Genetic variation inheritance, and quaternary structure of malic enzyme in brook trout, *Biochem. Genet.* **17**:599–619.

Svardson, G., 1945, Chromosome studies of salmonidae, *Rep. Swed. State Inst. Fresh-Water Fish. Res.* **23**:1–151.

Swaminathan, M. S., and Sulbha, K., 1959, Multivalent frequency and seed fertility in raw and evolved tetraploids of *Brassica campestris* var. *toria, Z. Vererbungsl.* **90**:385–392.

Swanson, C. P., Merz, T., and Young., W. J., 1981, *Cytogenetics: The Chromosome in Division, Inheritance and Evolution,* 2nd ed., Prentice-Hall, Englewood Cliffs, New Jersey.

Swarup, H., 1959, Production of triploidy in *Gasterosteus aculeatus* (L.), *J. Genet.* **56**:129–142.

Sybenga, J., 1972, *General Cytogenetics,* American Elsevier, New York.

Thorgaard, G. H., 1976, Robertsonian polymorphism and constitutive heterochromatin distribution in chromosomes of the rainbow trout *Salmo gairdneri, Cytogenet. Cell Genet.* **17**:174–184.

Thorgaard, G. H., 1977, Heteromorphic sex chromosomes in male rainbow trout, *Science* **196**:900–902.

Thorgaard, G. H., 1978, Sex chromosomes in the sockeye salmon: A Y-autosome fusion, *Can. J. Genet. Cytol.* **20**:349–354.

Thorgaard, G. H., and Gall, G. A. E., 1979, Adult triploids in a rainbow trout family, *Genetics* **93**:961–973.

Thorgaard, G. H., Jazwin, M. E., and Stier, A. R., 1981, Polyploidy induced by heat shock in rainbow trout, *Trans. Am. Fish. Soc.* **110**:546–550.

Utter, F. M., Allendorf, F. W., and Hodgins, H. O., 1973, Genetic variability and relationships in Pacific salmon and related trout based on protein variations, *Syst. Zool.* **22**:257–270.

Uyeno, T., 1972, Chromosomes of offspring resulting from crossing coho salmon and brook trout, *Jpn. J. Ichthyol.* **19**:166–171.

Uyeno, T., and Smith, G. R., 1972, Tetraploid origin of the karyotype of catostomid fishes, *Science* **175**:644–646.

Valenti, R. J., 1975, Induced polyploidy in *Tilapia aurea* (Steindachner) by means of temperature shock treatment, *J. Fish Biol.* **7**:519–528.

Valentine, J. W., and Campbell, C. A., 1975, Genetic regulation and the fossil record, *Am. Sci.* **63**:673–680.

Vasilev, V. P., Makeeva, A. P., and Ryabov, I. N., 1975, Triploidy of hybrids of carp with other representatives of the family Cyprinidae, *Genetika* **11**:49–56.

Vervoort, A., 1980, Tetraploidy in *Protopterus* (Dipnoi), *Experientia* **36**:294–295.

Wallace, B., 1963, The annual invitation lecture. Genetic diversity, genetic uniformity, and heterosis, *Can. J. Genet. Cytol.* **5**:239–253.

Walters, M. S., and Gerstel, D. V., 1948, A cytological investigation of a tetraploid *Rhoeo discolor, Am. J. Bot.* **35**:141–150.

White, M. J. D., 1973, *Animal Cytology and Evolution,* Cambridge University Press, London.

White, M. J. D., 1978, *Modes of Speciation,* W. H. Freeman, San Francisco.

Whitt, G. S., 1981, Developmental genetics of fishes: Isozymic analyses of differential gene expression, *Am. Zool.* **21:**549–572.

Wilson, A. C., 1976, Gene regulation in evolution, in: *Molecular Evolution* (F. J. Ayala, ed.), Sinauer, Sunderland, Massachusetts, pp. 225–234.

Wolf, U., Ritter, H., Atkin, N. B., and Ohno, S., 1969, Polyploidization in the fish family Cyprinidae, order Cypriniformes. I. DNA-content and chromosome sets in various species of Cyprinidae, *Humangenetik* **7:**240–244.

Wolf, U., Engel, W., and Faust, J., 1970, The mechanism of diploidization in vertebrate evolution: Coexistence of tetrasomic and disomic gene loci for the isocitrate dehydrogenases in trout (*Salmo irideus*), *Humangenetik* **9:**150–156.

Wolters, W. R., Libey, G. S., and Chrisman, C. L., 1981, Induction of triploidy in channel catfish, *Trans. Am. Fish. Soc.* **110:**310–312.

Wright, J. E., 1972, The palomino rainbow trout, *Pa. Angler Mag.* **41:**8–9.

Wright, J. E., Heckman, J. R., and Atherton, L. M., 1975 Genetic and developmental analyses of LDH isozymes in trout in: *Isozymes III: Developmental Biology* (C. L. Markert, ed.) Academic Press, New York, pp. 375–399.

Wright, J. E., May, B., Stoneking, M., and Lee, G. M., 1980, Pseudolinkage of the duplicate loci for supernatant aspartate aminotransferase in brook trout, *Salvelinus fontinalis, J. Hered.* **71:**223–228.

Wright, J. E., Johnson, K., Hollister, A., and May, B., 1983, Meiotic models to explain classical linkage, pseudolinkage and chromosome pairing in tetraploid derivative salmonids, in: *Isozymes: Current Topics in Biological and Medical Research,* Vol. 10 (M. C. Rattazzi, J. G. Scandalios and G. S. Whitt, eds.), Alan R. Liss, New York, pp. 239–260.

Wright, S., 1951, The genetical structure of populations, *Ann. Eugen.* **15:**323–354.

Zenzes, M. T., and Voiculescu, I., 1975, C-Banding patterns in *Salmo trutta,* a species of tetraploid origin, *Genetica* **45:**531–536.

CHAPTER 2

Tetraploidy and the Evolution of the Catostomid Fishes

STEPHEN D. FERRIS

1. Introduction

1.1. The Role of Gene Duplication in Evolution

The catostomid fishes provide an excellent opportunity to follow the evolutionary fates of duplicate genes. The genomes of these fish are still in a state of flux owing to polyploidization some 50 million years ago. The time of origin of these fish is advantageous in understanding the overall picture of genome evolution in eukaryotes. Catostomids, or the suckers, are not so recent in origin that duplicate genes are identical in structure and regulation, yet not so old that most homologies have been erased by the passage of time. This chapter will explore the many facets of tetraploidy as they effect the evolution in the catostomids, and further, to view the catostomids as a model for genome evolution in other taxa, including the early vertebrates. Finally, I will outline directions that promise to be fruitful in extending the model in the future.

The amplification of genetic material has been intimately connected with the evolution of multicellularity and complex biochemical organization. Studies of the DNA content among animals show a general increase of cellular DNA with organismal complexity (Hinegardner, 1976). This increase can occur by two major mechanisms: regional gene duplication and polyploidy. Ohno (1970) has argued that it is the latter mechanism that has been the really creative force in major bursts of evolution.

STEPHEN D. FERRIS • Department of Biochemistry, University of California, Berkeley, California 94720.

As organisms acquired greater numbers of specialized cell types, gene duplication facilitated the formation of quite different enzymes tailored for these cells, as well as increased regulatory specificity in their time of expression in development (Markert and Ursprung, 1971; Markert *et al.*, 1975; Zuckerkandl, 1978). New enzymes can be formed since the presence of redundant copies allows one sequence to mutate freely and acquire new catalytic functions. The wide variety of dehydrogenases, for instance, probably formed in this way (Eventhoff and Rossman, 1975).

The differential regulation of duplicate loci is theoretically easier to accomplish through polyploidization than tandem duplication because regulatory elements are always doubled. Thus, the evolution of tissue specialization is promoted when the elements diverge. Regional duplications may not always include regulatory elements, and so the scope of differential regulation may be limited (Ohno, 1970). Nevertheless, tandem duplication of genes is widespread (Jeffreys, 1981) and many instances are known where duplicates are expressed at various times in development, e.g., the globin genes (Dickerson and Geis, 1969). There is also growing evidence that duplicate genes may achieve differential expression through exotic mechanisms such as rearrangements or with transposable elements (Haber and Rogers, 1982).

Polyploidy, while a major force in the evolution of higher plants (Stebbins, 1971), is now largely restricted among the animals to lower vertebrates. Higher vertebrates have evolved delicate sex determination based on heteromorphic sex chromosomes, which preclude successful polyploidization (Ohno, 1970). Polyploidy is still common among fish, occurring in some paddlefish (Dingerkus and Howell, 1976) and sturgeons (Nikolsky, 1976), all salmonids (Ohno, 1970), some cyprinids (Makino, 1939; Wolf *et al.*, 1969) and cobitids (Muramato *et al.*, 1967; Ferris and Whitt, 1977*a*), the catostomids (Uyeno and Smith, 1972), and a few poeciliids (Schultz, 1969).

Polyploidization was probably an important factor in the formation of the early chordates (Ohno, 1970, 1974). Consistent with this hypothesis, a number of multilocus isozymes are unlinked in vertebrates (Wheat *et al.*, 1973; Kucherlapati *et al.*, 1974; Whitt *et al.*, 1976). The strongest evidence for an increase in the number of functioning genes comes from studies of specific isozyme loci among fish of increasing evolutionary grades. A striking trend for increased numbers of loci is seen for many enzymes, including lactate dehydrogenase (Markert *et al.*, 1975), creatine kinase (Fisher and Whitt, 1978), glucosephosphate isomerase (Avise and Kitto, 1973), and others (Fisher *et al.*, 1980). Furthermore, these studies reveal a trend toward increasing tissue specificity of the multiple isozymes among fishes of increasing evolutionary specialization. A classic example of loci formed by duplication over 500 million years ago are those for lactate dehydrogenase (LDH) A and B. They are found in essentially all fish and tetrapods. LDH-A is predominantly expressed in muscle and LDH-B in the heart. This widespread "pattern" in phylogeny is evidence that the discrimination of the loci by regulatory genes began early and was preserved in all descendant lineages. In addition, LDH isozymes have evolved characteristically different affinities for lactate and pyruvate, further evidence for early adaptation to different metabolic niches.

The fishes are a valuable model for studying the evolutionary paths of gene duplication. They still retain much "genomic plasticity" as evidenced by their ability to undergo polyploidization (Ohno, 1974) and to hybridize readily between different genera (Whitt, 1981*a,b*). The catostomids allow us to view the short-term consequences of gene duplication, as opposed to the long-established duplications illustrated by LDH-A and B. Catostomids may give us a hint about what happened in our early chordate ancestors. Finally, since there are now many genera of catostomids, it is possible to get a broad evolutionary picture of the changes following polyploidization.

1.2. Origin of the Catostomidae

The catostomids are a coherent family of freshwater fishes, comprised of 11 genera with over 60 species (Miller, 1959). Ten genera are native to North America, and the 11th, the monotypic *Myxocyprinus,* survives as a relic in China. While it is likely that the family originated in Asia, they did not radiate there, possibly because of competition from the diverse array of cypriniforms. Some of these cypriniforms also had evolved polyploidy (the carp and loaches). An evolutionary tree of the catostomid genera based on their morphology is given by Ferris and Whitt (1980). The greatest number of species occur in *Moxostoma* and *Catostomus,* with over 20 species in each. These genera, and *Hypentelium, Xyrauchen,* and *Chasmistes,* are morphologically advanced (Miller, 1959). *Minytrema* and *Erimyzon* are considered to be more intermediate in morphological specialization. These seven genera belong to the subfamily Catostominae. The less specialized genera are *Cycleptus* (Cycleptinae) and *Ictiobus* and *Carpiodes* (Ictiobinae).

The chromosomes of suckers were first described by Uyeno and Smith (1972). Representatives of various genera had 100 chromosomes, or twice the number of most diploid cypriniforms. Uyeno and Smith (1972) concluded that the family descended from a single tetraploidization event, and, based on several fossils, estimated this event at about 50 million years ago.

The initial genome doubling was apparently the result of allopolyploidization, i.e., tetraploidization after species hybridization. Hybridization is a common phenomenon among cypriniform fishes (Hubbs, 1955) and is known in the catostomids (Hubbs *et al.*, 1943; Dauble and Buschbom, 1981). Allopolyploidy has been invoked in the formation of some tetraploid cyprinids (Ohno, 1974). Also consistent with this mode of origin is the absence of multivalents in meiotic spreads of *Erimyzon* (Uyeno and Smith, 1972). Multivalents are usually associated with autotetraploids. Salmonids (Ohno, 1974) and odontophrynid frogs (M. L. Becak *et al.*, 1966) are good examples of presumptive autotetraploids with multivalents at meiosis. For a further discussion of the different modes or origin of catostomids and salmonids the reader is referred to Stoneking *et al.* (1981).

In addition to the chromosome evidence for tetraploidization in catostomids, there is evidence from cellular DNA content. The amount of DNA in cells of *Catostomus commersoni* is twice that of diploid cypriniforms (Uyeno and Smith, 1972). Ferris and Whitt (unpublished) have also observed this high DNA content in other catostomids. The results of fluoremetric assays in a cell sorter are given in Table I. Catostomids have 53–79% of the DNA content of mammals, compared to values from the literature of 20–40% for diploid cyprinids (Ohno, 1970). The covariances of the estimates are too large, however, to detect significant differences among the four species studied.

1.3. Disomic versus Tetrasomic Inheritance

Duplicate genes cannot become fixed in a tetraploid and begin diverging in sequence until they are inherited disomically. In allopolyploids, many chromosomes will already have diverged in structure and thus their genes will be inherited disomically. In autotetraploids, a long time may be required until all chromosomes segregate disomically (Li, 1980). Chromosomal "diploidization" is still occurring in salmonids, apparently through

Table I
Relative DNA Contents of Catostomid and Hamster Cells

Species	Channel number[a]		Percent DNA content of hamster
	Catostomid	Hamster	
Carpiodes cyprinus	39 (6.2)	61 (4.3)	64
Erimyzon oblongus	33 (10.9)	62 (4.0)	53
Moxostoma erythrurum	42 (5.7)	69 (3.2)	61
Catostomus commersoni	48 (7.5)	61 (4.5)	79

[a]Covariances of channel numbers are in parentheses.

numerous rearrangements (Ohno, 1970). The interesting phenomenon of pseudolinkage in salmonids is likely to be a remnant of homeologous chromosome pairing (Wright *et al.*, 1980). Li (1980) has summarized the widely held view that the unequal rate of transition to disomy among chromosomes will be a complicating factor in interpreting the range of functional and regulatory divergence among duplicate genes. This complication is more pertinent to the salmonids than the catostomids. Clearly, more research is needed in both groups regarding the rate of chromosomal diploidization.

For catostomids, available information is consistent with a rapid switch to disomy at most loci. In addition to the constancy of chromosome complements among species and the absence of multivalents, the isozymes themselves furnish evidence of disomic inheritance. Although detailed inheritance studies in catostomids must await the future, much may be inferred about inheritance by studying electrophoretically the ratios of activities in isozyme bands (Allendorf *et al.*, 1975). Twenty loci were examined among catostomids, and none showed evidence for tetrasomic inheritance using the criteria of Allendorf *et al.* (1975). In the vast majority of cases, homologous duplicates in species of different genera also had nonidentical mobilities.* Also consistent with a long period of structural gene divergence is that within each species, duplicate isozymes are resolved electrophoretically (Ferris and Whitt, 1978*a*).

1.4. Early Biochemical Studies

Before reviewing studies on gene duplication, I will briefly recount the initial biochemical studies on catostomids. These studies were done largely before it was known that the catostomids are tetraploids.

Tsuyuki *et al.* (1967) reported allozyme variation at several loci in various common suckers. The most prominent allozyme was in muscle, encoded by a single locus with two alleles. The allozymes are now known to be creatine kinase (Gosselin-Ray and Gerday, 1970; Buth, 1979*a*). Huntsman (1970) looked at serum and muscle proteins in an attempt to find diagnostic markers for taxa. The objective was difficult to achieve, in part because of the disc gel method in use at the time. Nevertheless,

*A minimum estimate of the genetic distance between species may be made even though precise orthologies of duplicates is not known. It is certain that a minimum of one codon substitution has occurred at each locus because no alleles have the same mobility. The Nei (1972) distance between *Hypentelium* and *Minytrema* is 1.13 (Ferris *et al.*, 1979). This distance may be calibrated to about 35 million years (Sarich, 1977). Since these genera belong to one subfamily, an estimate that the entire family originated over 50 million years ago is not unreasonable.

he showed that many loci were polymorphic and that there was a great similarity between *Ictiobus* and *Carpiodes*. Koehn (1966) and Koehn and Johnson (1967) described polymorphic serum esterases, and Koehn (1969*a*) found a cline in esterase allele frequencies in *Catostomus clarkii*. He attributed this cline to temperature selection in headwater verus downstream habitats. A hint that catostomids might have unusual genetics came from serum esterase studies in the subgenera *Pantosteus* and *Catostomus*. Species in *Pantosteus* had a polymorphism for a single esterase-1 locus, whereas a "fixed heterozygote" pattern was found in *Catostomus* (Koehn and Rasmussen, 1967). Koehn and Rasmussen interpreted the data in *Catostomus* as evidence of a tandem duplication. Since the silencing of genes is very frequent in catostomids (Ferris and Whitt, 1977*b*), it is more probable that *Pantosteus* has lost the expression of one duplicate.

A striking array of hemoglobins was observed in species of the genus *Catostomus* (Koehn, 1969*b*), although it was not suspected that the diversity in part might be due to polyploidy. A better understanding of globin complexity resulted from the fine sequencing work of Powers and Edmundson (1972). Although duplicate α-globins were discovered, we cannot now be certain whether their origin was through tetraploidy or unequal crossing over. The latter mechanism is prevalent among multigene families (Zimmer *et al.*, 1980).

One systematic effort deserves mention. Smith and Koehn (1971) were pioneers in the sense that they combined morphological and biochemical traits in one analysis. They studied the systematics of *Catostomus*. One problem with the biochemical data is that of determining the most direct homology among paralogous loci, or in fact, which ones are duplicated and which are not. We will examine these issues in more depth in the following section.

2. Experimental and Theoretical Approaches

2.1. Starch Gel Electrophoresis and Activity Staining

By far the most commonly used technique for comparing gene expression in many species is starch gel electrophoresis. Since the preparation of extracts, buffer and gel conditions, and staining recipes are published elsewhere (Ferris and Whitt, 1977*a*; Buth, 1978; Shaw and Prasad, 1970), they will not be given here.

The majority of proteins used to study gene duplication in the catostomids are glycolytic enzymes and kinases. These proteins were selected because their genetic bases were well known in diploid fishes (Ferris, 1978). This is an important consideration when trying to unravel the complexity of duplicate gene expression [e.g., see Fig. 4 of Ferris (1978)]. The enzymes have also been used in studies of variation in other vertebrates, thus facilitating comparisons between species.

Most of the proteins are visualized by activity staining with specific substrates. Creatine kinase of muscle, on the other hand, may be visualized with a general protein stain, because it constitutes 20% of the soluble protein in that tissue (Gosselin Ray and Gerday, 1970). In the next section I will outline some of the methods used by Ferris and Whitt (1978*a*) in determining the number of genes encoding isozymes in the Catostomidae.

2.2. Determination of the Number of Functional Gene Copies

The ideal way to find the number of loci expressed is through controlled inheritance studies. This has been elegantly done for several salmonid species by Utter *et al.* (1979), Allendorf *et al.* (1975), and Wright *et al.* (1980). A drawback of this approach is that it is time-consuming and not feasible for a broad survey of many species. Nevertheless, the approach seems to be necessary for salmonids because many duplicate genes still have identically migrating gene products. This, as I have mentioned, is not a significant problem in catostomids. Ferris and Whitt (1978*a*) have developed a number of criteria that allow the determination of the number of loci expressed with a high degree of confidence. Four of the major ones are as follows:

1. "Fixed heterozygote" patterns: One of the best ways to tell whether one or two loci are expressed is to examine a sufficiently large number of individuals. If a heterozygote pattern is observed in 10 or more individuals derived from several localities, than the chances are very small that one is observing a large number of heterozygotes at a single locus. Two loci are likely. Comparison of the patterns with diploids should further clarify the number of functioning genes.
2. Ratio of isozyme activities: To use this criterion, one must have a good understanding of the subunit structure of the enzymes employed. For example, a dimeric protein with two alleles will show a 1:2:1 ratio of activity in the heterozygote. Should the duplicate have allelic products with the same mobilities, then "skewed" ratios will result. Allendorf *et al.* (1975) have described the expected ratios under various models of inheritance.
3. Phylogenetic analysis: Comparison of the mobility of an allozyme in a variety of genera can provide indirect evidence on the number of genes. For example, Ferris and Whitt (1978*a*) found that creatine kinase of muscle (CK-A) displayed a single band in many

catostomid species, yet the mobility usually differed between species. Although it is possible that two *Ck-A* loci are encoding identical alleles, it is very unlikely that they would change mobilities concordantly in many lineages. The simplest hypothesis, therefore, is the silencing of a duplicate early in phylogeny.

4. Tissue specificity: The observation of varying ratios of activities of multiple isozymes among tissues is strong evidence for gene duplication. Alleles at a single locus would rarely exhibit this behavior. The criterion of differential tissue expression has been successfully used to ascertain many catostomid duplications (Ferris and Whitt, 1978*a*) and has been used in studies of other tetraploid fishes (Ohno, 1970; Bailey *et al.*, 1976; Crabtree and Buth, 1981).

2.3. Determination of Locus Homologies

Duplicate genes may have two kinds of homology: orthology and paralogy. Those duplicates related by speciation events are orthologous. Genes related by regional events or duplication of a single gene through ploidy changes are said to be paralogous. I extend the logical notation of Engel *et al.* (1973) to keep the relationships clear. Loci denoted by capital letters, e.g., *Ldh-A* and *B*, are related by ancient duplications, perhaps over 500 million years old. They are so uniquely diverged in tissue specificity that it is not too difficult to distinguish their homologies. Recent duplications, namely those of 50 or so million years old and due to polyploidy, are denoted by superscripts, e.g., $Ldh\text{-}A^1$ and A^2. Alleles at a locus may be given lower case letter superscripts, e.g., $Ldh\text{-}A^{2a}$ and A^{2b}. Subunit composition labels on a gel will often have subscripts to signify the number of subunits, such as the homotetramer A_4^{2a}. With this system, the A^2 of species X and Y are orthologous to each other, and the A^1 of X and A^2 of Y are paralogous; A^1 and A^2 of X are also said to be paralogous. Some of the more important means of finding homologies between genes are as follows:

1. Tissue specificity: Loci in fish that are related by ancient duplications usually exhibit a characteristic tissue pattern of enzyme expression (Avise and Kitto, 1973; Shaklee and Whitt, 1981; Fisher and Whitt, 1978). The loci tend to be expressed in the same tissues over broad taxonomic categories, and this is most certainly the case within a family (Shaklee and Whitt, 1981).
2. Immunology: By using antibodies to LDH-A, for example, and taking advantage of its cross-reactivity with related A subunits, one can demonstrate homologies of duplicates quite cleanly. This

has been elegantly done for LDH-B duplicates in the chubsucker (Champion *et al.*, 1974).

3. Relative electrophoretic mobilities: This method is useful for surveys of many species within different genera. For example, at *p*H 7, mitochondrial malate dehydrogenase (M-MDH) almost always migrates cathodally to the cytosolic form. Homology of the gene sets is further clarified because cytosolic and mitochondrial MDH subunits do not form heterodimers.

   Determining orthologies of genes among species is more difficult. In the genera of the advanced catostomids including *Minytrema,* orthologies of *Gpi-A* duplicates could be inferred by statistical analysis of the variance of electrophoretic mobilities of the A^1 and A^2 products, as well as the diminished activity of A^2 (Ferris and Whitt, 1978*a*). For other loci, such as superoxide dismutase (SOD), the duplicate isozymes vary widely in mobilities and occasionally comigrate (within species) such that the correct ascertainment of orthology between species is not possible.

4. Silenced genes: As might be inferred from the previous paragraph, it would be equally difficult to determine which specific duplicate (relative to its ortholog) had become fixed for a null allele in one of two species. This is the rule in catostomids. Buth (1979*b*) has an example where it may indeed be evident which gene was silenced in two closely related species of *Moxostoma. Moxostoma macrolepidotum* and *M. pappillosum* have an adenylate kinase-A isozyme with the same mobility, called A^2. The former has an anodal A^1 in all individuals surveyed, whereas the latter lacks it. Thus *M. pappillosum* is probably fixed for a null allele.

5. Other methods: The homologies of duplicates may also be inferred from measurements of heat stability, *p*H optima, and other kinetic and physical properties. These are more time-consuming and should be confined to closely related species, as it is not known how these traits vary between distant groups.

3. Pathways of Duplicate Gene Evolution

3.1. Gene Silencing

The phenomenon of gene silencing has been amply demonstrated in the catostomid fishes. Over 30 species have been surveyed in North America for gene expression at 20 loci relative to the diploid minnow *Notropis stramineus*. On the average, catostomids express about 50% of their loci in duplicate (range 35–65%). The remaining loci have been fixed for a null allele, and no activity is detected on gels. The number of doubly and singly expressed genes in catostomids has been published elsewhere (Ferris and Whitt, 1977*b*), as have the criteria for resolving the number of loci (Ferris and Whitt, 1978*a*). The level of functioning genes in these fish is comparable to that observed in other tetraploid fish lineages of apparently similar age, for example, the salmonids (Allendorf *et al.*, 1975), and the carps (Ohno, 1974; Ferris and Whitt, 1977*c*).

There is a significant difference in the number of duplicates expressed among morphologically primitive and advanced catostomids. The primitive taxa average 59% retention of duplicate loci, whereas the advanced have 41%. The loss of duplicate gene expression is most extreme in derived species within *Moxostoma* and *Catostomus*. Buth (1979*b*), using a slightly different array of enzymes, found 33% duplication in *M. pappilosum*. Ferris *et al.*, (1982) observed 35% duplication in *Catostomus plebeius*. The range in duplications between the most morphologically primitive *Cycleptus elongatus* and *M. pappilosum* is exactly twofold. Buth's (1979*b*) expanded allozyme survey of a large number of *Moxostoma* species and a large number of loci (41) for two species of *Catostomus* (Crabtree and Buth, 1981) confirms the original estimates of Ferris and Whitt (1977*b*) that both of these genera express about 40% of their genes in duplicate.

The explanation of these trends among primitive and derived species is undoubtedly complex. Is it a coincidence that retention of morphological traits is paralleled by retention of duplicate gene expression? Ohno (1970) has suggested that morphological change could be facilitated by divergence of newly formed regulatory genes after duplication. Evolutionary change in catostomid morphology has not been quantitated rigorously, but consists in part of divergences in the number of fin rays, scale rows, or bones in the skull and Weberian apparatus (Ramaswami, 1957; Nelson, 1948). Clearly it would be valuable to know whether catostomids have an unusually higher rate of morphological divergence. We may be witnessing another facet of regulation by observing changes in duplicate gene expression. As we show later, there is ample evidence for evolution in presumptive regulatory genes. Wilson *et al.* (1977) have speculated that changes in regulatory genes could account for rapid morphological change in vertebrates. Are we seeing a manifestation of rapid regulatory change at the level of gene expression in catostomids of advanced lineages?

Population factors must also be considered in gene silencing. The primitive species, e.g., *Ictiobus*, tend to inhabit larger rivers, which should provide a stabilizing influence on population size over evolutionary time Neutral null alleles would be hard to fix in a large, stable population. The advanced species, on the other hand, may have undergone more bottlenecks as they invaded more specialized niches, such as mountain streams. Vicariant speciation, mediated by stream capture, is common in freshwater fish. Small numbers of founders would be predicted. The fixation of neutral or even slightly deleterious nulls might be accelerated. This topic has been discussed at length by Ferris *et al.* (1979). Accelerated genetic change has indeed been documented among mountain species of *Moxostoma* (Buth, 1979*c*). The change was in the form of allozymically detected charge alterations, not null alleles. However, the behavior of many nulls might be similar.

Is there a selective advantage to null alleles? One may envision a selective advantage to the organism in ridding itself of "excess" copies of genes encoding isozymes that have failed to diverge in structure or function to fit new metabolic niches. Selection for efficiency of protein synthesis might be stronger in fish colonizing marginal habitats.* But until we know more about the energetic cost of null mutations the question will have to remain unanswered.

Much has been discussed about the potential roles of subunit structure and enzyme function in allozyme variation. Insofar as we are examining variation of null alleles, do these factors play a role in catostomids? In the first instance, no correlation has been found between the extent of silencing of a locus among species and subunit size (Ferris and Whitt, 1980). In the second case, we note first that studies of enzyme polymorphism in other vertebrates reveal a correlation between mean heterozygosity and enzyme function (Johnson, 1974). In catostomids, the loci examined may be divided into those encoding glucose-metabolizing enzymes and those encoding other enzymes. With these divisions, there is no correspondence between enzyme type and extent of silencing among species. In fact, all enzymes except two, CK-B and SOD, have had one copy eliminated in at least one species. Even CK-B and SOD are not immune to silencing, as they exist in the single state in either tetraploid cyprinids (Ferris and Whitt, 1977*c*) or salmonids (Allendorf *et al.*, 1975). Thus no enzyme has an absolute requirement to remain duplicated in tetraploids (cf. Li, 1980).

Finally, the loss of duplicate gene expression has been a continuous process in the phylogeny of the catostomids. We know that loss began almost immediately after polyploidization, since four loci are expressed singly in all species (Ferris and Whitt, 1977*a*). In some cases, species of

*Selection for ridding the excess DNA itself is a different matter. It is likely to be weak at best since most eukaryotes carry tremendous amounts of selfish DNA (Doolittle and Sapienza, 1980).

a single genus have lost extra genes, such as the *Gpi-A*[2] in *Erimyzon* (Ferris and Whitt, 1978*a*). Still more recent losses are indicated by the loss of *Gpi-B* in *Pantosteus* (Ferris and Whitt, 1978*a;* Crabtree and Buth, 1981), and probably transferrin (Koehn and Rasmussen, 1967).

3.2. Molecular Basis of Gene Silencing

The silencing of duplicate genes may occur at any of a variety places in the genetic apparatus. Nulls may be traced to proteins or nucleic acids. Each will be considered in turn.

3.2.1. Nulls Detectable at the Protein Level

In this section I will consider two broad categories of nulls, those due to protein defects, and those in which proteins are made but are regulated or altered in such a way as to escape detection. They are in a sense "false nulls."

We would score a false null if a duplicate had diverged such that its encoded protein no longer operated on the normal substrate. Ferris (1978) provides arguments that this is unlikely for the conservative glycolytic and energy metabolizing enzymes employed. Although 50 million years is long enough to accumulate sufficient numbers of mutations for charge differences, it is unlikely that complete changes in substrate specificity would happen in this length of time. Changes of this nature might be possible in rapidly evolving esterases, however. I attempted to address the question of substrate affinity by staining LDH with a substrate analogue, α-hydroxybutyrate, and alcohol dehydrogenase (ADH) with a variety of higher alcohols (Ferris, 1978). For those instances where a single locus was expressed, no new isozymes appeared with the analogues; in those examples with duplicate isozymes, only minor shifts in the ratios of their activities are seen.

A second outcome would be that an isozyme of a duplicate pair is active *in vivo,* but its stability *in vitro* is so brief that in the hour between extraction and gel loading it has lost all activity. Since the extracts are always kept cold, this possibility seems slight, yet there is one instance where long periods of time *in vitro* at an elevated temperature results in the loss of activity of a duplicate gene product (see Section 3.3).

The isozymes may have evolved extreme differences in tissue expression and one of them no longer appears in the usual tissue. This source of "error" is minimized since Ferris and Whitt (1979) have generally surveyed 10 tissues per enzyme. We feel it is likely that an extremely diverged duplicate would have been detected if it had shifted its expression to another tissue. A more troublesome possibility is that we are missing some duplicates that are transiently expressed in early development. Champion *et al.* (1974) have one instance where this appears to have happened for GPI. More research is needed before we can estimate this fate of gene expression.

A duplicate locus could be expressed at such a low level within a cell that its activity is below the threshold of detection by activity staining. If this is true, we must question whether the isozyme plays any significant role in metabolism, and is not in fact equivalent to being silenced.

We are able to draw stronger conclusions regarding the mechanism of silencing involving protein inactivation. If a point mutation destroys the active site, for example, nonfunctional protein (or CRM) would be produced. Small deletions or frameshifts would also yield CRM. In many events, the inactive proteins will retain the capacity to form heteropolymers with active proteins (Schwartz and Sofer, 1976). This feature should help us recognize null alleles. Seventeen of the proteins studied by Ferris and Whitt (1977*b*) were multimeric. Over 1000 individuals have now been surveyed by Ferris and Whitt (1978*a*), Buth (1978), Crabtree and Buth (1981), and Ferris *et al.* (1982). In only one species is there solid evidence for a null allele, and the encoded protein can still form a heterodimer (see Section 4.2). To this may be added the LDH-B null described in carp (Engel *et al.*, 1973). This protein apparently has no ability to form heteropolymers with active subunits. Finally, there is creatine kinase-A (CK-A) in catostomids. In catostomids with two CK-A isozymes visualized by activity staining, there are two identical bands after staining the gel with general protein (recall that CK is very abundant in muscle). Interestingly, in those species with one active band, there is but one corresponding protein stained band, never two (Ferris and Whitt, 1978*a,b*). These species belong to the Catostominae, and the simplest explanation is that CRM was eliminated prior to the radiation of this subfamily over 35 million years ago.

Many of the mutations causing silencing undoubtedly are occurring at regulatory loci. We do not know what fraction of all mutations originate in these regions. We can speculate that regulatory silencing must be rather frequent, based on evidence from other organisms. Myomorph rodents lack LDH-B activity in erythrocytes, whereas other rodents have activity in this tissue. Thus, a limited silencing mechanism is operating here (Shows *et al.*, 1969). Some loci in salmonids have been singled out as being far along in the process of silencing. The LDH-A duplicate, which is weakly expressed in all tissues (Lim and Bailey, 1977), is a good candidate. Measurements of subunit specific activity reveal that the reduced activity of certain isozymes on gels is due to reduced levels of subunits in the cytoplasm (a regulatory change) rather than reduction in activity of the subunits themselves. We may also gain some insight from recent work on mammalian pseudogenes. These genes are silenced duplicates that have arisen from regional unequal crossing-over events. Mutations in both regulatory and structural cistrons are known (Ottolenghi *et al.*, 1982) in thalassemias.

What is the likely course of events in null fixation? First, new mutations will produce nonfunctional protein. In view of the smallness of the active site as a "target," inactivation will only occasionally affect this region (and heteropolymers could still form). More frequently, mutations will affect the broadly distributed positions controlling protein tertiary and quaternary structure (reducing or eliminating the ability of the subunit to participate in the formation of heteropolymers). DNA rearrangements and regulatory mutations would soon act to eliminate CRM altogether. Ultimately, over time in excess of 50 million years, the DNA would gradually be eliminated or converted to some entirely new function.

3.2.2. Nulls Detectable at the Nucleic Acid Level

A variety of events could cause null expression to begin with mRNA. These include the production of incomplete message, message that cannot be bound to the ribosome, or messages bearing mutations that cause suppression of the normal stop codons. A recently discovered mechanism is the failure of correct splicing of intervening sequences. Mutations in globin genes are known with this defect (Busslinger *et al.*, 1981).

At the DNA level, nonsense mutations could result in chain termination prematurely. The DNA may be entirely deleted, partially deleted, or rearranged. The latter events are frequent in globin gene evolution (Zimmer *et al.*, 1980). The former event is not thought to be common in catostomids, because their DNA contents are at tetraploid levels. DNA loss from tetraploid frogs of recent origin is known, however, and the amount was 10% (Bachmann and Bogart, 1975). The loss may have easily arisen in satellite or repeated sequences. We note that large changes in DNA content of some amphibians has not resulted in changes at the isozyme level of expression (Comings and Berger, 1969).

Before reviewing evidence for substantial divergence of regulatory genes in catostomids, I will digress briefly to view the evidence for divergence in sequence at protein-coding loci.

3.3. Structural Divergence of Proteins

The most striking evidence for protein divergence of considerable extent in catostomids is that most duplicate isozymes have acquired charge substitutions with different electrophoretic mobilities. In fact, this has been exploited to gain an independent estimate of the age of the Catostomidae of about 50 million years. By contrast, many duplicates in salmonids still encode alleles with the same mobility, possibly because the loci have reached disomy only recently. Most of our knowledge of the evolution of enzyme kinetics in polyploids comes from studies of salmonids. These studies suggest that protein divergence can follow several routes. The elegant work of Bailey and co-workers (e.g., Lim *et al.*, 1975) has shown that duplicate LDH-B loci have diverged considerably in Michaelis constant K_m for lactate. Similar data for MDH-B duplicates fail to show any evidence of such divergence. Surprisingly, the substantial shifts in kinetic properties of LDH-B's have occurred with only minor changes in amino acid composition (Lim *et al.*, 1975) compared to the large divergence in amino acid composition of LDH-A and B (Pesce *et al.*, 1967).

There is definite evidence of structural divergence of proteins in catostomids from analysis of *in vitro* stability of subunits. As Fig. 1 shows, in *Hypentelium nigricans* M-MDH, the less anodal form gradually loses activity upon incubation at 37°C. After 14 h, it has lost all activity, while the anodal form is nearly unchanged! I hypothesize that the slower form, presumably a dispensible copy, has acquired a mutation causing reduced binding between subunits. Thus, the protein denatures more readily.

3.4. Evolution of the Regulation of Duplicate Genes

We have seen that structural genes have diverged subsequent to the polyploidization. In a tetraploid event, of course, regulatory genes are

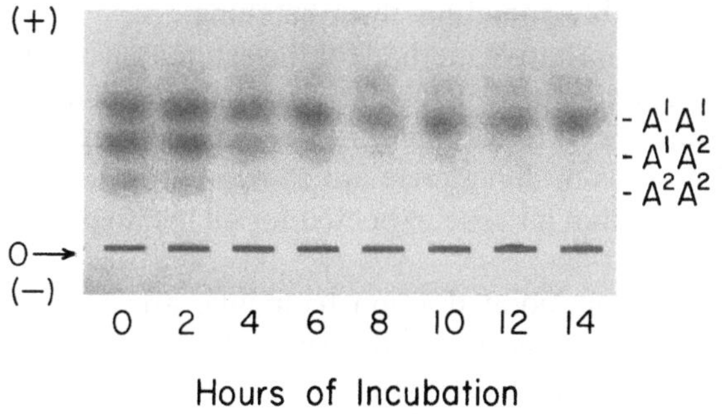

Figure 1. Differential *in vitro* stability of duplicate mitochondrial malate dehydrogenase isozymes. Crude extract was prepared from brain tissue of *Hypentelium nigricans* and then incubated from 0 to 14 h at 27°C and placed on ice. The A^2 isozymes have lost all activity by 14 h.

duplicated as well and will begin diverging. One of the most exciting findings in the catostomids is that there are dramatic and extensive changes in relative expression of duplicates both within and among tissues. For the first time, these changes were quantitated, and in many genera, allowing a phylogenetic picture to emerge.

The differential expression of duplicates is basically a developmental phenomenon. Most of the differences seen among isozyme ratios in the adult tissues studied by Ferris and Whitt (1979) must have been established sometime in the early ontogeny of the fish. This problem was addressed by Champion *et al.* (1974) in following the ontogeny of expression in the lake chubsucker (*Erimyzon sucetta*). Their research was among the first to recognize gene duplications in the catostomids. Using antibodies for LDH-A and B subunits, they showed that the unusual complexity of isozymes of LDH was attributable to duplicate LDH-B loci and a single LDH-A locus (Champion *et al.*, 1974). Extracts were prepared from various developmental stages of lake chubsuckers. The changes in levels of different isozymes at different developmental periods were then observed after electrophoresis of the extracts. Champion *et al.* (1974) were mainly concerned with the differential expression of ancient duplicates such as MDH-A and B or LDH-A and B. Resolution of the buffer systems at that time was not adequate to resolve some loci that are now known to be duplicated, e.g., CK-B; nevertheless, two sets of duplicates could be clearly followed in development. These were LDH-B and GPI-B. In the first instance, duplicates were expressed nearly equally in the embryos. Duplicates are also expressed in adults, but in less than equal amounts (Champion *et al.*, 1974; Ferris and Whitt, 1979). In the case of GPI-B, embryos express both loci equivalently, but adult tissues show only one isozyme in muscle. Thus, there is good evidence that one of the genes is switched off, probably sometime after hatching.

A rapid and quantitative method for the estimation of relative isozyme activities was developed by Klebe (1975). This method was utilized by Ferris and Whitt (1979) to estimate the degree of divergence of duplicate isozymes. The ratios of divergence are compared to the ideal ratio of 1:1 gene expression within a tissue expected for all loci immediately after the polyploidization.

Details of the endpoint dilution technique are described by Klebe (1975). The assumption is made that subunit activity is proportional to the amount of subunit, and not due to catalytically more efficient proteins. Evidence that this applies widely is inferred from Lucchesi and Rawls (1973), Lim and Bailey (1977), and Nadal-Ginard (1978). Although the assumption is well founded, it must be left for the future to elucidate what proportion of weakly staining duplicate isozymes are in fact due to less efficient proteins, and not to lowered subunit numbers (the two are not mutually exclusive).

Ferris and Whitt (1979) surveyed 15 representative catostomids with the aim of understanding the nature of differential expression of duplicate genes. The key result was that the majority of duplicates are differentially expressed within and among tissues. Thus, they have come to be discriminated by regulatory genes. First let us consider the qualitative patterns of expression among tissues, and then the ratios within tissues.

3.4.1. Patterns of Expression

The kinds of patterns found are classified into four arbitrary but operationally useful categories. "Nondivergence," the primitive condition, is when equal activities of duplicates prevail in all tissues. An example is shown in Fig. 2A. In the "unidirectional" case, one of the isozymes consistently predominates in activity in all tissues in which the

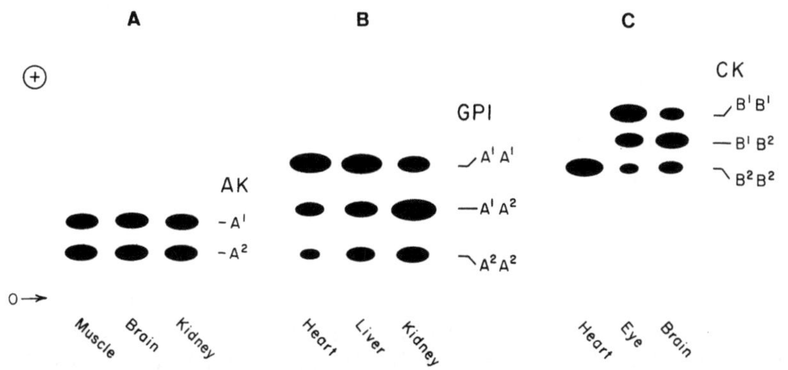

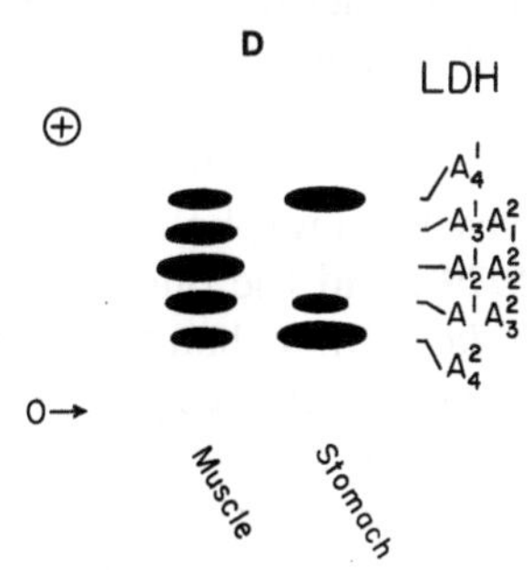

Figure 2. Four major patterns of gene expression in the Catostomidae. (A) Nondivergent: adenylate kinase-A isozymes from tissues of *Ictiobus bubalus*. (B) Unidirectional divergence: glucosephosphate isomerase-A isozymes of *Hypentelium nigricans*. The A^1 isozymes usually predominate over the expression of the A^2 isozymes. (C) Bidirectional divergence: creatine kinase-B expression in *Carpiodes cyprinus*. The B^1 isozyme predominates in the eye, and the B^2 is exclusively expressed in heart. (D) Complex pattern of lactate dehydrogenase-A expression in *Cycleptus elongatus*. Random assembly of A subunits occurs in muscle, but in stomach the subunits appear to be spatially isolated, since heterotetramers do not form to a significant extent. Redrawn with permission from Ferris and Whitt (1979).

two are unequally expressed. An example of this kind of divergence is given in Fig. 2B. The third, or "bidirectional" divergence, is a more sophisticated kind of regulatory pattern where one isozyme predominates in some tissues and the duplicate predominates in other tissues. An example is in Fig. 2C. The last category, "complex," is known from only one example in catostomids. It is a bidirectional pattern with restricted heteropolymer formation. This is seen in Fig. 2D. The enzyme is LDH-A, and in muscle the duplicates A^1 and A^2 are about equally expressed and the subunits have random access to one another. All heterotetramers are formed as expected. In stomach (Fig. 2D, right), the heteropolymers are nearly all absent, yet the homotetramers are active (Ferris and Whitt, 1979).

About 19% of a total of 140 patterns were of the nondivergent type, 64% unidirectional, and 21% bidirectional. The regulatory implications of these patterns are discussed in detail by Ferris and Whitt (1979). The complex LDH-A's are possibly a kind of regulation where the duplicates are now restricted to different cell types within stomach. Similar restricted tissue expression has been found for LDH-A and B in mouse kidney cells (Nadal-Ginard, 1978). Lastly, it should be noted that the lack of heteropolymer formation between duplicate CK-A's in primitive catostomids is not a complex pattern that evolved within the catostomids. A survey of many teleost families indicates that the failure to form heterodimers is widespread, and thus arose early in phylogeny (Ferris and Whitt, 1978).

3.4.2. Quantitation of Isozyme Regulation

Based on the method of Klebe, we know the ratios of activities of duplicate isozymes in tissues from 15 species. The distribution of the ratios in 864 tissues is given in Table II. The majority of tissues, 59%, have unequal ratios of subunit activities. The average ratio was 1:2. Let us consider in turn how these ratios vary over different species, enzymes, and tissues.

Species: The species with the least amount of isozyme divergence was the morphologically primitive *Ictiobus bubalus* (Ferris and Whitt, 1979), and the one with the most was the advanced *Catostomus discobolus*. Other than these extremes, there was no correspondance between primitiveness (as judged by retention of duplicates) and the retention of primitive patterns of isozyme regulation (i.e., close to 1:1 ratios).

The results of studies from the Catostomidae are shown in Table III. Interestingly, the lack of correspondence between the factors above is contrary to that seen by Fisher and Whitt (1978), where primitive fishes

Table II
Distribution of Subunit Activity Ratios for Duplicate Genes for 15 Enzymes and 15 Species[a]

Ratio of subunit activities	Number of tissues with ratio
1.0	354
1.01–1.5	120
1.51–2.00	123
2.01–3.00	87
3.01–4.00	91
4.01–6.00	41
6.01–8.00	17
8.01–16.00	30
16.01–32.00	1
Total	864

[a]Reprinted with permission from Ferris and Whitt (1979).

also retained primitive patterns of gene expression. The discrepancy may be due to the sampling of taxa; those studied by Fisher and Whitt (1978) were on the average older and much more divergent morphologically.

Enzymes: The enzymes were ranked according to their degree of departure from 1:1 ratios in tissues. Creatine kinase-A and acid phosphatase-A were the most similar to 1:1, while LDH-A and GPI-A were most divergent. Their ratios were nearly 1:3 (Ferris and Whitt, 1979). The differences among the enzymes may reflect differences in the receptivity of regulatory genes for these loci to evolutionary change.

Tissues: There is some evidence for a "tissue effect" in the ratios of duplicate genes expressed among tissues. Brain tends to have closest to the 1:1 ratio, i.e., 1:1.5, whereas liver is the most divergent, with 1:2.3. We do not know why these differences exist. Perhaps the regulatory genes or their products are sensitive to different levels of effector molecules in the various cell types. The fact that protein turnover rates are low in brain and high in liver (Ferris and Whitt, 1979) may also play a role. Slight shifts in protein stability might be exaggerated in a tissue such as liver, thus resulting in a more divergent ratio of isozymes. As Ferris and Whitt (1979) have argued from examples elsewhere in the animal kingdom, this mechanism pobably does not play a major role. More research is needed on the factors affecting tissue ratios in catostomids, and how the ratios change throughout development. Finally, the ratios of isozyme divergence

Table III
Average Divergence of Duplicate Gene Expression for All Enzymes and Tissues in 15 Species of Catostomids[a]

Species	Percent duplicate genes expressed	Ratio of divergence of duplicate genes
Cycleptinae		
Cycleptus elongatus	65	2.03
Ictiobinae		
Ictiobus bubalus	60	1.18
Carpiodes cyprinus	55	2.02
Catostominae		
Erimyzon sucetta	45	1.88
Erimyzon oblongus	45	1.70
Erimyzon tenuis	35	1.52
Minytrema melanops	35	1.64
Moxostoma duquesnei	35	1.99
Moxostoma erythrurum	35	1.55
Hypentelium nigricans	35	1.79
Catostomus commersoni	50	2.08
Catostomus catostomus	45	2.12
Catostomus discobolus	40	2.37
Chasmistes brevirostris	45	2.01
Xyrauchen texanus	50	1.80

[a]Reprinted with permission from Ferris and Whitt (1979).

within tissues are not correlated with subunit size or metabolic type of the enzyme [i.e., glucose- versus nonglucose-metabolizing (Ferris and Whitt, 1979)].

All the above evidence points to a predominantly random pattern of change among presumed regulatory genes. But is it really random? Bailey *et al.* (1976) speculate that the less active duplicate of LDH-A in salmonids is well on the way to being eliminated. This is certainly a plausable idea. If all gene silencing begins by mutations in isozymes that are already weakly expressed, then one would expect more divergent ratios in species that have already silenced many genes. Ferris and Whitt (1979) observed no such trend. While the proposal may apply to some duplicates, especially at this relatively late stage in catostomid phylogeny [see also Zuckerland (1978)], it probably is not universally applicable. For most of catostomid evolution, nevertheless, we may say that a silencing mutation can strike in any species, at any time, and at almost any locus. Some of the constraints on silencing, such as tissue differentiation or regulation of dosage levels, will be taken up subsequently.

3.4.3. The Time of Appearance of Regulatory Patterns in Evolution

Silencing has been going on throughout much of catostomid evolution. In a few instances we may infer the relative time of appearance of specific patterns. The tissue-specific pattern of CK-B in Fig. 2C is found in almost all catostomids and therefore must have appeared very soon after polyploidization. The rapid evolution of duplicate gene regulation has also been inferred in recent tetraploid loaches (Ferris and Whitt, 1977*a*). The GPI-A pattern, in which the less anodal duplicate isozyme is very weakly expressed, or is even absent in *Erimyzon,* is common to all the species of the Catostominae, and so presumably arose somewhat later in phylogeny. Other patterns have arisen more recently and are restricted to certain genera or species (Ferris and Whitt, 1979). Thus, "regulatory evolution" of duplicates, whether the result of selection or not, has been proceeding from ancient to recent times.

May silenced genes ever be reactivated? Koch (1972) presents theoretical arguments that genes may be silenced for a short time and then, after accumulating "forbidden" mutations, may reemerge with a new function. Li (1980) has argued that once a gene is silenced, the high neutral mutation rate of DNA would very quickly render sequences meaningless. There is indeed evidence for this among mammalian pseudogenes (Li *et al.,* 1981).

At the experimental level, we have slight evidence for reversals in silencing. A parsimony ("cladistic") analysis of silencing events in catostomids shows that at least one locus may have been reactivated after silencing in *Minytrema* (Ferris and Whitt, 1978*a*). Similarly, Buth (1982) has good evidence through similar analyses for a duplicate gene in an otherwise highly derived lineage, *Moxostoma lachneri.* The gene for GPI-B is silenced in all of its close relatives, and thus this species may have reactivated it. A problem in interpreting these kinds of data is possible confusion with recent tandem duplications of singly expressed loci. Ferris and Whitt (1978*a*) have argued strongly against this being a common phenomenon in catostomids, since three, four, or more gene states were not encountered. Nonetheless, tandem duplications occur sporadically and are known in catostomids. Crabtree and Buth (1981) and Buth (1979*b*) have uncovered a tandem duplication at one of the duplicated mitochondrial aspartate aminotransferase loci (not included in the Ferris and Whitt surveys) in *Catostomus* and *Moxostoma,* respectively. These two genera may therefore share an ancestral regional duplication.

3.4.4. The Role of Regulatory Mutations

The unraveling of the roles of structural and regulatory mutations in gene silencing is very complex, and the following observation illustrates the fact that much speculation is involved, a situation likely to persist until more detailed followup studies are performed. Ferris and Whitt (1979) observed that in catostomid tissues having nonequal expression of duplicates, 63% of the tissues had weaker expression of the less anodally migrating (less negative or more "acidic") isozyme. Lim and Bailey (1977) have noted this in salmonids and Danzmann and Down (1982) in carp. In mammals, acidic proteins are less stable (Dice and Goldberg, 1975). The difference in stability has been attributed to differential susceptibility to proteases. Ferris and Whitt (1979) suggested that a structural gene mutation was fixed by chance in some lineages early in phylogeny. Although this may be true, we must reexamine this idea in the light of the above findings in other species. From Lim and Bailey's (1977) studies on protein-specific activity of LDH-B duplicates, it is clear that the reason for the lowered activity in the slower band is because there is less protein in the cell. Probably there is less message made, and the control of the isozyme ratios is at the regulatory level. But what are we to conclude from the instability of the slow M-MDH in Fig. 1? Perhaps net charge is selected because it somehow affects the overall stability of tertiary and quaternary structures. If this is the case, regulatory mutations that reduce the levels of the unstable slow forms would be selected for.

What other evidence is there that mutations are often of a regulatory nature in catostomids? Some of it is from inference from work on other polyploid species. Regulatory discrimination has been postulated for both salmonids and cyprinids (Engel *et al.*, 1975; Bailey *et al.*, 1976; Allendorf, 1979; Ohno, 1970, 1974). The ancient duplicates *Ldh-A* and *B*, for instance, have very different ratios in mouse kidneys. Nadal-Ginard (1978) demonstrated that the differences are brought about by unequal rates of message synthesis, not differential subunit degradation. Furthermore, the non-binomial assembly of subunits in this tissue was found to be the consequence of spatial isolation of subunit synthesis, namely the expression of the subunits in two kinds of cell types (cf. the example in Fig. 2D).

Will gene silencing continue in catostomid evolution? Some isozyme patterns must be under strong selection because of their maintenance in many lineages. The most notable is CK-B. As first proposed by Ferris and Whitt (1979) and later emphasized by others (Allendorf, 1979; Li, 1980), the evolution of strong tissue specificity is expected to play a retarding role in the future fixation of nulls. For example, only one duplicate *Ck-B* locus is expressed in the heart of most catostomids, whereas both loci are active in brain. A null fixed at the heart-predominant locus would possibly be detrimental, since no CK activity would be left in that tissue. In the many instances of unidirectional divergence, especially the ones with extreme ratios, a null fixed at the predominant locus could lead to severe problems of dosage. Approximately 20% of the patterns were bidirectional and thus most differentiated with respect to regulation. Perhaps this 20% is a lower limit for the retention of duplicate genes in the future. As mentioned earlier, loss of duplicates appears to be random in phylogeny. Noting the rarity of nulls now, I hypothesize that random mutations will continue to eliminate the expression of duplicates, but at a slower rate than in the past. This is because some of the duplicates are acquiring different kinetic properties and tissue specificities (Zuckerland, 1978). A second retarding factor is likely to be dosage related. It may be that some duplicates are retained because they help to maintain an enzyme balance in finely integrated biochemical pathways (Ferris and Whitt, 1980). Conversely, the loss of a duplicate now might begin a cascade of further losses, until new balances are again achieved. Thus a complex set of losses might appear random and yet in fact have been highly selected for.

Bailey *et al.* (1978) have argued that randomness of loss among species means that regulatory mutations also are random. We do not, in fact, know at this stage just what fraction of changes in catostomid gene expression are neutral and what fraction are mediated by natural selection. It should be pointed out that if silencing were a totally neutral affair, and especially in view of the high intrinsic rate of DNA mutation, all duplicates would have been lost by now. Yet, 50% are retained after 50 million years, strongly implying some sort of selection to maintain them. We would still like to know how much selection is involved, and whether selection for a new null allele is positive or negative.

3.4.5. Dosage

The question of dosage is an important one, and much further work is needed in this regard. Dosage adjustments have been reported in other tetraploid organisms, including frogs (W. Becak and Pueyo, 1970) and cyprinids (Schmidtke *et al.*, 1976). These authors report that the overall level of isozyme activity, e.g., for LDH and GPI, is comparable to those levels in diploid relatives. It is apparantly achieved by lowering the effective dosage of each duplicate locus. Loss of ribosomal genes is not involved, according to Schmidtke and Engel (1976).

Some further tanatalizing insights have emerged from recent DNA studies. The globin gene clusters in a tetraploid frog, *Xenopus laevis,* have been examined by restriction enzyme cleavage mapping of DNA (Jeffreys *et al.*, 1980). In the 30 or so million years since this polyploidization event, there has been substantial DNA divergence based on many changes in restriction sites [for a further discussion of restriction mapping, see Ferris *et al.* (1981*a,b*)]. In *Xenopus,* the globin gene cluster has been duplicated;

however, the peptides encoded fall into strong and weak activity groups. Jeffreys *et al.* (1980) conclude that silencing is occurring primarily in one gene cluster, and probably involves DNA rearrangements as well as point mutations.

4. Population Genetics

4.1. Genetic Variability

An understanding of the gene silencing process must include knowledge of the genetic variation present in tetraploid fishes. Null alleles first appear as a polymorphism, and then may become fixed at one of the duplicate loci, with the result that the duplicate is no longer expressed. In this section I will describe some of the basic observations on levels of genetic variation both at singly and duplicately expressed loci, including null alleles, and then relate these findings to mathematical models of gene silencing that have recently been developed.

The first report of variation in tetraploid fish of major scope was that by Allendorf (1975) for a large number of *Salmo gairdneri*. This species has slightly higher than average per-locus heterozygosity *H*, about 10%. Most fish have *H* values around 5–10% (Powell, 1975), so salmonids are not significantly different. *A priori,* one might predict that mutant alleles could accumulate at one of the duplicate genes, especially if there was reduced selection, and thus very high frequencies of polymorphism would ensue. This appears to be the case for recent tetraploid frogs (Jeffreys *et al.*, 1980), but not for the older salmonids. But what of the suckers, with their different mode of tetraploidization than salmonids, and different life style?

The mean per-locus heterozygosity for catostomids is 5% (Ferris and Whitt, 1980). This figure is not demonstrably different from values in other fish or higher vertebrates (Powell, 1975). However, the story is not so simple. The variation has been partitioned according to those loci that remain duplicated and those that have been silenced. Among the former, the mean *H* is 7.6%, while it is 4.3% for the latter. This difference is significant at the $p = 2\%$ level (Ferris and Whitt, 1980).

The elevated *H* at singly expressed loci is consistent with a recent model by Allendorf (1978). He argues that the singly expressed loci should have higher *H* than duplicate loci. The basis of the model is that singly expressed loci have historically had a higher rate of null fixation due to high intrinsic mutation rates of these loci. The model has been contested by Li (1980), on the grounds that inclusion of additional species and reanalysis of the data reduces the correlation substantially.

One piece of evidence does not seem to fit the model at all. Ferris and Whitt (1980) considered all loci that exist singly in catostomids as one set, and those that exist in duplicate as a second set, providing that each locus has a homolog in the other set. The average heterozygosities were determined for each kind of locus. Surprisingly, the heterozygosity was twice as large for a locus when it existed in the single state among species than when it existed in the duplicate state (Ferris and Whitt, 1980). The difference was also highly significant. If Allendorf's model were correct, we would predict no difference in per-locus heterozygosity for the *same* gene whether it was expressed singly or in duplicate. Perhaps there is an upper limit to the amount of heterozygosity that is tolerated in the organism (Allendorf, 1979; Ferris and Whitt, 1980).

By examining many genera of catostomids, we begin to have an evolutionary picture of the allelic variation. The mean heterozygosity and number of loci polymorphic was by far the greatest in the primitive genera *Ictiobus* and *Carpiodes*. These groups had an average heterozygosity of 9.2%, over twice that of the other genera (Ferris and Whitt, 1980). Some of this variation may be linked to frequent hybridization among the primitive species. The trend for high variation is not a general feature of primitive catostomids, since *Cycleptus* had low levels of heterozygosity. At this time we do not know whether mean heterozygosity relates mainly to primitiveness, population size, life style, or even frequency of hybridization. The tendency for low heterozygosity in advanced catostomids like *Catostomus, Hypentelium,* and *Moxostoma* has been well documented (Buth, 1977*b*, 1978, 1979*a;* Buth and Crabtree, 1982), and values of $H = 3\%$ are typical. Buth (personal communication) hypothesizes that in many cases, levels of polymorphism are a species-specific trait, owing to unidentified parameters of selection or history of the population. For instance, *M. macrolepidotum* is peculiar for the genus in having $H = 10\%$ (like *Ictiobus*), and *M. pappillosum* has zero heterozygosity.

A more quantifiable parameter is that of species per genus. Avise (1977) has proposed that speciose fish taxa might have more genetic variation then depauperate ones. He finds no difference in levels of H among cyprinids and centrarchids. In the catostomids, the primitive and moderately primitive taxa have two to three species each and are therefore "depauperate." The advanced taxa, as mentioned earlier, are quite speciose. Table IV summarizes the mean heterozygosities in species of the two kinds of lineages. Clearly, there is more heterozygosity in the primitive, depauperate taxa than in the speciose ones! The catostomids are admittedly not a perfect test of the model, because of the complicating factor of duplicate gene differences. More teleosts must be studied before we can make any judgement on the model.

Table IV
Mean Heterozygosity in Speciose and Depauperate Genera of Catostomidae[a]

Depauperate genera		Speciose genera	
Species	Heterozygosity	Species	Heterozygosity
Ictiobus		*Moxostoma*	
I. bubalus	0.103	*M. erythrurum*	0.034
I. cyprinellus	0.102	*M. duquesnei*	0.015
		M. macrolepidotum	0.075
		M. cervinum	0.044
Carpiodes		*Catostomus*	
C. carpio	0.089	*C. commersoni*	0.035
C. velifer	0.081	*C. catostomus*	0.038
C. cyprinus	0.083	*C. columbianus*	0.048
		C. plebeius	0.019
Cycleptus			
C. elongatus	0.050		
Erimyzon			
E. sucetta	0.058		
E. oblongus	0.043		
E. tenuis	0.047		
Minytema			
M. melanops	0.059		
Hypentelium			
H. nigricans	0.024		
Mean	0.067		0.038
Standard deviation	0.026		0.019

[a]Modified from Ferris and Whitt (1980). Reprinted with permission of *American Naturalist*, University of Chicago Press.

4.2. Mathematical Models of the Rate of Gene Silencing

A lively interest in modeling gene loss in polyploids has developed in the last several years. How is it possible to maintain a 50% level of functioning genes after 50 million years? All models of silencing must at some point deal with the frequencies of null polymorphism in populations. Some models predict rather high frequencies (Bailey *et al.*, 1978) and others low frequencies (Takahata and Maruyama, 1979; Li, 1980). None of the existing models take into account the different mode of origin of tetraploidy in salmonids and catostomids. Stoneking *et al.* (1981) argue

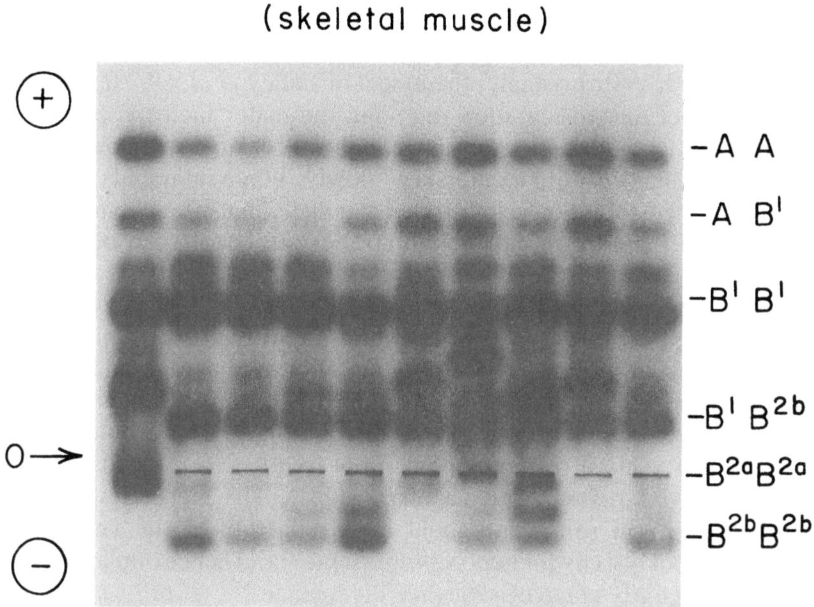

Figure 3. Polymorphism for one or more null alleles at the glucosephosphate isomerase-B loci in *Erimyzon oblongus*. Lane 1 shows a B^{2a} homozygote, lanes 2–5 and 10 a B^{2b} homozygote. Lane 8 contains a B^{2a}/B^{2b} heterozygote. Lanes 6 and 9 show presumptive nulls. One null forms active heterodimers with the B^{1} product, as is evident from the $B^{1}B^{2b}$ band. Another null with mobility close to that of B^{2a}, forms an active heterodimer with the B^{1} and is just anodal to the $B^{1}B^{2b}$ band.

that this could lead to rather different predictions on levels of null polymorphisms. What are the data? Stoneking *et al.* (1981) have solid evidence for a null in a salmonid and further indicate that nulls are fairly common in the group. On the other hand, in over 50 species of catostomids examined electrophoretically, in only one is there reasonable evidence for a null allele segregating at a locus. The null at *Gpi-B* is illustrated in Fig. 3 for *Erimyzon oblongus*. The null product apparently may form heterodimers with the active form. Certainly many nulls will be difficult to detect, especially if they do not form heteropolymers, yet even if we are observing only a fourth of them, the levels of null polymorphism are significantly below those predicted by neutral drift models such as that of Bailey *et al.* (1978).*

*The model could be valid for salmonids, since nulls are segregating at higher frequencies in this group (Stoneking *et al.*, 1981).

As Lim *et al.* (1975) point out, the high neutral mutation rate should have resulted in the silencing of all but a handful of duplicate loci after 50 million years. Surprisingly, the model of Bailey *et al.* (1978) finds only a minor effect due to mutation rate, and the major factor is population size. Because population sizes are thought to be large in salmonids and catostomids, nulls drift to fixation very slowly. While salmonids may have large effective population sizes, it seems unlikely that effective population sizes for the majority of catostomid species are larger than one million or have remained this large over evolutionary time (Ferris *et al.*, 1979; Li, 1980). The prediction of the above model that high levels of nulls be present even now does not appear to be true for the catostomids.

More recent models invoke both mutation rate and population size. Takahata and Maruyama (1979) developed a diffusion model for gene silencing where nulls are selected against strongly. This would account for their rarity in catostomids now. Low frequencies of nulls are also predicted by Maruyama and Takahata (1981). The higher level of null polymorphisms in salmonids could be a reflection of the very recent achievement of disomy for many chromosomes, and hence duplicate genes are in a very early stage of sequence divergence.

Allendorf (1979) recently expanded on the selection models in suggesting that gene regulation is now the main reason for the infrequency of null alleles. I agree that this process is important now, but it is unlikely to have played a major role in the past, when only a few loci were differentially regulated. Li (1980) also stresses this idea. In summary, we do not know the mechanism for maintenance of nulls; no doubt some are selected for and some are neutral.

Li (1980) has proposed the most comprehensive model of gene silencing to date. In this model, changes in population size and the mutation rate are both factors. Three phases are recognized in tetraploid gene evolution: (1) Chromosomes begin the process of diploidization, and at varying rates, some silencing begins. (2) Loss of gene expression is rapid for those loci with alleles with no deleterious effects, and occurs at loci with high mutation rates, providing that the populations sizes are small; if the population size is large in some species and the mutation rate low, duplicates will remain functional for a long time. (3) Loss of expression slows down in the terminal phase, primarily due to the evolution of differential expression among tissues. This latter point fits well with ideas already discussed by Ferris and Whitt (1979) and Allendorf (1979) that silencing in the bidirectional category will be slowed down, and that some slowing is expected even within the unidirectional category, since dosage problems may occur at some loci.

In summary, allelic polymorphisms are at levels comparable to those in diploid teleosts, although there are higher levels at duplicate loci than single loci in catostomids. Null polymorphisms are rare, suggesting that they are deleterious. Polymorphism of nulls and nonnulls was probably higher in the past, however. Population size appears to play an important role in the rate of silencing.

5. Systematics

5.1. Gene Duplication Analyses and Allozyme Approaches

The catostomids are an excellent testing ground for systematics in all areas, including morphology, allozyme variation, gene duplications, and, eventually, DNA studies. The combination of data from more than one of the above areas, such as done by Smith and Koehn (1971), is sure to be a fruitful approach. The cladistic branching sequence is important in the catostomids so that we can understand rates of silencing, identify reactivation events, look for accelerated morphological change, and correlate morphological change with biochemical change.

The systematics of catostomids has been examined morphologically by a variety of workers (Ramaswami, 1957; Nelson, 1948; Miller, 1959; Smith and Koehn, 1971; Jenkins, 1980). From these efforts there is a basis of comparison for phylogenies based on biochemical and morphological data. Ferris and Whitt (1978*a*) examined the systematics of catostomids, but instead of using traditional allozyme methods, they used gene duplication states. With duplicate genes, primitive and derived states are relatively clear-cut. Retention of two copies is an ancestral state, and the loss of a functional copy is a derived state. This "polarity" was exploited by Ferris and Whitt (1978*a*) in a construction of a phylogeny of the North American genera. A Wagner analysis of the gene states resulted in a tree with a topology very similar to those based on morphology. The only notable difference was the placement of *Cycleptus*. Ferris and Whitt (1978*a*) argue that this genus may be a basal stock of the Catostominae, as have others based on dietary and morphological features [see references in Ferris and Whitt (1978*a*)]. The occasional placement of a member of one species amidst members of another is almost certainly a reflection of parallel changes and the relatively small number of gene states studied. Buth (1982) has discussed the very real possibility that occasional reduplication of singly expressed genes can confound efforts at phylogenetic reconstruction.

Buth (1977*a*, 1978, 1979*c*, 1980) has carried out in-depth phylogenetic reconstruction of the tribe Moxostomatini, using both gene duplications and allelic polymorphisms. In general, his work validates the hypothesis that two genes are primitive and one gene is derived. Increasing the number of loci by a factor of two (Crabtree and Buth, 1981) gives the same overall percent gene duplication. There is also more information for tree-building. In my opinion, however, gene duplication data will not be particularly useful for high-resolution trees within a genus unless hundreds of loci are sampled. Rather, these data will be best suited for resolving higher taxonomic categories. A few problems should be noted with this method. One is that we are oversimplifying the picture by assuming all losses are orthologous. A set of silenced genes may actually bear a paralogous relationship to one another, in which case computer analyses that treat the single states as shared derived may sometimes be in error. Furthermore, orthologous duplicates may sometimes have been silenced independently in two lineages. Another problem is estimating the frequency of reversals, or reactivation events. We should recall Buth's (1982) finding of a reversal in *M. lachneri*. A few of these reversals in a large data set should not adversely affect the reliability of trees; character reversals in fact are inherent in all data sets. Li (1980) gives additional reasons why reversals of tetraploid gene expression should be rare.

Allozyme analyses for reconstructing catostomid history have been most persuasively used by Buth and his colleagues. His use of the cladistic approach, especially the assignment of ancestral alleles from comparisons of outgroups, is the method of choice. Very detailed phylogenies of *Moxostoma* and *Hypentelium* have resulted (Buth, 1978, 1979*c*, 1980).

5.2. Species Hybridization

One area that impinges upon systematics is the phenomenon of hybridization. Hybrids between catostomids are known in nature. An interesting question is what happens when one species has a duplication for a locus and the other does not. Do the hybrids express one gene or two? So far, the hybridizing species that have been surveyed electrophoretically have not had detectable differences in the number of duplicate genes (Buth, personal communication). We know that hybrids between distantly related diploid species can have aberrant expression of maternal or paternal alleles (Whitt *et al.*, 1977). It would be exciting to find a reactivated silenced gene as the result of catostomid hybridization. We can get further insights from recent studies of hybrids between tetraploid carp and goldfish (Danzmann and Down, 1982). Although there are no duplication differences separating the species, the F_1 hybrids are sterile. The authors speculate that regulatory incompatabilities may play a role. Esterase and MDH-A loci show unidirectional expression of duplicates in both species. The unique migrating carp isozymes appear to predominate over the goldfish isozymes in the hybrid animals. Further research into the behavior of duplicates in these and other hybrids should shed much light on the mechanism of regulation of duplicate genes.

6. Speculations on Catostomid Evolution and Directions for Future Research

As I hope to have demonstrated, the catostomids are an unusually rich system for testing hypotheses about genic and organismal evolution. In this section I will speculate on some of the factors that have promoted the retention of multiple genes since that genome doubling in their distant past. Lastly, I will outline several potentially rewarding avenues for future research on catostomids and other tetraploid organisms.

6.1. The Advantages of Polyploidy

We know from hindsight that many of the duplicates have diverged in structure and regulation. A large array of multiple isozymes, as well as new ways of expressing them, may have contributed to the successful radiation of the group. This diversity would have been much smaller in the initial period subsequent to polyploidization but may still have an advantage represented.

The polyploidization must have had profound effects on the organism. A new species was created in an instant. The niche of this new species was probably very similar to that of its parental species. Yet, it was perhaps new combinations of alleles or ways of regulating old alleles for a few loci that proved to be a distinct advantage. It is possible that entirely new alleles arose by "hybrid dysgenesis" (intragenic recombination?). In fact, completely new proteins have been noted in other allotetraploid vertebrates (Sage and Selander, 1979). If the event actually was of the allopolyploid type, a "fixed heterozygote" advantage would be realized at a number of loci. This advantage has been discussed in detail elsewhere (Spofford, 1969; Wheat *et al.*, 1973; Lewontin, 1974).

Niche differentiation in the new polyploid would have been facilitated by various changes accompanying polyploidization in addition to the allozymes discussed above. For one thing, the fish have larger size and larger growth rates than diploid relatives (Uyeno and Smtth, 1972). Furthermore, because cell volume in animals is proportional to the DNA content (Hinegardner, 1976), catostomids with their larger cell volume and more favorable surface to volume ratio in the cells for the exchange of metabolites may have had an advantage in oxygen-poor environments. Catostomids often thrive to this day in slow-moving or stagnant water with low oxygen. This rationale does not account convincingly for the success of the salmonids, which require oxygen-rich habitats. And lastly, the gene doubling would have increased the dosage of a number of physiological key proteins, such as hemoglobin and digestive enzymes.

Catostomids were never very successful in Asia, their place of origin. By the time the ancestor of North American catostomids reached this continent, gene divergence and regulation was already well underway and could have led to a competitive edge. The assumed absence of North American polyploid cyprinids prior to their arrival no doubt also favored their unhindered radiation.

What will happen to the genome of the catostomids in the future? We expect a continued silencing of duplicates perhaps to 20 or 25% retention. Those retained will continue to diverge in structure and tissue regulation. Eventually, many will have become as distinct in their properties as LDH-A and B are now. Duplicates now expressed equally in many tissues, such as adenylate kinase-A, should show ever-increasing tissue specificity of one of the duplicates. Some will even narrow to expression in a few cell types within a tissue, much like LDH-A duplicates in *Cycleptus*. At the DNA level, there will be gradual loss of genetic material, following the trend in fish evolution seen by Hinegardner (1976). On the other hand, it is not inconceivable that yet another round of ploidy could occur in catostomids. Indeed, a triploid goldfish has been found (Kobayasi *et al.*, 1970).

6.2. Future Research

We need to know more about the physical and chemical changes that have arisen in duplicate isozymes, along the lines of research recently carried out by DiMichelle and Powers (1982) for the kinetic differences of LDH alleles in *Fundulus*. They show how structural differences between the alleles actually cause changes in developmental rates. How do catostomid isozymes differ in K_m, *p*H optima, and amino acid sequence? Can we demonstrate greater differences in these properties among duplicates that have evolved more disparate ratios of activities within and between tissues? What is the molecular mechanism(s) for the departure from the ancestral 1:1 ratio?

At the population level, more information is needed on the frequencies of null polymorphisms. And when nulls first appear, are they primarily "CRM" or is there no protein at all?

This leads us to what could be one of the more exciting areas of research in coming years, the elucidation of the molecular mechanisms of gene silencing. There is already considerable progress in understanding the nature of pseudogenes of globins (Martin *et al.*, 1980), a multigene family. The nature of silencing has already been examined at the DNA level in tetraploid frogs (Jeffreys *et al.*, 1980), but we are still a long way from knowing the mechanism of silencing for "household function" enzymes like MDH.

Another consequence of the DNA studies should be a better understanding of what fraction of mutations occurs in regulatory versus structural genes. With DNA hybridization (the Southern method) and recombinant DNA technologies, it shall be possible to make probes for specific genes. A good candidate is creatine kinase of muscle. Its abundance in that tissue is probably paralleled by an abundance of message. This message could be isolated and its cDNA cloned. The gene could be sequenced or used as a probe to visualize other CK genes. From restriction mapping and DNA sequencing it will be possible to determine the precise number and locations of mutations, not to mention rearrangements. How many mutations are in intervening sequences versus coding sequences? Have introns been lost, as in recently discovered silenced globin genes (Leuders *et al.*, 1982)?

It is hoped that more polyploids will be discovered that are of different ages than catostomids. Then it will be possible to understand how the rate of gene silencing changes, as well as rates of regulatory divergence and chromosomal "diploidization."

Systematics, too, will continue to be a valuable avenue of research. An accurate phylogeny will allow us to pinpoint the origins of specific regulatory patterns. Protein electrophoresis will continue to be instrumental as a source of information in phylogenetic inference. New techniques, such as DNA and mitochondrial DNA restriction analysis, will add information on systematics. Mitochondrial DNA has already been purified from salmonids (Berg and Ferris, 1984, B. J. Turner, personal communication). This type of DNA evolves fast, and thus will provide a magnified view of the evolution of populations and closely related species (Ferris *et al.*, 1981*a,b*). It offers the further advantage of being relatively unaffected by nuclear genome doubling and its concomitant problems in ascertaining homologies among chromosomal genes.

The catostomids and other tetraploids should continue to reveal insights into gene evolution and gene regulation in a much more forceful manner than possible by studying diploid fishes. Just as important, detailed studies along these lines will tell us more about how tetraploidy has changed the biology of the catostomids.

ACKNOWLEDGMENTS. I thank Drs. Donald Buth and Mark Stoneking for helpful comments, and especially Dr. Gregory Whitt for the inspiration and resources to carry out much of the research described in this chapter. I also thank Dr. Anthony Carrano for assistance with the DNA content measurements. The research was supported by grants PCM 76-08383 to Gregory Whitt and a NIH predoctoral traineeship to the author.

References

Allendorf, F. W., 1975, Genetic variability in a species possessing extensive gene duplication: "Genetic interpretation of duplicate loci and examination of variation in populations of rainbow trout", Ph.D. dissertation, University of Washington.

Allendorf, F.W., 1978, Protein polymorphism and the rate of loss of duplicate gene expression, *Nature* **272**:76–78.

Allendorf, F. W., 1979, Rapid loss of duplicate gene expression by natural selection, *Heredity* **43**:247–258.

Allendorf, F. W., Utter, F. M., and May, B. P., 1975. Gene duplication within the family Salmonidae: II. Detection and determination of the genetic control of duplicate loci through inheritance studies and the examination of populations. in: *Isozymes IV: Genetics and Evolution* (C. L. Markert, ed.), Academic Press, New York, pp. 415–432.

Avise, J. C., 1977, Genic heterozygosity and rate of speciation, *Paleontology* **3**:422–432.

Avise, J. C., and Kitto, B. G., 1973, Phosphoglucose isomerase gene duplication in the bony fishes, an evolutionary history, *Biochem. Genet.* **8**:113–132.

Bachman, K., and Bogart, J. P., 1975, Comparative cytochemical measurements in diploid-tetraploid species pair of hylid frogs *Hyla chrysocelis* and *H. versicolor, Cytogenet. Cell Genet.* **15**:186–194.

Bailey, G. S., Tsuyuki H., and Wilson, A. C., 1976, The number of genes for lactate dehydrogenase in salmonid fishes, *J. Fish. Res. Board Can.* **33**:760–767.

Bailey, G. S., Poulter, R. T. M., and Stockwell, P. A., 1978, Gene duplication in tetraploid fish. Model for gene silencing at unlinked duplicate loci, *Proc. Natl. Acad. Sci. USA* **75**:5575–5579.

Becak, M. L., Becak, W., and Rabello, M. N., 1966, Cytological evidence of constant tetraploidy in the bisexual South American frog *Odontophrynus americanus, Chromosoma* **19**:188–193.

Becak, W., and Pueyo, M. T., 1970. Gene regulation in the polyploid amphibian *Odontophrynus americanus, Exp. Cell Res.* **63**:448–451.

Berg, W. J., and Ferris, S. D., 1984, Restriction endonuclease analysis of salmonid mitochondrial DNA, *Can. J. Fish Aquat. Sci.*, in press.

Busslinger, M., Moschonas, N., and Flavell, R. A., 1981, β^+ Thalassemia: Aberrant splicing results from a single point mutation in an intron, *Cell* **27**:289–298.

Buth, D. G., 1977*a*, Biochemical identification of *Moxostoma rhothoecum* and *M. hamiltoni, Biochem. Syst. Ecol.* **5**:57–60.

Buth, D. G., 1977*b*, Alcohol dehydrogenase variability in *Hypentelium nigricans, Biochem. Syst. Ecol.* **5**:61–63.

Buth, D. G., 1978, Biochemical systematics of the Moxostomatini (Cypriniformes, Catostomidae), Ph.D. dissertation, University of Illinois, Urbana.

Buth, D. G., 1979*a*, Creatine kinase variability in *Moxostoma macrolepidotum* (Cypriniformes, Catostomidae), *Copeia* **1979**:152–154.

Buth, D. G., 1979*b*, Duplicate gene expression in tetraploid fishes of the tribe Moxostomatini (Cypriniformes, Catostomidae), *Comp. Biochem. Physiol.* **63B**:7–12.

Buth, D. G., 1979*c*, Genetic relationships among the torrent suckers, genus *Thoburnia*, *Biochem. Syst. Ecol.* **7**:311–316.

Buth, D. G., 1980, Evolutionary genetics and systematic relationships in the catostomid genus *Hypentelium*, *Copeia* **1980**:280–290.

Buth, D. G., 1982, Glucosephosphate-isomerase expression in the tetraploid fish, *Moxostoma lachneri* (Cypriniformes, Catostomidae): Evidence for a "retetraploidization"? *Genetica* **57**:171–175.

Buth, D. G., and Crabtree, C. B., 1982, Genetic variability and population structure of *Catostomus santaanae* in the Santa Clara drainage, *Copeia* **1982**:439–444.

Champion, M. J., Shaklee, J. B., and Whitt, G. S., 1974, Developmental genetics of teleosts: A biochemical analysis of lake chubsucker ontogeny, *Dev. Biol.* **38**:356–382.

Comings, D. E., and Berger, R. O., 1969, Gene products of Amphiuma: An amphibian with an excessive amount of DNA, *Biochem. Genet.* **2**:319–333.

Crabtree, C. B., and Buth, D. G., 1981, Gene duplication and diploidization in tetraploid catostomid fishes *Catostomus fumeiventris* and *C. santaanae*, *Copeia* **1981**:705–708.

Danzmann, R. G., and Down, N. E., 1982, Isozyme expression in F_1 hybrids between carp and goldfish, *Biochem. Genet.* **20**:1–15.

Dauble, D. D., and Buschbom, R. L., 1981, Estimates of hybridization between two species of catostomids in the Columbia River, *Copeia* **1981**:802–810.

Dice, F. J., and Goldberg, A. L., 1975, Relationships between *in vivo* degradative rates and isoelectric points of proteins, *Proc. Natl. Acad. Sci. USA* **72**:3893–3897.

Dickerson, R. E., and Geis, I., 1969, *The Structure and Action of Proteins*, Benjamin, Menlo Park, California.

DiMichele, L., and Powers, D. A., 1982, LDH-B genotype-specific hatching times of *Fundulus heteroclitus* embryos, *Nature* **296**:563–564.

Dingerkus, G., and Howell, W. M., 1976, Karyotypic analysis and evidence of tetraploidy in North American paddlefish, *Polyodon spathula*, *Science* **194**:842–844.

Doolittle, W. F., and Sapienza, C., 1980, Selfish genes, the phenotype paradigm, and genome evolution, *Nature* **284**:601–603.

Engel, W., Schmidtke, J., Vogel, W., and Wolf, U., 1973, Genetic polymorphism of lactate dehydrogenase isoenzymes in the carp, *Cyprinus carpio*, apparently due to a "null allele," *Biochem. Genet.* **8**:281–289.

Engel, W., Schmidtke, J., and Wolf, U., 1975, Diploid–tetraploid relationships in teleostean fishes in: *Isozymes IV: Genetics and Evolution* (C. L. Markert, ed.), Academic Press, New York, pp. 449–462.

Eventhoff, W., and Rossman, M. G., 1975, The evolution of the dehydrogenases and kinases, *CRC Crit. Rev. Biochem.* **3**:111–140.

Ferris, S. D., 1978. Evolution of duplicate gene expression after polyploidization, Ph.D. dissertation, University of Illinois, Urbana.

Ferris, S. D., and Whitt, G. S., 1977*a*, Duplicate gene expression in diploid and tetraploid loaches (Cypriniformes, Cobitidae), *Biochem Genet.* **15**:1097–1111.

Ferris, S. D., and Whitt, G. S., 1977*b*, Loss of duplicate gene expression after polyploidisation, *Nature* **265**:258–260.

Ferris, S. D., and Whitt, G. S., 1977*c*, The evolution of duplicate gene expression in the carp (*Cyprinus carpio*), *Experientia* **33**:1299–1301.

Ferris, S. D., and Whitt, G. S., 1978*a*, Phylogeny of the tetraploid catostomid fishes based on the loss of duplicate gene expression, *Syst. Zool.* **27**:189–206.

Ferris, S. D., and Whitt, G. S., 1978*b*, Genetic and molecular analysis of the nonrandom dimer assembly of the creatine kinase isozymes of fishes, *Biochem. Genet.* **16**:811–829.

Ferris, S. D., and Whitt, G. S., 1979. Evolution of the differential regulation of duplicate genes after polyploidization, *J. Mol. Evol.* **12:**267–317.
Ferris, S. D., and Whitt, G. S., 1980, Genetic variability in species with extensive gene duplication: The tetraploid catostomid fishes, *Am. Nat.* **115:**650–666.
Ferris, S. D., Portnoy, S., and Whitt, G. S., 1979, The roles of speciation and divergence times in the loss of duplicate gene expression, *Theor. Popul. Biol.* **15:**114–139.
Ferris, S. D., Wilson, A. C., and Brown, W. M., 1981*a,* Evolutionary tree for humans and apes based on cleavage maps of mitochondrial DNA, *Proc. Natl. Acad. Sci. USA* **78:**2432–2436.
Ferris, S. D., Brown, W. M., Davidson, W. S., and Wilson, A. C., 1981*b,* Extensive polymorphism in the mitochondrial DNA of apes, *Proc. Natl. Acad. Sci. USA* **78:**6319–6323.
Ferris, S. D., Buth, D. G., and Whitt, G. S., 1982, Substantial genetic differentiation in populations of *Catostomus plebeius, Copeia* **1982:**444–449.
Fisher, S. E., and Whitt, G. S., 1978, Evolution of isozyme loci and their differential tissue expression: Creatine kinase as a model system, *J. Mol. Evol.* **12:**25–55.
Fisher, S. E., Shaklee, J. B., Ferris, S. D., and Whitt, G. S., 1980, Evolution of five multilocus isozyme systems in the chordates, *Genetica* **52/53:**73–85.
Gosselin-Ray, C., and Gerday, C., 1970, Isolation and molecular properties of creatine kinase from carp white muscle, *Biochim. Biophys. Acta* **221:**241–254.
Haber, E. J., and Rogers, T. D., 1982, Transposition of a tandem duplication of yeast mating type genes, *Nature* **296:**768–770.
Hinegardner, R., 1976, Evolution of genome size, in: *Molecular Evolution* (F. J. Ayala, ed.), Sinauer, Sunderland, Massachusetts, pp. 179–199.
Hubbs, C. L., 1955, Hybridization between fish species in nature, *Syst. Zool.* **4:**1–20.
Hubbs, C. L., Hubbs, L. C., and Johnson, R. E., 1943, Hybridization in nature between species of catostomid fishes, *Contrib. Lab. Vertebr. Biol. Univ. Mich.* **1943**(22)**:**1–76.
Huntsman, G. R., 1970, Disc electrophoresis of blood sera and muscle extracts from some catostomid fishes, *Copeia* **1970:**457–467.
Jeffreys, A. J., 1981, Recent studies of gene evolution using recombinant DNA technology. in: *Genetic Engineering 2,* (R. Williamson, ed.), Academic Press, New York, pp. 1–48.
Jeffreys, A. J., Wilson, V., Wood, D., and Simons, J. P., 1980, Linkage of adult α- and β-globin genes in *X. laevis* and gene duplication by tetraploidization, *Cell* **21:**555–564.
Jenkins, R. E., 1970, Systematic studies of the catostomid fish tribe Moxostomatini, Ph.D. dissertation, Cornell University, Ithaca, New York.
Johnson, G. B., 1974, Enzyme polymorphism and metabolism, *Science* **184:**28–37.
Klebe, R. J., 1975, A simple method for the quantitation of isozyme patterns, *Biochem. Genet.* **13:**805–812.
Kobayasi, H., Kawashima, Y., and Takeuchi, N., 1970. Comparative chromosome studies in the genus *Carassius,* especially with a finding of polyploidy in the ginbuna (*C. auratus langsdorfi*), *Jpn. J. Ichthyol.* **17:**153–160.
Koch, A. L., 1972, Enzyme evolution: I. The importance of untranslatable intermediates, *Genetics* **72:**297–316.
Koehn, R. K., 1966, Serum haptoglobins in some North American catostomid fishes, *Comp. Biochem. Physiol.* **17:**349–352.
Koehn, R. K., 1969*a,* Esterase heterogeneity: Dynamics of a polymorphism, *Science* **163:**943–944.
Koehn, R. K., 1969*b,* Hemoglobins of fishes of the genus *Catostomus* from western North America, *Copeia* **1969:**21–30.
Koehn, R. K., and Johnson, D. W., 1967, Serum transferrin and serum esterase polymor-

phisms in an introduced population of bigmouth buffalofish, *Ictiobus cyprinellus, Copeia* **1967**:805–808.
Koehn, R. K., and Rasmussen, D. I., 1967, Polymorphic and monomorphic serum esterases in catostomid fish populations, *Biochem Genet.* **1**:131–134.
Kucherlapati, R. S., Cregan, R. P., and Ruddle, F. H., 1974, Progress in human gene mapping by somatic cell hybridization, in: *The Cell Nucleus* (H. Busch, ed.), Academic Press, New York, pp. 209–222.
Leuders, K., Leder, A., Leder, P., and Kuff, E., 1982. Association between a transposed α-globin pseudogene and retrovirus-like elements in the BALB/c mouse genome, *Nature* **295**:426–428.
Lewontin, R. C., 1974, *The Genetic Basis of Evolutionary Change,* Columbia University Press, New York.
Li, W.-H., 1980, Rate of gene silencing at duplicate loci: A theoretical study and interpretation of data from tetraploid fishes, *Genetics* **95**:237–258.
Li, W.-H, Gojobori, T., and Nei, M., 1981, Pseudogenes as a paradigm of neutral evolution, *Nature* **292**:237–239.
Lim, S. T., and Bailey, G. S., 1977, Gene duplication in salmonid fishes: Evidence for duplicated but catalytically equivalent A_4 lactate dehydrogenase, *Biochem. Genet.* **15**:707–721.
Lim, S. T., Kay, R. M., and Bailey, G. S., 1975, Lactate dehydrogenase isozymes of salmonid fish. Evidence for unique and rapid functional divergence of duplicated H_4 lactate dehydrogenases, *J. Biol. Chem.* **250**:1790–1800.
Lucchesi, J. C., and Rawls, J. M., Jr., 1973, Regulation of gene function: A comparison of enzyme activity levels in relation to gene dosage in diploids and triploids of *Drosophila melanogastor, Biochem. Genet.* **9**:41–51.
Martin, S. L., Zimmer, E. A., Kan, Y. W., and Wilson, A. C., 1980, Silenced α-globin gene in Old World monkeys, *Proc. Natl. Acad. Sci. USA* **77**:3563–3566.
Makino, S., 1939, The chromosomes of the carp, *Cyprinus carpio,* including those of some related species of Cyprinidae for comparison, *Cytologia* **9**:430–440.
Markert, C. L., and Ursprung, H., 1971, *Developmental Genetics,* Prentice-Hall, Englewood Cliffs, New Jersey.
Markert, C. L., Shaklee, J. B., and Whitt, G. S., 1975, Evolution of a gene, *Science* **189**:102–114.
Maruyama, T., and Takahata, N., 1981, Numerical studies of the frequency trajectory in the process of fixation of null genes at duplicated loci, *Heredity* **44**:49–57.
Miller, R. R., 1959, Origin and affinities of the freshwater fish fauna of Western North America, in: *Zoogeography, American Association for the Advancement of Science,* Washington, D. C., pp. 187–222.
Muramoto, J. C., Ohno, S., and Atkin, N. B., 1967, On the diploid state of the order Ostariophysi, *Chromosoma* **24**:59–66.
Nadal-Ginard, B., 1978, Regulation of lactate dehydrogenase levels in the mouse, *J. Biol. Chem.* **253**:170–177.
Nei, M., Genetic distance between populations, *Am. Nat.* **106**:283–292.
Nelson, E. M., 1948, The comparative morphology of the Weberian apparatus of the Catostomidae and its significance in systematics, *J. Morphol.* **83**:225–251.
Nikolsky, G., 1976, The interrelation between variability of characters, effectiveness of energy utilization, and karyotypic structure in fishes, *Evolution* **30**:180–185.
Ohno, S., 1970, *Evolution by Gene Duplication,* Springer-Verlag, New York.
Ohno, S., 1974, *Animal Cytogenetics: Protochordata, Cyclostomata, and Pisces,* Vol. 4, *Chordata 1* (B. John, ed.), Gebrüder-Bornträger, Berlin.

Ohno, S., 1974, *Animal Cytogenetics: Protochordata, Cyclostomata, and Pisces,* Vol. 4, *Chordata 1* (B. John, ed.), Gebrüder-Bornträger, Berlin.
Ottolenghi, S., Giglioni, B., Taramelli, R., Comi, P., Mazza, U., Saglio, G., Camaschella, C., Izzo, P., Cao, A., Galanello, R., Gimferrer, E., Baiget, M., and Gianni, A. M., 1982, Molecular comparison of $\delta\beta$-thalassemia and hereditary persistence of fetal hemoglobin DNA: Evidence of a regulatory area? *Proc. Natl. Acad. Sci. USA* **79:**2347–2351.
Pesce, A., Fondy, T. P., Stolzenbach, F., Castillo, F., and Kaplan, N. O., 1967, The comparative enzymology of lactate dehydrogenase, *J. Biol Chem.* **242:**2151–2167.
Powell, R. J., 1975, Protein variation in natural populations of animals, *Evol. Biol.* **8:**79–119.
Powers, D. A., and Edmundson, A. B., 1972, Multiple hemoglobins from catostomid fish I. Isolation and characterization of the isohemoglobins from *Catostomus clarkii, J. Biol. Chem.* **247:**6686–6693.
Ramaswami, L. S., 1957, Skeleton of Cyprinoid fishes in relation to phylogenetic studies. 8. The skull and Weberian ossicles of Catostomidae, in: *H. Mookerjee Memorial Volume, Proceedings of the Zoological Society of Calcutta* (J. L. Bhaduri, B. Biswas, and S. P. Ray-Chaudhury, eds.), pp. 293–303.
Sage, R. D., and Selander, R. K., 1979, Hybridization between species of the *Rana pipiens* complex in central Texas, *Evolution* **33:**1069–1088.
Sarich, V. M., 1977, Rates, sample sizes, and the neutrality hypothesis for electrophoresis in evolutionary studies, *Nature* **265:**24–28.
Schmidtke, J., and Engel, W., 1976, Gene action in fish of tetraploid origin III. Ribosomal DNA amount in cyprinid fish, *Biochem. Genet.* **14:**19–26.
Schmidtke, J., Schulte, B., Kuhl, P., and Engel, W., 1976, Gene action in fish of tetraploid origin. V. Cellular RNA and protein content and enzyme activities in cyprinid, clupeoid, and salmonoid species, *Biochem. Genet.* **14:**975–980.
Schultz, J. R., 1969, Hybridization, unisexuality, and polyploidy in the teleost poeciliopsis (Poeciliidae) and other vertebrates, *Am. Nat.* **103:**605–619.
Schwartz, M., and Sofer, W., 1976, Alcohol dehydrogenase-negative mutants in *Drosophila:* Defects at the structural locus? *Genetics* **83:**125–136.
Shaw, C. R., and Prasad, R., 1970, Starch gel electrophoresis of enzymes—A compilation of recipes, *Biochem. Genet.* **4:**297–320.
Shaklee, J. B., and Whitt, G. S., 1981, Lactate dehydrogenase isozymes of Gadiform fishes: Divergent patterns of gene expression indicate a heterogeneous taxon, *Copeia* **1981:**563–578.
Shows, T. B., Massaro, E. J., and Ruddle, F. H., 1969, Evolutionary evidence for a regulator gene controlling the LDH-B gene in rodent erythrocytes, *Biochem. Genet.* **3:**525–536.
Smith, G. R., and Koehn, R. K., 1971, Phenetic and cladistic studies of biochemical and morphological characteristics of *Catostomus, Syst. Zool.* **20:**282–297.
Spofford, J. B., 1969, Heterosis and the evolution of duplications, *Am. Nat.* **103:**407–432.
Stebbins, G. L., 1971, *Chromosomal Evolution in Higher Plants,* Columbia University Press, New York.
Stoneking, M., May, B., and Wright, J. E., 1981, Loss of duplicate gene expression in salmonids: Evidence for a null allele polymorphism at the duplicate aspartate amino transferase loci in brook trout *(Salvelinus fontinalis), Biochem. Genet.* **19:**1063–1077.
Takahata, N., and Maruyama, T., 1979, Polymorphism and loss of duplicate gene expression: A theoretical study with application to tetraploid fish, *Proc. Natl. Acad. Sci. USA* **76:**4521–4525.
Tsuyuki, H., Roberts, E., Kerr, R. H., Uthe, J. F., and Clarke, L. W., 1967, Comparative electropherograms in the family Catostomidae, *J. Fish, Res. Board Can.* **24:**299–304.
Utter, F. M., Allendorf, F. W., and May, B., 1979, Genetic basis of creatine kinase isozymes in skeletal muscle of salmonid fishes, *Biochem. Genet.* **17:**1079–1091.

Uyeno, T., and Smith, G. R. 1972, Tetraploid origin of the karyotype of catostomid fishes, *Science* **175**:644–646.

Wheat, T. E., Whitt, G. S., and Childers, W. F., 1973, Linkage relationships of six enzyme loci in interspecific sunfish hybrids (genus *Lepomis*), *Genetics* 74:343–350.

Whitt, G. S., 1981*a*, Developmental genetics of fishes: Isozymic analysis of differential gene expression, *Am. Zool.* **21**:549–572.

Whitt, G. S., 1981*b*, Evolution of isozyme loci and their differential regulation, in: *Evolution Today* (G. G. Scudder and J. L. Reveal, eds), University of British Colombia, Vancouver, pp. 271–289.

Whitt, G. S., Childers, W. F., Shaklee, J. B., and Matsumoto, J., 1976, Linkage analysis of the multilocus glucosephosphate isomerase isozyme system in sunfish (Centrarchidae, Teleostii), *Genetics* **82**:35–42.

Whitt, G. S., Phillip, D. P., and Childers, W. F., 1977, Aberrant gene expression during development of hybrid sunfishes (Perciformes, Teleosti), *Differentiation* **9**:97–109.

Wilson, A. C., Carlson, S. S., and White, T. J., 1977, Biochemical evolution, *Annu. Rev. Biochem* **46**:573–639.

Wolf, U., Ritter, H., Atkin, N. B., and, Ohno, S., 1969, Polyploidization in the fish family Cyprinidae, order Cypriniformes. I. DNA content and chromosome sets in various species of Cyprinidae, *Humangenetik* **7**:240–244.

Wright, J. E., May, B., Stoneking, M., and Lee, G. M., 1980, Pseudolinkage for the duplicate supernatant aspartate aminotransferase in brook trout (*Salvelinus fontinalis*), *J. Hered.* **71**:223–228.

Zimmer, E. A., Martin, S. L., Beverly, S. M., Kan, Y. W., and Wilson, A. C., 1980, Rapid duplication and loss of genes coding for the α chains of hemoglobins, *Proc. Natl. Acad. Sci. USA* **77**:2158–2162.

Zuckerkandl, E., 1978, Multilocus isozyme systems, gene regulation, and genetic sufficiency, *J. Mol. Evol.* **12**:57–89.

CHAPTER 3

A New Look at Sex Determination in Poeciliid Fishes

KLAUS D. KALLMAN

1. Introduction

The recent increased interest in sex-determining mechanisms can be traced directly to Wachtel's concept that the H-Y antigen is the primary testis organizer. There have been several comprehensive reviews dealing with genetic mechanisms of sex determination and sex differentiation, but all of these deal largely with the mammalian literature and pay only lip service to the other classes of vertebrates (J.W. Gordon and Ruddle, 1981; Haseltine and Ohno, 1981; McCarrey and Abbot, 1979; Ohno, 1979; Wachtel and Ohno, 1979). Of particular interest to geneticists are several cases of atypical sex determination in mammals, including humans, in which an autosomal gene causes the testicular differentiation of XX individuals. This has raised the question as to the location of the male-determining gene.

Similar exceptional cases of XX males and XY females have occurred in several species of teleost fishes. These species are unusually amenable to genetic experimentation, and their study has led to a better understanding of the XX male and XY female conditions. In this chapter, I review the literature on the subject and summarize many of my still unpublished results. I come to the conclusion that most cases of atypical sex determination are due to the interaction of a single autosomal locus with a specific sex chromosome, the autosomal gene modifying the action of the sex-determining gene located on the gonosome.

KLAUS D. KALLMAN • Genetics Laboratory, Osborn Laboratories of Marine Sciences, New York Aquarium, New York Zoological Society, Brooklyn, New York 11224.

2. Polygenic Sex Determination in Fishes

Sex determination in fishes is said to be polyfactorial (polygenic), even in those species in which a sex chromosome mechanism has been demonstrated. This polygenic theory, which can be traced to Winge (1934) and Kosswig (1935), has been discussed in great detail by Kosswig (1964). According to this theory, sex determination in fishes is governed by a large number of sex genes that are located on many chromosomes. It is the ratio of male to female factors that ultimately determines the sex into which a given individual will develop. The sex chromosomes carry superior sex genes, i.e., male (*M*) and female (*F*) determinants with greater potencies than those of the autosomes. Under ordinary conditions the sum of the potencies of the *M* and *F* factors on the autosomes cancels each other out, so that sex determination proceeds according to the switch mechanism inherent in the sex chromosomes. Kosswig (1964) pointed out that it is not known whether there are separate loci for *M* and *F* determinants with different valences or whether a single locus may exhibit *M* and *F* alleles or alleles with neither *M* nor *F* valences. Inasmuch as sex is determined by the ratio of the sum of *M* to *F* valences and not by the absolute number of sex factors, individual males or females may have very different genotypes with respect to sex determination. To my knowledge, this theory has never been critically tested; nevertheless it has been accepted by most investigators and authors in this field. Only Bull and Charnov (1977) expressed some doubts about the validity of Winge's and Kosswig's model. Kallman (1968) pointed out that the autosomal factors in the southern platyfish (*Xiphophorus maculatus*) cannot be sex genes *per se,* because they had to interact with specific, independently segregating sex chromosomes to bring about their effect. In 1968, I referred to them as "transformer genes for sex," but today I would refer to them as regulatory genes. Similarly, Avtalion and Hammerman (1978) have proposed a model of sex determination for *Sarotherodon* (*Tilapia*) that involves a pair of gonosomes and a single autosomal locus. Each species is homozygous for the autosomal factor, but different alleles have become fixed in species with male or female heterogamety. The effect of the autosomal locus on sex determination becomes apparent only in species hybrids.

3. The H-Y Locus

Originally it was thought that the H-Y structural gene in mammals was located on the Y chromosome, because of the high (but not perfect) correlation between the presence of the Y chromosome and H-Y antigen

(Wachtel *et al.*, 1975*a*) and the increased levels of antigen in individuals with two Y chromosomes (Wachtel *et al.*, 1975*b*). Subsequently, it also was shown that YY males of the teleost *Xiphophorus maculatus* absorbed more H-Y antigen than did XY males (Pechan *et al.*, 1979). However, the occurrence in mammals of XX individuals that are H-Y positive and male and XY individuals that are H-Y negative and female has raised doubts about the location of the H-Y structural gene and has implicated several sexlinked or autosomal factors in its regulation (Ohno, 1979; Wachtel and Ohno, 1979; Wachtel, 1981).

An even better correlation exists in mammals between the presence of H-Y antigen and testis differentiation. The most direct evidence that in mammals the H-Y antigen causes the differentiation of the gonad primordium into a testis comes from reaggregation experiments on dispersed gonadal cells. Dispersed testicular cells of rat and mouse reorganized into a structure that resembled testicular tubules, but the cells developed an ovarian follicle-like organization after they had been stripped of H-Y antigen (Ohno *et al.*, 1978; Zenzes *et al.*, 1978*a*). Conversely, in the presence of H-Y antigen, dissociated rat ovarian cells and bovine embryonic indifferent gonads from XX individuals reorganized into testicularlike structures (Zenzes *et al.*, 1978*b;* Nagai *et al.*, 1979).

Wachtel *et al.* (1975*c*, 1980*a*) suggested that the H-Y antigen has been preserved throughout vertebrate evolution and that it has functioned as the primary inducer of the heterogametic gonad in all vertebrates. Thus, in mammals and in those amphibians with an XX ♀♀–XY ♂♂ mechanism the males are H-Y positive, while in birds, reptiles and in amphibians with female heterogamety, H-W (Y) antigen activity is restricted to females (Wachtel, 1983). Recently, these ideas have been extended to teleosts and they have been supported by the discoveries of H-Y activity in males but not in females in *Oryzias latipes, Xiphophorus maculatus,* and *Haplochromis burtoni* (Pechan *et al.*, 1979), *Poecilia reticulata* (Múller and Wolf, 1979; Shalev and Huebner, 1980), and *X. helleri* (Múller and Wolf, 1979). The Eichwald–Silmser phenomenon has also been reported for an inbred strain of *X. maculatus* (XX ♀♀–XY ♂♂), using scale grafts (Miller, 1962). The sex-determining mechanisms of *X. helleri* and *H. burtoni* are unknown, but in *O. latipes* and *P. reticulata* males represent the heterogametic sex, whereas in *X. maculatus* males are either XY or YY (see Section 4). But in the European eel, *Anguilla anguilla,* which shows female heterogamety (Park and Grimm, 1981), H-W (Y) activity is restricted to females (Wiberg, 1982).

In a number of species of *Salvelinus, Salmo, Rutilus, Carassius,* and *Barbus,* however, no sex difference in the level of anti-H-Y antiserum absorption could be detected (Müller and Wolf, 1979). These authors pointed out that cross-reacting antigen resembling H-Y may not be sex-limited in all lower vertebrates, although in the more advanced teleosts, H-Y activity, as far as is known, is restricted to the heterogametic sex. According to Shalev and Huebner (1980), the fact that in some species no significant differences in H-Y activity can be detected between males and females as well as the relative ease with which XX and XY embryos of certain species of teleosts can be made to differentiate into the opposite sex by exposing them to suitable extrinsic agents (Clemens and Inslee, 1968; Dzwillo, 1962; Yamamoto, 1958, 1959; Yamamoto and Kajishima, 1968) suggests that the H-Y cross-reactive antigen has only a limited role in gonad differentiation. Before this view can be accepted, however, it must be determined whether the external agents (steroid hormones, temperature) do not induce H-Y antigen first and subsequently testicular differentiation. Shalev and Huebner (1980) also suggested that present-day fishes may represent an early stage in the evolution of the gonad-determining function of H-Y. If this is the case, the H-Y antigen must have evolved its induction of heterogametic gonads independently in the different lineages exhibiting heterogamety. The function of the H-Y antigen may indeed be an ancestral one, but its regulation may be quite diverse in different groups.

4. Polygenic Sex Determination in Mammals

There is some evidence that in mammals other sex-linked genes besides the H-Y locus are involved in sex determination. If the transcription of a Y-linked H-Y locus were dependent upon the activity of an X-linked (X^+) locus, its presence could not be detected unless a mutation had occurred (X^*) that exhibited an altered interaction with the H-Y locus. Such an X-linked gene could be referred to as the activator of the H-Y locus on the Y chromosome. In this scheme, in the presence of X^+Y, the H-Y locus would be normally transcribed and would result in the testicular differentiation of the gonad primordium, whereas in X^*Y individuals the H-Y locus would not be active and in the absence of H-Y antigen, the undifferentiated gonad primordium would develop into an ovary. The best evidence for the existence of such an X-linked gene can be found in two species of lemming, *Myopus schisticolor* and *Dicrostonyx torquatus* (Fredga *et al.*, 1976; Gileva and Chebotar, 1979). It has been shown that *M. schisticolor* is polymorphic for an X-linked factor that suppressed H-Y activity (Wachtel *et al.*, 1976). Three kinds of females were identified in *M. schisticolor,* X^+X^+, X^+X^*, and X^*Y, and one kind of male, X^+Y. The short arm of the X and X^* chromosomes differ in length and G-band staining, suggesting that a chromosomal rearrangement was involved in the origin of the X* (Herbst *et al.*, 1978). Females of genotype X^+X^+ produced males and females in equal proportions, X^+X^* females gave rise to males and females in a ratio of 1:3, and X*Y females resulted in all-female progeny. No YY zygotes were produced by the X*Y females, because the Y chromosome is selectively eliminated in the female germ line, and the X* chromosome duplicates itself to yield X*X* oogonia (Fredga *et al.*, 1977).

There are two additional well-established cases in mammals in which sex differentiation does not proceed according to sex-chromosome constitution. The *Sxr* factor causes XX mice to develop into males (Cattanach *et al.*, 1971) and these males are H-Y positive (Bennet *et al.*, 1977; Wachtel *et al.*, 1977). Clearly, the H-Y locus cannot be restricted to the Y chromosome. Cattanach *et al.* (1971) originally favored the view that the Sxr condition is caused by an autosomal dominant gene mutation and they also suggested that the Sxr locus is similar to other genes that cause male differentiation in XX individuals of other mammalian species.

As an alternate explanation it has been suggested that there is translocation of the H-Y locus from the Y chromosome to an autosome or, if the H-Y gene exists in multiple copies on the Y, the translocation of only some of these copies (Ohno, 1979). There is, however, no convincing cytological evidence for such a translocation, as Cattanach has pointed out. Alternately, Wachtel (1977) proposed that the H-Y locus may be an autosomal factor that, under normal conditions, is induced by an activator on the Y chromosome. The Sxr factor would then represent a constitutive mutation of the H-Y locus. It is difficult to see why in one species (lemming) the H-Y locus should be sex-linked and in another (mice) autosomal. If the H-Y locus is autosomal, a hierarchy of regulating genes would have to exist. The activator on the Y would control the autosomal H-Y locus, while the activator itself would be controlled by the X-linked regulator. Recently, Singh and Jones (1982) showed, using DNA hybridization, that the XX-Sxr male condition arises from a nonreciprocal crossover between an aberrant Y chromosome and the X. As a result, the X chromosome gains at its distal end the male sex determiner. According to a slightly different interpretation, only one chromatid of the aberrant Y chromosome is involved in the exchange (Hansmann, 1982).

The second experimental situation involves the autosomal *Po* locus in goats. The dominant *Po* factor causes hornlessness with no effect on sex determination in the *Po* heterozygotes. The XX *PoPo* homozygotes, however, develop into male pseudohermaphrodites (Hamerton *et al.*, 1969). These animals are H-Y positive, but the amount of H-Y antigen is quite variable and less than in normal XY males (Shalev *et al.*, 1980). These authors suggested that *Po,* or a factor closely linked to it, is a mutant regulatory gene that mimics the Y-linked regulator of the X-linked H-Y locus.

Additional evidence that both sex-linked and autosomal loci are involved in sex determination and that the H-Y structural gene cannot be restricted to the Y chromosome comes from the XX male and XY female condition in humans. Although the XX male condition results from heterogeneous causes, the common denominator is that the individuals are H-Y positive (Wachtel *et al.*, 1981, 1980*b*). Wachtel *et al.* (1980*b*), following Hamerton *et al.* (1969), pointed out that the condition inherited as an X-linked trait is consistent with the X-linkage of the H-Y structural gene, which is normally activated by a regulatory gene on the Y. As a result of a change on the X chromosome, the H-Y locus might become constitutive. The XY female condition with dysgenetic ovaries, which is H-Y negative, is inherited as an X-linked trait and there is some evidence for a structural change on the X chromosome involved (Bernstein *et al.*, 1980). Wolf (1979) suggested that the H-Y structural gene is autosomal and that regulatory genes are located on both X and Y chromosomes. The Y-linked factor interacts with (that is, blocks) the X-linked regulator and causes the H-Y locus to become active. This idea is supported by the observations that patients with Turner's syndrome (XO) and XO mice are H-Y^+ but that their level of antigen is significantly less than in normal males (Wolf *et al.*, 1980; Engel *et al.*, 1981*a,b*). A mutation of the X-linked locus could render the H-Y gene inactive even in the presence of a Y. The XY female condition that is inherited as an autosomal trait was attributed by Wolf (1979) to a mutation at the H-Y structural locus.

These different and often contradictory interpretations share in common the idea that the sex-determining mechanism does not consist only of a single locus (H-Y), but also of one or more regulatory genes. The problem is that no critical experiment that could pinpoint their location has yet been designed. Implicit in these ideas, although expressed in detail only by Thompson (1978), is the assumption that the H-Y locus consists of two units, an operatorlike control element and the adjacent structural gene *H-Y*.

5. The Sex-Determining Mechanism of the Platyfish, *Xiphophorus maculatus*

There are three sex chromosomes in *X. maculatus*, W, X, and Y. Three combinations of sex chromosomes, WY, WX, and XX, cause female differentiation, and two combinations, XY and YY, give rise to males. In wild populations, six kinds of matings occur (Table I), of which

Table I
Sex Ratio of the Platyfish, *Xiphophorus maculatus*

Mating ♀	Mating × ♂	Sex ratio ♀♀	Sex ratio : ♂♂	Genotypes of offspring ♀♀	Genotypes of offspring ♂♂
WY	XY	1	1	WX, WY	XY, YY
WY	YY	1	1	WY	YY
WX	XY	3	1	WX, WY, XX	XY
WX	YY	1	1	WY	XY
XX	XY	1	1	XX	XY
XX	YY	0	1	—	XY

four result in an even sex ratio, one gives rise to a ratio of three females to one male, and one yields all male progeny (Kallman, 1965, 1973). The genotype, WW, is not encountered under natural conditions, but can be produced in the laboratory by selective breeding (see Section 8). Extensive sampling of natural populations has revealed the occurrence of W and X chromosomes and XY and YY males in most river systems except the Río Jamapa and Río Papaloapan, Vera Cruz, Mexico, in which the W chromosome has not been demonstrated (Kallman, 1973). This is a sex-determining mechanism that closely resembles the condition found in the lemming, *M. schisticolor* (see Section 4). In both species, three kinds of females are known: X^+X^+, $W(=X^*)X^+$, and $W(=X^*)Y$. The major difference between the platyfish and lemming mechanisms is that the former results not only in X^+Y males but also in YY males. From a genetic and evolutionary point of view, this difference can be readily accounted for, because in *X. maculatus* there is convincing evidence that X-linked loci are also present on the Y chromosome and that YY individuals do not lack any loci that are essential for viability. In contrast, chromosomal evolution in lemmings, as in other mammals, has resulted in a Y chromosome that is largely inert. The YY condition, therefore, lacks many genes that are necessary for normal development and viability. This in turn probably led to the evolution of the modified meiotic mechanism in X^*Y female lemmings, which selectively eliminates the Y chromosome and prevents the waste of YY zygotes.

If this view is correct, the evolution of the sex-chromosome mechanism of *X. maculatus* becomes less difficult to envisage. The notion that the sex-chromosome mechanism of different *X. maculatus* populations arose independently from a condition in which the 24 pairs of chromosomes were undifferentiated as far as sex determination was concerned (Anders and Anders, 1963; Dzwillo and Zander, 1967), leading to the XX♀♀–XY♂♂ and the WY ♀♀–YY♂♂ systems, can now be discarded.

This old idea was based on a few fish collected in the Río Jamapa (XX–XY) and on an equally small sample from Belize in which all the females happened to be heterogametic, WY, and males homogametic, YY (Kallman, 1973). Kallman (1965, 1983) had previously pointed out that the W, X, and Y chromosomes of *X. maculatus* are homologous with the X and Y chromosomes of several other species of *Xiphophorus* (*X. xiphidium, X. variatus, X. milleri, X. pygmaeus, X. nigrensis, X. montezumae*), and that it is highly unlikely the same undifferentiated pair of autosomes could have evolved into a sex chromosome in all these species. Moreover, similar linkage groups are present on the sex chromosomes of these species. These observations strongly indicate that the ancestral species of *Xiphophorus* already exhibited the XX–YY mechanism and that the W–X–Y system of *X. maculatus* evolved from it as a result of a gene mutation or chromosomal rearrangement.

Although the differentiation of W(=X*)Y individuals in the platyfish, *X. maculatus,* and in the lemming, *M. schisticolor,* into functional females could be regarded as evidence for the existence of an X-linked activator gene for a Y-linked H-Y locus, the function of the normal allele at the activator locus (X^+) is problematic, because in *X. maculatus,* YY individuals differentiate into males. This is a point that can never be tested in the lemming or in any other mammals, because the YY condition is nonviable. The same consideration also holds true for several other species of atheriniform fishes with the XX–XY mechanism, species in which the YY genotype, in contrast to *X. maculatus,* arises only under abnormal conditions. In these species, YY progeny can be produced by mating exceptional XY females with normal XY males. The viability of such YY fish and their differentiation into males have been demonstrated for *X. milleri* (Kallman and Borowsky, 1972), *X. montezumae* (Kallman, 1983), *Oryzias latipes* (Aida, 1936; Yamamoto, 1964*a*, 1967) and *Poecilia reticulata* (Winge, 1934; Haskins *et al.,* 1970). In the two latter species, however, YY males were viable only if they were heterozygous for certain Y-linked pigment genes (Yamamoto, 1964*b*; Haskins *et al.,* 1970; Winge and Ditlevsen, 1938). The same set of observations also refutes the idea that the H-Y locus in these fish is restricted to the X chromosome and that the Y chromosome carries an activator gene.

According to these observations, the male-determining gene must be on the Y chromosome. A gene that blocks the male-determining gene or the H-Y locus must be present on the W chromosome of *X. maculatus* and on the X* of *M. schisticolor,* but no function can be assigned to its corresponding allele on the X chromosome. Nevertheless, in the presence of certain autosomal regulatory genes (see Section 8), XX, WY, WX, and WW platyfish can develop into functional males. To reconcile these apparent paradoxical cases of atypical sex determination, I propose, following Thompson (1978), that the male-determining gene is present on all sex chromosomes. In species with an XX–XY mechanism, it is a constitutive locus on the Y, but a facultative locus on the X (and W), where it has to be induced by appropriate signals. It follows that the operator-like control elements, O^1, O^2, or O^3, adjacent to the male-determining gene on the X, W, and Y chromosomes cannot be identical (Fig. 1). In this model the control loci play the key role in sex determination. The male-determining gene on the sex chromosome could be the H-Y locus. But even if the location of the H-Y locus is autosomal, my theory would not require changes in any great detail. In this case, there would merely be one additional step in the hierarchy of control genes that determine whether or not the H-Y locus becomes active. The male-determining gene on the sex chromosome would then become an activator for the autosomal H-Y locus. I attribute the differentiation of XX, WX, and WW platyfish into functional males (see Section 8) to the effect of mutant autosomal regulatory genes that activate the male-determining gene on the X or W chromosome. It should be remembered that XX males are also known from other species of fish.

Crossing over between the sex chromosomes of *X. maculatus* occurs at a frequency of 0.2–0.3% (Bellamy and Queal, 1951; Kallman, 1965). All crossover data indicate that the sex-determining locus is located at the proximal end of the chromosome near the centromere (Anders *et al.*, 1973, Kallman, 1975). Similar conclusions have been reached for the guppy, *P. reticulata* (Winge, 1934), and the medaka, *O. latipes* (Yamamoto, 1964*a*). These observations suggest that the location of the sex-determining gene

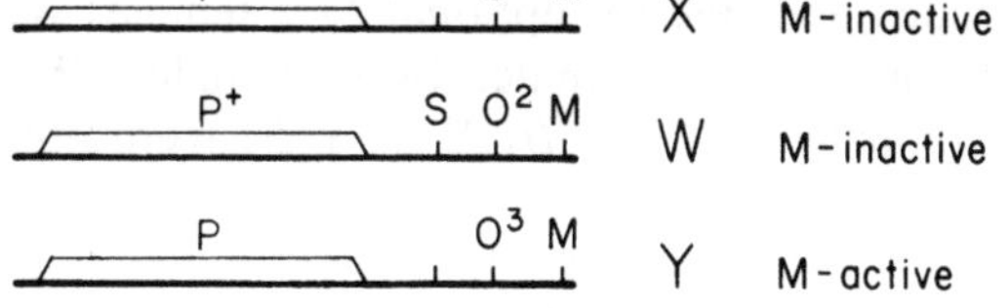

Figure 1. The male-determining gene *M* is located on the X, W, and Y chromosomes. Adjacent to *M* is an operator-like control gene *O*, which controls its transcription. The control elements on the three sex chromosomes are different. The interaction of a regulatory protein of an autosomal factor with O^1 and O^2 prevents the transcription of *M* on the X and W chromosomes. The control gene O^3 on the Y chromosome is not blocked and *M* is transcribed. A suppressor gene *S* is located on the W chromosome and inhibits the transcription of *M* on the Y. The region of the pigment loci is denoted as *P*. The P^+ denotes the pigment loci on the W that carry the wild-type alleles.

near the centromere may be an ancestral condition in these fishes and may have been preserved in different descendant lineages.

Crossing over between the W and Y chromosomes also provides insight into the location of the suppressor for the male-determining gene on the Y chromosome. If the suppressor is located proximal to the pigment genes and close to the sex-determining locus, the recombinant chromosome with *S* would remain a W but would acquire the pigment factors, and the Y chromosome would remain a Y but would acquire the wild-type allele at the pigment loci (Fig. 1). If the suppressor is located distally to the group of pigment genes, however, a crossover between *S* and the pigment loci would result in two chromosomes on which the suppressor and control genes are differently arranged than on the original two chromosomes. One crossover product would have the sequence $S\text{-}P\text{-}O^3\text{-}M$ and presumably this chromosome would continue to function as a W. The reciprocal crossover product with the arrangement $+\text{-}P^+\text{-}O^2\text{-}M$ would be expected to act like an X, because *M* linked to either O^1 or O^2 is suppressed. One could, therefore, expect the existence of at least two kinds of X chromosomes in *X. maculatus* and three kinds of W chromosomes. I know of no easy method to distinguish between the different W and X chromosomes, but whether the crossover product without the pigment factor remains a Y or becomes an X can easily be tested, although to my knowledge this has not yet been done.

Six well-documented instances of crossing over between the W and Y chromosomes have come to my attention and in each case the crossover product remained a W [Fraser and Gordon (1929), two cases; Kosswig (1937), one case; Kallman (1965), two cases]. The sixth crossover occurred recently in the Belize stock and involved a cross between a W-+ Y-*Br* ♀ and a Y-*Ir* Y-*Br* ♂ (*Ir* and *Br* are dominant pigment genes). The progeny (pedigree 3492) consisted of 13 *Ir* and 12 *Br* females and 12 *Br* and six *BrIr* males and one exceptional *BrIr* female. When this female was bred with an X-+ Y-+ male, 18 *Br* females (W-*Br* X-+, W-*Br* Y-+), and 16 *Ir* males (X-+ Y-*Ir*, Y-+ Y-*Ir*) were obtained (pedigree 3572).

The overall incidence of atypical sex determination (ASD), based on 14,127 platyfish raised in this laboratory, was 0.0085 (Table II). These fish comprise the offspring of hundreds of matings within and between stocks, and the details of these crosses have been published previously (Kallman, 1965, 1970). The fish with atypical sex determination were concentrated in just a few pedigrees or stocks; the familial nature of their occurrence clearly points to a genetic origin. Thirteen of the 17 known WY males came from a single pedigree with 16 WY offspring [pedigree 1952 (Kallman, 1970)]. The 27 WX or WY males were found among the 50 WX or WY offspring of pedigree 27 involving New River (Np) ancestry

Table II
Sex Chromosome Genotypes of *Xiphophorus maculatus*

	WY	WX	WY or WX	XX	XY	YY	XY or YY	WY or YY
♀♀[a]	1243	462	436	3033	44[c]	0	0	
♂♂[a]	17[c]	0	27[c]	30[c]	7298	1051	445	41[d]
♀♀[b]	2692	—	—	—	—	—	—	—
♂♂[b]	—	—	—	—	—	2734	—	14[e]

[a]Data from Kallman (1965, 1970).
[b]Data from Table I, Bellamy and Queal (1951).
[c]Instances of atypical sex determination.
[d]The exact number of WY males is not known.
[e]According to Bellamy and Queal's data, half of these males were WY, half YY.

(Kallman, 1965; see also Kallman, 1968). The 41 males listed as either WY or YY also belonged to the Np stock. Additional males with either WY or YY genotypes were subsequently obtained in the Np stock (Kallman, 1968). Two of these were tested and proved to possess the WY genotype. All 20 XX males belonged to the Lake Petén stock or were derived from it. Forty of the 44 XY females occurred in the Jamapa (Jp) stocks 163A and B. Another instance of the familial occurrence of XY females will be analyzed in Section 8.

A similar estimate of the incidence of exceptional WY males has been provided by Bellamy and Queal (1951), who reported their compositè observations on a domesticated stock of *X. maculatus* maintained for over 20 years. From crosses of the type W-+ Y-*C* ♀♀ × *Y*-+ *Y*-+ ♂♂ (where *C* stands for a dominant pigment gene), wild-type female offspring W-+ Y-+ and males Y-*C* Y-+ with the maternal pigment patterns were expected.

Bellamy and Queal obtained 2679 + females and 2734 *C* males besides 13 exceptional *C* females and 14 + males. Since I know of no report of YY females (but see Section 8) in *X. maculatus* (Kallman, 1973), it is certain that the females exhibiting the maternal pigment pattern were crossover products, W-*C* Y-+. Thus, they do not have an exceptional sex chromosome constitution. However, when some of the wild-type males were tested, one-half of them turned out to be the reciprocal crossover products, Y-+ Y-+, whereas the other half possessed the sex genotype W-+ Y-+, which is exceptional for males.

The incidence of atypical sex determination in *O. latipes* is equally low. Only 0.38% of the XY fish developed into females (Yamamoto, 1963) and only 0.11% of the XX genotype differentiated into males (Aida, 1930). Precise figures are difficult to obtain for the guppy, but it is apparent from

Winge's (1930) work that the spontaneous occurrence of XX males and XY females is a rather rare event. Haskins *et al.* (1961) reported the capture of an XX male in the Paria River of Trinidad, one of the few examples documenting the occurrence of atypical sex determination within natural populations.

6. The Sex Ratio in the Swordtail, *Xiphophorus helleri*

The sex-determining mechanism of *X. maculatus* may be considered a paradigm of stability, but the same certainly cannot be said for another member of this genus, *X. helleri,* the swordtail. There are many reports in the early literature, reviewed by Breider (1935*a*), Atz (1964), and Peters (1964), that *X. helleri* exhibits highly varying sex ratios, that males may mature at widely different ages and sizes, that females occasionally undergo sex inversion and become functional males, and that often females exhibit arrhenoidy. These reports are suspect, however, because most of the fish were not *X. helleri* but hybrids between *X. helleri* and *X. maculatus* backcrossed to *X. helleri* (M. Gordon, 1937, 1938). It was not apparent, therefore, whether these phenomena were an effect of hybridization or were characteristic of *X. helleri* alone. As will be pointed out below, all these phenomena apply to the hybrids as well as to the species.

Some of the best documentation that the sex ratios exhibited by *X. helleri* deviate significantly from unity was provided by Peters (1964), who reported that in one stock of *X. h. helleri,* males ($N = 459$) greatly outnumbered females ($N = 164$) and that in a second stock of *X. h. guentheri* from the Río Lancetilla, Honduras, the frequencies of males in individual crosses ranged from 18% to 70%. She also discovered that in her stocks two kinds of males were present, which became sexually mature at different ages and sizes. Furthermore, a correlation was established between the size and age of the males at the onset of sexual maturity and the sex ratio of their progeny. The small, early-maturing males sired significantly more male than female offspring, the percentage of males in the different crosses ranging from 45% to 82%. The large, late-maturing males, on the other hand, produced a significant excess of females; the percentage of males ranged from 18% to 47%, depending upon the cross. Altogether, the small males produced 423 ♀♀ and 711 ♂♂ and the large males 1329 ♀♀ and 897 ♂♂. By mating the same male with more than one female, Peters could also demonstrate that the sex ratio of the offspring depended in part upon the genotype of the female.

Peters (1964), following Kosswig (1964), explained her observations by stating that sex determination in swordtails has a polyfactorial basis and that in late-maturing males the sum of male valences barely predominates over the sum of female valences, whereas in early-maturing males the ratio of male-to-female valence is much larger than one.

Implied in Peters' publication, but not spelled out in detail, was the idea that the same genes involved in sex determination also control the onset (age or size) of sexual maturity.

In fish, as in other vertebrates, gonad maturation is mediated by the appropriate signal from the hypothalamus to the gonadotrops of the pituitary gland, which in turn secrete gonadotropin, which acts upon the gonad. Just how a polyfactorial system of sex determination can effect the functioning of the hypothalamus is not clear. Peters' conclusions were also based upon very limited material involving only four large, late-maturing males and four small, early-maturing males. Moreover, all these males, and also the females to which they were mated, were related to each other, decreasing even further the effective number of the experimental animals.

Lodi (1980) has recently made similar claims for a domesticated stock of *X. helleri* without giving any detailed evidence. But the data are much too limited to indicate whether they are of general significance and might be applied to other species of *Xiphophorus* as well. In order to ascertain whether the same genetic condition is responsible for late maturation and large size on the one hand and for an excess of female offspring on the other, the late-maturing males of Peters' and other stocks must be outcrossed to other, unrelated strains. It is difficult to escape the conclusion that because of inbreeding there was actually selection for both traits in Peters' experiment. No such correlation exists in *X. cortezi* (Zander, 1965) and, as will be shown below, a sex ratio biased in favor of females is associated with small size and early maturation in *X. nigrensis*.

The unequal sex ratios of *X. helleri* are not a result of inbreeding. In Table III I have listed the overall sex ratios of several laboratory stocks of six species of *Xiphophorus*, all without known sex-linked traits and therefore without any known sex-determining mechanism. For some of these stocks, the early records are not extant, but for others their entire genetic history is known. These stocks are maintained either by full-sib matings, each generation being produced by a single male and female, or, occasionally, by cousin matings. The complete data for some of these stocks are listed in the Appendix. Most of the uneven sex ratios occurred in different stocks of *X. helleri* and *X. signum*, in which they must be considered the rule in four out of the seven strains listed. These uneven sex ratios were never associated with differences in size or age at maturation of the male parent. Only one of the *X. helleri* stocks had sex ratios that always conformed to an expectancy of unity. On the other hand, of 156 crosses involving *X. clemenciae*, *X. alvarezi*, *X. cortezi*, and

Table III

Sex Ratios of Six Species of *Xiphophorus* from Which No Sex-Linked Traits Are Known[a]

Taxon, code of stock, and drainage	Total number of ♀♀	Total number of ♂♂	Generation	Number of crosses	Number of crosses[b] with significant ($P \leq 0.01$) excess of: ♀♀	Number of crosses[b] with significant ($P \leq 0.01$) excess of: ♂♂
X. helleri helleri, Cd, Río Jamapa–Atoyac system, Mexico	406	812	8–35	35	4	16
X. helleri strigatus, 3B, Río Papa-loapan system, Mexico	442	496	17–35	28	1	7
X. helleri strigatus, Río Sarabia, Rio Coatzacoalcos basin, Mexico	370	382	1–16	27	0	0
X. helleri guentheri, Gx, Río Grijalva system, Mexico	551	508	8–35	37	1	1
X. helleri guentheri, Río de la Pasión, Rio Usumacinta basin, Mexico	128	166	1–4	7	0	1
X. helleri guentheri, Bx, Belize River drainage, Belize	641	913	2–33	43	4	19
X. signum, Rio Semococh, Río Chajmaic drainage, Río Usumacinta basin, Guatemala	550	267	1–16	26	12	1

X. alverezi, Río San Ramon, Río Lacantun drainage, Río Usumacinta basin, Guatemala	48	151	1–6	6	0	3
X. alvarezi, Río Candelaria–Yalicar, Rio Usumacinta basin, Guatemala	102	113	1–5	6	0	0
X. alvarezi, Río Dolores, Río Salinas drainage, Río Usumacinta basin, Guatemala	209	174	1–10	11	0	0
X. clemenciae, Río Sarabia, Río Coatzacoalcos basin, Mexico	245	279	1–15	16	0	0
X. cortezi,[c] Río Axtla, Río Panuco basin, Mexico	302	295	—	21	0	0
X. cortezi,[d] (Arroyo Palitla x Axtla)	300	286	—	16	0	0
X. cortezi (strain 38), Río Axtla, Mexico	406	409	1–32	38	0	0
X. couchianus, Río Santa Catarina, Rio Grande basin, Mexico	473	495	1–43	41	0	1

[a]All stocks maintained at the Genetics Laboratory of the New York Aquarium.
[b]Crosses with fewer than 10 offspring have been omitted.
[c]Summarized from Kallman and Atz (1966).
[d]Summarized from Kallman (1972).

X. couchianus, only four deviated significantly from unity. If larger samples of these species are eventually tested, sex ratios that deviate significantly from unity may also be discovered in these species. There is already convincing evidence that this is the case for *X. cortezi* (Zander, 1965).

Uneven sex ratios in the Bx stock first occurred in the third laboratory generation, in which two females and 16 males were obtained. In the fourth generation of 31 offspring, only one was a female. The swordtail stock from the Río de la Pasíon can be traced to two gravid wild-caught females. One of these gave rise to 40 offspring, all males, while the other produced 10 females and 10 males. A highly biased sex ratio in favor of females was present from the very beginning in the *X. signum* stock. This stock was derived from a single gravid female, which produced 67 females and one male, and in the following generation, 26 females and five males were obtained. The two crosses comprising the third generation yielded 28 females on the one hand and 19 females and eight males on the other.

All stocks of *Xiphophorus* are maintained at the Genetics Laboratory of the New York Aquarium under similar conditions. The highly distinctive sex ratios that have been observed over many generations clearly indicate that the sex ratio is under genetic control. Significant genetic differences with respect to sex determination must exist among the various stocks of *X. helleri.* Although Winge (1934) and Aida (1936) thought that the season had some effect on the frequency of XX males in the guppy and the medaka, respectively, Breider (1935*a*) did not obtain any indication that an environmental factor had any effect on the sex ratio of *X. helleri.* Breider concluded that the differences in the sex ratio must have a genetic basis, and Zander (1965) came to the same conclusion for *X. cortezi.*

However, sex ratios that deviate significantly from unity cannot automatically be attributed to a polyfactorial system of sex-determining genes without corroborative evidence. There is one example in *X. maculatus* in which uneven sex ratios are clearly due to other causes. Crosses between WY females from the Río Hondo and XY males of other stocks produced a significant excess of females (627 ♀♀, 472 ♂♂) (Kallman, 1965). Inasmuch as the sex chromosomes carried suitable marker genes, it could be determined that the differentiation of any XY or YY fish into females was most unlikely.

7. Do Swordtails Change Sex?

There are four scientific reports (Peters lists a few additional references from the popular literature) that describe how female *X. helleri* transformed into functional males (Essenberg, 1926; Harms, 1929; Schmidt,

1930; Lodi, 1980). As already pointed out by M. Gordon (1956), Atz (1964), and Peters (1964), the first three reports are so poorly documented as to raise serious doubts whether any "sex reversals" really occurred. Only Peters allows two of Essenberg's females as possible candidates for sex reversals.

Referring to Essenberg's alleged sex reversals, M. Gordon (1956) wrote, "From the thousands of swordtails and platyfish that have been reared in various laboratories for more than a quarter of a century since 1926, no additional cases of spontaneous and functional sex reversals have been reported and substantiated during the last 25 years." During the following 25 years not a single sex reversal has been observed in this laboratory and a similar absence of sex reversals in *X. helleri* has been noted by other researchers (Vallowe, 1957; Peters, 1964; Gomelsky and Fetisov, 1977), all of whom agree that differentiation in the swordtail is stable. The prominence that has been given in the literature to these alleged cases of sex reversal is totally unwarranted.

Lodi (1980) reported that three females that previously had given rise to broods eventually transformed into functional males, which, when mated to six females, sired a total of 569 offspring, of which 447 were females. However, the evidence for functional sex reversal as presented by Lodi is again not as clear-cut as one might wish. The females that were believed to have transformed into males were kept together with other females that had previously given rise to young. Although these fish were said to have been kept under close observation, Lodi provided no information as to the number of females maintained in the tank, the age of the females, the number of broods and young before transformation, and the number of months between the last brood and the first appearance of male secondary sex characters.

Lodi stated that before the development of their male secondary sex characters, the late-maturing males were not only large, but that they also looked like females. However, as already pointed out by Kosswig (1941), the anal fin of late-maturing males at 5–7 weeks of age is already clearly different in shape from that of females. It is difficult to avoid the suspicion that the alleged functional sex reversals were really late-maturing males and that Lodi had difficulty in distinguishing between females and late-maturing males. More convincing proof that functional sex reversals had indeed occurred would have been an illustration of the axial skeleton of the hermaphrodites in question. As discussed in detail by M. Gordon (1956) and M. Gordon and Benzer (1945), it is well known that in *Xiphophorus* the first three hemal spines of females undergo histolysis and eventually disappear altogether. In males, however, the first hemal spine becomes detached and moves anteriorly to form the ligastyle, whereas the second through fourth hemal spines become strengthened through additional bone growth to form the gonapophyses (Rosen and Kallman,

1959), to which the gonactinosts of the gonopodium become anchored. When adult females of *Xiphophorus* were treated with anhydrooxyprogesterone they developed a single gonapophysis, whereas immature females treated similarly developed all three (Vivien and Mohsen, 1952). "Sex-reversed" males of *Xiphophorus,* therefore, can possess only a single gonapophysis.

It is also not clear how the sex-reversed females could have formed a normal gonopodium. No illustration of the transformed anal fins was provided. Both Turner (1942) and Grobstein (1942) demonstrated that the anal fin of mature *Gambusia a. affinis* and *X. maculatus* females are incapable of transforming into a typical gonopodium. The anal fin of adult females has lost the ability to elongate and under the influence of androgenic hormone it merely produces some of the specialized distal segments of the 3, 4, 5 complex in an aberrant fashion. More typical gonopodia can be induced in the regenerating fin of adult females (Grobstein, 1942).

Lodi (1980) described the testes of normal males as a deeply bilobed structure with two ducts, while those of the alleged sex-reversed males were oval in shape with one central cavity acting as the sperm duct. The testes of male *Xiphophorus* often appear ovoid in shape, however, especially when they are sectioned slightly lateral to their exact center. The central sperm duct referred to may actually represent a composite of various spermatogenic cysts, efferent ducts, and connective tissue that fill the spaces between the true ducts. The two sperm ducts do not always run in a straight line and sometimes part of one may stray toward the other side. Two ducts to the right and left of the central cavity in the lower half of his Fig. 2 appear to be the main sperm ducts. Moreover, no trace of ovarian tissue could be detected in the gonads of these alleged sex reversals and histologically they looked indistinguishable from normal testes.

On the basis of histological observations, Vallowe (1957) and Peters (1964) pointed out that sex differentiation is stable in *X. helleri* and that the testes of early- and late-maturing males are identical in structure.

Two phenomena may have contributed to the repeated claims of functional sex reversal in *X. helleri* (Atz, 1964; Peters, 1964). There is no doubt that both small, early-maturing and large, late-maturing males exist in *X. helleri* and, as will be shown, these differences have a genetic basis. In *Xiphophorus* at the onset of sexual maturity, the growth rate of males decreases significantly, whereas females continue to grow. For this reason, late-maturing males continue to grow for a longer period of time than do early-maturing ones. As long as the late-maturing males remain immature, they have the general habitus of females and unless one is thoroughly familiar with this species, these males could be mistaken for females. In Peters' (1964) opinion many of the "transforming" males of early authors were in reality late-maturing males.

The second phenomenon concerns the assumption of male secondary sex characters by females. Externally, it involves the elongation of the ventral rays of the caudal fin into a sword, the appearance of the melanophores that line the ventral and dorsal margins of the sword, the thickening of the third, fourth, and fifth anal fin rays, and the development of the specialized distal segments on them. In some fish the so-called "gravidity" spot may disappear. This problem of arrhenoidy has been discussed by Atz (1964) and Peters (1964), who pointed out that a variety of conditions and agents, including old age, appear to be able to initiate masculinization in female swordtails, perhaps as a result of an altered metabolism. During the last 15 years the development of male secondary sex characters in female swordtails has been observed in every stock of *X. helleri* and also in *X. signum* and *X. alvarezi,* but never in any of the other 11 species of *Xiphophorus* maintained in this laboratory. Only Cohen *et al.* (1941) reported the spontaneous development of masuclinized anal fins in female *X. maculatus* belonging to a single pedigree. Upon histological examination of some of these fish, typical ovaries were found. At least in this laboratory, arrhenoidy is not dependent upon age, because females as young as 4 months have exhibited masculinization. None of these females ever transformed into a male, even when kept under observation for a period of more than 1 year. Histological examination of many of these maculinized females showed that they possessed ripe ovaries and that none of them showed any trace of testicular tissue. In general, these females were poor breeders, but three of them gave rise to small broods. Often more than one-half of an entire generation of females has become masculinized. Some observations in this laboratory suggest that arrhenoidy is caused by an extrinsic agent. When the females of a given brood are raised in two separate tanks, all the females in one may become masculinized whereas the females in the second remain unmodified. Transfer of some of these females into the first tank has caused the appearance of male secondary sex characters within 4 weeks.

Xiphophorus helleri, X. signum, and *X. alvarezi* are more closely related to one another than to any other species of *Xiphophorus* (Rosen, 1979). The occurrence of arrhenoidy in these three species and its absence from all other *Xiphophorus* suggests that this condition has a genetic basis, which is shared by the three taxa. Arrhenoidy in these fishes should be looked upon as a genetically controlled response to an extrinsic agent.

8. Taxonomy and the Induction of the Heterogamic Gonad by H-Y (H-W)

It should be recalled that Wachtel *et al.* (1975*c*) proposed that H-Y (H-W) antigen functions as the inducer of the gonad of the heterogametic sex. This statement holds true when certain taxonomic categories (Aves, Mammalia) are considered, but no one has yet explored how it may be applied to more closely related taxa with opposing sex-determining mechanisms or to species in which either sex can be heterogametic. The restriction of the H-Y antigen to the heterogametic males (XY) in the lemming, *M. schisticolor,* and its absence from the X*Y females finds its explanation in the evolutionary history of this form. There can be little doubt that *M. schisticolor* arose recently from a form with an XX–XY mechanism. The pattern of H-Y activity in this lemming is still in accordance with its recent evolutionary past. [Recently, it was reported, in contradiction to Wachtel *et al.* (1976), that the female X*Y wood lemming is H-Y positive (Wiberg *et al.,* 1982). The reasons for these divergent results are not apparent at this time.] The same explanation applies to the teleost *X. maculatus* and provides another indication that its W–X–Y system has arisen from an XX–XY mechanism. Does this also mean that *X. helleri,* in which only males are H-Y positive (Müller and Wolf, 1979), possesses an XX–XY mechanism or has recently evolved from a form with such a system?

9. Atypical Sex Determination in Fishes

9.1. XX Males

There exist a number of instances of atypical sex determination (ASD) in fishes and I shall now examine how well these fit into a proposed scheme. Öktay (1959*a,b*) crossed an XX platyfish female of the Jp stock with a YY male of a domesticated stock of unknown origin. As expected, all offspring ($N = 205$) were males. The X chromosome was marked by *Sp* and the Y chromosome by either *Fu* or *R* (*Sp, Sd, Fu,* and *R* are dominant pigment genes). When one of the males was backcrossed to a Jp X-*Sp* X-*Sd* female, all but one of the 29 X-*Sp* X-*Sd* offspring were females. The exceptional male, X-*Sp* X-*Sd,* was in turn backcrossed to a Jp X-*Sd* X-*Sd* female. There were three exceptional males, X-*Sd* X-*Sd,* among the 93 offspring. When these three males were inbred, the incidence of XX males (24.3%) increased dramatically in just one generation. From 43 additional crosses over a span of nine generations of inbreeding, a total of 1858 ♀♀ and 865 ♂♂ were obtained. The percentage of males among the individual crosses varied from 4.5 to 69.5, but the frequency of XX males was never significantly higher than 50%. In two crosses the excess of males was of borderline significance ($P = 0.03$). However, when the XX males were outcrossed to Jp females the incidence of XX males decreased rapidly. From 15 such crosses, 854 females and 34 males were obtained.

The XX males occurred among the X-*Sp* X-*Sp*, X-*Sp* X-*Sd*, and X-*Sd* X-*Sd* classes, clearly indicating that the differently marked X chromosomes, all derived from the Jp stock, were equal as far as sex determination was concerned (Öktay, 1959*b*). In this case a mutation of the control element adjacent to the male-determining gene cannot explain the occurrence of the XX males. The genetic causes, therefore, must be autosomal. The origin of the XX males in Öktay's strain best fits the model of a mutant autosomal regulator gene, which, when present in homozygous condition, results in the activation of the male-determining gene on the X (Table IV). If a single mutant autosomal regulatory gene *a* were involved, then the heterozygous condition *Aa* must result primarily in females because of the consistent low incidence of XX males in outcrosses. However, most of the *aa* individuals must differentiate into males. Inbreeding would result in the following types of matings: *AA* ♀ × *Aa* ♂, *AA* ♀ × *aa* ♂, *Aa* ♀ × *Aa* ♂, and *Aa* ♀ × *aa* ♂. The frequency of XX male progeny would be less than 5% in the first cross and close to 50% in the last. When both parents are heterozygous, only 25% males are expected. Because most *aA* fish will be females, crosses involving *aA* males will be relatively uncommon.

According to this explanation, the original XX male was *Aa* at the autosomal locus. This male was backcrossed to a Jp female, XX *AA*. The offspring consists of two classes, XX *AA* and XX *Aa*. All *AA* and most of the *Aa* individuals would have been females. The actual result was 90 females and three males. The three males were in turn inbred and these crosses could have involved *Aa* females. This would lead to 25% *aa* offspring in the first inbred generation and the actual results fit this expectation rather well (187 ♀♀, 60 ♂♂).

A second instance of XX males in *X. maculatus* was enountered among the descendants of two fish collected in Lake Petén (Pp). Both these fish and their descendants were crossed to the Jamapa stocks and XX males occurred with varying frequency in the F_1, first Bc, and second Bc generations (Table V). Although the genetic basis for the development of these XX fish into males was not studied in any detail, it is obvious that it is independent of the different origins of the X chromosomes. In

Table IV
Proposed Interaction of the Autosomal *A* Locus with the X Chromosome

	Autosomal genotype		
Sex genotype	*AA*	*Aa*	*aa*
		Phenotype	
XX	♀♀	95% ♀♀	♂♂

Table V
Occurrence of XX Male Progeny in the F_1 and Bc generations of Jamapa (Jp) × Petén (Pp) *Xiphophorus maculatus*[a]

Pedigree	Generation	Sex of XX offspring ♀♀	Sex of XX offspring ♂♂	Origin of parents ♀♀	Origin of parents ♂♂
From Pp ♂-16					
1413	F_1	12	6	Jp 163 B	Pp-16
1580	First Bc	53	0	Jp 163 B	1413-11
1595	First Bc	62	0	Jp 163 B	1413-12
From Pp ♀-1					
1562	F_1	61	1	1382a-1[b]	Jp 163 B
1490	F_1	22	1	Jp 163 A	1382a-12[b]
1685	First Bc	47	10	Jp 163 B	1562-11[c]
1570	First Bc	71	2	Jp 163 A	1490-11
1686	Second Bc	53	0	Jp 163 A	1570-11
1687	Second Bc	71	10	Jp 163 A	1570-12[d]

[a]Modified from Kallman (1965).
[b]Two offspring of Pp ♀-1.
[c]Male parent was X-*Sp* X-*N*, with the X-*N* from Pp. Of 27 X-*Sp* X-*Sp* progeny, seven were males; of 30 X-*Sp* X-*N* progeny, three were males.
[d]Male parent was X-*Sd* X-*N*, with the X-*N* from Pp. Of 49 X-*Sd* X-*Sd* progeny, six were males; of 32 X-*Sd* X-*N*, four were males.

some of the males, both X chromosomes were derived from Jp, in others one of them came from Pp (Table V). It is also apparent that no fortuitous combination of "*M*" factors, as postulated by the polygenic theory of sex determination of Kosswig (1964), can be responsible for the occurrence of the XX males, because their frequency was approximately the same in the F_1 as in some of the second Bc generations. Moreover, no XX males are known to have occurred among the Jp stocks. Once again, the occurrence of XX males has to be attributed to a single autosomal factor *a*. As in Öktay's experiments, the Jp fish are *AA*, but some of the Pp fish fish must have been *Aa*, and apparently a small percentage of XX *Aa* fish again differentiated into males. This explanation readily accounts for the fact that the frequency of XX males did not diminish during two generations of backcrossing to Jp. None of the Jp × Pp fish were inbred; it is therefore not known whether the *aa* condition would have resulted in male differentiation.

The phenomenon of XX males has also been observed in one of the pygmy swordtails, *X. nigrensis* (see Section 10), in the guppy, *P. reticulata* (Winge, 1934), and in the medaka, *O. latipes* (Aida, 1936). The pattern of occurrence of XX males in these three species is in agreement with the above theory. The original three XX males of *P. reticulata* were mated to their own sisters and one of these males was also mated to one female in each of the F_1 and F_2 generations. All together, 508 females and two males were obtained from these five crosses. The cross of an F_2 female with the single F_3 male yielded males and females in approximately equal proportion, but no actual figures were reported. According to Winge's report, more XX males were obtained when these fish were inbred, but outbreeding resulted only in females (Winge and Ditlevson, 1948).

Aida (1936) obtained seven exceptional XX males among more than 5000 XX medaka. When these males and their descendants were in turn bred to related XX females, additional exceptional XX males were produced. An analysis of his data is difficult, because most of the males were mated simultaneously to three or five females and their offspring were combined. It is also not clear just how closely the females were related to the males. There were five sets of crosses in which the females were siblings of the males, but in other crosses the females apparently belonged to the same colony as the original males, although not directly related to the XX males. Inbreeding led to a consistently high frequency of male offspring, but when the same males were outbred to females of an unrelated stock, only female progeny were produced. The X chromosomes of Aida's stock were marked by two pigment factors, *R* and *r;* XX males occurred among the *RR, Rr,* and *rr* classes. With respect to sex determination, the two X chromosomes were equivalent.

In the absence of detailed knowledge concerning gene regulation in higher forms and of the sex-determining gene in particular, little can be said about the function of the common or "normal" allele at this autosomal locus. The *A* allele might be concerned with the suppression of the male-determining gene on the X chromosome, whereas the *a* allele has no such effect. Apparently *a* has no effect on the Y chromosome, but this can only be ascertained critically if YY *aa* individuals are produced.

9.2. WW, WX, and WY Males in the Platyfish

A very different genetic condition is responsible for the differentiation of WW, WY, and WX platyfish into males. The occurrence of these exceptional males is largely restricted to hybrids between the New River (Np) (WY ♀♀–YY ♂♂) and Coatzacoalcos (Cp) (XX ♀♀–XY ♂♂) strains and to fish of Np × Cp ancestry (Kallman, 1968).

There were two crosses between XY Np females and XY males of the Cp stock (pedigrees 627, 1571) and two additional crosses between Np females and exceptional WY males of Cp ancestry (pedigrees 1607,

1612). The chromosomes were marked by different pigment genes, so that the different chromosome classes could be identified without difficulty (Table VI). In pedigrees 627 and 1571, there were no appreciable differences in the frequencies of males between the WY and WX classes (total frequency of males 39%). This result suggests that the autosomal factor interacts specifically with the W chromosome in bringing about male determination and that it is unimportant whether the second gonosome is an X or a Y. In pedigrees 1607 and 1612, the overall frequency of males in the four WY classes was 59% (range 32–72), but only 18% of the WW fish differentiated into males. The significantly lower frequency of males in the WW class suggests that the autosomal factor(s) interacts independently with both W chromosomes to cause male determination.

At least some of the WW females transmitted the genetic conditions that result in atypical sex determination. Three WW females were mated with YY males of the Hp-1 stock; from two of these crosses 33 and 40% WY males were obtained (pedigrees 1735a, b), but the third cross resulted in virtually all-female progeny (pedigree 1759).

There were seven crosses between Jp XX females and some of the WX males (Tables VI and VII). Five of these matings represented the second backcross generation to Jp. Of the 382 WX progeny, 163 (43%) were males, whereas of 373 XX offspring, only nine (3%) were males. These results indicate very clearly that the autosomal factor interacts specifically with the W chromosome and not with the X to cause WX individuals to develop into males.

These observations differ fundamentally from those of Öktay, Aida, and Winge and my own on *X. nigrensis* (see Section 10) in that the incidence of atypical sex determination did not decrease after outcrossing. The autosomal factor responsible for the differentiation of WW, WX, and W zygotes into males must be effective in heterozygous condition.

Of special significance is the fact that the autosomal regulatory gene interacts epistatically with the W chromosome. The product of the rare autosomal factor, by nature of its interaction with the W chromosome, causes the transcription of the male-determining gene in 86% of all WX individuals (Table VII), but the interaction with the X chromosome is different and results only rarely (6%) in activation of the male-determining gene. This suggests a similarity of the sites on both W and X chromosomes with which the product of the autosomal factor interacts. Its product may have a differential affinity for the control elements of the W- and X-linked male-determining gene. In other words, the control elements cannot be identical (Fig. 1). Thus two differences can be recognized between W and X chromosomes: (1) As pointed out earlier, the former possesses a suppressor gene for the Y-linked male-determining gene, and (2) the control

Table VI
The Occurrence of WX, WY, and WW Males in *Xiphophorus maculatus*[a]

Pedigree	Parents[b] ♀	Parents[b] ♂	Progeny WW ♀♀	WW ♂♂	WY ♀♀	WY ♂♂	WY ♀♀	WY ♂♂	WX ♀♀	WX ♂♂	XX ♀♀	XX ♂♂	YY ♂♂	XY ♂♂
627	WY-*Sd*	X-*Sp* Y-+			+: 13	7			*Sp*: 17	7			*Sd*: 19	*SpSd*: 11
1571	WY-*Sr*	X-*Sp* Y-*Sd*			*Sd*: 11	10			*Sp*: 10	8			*SdSr*: 33	*SpSr*: 27
1607	WY-*N*	WY-*Sd*	22	1	*Sd*: 15	7	*N*: 8	10					*SdN*: 18	
1612	WY-*N*	WY-*Sd*	29	10	*Sd*: 21	32	*N*: 12	32						
1735a	WW	YY-*Sd*			*Sd*: 16	10	+: 35	15						
1735b	WW	YY-*Sd*			*Sd*: 8	11	+: 8	4						
1759	WW	YY			+: 75	1								
	X-*Dr* X-*Dr*	WX-*Sp*[c]							*Dr*: 188	130	*DrSp*: 313	7		
	X-*Sp* X-*Sp*	WX-*Dr*							*Sp*: 31	33	*DrSp*: 51	2		

[a]From Kallman (1968).
[b]WY ♀♀ taken from the Np stock. XY ♂♂ belonged to the Cp stock. WY ♂♂ were of (Up × Jp) × Cp ancestry. WW ♀♀ and ♂♂ derived from pedigree 1607 or 1612. The YY ♂♂ belonged to the Hp stock. XX ♀♀ came from the Jp stocks.
[c]A summary of six crosses. Kallman (1968) listed the offspring of one of these pedigrees, 1839, incorrectly. Pedigree 1839 consisted of 31 WX ♀♀, 18 WX ♂♂, and 47 XX ♀♀.

Table VII
The Percentage of WX Male Progeny in Crosses between Jp XX Females and WY, WX, or WW Males[a]

Pedigree of offspring	Progeny ♀♀	Progeny ♂♂	Pedigree and sex chromosomes of male parent	Value of *P* for the expectation of[b]: 25% ♂♂	50% ♂♂
1701	45	14	1571[c] WY	NS	S
1704	37	18	1571[c] WX	NS	S
1771[d]	31	33	1704 WX	S	NS
1773	11	1	1607[c] WY	NS	S
1861	24	16	1607[c] WY	0.04	NS
1787	26	3	1612[c] WY	NS	S
1756	27	4	1612[c] WY	NS	S
1849	78	24	1612[c] WW	NS	S
1772	40	16	1612[c] WW	NS	S
1834[d]	17	17	1772 WX	S	NS
1836[d]	30	29	1772 WX	S	NS
1837[d]	29	33	1772 WX	S	NS
1839[d]	31	18	1772 WX	NS	NS
1840[d]	44	15	1772 WX	NS	S

[a]From Kallman (1968).
[b]Not significant (NS): $P > 0.05$; significant (S): $P < 0.01$. Twenty five percent WX males are expected under the assumption that the presence of two autosomal factors is required; 50% WX males are expected under the assumption that only one is required.
[c]Pedigree from Table VI.
[d]Second backcross to Jp.

elements for the male-determining gene on the two chromosomes are not identical.

This explanation poses a problem with respect to the suppressor on the W chromosome that blocks the action of the Y-linked male-determining gene. Is it not reasonable to assume that the suppressor has the same effect on the male-determining gene on the W after it has been induced by the autosomal regulator? Crossover data are in agreement with the view that the suppressor gene and the sex locus including the control element are closely linked to each other. The observation that WW females transmit the condition for atypical sex determination and that significantly fewer males occur in the WW class than in the WX and WY classes can be taken as evidence that WW zygotes develop into males only if the suppressor genes on both chromosomes are blocked at the same time as the male-determining gene is induced. This means that the product of the autosomal regulator has a dual function: it blocks the suppressor and at the same time it induces the male-determining gene.

This can only take place if all three genes are closely linked to one another and form a functional unit.

When Np fish were crossed with stocks of *X. maculatus* other than Cp, or Cp fish were crossed with stocks other than Np, virtually all WY and WX offspring were females (Kallman, 1968). The high incidence of WY and WX males were restricted to fish of Np × Cp ancestry. These observations could be taken as evidence that the interaction of two autosomal factors was required to cause atypical sex determination, although in the discussion so far this phenomenon has been treated as if it were dependent upon a single autosomal mutant gene. If two unlinked factors *b* and *c* were involved, it would not essentially change my interpretation, but it would render a genetic analysis even more difficult. ("*B*" and "*C*" would represent the common alleles and *b* and *c* the rare ones at the two loci. No implication has been made as to the degree of dominance or recessiveness.) One of the autosomal factors, *b*, would be concerned with blocking the suppressor on the W chromosome whereas the other, *c*, would be involved in the control of the adjacent male-determining gene. According to this scheme, WW and WX zygotes could develop into males only in the presence of both *b* and *c*, but WY fish also could become males in the presence of only *b* (Table VIII). Activation of the male-determining gene is unnecessary in the WY *Bb CC* genotype because a constitutive male-determining gene is already present on the Y chromosome.

The detection of *b* and *c* poses considerable difficulties. The factor *c* can only be recognized in WX and WW individuals in which the factor *b* is also present. The factor *b* can be detected in strains with the WY–YY mechanism, but goes undetected in XX–XY stocks (Table VIII). Assuming that two factors were involved, there is the possibility that the factor *b* was introduced by the Np stock (WY–YY). The sex ratio of this stock

Table VIII

Atypical Sex Determination in *Xiphophorus maculatus:* Proposed Interaction of Two Autosomal Loci with the W Chromosome

Sex chromosome genotype	Autosomal genotype[a]			
	BB CC	*Bb CC*	*BB Cc*	*Bb Cc*
	Phenotype			
WY	♀♀	♂♂	♀♀	♂♂
WX	♀♀	♀♀	♀♀	♂♂
WW	♀♀	♀♀	♀♀	♂♂

[a]The factor *b* blocks the W-linked suppressor of the male-determining gene. The factor *c* causes transcription of the male-determining gene on the W chromosome, but has no effect of the X chromosome.

over 17 generations was 331 ♀♀:393 ♂♂, a significant excess of males ($P = 0.02$). Each of the three Y chromosomes was marked by a different pigment gene, whereas the W chromosome was unmarked. Consequently, each male should exhibit two patterns but each female only one (Kallman, 1965, 1968). There were a number of males, however, that exhibited only a paternal pattern. Two of these fish were mated and identified as WY males. Nevertheless, one cannot assign the WY genotype to all the males with only one pattern, because in this stock the pigment genes do not exhibit 100% penetrance. The upper limit for the frequency of WY males in this stock has been estimated to be 0.13 (Kallman, 1968). The other factor, *c*, then must have been introduced by the Cp stock (XX–XY), but there is no independent confirmation for this. The factor *c* cannot be identical with *a*, which causes XX *aa* individuals to develop into males, because no XX males have ever been detected in the inbred Cp stock (Kallman, 1965).

It follows that crosses between XX *BB CC* females and males of genotypes WY, WX, or WW, all of which are *BbCc*, should result in WX progeny of which never more than 25% should become males (Table VIII). The results of 14 such experimental crosses were ambiguous (Table VII). Nine crosses were in excellent agreement with the expectation ($P > 0.05$), and one cross was marginally acceptable ($P = 0.04$), but four crosses, in which equal numbers of WX males and females were obtained, did not fit the expectation ($P < 0.01$). Curiously, all four of these crosses belonged to the second backcross generation to Jp.

The idea that an autosomal regulator exists that merely blocks the W-linked suppressor without inducing the male-determining gene is not without some experimental evidence. There is one cross (pedigree 1952; Kallman, 1970) in which a Bp WY female was crossed to a Grijalva (Gp) XY male. The sex chromosomes of the two fish were marked by different pigment genes, ensuring an easy identification of the four chromosome classes among the offspring:

W-+ Y-*N* ♀ × X-*Sp* Y-*Sd* ♂

progeny	*Sp*: ♀♀ 24	*N Sp*: ♂♂ 9
(pedigree 1952)	*Sd*: ♀♀ 3 ♂♂ 13	*N Sd*: ♂♂ 12

Most pertinent is the observation that all but three of the *Sd* offspring (WY) differentiated into males, whereas all *Sp* (WX) offspring were females. The absence of any males in the WX class and the high frequency of their occurrence among the WY fish is statistically highly significant ($P < 0.0001$). This result should be contrasted with the crosses listed in Table VI, which yielded males in both WX and WY classes. One of the

W-+ Y-*Sd* females of pedigree 1952 was in turn crossed with a Jp X-*Sp* Y-*Sr* male. All 36 WX (*Sp*) offspring were again females, but nine of 37 WY progeny were males (pedigree 2071). This pedigree demonstrates that the genetic condition for atypical sex determination of WY fish can be transmitted by WY females, i.e., it exhibits less than 100% penetrance, and that the genetic factor present in pedigrees 1952 and 2071 is different from the condition present in the pedigrees listed in Table VI. The Bp WY female was a wild-caught fish, but the Gp XY male had been taken from the highly inbred Gp stock. Since the WY offspring of three other crosses between Bp WY females and Gp males (pedigrees 1467, Kallman, 1965; pedigrees 1933, 1917b, Kallman, 1970) differentiated into females, the mutant autosomal gene must have been introduced by the Bp female into pedigree 1952.

The Bp female could also have been homozygous for the autosomal gene, because virtually all of her WY offspring were either males or carriers. This would also account for the lower incidence of WY males in pedigree 2071, in which only half of the fish would have inherited this autosomal gene. Fourteen other wild-caught Bp females from the same geographic location were tested in the laboratory, but none of their WY offspring differentiated into males.

Little can be said about the function of the common allele at the *B* locus. Is its presence required for the activity of the W-linked suppressor of the male-determining gene? If this is the case, it must also be present in the Jamapa population of *X. maculatus*, in which the W chromosome is absent (Kallman, 1973). This chromosome has always continued to function like a "W" after introduction into the Jamapa stocks, provided the fish were not of Np × Cp ancestry (Kallman, unpublished). The factor *c* resembles *a* in that a small percentage of heterozygous XX fish (*Cc*) develop into males. No XX fish have been made homozygous for *c* and no WX fish have been made homozygous or heterozygous for *a;* it cannot be determined, therefore, whether "*a*" and "*c*" are allelic.

9.3. XY and YY Females in *Xiphophorus maculatus*

Females of genotype XY have arisen spontaneously in *X. maculatus* (MacIntyre, 1961; Anders and Anders, 1963; Kallman, 1965), *X. milleri* (Kallman and Borowsky, 1972), *X. montezumae* (Kallman, 1983), *P. reticulata* (Winge, 1934), and *O. latipes* (Aida, 1936; Yamamoto, 1963). Virtually all published cases of XY females in *X. maculatus* occurred in the Jamapa stocks. All nine XY females in stock Jp 163 B arose in a single mating of the 12th inbred generation. This stock is now in the 57th inbred generation. XY females arose also in strain Jp 163 A between the 14th and 31st generations, but since then no such females have been encountered. This strain is now in its 61st inbred generation. When such XY females were mated to XY males, sex determination returned to normal; all XY progeny were males (MacIntyre, 1961). Identical observations were made on *O. latipes* and *P. reticulata*.

An offshoot of the Jamapa stock was brought to Europe in 1950 by the late Dr. M. Gordon, and in 1959 Anders and Anders (1963) obtained four all-female broods in which all XY fish had differentiated into females. As in the case of MacIntyre's experiments, sex determination returned to normal when some of these females ($N=13$) were mated to Jp XY males. Of 358 XY progeny all but two were males. Through subsequent selection, a strain was eventually created in which approximately 25% of all females possessed the XY genotype; however, no data to support this claim were provided. It would, therefore, seem that the XY female condition has a genetic basis, although the details remain obscure. Totally unexplained is why, without selection, both MacIntyre (1961) and Anders and Anders (1963) obtained broods in which all XY fish became females.

Anders and Anders (1963) explained these results on the basis of Winge's theory of polygenic sex determination. Without explanation, Anders *et al.* (1969) subsequently changed their interpretation and now attributed the occurrence of XY females in the Jp stock to unspecified environmental factors. MacIntyre (1961) and Atz (1964) also thought, following Winge's polygenic theory, that genes with a male and female tendency had become relatively closely balanced as a result of inbreeding, thus making certain individuals susceptible to the influence of exogenous factors. Although no specific exogenous factors with effects on sex determination could be discerned, this idea should not be rejected, in view of the known influence of temperature on sex determination in the teleosts *Rivulus marmoratus* (Harrington, 1967) and *Menidia menidia* (Conover and Kynard, 1981) and in some reptiles (Bull, 1980). Both Aida (1936) and Winge (1934) obtained some evidence that in their artificial strains, the incidence of XX males increased during the summer months. Alternately, MacIntyre and Atz proposed that this phenomenon may be a reflection of the poorly canalized development often found in inbred organisms. However, this explanation appears to be ruled out by the absence of XY females in subsequent inbred generations.

A second instance of XY females, in *X. maculatus* unrelated to the Jamapa fish, has recently been discovered in this laboratory and has been analyzed genetically in great detail. This XY female condition eventually led to the production of YY females. Females of genotype XY were initially observed among the offspring of an *ArMr* male and a Jp XX female. The color patterns Ar and Mr are known to be sex-linked (Kallman, 1975). This male was the offspring of a gravid female collected near Bermudian Landing, Belize. This cross, which was part of a larger series, was performed to identify the sex-chromosome constitution of the male. If its genotype had been YY, all-male progeny would have been expected; but if the male had been XY, its progeny should have consisted of males and females with each sex inheriting a different sex-linked pigment pattern. Test crosses of this type were described by Kallman (1965, 1970).

Neither expectation was realized among the offspring of this cross. The 13 *Mr* offspring were all males, but the *Ar* fish were seven females and eight males (Table IX, cross 1). These *Ar* females and some of their *Ar* progeny as well were repeatedly backcrossed to Jp X-*Sp* Y-*Sr* males for 13 generations. All together, there were 34 crosses (Table IX, series A), which resulted in four pigment classes: *SpSp, SpAr, SpSr,* and *SrAr*. All 581 *Sp* offspring were females, all 590 *SpSr* and 613 *SrAr* were males. Of the *SpAr* progeny, 216 (35.9%) were females and 385 were males. A second series of 13 crosses (Table IX, series B) involved Jp XX females and *Ar* males, seven of which came from the original cross (cross 1), but the others can be traced to one male of series A. Again all *SpSp* progeny (N=254) differentiated into females, but in contrast to series A, most of the *SpAr* fish were males (four ♀♀, 245 ♂♂). The four females occurred among the offspring of only two of the crosses.

The interpretation of these observations assumes that the original *ArMr* male possessed two Y chromosomes that were qualitatively different, Y′-*Ar* Y-*Mr*. The symbol Y′ is introduced here to differentiate Y-*Ar* from the more usual Y chromosome, which, when combined with an X, always gives rise to males.

Under certain conditions, X-*Sp* Y′-*Ar* fish differentiate into females. Because of the significant difference in the frequency of *Ar* males that exists between series A (64%, N=601) and series B (98.4%, N=249), the existence of an autosomal factor *T* must be postulated, which interacts with the Y′ to produce XY′ females. Presumably, *T* is absent from series B, and this will be discussed further below. The point being made here is that series A provides critical evidence that the two Y chromosomes are different inasmuch as all 590 XY-*Sr* males differentiated into males, while 216 of 601 XY′-Ar fish were females. Because the two classes occurred within the same crosses, differences in genetic background cannot be responsible for the presence of females in the *Ar* class and their absence from the *Sr* class.

Subsequent crosses with the XY-*Mr* males (Table IX, series C) demonstrated once more that the Y chromosome marked by *Mr* is the common Y chromosome, which, in association with an X, always results in male differentiation.

Table IX
Sex Determination in *Xiphophorus maculatus* with Y′

Cross number or series	X-*Sp* X-*Sp* ♀♀	X-*Sp* Y′-*Ar* ♀♀	X-*Sp* Y′-*Ar* ♂♂	X-*Sp* Y-*Sr* ♂♂	Y′-*Ar* Y-*Sr* ♂♂	Y′-*Ar* Y′-*Ar* ♀♀	Y′-*Ar* Y′-*Ar* ♂♂	X-*Sp* Y-*Mr* ♂♂
1	Jp X-*Sp* X-*Sp* ♀ × Y′-*Ar* Y-*Mr* ♂							
		8	7					13
Series A[a]	X-*Sp* Y′-*Ar* ♀ × Jp X-*Sp* Y-*Sr* ♂							
	581	216	385	590	613			
Series B[b]	Jp X-*Sp* X-*Sp* ♀ × X-*Sp* Y′-*Ar* ♂							
	254	4	245					
Series C[c]	Jp X-*Sp* X-*Sp* ♀ × X-*Sp* Y-*Mr* ♂							
	117							123
2[d]	Jp X-*Sp* X-*Sp* ♀ × Y-*Sr* Y′-*Ar* ♂							
		33	151	169				
3[e]	X-*Sp* Y′-*Ar* ♀ × X-*Sp* Y′-*Ar* ♂							
	75	65	84			2	101	
4[f]	X-*Sp* Y′-*Ar* ♀ × Y-*Sr* Y′-*Ar* ♂							
		18	17	47	43	4	40	
5[g]	Y′-*Ar* Y′-*Ar* ♀ × Y′-*Ar* Y′-*Ar* ♂							
						215	203	
	Y′-*Ar* Y′-*Ar* ♀ × Jp X-*Sp* Y-*Sr* ♂							
6[h]		61	51		121			
7[i]		12	31		31			

[a]A summary of 34 crosses.
[b]A summary of 13 crosses.
[c]A summary of 10 crosses.
[d]P_1 ♂ from series A.
[e]P_1 ♀ from series A, P_1 ♂ from series B.
[f]P_1 ♀ and ♂ from cross 2.
[g]P_1 ♀ and ♂ from cross 4.
[h]P_1 ♀ from cross 4.
[i]P_1 ♀ from cross 3.

The combination XY′ was not sufficient by itself to cause the differentiation of these fish into females. In addition there must exist an autosomal regulatory locus with two alleles, *T* and *t*. Allele *T* interacts specifically with Y′ to cause female differentiation, but *t* has no detectable effect on either X, Y, or Y′. Moreover, the presence of a Y chromosome shelters Y′ from the effect of *T*, because all 808 Y-*Sr* Y′-*Ar* fish were males (Table IX). According to my interpretation, the control elements adjacent to *M* on the Y and Y′ chromosomes are not identical. The factor O^3 is located on the Y, but O^4 is present on Y′. There is a certain probability that the product of *T* interacts with O^4 to block the transcription of *M* and this causes XY′ individuals to differentiate into females. The factor *T* does not block O^3 on the Y chromosome; consequently, all *tT* YY′ or *TT* YY′ fish will be males (Table X). The Y′ chromosome has been introduced into the Jamapa stock through a series of backcrosses. The Jamapa stocks possess the *A* factor, which, as was pointed out, blocks O^1 on the X but not O^3 on the Y. According to series B (Table IX) the factor *A* also does not block O^4 on Y′. The possibility cannot be ruled out that the factor *T* represents a third allele at the *A* locus that specifically interacts with O^4, but does not differ from *A* in any other way.

Because the XY′ females were first discovered in Jp × Bp hybrids, it is not readily apparent from which strain or population *T* was derived. If the origin of *T* was the Jp stock, which has been inbred by brother-to-sister matings for more than 60 generations, *T* must be present in homozygous condition. Then, however, the frequency of only 1.5% XY′ females of series B as compared with 35.9% in series A cannot be explained. Note that both series involved backcrosses to the Jp stock! The second possibility is that *T* was introduced by the *ArMr* male of the Belize population, for which it was heterozygous. Then *T* was passed on to series A through the XY′-*Ar* females. Apparently *T* was not introduced by the XY′-*Ar* males into series B.

The Ar females of the F_1 generation, Jp × Bp (Table IX, cross 1), must have been heterozygous for *T*. The matings of series A represent a series of backcrosses to Jp for up to 13 generations. During this period, the frequency of *Ar* females remained rather constant, with an overall frequency, combining the results of all crosses, of 35.9%. If the additive effects of a number of autosomal loci were responsible for the differentiation of XY′-*Ar* fish into females, a decrease in the number of such

Table X
Proposed Interaction of the Autosomal *T* Locus with Y′

Sex chromosomes	Autosomal genotype *TT*	*Tt*	*tt*
		Phenotype	
X Y′	♀♀	72% ♀♀	♂♂
X Y	♂♂	♂♂	♂♂
Y′Y′	♀♀	5% ♀♀	♂♂
Y′Y	♂♂	♂♂	♂♂

females would be expected in succeeding generations. In the event that only two such autosomal factors were required, not more than 25% XY′-*Ar* females should have occurred in each generation, and in the case of three autosomal factors, not more than 12.5%. The actual percentage of females was signficantly higher and this argues strongly against the possibility that more than one autosomal factor was involved.

There is, however, one serious difficulty with the assumption of a single autosomal factor *T*. The genotype that must be assigned to the XY′-*Ar* females of series A is XY′-*Ar Tt,* and in each backcross generation one-half of the Ar fish would have been *Tt*, the other half *tt*. The frequency of females should therefore be 50%, but clearly the actual percentage (35.9%) is significantly lower. It must, therefore, be assumed that the penetrance of *T* is incomplete. According to series A, the penetrance of *T* in the XY′ *Tt* genotype must be about 72% [one-half of 601 *SpAr* fish are *Tt,* of which 216 (71.9%) were females]. The frequency of *SpAr* females in only three of these crosses differed significantly from the average of 35.9% *SpAr* females ($P = 0.03$; $P = 0.05$; $P = 0.01$).

The observation that the vast majority of the 249 XY′-*Ar* offspring of series B differentiated into males can readily be accounted for by the hypothesis. The P_1 males of these crosses possessed, with a single exception, the genotype XY′-*Ar tt*. For that reason the original *Ar Mr* male must have been heterozygous for *T*. The occurrence of four females in series B does appear to represent an embarrassment for the hypothesis, but three of these females can easily be explained. The results of series A indicated that the penetrance of *T* is only 72%; the *Ar* males of cross 1 and series A were in fact of two kind; that cannot be distinguished phenotypically, X-*Sp* Y′-*Ar Tt* (22%), and X-*Sp* Y-*Ar tt* (78%). Series B represents the combined results of 13 matings involving all seven Ar males of cross 1. The three females occurred in only one of these crosses (17 + ♀♀, three *Ar* ♀♀, 11 *Ar* ♂♂). In my interpretation, the male that sired this brood was *Tt*. The presence of three out of 14 *Ar* fish also agrees statistically with the expected average of 35.9% females ($P = 0.40$). All other males of series B were *tt*.

From these considerations it also follows that the Y-*Sr* Y′-*Ar* males produced in series A were either *tt* or *Tt*. Apparently a male of the *Tt* genotype was used in cross 2, because 18% of the XY′-*Ar* progeny were females. However, the frequency of females deviated significantly ($P < 0.0001$) from the expectation of 35.9%. This result represents one of two instances that cannot easily be reconciled with the hypothesis.

Cross 3 involved a mating in which both parents possessed the XY′-*Ar* constitution. The female was taken from series B and therefore was *Tt,* but the genotype *tt* has to be assigned to the male, because it was derived from series B. The expectation of 35.9% XY′-*Ar* females fits the actual result rather well ($P = 0.24$). Among the offspring of this cross, fish homozygous for Y′-*Ar* occurred for the first time. One-half of these fish should have been *tt*, the others *Tt*. Since all but two of 103 homozygous *Ar* offspring were males, it must be concluded that most Y′-*Ar* Y′-*Ar* *Tt* fish differentiated into males. Thus the penetrance of *T* in the heterozygous condition in the Y′-*Ar* Y′-*Ar* genotype is significantly lower than in XY′-*Ar* fish. This is analogous to the interaction of the autosomal regulatory gene with the W chromosome discussed earlier: the frequency of males was significantly higher in the WX than in the WW class (Table V).

In cross 4 the genotype XY′ *Tt* must be assigned to the female parent, but the YY′ male could have been either *tt* or *Tt*. For the first possibility, the expectation is again 35.9% female XY′ progeny and the actual result is in agreement with this expectation ($P = 0.08$). But based upon the results of cross 3, only a small percentage of the Y′Y′ *Tt* fish should have differentiated into females. Inasmuch as one-half of the *ArAr* males must be assigned the genotype *tt*, only about 2% of the *ArAr* class would be expected to be females. The actual result of 9.1% females does not fit this expectation very well ($P = 0.005$).

The second possibility assumes that the P_1 male of cross 4 was heterozygous, *Tt*. Each pigment class among the progeny consequently would consist of *TT*, *Tt*, and *tt* individuals in ratio of 1:2:1. None of the crosses discussed so far resulted in XY′-*Ar* *TT* fish and it is therefore not known what proportion of such a genotype would become females. According to cross 5, which is closely related to cross 4, the Y′-*Ar* Y′-*Ar* *TT* genotype develops into females. If this is the case, it would be expected that XY′-*Ar* *TT* fish would also be female. The expected frequency of XY′-*Ar* females of cross 4 consequently becomes 0.61 (25% of the XY′-*Ar* fish are *TT* and these will be females, plus 72% of the *Tt* class, which comprises half of the XY′-*Ar* fish). The observed result fits this expectation rather well ($P = 0.33$). Assuming that all Y-*Ar* Y-*Ar* *TT* fish differentiate into females, we find the expected frequency of *ArAr* ♀♀ in cross 4 to be 0.27 (25% of the *TT* class and 4% of the *Tt* class, which make up 50% of all *Ar* fish). The actual result shows a deficiency of females ($P = 0.01$), but provides a better fit than under the assumption that the P_1 male of cross 4 was *tt*.

Cross 5 is derived from cross 4 and both parents were homozygous for *Ar*. Highly significant is the sharp increase in the frequency of homozygous *Ar* females from 10% in cross 4 to over 50% in cross 5. The female parent of cross 5 could have been either *TT* or *Tt*, and the male either *Tt* or *tt*. Of the four possible types of matings, *Tt* ♀ × *tt* ♂ and *TT* ♀ × *tt* ♂ can be eliminated immediately, inasmuch as only a small percentage of *ArArTt* fish should have differentiated into females according to cross 3. If both parents were heterozygous for *T*, the frequency of female progeny should have been 0.27, assuming that all *TT* fish are females and that, in accordance with the crosses discussed above, only 4% of *Tt* fish are females. The observed frequency of 0.52 female progeny ($P < 0.0001$) shows decisively that both parents could not have been heterozygous for *T*.

The final possibility is that the female parent was homozygous and the male heterozygous for *T*. One-half of the offspring then would have been *TT* and expected to be females and the remainder *Tt*, of which only about 4% are females. The expected frequency of females thus comes to 0.52 and the actual result of 0.516 fits this expectation perfectly (P = 0.84). The discussion of cross 4 indicated that the genotype of the *ArAr* males was most likely *Tt* and cross 5 seems to bear this out.

Although the result of cross 5 is in good agreement with the assumption that all Y′Y′ *TT* fish differentiate into females and most Y′Y′ *Tt* fish into males, there is no independent genetic confirmation that this is indeed the case. The results can also be explained by assuming that only about 90% of the Y′Y′ *TT* fish are females and that about 10% of the Y′Y′ *Tt* also become females. Under this assumption, a somewhat better fit is obtained for cross 4. Incomplete penetrance of the autosomal regulatory genes was also observed in the XX, WX, and WY male conditions and this seems to be a common feature of atypical sex determination.

Two crosses in Table IX remain to be discussed. Both crosses involve *ArAr* females and Jp X-*Sp* Y-*Sr* males. The female parent of cross 6 was derived from cross 4, in which the *TT* genotype has been assigned to the *ArAr* females. Inasmuch as Jp fish are *tt*, all progeny of cross 6 are *Tt*. According to series A, 72% of the X-*Sp* Y′-*Ar Tt* fish can be expected to be females. The results of cross 6, in which only 54% of this class were females, deviate from this expectation ($P < 0.0001$). This represents one of the few results that differ significantly from expectation and are not in agreement with the hypothesis. There is a small possibility that some of the *Ar* females of cross 4 were actually *Tt*. If this is the case, only one-half of the progeny of cross 6 are *Tt*, with the other half *tt*. The expected frequency of females then becomes 36%, significantly lower than the observed frequency of 0.54 ($P < 0.001$). In cross 7, the *Ar* female was derived from cross 3. To these females the genotype *Tt* had been assigned. The expected frequency of 36% females is in excellent agreement with the actual result (P = 0.42).

One additional report points to the existence of two Y chromosomes in *X. maculatus* that are qualitatively different with respect to sex determination. Kosswig (1939) crossed a Y-*Dr* Y-*N* male of *X. maculatus* with a female of *X. helleri*. Among the *N* offspring, there were eight females and 67 males, but among the *Dr* class, there were 42 females and 33 males. In his discussion, Kosswig (1939) stated that either a gene locus that affects sex determination is closely linked to the pigment genes and different alleles at this locus are present on the two Y chromosomes, or the pigment genes themselves act as relative sex realizers. Kosswig preferred the latter explanation.

9.4. XY Females in *Xiphophorus montezumae* and *Xiphophorus milleri*

The existence of two Y chromosomes not identical with respect to sex determination has also been discovered in the Montezuma swordtail, *X. montezumae,* and in the Catemaco livebearer, *X. milleri.* Like several other species of this genus, *X. montezumae* and *X. milleri* exhibit X and Y chromosomes. In most instances females are XX and males XY, but occasionally some females that are XY have been obtained. The mating of an XY female with an XY male has produced male progeny with two Y chromosomes.

In *X. montezumae,* the evidence for this comes from a series of experiments, the details of which will be presented elsewhere (Kallman, 1983). Only a few examples will be given here. Crosses 1–4 (Table XI) are in agreement with the XX ♀♀–XY ♂♂ scheme of sex determination with the macromelanophore factor "*Sp*" being X-linked. Crosses 1 and 2 are especially compelling. However, cross 5 of a heterozygous *Sp* female (X-+ X-*Sp*) with an X-*Sp* Y-+ male does not fit easily into this scheme. Instead of the offspring consisting of *Sp* females and + and *Sp* males, some female progeny were also +. Presumably these females represent instances of atypical sex determination, and the sex-chromosome genotype that has to be assigned to them is X-+ Y-+. The correctness of this view will be discussed below (series III).

The *Sp* factor of the crosses listed in series II is obviously Y-linked. All females in crosses 6–9 are X-+ X-+ and males X-+ Y-*Sp*.

Crosses 10 and 11 of series III represented matings between the exceptional + females (from cross 5) and X-+ Y-*Sp* males of series II. The results of both crosses are in excellent agreement with the assumption of the X-+ Y-+ genotype of the female parents. The wild-type female progeny of both crosses are X-+ X-+, the wild-type males X-+ Y-+, the *Sp* males being of two genotypes, X-+ Y-*Sp* and Y-+ Y-*Sp,* that cannot be told apart phenotypically. The genotype of the *Sp* males, however, could be tested by mating them to X-+ X-+ females. There can be little doubt that crosses 12 and 13 are sired by YY males, because all but one of the 41 offspring developed into males. On the other hand, cross

Table XI
Sex Determination in *Xiphophorus montezumae*

Cross number	Genotype of parents ♀	Genotype of parents ♂	Pattern of progeny + ♀♀	+ ♂♂	*Sp* ♀♀	*Sp* ♂♂
Series I						
1	X-+ X-+	X-*Sp* Y-+		28	25	
2	X-+ X-+	X-*Sp* Y-+		10	7	
3	X-+ X-*Sp*	X-+ Y-+	7	5	4	3
4	X-+ X-*Sp*	X-+ Y-+	5	4	3	7
5	X-+ X-*Sp*	X-*Sp* Y-+	4	3	16	6
Series II						
6	X-+ X-+	X-+ Y-*Sp*	7			9
7	X-+ X-+	X-+ Y-*Sp*	15			10
8	X-+ X-+	X-+ Y-*Sp*	22			33
9	X-+ X-+	X-+ Y-*Sp*	5			3
Series III[a]						
10	X-+ Y-+[b]	X-+ Y-*Sp*	16	7		19
11	X-+ Y-+[b]	X-+ Y-*Sp*	7	5		14
12	X-+ X-+[c]	Y-+ Y-*Sp*		12		7
13	X-+ X-+[d]	Y-+ Y-*Sp*	1	12		9
14	X-+ X-+[d]	X-+ Y-*Sp*	3			4
Series IV[a]						
15	X-+ Y-+[e]	X-+ Y-*Sp*	2	3		7

[a]Parental males with *Sp* from series II.
[b]The + females from cross 5.
[c]From series II.
[d]From cross 4.
[e]Females from cross 13.

14 must have been sired by an X-+ Y-*Sp* male. Again, the single female of cross 13 must have been XY, as indicated by the result of cross 15.

The control elements adjacent to the male-determining gene on the gonosomes Y-+ and Y-*Sp* cannot be identical. All examples of atypical sex determination in this stock possessed the XY-+ genotype. In all, 246 progeny known to be X-+ Y-+ were raised (e.g., from crosses of the X-+ X-+ ♀♀ × X-*Sp* Y-+ ♂♂) and of these, 24 (9.8%) differentiated into females, whereas all 231 X-+ Y-*Sp* progeny were males (Kallman, 1983). This genetic mechanism is analogous to the situation in *X. maculatus* with the Y′ chromosome. Inasmuch as the XY-+ females occurred in only 10 out of 16 such crosses, it appears that the male-determining gene on Y-+ (=Y′) is blocked only in the presence of a certain autosomal regulatory factor for which the *X. montezumae* stock must be heterozygous. No crosses have yet been performed to obtain a pedigree of *X. montezumae* homozygous for this autosomal factor.

The two Y chromosomes in the stock of *X. milleri* were marked by either *Sv* or *Gn,* two genes controlling melanophore pigmentation. Females of the genotype XY always possessed the Y (=Y′) chromosome marked by *Sv* (Kallman and Borowsky, 1972). Males of genotype Y′-*Sv* Y-Gn were obtained by mating X-+ Y′-*Sv* females with X-+ Y-*Gn* males. From a mating of a X-+ X-+ female with a *SvGn* male, 37 X-+ Y′-*Sv* progeny were obtained, of which 10 were females (27%), and all 30 X-+ Y-*Gn* offspring were males. With respect to sex determination, the two Y chromosomes cannot be identical. Since XY-*Sv* females did not occur in every pedigree (Kallman and Borowsky, 1972), it must be assumed that, as in *X. maculatus,* the presence of an additional autosomal factor is necessary for this phenomenon to arise.

10. The Relationship between Atypical Sex Determination, Sex Ratio, Age at Maturity, and Adult Size

Reference has already been made to some instances of biased sex ratios in *Xiphophorus* that were associated with late maturation (Peters, 1964; Lodi, 1980). Also, Anders and Anders (1963) reported that in their Jamapa stock of *X. maculatus,* the exceptional YY males were larger and became sexually mature 3–4 weeks later than XY males. They suggested that a gene for "normal" onset of sexual maturation was located on the X chromosome.

It is now known that a sex-linked multiple-allelic series at the *P* locus controls the onset of sexual maturity in *X. maculatus* (Kallman and Borkoski, 1978), *X. montezumae* (Kallman, 1983), *X. milleri* (Kallman and Borowsky, 1972), and *X. nigrensis* (Kallman, unpublished). The action of this gene has been traced histologically to the gonodotrops of the pituitary gland (Kallman and Schreibman, 1973; Schreibman and Kallman, 1977). More recent information suggests that the *P* gene controls the fate of luteinizing hormone-releasing hormone (Bao and Kallman, 1982). Differences in the presence of ir-LHRH have been demonstrated at 4 weeks of age in the nucleus olfactoretinalis and nucleus praeopticus periventricularis between P^1P^2 (early maturing) and P^2P^5 (late maturing) fish (Halpern-Sebold and Schreibman, 1983). These observations suggest that the two centers are involved in regulating the sexual maturation process and that they may represent the site of action of the *P* gene. At least five different alleles exist at the *P* locus of *X. maculatus*. Under routine laboratory conditions, depending upon genotype, platyfish may mature as early as 8 weeks or still may be immature at 2 years. One *P* allele (P^5) has been identified, which in homozygous condition results in arrested gonadal development in some fish, and the P^5P^5 females that do become sexually mature produce a smaller number of eggs than females of other *P* genotypes, but of equal size or age. The arrested gonadal development of fish homozygous for P^5 apparently results from an abnormality of the hypothalamo-pituitary axis and is analogous to certain forms of secondary hypogonadism (hypothalamic insufficiency) in man (Rimoin and Schimke, 1971; Mortimer *et al.*, 1974).

Although actual figures cannot be provided, all observations indicate that the female-determining chromosomes W and X mostly carry *P* factors that determine maturation at a small size and at relatively early age. The only exception to this rule is the X chromosome marked by P^5, which was isolated from the Belize population of *X. maculatus*. It is noteworthy that in this population, the frequency of the X chromosome is only 0.045 (Kallman, 1973) and that X-P^5 is rarer still. Hence, most females are WY and males YY and the occurrence of XX females homozygous for P^5 is a most unlikely event. It is a consequence of the W–X–Y system of *X. maculatus* that females may inherit a late Y-linked *P* factor from their male parent and that, therefore, not all females of a given brood necessarily mature at the same size or age. Some males will be homozygous for late maturation, however, and they attain sexual maturity at a larger size and greater age than does any female. As I shall show below, the *P* gene polymorphism is restricted to the Y chromosome, and thus to males, in *X. montezumae, X. milleri,* and *X. nigrensis*.

Because the *P* gene polymorphism is of common occurrence in several species of *Xiphophorus*, it often is associated with the genetic condition for atypical sex determination. There is, however, no causal relationship between the two. In the experiments of Anders and Anders (1963) P^1 was linked to the X and P^2 to the Y. Males of genotype XY were P^1P^2 and matured earlier and at a smaller size than did YY males homozygous for P^2. When mated to XX females, the smaller XY males produced equal numbers of males and females, whereas the larger YY males sired all-male offspring.

In the ASD stock of *X. maculatus* with Y′-*Ar* the *P* factors on the X, Y, and Y′ chromosomes are not identical. Factor P^1 is located on the X chromosome, P^2 is present on the Y, and a factor apparently identical with P^3 is located on the Y′ (Kallman, unpublished). The XY and XY′ males matured earlier and at smaller sizes than males of the genotypes YY′ and Y′Y′ (Table XII). The effects of P^2 and P^3 in the ASD stock were similar to each other and this agrees with the earlier report of Kallman and Borkoski (1978), which showed that the main difference between P^2 and P^3 concerns their interaction with P^5.

In *X. montezumae,* males with Y-*Sp* matured later and at a larger size than did males with Y-+ (Table XIII). No differential effect on size at maturation could be ascribed to the X-+ or X-*Sp* chromosome. Especially telling are the results of series III, in which XY-+ and XY-*Sp* males were obtained in the same cross. Thus both kinds of males were raised together under identical conditions, ruling out conclusively the possibility that environmental conditions could be responsible for the size differences. The females of this stock matured at an age and size that corresponded closely with those of the XY-+ males (Kallman, 1983).

My interpretation of these observations is that *X. montezumae* also possesses a sex-linked *P* locus and that at least two alleles, *a* and *b,* are segregating in this stock. The *a* allele, which determines maturation relatively early and at a small size, is located on the X chromosomes and on Y-+ (Y′), and the *b* allele for maturation at a large size is present on Y-*Sp*.

A similar explanation can be applied to the stock of *X. milleri,* in which XY′ (*Sv*) and XY (*Gn*) males differed significantly in size as adults (Kallman and Borowsky, 1972). The *Gn* males were always significantly larger than *Sv* males. The difference in size between the two genotypes of males had nothing to do with the process of sex determination. It merely signifies that the allele at the *P* locus on Y′-*Sv* is not identical with the one on Y-*Gn,* the *P* factor on Y′ resulting in maturation at a smaller size than the *P* factor on the Y chromosome.

It must also be pointed out that whereas XY′ males were larger than XY males in *X. maculatus,* this situation was reversed in *X. montezumae* and *X. milleri.*

11. Most Small Males of *Xiphophorus nigrensis* (Río Choy) Are XX

A striking example of the association of unequal sex ratio with adult male size comes from the Río Choy population of *X. nigrensis*. In this population, the size of adult males ranges from 22 to 40 mm and can be attributed to four alleles at the sex-linked *P* locus. To differentiate these alleles from those of *X. maculatus* and *X. montezumae,* they are referred to as *S* (small), *I* (intermediate size), and *L* (large).

Table XII

Age (weeks) and Size (mm) at Sexual Maturity of Male *Xiphophorus maculatus* with and without Y′[a]

		N	XY P^1P^2 *SpSr*	N	XY′ P^1P^3 *SpAr*	N	YY′ P^2P^3 *SrAr*	N	Y′Y′ P^3P^3 *ArAr*
4167	Size[b]	34	23.4 ± 0.4	25	24.5 ± 0.4	23	26.5 ± 0.6		
	Age[c]		17.2 ± 0.8		19.0 ± 1.0		23.3 ± 1.1		
4252	Size[d]	20	23.8 ± 0.3	14	24.0 ± 0.6	19	26.4 ± 0.8	24	27.9 ± 0.3
	Age[e]		16.8 ± 1.1		17.5 ± 1.1		21.0 ± 0.8		23.0 ± 0.8

[a]Means ± SE. The underlined means do not differ significantly from each other (Newman–Keuls test, $P > 0.05$).
[b]Trt. Df 2, Ms 66.2; error Df 79, Ms 4.4, $F = 15.2$, $P < 0.01$.
[c]Trt. Df 2, Ms 252.9; error Df 79, Ms 24.4, $F = 10.4$, $P < 0.01$.
[d]Trt. Df 3, Ms 79.9; error Df 73, Ms 3.1, $F = 25.5$, $P < 0.01$.
[e]Trt. Df 3, Ms 181.1; error Df 73, Ms 15.9, $F = 11.4$, $P < 0.01$.

Table XIII
Adult Size (mm) of Male *Xiphophorus montezumae*[a]

Cross number	+ ♂♂	♂♂ *Sp* is X-linked	♂♂ *Sp* is Y-linked
Series I			
1	34.5±1.03		
2	34.4±0.63		
3	34.0±1.22	35.3±1.02	
4	34.8±0.67	36.0±0.47	
5	31.0±0.71	30.2±0.77	
Series II			
6			45.4±0.89
7			44.0±0.83
8			41.5±0.37
9			42.0±1.22
Series III			
10		33.5±1.50	43.0±0.78
11		31.8±1.39	40.3±0.93
12		33.3±0.63	42.3±1.22
13		32.2±0.28	43.1±1.38
14			40.8±1.91
Series IV			
15		30.0±1.80	44.6±0.60

[a]Means ± SE.

In the laboratory stock, males of intermediate size always exhibit a yellow caudal fin pattern. From 15 matings involving *I* males (comprising six generations), 161 *I* males with yellow coloration, ranging in size from 25 to 32 mm (28.9±0.13), were obtained (footnotes *c* and *d,* Table XIV). In addition there were nine *S* male progeny 22–24 mm long without yellow coloration and 136 females. All individual crosses conformed to a statistical sex ratio of unity.

From 35 crosses (footnotes *a* and *b,* Table XIV) involving *L* males spanning five generations, 295 ♀♀ and 343 ♂♂ were obtained, a slight excess of males ($P = 0.06$), with all crosses again exhibiting a sex ratio that did not deviate significantly from unity. The male progeny fell into two nonoverlapping size classes. There were 327 *L* males ranging in size from 30 to 40 mm (35±0.10) and 16 *S* males from 22 to 24 mm.

The *S* males that were generated by the *I* or *L* males or that were collected in the Río Choy were either inbred or mated to females of the *L* line. All male progeny ranged in size from 22 to 24 mm (23.1±0.06).

Table XIV
Inheritance of Sex Ratio and Size in *Xiphophorus nigrensis* (Río Choy population)

Pedigree	Phenotype of offspring XX ♀♀	XX *S* ♂♂	XY *I* ♂♂	XY *L* ♂♂
3280[a]	26	1		28
3267[a]	15	5		13
3333[a]	14	5		32
3702 A[a]	4	2		6
3702 B[a]	6	2		5
3508[a]	3	1		1
Many[a,b]	227	0		242
3415[c]	9	5	11	
3574[c]	11	2	11	
3834[c]	13	2	14	
Many[c,d]	103	0	125	
3936[e]	8	2		5
3938[e]	8	6		16
3946[e]	3	1		5
3948[e]	0	1		
3955[e]	19	3	25	
Many[f]	205	181		
Many[g]	285	12		

[a]Crosses between X-*S* X-*S* ♀♀ and X-*S* Y-*L* ♂♂.
[b]Summary of 29 crosses.
[c]Crosses between X-*S* X-*S* ♀♀ and X-*S* Y-*I* ♂♂.
[d]Summary of 12 crosses.
[e]Progeny of gravid females collected in the Río Choy.
[f]Summary of 17 crosses between X-*S* X-*S* females and X-*S* X-*S* males. The sex ratio of each of the crosses conformed to a statistical expectation of unity. Males and females were siblings or closely related to each other.
[g]Summary of 17 crosses between X-*S* X-*S* females and X-*S* X-*S* males. Some of these males were the same as in footnote *f*. Male and female parents were unrelated. The 13 male offspring occurred in six of the crosses.

Not a single *I* or *L* male was produced. The sex ratio depended in part upon the genotype of the female parent. This could be shown by a number of crosses in which the same male was mated to more than one female. With some females, these males sired all-female progeny, but with others an equal number of males and females. Significantly, the sex ratio always conformed to unity when the male and female parents were closely related.

There were all together 34 such matings, comprising some of the original *S* males and some of their *S* male descendants. Eleven of these crosses resulted in all-female progeny (N = 191) and a significant excess of females (94 ♀♀, 12 *S* ♂♂) was obtained in six other crosses (footnote *g*, Table XIV). The remaining 17 crosses, however, gave rise to sex ratios (205 ♀♀, 181 ♂♂) that never deviated significantly from unity (footnote *f*, Table XIV).

In my interpretation, the sex chromosome mechanism of *X. nigrensis* is XX–XY. An autosomal locus with two alleles, *A* and *a*, that affect sex determination also exists. The genotype XX *AA* is invariably females, but XX *aa* zygotes become males. All fish with a Y chromosome become males regardless of autosomal genotype. This explanation is identical with the one previously proposed to account for Öktay's XX males in *X. maculatus* and Aida's and Winge's XX males in *O. latipes* and *P. reticulata*. The *S* allele (maturation at a size smaller than 24 mm) is located at the *P* locus of all X chromosomes, but the Y chromosomes possess either the *I* or *L* allele. If this interpretation is correct, all offspring of the cross XX *AA* ♀ × XX *aa* ♂ possess the genotype XX *Aa*. The 17 crosses involving *S* males that resulted in predominantly female offspring appear to have been of this kind (Table XIV). From this result, it must also be concluded that about 4.4% of the XX *Aa* progeny differentiate males. The 17 crosses that resulted in a 1:1 sex ratio must have involved X-*S* X-*S* *Aa* females and X-*S* X-*S* *aa* males. Note that most crosses involving small males give rise to either 50% males and females (footnote *g*, Table XIV) or to virtually all-female offspring (footnote *f*, Table XIV). It is only when a rare X-*S* X-*S* *Aa* male is mated to an X-*S* X-*S* *Aa* female that one can expect female and male progeny in a ratio of 3:1.

This scheme of sex determination also readily accounts for the occurrence of small XX males among the progeny sired by X-*S* Y-*I* and X-*S* Y-*L* males. There were 18 such crosses (Table XIV). The genotypes that must be assigned to the P_1 females is X-*S* X-*S* *Aa* and to the male parents X-*S* Y- *Aa* or *aa*, the Y chromosomes being marked by either *I* or *L*. Males that are autosomally *aa* give rise to X-*S* X-*S* progeny, of which 50% are small males. On the other hand, X-*S* Y- *Aa* males produce females and small males in a ratio of 3:1. With one exception (pedigree 3280), the frequency of small males in the 13 crosses agrees with the expectations (Table XIV). The single small male of pedigree 3280 may represent a rare X-*S* X-*S* *Aa* male.

Not all small males of the Río Choy are XX. A small percentage are XY and these males produce even sex ratios. One such male was identified and a stock of small XY males was maintained in the laboratory for three generations. The Y chromosome was marked by a gene for a yellow caudal fin pattern, which permitted easy differentiation from possible XX males. The size of these XY males ranged from 22 to 26 mm (24.1±0.10), and the difference between the mean size of XX and XY males was significant ($P < 0.001$). This raised the possibility that the *S* on the X chromosome is not identical with the Y-linked *P* allele that also determines small adult size.

12. The Frequency of the Autosomal Factors for Atypical Sex Determination in Natural Populations of *Xiphophorus*

The frequencies of the Y′ chromosome and the autosomal regulatory factors involved in atypical sex determination are surprisingly high in certain natural populations. Among hundreds of crosses involving *X. maculatus* from all river systems (Kallman, 1965, 1970), XX males have been encountered only among the descendants of two fish collected in Lake Petén, part of a sample of 14 fish. These observations suggest that the factor *a* is relatively common in this local population, but rare or absent elsewhere.

The calculation of the frequency of *a* in the Río Choy population of *X. nigrensis* is based upon the progeny of 24 gravid females collected in the Río Choy in 1979. Five of these females produced small XX males. The offspring of these 24 females consisted of 180 X-*S* X-*S* females, 13 X-*S* X-*S* males, and 168 X-*S* Y-*I* or X-*S* Y-*L* males. Inasmuch as 6.7% of the XX class were males, *aa,* the frequency of *a* in this population is 0.26. Thus *a,* the autosomal regulator that in homozygous condition determines male differentiation in XX individuals, is a rather common component of the Río Choy gene pool. This is also indicated by our field studies, because X-*S* X-*S* males were present in both the small samples taken in 1972 and 1974.

The high frequency of *a* in the Río Choy population of *X. nigrensis* argues strongly that it is maintained by natural selection. Because most XY males of this population possess the *I* and *L* alleles on the Y chromosome, they mature at a larger size and greater age than do the X-*S* X-*S* females and X-*S* X-*S* *aa* males. If reproduction in this species is seasonal, these males at the beginning of the reproductive season will have no competition from the larger XY males. In this connection it should be noted that the factor *a* is absent from the Río Coy population, in which many Y chromosomes carry the *S* allele.

The detection of Y′ and *T* within the Belize (Bp) population of *X. maculatus* is difficult, because neither Y′ nor *T* by itself has a detectable effect. Because the frequency of the X chromosome in the Belize population is only 0.045 (Kallman, 1973), most males will be either YY, YY′, or Y′Y′. According to Table IX, only Y′Y′ fish that are also homozygous for *T* will differentiate into females, and these are the only fish in which Y′ and *T* may be recognized. In the experimental crosses, however, it was possible to refine the detection of Y′ and *T* by pairing each Bp sex chromosome with an X of Jp, inasmuch as XY′ *Tt* fish also differentiate into females. In other words, in hybrids with Jp, the Y′-*T* combination can also be recognized when present in heterozygous condition.

The matings listed in Table XV, part A, were part of a larger series of crosses originally designed to determine the frequency of the X chromosome in the Bp population. Seventy-two of these crosses between Bp males and XX females of the Jamapa (Jp) and Coatzacoalcos (Cp) stocks and 10 crosses between Bp females and Jp XY-*Sr* males can also be used to detect the presence of the Y′*T* combination. These crosses are arranged according to sex ratio (A versus B) and according to whether the progeny consisted of one or two classes of males and females (B1 versus B2).

The Bp fish were obtained in 1969 from the Belize River drainage near Bermudian Landing. The collections were made at five locations along the road from Bermudian Landing to Rancho Dolores, and at four points along the road to Lemonal. The 72 Bp males listed in Table XV represent 18 wild-caught males and one offspring each of 54 wild-caught nonvirgin females.

Sixty-one of these males gave rise to all-male broods ($N = 1669$). The 122 sex chromosomes of these Bp males were either Y or Y′. Little can be said about the presence of *T*, except that the Y′ *T* combination did not occur in any of these fish. Five Bp males produced male and female offspring, the males and females of each cross inheriting different pigment patterns. These Bp males were obviously XY or XY′, the gonosomes being marked by different pigment genes. The combination Y′ *T* did not occur in these five crosses.

Five other Bp males also gave rise to male and female offspring (Table XV, part B2), but one pigment class consisted of males only, whereas the other class consisted of both sexes. The genotypes of these five Bp males were designated as YY′ *T*, with the Y and Y′ chromosomes carrying different pigment genes. Because, as was shown above, the penetrance of *T* in the XY′ *Tt* genotype is only 72%, the Y′*T* progeny differentiated into both males and females, while the progeny that inherited Y and *T* were all males. Whether the five Bp males were homozygous or heterozygous for *T* cannot be determined from these data. The relatively high proportion of males among the Y′ class in two of these crosses suggests that they were heterozygous.

The progeny of the remaining Bp male consisted of two classes of

Table XV

Results of 72 Crosses of Belize (Bp) Males with XX Females and Ten Crosses of Bp Females with Jp XY-*Sr* Males[a]

Sex-linked pigment pattern of Bp parent	♀♀	♂♂	Possible genotypes of Bp parent	Presence (+) or absence (−) of Y′*T* among progeny
A. Summary of 61 crosses of Bp ♂♂ resulting in all-male progeny				
	0	1669	YY or Y′Y	−
B. Eleven crosses with Bp ♂♂ resulting in male and female progeny				
1. Males and females occur in different pigment classes				
CPo	17 *CPo*	19 +	XY or XY′	−
IyCPoN	19 *CPoN*	21 *IyCPo*	XY or XY′	−
IyTr	7 +	14 *IyTr*	XY or XY′	−
Br	15 *Br*	15 +	XY or XY′	−
IyVoSp[8]	18 *Sp*[8b]	15 *Iy*	XY or XY′	−

2. One sex occurs in one pigment class, the other sex in two classes						
Ar	3 *Ar*		24 +	8 *Ar*	Y′Y	+
ArMrSd	8 *Ar*		13 *MrSd*	7 *Ar*	Y′Y	+
DrSdIy	8 *DrSd*		13 *Iy*	9 *DrSd*	Y′Y	+
DrSd	10 *DrSd*		9 +	2 *DrSd*	Y′Y	+
CPyVoSp^8	2 +		9 *CPyVoSp*^{8}	10 +	Y′Y	+
ArAy	3 *Ar*	12 *Ay*	2 *Ar*		XY′	+
C. Summary of 10 crosses involving Bp females						
"*M*"[c]	113	85 *Sr*	91 "*M*"	87 "*M*" *Sr*	WY or WY′	–

[a]The 122 sex chromosomes were marked as follows: + 42, *Ay* 18, *Iy* 12, *Ir* nine, *CPo* seven, *Dr* six, *N* four, *Sr* three, *Mr* two, *Ar* two, *Nr* one, *Br* one, *r-Sr* (red horizontal stripes) one, *Tr* one, *VoSp*^{8} four, *SrCPo* two, *MrSd* two, *CPySr* one, *TrSr* one, *IyAy* one, *IySr* one, *ArIySrAy* one [for symbols, see Kallman (1975)].

[b]The pattern *Vo* is a secondary sex character that is expressed in males only. When these females, in turn, were crossed to Jp XY males, all *Sp*^{8} male progeny exhibited *Vo*.

[c]The sex chromosome constitution of these females was WY. The 10 Y chromosomes were marked by the following pigment factors "*M*": *Ay* (three), *Ir* (two), *Dr* (two), *Iy* (one), *Iy Ay* (one), *VoSp*^{8} (one).

females and one class of males. My interpretation of this cross is that the genotype of this Bp male was XY′ *T*-, the factor *Ay* linked to the X and *Ar* to Y′. Again the question of whether this Bp male was homozygous or heterozygous for *T* remains unanswered.

The 10 Bp females were identified as either WY or WY′, because all their male offspring inherited the color pattern "M" of the female parent and either the X or the Y-*Sr* chromosome of the Jp male. The female offspring were of two genotypes, WX and WY-*Sr*. The combination Y′ *T* could not be detected.

The six males with Y′ *T* were not randomly distributed among the nine collections. The 148 Y or Y′ chromosomes of the Bp fish (Table XV) were marked by 25 different pigment genes or combinations of pigment genes. Five of the gonosomes were marked by *Ar* and of them three were identified as Y′. Two of these Y′-*Ar* chromosomes were derived from the Willows Bank Creek collection, where they were present in one wild-caught male and also among the progeny of one wild-caught nonvirgin female. It is, therefore, possible that these *Ar* fish were closely related to each other. The third Y′ chromosome, also marked by *Ar,* came from Governors Creek, 8 km north of the previous location. A fourth Y′, unmarked, was obtained 1.9 km west of Willows Bank Creek. The two remaining Y′ chromosomes were marked by *DrSd,* the only two chromosomes in the sample with this combination. They were present in two females collected 3.5 km west of Willows Bank Creek, and again the presumption is strong that the two fish were closely related and had inherited Y′-*DrSd* and *T* from an immediate common ancestor.

As was pointed out above, the combination XY′*Tt* results in the production of both males and females. A particular gonosome was designated as Y′ in the crosses in which both sexes of the offspring were observed to inherit the same color pattern (Table XV, part B2). The detection of Y′ must then be partially dependent upon the number of offspring. In one cross, for example, 10 *DrSd* females and two *DrSd* males were obtained. Had there been fewer offspring, no such males might have occurred and in this case the chromosome marked by *DrSd* would have been classified as an X. Similarly, in the cross resulting in two + females and 10 + males, a smaller number of offspring could have easily resulted in all wild-type male progeny, and in this case the genotype of the fish would have been designated as either Y *T* or *t,* or Y′ *t*. These are methodological limitations that cannot be avoided. On this basis the very presence of the X chromosome in the Belize population must be questioned. The existence of the X chromosome in this population has been demonstrated in the past (Kallman, 1963, 1970), however, and fish listed with the X chromosome in Table XV were used in other matings. The results of these crosses showed without exception that indeed these chromosomes were X. The most extensive series of crosses involved the X chromosome marked by *CPoN,* which has been maintained in the laboratory for over 10 generations. This is the chromosome that is also marked by P^5 (Kallman and Borkoski, 1978), and without exception, all XX-*CPoN* or X-*CPoN* X-*CPoN* fish developed into females.

The occurrence of Y′ as well as *T* therefore cannot be considered to be an isolated, localized event. At least within the study area, Y′ and *T* are widespread. If Y′ originally arose as a single mutation, this must have occurred many generations ago, permitting Y′ not only to spread throughout the population, but also to become linked to a number of pigment genes in spite of the low frequency of crossing over between the sex chromosomes of *X. maculatus* (Kallman, 1965). Apparently, the limited dispersal of these fish in small peripheral waters and the resulting inbreeding increases the chances of encountering fish with the same genotype within a given collection.

Although the exact frequencies of Y′ and *T* are not yet known, a few simple calculations show that neither one is particularly rare. It is known from Table XV that there were 148 Y or Y′ chromosomes among the 72 Bp males and the 10 Bp females. Six of these chromosomes (4.05%) were definitely identified as Y′. Assuming arbitrarily the unlikely event that *T* has become fixed in the Belize population, then every Y′ in our sample would have been manifested. This sets the lower limit for the frequency of Y′ at 0.04. It is virtually certain that this assumption is unreasonable. At least one of the Y′ *T* males was heterozygous, *Tt,* as was discussed in detail when the crosses of Table VIII were analyzed. No information is available for the other five males. The presence of male progeny among the Y′ classes cannot be taken as evidence for the absence of *t,* because the penetrance of T in the XY′ *Tt* genotype is only 71%.

The upper limit for the frequency of Y′ has to be set at about 0.50, because none of the males identified as Y′ was homozygous for this factor. Inasmuch as the frequency of males with both Y′ and *T* was 0.076 (five of 66 YY, YY′, Y′Y′ males, Table XVI), a value of 0.50 for Y′ corresponds to a frequency of 0.052 for *T,* which is its lower limit. Values for *T* for

Table XVI

Calculated Values for *T* for Frequencies of Y′ between 0.04 and 0.50 in a Local Population of *Xiphophorus maculatus* from Belize

Y′	0.04	0.05	0.10	0.20	0.30	0.40
T	0.79	0.49	0.22	0.12	0.07	0.06

intermediate frequencies of Y′ between 0.04 and 0.50 are listed in Table XVI.

The Y′ *T* combination of *X. maculatus* has only been encountered in one local population from Belize. No evidence for the existence of Y′ *T* was obtained in any of the other populations studied, including two earlier samples from Belize, one of which came from within 8 km of the Y′ *T* location (Kallman, 1965, 1970).

The Y′ chromosomes of *X. montezumae* and *X. milleri* appear to be relatively common. The Y′ chromosomes were present in our only live sample of *X. montezumae,* comprising four males, and in one of two equally small samples of *X. milleri*.

13. The Effect of Extrinsic Factors on Sex Determination in Fishes

All instances of atypical sex determination discussed here share the phenomenon of incomplete penetrance, i.e., individuals with the same genotype for ASD may develop into either male or female. The physiological basis for this labile determination of the undifferentiated gonad is not known. It should be determined whether fishes with genotypes for ASD are susceptible to still unrecognized environmental influences on sex determination. So far, there is little evidence for this. Only Winge (1934) reported that in the XX ♀♀–XX ♂♂ stock of *P. reticulata* the frequency of males in some crosses were significantly higher during the winter than during the summer months. No information was provided, however, as to the conditions under which the fish were reared.

In view of the recent discovery that temperature has a decisive effect on sex determination in certain reptiles (Bull, 1980), it is somewhat surprising that little information is available concerning the effect of extrinsic factors on sex determination in fishes. The differentiation of XX embryos into functional males and of XY embryos into functional females after exposure to heterotypic sex hormones has already been discussed. Anders *et al.* (1969) reported briefly that gravid females of *X. maculatus* (Jp) exposed to 1500–2000 R gave rise to XY progeny, of which about 40% developed into functional females. Subsequent crosses showed that this condition was not inherited. This experiment was poorly controlled and the spontaneous occurrence of XY females in the same stock, which had been attributed in 1963 to a polyfactorial system of sex genes on both gonosomes and autosomes (Anders and Anders, 1963), was now attributed to unspecified environmental factors, notwithstanding the earlier report that through selection a stock had been created in which 25% of all XY individuals were females.

Breider (1935*a*) was unable to establish a correlation between the sex ratio of *X. helleri* and a number of environmental variables, such as season, temperature, poor food conditions, crowding, age of parents, age of spermatozoa stored in the female genital tract, and size of brood, and Schröder (1964) came to a similar conclusion for *Poecilia sphenops*. Most laboratories raise tropical or subtropical species of poeciliid fishes under a temperature regimen that resembles that of the natural environment. Under these conditions, sex determination is according to the sex chromosome constitution. Apparently no attempt has been made to breed these species at temperatures below or well above those usually encountered under natural conditions.

The best documented example of the effect of temperature on sex determination in fish is Harrington's (1967, 1971) work on *Rivulus marmoratus,* a self-fertilizing hermaphrodite consisting, at least near the northern limit of its range in Florida, of numerous homozygous clones. About 75% of the embryos reared at 20°C during a phenocritical period develop into primary males, whereas those exposed to higher temperatures develop into hermaphrodites. Temperatured as low as 20°C, however, are probably never encountered by the embryos under natural conditions (Harrington, 1971). The effect of low temperature is on the undetermined gonad of the embryo. On the other hand, high posthatching temperatures (30°C) for a period of 4–5 months affect the amphisexual gonad and hastens the eventual involution of the ovarian tissue that changes the functional hermaphrodite into a secondary male. The involution of the ovarian component is triggered by short day length. The two clones studied differed in the timing of this event. In one clone the change to secondary males occurred during the first two short photoperiods, whereas the hermaphrodites of the other clone did not do so until the third and fourth short photoperiod. Since all other variables were controlled, the different response of the two clones to high posthatching temperatures must be genetic.

An interaction of genotype and temperature on sex determination has also been recorded for *Menidia menidia* (Conover and Kynard, 1981). The proportion of females derived from egg clutches reared under low temperatures during a critical developmental period was significantly higher in five out of six cases than when reared under high temperatures. Since the proportion of female progeny derived from different female parents but the same male differed significantly under both temperature conditions, the population studied appears to have been variable for genotypes that govern the response to temperature.

14. Sex Ratios and Sex Determination in Species Hybrids

Avtalion and Hammerman (1978) proposed a genetic mechanism of sex determination that consists of one pair of gonosomes and a single autosomal locus. This mechanism was proposed to account for peculiar sex ratios among hybrids of different species of *Sarotherodon* (*Tilapia*). According to these authors, each species is homozygous for a certain allele at the autosomal locus, but different alleles have become fixed in different species. Those species with male heterogamety were assigned the genotype XX *AA* ♀♀–XY *AA* ♂♂ and the species with female heterogamety the genotype WY *aa* ♀♀–YY *aa* ♂♂. *Interspecific* crosses of XX ♀♀ with YY ♂♂ resulted in all-male progeny, XY *Aa*, but the reciprocal crosses resulted in a ratio of one ♀♀ to three ♂♂ (WX *Aa* ♀, WY *Aa* ♂, XY *Aa* ♀, YY *Aa* ♀). This ratio differs from the one obtained in crosses involving heterozygous male (XY) and female (WY) *X. maculatus*. This *intraspecific* cross always results in a 1:1 sex ratio (WY, WX ♀♀; YY, XY ♂♂).

Based upon the sex ratios of subsequent crosses (F_2, Bc generations), Avtalion and Hammerman (1978) worked out theoretically the sex phenotypes of the 18 possible genotypes (Table XVII), but they did not go into any detail of just how their autosomal locus affects sex determination. Because no sex-linked marker genes are known in these cichlids, the proposed sex chromosome constitutions must be considered as hypothetical. The Y chromosomes of the different *Tilapia* species appear to be identical. The factor *A* appears to be involved in suppressing the female-determining action of the W chromosome, hence the male differentiation of WY *AA*, WY *Aa*, and WW *AA* individuals. However, female differ-

Table XVII
Proposed Influence of Sex Chromosomes and Autosomal Locus on Sex Determination in *Sarotherodon*[a]

	Autosomal factors		
Sex chromosomes	*AA*	*Aa*	*aa*
	Phenotype		
YY	♂	♂	♂
WY	♂	♂	♀
XY	♂	♂	♀
WW	♂	♀	♀
WX	♀	♀	♀
XX	♀	♀	♀

[a]From Avtalion and Hammerman (1978).

entiation of the WY *Aa* genotype remains unexplained. Another problem is presented by the female differentiation of XY *Aa* fish. The XX *aa* fish become females and YY *aa* fish become males, indicating that the factor *a* has no detectable effects on either the X or Y chromosomes. The female differentiation of XY *aa* individuals, therefore, is unexpected.

My theory of sex determination in fish also assumes the presence of at least one autosomal locus involved in sex determination, and evidence for intraspecific heterozygosity has been presented. The theory of Avtalion and Hammerman can be expanded and there is the possibility that each species has not only species-specific alleles at an autosomal locus, but also species-specific control elements on the sex chromosome. If this is true, the sex-determining mechanism within a genus or a family would be the same and the sex chromosomes and the autosomal loci of the different species would be homologous, although the individual components (alleles at the regulatory and control loci) of the system would be different. This situation brings up the problem of whether changes in the regulatory systems are more important for evolution than changes at the structural loci.

If the alleles at the loci involved in sex determination are indeed species-specific, one could expect significant differences between the sex ratio of the hybrids and the parental forms. The best evidence for such species specificity would be the demonstration that certain genetically *marked* sex chromosomes of one species when introduced into another would become associated with atypical sex determination. The evidence for or against species specificity of such regulatory and operator genes can be examined critically in four series of relatively extensive crosses between species of poeciliid fishes.

Schröder (1964) found little evidence for atypical sex determination when different species of *Poecilia* (subgenus *Mollienesia*) were hybridized with each other. From the sex ratios of various inter- and intraspecific crosses, Schröder concluded that *P. velifera, P. latipinna,* and one domesticated stock of *P. sphenops* ("black molly") exhibited the XX♀♀–XY ♂♂ mechanism, but that a second stock of *P. sphenops* ("liberty molly") was WY ♀♀ and YY ♂♂. The sex ratios of the F_1, F_2, and backcross hybrids agreed with expectations, and Schröder concluded that the sex chromosomes of the three species could replace one another. The only instances of abnormal sex ratios occurred in some crosses between the two domesticated stocks of *P. sphenops*. A different result was obtained when *Poecilia reticulata,* a species more distantly related to the subgenus *Mollienesia,* was crossed with *P. velifera* and *P. latipinna*. All 39 offspring were males, suggesting that some XX fish had differentiated into males contrary to their sex-chromosome constitution. On the basis of these observations, it appears that closely related species possess regulatory and sex-determining genes that can adequately function with each other. In contrast, the regulatory genes of more distantly related species may be distinctly different.

Breider (1935*b*) carried out a series of crosses with *P. nigrofasciata, P. caudofasciata,* and *P. vittata,* species from another section (subgenus *Limia*) of the genus *Poecilia*. None of these fishes has any known sex-linked marker genes, and even the sex ratios of their hybrids provides no clue as to the sex-determining mechanism of these forms. The data, therefore, are of somewhat limited use and the analysis must be restricted to whether the sex ratio of the hybrids is significantly different from that of the parental species. However, even this is difficult, because few data were provided concerning the sex ratio of the pure species. Moreover, each stock was derived from but a single male and female.

The sex ratio of adult *P. nigrofasciata* and *P. caudofasciata* appears to be 1:1. Breider (1935*b*) also reported that most juveniles of *P. caudofasciata* were females and concluded that many of these fish must have transformed into males. This "transformation" is very poorly documented, however, and I suggest that the immature fish were incorrectly identified as to sex. As Atz (1964) noted, the notion that *P. caudofasciata* and *P. vittata* (Breider, 1935*b*) and *X. helleri* (Kosswig, 1932; Breider, 1937) were "latent" hermaphrodites was based in part on faulty histological interpretation of the gonads of new-born fish. Broods from three pairs of crosses of *P. vittata* exhibited highly varying sex ratios (50 ♀♀–257 ♂♂, 167 ♀♀–65 ♂♂, 43 ♀♀–32♂♂), results that recall the condition in *X. helleri*.

The crosses between *P. caudofasciata* ♀♀ and *P. nigrofasciata* ♂♂ (F_1, F_2, Bc to *nigrofasciata* ♀ and to *caudofasciata* ♀) resulted in a 1:1 sex ratio, but in the reciprocal crosses, the F_1 generation consisted exclusively of females, and in the backcross to *caudofasciata* a significant excess of females was obtained (five crosses: 385 ♀♀, 110 ♂♂). It appears the *P. nigrofasciata* has an XX–XY mechanism and that its Y chromosome maintains its male-determining function in the hybrids. The excess of females in the reciprocal hybrids suggests the suppression of the male-determining gene of *P. caudofasciata*. The crosses of *P. nigrofasciata* or *P. caudofasciata* with *P. vittata* gave rise either to a 1:1 sex ratio or to a large excess of females. These results are not informative, however, inasmuch as *P. vittata* itself had pedigrees in which females sometimes outnumbered males in significant numbers.

Within the individual species of *Poeciliopsis*, sex ratios that deviate from unity are uncommon, but in hybrids, depending upon the species involved, sex ratios ranged from all female to all male. In some interspecies crosses, the sex ratio depended upon the precise geographic origin of the fish (Schultz, 1961, 1973, 1977; Vrijenhoek and Schultz, 1974). Particularly interesting are the crosses between females of *P. monacha* and males of *P. lucida* or *P. occidentalis,* because they invariably result in all-female offspring. The meiotic divisions of these hybrids are highly modified and they result in the elimination of all paternally derived chromosomes (hybridogenesis), in effect reconstituting a *P. monacha* genome (Schultz, 1969; Cimino, 1972). Under natural conditions, the hybrids have formed an all-female breeding complex that has been maintained over uncounted generations by repeated backcrosses to *P. lucida* or *P. occidentalis,* and the *monacha* genome has extended far beyond the natural range of the species itself. The preconditions for the evolution of this all-female complex must include not only a genotype that ensures the establishment of the highly modified meiotic divisions, but also a genetic mechanism that never fails to block the action of the male-determining gene of *P. lucida* or *P. occidentalis*. One possible explanation is that *P. monacha* is homozygous for a certain allele at an autosomal locus that blocks the male-determining gene of *P. lucida* and *P. occidentalis*. It follows that the alleles at the homologous loci of these two species cannot be identical with the factor of *P. monacha.*

The *P. monacha* genome has also adopted *P. latidens* as a sexual host species (Schultz, 1977). The sex ratio of the *P. monacha-latidens* complex depends in part upon the geographic origin of both parents. Since the results of individual crosses were seldom reported, these data are difficult to analyze and few details were provided. The *monacha-latidens* × *latidens* crosses of fishes from the Río Fuerte resulted in all-female offspring, but the *monacha-lucida* × *latidens* crosses resulted in all-male progeny. Obviously, either the two *monacha* genomes involved cannot have been identical or *P. latidens* must be heterozygous for the male-determining gene, a situation analogous to the Y and Y′ of *X. maculatus, X. milleri,* and *X. montezumae*. Some crosses between *monacha-latidens* × *latidens* from the Río Mocorito resulted in all-female offspring, but others produced both sexes. Moreover, there is some evidence that the "*monacha*" genome found in the Río Mocorito contains some genes derived from still another species, *P. viriosa.*

Schultz (1977) did not discuss a mechanism to account for the abnormal sex ratios, but pointed out that in hybrids unbalanced sex ratios are of common occurrence, which he attributed to the different "strength" of the sex-determining mechanisms of the species involved. The interspecies hybrids of *Poeciliopsis* lend support to the idea that different species possess different alleles at the regulatory loci for the sex-determining gene, but any further interpretation has to await more detailed information concerning the various pedigrees.

Among the interspecific hybrids of *Xiphophorus*, sex ratios that deviate significantly from unity are primarily observed in hybrids involving *X. helleri*. For a summary of an extensive list of references to *X. helleri* × *X. maculatus* crosses the reader is referred to Kallman (1965), and additional cases have been discussed by Zander (1964). *Xiphophorus helleri* × *X. xiphidium* crosses have been listed by Öktay (1963) and Zander (1964) and *X. helleri* × *X. variatus* crosses by Kosswig (1959) and Zander (1964). Zander demonstrated that the sex ratio of the Xx′ and Yx′ classes of the F_1 generation (x′ denotes the chromosome of *X. helleri* that is homologous with the gonosomes of the other species) depends both on the stock of *X. helleri* and the species identity of the other parent. Males were especially common in the *X. helleri* × *X. xiphidium* crosses and least common in the hybrids with *X. maculatus*. The frequency of males in the Xx′ classes ranged from 0% to 70% and in the Yx′ classes from 27% to 100%. The results of these crosses are not very informative because the role of the x′ chromosome in sex determination within *X. helleri* is not known and also because *X. helleri* itself has sex ratios that often deviate from unity (Table III).

Apparently only three crosses between heterozygous *X. maculatus* females, WY, and *X. helleri* males have ever been reported (Kosswig, 1928; Kosswig and Öktay, 1955; Sengün, 1941). Virtually all Wx′ offspring of the F_1 and the backcross generations to *X. helleri* differentiated into females, indicating that *X. helleri* has no regulatory genes that activate the male-determining gene on the W.

The problem of whether related species possess different alleles at the homologous autosomal regulatory loci can only be examined in hybrids between two taxa with a known sex chromosome mechanism or in *X. helleri* hybrids with no x′ chromosome (e.g., F_2 or backcross generations to species with marked sex chromosomes). There is some evidence that within *Xiphophorus* the juxtaposition of the sex chromosomes of one species with the regulatory genes of another results in atypical sex determination. ASD is observed in certain *X. maculatus* × *X. helleri*, *X. maculatus* × *X. milleri* and *X. milleri* × *X. nigrensis* hybrids.

The Y chromosome of *X. xiphidium* maintains its male-determining function after introgression into *X. cortezi* and *X. maculatus*, and the X chromosomes of *X. maculatus* and *X. xiphidium* can replace each other (Kosswig, 1959; Öktay, 1963). All F_1 hybrids with one X chromosome of *X. maculatus* and the other from *X. xiphidium* were females (Öktay, 1963). Similarly, the male-determining action of the Y chromosome of *X. ni-*

grensis remains unchanged after introgression into *X. maculatus* (Zander, 1968; Bao and Kallman, 1982) and *X. milleri* (Zander, 1968), but a high percentage of *X. milleri* × *X. nigrensis* backcross hybrids with the X chromosomes derived from *X. milleri* developed into males (Zander, 1968). The XX F_1 hybrids between *X. maculatus* and *X. nigrensis* were all females. The sex chromosomes of *X. maculatus* and *X. variatus* can fully replace each other and sex differentiation of their hybrids is in accordance with their sex chromosome constitution (Kosswig, 1935; Bellamy, 1936).

The Y chromosome of *X. maculatus* can replace the Y chromosome of *X. milleri* (no backcross to *X. maculatus* involving the Y of *X. milleri* was performed), but some XX hybrids that possess at least one X^{ma} chromosome develop into males (Kallman and Atz, 1966). This phenomenon will be discussed in detail below. According to two rather well-documented crosses involving *X. helleri* (3 B) and *X. maculatus* (Jp), all $X^{ma}X^{ma}$ hybrids of the F_2 ($N=82$) and backcross generations ($N=69$) were females (Berg and Gordon, 1953). In these crosses, at least, there is no evidence that *X. helleri* has regulatory genes that cause ASD when brought together with the sex chromosomes of *X. maculatus*. But in other crosses involving domesticated stocks of *X. helleri* and *X. maculatus* some females occured in the $Y^{ma}Y^{ma}$ classes of the F_2 and backcross generations and some males were present among the $W^{ma}Y^{ma}$ fish (Kosswig, 1939). In still another series of crosses between *X. helleri* and *X. maculatus* (Jp) about 25% of the $Y^{ma}Y^{ma}$ class of the third and fourth backcross generations to *X. maculatus* differentiated into functional females (Anders *et al.*, 1969). These results imply that various stocks of *X. helleri* carry different alleles at the autosomal locus for ASD.

Atypical sex determination has been observed in hybrids that involve at least four species (*X. helleri, X. variatus, X. xiphidium, X. maculatus*). Some of the $W^{ma}X^{xi}$ and $X^{ma}X^{va}$ fish developed into both sexes (Kosswig, 1937), but because of the complicated origin of the hybrids it cannot be determined from which species the autosomal regulators for ASD were derived.

The X chromosome of *X. maculatus* and *X. xiphidium* also behave identically in the presence of the rare autosomal factor *a* of *X. maculatus* that, in homozygous condition, causes XX fish of *X. maculatus* to develop into males (see page 115). All F_1 X^{ma} X^{xi} hybrids of *X. xiphidium* (XX) ♀ × *X. maculatus* (XX) ♂ (from Öktay's XX ♀♀ – XX ♂♂ stock) were females. This result is in perfect agreement with the observation that, in intraspecific crosses, XX males give rise to all-female or mostly female progeny when outcrossed to unrelated strains. When one of the F_1 *xiphidium* × *maculatus* females was backcrossed to an XX *maculatus* male, one-half both of the $X^{ma}X^{xi}$ and $X^{ma}X^{ma}$ classes developed into males

(Öktay, 1963). My explanation for these crosses is that the *maculatus* male was *aa* at the autosomal locus and the female of *X. xiphidium* was *AA*. According to these observations, there is no detectable diference between the "*A*" allele of the two species. The combination $X^{xi}X^{xi}$ aa was not tested.

The best example of a regulatory factor from one species interacting with a sex chromosome of another to cause atypical sex determination concerns certain crosses of *X. maculatus* with *X. milleri* (Kallman and Atz, 1966). Some of these $X^{ma}X^{mi}$ hybrids were males (Table XVIII). Inasmuch as the two *X. maculatus* stocks involved, Hp-2 and Gp (Kallman, 1965), and *X. milleri* (Kallman and Atz, 1966; Kallman and Borowsky, 1972) have no history of any XX males, one must conclude that the autosomal factor was derived from *X. milleri* and that it caused the activation of the male-determining gene on the X chromosome of *X. maculatus*. Not every cross involving these two species resulted in $X^{ma}X^{mi}$ males (Kallman and Atz, 1966); the *X. milleri* stock thus appears to have been heterozygous at an autosomal regulatory locus, E^{mi-1} E^{mi-2}.

The autosomal genotype assigned to the 13 $X^{ma}X^{mi}$ females is $E^{ma}E^{mi-1}$ and to the 20 $X^{ma}X^{mi}$ males is $E^{ma}E^{mi-2}$ (pedigree 1532). The backcross of an XX male to an *X. maculatus* female must have resulted in four classes of offspring (pedigree 1605):

1. $X^{ma}X^{ma}E^{ma}E^{ma}$
2. $X^{ma}X^{ma}E^{ma}E^{mi-2}$
3. $X^{ma}X^{mi}E^{ma}E^{ma}$
4. $X^{ma}X^{mi}E^{ma}E^{mi-2}$

All the offspring are XX, but only classes 2 and 4, which exhibit the $X^{ma}E^{mi-2}$ combination and which make up 50% of the total progeny, are expected to develop into males. The actual result (pedigree 1605) fits this expectation well (χ^2 = 0.68, df = 1, P = 0.41).

In another backcross mating to *X. maculatus* (pedigree 1603), an F_1 hybrid male of genotype XY-*Sd* was used. Since the female was X-*Sp* X-*Sd,* all the *Sp* offspring are $X^{ma}X^{mi}$, all the *SpSd* fish are $X^{ma}Y^{ma}$, but the *Sd* progeny consist of two sex-chromosome genotypes, $X^{ma}X^{mi}$ and $X^{ma}Y^{ma}$. If the F_1 hybrid male also carried the E^{mi-2} allele, one-half of each sex chromosome class would have been $X^{ma}E^{mi-2}$. Consequently, one-half of all the XX fish should have developed into males. The frequency of *Sp* males does not differ from this expectation at the 0.05 level of significance (χ^2 = 3.7, df = 1, P = 0.05). The *Sd* class should have consisted of 75% males (all XY fish and one-half of the XX class) and the observed result is in perfect agreement with this explanation (χ^2 = 0.07, df = 1, P = 0.79).

Table XVIII

Atypical Sex Determination in Hybrids between *Xiphophorus maculatus* (*ma*) and *Xiphophorus milleri* (*mi*)

Parents		Phenotype of progeny					
♀♀	♂♂	Pedigree	♀♀		♂♂		
(*mi*) X-+ X-+	(*ma*)[a] X-+ Y-*Sd*	1532	+: 13		+: 20[c]	*Sd*: 42	
Backcross to *maculatus*							
(*ma*)[b] X-*Sd* X-*Sd*	1532 X-+ X-+	1605	*Sd*: 51		*Sd*: 43[c]		
(*ma*)[b] X-*Sp* X-*Sd*	1532 X-+ Y-*Sd*	1603	*Sd*: 10	*Sp*: 8	*Sd*: 33[d]	*Sp*: 19[c]	*SpSd*: 20
Backcross to *milleri*							
1532 X-+ X-+	(*mi*) X-+ Y-*Sv*	1587	+: 24		+: 4[c]	*Sv*: 37	
(*mi*) X-+ X-+	1532 X-+ X-+	1606	+: 50		+: 23[c]		

[a]From Hp-2 stock.
[b]From Gp stock.
[c]These fish are XX males.
[d]Some of these fish are XX males.

The reciprocal backcross to *X. milleri* provides an opportunity to ascertain whether E^{mi-2} interacts with X^{ma}, but not with X^{mi}, to cause ASD. One-half of the XX offspring of this backcross will be $X^{mi}X^{mi}$, the other half $X^{ma}X^{mi}$. Unfortunately, none of the X chromosomes carried marker genes; nevertheless, if only the $X^{ma}E^{mi-2}$ combination results in ASD, 25% of the offspring, not 50% as in the backcross to *X. maculatus*, should be males. The results of pedigree 1606 ($\chi^2 = 1.65$, df = 1, $P = 0.20$) and pedigree 1587 ($\chi^2 = 1.79$, df = 1, $P = 0.18$) are in agreement with this expectation, but differ significantly from an expectation of 50% ($P < 0.001$ and $P < 0.0001$, respectively).

The XX male condition in *maculatus* × *milleri* hybrids is caused by a dominant autosomal gene E^{mi-2}, whereas XX males within species are caused by the recessive autosomal factor *a*. No genetic data are available to determine whether these factors are allelic or whether they represent different loci. It was suggested above that the product of the common allele at the *A* locus is involved in the suppression of the male-determining gene *M* on the X, and that the rare allele *a* at this locus does not possess the ability to do so. Hence all XX *AA* and most XX *Aa* fish become females, but XX *aa* individuals develop into males. But this interpretation poses a problem for the male XX hybrids, $X^{ma}X^{mi}$ A^{ma}-E^{mi-2}-. Why does not A^{ma} block the transcription of *M* on X^{ma}? This observation may indicate that *A* and *E* are not allelic and that the product of E^{mi-2} inactivates the gene product of *A* or blocks its transcription. Since E^{mi-2} has no such effect within *X. milleri,* it follows that A^{ma} and A^{mi} cannot be identical and this raises the possibility that the control element *O* on X^{ma} and X^{mi} are also distinct.

In a purely formal way, the sex-determining mechanism of a female homogametic–male heterogametic system may be written as O^1M O^1M *AA* ♀♀ and O^1M O^2M *AA* ♂♂, in which O^1M represents the X and O^2M the Y chromosome (Fig. 1) and *A* is the autosomal regulatory locus that blocks O^1 and prevents the transcription of the adjacent *M* gene. Let us assume that *A* mutates to *a* and that *a* recognizes neither O^1 nor O^2. Therefore, O^1M O^1 *aa* individuals (XX) will develop into males. Bull and Charnov (1977) have studied the evolutionary consequences of several two-locus mechanisms of sex determination. One of their models, 2B, corresponds closely to the genetic explanation proposed above to account for the occurrence of XX males. According to Bull and Charnov, the Y chromosome will be lost if the fitness of the XX *Aa* (♀♀) or XX *aa* (♂♂) genotypes is greater than that of XX *AA* or XY *AA*. A stable multigenic system will be established if the viability of one of the nonrecurrent genotypes XY *Aa* or XY *aa* is greater than that of any other. The two-locus system has the potential to evolve from a homogametic female–heterogametic male mechanism to a heterogametic female–homogametic male one (Bull and Charnov, 1977). During this change sex determination will be taken over by the pair of autosomes on which the *A* gene is located and the ancestral gonosome X will show autosomal inheritance. Consequently, in related species the sex chromosomes may not be homologous to each other. The evolution of an XX *Aa* ♀♀–XX *aa* ♂♂ system will have important consequences for *Xiphophorus*. It would immediately abolish sex linkage for all traits governed by loci on the ancestral gonosomes and do away with the marked sexual dimorphism related to pigmentation and size and age at sexual maturity. In *X. nigrensis* it would eliminate the large male morphs.

There is no evidence that such a process has occurred in *Xiphophorus*. The sex chromosomes of the seven species that on the basis of the inheritance of certain traits are known to be homogametic in females and heterogametic in males are homologous to each other, i.e., during meiosis in species hybrids their gonosomes segregate from each other. Five of these species, e.g., *X. maculatus*, possess a sex-linked macromelanophore locus. In an eighth species, *X. helleri*, with a still unknown sex-determining mechanism, the macromelanophore locus is on a chromosome to which no role in sex determination can be ascribed (Kallman and Atz, 1966). If at one time this chromosome had a sex-determining function, but if this function was eventually taken over by another chromosome, or if *X. helleri* has a two-locus system, in which the sex-determining function of a particular chromosome cannot easily be recognized, then the chromosome with the macromelanophore locus of *X. helleri* should be homologous to the gonosomes of *X. maculatus*. Experimental evidence does not support this possibility. In the hybrids the two chromosomes segregate independently from each other (Kallman and Atz, 1966).

A similar and more complicated problem is presented by *X. cortezi*, which has three unlinked macromelanophore loci, *Sc*, *At* [= *Sp* of Kosswig (1959) and Zander (1965, 1969)], and *Cam* (Kallman, 1971). The inheritance of the three color patterns does not indicate sex linkage within *X. cortezi* (Kallman, 1971; Zander, 1965), but in a species hybrid, *Sc* segregated from a sex-linked pattern of *X. maculatus* [discussed by Kallman and Atz (1966)], suggesting that the pair of chromosomes of *X. cortezi* marked by *Sc* is homologous to the gonosomes of *X. maculatus*. This report may have provided the basis for Zander's (1969) statement that *Sc* is X- or Y-linked, although no new data were offered. The situation in *X. cortezi* could represent an example in which a pair of chromosomes has no role in sex determination in one species but has a sex-determining function in another. In this case *X. cortezi* could possess only X chromosomes, females would be heterozygous for *A* and *a*, and males homozygous for *a*. Whether this is the case in *X. cortezi* can be examined in hybrids between males of *X. cortezi* and females of species with the XX ♀♀–XY ♂♂ mechanism. Zander (1965) performed crosses between 16 *X. cortezi* males and 26 females of *X. maculatus, X. variatus,* and *X. xiphidium*. Six of these males gave rise to 379 female and 20 male progeny and 10 males sired 224 female and 221 male progeny. On the basis of these results it appears that there exist two kinds of males in *X. cortezi* and these data do not fit a XX *AA* ♀♀–XX *aa* ♂♂ model.

When the male F_1 hybrids of *maculatus* ♀♀ × *cortezi* ♂♂ were backcrossed to *X. maculatus* for two generations, the $X^{ma}X^{ma}$ progeny consisted almost exclusively of females (first Bc: 53 ♀♀, one ♂; second Bc: 45 ♀♀), whereas the progeny with one X^{ma} and the homologous chromosome of *X. cortezi* consisted of both sexes (first Bc: 43 ♀♀; 23 ♂♂; second Bc: 35 ♀♀, 24 ♂♂).

Unfortunately, the sex-determining mechanism of *X. cortezi* is not well known. Although in our laboratory *X. cortezi* gives rise to both sexes in equal numbers (Table III), Zander (1965) obtained several all-female broods. Based exclusively on sex ratio data, Zander concluded that females were homogametic (XX), but that males were either homogametic (XX) or heterogametic (XY). Of 18 males tested (the 16 males tested above plus two additional ones), seven *Sc* and two wild-type individuals were identified as being heterogametic and five *AtSc* and two *At* fish were thought to be homogametic. The significance of the association of *At* with homogamety is not clear. At least two of Zander's homogametic males sired with three females of *X. cortezi* 286 ♀♀ and three males, but with a fourth female one of the males produced 15 female and 20 male offspring and the other produced 24 female and 31 male progeny. Thus as in the ASD stocks of *X. maculatus* and *X. nigrensis* the sex ratio of *X. cortezi* depends upon the genotypes of both the male and female parents. The entire problem of sex determination in *X. cortezi* must be reinvestigated.

15. The Sex-Determining System of *Xiphophorus helleri*

The alleged polyfactorial system of sex determination in fishes has generally been regarded as a primitive mechanism (Anders and Anders, 1963; Chan,1970; Kosswig, 1964; Ohno, 1967; Öktay, 1959*a,b;* Schröder, 1964, White, 1973; Zander, 1965). Schröder (1964) looked upon the sex chromosomes and sex-determining mechanisms of poeciliid fishes as being in "*in statu nascendi*" from an ancestral autosomal-polygenic system to a monofactorial one. Similar ideas concerning *X. cortezi* were expressed by Zander (1965), who thought that this species had reached an intermediate evolutionary stage between an original purely polyfactorial (*X. helleri*) and a derived, strictly gonosomal (*X. maculatus*) sex-determining mechanism.

Reference has been made previously to several species of *Xiphophorus* of different lineages (Rosen, 1979) that possess the XX–XY system and that the sex chromosomes of these species are homologous with one another. Inasmuch as the odds are enormously against the possibility that the same pair of chromosomes out of 24 pairs could have evolved independently into gonosomes several times, the most parsimonious explanation is that in the ancestral *Xiphophorus* this chromosome pair already possessed a sex-determining function. The present genetic system of sex determination of *X. helleri* cannot represent the primitive, undifferentiated condition. The sex-determining system of *X. helleri* may closely resemble that of the other species in the genus, the only difference being that in *X. helleri* the frequencies of the ASD factors are relatively higher.

Witschi (1959) noted that the ubiquitous occurrence of the XX–XY and WZ–ZZ mechanisms among mammals on the one hand and birds and reptiles on the other suggests that the two mechanisms had evolved during Jurassic times or earlier and were already present in the ancestors of these lineages. More evidence for the correctness of this view comes from Ohno's (1967) determination that similar linkage groups are present on the X chromosome of diverse mammalian species. Ohno proposed that the X and Z chromosomes have been conserved throughout mammalian and avian evolution. It would be surprising if the situation in fishes were different and that sex chromosomes had only recently arisen in the teleosts. I find it unreasonable to assume that the sex chromosomes in the different genera of poeciliid fishes have independently evolved. Although evidence is still not available, because suitable biochemical marker genes are absent, it is quite possible that the sex chromosomes of the different genera are homologous. Of the more than 130 known species of poeciliid fishes (Rosen and Bailey, 1963), the sex-determining mechanism of hardly more than a dozen species has been investigated. In this small sample, most species are XX–XY, but two species, *X. maculatus* and *P. sphenops,* possess W as well as X and Y chromosomes and this suggests that the evolution from X or Y to W has occurred at least twice and that this change is a relatively simple step. Female heterogamety has also been demonstrated in *Gambusia* on the basis of karyological evidence (Chen and Ebeling, 1968; Campos and Hubbs, 1971; Black and Howell, 1979) but whether female heterogamety is characteristic for this poeciliid genus is not yet known.

In *X. maculatus* and *X. nigrensis,* sex ratios have been obtained that deviate from unity as much as those of *X. helleri.* Nevertheless, the existence of sex chromosomes in *X. maculatus* and *X. nigrensis* cannot be in doubt inasmuch as their inheritance can be traced through dominant marker genes. Had such markers been absent, however, the demonstration of W, X, Y, and Y′ chromosomes, as well as specific autosomal regulatory genes that interact with just one of these sex chromosomes, would have been indeed difficult. Additional difficulties arise from the fact that the ASD genotypes exhibit incomplete penetrance. Based on the sex-determining systems of these species, uneven sex ratios are not indicative of the absence of gonosomes. Although no sex-linked trait has been discovered in *X. helleri,* this cannot be taken as evidence that sex chromosomes are absent. On the contrary, *X. helleri* may possess sex chromosomes just as do the other species in this genus. Moreover, alleles for atypical sex determination must occur at an autosomal regulatory locus with high frequency in order to account for the prevalence of uneven sex ratios. Such factors may become rapidly fixed in laboratory strains, which are usually derived from few animals. It should be recalled that the factor *a* that causes XX *aa* individuals to develop into males is present at a relatively high frequency of 0.26 in the Río Choy population of *X. nigrensis*.

The phenomenon of early- and late-maturing males in *X. helleri* also has its counterpart in many other species of *Xiphophorus,* in which it is controlled by a sex-linked gene. The existence of small, early-maturing and large, late-maturing males in *X. helleri* raises not only the strong possibility that the *P* gene also exists in this species, but that it is also located on a chromosome that is homologous with the gonosome of the other species. There is even a hint that it may have a sex-determining function. The association of late maturation and a sex ratio biased in favor of females in two stocks of *X. helleri* (Peters, 1964) is in accordance with the view that the *P* factor is located on a chromosome that is involved in atypical sex determination, analogous to the X chromosome of *X. nigrensis* associated with small size and Y′ of *X. maculatus* associated with large size.

16. Summary

Sex determination in several species of poeciliid fishes is controlled by a sex chromosome mechanism. A number of species of *Xiphophorus* possess the XX–XY system, and one form, *X. maculatus,* has three sex chromosomes, W, X, and Y. In natural populations, females are either XX, WX, or WY and males are XY or YY. The sex chromosomes of the different species of *Xiphophorus* are homologous with one another and, presumably, the same pair of chromosomes already had a sex-determining function in the ancestral *Xiphophorus*. In all species sporadic instances of atypical sex determination occur in which individuals with the sex chromosome constitution typical of females, XX, WX, or WY, differentiate into functional males, and XY individuals develop into functional females. The phenomenon of atypical sex determination has a genetic basis and each instance is caused by the interaction of an autosomal factor with a specific sex chromosome. When exceptional WX or WY males are mated with WY females, WW progeny are obtained that can develop into either males or females, and when exceptional XY females are mated to XY or YY males, the YY offspring may differentiate into either sex.

A theory of sex determination has been proposed according to which the male-determining gene *M* specifies whether the undifferentiated gonad primordium develops into an ovary or a testis. The *M* gene could be identical with the H-Y locus. The *M* gene is present on all sex chromosomes, W, X, and Y. Closely linked to *M* is an operatorlike control element that determines whether or not *M* is transcribed. The control elements on the three sex chromosomes are different and by their interaction with a regulatory protein of an autosomal locus, the control elements on the X and W chromosomes prevent the transcription of the adjacent male-determining gene *M*. In addition, a suppressor gene *S* has to be postulated on the W chromosome that prevents transcription of *M* on the Y chromosome.

The XX male condition is caused by a recessive autosomal factor *a*, which, when present in homozygous condition, causes transcription of *M* on the X chromosome. The normal or common allele *A* of this locus may be involved in blocking the control element on the X. The allele *A* does not exhibit complete dominance, because a small percentage of *Aa* individuals also develop into males.

Another autosomal factor apparently blocks the suppressor *S* on the W chromosome and is responsible for the differentiation of WY, but not WX, individuals into males. This factor does not cause the transcription of *M* on the W chromosome. A third autosomal factor is responsible for the development of WX, WY, and WW fish into males. This factor blocks *S* at the same time as it causes the transcription of *M* on the W chromosome. This factor has only a very minor effect on the X chromosome.

The XY and YY female condition can be attributed to a dominant autosomal factor *T* with incomplete penetrance that interacts with a mutant control element on the Y and prevents the transcription of *M*. The Y chromosome with the mutant control element is denoted as Y′, and Y′ chromosomes have been identified in *X. maculatus, X. montezumae,* and *X. milleri*. The factor *T* has no effect on the common Y chromosome, and the common allele *t* at the autosomal locus has no detectable effect on Y′. Hence Y′Y′ *TT* fish are females and YY′ *TT* or Y′Y′ *tt* individuals are males.

Although the sex chromosomes of the different species of *Xiphophorus* are homologous with one another and the male-determining gene *M* may be identical in all species, there exists the possibility that the control factors on the X and Y chromosomes of the different species are not identical. Similarly, the autosomal regulatory loci may be the same in all species, but different alleles may be present. The interaction of the control elements on the sex chromosome of one species with a foreign autosomal regulatory factor in species hybrids may lead to the activation of *M* on the X or W chromosomes or may block the transcription of *M* on the Y. In certain interspecies hybrids, sex determination is completely normal, but in other hybrids there is an increased frequency of atypical sex determination. Certain crosses between *X. maculatus* and *X. milleri* indicate that in the presence of a dominant autosomal factor of *X. milleri* the *M* gene on the X chromosome of *X. maculatus* is transcribed and such hybrids with one or two X chromosomes of *X. maculatus* develop into males. This autosomal factor and its homologous counterpart in *X. maculatus* have no detectable effect on the X chromosome of *X. milleri*. All hybrids with two X chromosomes of *X. milleri* develop into females.

According to this theory, the control elements on the sex chromosomes must be regarded as the true sex-determining genes, because they represent the only consistent genetic difference between male and female under ordinary conditions.

ACKNOWLEDGMENTS. I thank Dr. J. W. Atz for many editorial comments and bibliographical references, and Dr. S. S. Wachtel for his interest in this study and several interesting discussions. The research in the Genetics Laboratory is supported in part by a U.S. Public Health Service Grant, GM 19934, which is gratefully acknowledged.

Appendix. Sex Ratio Data for Various *Xiphophorus* Stocks

Table XIX

Sex Ratio in *Xiphophorus helleri helleri*, Cd Stock[a]

	Progeny		Parental		Progeny		Parental
Generation	♀♀	♂♂	generation	Generation	♀♀	♂♂	generation
8[b]	9	44		21B	5	8	
9	13	20		22	2	35	21A
10A	12	20		23	7	18	
10B	1	23		24	13	33	
11A	11	13	10A	25	17	46	
11B	0	10	10A	26A	10	22	
12	15	10	11A	26B	0	9	

Table XIX (*Continued*)

Generation	Progeny ♀♀	Progeny ♂♂	Parental generation	Generation	Progeny ♀♀	Progeny ♂♂	Parental generation
13	25	2		27	5	16	26A,B
14	13	22		28A	8	46	
15A	15	2		28B	3	27	
15B	2	3		29A	3	32	28B,A
16A	24	3	15B	29B	12	28	28B
16B	7	0	15A	30	3	30	29B,A
17	34	5	16A	31	26	50	
18A	7	15		32	9	22	
18B	8	5		33	15	64	
19A	32	30	18A	34A	1	20	
19B	8	1	18B	34B	2	33	
20	17	11	19A	34C	4	6	
21A	5	28		35	3	27	34A

[a]This stock is derived from fish collected near Cordoba, Mexico, by Dr. C. L. Turner in the early 1930s. This stock was maintained at the University of Michigan until 1948.
[b]This represents the eighth generation after the stock was transferred to the Genetics Laboratory of the New York Aquarium.

Table XX
Sex Ratio in *Xiphophorus helleri strigatus*, 3B Stock[a]

Generation	Progeny ♀♀	Progeny ♂♂	Parental generation	Generation	Progeny ♀♀	Progeny ♂♂	Parental generation
17[b]	13	53		26A	2	22	
18A	56	72		26B	8	0	
18B	10	11		27	5	23	26A
19A	28	20	18A	28	10	24	
19B	7	3	[c]	29	20	32	
20A	30	18	[c]	30	17	19	
20B	25	13	[c]	31A	6	1	
21	49	9	20A	31B	14	3	
22	14	6		32	8	9	31A
23A	20	11		33A	8	12	♀31A, ♂32
23B	12	11		33B	2	1	
23C	10	5		34A	7	5	33A,B
23D	5	0		34B	10	4	33A
24A	2	11	23B	34C	1	2	33A
24B	1	14	23A	35A	10	4	34A
24C	2	17	23A, same ♂ as in 24B	35B	1	7	34A
25	27	28	24A	35C	6	13	34A,B

[a]This strain is descended from fish collected in Arroyo Zacatispan, Río Papaloapan drainage, Oaxaca, Mexico, in 1939.
[b]Records for first 16 generations not available.
[c]Parents from preceding generations, but not recorded whether they were taken from A or B.

Table XXI
Sex Ratio in *Xiphophorus helleri guentheri*, Bx Stock[a]

Generation	Progeny ♀♀	Progeny ♂♂	Parental generation	Generation	Progeny ♀♀	Progeny ♂♂	Parental generation
2A	15	15		19A	7	18	18B
2B	31	45		19B	1	20	18B
3	2	16	2A or B	20	5	40	19A
4	1	30		21A	3	19	
	Not recorded			21B	43	7	
5							
6	2	8		22	25	17	21B
7	14	38		23	11	30	
8	21	42		24	13	16	
9	30	19		25	25	17	
10A	0	11		26	29	22	
10B	4	8		27	36	15	
10C	2	16	Same ♂ as in 10B	28	16	25	
10D	6	21	Same ♂ as in 10A	29A	30	21	
11A	7	7	10C	29B	22	8	
11B	6	10	10C	30A	4	2	29A
12	23	26	11A	30B	11	10	29B
13	14	32		30C	39	35	29A
14A	33	4		30D	9	7	29B
14B	3	4		31A	6	24	30C
15	4	16	14A	31B	13	13	30C
16	10	33		32A	12	38	31B
17	7	43		32B	28	7	31B Same ♂ as in 32A
18A	0	3		33A	13	25	32A
18B	3	23		33B	2	7	32A

[a]This stock originated from fish collected in the Belize River, Belize, in 1949.

Table XXII
Sex Ratio in a Strain of *Xiphophorus signum*[a]

Generation	Progeny ♀♀	Progeny ♂♂	Parental generation	Generation	Progeny ♀♀	Progeny ♂♂	Parental generation
1	67	1		8A	27	36	7B,A
2	26	5		8B	6	6	7B,A
3A	28	0		9	17	20	8A,B
3B	19	8		10A	13	15	
4A	19	1	3B	10B	26	3	
4B	23	2	3B	11	8	19	10A
4C	23	6	3B	12	23	26	

Table XXII (*Continued*)

Generation	Progeny ♀♀	Progeny ♂♂	Parental generation	Generation	Progeny ♀♀	Progeny ♂♂	Parental generation
5A	41	1	4C	13	9	2	
5B	9	6	4B,C	14	12	21	
6A	8	15	5A	15A	27	8	
6B	37	13	5A,B	15B	15	4	
7A	33	18	6A,B	16A	17	0	15B
7B	13	1	6B,A	16B	4	30	15A,B

[a]This strain originated from a single gravid female collected in the Río Semococh, Río Chajmaic system, Alta Verapaz, Guatemala, in 1963.

Table XXIII
Sex Ratio in a Stock of *Xiphophorus alvarezi*[a]

Generation	Progeny ♀♀	Progeny ♂♂
1	12	15
2	17	19
3	5	24
4	5	63
5	3	35
6	8	3

[a]This stock originated from a single gravid female collected in the Río San Ramón, Río Lacantún drainage, Huehuetenango, Guatemala, in 1976.

References

Aida, T., 1930, Further genetical studies of *Aplocheilus latipes, Genetics* **15:**1–16.

Aida, T., 1936, Sex reversal in *Aplocheilus latipes* and a new explanation of sex differentiation, *Genetics* **21:**136–153.

Anders, A., and Anders, F., 1963, Genetisch bedingte XX-♀♀ und XY- und YY-♂♂ beim wilden *Platypoecilus maculatus* aus Mexico, *Z. Vererbungsl.* **94:**1–18.

Anders, A., Anders, F., Förster, W., Klinke, K., und Rase, S., 1969, XX-, XY-, YY-♀♀ und XX-, XY-, YY-♂♂ bei *Platypoecilus maculatus* (Poeciliidae), *Zool. Anz. Suppl.* **33:**333–337.

Anders, A., Anders, F., and Klinke, K., 1973, II. The arrangement of chromatophore determining loci and regulating elements in the sex chromosomes of xiphophorin fish, *Platypoecilus maculatus* and *Platypoecilus variatus,* in: *Genetics and Mutagenesis of Fish* (J. H. Schröder, ed.), Springer-Verlag, New York, pp. 33–63.

Atz, J. W., 1964, Intersexuality in fishes, in: *Intersexuality in Vertebrates Including Man* (A. J. Marshall and C. N. Armstrong, eds.), Academic Press, New York, pp. 145–232.

Avtalion, R. R., and Hammerman, I. S., 1978, Sex determination in *Sarotherodon* (Tilapia). I. Introduction to a theory of autosomal influence, *Bamidgeh* **30:**110–115.

Bao, I. Y., and Kallman, K. D., 1982, Genetic control of the hypothalamo-pituitary axis and the effect of hybridization on sexual maturation (*Xiphophorus*, Pisces, Poeciliidae), *J. Exp. Zool.* **220**:297–309.

Bellamy, W. A., 1936, Inter-specific hybrids in *Platypoecilus:* one species ZZ–WZ; the other XY–XX, *Proc. Natl. Acad. Sci. USA* **22**:531–536.

Bellamy, A. W., and Queal, M. L., 1951, Heterosomal inheritance and sex determination in *Platypoecilus maculatus, Genetics* **36**:93–107.

Bennet, D., Mathieson, B. J., Scheid, M., Yanagisawa, K., Boyse, E. A., Wachtel, S. S., and Cattanach, B. M., 1977, Serological evidence for H-Y antigen in *Sxr,* XX sex-reversed phenotypic males, *Nature* **265**:255–257.

Berg, O., and Gordon, M., 1953, Relationship of atypical pigment cell growth to gonadal development in hybrid fishes, in: *Pigment Cell Growth* (M. Gordon, ed.), Academic Press, New York, pp. 43–72.

Bernstein, R., Koo, G. C., and Wachtel, S. S., 1980, Abnormality of the X chromosome in human 46, X Y female siblings with dysgenetic ovaries, *Science* **207**:768–769.

Black, D. and Howell, W. M., 1979, A unique case in sex chromosome evolution, *Copeia* **3**:509–513.

Breider, H., 1935*a*, Über Aussenfaktoren, die das Geschlechtsverhältnis bei *Xiphophorus helleri* Heckel kontrollieren sollen, *Z. Wiss. Zool.* **146**:383–416.

Breider, H., 1935*b*, Geschlechtsbestimmung und -differenzierung bei *Limia nigrofasciata, caudofasciata, vittata* und deren Artbastarden, *Z. Indukt. Abstammungs. Vererbungsl.* **68**:265–299.

Breider, H., 1937, Juveniles und adultes Geschlechtsverhaeltnis bei *Xiphophorus helleri, Z. Indukt. Abstammungs. Vererbungsl.* **73**:371–475.

Bull, J. J., 1980, Sex determination in reptiles, *Quart. Rev. Biol.* **55**:3–21.

Bull, J. J., and Charnov, E. L., 1977, Changes in the heterogametic mechanism of sex determination, *Heredity* **39**:1–14.

Campos, H. H. and Hubbs, C., 1971, Cytomorphology of six species of gambusiine fishes, *Copeia* **3**:566–569.

Cattanach, B. M., Pollard, C. E., and Hawkwes, S. G., 1971, Sex-reversed mice: X X and X O males, *Cytogenetics* **10**:318–337.

Chan, S. T. H., 1980, Natural sex reversal in vertebrates, *Phil. Trans. R. Soc. Lond. B* **259**:59–71.

Chen, T. R., and Ebeling, A. W., 1968, Karyological evidence for female heterogamety in the mosquitofish, *Gambusia affinis, Copeia* **1**:70–75.

Cimino, M. C., 1972, Egg production, polyploidization and evolution in a diploid all-female fish of the genus *Poeciliopsis, Evolution* **26**:294–306.

Clemens, H. P., and Inslee, T., 1968, The production of unisexual broods of *Tilapia mossambica* sex-reversed with methyl testosterone, *Trans. Am. Fish. Soc.* **67**:18–21.

Cohen, H., Gordon, M., and Nigrelli, R. F., 1941, Spontaneous development of gonopods in females of *Platypoecilus maculatus, Anat. Rec.* **81**(suppl. 1):89.

Conover, D. O., and Kynard, B. E., 1981, Environmental sex determination: Interaction of temperature and genotype in a fish, *Science* **213**:577–579.

Dzwillo, M., 1962, Über künstliche Erzeugung funktioneller Männchen weiblichen Genotyps bei *Lebistes reticulatus, Biol. Zentralbl.* **81**:575–584.

Dzwillo, M., and Zander, C. D., 1967, Geschlechtsbestimmung und Geschlechtsumstimmung bei Zahnkarpfen (Pisces), *Mitt. Hamb. Zool. Mus. Inst.* **64**:147–162.

Engel, W., Klemme, B., Probeck, H. D., and Hansmann, I., 1981*a*, H-Y antigen in Turner's syndrome patients with different sex chromosome constitutions, *Hum. Genet.* **59**:333–336.

Engel, W., Klemme, B., and Ebrecht, A., 1981*b*, Serological evidence for H-Y antigen in X O female mice, *Hum. Genet.* **57**:68–70.

Essenberg, J. M., 1926, Complete sex reversal in the viviparous teleost, *Xiphophorus helleri, Biol. Bull.* **51:**98–111.

Evans, E. P., M. D. Burtenshaw, and Cattanach, B. M., 1982, Meiotic crossing-over between the X and Y chromosomes of male mice carrying the sex-reversed (*Sxr*) factor, *Nature* **300:**443–445.

Fraser, A. C., and Gordon, M., 1929, The genetics of *Platypoecilus*. II. The linkage of two sex-linked characters, *Genetics* **14:**160–179.

Fredga, K., Gropp, H., Winking, H., and Frank, F., 1976, Fertile XX- and XY-type females in the wood lemming *Myopus schisticolor, Nature* **261:**225–227.

Fredga, K., Gropp, A., Winking, H., and Frank, F., 1977, A hypothesis explaining the exceptional sex ratio in the wood lemming (*Myopus schisticolor*), *Hereditas* **85:**101–104.

Gileva, E. A., and Chebotar, N. A., 1979, Fertile XO males and females in the varying lemming, *Dicrostonyx torquatus* Pall (1779). A unique genetic system of sex determination, *Heredity* **42:**67–77.

Gomelsky, B. I., and Fetisov, A. N., 1977, Sex ratio in the swordtail *Xiphophorus helleri* Heckel (Cyprinodontiformes, Poeciliidae) at different stages of ontogenesis in connection with the problem of sex differentiation, *Vopr. Ikhtiol.* **17:**767–769.

Gordon, J. W., and Ruddle, F. H., 1981, Mammalian gonadal determination and gametogenesis, *Science* **211:**1265–1271.

Gordon, M., 1937, Heritable color variations in the Mexican swordtail fish, *J. Hered.* **28:**222–230.

Gordon, M., 1938, The genetics of *Xiphophorus hellerii:* Heredity in Montezuma, a Mexican swordtail fish, *Copeia* **1:**19–29.

Gordon, M., 1956, Evidence for complete sex reveral in fishes, *Aquarist Pondkeeper* **21:**66–69.

Gordon, M., and Benzer, P., 1945, Sexual dimorphism in the skeletal elements of the gonopodial suspensoria in xiphophorin fishes, *Zoologica* (*N. Y.*) **30:**57–72.

Grobstein, C., 1942, Endocrine and developmental studies of gonopod differentiation in certain poeciliid fishes, *J. Exp. Zool.* **89:**305–328.

Halpern-Sebold, L. R. and Schreibman, M. P., 1983, Ontogeny of centers containing luteinizing hormone-releasing hormone in the brain of platyfish (*Xiphophorus maculatus*) as determined by immunocytochemistry, *Cell Tissue Res.* **229:**75–84.

Hamerton, J. L., Dickson, J., Pollard, C. E., Grieves, S. A., and Short, R. V., 1969, Genetic intersexuality in goats, *J. Reprod. Fertil. Suppl.* **7:**25–51.

Hansmann, I., 1982, Sex reversal in the mouse, *Cell* **30:**331–332.

Harms, J. W., 1929, Realisation von Genen und die consecutive Adaption. I. Phasen in der Differenzierung der Anlagekomplexe und die Frage der Landtierwerdung, *Z. Wiss. Zool.* **133:**211–397.

Harrington, R. W., Jr., 1967, Environmentally controlled induction of primary male gonochorists from eggs of the self-fertilizing hermaphroditic fish, *Rivulus marmoratus* Poey, *Biol. Bull.* **131:**174–199.

Harrington, R. W., Jr., 1971, How ecological and genetic factors interact to determine when self-fertilizing hermaphrodites of *Rivulus marmoratus* change into functional secondary males, with a reappraisal of the modes of intersexuality among fishes, *Copeia* **3:**389–432.

Haseltine, F. P., and Ohno, S., 1981, Mechanisms of gonadal differentiation, *Science* **211:**1272–1278.

Haskins, C. P., Haskins, E. F., McLaughlin, J. J. A., and Hewitt, R. E., 1961, Polymorphism and population structure in *Lebistes reticulatus,* an ecological study, in: *Vertebrate Speciation* (W. F. Blair, ed.), University of Texas Press, Austin, Texas, pp. 320–395.

Haskins, C. P., Young, P., Hewitt, R. E., and Haskins, E. F., 1970, Stabilized heterozygosis of supergenes mediating certain Y-linked colour patterns in populations of *Lebistes reticulatus, Heredity* **25:**575–589.

Herbst, E. W., Fredga, K., Frank, F., and Winking, H., 1978, Cytogenetic identification of two X-chromosome types in the wood lemming (*Myopus schisticolor*), *Chromosoma* **69:**185–191.

Kallman, K. D., 1965, Genetics and geography of sex determination in the poeciliid fish, *Xiphophorus maculatus, Zoologica* (*N. Y.*) **50:**151–190.

Kallman, K. D., 1968, Evidence for the existence of transformer genes for sex in the teleost *Xiphophorus maculatus, Genetics* **60:**811–828.

Kallman, K. D., 1970, Sex determination and the restriction of pigment patterns to the X and Y chromosomes in populations of the poeciliid fish, *Xiphophorus maculatus,* from the Belize and Sibun Rivers of British Honduras, *Zoologica* (*N. Y.*) **55:**1–16.

Kallman, K. D., 1973, The sex-determining mechanism of the platyfish, *Xiphophorus maculatus,* in: *Genetics and Mutagenesis of Fish,* (J. H. Schröder, ed.), Springer-Verlag, New York, pp. 19–28.

Kallman, K. D., 1975, The platyfish, *Xiphophorus maculatus,* in: *Handbook of Genetics,* Vol. 4 (R. C. King, ed.), Plenum Press, New York, pp. 81–132.

Kallman, K. D., 1983, The sex-determining mechanism of the poeciliid fish, *Xiphophorus montezumae* Jordan and Snyder and the genetic control of the sexual maturation process and adult size, *Copeia,* **3:**755–769.

Kallman, K. D., and Atz, J. W., 1966, Gene and chromosome homology in fishes of the genus *Xiphophorus, Zoologica* (*N. Y.*) **51:**107–135.

Kallman, K. D., and Borkoski, V., 1978, A sex-linked gene controlling the onset of sexual maturity in female and male platyfish (*Xiphophorus maculatus*), fecundity in females and adult size in males, *Genetics* **89:**79–119.

Kallman, K. D., and Borowsky, R., 1972, The genetics of gonopodial polymorphism in two species of poeciliid fish, *Heredity* **28:**297–310.

Kallman, K. D., and Schreibman, M. P., 1973, A sex-linked gene controlling gonadotrop differentiation and its significance in determining the age of sexual maturation and size of the platyfish, *Xiphophorus maculatus, Gen. Comp. Endocrinol.* **21:**287–304.

Kosswig, C., 1928, Über Kreuzungen zwischen den Teleostiern *Xiphophorus helleri* und *Platypoecilus maculatus, Z. Indukt. Abstammungs. Vererbungsl.* **52:**114–120.

Kosswig, C., 1932, Hermaphroditismus im Tierreich vom genetischen Standpunkt, *Züchter* **4:**22–32.

Kosswig, C., 1935, Die Kreuzung zweier XX- bzw. XY-Geschlechter miteinander und der Ersatz eines Y-Chromosoms einer Art durch das X-Chromosom einer anderen, *Züchter* **7:**40–48.

Kosswig, C., 1937, Genotypische und phänotypische Geschlechtsbestimmung bei Zahnkarpfen. VII. (Kreuzungen mit *Platypoecilus xiphidium*), *Wilhelm Roux' Arch. Entwicklungs. Org.* **136:**491–528.

Kosswig, C., 1939, Die Geschlechtsbestimmung in Kreuzungen zwischen *Xiphophorus* und *Platypoecilus, Rev. Fac. Sci. Univ. Istanbul Ser. B Sci. Nat.* **4:**91–144.

Kosswig, C., 1941, Mitteilungen zum Geschlechtsbestimmungs-problem bei Zahnkarpfen, *Rev. Fac. Sci. Univ. Istanbul Ser. B Sci. Nat.* **6:**1–32.

Kosswig, C., 1959, Beiträge zur genetischen Analyse xiphophoriner Zahnkarpfen, *Biol. Zentralbl.* **78:**711–718.

Kosswig, C., 1964, Polygenic sex determination, *Experientia* **20:**190–199.

Kosswig, C., und Öktay, M., 1955, Die Geschlechtsbestimmung bei den Xiphophorini (Neue Tatsachen und neue Deutungen), *Istanbul Univ. Fen Fak. Hidrobiol. B* **2:**133–156.

Lodi, E., 1980, Sex inversion in domesticated strains of the swordtail *Xiphophorus helleri* Heckel (Pisces, Osteichthyes), *Bull. Zool.* **47:**1–8.

MacIntyre, P. A., 1961, Spontaneous sex reversals of genotypic males in the platyfish (*Xiphophorus maculatus*), *Genetics* **46:**575–580.

McCarrey, J. R., and Abbot, V. K., 1979, Mechanisms of genetic sex determination, gonadal sex differentiation and germ-cell development in animals, *Adv. Genet.* **20**:217–290.

Miller, L., 1962, The Eichwald–Silmser phenomenon in an inbred strain of platyfish, *Transplant. Bull.* **30**:147.

Mortimer, C. H., McNeilly, A. S., Murray, M. A. F., Fisher, R. A. F., and Besser, G. M., 1974, Gonadotropin releasing hormone therapy in hypogonadal males with hypothalamic pituitary dysfunction, *Br. Med. J.* **4**:617–621.

Müller, U., and Wolf, U., 1979, Cross-reactivity to mammalian anti-H-Y antiserum in teleostean fish, *Differentiation* **14**:185–187.

Nagai, Y., Ciccarese, S., and Ohno, S., 1979, The identification of human H-Y antigen and testicular transformation induced by its interaction with the receptor site of bovine fetal ovarian cells, *Differentiation* **13**:155–164.

Ohno, S., 1967, Sex chromosomes and sex-linked genes, in: *Monographs in Endocrinology*, Vol. 1, Springer-Verlag, New York, p. 192.

Ohno, S., 1979, Major sex-determining genes, in: *Monographs in Endocrinology*, Vol. 11, Springer-Verlag, New York, p. 142.

Ohno, S., Nagai, Y., and Ciccarese, S., 1978, Testicular cells lysostripped of H-Y antigen organize ovarian follicle-like aggregates, *Cytogenet. Cell Genet.* **20**:351–364.

Öktay, M., 1959*a*, Über Ausnahmemaennchen bei *Platypoecilus maculatus* und eine neue Sippe mit XX-Maennchen und XX-Weibchen, *Rev. Fac. Sci. Univ. Istanbul Ser. B Sci. Nat.* **24**:75–92.

Öktay, M., 1959*b*, Weitere Untersuchungen über eine Ausnahme (XX-) Sippe des *Platypoecilus maculatus* mit polygener Geschlechtsbestimmung, *Rev. Fac. Sci. Istanbul Ser. B Sci. Nat.* **24**:225–233.

Öktay, M., 1963, Die Rolle artfremder Gonosomen bei der Geschlechtsbestimmung von Bastarden mit *Platypoecilus xiphidium, Istanbul Univ. Fen Fak. Hidrobiol. B* **4**:1–13.

Park, E. H. and Grimm, H., 1981, Distribution of C-band heterochromatin in the ZW sex chromosomes of European and American eels (Anguillidae, Teleostomi), *Cytogenet. Cell Genet.* **31**:167–174.

Pechan, P., Wachtel, S. S., and Reinboth, R., 1979, H-Y antigen in the teleost, *Differentiation* **14**:189–192.

Peters, G., 1964, Vergleichende Untersuchungen an drei Subspecies von *Xiphophorus helleri* Heckel (Pisces), *Z. Zool. Syst. Evolutionsforsch.* **2**:185–271.

Rimoin, D. L., and Schimke, R. N., 1971, *Genetic Disorders of the Endocrine Gland,* V. C. Mosby, St. Louis, Missouri.

Rosen, D. E., 1979, Fishes from the uplands and intermontane basins on Guatemala: Revisionary studies and comparative geography, *Bull. Am. Mus. Nat. Hist.* **162**(5):267–376.

Rosen, D. E., and Bailey, R. M., 1963, The poeciliid fishes (Cyprinodontiformes), their structure, zoogeography, and systematics, *Bull. Am. Mus. Nat. Hist.* **126**:1–176.

Rosen, D. E., and Kallman, K. D., 1959, Development and evolution of skeletal deletions in a family of viviparous fishes (Cyprinodontiformes, Poeciliidae), *Q. J. Fla. Acad. Sci.* **22**:169–190.

Schmidt, H., 1930, Geschlechtsumwandlungen bei tropischen Zierfischen, *Züchter* **2**:297–305.

Schreibman, M. P., and Kallman, K. D., 1977, The genetic control of the pituitary-gonadal axis in the platyfish, *Xiphophorus maculatus, J. Exp. Zool.* **200**:277–294.

Schröder, J. H., 1964, Genetische Untersuchungen an domestizierten Stämmen der Gattung *Mollienesia* (Poeciliidae), *Zool. Beitr.* **10**:369–463.

Schultz, R. J., 1961, Reproductive mechanisms of unisexual and bisexual strains of the viviparous fish *Poeciliopsis, Evolution* **15**:302–325.

Schultz, R. J., 1969, Hybridization, unisexuality, and polyploidy in the teleost *Poeciliopsis* (Poeciliidae) and other vertebrates, *Am. Nat.* **103**:605–619.

Schultz, R. J., 1973, Origin and synthesis of a unisexual fish, in: *Genetics and Mutagenesis of Fish* (J. H. Schröder, ed.). Springer-Verlag, New York, pp. 207–211.
Schultz, R. J., 1977, Evolution and ecology of unisexual fishes. in: *Evolutionary Biology,* Vol. 10 (M. K. Hecht, W. C. Steere, and B. Wallace, eds.), Plenum Press, New York, pp. 277–331.
Sengün, A., 1941, Ein Beitrag zur Geschlechtsbestimmung bei *Platypoecilus maculatus* und *Xiphophorus helleri, Istanbul Univ. Fen Fak. Mecm. Seri B* **4:**33–48.
Shalev, A., and Huebner, E., 1980, Expression of H-Y antigen in the guppy (*Lebistes reticulatus*), *Differentiation* **16:**81–83.
Shalev, A., Short, R. V., and Hamerton, J. L., 1980, Immunogenetics of sex determination in the polled goat, *Cytogenet. Cell Genet.* **28:**195–202.
Singh, L., and Jones, K. W., 1982, Sex reversal in the mouse (*Mus musculus*) is caused by a recurrent nonreciprocal crossover involving the X and an aberrant Y chromosome, *Cell* **28:**205–216.
Thompson, F. H., 1978, H-Y antigen loci, *Science* **201:**842.
Turner, C. L., 1942, Gonopodial characteristics produced in the anal fins of females of *Gambusia affinis affinis* by treatment with ethinyl testosterone, *Biol. Bull.* **80:**371–383.
Vallowe, H. H., 1957, Sex differentiation in the teleost fish, *Xiphophorus helleri,* as modified by experimental treatment, *Biol. Bull.* **112:**422–429.
Vivien, J., and Mohsen, T., 1952, Action de l'anhydrooxyprogesterone sur les caracteres sexuels squelettiques du xiphophore, *C. R. Soc. Biol. Paris* **146:**773–777.
Vrijenhoek, R. C., and Schultz, R. J., 1974, Evolution of a trihybrid unisexual fish (*Poeciliopsis,* Poeciliidae), *Evolution* **28:**306–319.
Wachtel, S. S., 1977, H-Y antigen and the genetics of sex determination, *Science* **198:**797–799.
Wachtel, S. S., 1981, Induction of the heterogametic gonad, in: *Levels of Genetic Control in Development* (S. Subtelny and U. K. Abbot, eds.), A. R. Liss, New York, pp. 219–234.
Wachtel, S. S., 1983, *H-Y antigen and the biology of sex determination,* Grune-Stratton, p. 302.
Wachtel, S. S., and Ohno, S., 1979, The immunogenetics of sexual development, *Prog. Med. Genet.* **3:**109–142.
Wachtel, S. S., Ohno, S., Koo, G. C., and Boyce, E. A., 1975*a,* Possible role for H-Y antigen in the primary determination of sex, *Nature* **257:**235–236.
Wachtel, S. S., Koo, G. C., Breg, W. R., Elias, S., Boyse, E. A., and Miller, O. J., 1975*b,* Expression of H-Y antigen in human males with two Y chromosomes, *N. Engl. J. Med.* **293:**1070–1072.
Wachtel, S. S., Koo, G. C., and Boyce, E. A., 1975*c,* Evolutionary conservation of H-Y ('male') antigen, *Nature* **254:**270–272.
Wachtel, S. S., Koo, G. C., Ohno, S., Gropp, A., Dev, V. G., Tantravahi, R., Miller, D. A., and Miller, O. J., 1976, H-Y antigen and the origin of XY female wood lemmings (*Myopus schisticolor*), *Nature* **264:**638–639.
Wachtel, S. S., Koo, G. C., and Ohno, S., 1977, H-Y antigen and male development, in: *The Testis in Normal and Infertile Men* (P. Troen and H. R. Nankin, eds.), Raven Press, New York, pp. 35–43.
Wachtel, S. S., Bresler, P. A., and Koide, S. S., 1980*a,* Does H-Y antigen induce the heterogametic ovary?, *Cell* **20:**859–864.
Wachtel, S. S., Koo, G. C., Breg, W. R., and Genel, M., 1980*b,* H-Y antigen in X,i (Xq) gonadal dysgenesis: Evidence of X-linked genes in testicular differentiation, *Hum. Genet.* **56:**183–187.
Wachtel, S. S., Hall, J. L., and Cahill, L. T., 1981, H-Y antigen in primary sex determination.

in: *Bioregulators of Reproduction* (G. Jagiello and H. J. Vogel, eds.), Academic Press, New York, pp. 9–24.

White, M. J. D., 1973, *Animal Cytology and Evolution,* Cambridge University Press, New York.

Wiberg, U., 1982, Serological cross-reactivity to rat anti H-Y antiserum in the female European eel (*Anguilla anguilla*), *Differentiation* **21**:206–208.

Wiberg, U., Mayerova, A., Müller, U., Fredga, K., and Wolf, U., 1982, X-linked genes of the H-Y antigen system in the wood lemming (*Myopus schisticolor*), *Hum. Genet.* **60**:163–166.

Winge, O., 1930, On the occurrence of XX males in *Lebistes,* with remarks on Aida's so-called "Non-Disjunctional" males in Aplocheilus, *J. Genet.* **23**:69–76.

Winge, O., 1934, The experimental alteration of sex chromosomes into autosomes and vice versa, as illustrated by *Lebistes. C. R. Trav. Lab. Carlsberg Ser. Physiol.* **21**:1–49.

Winge, O., and Ditlevsen, E., 1938, A lethal gene in the Y chromosome of *Lebistes, C. R. Trav. Carlsberg Ser. Physiol.* **22**:203–210.

Winge, O., and Ditlevsen, E., 1948, Colour inheritance and sex determination in *Lebistes, C. R. Trav. Lab. Carlsberg Ser. Physiol.* **24**:227–248.

Witschi, E., 1959, Age of sex-determining mechanisms in vertebrates, *Science* **130**:372–375.

Wolf, U., 1979, XY gonadal dysgenesis and the H-Y antigen, *Hum. Genet.* **47**:269–277.

Wolf, U., Fraccaro, M., Mayerova, A., Hecht, T., Zuffardi, O., and Hameister, H., 1980, Turner syndrome patients are H-Y positive, *Hum. Genet.* **54**:315–318.

Yamamoto, T., 1958, Artificial induction of functional sex-reversal in genotypic females of the medaka (*Oryzias latipes*), *J. Exp. Zool.* **137**:227–264.

Yamamoto, T., 1959, A further study on induction of functional sex-reversal in genotypic males of the medaka (*Oryzias latipes*) and progenies of sex-reversals, *Genetics* **44**:739–757.

Yamamoto, T., 1963, Induction of reversal in sex differentiation of YY zygotes in the medaka, *Oryzias latipes, Genetics* **48**:293–306.

Yamamoto, T., 1964*a,* The problem of viability of YY zygotes in the medaka, *Oryzias latipes, Genetics* **50**:45–58.

Yamamoto, T., 1964*b,* Linkage map of sex chromosomes in the medaka, *Oryzias latipes, Genetics* **50**:59–64.

Yamamoto, T., 1967, Estrone-induced white YY females and mass production of white YY males in the medaka, *Oryzias latipes, Genetics* **55**:329–336.

Yamamoto, T., and Kajishima, T., 1968, Sex hormone induction of sex reversal in the goldfish and evidence for male heterogamety, *J. Exp. Zool.* **168**:215–221.

Zander, C. D., 1964, Physiologische und genetische Untersuchungen zur Systematik xiphophoriner Zahnkarpfen, *Mitt. Hamb. Zool. Mus. Inst. Kosswig-Festschrift* **1964**:333–348.

Zander, C. D., 1965, Die Geschlechtsbestimmung bei *Xiphophorus montezumae cortezi* Rosen (Pisces), *Z. Vererbungsl.* **96**:128–141.

Zander, C. D., 1968, Über die Vererbung von Y-gebundenen Farbgenen des *Xiphophorus pygmaeus nigrensis* Rosen, *Mol. Gen. Genet.* **101**:29–42.

Zander, C. D., 1969, Über die Entstehung und Veränderung von Farbmustern in der Gattung *Xiphophorus* (Pisces). I. Qualitative Veränderungen nach Artkreuzung, *Mitt. Hamb. Zool. Mus. Inst.* **66**:241–271.

Zenzes, M. T., Wolf, U., Günther, E., and Engel, W., 1978*a,* Studies on the function of H-Y antigen: Dissociation and reorganization experiments on rat gonadal tissue, *Cytogenet. Cell Genet.* **20**:365–371.

Zenzes, M. T., Wolf, U., and Engel, W., 1978*b,* Organization *in vitro* of ovarian cells into testicular structures, *Hum. Genet.* **44**:333–338.

CHAPTER 4

Gene Mapping in Fishes and Other Vertebrates

DONALD C. MORIZOT and MICHAEL J. SICILIANO

1. Introduction

1.1. Evolutionary Stability of Linkage Groups

Assessments of genetic divergence during vertebrate evolution have been studied using a variety of methods during the past several decades. A question that remains in large degree unanswered has provided an underlying theme to this area of research: what kinds of genetic change are associated with speciation events and major adaptive radiations? The pursuit of general correlates to morphological divergence has generated a wealth of data relating to variability and evolutionary rates of change of: (1) protein structure, as assessed indirectly by electrophoretic and immunological methods or more directly by amino acid sequencing (2) chromosome organization, as revealed by karyotypic analyses and DNA hybridization techniques; and (3) most recently, DNA and RNA structure determined directly by nucleotide sequencing.

The rapid expansion of gene maps of vertebrate species provides a starting point for the assessment of yet another level of genetic organization, the rate of evolutionary change in linkage relationships of homologous loci. In particular, a question might be phrased: do groups of homologous genes tend to remain linked over long periods of evolutionary

DONALD C. MORIZOT • University of Texas Science Park, Research Division, Smithville, Texas 78957. *MICHAEL J. SICILIANO* • Department of Genetics, University of Texas System Cancer Center, M.D. Anderson Hospital and Tumor Institute, Houston, Texas 77030.

time, even when karyotypic divergence has been large? An answer of yes implies a functional significance of such conserved gene arrangements, perhaps relating to chromosome structure or to gene regulation.

The development of somatic cell hybridization methods provided a major impetus for determining linkage relationships of loci phenotypically expressed by somatic cells in culture. The most obvious candidate loci were those that code for enzymes necessary for cell metabolism and thus for survival. Gel electrophoresis methods provided a sensitive means for separation of homologous enzymes with differences in charged amino acids, and specific histochemical stains were available for a large number of such enzymes. Thus, the stage was set, at least in mammals, for the mapping of a large number of loci whose homology can be presumed even in distantly related organisms, since the possibility of producing a protein as complex as an enzyme by chance convergence of nucleotide sequences is without question very low.

Given a wealth of new data, it can be asked whether any evidence has been amassed for long-term evolutionary stability of arrangements of enzyme-coding genes. The answer is an unqualified yes, at least among mammals. A striking example is the conservation, almost *in toto,* of the mammalian X chromosome in most species, both in chromosome structure (Pathak and Stock, 1974) and in gene content. Glucose-6-phosphate dehydrogenase has been demonstrated to be X-linked in man (Childs *et al.*, 1958; Motulsky and Yoshida, 1969), great apes and cercopithecoid monkeys (Garver *et al.*, 1978; Mathai *et al.*, 1966), rabbit and wild hare (Ohno *et al.*, 1965; Echard *et al.*, 1981), mouse (Epstein, 1969), rat (Yoshida, 1978), field vole (Cook, 1975), Chinese hamster (Westerveld *et al.*, 1972), muntjac deer (Shows *et al.*, 1976), horse and Donkey (Trujillo *et al.*, 1965), dog (Pearson and Roderick, 1978), and marsupials (Richardson *et al.*, 1971; J. A. Donald and Hope, 1981). Similarly, four other loci shown to be X-linked in humans, α-galactosidase (Opitz *et al.*, 1965), 3-phosphoglycerate kinase (PGK) (Chen *et al.*, 1971), hypoxanthine-guanine phosphoribosyl transerase (HPRT) (Nyhan, 1968), and phosphorylase kinase (L. J. Donald and Hamerton, 1978) are also X-linked in other mammalian species: phosphorylase kinase in the mouse (M. T. Davisson and Roderick, 1978); HPRT in the mouse (Epstein, 1972), Chinese hamster (Westerveld *et al.*, 1972), rabbit (Echard *et al.*, 1981), rat (Yoshida, 1978), cattle (Shimizu *et al.*, 1981), horse (Ohno, 1973), field vole (Cook, 1975), muntjac deer (Shows *et al.*, 1976), dog (Pearson and Roderick, 1978), and many primates (Garver *et al.*, 1978); PGK in mouse (Chapman and Shows, 1976), Chinese hamster (Westerveld *et al.*, 1972), rabbit (Echard *et al.*, 1981), cattle (Shimizu *et al.*, 1981), rat (Yoshida, 1978), and African green monkey (Garver *et al.*, 1978); and α-galactosidase in cattle (Shimizu *et al.*, 1981), rabbit (Echard *et al.*, 1981), mouse, Chinese hamster, horse, and primates (Pearson and Roderick, 1978).

Much evidence for autosomal linkage group conservation in mammals also has been adduced. Comparative mapping within the primates has revealed an extremely strong homology between ape and human gene maps. For example, five loci have been mapped to homologous chromosomes in man, chimpanzee, gorilla, and orangutan (Pearson and Roderick, 1978). Other great ape linkage relationships presumably homologous to human gene arrangements are found in markers carried on human chromsomes 2–15, 17–19, 21, and 22 (Pearson and Roderick, 1978). Such comparisons have been expanded to include rhesus monkeys, African green monkeys, and baboons. In at least one of the three species, homology with human gene assignments has been indicated for human chromosomes 1, 3–6, 8, 9, 11, 12, 15, 17, and 19 (Pearson and Roderick, 1978). It is apparent from the primate studies that, although chromosomal rearrangements have occurred, the linkage relationships of many, if not most, genes on the same chromosome arm remain in large degree undisturbed.

Evidence suggestive of longer term evolutionary stability of linkage relationships in mammals has accumulated more slowly, but an impressive body of data is beginning to emerge. Womack and Sharp (1976) demonstrated striking homology between clusters of esterase loci in mouse and rat; further, they suggested that two linked erythrocyte esterase loci in the rabbit (Schiff and Stormont, 1970) and two linked plasma esterase loci in voles (Semenoff, 1972) might indicate further conservation of the autosomal linkage group in mammals. Considerable comparative data between man and mouse have indicated many apparently conserved linkages where two loci are relatively tightly linked; such conservation would have been preserved for some 80 million years (Lalley *et al.*, 1978*c*). Mouse chromosome 4 contains five loci assigned to human chromosome lp; the only two loci of this set tested in the rat are syntenic on chromosome 5 (Pearson and Roderick, 1978). Similarly, genes coding for pyruvate kinase (*M2*) and mannosephosphate isomerase (Lalley *et al.*, 1978*d*) are linked in man and mouse. Two other pairs apparently conserved between man and mouse are the loci for peptidase-S and a phosphoglucomutase locus (Lalley *et al.*, 1978*b*) and for glucosephosphate isomerase and peptidase-D (Lalley *et al.*, 1978*c*). There is no need to list the nonconserved linkage pairs that differ in man and mouse. Taking the data as a whole, Pearson and Roderick (1978) speculate that linked mouse loci more than 30 map units apart will usually be unlinked in man, but linked loci less than 25 map units apart in the mouse may in man be unlinked; any pair of loci in the mouse less than one map unit apart will probably always be linked in man.

Some evidence from the dog also indicates probable long-term evolutionary conservation. Bruns *et al.* (1978) report the synteny of glyceraldehyde-3-phosphate dehydrogenase and triosephosphate isomerase loci; the same is true in man and other primates. The linkage between phosphoglucomutase-3 and the major histocompatibility complex in the dog yields remarkably similar sex-specific recombination frequency estimates to those observed in man (Meera Khan *et al.*, 1978).

Though the comparative data cited above are far from exhaustive, it should be clear that long-term evolutionary conservation of gene arrangements is a phenomenon documented by undisputable evidence.

1.2. A Perspective on Gene Mapping

Perhaps the best way to instill appreciation of the present state of the art and the near-future promise in developing genetic maps of a large number of vertebrate species is to explore briefly the development of gene mapping methods and their application to vertebrate species. Two questions are salient in such a review: what kinds of genes have been mapped and what species have been studied?

With respect to the kinds of genes that have been mapped, the vast majority of mapped genes can be assigned to one of four classes: (1) genes producing inherited diseases or defects; (2) genes producing "mutant" phenotypes, though not necessarily affecting viability; (3) genes producing distinctive color patterns; and (4) biochemical loci, coding for enzymes and other proteins. These classes, of course, are not mutally exclusive: albinism, e.g., can be inserted in any of the four categories, given enough information. The point that needs to be made here is one relating to homology: in the first three classes, establishment of homology between species is usually difficult. Even within a single genus of fishes, *Xiphophorus* (platyfishes and swordtails of the family Poeciliidae), very similar patterns of lateral black spots comprised of macromelanophore cells are coded by quite different, almost certainly nonhomologous genes (Kallman, 1975). Since the mapping of most biochemical loci has been accomplished only during the last decade or so, it is easy to appreciate that comparative studies of gene arrangements are still in their infancy.

Turning to the question of which vertebrate species have been studied, three classes can easily be identified: (1) man and his primate relatives; (2) laboratory experimental animals and their relatives; and (3) domestic animals and other species of economic importance. The result of such limitations on vertebrate species studied genetically is that for many years only for man, mouse, and chicken were more than rudimentary genetic maps established (Robinson, 1971; Warren, 1949; Abbott and Yee, 1975).

In poeciliid fishes a few species characterized by unusual diversity of pigment pattern polymorphisms received early genetic study resulting in considerable linkage data. In the guppy, *Poecilia reticulata,* at least eight autosomal loci have been identified, as well as 10 sex-linked loci expressed either as color patterns or variations in fin shape (Schroder, 1976). An even larger number of color pattern loci, both sex-linked and autosomal, have been described, and in many cases carefully studied, in species of the genus *Xiphophorus*. These loci control black pigment patterns comprised of macromelanophore or micromelanophore pigment cells and red and yellow patterns comprised of xanthoerythrophores (Kallman and Atz, 1966; Kallman, 1975).

Having considered the types of genes most often mapped and the vertebrate species most often studied, a brief description of the methods available for assigning genes to chromosomes or linkage groups is pertinent. In what has become known as the classical method of detecting linkage between two loci, genetic crosses are utilized to generate an individual that is heterozygous for two or more loci of interest. The most efficient cross for detection of linkage is the backcross, where a doubly heterozygous individual is crossed to a double homozygote for two loci (a double recessive where dominance is involved). A significant excess of parental-type individuals in the backcross progeny constitutes evidence for linkage; the frequency of nonparental types provides an estimate of recombination between the loci that can often be related to the actual separation distance on a chromosome. For a detailed discussion of cross design and statistical analysis using classical linkage methods, the reader is directed to the treatises of Mather (1957), Bailey (1961), and Elandt-Johnson (1971). The strengths of the classical method are that it can be used for any gene resulting in relatively unambiguous phenotypes, and that the method yields estimates of recombination, thus allowing detection of sex or population differences reflecting either physically different gene locations or differences in regulation. The major weakness of the classical method is that genes can be assigned only to linkage groups in the absence of other data concerning chromosomal location.

The second major gene mapping method, which has produced a quantum expansion in gene map assignments, involves the use of somatic cell hybrids [see Ruddle, (1969) for detailed discussion]. The power of such methods lies in the use of cells from distantly related species; in such cases, a large proportion of enzymes that can be analyzed exhibit electrophoretic differences due to amino acid sequence divergence. The disadvantage of somatic cell method is that, in the absence of detected deletions, insertions, or translocations, no information is obtained concerning the positions of loci relative to each other on the chromosome.

The last and most recently developed method of gene mapping with general utility is the mapping of genes to chromosomes by hybridizing DNA. The process seems quite rigorous, but is proving to be technically feasible. Though there doubtless will prove to be technical limitations to such methods, in principle the chromosomal location of the structural gene for any protein could be determined. For a summary of the use of *in situ* hybridization methods, see Evans and Atwood (1978).

Other methods of mapping genes have been utilized, but these three are perhaps the most powerful and of the most general utility in assessing linkage relationships of protein-coding loci.

1.3. Genetic Maps of Protein-Coding Loci in Vertebrates

A brief survey of the available mapped loci in vertebrates will be presented as introductory to a more detailed assessment of enzyme gene mapping progress in fishes. For several species, no adequate review of the linkage data has yet appeared, and reference citations will be deferred and presented in the more detailed reviews that follow.

1.3.1. Fishes

Only three families of fishes have been utilized at all extensively in genetic mapping studies: Salmonidae (trouts and salmon), Centrarchidae (freshwater sunfishes), and Poeciliidae (platyfishes and swordtails, *Xiphophorus; Poeciliopsis,* including sexually and clonally reproducing forms; and *Poecilia reticulata,* the guppy).

Linkage data for pairs of biochemical loci have been collected in seven species of salmonids, representing three genera: *Salmo gairdneri,* rainbow trout, *S. trutta,* brown trout, and *S. clarki,* cutthroat trout; *Salvelinus fontinalis,* brook trout, and *S. namaycush,* lake trout; and *Oncorhynchus gorbuscha,* pink salmon, and *O. keta,* chum salmon. The recent reviews of May *et al.* (1979*b*, 1980, 1982) and J. E. Wright *et al.* (1983) summarize the linkage data for over 30 enzyme loci. Due primarily to an ancestral tetraploidization event in salmonids (Ohno *et al.*, 1969; Ohno, 1970), and the ongoing, apparently incomplete diploidization process (May *et al.*, 1979*b*), workers in the area of salmonid gene mapping wisely have refrained from designating linkage groups. Comparative data from several species are needed to pinpoint linked genes to specific chromosome arms, since within the family, arm number is much more conservative than chromosome number (Gold, 1979).

The more meager linkage data from centrarchids were derived from fertile interspecific hybrids, *Lepomis gulosus* × *L. cyanellus* (warmouth and green sunfish, respectively). In total, eight enzyme loci have been analyzed for linkage.

In *Poecilia reticulata,* the guppy, only four enzyme loci have been assessed for linkage in a single study (Shami and Beardmore, 1978). Even for these few loci, sample sizes are very inadequate, only 16 individuals being surveyed for three locus pairs.

Intra- and interspecific crosses involving *Poeciliopsis monacha, P. viriosa, P. lucida,* and *P. occidentalis* have yielded linkage data for 17 enzyme loci. One linkage group including five loci has been identified along with a possible linkage of two others.

Our studies of crosses within and among five species of *Xiphophorus* (*X. maculatus, X. helleri, X. clemenciae, X. milleri,* and *X. couchianus*) have assessed the linkage relationships of 41 enzyme loci and two pigment pattern genes. These results will be presented in detail below. Briefly summarizing, six multipoint independently assorting linkage groups comprised of 19 or 20 loci have thus far been identified. Two other two-point linkage groups have been demonstrated, but sample sizes must be increased before independence can be presumed. With the addition of the remaining 18 loci that assort independently within the limits of the data, each of the 24 pairs of chromosomes could be marked by at least one gene.

1.3.2. Amphibians

To our knowledge, only two genera of anurans have contributed to knowledge of linkage of enzyme loci in amphibians. Hybrids between two species of toads of the genus *Bombina* have been studied for linkage of five enzyme loci; no linkages were detected (Szymura and Farana, 1978). Much more extensive data have been amassed for crosses involving frogs of the genus *Rana* (D. A. Wright, 1975; D. A. Wright and Richards, 1980; D. A. Wright *et al.,* 1980). Thus far, at least 20 loci have provided data; four groups of two or more linked biochemical loci have been identified, including 12 loci.

1.3.3. Reptiles

We know of no linkage studies of biochemical loci in reptiles.

1.3.4. Birds

The avian species most extensively studied for linkage relationships of protein-coding loci is the domestic chicken, *Gallus domesticus.* Only a few enzyme loci have been mapped (including an acid phosphatase locus and two plasma esterase loci), but many blood and egg proteins (blood groups, albumins, transferrin, etc.) have been studied (Abbott and Yee, 1975).

1.3.5. Mammals

The genetic maps of the laboratory mouse (*Mus musculus*) and man continue to be the most comprehensive among vertebrate species; in both species, at least one locus has been associated with each chromosome. The growing interest in comparative gene mapping has resulted in extensive map assignments in other rodent and primate species.

In a recent review of the linkage map of the mouse (M. T. Davisson and Roderick, 1978), more than 50 enzyme loci are included among the 350 or so loci that have been mapped. The assignment of enzyme loci is progressing rapidly, as evidenced by the reports of the mapping of seven loci at the 1977 Winnipeg Conference on human gene mapping (Lalley *et al.*, 1978*a–d*).

More than 200 genes have been mapped in man, including more than 75 enzyme loci (Shows and McAlpine, 1978). The number of mapped genes more than tripled in the years 1972–1978 (L. J. Donald and Hamerton, 1978), due largely to the use of somatic cell hybridization mapping methods.

The development of extensive linkage maps of protein-coding loci promises to allow important conclusions concerning the evolutionary stability of gene arrangements. In the following sections we will review in some detail the linkage maps developed thus far in fishes, and will attempt to assess the extent of linkage group homology within fishes and between fishes and other vertebrate species.

2. Linkage Relationships of Protein-Coding Loci in Fishes

2.1. *Xiphophorus*, Poeciliidae

Since platyfishes and swordtails of the genus *Xiphophorus* are the most extensively analyzed fishes with respect to linkage relationships of protein-coding loci, we will introduce our treatment of linkage studies in fishes with a detailed review of current progress in linkage analysis in *Xiphophorus*.

The vast majority of linkage data in *Xiphophorus* are derived from backcrosses produced using an interspecific F_1 hybrid. *Xiphophorus helleri* × *X. maculatus* crosses have been predominant in our analyses, but F_1 hybrids have been utilized from *X. maculatus* × *X. milleri, X. ma-*

culatus × *X. clemenciae,* and *X. maculatus* × *X. couchianus* matings. We are beginning large-scale production of backcross hybrids from *X. variatus* × *X. helleri* matings. Additionally, some intraspecific matings between various populations of *X. maculatus, X. helleri,* and *X. pygmaeus* have been produced and yield linkage data for a substantial number of locus pairs.

To date, some 100 protein-coding loci have been characterized in *Xiphophorus;* electrophoretic variants in the genus have been identified for 63 of these proteins (Table I). Of these 63, 41 have been assessed at least to some degree for segregation and linkage relationships (Tables II and III). While sample sizes fluctuate widely among pairwise comparisons for linkage detection (from a low of two to a high of over 700 individuals), some data have been accumulated for over 70% of the 780 possible pairwise comparisons of these protein-coding loci. Such a high rate of completion with respect to pairwise analyses stems from the substantial variability between *Xiphophorus* species, while preserving the capability of fertile interspecific hybrid production. Protein variability among inbred strains of *Xiphophorus* has been reviewed by Morizot and Siciliano (1982*a*): it is not unusual for two species to differ in their allozymes at 20 enzyme loci, resulting in 190 possible pairwise comparisons in a single backcross.

Since all linkage tests are to some degree contingent upon normal Mendelian segregation of alleles, it is important to assess critically the segregation of allozyme phenotypes and thus of presumed genotypes under a codominant inheritance model, particularly in interspecific crosses. The segregations of 41 protein-coding loci and two pigment pattern genes of *Xiphophorus* are summarized in Table II. The salient conclusion evidenced by these data is that most, if not all, of the biochemical loci segregate in backcrosses and intercrosses in excellent agreement with the Mendelian expectation of 50% homozygotes and 50% heterozygotes. No large-scale heterotic effects, or conversely, excessive homozygosity, are observed. Such a result is of paramount importance in interspecific crosses, where chromosomal incompatibility could confer reduced viability upon some hybrid classes with accompanying segregation distortion. Such a situation is not merely hypothetical: in a pond-release experiment (with possible subsequent predation), Whitt *et al.* (1973) observed F_2 interspecific sunfish hybrids that were morphologically similar to the F_1 generation and were predominantly heterozygous at three enzyme loci exhibiting variation between the parental stocks.

A more detailed analysis of the most deviant segregations in Table II is presented to ascertain whether any reasons are evident to reject the hypothesis of sample error. Only two loci exhibit segregations significantly different from 50% homozygotes at the 0.05 level, *IDH-2* and *MP-1.* It

Table I

Proteins Resolved Electrophoretically in *Xiphophorus* Fishes, Number of Loci Coding for Each Protein, and Number of Loci for Which Electrophoretic Variants Have Been Identified

Protein	EC Number[a]	Number of loci	Number of variant loci
Acid phosphatase (ACP)	3.1.3.2	1[b]	1
Aconitase (ACON)	4.2.1.3	2[b]	2
Adenine phosphoribosyltransferase (APRT)	2.4.2.7	1[b]	0
Adenosine deaminase (ADA)	3.5.4.4	1	1
Adenosine kinase (ADK)	2.7.1.20	1[b]	0
Adenylate kinase (AK)	2.7.4.3	1[b]	0
Alcohol dehydrogenase (ADH)	1.1.1.1	1	1
Aldolase (ALD)	4.1.2.13	1[b]	0
Amylase (AMY)	3.2.1.1	1[b]	1
α-Arabinosidase (αARA)	3.2.1.55	1[b]	0
Catalase (CAT)	1.11.1.6	1[b]	1
Creatine kinase (CK)	2.7.3.2	4	1
Diaphorase-NADH (DIA)	1.6.2.2	1[b]	0
Enolase (ENO)	4.2.1.11	2	0
Esterase (ES)	3.1.1.1	6[b]	6
Fructose-1,6-diphosphatase (FDP)	3.1.3.11	2	1
α-Fucosidase (αFUC)	3.2.1.51	1[b]	0
Fumarase (FUM)	4.2.1.2	1	1
Galactokinase (GALK)	2.7.1.6	1[b]	0
Galactose-1-phosphate uridyltransferase (GALT)	2.7.7.12	1[b]	1
α-Galactosidase (αGAL)	3.2.1.22	1[b]	1
β-Galactosidase (βGAL)	3.2.1.23	1	0
Glucose-6-phosphate dehydrogenase (G6PD)	1.1.1.49	1[b]	1
Glucosephosphate isomerase (GPI)	5.3.1.9	2	2
α-Glucosidase (αGLU)	3.2.1.20	1	0
β-Glucosidase (βGLU)	3.2.1.21	1	1
β-Glucuronidase (βGUS)	3.2.1.31	1[b]	0
Glutamate dehydrogenase (GLUD)	1.4.1.2	1	0
Glutamate-oxaloacetate transaminase (GOT)	2.6.1.1	3	2
Glutamate-pyruvate transaminase (GPT)	2.6.1.2	1[b]	1
Glutamine synthetase (GS)	6.3.1.2	1	1
Glutathione reductase (GSR)	1.6.4.2	1[b]	1
Glyceraldehyde-3-phosphate dehydrogenase (GAPD)	1.2.1.12	3	3
Glycerate-2-dehydrogenase (G2DH)	1.1.1.29	1[b]	1
α-Glycerophosphate dehydrogenase (αGPD)	1.1.1.8	1[b]	1
Glyoxylase I (GLO)	4.4.1.5	1	0
Guanine deaminase (GDA)	3.5.4.3	1	1
Guanylate kinase (GUK)	2.7.4.8	2[b]	2
Hemoglobin (Hb)	—	2	1
Hexokinase (HK)	2.7.1.1	1[b]	0
Hexosaminidase (HA)	3.2.1.30	1[b]	0
Hexose-6-phosphate dehydrogenase (GDH)	1.1.1.47	1[b]	1
Hypoxanthine phosphoribosyltransferase (HPRT)	2.4.2.8	1	0

Table I (*Continued*)

Protein	EC Number[a]	Number of loci	Number of variant loci
Inosine triphosphatase (ITP)	3.6.1.19	1	1
Isocitrate dehydrogenase (IDH)	1.1.1.42	3	2
Lactate dehydrogenase (LDH)	1.1.1.27	3	3
Malate dehydrogenase (MDH)	1.1.1.37	3	2
Malic enzyme (ME)	1.1.1.40	1[b]	1
Mannosephosphate isomerase (MPI)	5.3.1.8	1	1
α-Mannosidase (αMAN)	3.2.1.24	1	0
Muscle proteins (MP)	—	4[c]	4[c]
Ornithine transcarbamylase (OTC)	2.1.3.3	1	0
Peptidase (PEP)	3.4.11	3	2
Peroxidase (PER)	1.11.1.7	1[b]	0
Phosphofructokinase (PFK)	2.2.1.11	1	0
Phosphoglucomutase (PGM)	2.7.5.1	1[b]	1
6-Phosphogluconate dehydrogenase (6PGD)	1.1.1.44	1	1
Phosphoglycerate kinase (PGK)	2.7.2.3	1[b]	1
Phosphoglycerate mutase (PGAM)	2.7.5.3	2	2
Inorganic pyrophosphatase (PP)	3.6.1.1	1[b]	0
Pyruvate kinase (PK)	2.7.1.40	2[b]	2
Sorbitol dehydrogenase (SDH)	1.1.1.14	1	0
Superoxide dismutase (SOD)	1.15.1.1	2	1
Thymidine kinase (TK)	2.7.1.21	1[b]	0
Transferrin (Tf)	—	1	1
Triosephosphate isomerase (TPI)	5.3.1.1	2	1
Uridine monophosphate kinase (UMPK)	2.7.4	1[b]	1
Xanthine dehydrogenase (XDH)	1.2.3.2	1	0
Total		101	64

[a]Numbers recommended by Commission on Biological Nomenclature (1978).
[b]Additional zones of activity sometimes observed, but not at this time meeting criteria for designation as separate locus products.

Table II
Segregation Analyses of 41 Protein-Coding Loci and Two Pigment Pattern Markers in *Xiphophorus* Crosses[a]

Locus	Species tested in crosses	Homozygotes	Heterozygotes	N	χ^2
ACP	M;H,M;M,Mi	242	217	459	1.36
ACON-1	H,M;C,M	177	167	344	0.29
ACON-2	M,Mi	11	5	16	2.25
ADA	M;H,M;C,M;M,Mi	317	310	627	0.08
ES-1	H,M	199	198	397	0.00
ES-2	M;H,M;Co,M	445	480	925	1.32
ES-3	M;H,M;C,M;M,Mi;Co,M	435	461	896	0.75

(*continued*)

Table II (*Continued*)

Locus	Species tested in crosses	Homozygotes	Heterozygotes	N	χ^2
ES-4	H,M	32	22	54	1.85
ES-5	H,M;M,Mi	49	51	100	0.04
GALT	H,M;M,Mi	54	60	114	0.32
αGAL	H,M;C,M	109	111	220	0.02
G6PD	H,M	327	340	667	0.25
GPI-1	M;H,M;M,Mi;Co,M	463	498	961	1.27
GS	M,Mi	12	15	27	0.33
GAPD-1	M;H,M;C,M	592	585	1177	0.04
GAPD-2	M	125	137	262	0.55
GUK-1	M,Mi;Co,M	22	20	42	0.10
GUK-2	H,M	32	48	80	3.20
IDH-1	H,M	255	276	531	0.83
IDH-2	H,M;C,M;M,Mi	321	374	695	4.04*
LDH-1	H,M	135	132	267	0.04
MDH-2	H,M;C,M	210	222	432	0.33
MPI	M;H,M;C,M;M,Mi	492	475	967	0.30
MP-1	M	154	117	271	5.05*
MP-2	H,M;M,Mi;Co,M	173	176	349	0.03
MP-3	C,M	7	10	17	0.53
MP-4	H,M;C,M	71	74	145	0.06
MP-5	H,M	117	119	236	0.02
PEP-2	H,M;C,M;M,Mi;Co,M	212	235	447	1.18
PEP-3	M;H,M;Co,M	152	160	312	0.21
PGM	H,M	302	322	624	0.64
6PGD	M;H,M;C,M;M,Mi	424	475	899	2.89
PGK	H,M	42	46	88	0.18
PGAM-1	H,M	28	32	60	0.27
PGAM-2	H,M	39	45	84	0.43
PK-1	H,M	37	52	89	2.53
PK-2	M,Mi	11	16	27	0.93
Sd[b]	M;H,M;C,M	288	322	610	1.90
SOD	M	57	43	100	1.96
Tailspot[c]	M	94	97	191	0.05
Tf	M;H,M;C,M;M,Mi	368	383	751	0.30
TPI-1	M;H,M;C,M;M,Mi	178	215	393	3.48
UMPK	H,M;M,Mi	46	49	95	0.09

[a]Abbreviations and symbols are as follows: N, total number of individuals; asterisk denotes the probability that a given χ^2 is due to chance, <0.05; M: intraspecific cross between *X. maculatus* populations; H,M: interspecific cross involving *X. helleri* × *X. maculatus* hybrids; C,M: *X. clemenciae* × *X. maculatus* crosses; M,Mi: *X. maculatus* × *X. milleri* crosses; Co,M: *X. couchianus* × *X. maculatus* crosses. Locus abbreviations as in Table I.

[b]*Sd*, Spotted dorsal, a sex-linked macromelanophore pigment pattern gene.

[c]*Tailspot*, A series of autosomal micromelanophore pigment pattern genes; interspecific crosses exhibiting variable expressivity have been excluded.

Table III
Summary of Pairwise Linkage Tests of 42 Loci in *Xiphophorus* Fishes[a]

Locus pair	Species tested in crosses	*N*	*P*	χ^2	Linkage group assignment
ACP–ACON-1	H,M	50	0.66	5.12*	
ACP–ACON-2	M	11	0.45	0.09	
ACP–ADA	M;H,M;M,Mi	119	0.50	0.01	
ACP–ES-1	H,M	50	0.42	1.28	
ACP–ES-2	M;H,M	326	0.50	0.01	
ACP–ES-3	M;H,M;M,Mi	327	0.48	0.52	
ACP–ES-5	H,M;M,Mi	38	0.47	0.11	
ACP–GALT	H,M;M,Mi	41	0.59	1.20	
ACP–αGAL	H,M	48	0.48	0.08	
ACP–G6PD	H,M	50	0.50	0.00	
ACP–GPI-1	M;H,M;M,Mi	135	0.50	0.01	
ACP–GS	M,Mi	27	0.59	0.93	
ACP–GAPD-1	M;H,M	310	0.45	2.53	
ACP–GAPD-2	M	30	0.43	0.53	
ACP–GUK-1	M,Mi	27	0.56	0.33	
ACP–GUK-2	H,M	14	0.50	0.00	
ACP–IDH-1	H,M	36	0.50	0.00	
ACP–IDH-2	H,M;M,Mi	76	0.51	0.05	
ACP–MDH-2	H,M	50	0.48	0.08	
ACP–MPI	M;H,M;M,Mi	335	0.52	0.36	
ACP–MP-1	M	83	0.42	2.04	
ACP–MP-2	H,M;M,Mi	77	0.55	0.64	
ACP–MP-4	H,M	31	0.52	0.03	
ACP–PEP-2	H,M;M,Mi	77	0.55	0.64	
ACP–PEP-3	M;H,M	249	0.49	0.20	
ACP–PGM	H,M	49	0.49	0.02	
ACP–6PGD	M;H,M;M,Mi	79	0.52	0.11	
ACP–PGK	H,M	14	0.43	0.29	
ACP–PGAM-2	H,M	14	0.50	0.00	
ACP–PK-1	H,M	14	0.50	0.00	
ACP–PK-2	H,Mi	25	0.36	1.96	
ACP–Sd	M;H,M	256	0.52	0.56	
ACP–SOD	M	100	0.51	0.04	
ACP–Tailspot	M	132	0.50	0.00	
ACP–Tf	M;H,M;M,Mi	318	0.55	2.83	
ACP–TPI-1	M;H,M;M,Mi	196	0.50	0.00	
ACP–UMPK	H,M;M,Mi	39	0.54	0.23	
ACON-1–ADA	H,M;C,M	308	0.51	0.05	
ACON-1–ES-1	H,M	205	0.53	0.59	
ACON-1–ES-2	H,M	311	0.52	0.54	
ACON-1–ES-3	H,M;C,M	299	0.51	0.16	
ACON-1–ES-4	H,M	5	0.40	0.20	

[a]Abbreviations and symbols: *p*, maximum likelihood recombination frequency estimate; **, probability that a given chi-square is due to chance <0.01. Other symbols as in Table II.

(*continued*)

Table III (*Continued*)

Locus pair	Species tested in crosses	*N*	*P*	χ^2	Linkage group assignment
ACON-1–ES-5	H,M	76	0.49	0.05	
ACON-1–GALT	H,M	86	0.59	2.98	
ACON-1–αGAL	H,M	167	0.47	0.49	
ACON-1–G6PD	H,M	320	0.51	0.11	
ACON-1–GPI-1	H,M	290	0.45	3.10	
ACON-1–GAPD	H,M;C,M	314	0.53	0.82	
ACON-1–GUK-2	H,M	78	0.45	0.82	
ACON-1–IDH-1	H,M	269	0.48	0.63	
ACON-1–IDH-2	H,M;C,M	337	0.52	0.36	
ACON-1–LDH-1	H,M	128	0.52	0.13	
ACON-1–MDH-2	H,M;C,M	181	0.51	0.14	
ACON-1–MPI	H,M;C,M	225	0.50	0.00	
ACON-1–MP-2	H,M	192	0.55	1.69	
ACON-1–MP-3	C,M	6	0.50	0.00	
ACON-1–MP-4	H,M;C,M	76	0.49	0.05	
ACON-1–MP-5	H,M	128	0.54	0.78	
ACON-1–PEP-2	H,M;C,M	271	0.55	3.10	
ACON-1–PEP-3	H,M	51	0.45	0.49	
ACON-1–PGM	H,M	316	0.47	0.81	
ACON-1–6PGD	H,M;C,M	342	0.51	0.05	
ACON-1–PGK	H,M	86	0.48	0.19	
ACON-1–PGAM-1	H,M	60	0.48	0.07	
ACON-1–PGAM-2	H,M	81	0.58	2.09	
ACON-1–PK-1	H,M	86	0.44	1.16	
ACON-1–Sd	H,M;C,M	249	0.48	0.33	
ACON-1–Tf	H,M;C,M	317	0.52	0.38	
ACON-1–TPI-1	H,M;C,M	76	0.51	0.05	
ACON-1–UMPK	H,M	70	0.54	0.51	
ACON-2–ES-2	M	16	0.31	2.25	
ACON-2–ES-3	M	16	0.13	9.00**	II ?
ACON-2–GAPD-1	M	16	0.44	0.25	
ACON-2–MPI	M	16	0.44	0.25	
ACON-2–PEP-3	M	16	0.38	1.00	
ACON-2–Tf	M	16	0.69	2.25	
ACON-2–TPI-1	M	16	0.44	0.25	
ADA–ES-1	H,M	216	0.54	1.50	
ADA–ES-2	M;H,M	485	0.50	0.00	
ADA–ES-3	M;H,M;C,M;M,Mi	492	0.51	0.07	
ADA–ES-4	H,M	5	0.20	1.80	
ADA–ES-5	H,M;M,Mi	100	0.45	1.00	
ADA–GALT	H,M;M,Mi	114	0.50	0.00	
ADA–αGAL	H,M;C,M	178	0.51	0.09	
ADA–G6PD	H,M	452	0.13	243.86**	I
ADA–GPI-1	M;H,M;M,Mi	546	0.48	1.05	

Table III (*Continued*)

Locus pair	Species tested in crosses	N	P	χ^2	Linkage group assignment
ADA–GS	M,Mi	27	0.59	0.93	
ADA–GAPD-1	M;H,M;C,M	495	0.49	0.05	
ADA–GUK-1	M,Mi	27	0.41	0.93	
ADA–GUK-2	H,M	80	0.50	0.00	
ADA–IDH-1	H,M	341	0.47	1.29	
ADA–IDH-2	H,M;C,M;M,Mi	485	0.49	0.17	
ADA–LDH-1	H,M	239	0.45	2.62	
ADA–MDH-2	H,M;C,M	195	0.56	3.21	
ADA–MPI	M;H,M;C,M;M,Mi	446	0.45	3.96*	
ADA–MP-1	M	86	0.43	1.67	
ADA–MP-2	H,M;M,Mi	247	0.54	1.46	
ADA–MP-3	C,M	17	0.53	0.06	
ADA–MP-4	H,M;C,M	82	0.51	0.05	
ADA–MP-5	H,M	214	0.54	1.20	
ADA–PEP-2	H,M;C,M;M,Mi	350	0.50	0.01	
ADA–PEP-3	H,M	25	0.40	1.00	
ADA–PGM	H,M	437	0.50	0.00	
ADA–6PGD	H,M	457	0.30	71.69**	
	C,M	17	0.24	4.76*	
	M,Mi	27	0.30	4.48*	
	Total	501	0.30	80.64**	I
ADA–PGK	H,M	89	0.57	1.90	
ADA–PGAM-1	H,M	60	0.58	1.67	
ADA–PGAM-2	H,M	84	0.44	1.19	
ADA–PK-1	H,M	89	0.52	0.10	
ADA–PK-2	M,Mi	25	0.36	1.96	
ADA–Sd	M;H,M;C,M	423	0.50	0.02	
ADA–Tailspot	M	94	0.53	0.38	
ADA–Tf	M;H,M;C,M;M,Mi	522	0.47	1.72	
ADA–TPI-1	H,M;C,M;M,Mi	163	0.43	3.25	
ADA–UMPK	H,M;M,Mi	95	0.53	0.26	
ES-1–ES-2	H,M	277	0.51	0.09	
ES-1–ES-3	H,M	269	0.53	0.84	
ES-1–ES-4	H,M	54	0.06	42.67**	V
ES-1–ES-5	H,M	75	0.53	0.33	
ES-1–GALT	H,M	87	0.61	4.15*	
ES-1–αGAL	H,M	182	0.55	1.78	
ES-1–G6PD	H,M	304	0.51	0.21	
ES-1–GPI-1	H,M	380	0.50	0.00	
ES-1–GAPD-1	H,M	273	0.51	0.09	
ES-1–GUK-2	H,M	80	0.44	1.25	
ES-1–IDH-1	H,M	282	0.50	0.00	
ES-1–IDH-2	H,M	346	0.50	0.01	
ES-1–MDH-2	H,M	285	0.39	14.82**	V

(continued)

Table III (*Continued*)

Locus pair	Species tested in crosses	*N*	*P*	χ^2	Linkage group assignment
ES-1–MPI	H,M	226	0.52	0.28	
ES-1–MP-2	H,M	277	0.54	1.59	
ES-1–MP-4	H,M	125	0.48	0.20	
ES-1–PEP-2	H,M	213	0.45	2.48	
ES-1–PEP-3	H,M	55	0.55	0.45	
ES-1–PGM	H,M	329	0.47	1.10	
ES-1–6PGD	H,M	333	0.53	0.87	
ES-1–PGK	H,M	89	0.47	0.28	
ES-1–PGAM-1	H,M	60	0.47	0.27	
ES-1–PGAM-2	H,M	84	0.60	3.05	
ES-1–PK-1	H,M	89	0.53	0.28	
ES-1–Sd	H,M	262	0.54	1.85	
ES-1–Tf	H,M	214	0.56	3.16	
ES-1–TPI-1	H,M	14	0.57	0.29	
ES-1–UMPK	H,M	70	0.57	1.43	
ES-2–ES-3	M	287	0.23	83.71**	
	H,M	390	0.25	94.52**	
	M,Co	26	0.08	18.62**	
	Total	703	0.24	193.69**	II
ES-2–ES-4	H,M	27	0.63	1.81	
ES-2–ES-5	H,M	75	0.31	11.21**	II
ES-2–GALT	H,M	87	0.54	0.56	
ES-2–αGAL	H,M	183	0.49	0.05	
ES-2–G6PD	H,M	304	0.51	0.05	
ES-2–GPI-1	M;H,M	531	0.49	0.09	
ES-2–GAPD-1	M;H,M	708	0.47	1.83	
ES-2–GAPD-2	M	15	0.47	0.07	
ES-2–GUK-2	H,M	80	0.50	0.00	
ES-2–IDH-1	H,M	344	0.54	1.97	
ES-2–IDH-1	H,M	344	0.54	1.97	
ES-2–IDH-2	H,M	423	0.52	0.40	
ES-2–LDH-1	H,M	189	0.31	26.67**	II
ES-2–MDH-2	H,M	209	0.50	0.00	
ES-2–MPI	M	310	0.45	3.30	
	H,M	281	0.42	7.21**	
	Total	591	0.43	10.03**	II
ES-2–MP-1	M	136	0.54	0.74	
ES-2–MP-2	H,M	267	0.51	0.03	
ES-2–MP-4	H,M	102	0.51	0.04	
ES-2–MP-5	H,M	154	0.48	0.23	
ES-2–PEP-2	H,M	295	0.58	7.49**	
ES-2–PEP-3	M;H,M	255	0.46	1.73	
ES-2–PGM	H,M	453	0.51	0.11	
ES-2–6PGD	M;H,M	458	0.47	2.24	

Table III (*Continued*)

Locus pair	Species tested in crosses	N	P	χ^2	Linkage group assignment
ES-2–PGK	H,M	89	0.47	0.28	
ES-2–PGAM-1	H,M	60	0.53	0.27	
ES-2–PGAM-2	H,M	84	0.55	0.76	
ES-2–PK-1	H,M	89	0.55	0.91	
ES-2–Sd	M;H,M;Co,M	661	0.51	0.34	
ES-2–SOD	M	100	0.53	0.36	
ES-2–Tailspot	M	185	0.47	0.65	
ES-2–Tf	M;H,M	668	0.51	0.15	
ES-2–TPI-1	M;H,M	233	0.49	0.04	
ES-2–UMPK	H,M	70	0.60	2.80	
ES-3–ES-4	H,M	23	0.74	5.26*	
ES-3–ES-5	H,M	75	0.00	75.00**	
	M,Mi	25	0.00	25.00**	
	Total	100	0.00	100.00**	II
ES-3–GALT	H,M;M,Mi	114	0.54	0.56	
ES-3–αGAL	H,M;C,M	191	0.46	1.18	
ES-3–G6PD	H,M	420	0.52	0.61	
ES-3–GPI-1	M;H,M;M,Mi	528	0.50	0.01	
ES-3–GS	M,Mi	27	0.56	0.33	
ES-3–GAPD-1	M;H,M;C,M	668	0.49	0.22	
ES-3–GAPD-2	M	10	0.80	3.60	
ES-3–GUK-1	M,Mi	27	0.44	0.33	
ES-3–GUK-2	H,M	80	0.59	2.45	
ES-3–IDH-1	H,M	326	0.52	0.79	
ES-3–IDH-2	H,M;C,M;M,Mi	452	0.52	0.57	
ES-3–LDH-1	H,M	161	0.23	47.01**	II
ES-3–MDH-2	H,M;C,M	227	0.44	2.75	
ES-3–MPI	M	259	0.30	40.96**	
	H,M	225	0.26	52.80**	
	C,M	17	0.35	1.47	
	M,Mi	27	0.22	8.33**	
	Total	528	0.28	101.94**	II
ES-3–MP-1	M	113	0.48	0.22	
ES-3–MP-2	H,M;M,Mi	266	0.47	1.22	
ES-3–MP-3	C,M	17	0.41	0.53	
ES-3–MP-4	H,M;C,M	102	0.55	0.98	
ES-3–MP-5	H,M	163	0.53	0.74	
ES-3–PEP-2	H,M;C,M;M,Mi	330	0.50	0.00	
ES-3–PEP-3	M;H,M	208	0.50	0.00	
ES-3–PGM	H,M	412	0.50	0.04	
ES-3–6PGD	M;H,M;C,M;M,Mi	475	0.50	0.00	
ES-3–PGK	H,M	89	0.45	0.91	
ES-3–PGAM-1	H,M	60	0.50	0.00	
ES-3–PGAM-2	H,M	84	0.51	0.05	

(continued)

Table III (*Continued*)

Locus pair	Species tested in crosses	N	P	χ^2	Linkage group assignment
ES-3–PK-1	H,M	89	0.45	0.91	
ES-3–PK-2	M,Mi	25	0.08	17.64**	II ?
ES-3–Sd	M;H,M;C,M;Co,M	605	0.51	0.37	
ES-3–SOD	M	100	0.45	0.00	
ES-3–Tailspot	M	162	0.51	0.10	
ES-3–Tf	M;H,M;C,M;M,Mi	639	0.51	0.26	
ES-3–TPI-1	M,H,M;C,M;M,Mi	243	0.42	6.92**	?
ES-3–UMPK	H,M;M,Mi	95	0.57	1.78	
ES-4–G6PD	H,M	53	0.49	0.02	
ES-4–GPI-1	H,M	53	0.57	0.92	
ES-4–GAPD-1	H,M	54	0.46	0.30	
ES-4–IDH-1	H,M	46	0.54	0.35	
ES-4–IDH-2	H,M	51	0.45	0.49	
ES-4–MDH-2	H,M	12	0.33	1.33	V
ES-4–MPI	H,M	5	0.20	1.80	
ES-4–MP-2	H,M	18	0.50	0.00	
ES-4–MP-4	H,M	36	0.53	0.11	
ES-4–PEP-2	H,M	5	0.60	0.20	
ES-4–PGM	H,M	54	0.46	0.30	
ES-4–6PGD	H,M	54	0.50	0.00	
ES-4–Sd	H,M	18	0.44	0.22	
ES-4–Tf	H,M	4	1.00	4.00	
ES-5–GALT	H,M;M,Mi	100	0.51	0.04	
ES-5–αGAL	H,M	73	0.49	0.01	
ES-5–G6PD	H,M	75	0.44	1.08	
ES-5–GPI-1	H,M;M,Mi	101	0.47	0.49	
ES-5–GS	M,Mi	25	0.52	0.04	
ES-5–GAPD-1	H,M	62	0.53	0.26	
ES-5–GUK-1	M,Mi	25	0.40	1.00	
ES-5–GUK-2	H,M	68	0.54	0.53	
ES-5–IDH-1	H,M	55	0.38	3.07	
ES-5–IDH-2	H,M;M,Mi	98	0.52	0.16	
ES-5–MDH-2	H,M	75	0.41	2.25	
ES-5–MPI	H,M	62	0.19	23.29**	
	M,Mi	25	0.20	9.00**	
	Total	87	0.20	32.29**	II
ES-5–MP-2	H,M;M,Mi	91	0.44	1.33	
ES-5–PEP-2	H,M;M,Mi	100	0.52	0.16	
ES-5–PEP-3	H,M	13	0.54	0.08	
ES-5–PGM	H,M	70	0.47	0.23	
ES-5–6PGD	H,M;M,Mi	100	0.53	0.36	
ES-5–PGK	H,M	75	0.43	1.61	
ES-5–PGAM-1	H,M	55	0.49	0.02	
ES-5–PGAM-2	H,M	70	0.50	0.00	

Table III (*Continued*)

Locus pair	Species tested in crosses	N	P	χ^2	Linkage group assignment
ES-5–PK-1	H,M	75	0.44	1.08	
ES-5–PK-2	M,Mi	23	0.09	15.70**	II ?
ES-5–Sd	H,M	75	0.39	3.85*	
ES-5–Tf	H,M;M,Mi	98	0.53	0.37	
ES-5–TPI-1	M,Mi	24	0.25	6.00*	
ES-5–UMPK	H,M;M,Mi	89	0.57	1.90	
GALT–αGAL	H,M	80	0.45	0.80	
GALT–G6PD	H,M	87	0.49	0.01	
GALT–GPI-1	H,M;M,Mi	114	0.45	1.26	
GALT–GS	M,Mi	27	0.41	0.93	
GALT–GAPD-1	H,M	73	0.49	0.01	
GALT–GUK-1	M,Mi	27	0.52	0.04	
GALT–GUK-2	H,M	79	0.52	0.11	
GALT–IDH-1	H,M	64	0.42	1.56	
GALT–IDH-2	H,M;M,Mi	112	0.51	0.04	
GALT–MDH-2	H,M	87	0.46	0.56	
GALT–MPI	H,M;M,Mi	100	0.46	0.64	
GALT–MP-2	H,M;M,Mi	103	0.50	0.01	
GALT–PEP-2	H,M;M,Mi	114	0.42	2.84	
GALT–PEP-3	H,M	14	0.36	1.14	
GALT–PGM	H,M	82	0.52	0.20	
GALT–6PGD	H,M;M,Mi	114	0.53	0.32	
GALT–PGK	H,M	87	0.59	2.59	
GALT–PGAM-1	H,M	60	0.58	1.67	
GALT–PGAM-2	H,M	82	0.02	74.20**	Unassigned
GALT–PK-1	H,M	87	0.46	0.56	
GALT–PK-2	M,Mi	25	0.60	1.00	
GALT–Sd	H,M	87	0.45	0.93	
GALT–Tf	H,M;M,Mi	109	0.48	0.23	
GALT–TPI-1	H,M;M,Mi	40	0.55	0.40	
GALT–UMPK	H,M;M,Mi	95	0.52	0.09	
αGAL–G6PD	H,M	195	0.50	0.01	
αGAL–GPI-1	H,M	148	0.49	0.03	
αGAL–GAPD-1	H,M;C,M	187	0.55	1.55	
αGAL–GUK-2	H,M	73	0.62	3.96*	
αGAL–IDH-1	H,M	156	0.46	1.26	
αGAL–IDH-2	H,M;C,M	202	0.46	1.27	
αGAL–MDH-2	H,M;C,M	212	0.52	0.47	
αGAL–MPI	H,M;C,M	202	0.47	0.71	
αGAL–MP-2	H,M	190	0.56	3.03	
αGAL–MP-3	C,M	11	0.64	0.82	
αGAL–MP-4	H,M;C,M	51	0.53	0.18	
αGAL–PEP-2	H,M;C,M	188	0.52	0.34	
αGAL–PEP-3	H,M	47	0.51	0.02	

(*continued*)

Table III (*Continued*)

Locus pair	Species tested in crosses	*N*	*P*	χ^2	Linkage group assignment
αGAL–PGM	H,M	193	0.46	1.50	
αGAL–6PGD	H,M;C,M	210	0.54	1.54	
αGAL–PGK	H,M	80	0.45	0.80	
αGAL–PGAM-1	H,M	59	0.41	2.05	
αGAL–PGAM-2	H,M	76	0.46	0.47	
αGAL–PK-1	H,M	80	0.50	0.00	
αGAL–Sd	H,M;C,M	163	0.50	0.01	
αGAL–Tf	H,M;C,M	183	0.53	0.66	
αGAL–TPI-1	H,M;C,M	24	0.54	0.17	
αGAL–UMPK	H,M	69	0.48	0.13	
G6PD–GPI-1	H,M	462	0.52	0.55	
G6PD–GAPD-1	H,M	444	0.47	2.03	
G6PD–GUK-2	H,M	80	0.58	1.80	
G6PD–IDH-1	H,M	426	0.48	0.60	
G6PD–IDH-2	H,M	545	0.50	0.00	
G6PD–LDH-1	H,M	252	0.46	1.59	
G6PD–MDH-2	H,M	244	0.57	4.74*	
G6PD–MPI	H,M	334	0.46	2.02	
G6PD–MP-2	H,M	270	0.53	0.73	
G6PD–MP-4	H,M	111	0.49	0.08	
G6PD–MP-5	H,M	222	0.52	0.29	
G6PD–PEP-2	H,M	317	0.51	0.15	
G6PD–PEP-3	H,M	55	0.53	0.16	
G6PD–PGM	H,M	545	0.50	0.05	
G6PD–6PGD	H,M	577	0.17	256.87**	I
G6PD–PGK	H,M	89	0.55	0.91	
G6PD–PGAM-1	H,M	60	0.47	0.27	
G6PD–PGAM-2	H,M	84	0.50	0.00	
G6PD–PK-1	H,M	89	0.49	0.01	
G6PD–Sd	H,M	370	0.46	1.83	
G6PD–Tf	H,M	418	0.48	0.78	
G6PD–TPI-1	H,M	112	0.41	3.57	
G6PD–UMPK	H,M	70	0.53	0.23	
GPI-1–GS	M,Mi	27	0.48	0.04	
GPI-1–GAPD-1	M;H,M	588	0.48	0.82	
GPI-1–GAPD-2	M	37	0.65	3.27	
GPI-1–GUK-1	M,Mi	27	0.52	0.04	
GPI-1–GUK-2	H,M	80	0.40	3.20	
GPI-1–IDH-1	H,M	399	0.41	14.10**	IV
GPI-1–IDH-2	H,M;M,Mi	517	0.54	2.65	
GPI-1–LDH-1	H,M	252	0.46	1.59	
GPI-1–MDH-2	H,M	205	0.51	0.12	
GPI-1–MPI	M;H,M;M,Mi	426	0.49	0.23	
GPI-1–MP-1	M	139	0.50	0.01	

Table III (*Continued*)

Locus pair	Species tested in crosses	N	P	χ^2	Linkage group assignment
GPI-1–MP-2	H,M;M,Mi;Co,M	275	0.50	0.00	
GPI-1–MP-4	H,M	59	0.39	2.86	
GPI-1–MP-5	H,M	224	0.47	0.64	
GPI-1–PEP-2	H,M;M,Mi	324	0.48	0.31	
GPI-1–PEP-3	M;H,M;Co,M	85	0.53	0.29	
GPI-1–PGM	H,M	496	0.51	0.20	
GPI-1–6PGD	H,M;M,Mi	540	0.54	3.27	
GPI-1–PGK	H,M	89	0.54	0.55	
GPI-1–PGAM-1	H,M	60	0.62	3.27	
GPI-1–PGAM-2	H,M	84	0.48	0.19	
GPI-1–PK-1	H,M	89	0.10	56.64**	IV
GPI-1–PK-2	M,Mi	25	0.44	0.36	
GPI-1–Sd	M;H,M;Co,M	530	0.49	0.07	
GPI-1–Tailspot	M	138	0.46	0.72	
GPI-1–Tf	M;H,M;M,Mi	454	0.50	0.00	
GPI-1–TPI-1	H,M;M,Mi	137	0.48	0.18	
GPI-1–UMPK	H,M;M,Mi	95	0.56	1.27	
GS–GUK-1	M,Mi	27	0.41	0.93	
GS–IDH-2	M,Mi	27	0.44	0.33	
GS–MPI	M,Mi	27	0.48	0.04	
GS–MP-2	M,Mi	27	0.52	0.04	
GS–PEP-2	M,Mi	27	0.41	0.93	
GS–6PGD	M,Mi	27	0.44	0.33	
GS–PK-2	M,Mi	27	0.56	0.33	
GS–Tf	M,Mi	27	0.00	27.00**	VI
GS–TPI-1	M,Mi	26	0.54	0.17	
GS–UMPK	M,Mi	25	0.12	14.44**	VI
GAPD-1–GAPD-2	M	37	0.54	0.24	
GAPD-1–GUK-2	H,M	66	0.08	47.52**	III
GAPD-1–IDH-1	H,M	405	0.52	0.89	
GAPD-1–IDH-2	H,M;C,M	510	0.48	1.13	
GAPD-1–LDH-1	H,M	232	0.51	0.16	
GAPD-1–MDH-2	H,M;C,M	251	0.47	0.67	
GAPD-1–MPI	M;H,M;C,M	627	0.51	0.13	
GAPD-1–MP-1	M	97	0.48	0.09	
GAPD-1–MP-2	H,M	276	0.52	0.52	
GAPD-1–MP-3	C,M	17	0.59	0.53	
GAPD-1–MP-4	H,M;C,M	152	0.53	0.66	
GAPD-1–MP-5	H,M	178	0.49	0.02	
GAPD-1–PEP-2	H,M;C,M	298	0.54	1.93	
GAPD-1–PEP-3	M;H,M	250	0.53	0.78	
GAPD-1–PGM	H,M	511	0.53	1.88	
GAPD-1–6PGD	H,M;C,M	533	0.46	3.15	
GAPD-1–PGK	H,M	75	0.53	0.33	

(continued)

Table III (*Continued*)

Locus pair	Species tested in crosses	N	P	χ^2	Linkage group assignment
GAPD-1–PGAM-1	H,M	60	0.52	0.07	
GAPD-1–PGAM-2	H,M	70	0.50	0.00	
GAPD-1–PK-1	H,M	75	0.59	2.25	
GAPD-1–Sd	M;H,M;C,M	655	0.53	2.82	
GAPD-1–SOD	M	100	0.43	1.96	
GAPD-1–Tailspot	M	146	0.47	1.68	
GAPD-1–Tf	M;H,M;C,M	644	0.47	1.80	
GAPD-1–TPI-1	M;H,M;C,M	238	0.47	1.08	
GAPD-1–UMPK	H,M	56	0.52	0.07	
GAPD-2–MPI	M	86	0.49	0.05	
GAPD-2–MP-1	M	8	0.38	0.50	
GAPD-2–PEP-3	M	81	0.41	2.78	
GAPD-2–6PGD	M	14	0.64	1.14	
GAPD-2–Sd	M	16	0.25	4.00*	
GAPD-2–Tailspot	M	8	0.25	2.00	
GAPD-2–Tf	M	16	0.75	4.00*	
GUK-1–IDH-2	M,Mi	27	0.41	0.93	
GUK-1–MPI	M,Mi	27	0.37	1.81	
GUK-1–MP-2	M,Mi	27	0.48	0.04	
GUK-1–PEP-2	M,Mi;Co,M	42	0.36	3.43	
GUK-1–6PGD	M,Mi	27	0.52	0.04	
GUK-1–PK-2	M,Mi	27	0.44	0.33	
GUK-1–Sd	Co,M	15	0.33	1.67	
GUK-1–Tf	M,Mi	27	0.41	0.93	
GUK-1–TPI-1	M,Mi	26	0.46	0.15	
GUK-1–UMPK	M,Mi	25	0.44	0.36	
GUK-2–IDH-1	H,M	57	0.49	0.02	
GUK-2–IDH-2	H,M	77	0.51	0.01	
GUK-2–MDH-2	H,M	80	0.44	1.25	
GUK-2–MPI	H,M	66	0.48	0.06	
GUK-2–MP-2	H,M	71	0.55	0.69	
GUK-2–PEP-2	H,M	80	0.50	0.00	
GUK-2–PEP-3	H,M	14	0.36	1.14	
GUK-2–PGM	H,M	76	0.45	0.84	
GUK-2–6PGD	H,M	80	0.50	0.00	
GUK-2–PGK	H,M	80	0.50	0.00	
GUK-2–PGAM-1	H,M	53	0.47	0.17	
GUK-2–PGAM-2	H,M	75	0.52	0.12	
GUK-2–PK-1	H,M	80	0.43	1.80	
GUK-2–Sd	H,M	94	0.48	0.17	
GUK-2–Tf	H,M	76	0.47	0.21	
GUK-2–TPI-1	H,M	14	0.43	0.29	
GUK-2–UMPK	H,M	64	0.50	0.00	
IDH-1–IDH-2	H,M	448	0.49	0.08	

Table III (*Continued*)

Locus pair	Species tested in crosses	*N*	*P*	χ^2	Linkage group assignment
IDH-1–LDH-1	H,M	160	0.53	0.40	
IDH-1–MDH-2	H,M	198	0.47	0.51	
IDH-1–MPI	H,M	265	0.46	1.66	
IDH-1–MP-2	H,M	243	0.53	1.19	
IDH-1–MP-4	H,M	112	0.51	0.04	
IDH-1–MP-5	H,M	160	0.49	0.10	
IDH-1–PEP-2	H,M	251	0.49	0.10	
IDH-1–PEP-3	H,M	38	0.47	0.11	
IDH-1–PGM	H,M	428	0.52	0.46	
IDH-1–6PGD	H,M	440	0.51	0.08	
IDH-1–PGK	H,M	64	0.56	1.00	
IDH-1–PGAM-1	H,M	47	0.57	1.04	
IDH-1–PGAM-2	H,M	62	0.45	0.58	
IDH-1–PK-1	H,M	64	0.34	6.25*	IV
IDH-1–Sd	H,M	305	0.52	0.74	
IDH-1–Tf	H,M	326	0.51	0.20	
IDH-1–TPI-1	H,M	75	0.48	0.12	
IDH-1–UMPK	H,M	59	0.49	0.02	
IDH-2–LDH-1	H,M	232	0.51	0.16	
IDH-2–MDH-2	H,M;C,M	252	0.48	0.57	
IDH-2–MPI	H,M;C,M;M,Mi	387	0.48	0.58	
IDH-2–MP-2	H,M;M,Mi	310	0.48	0.63	
IDH-2–MP-3	C,M	17	0.47	0.06	
IDH-2–MP-4	H,M;C,M	146	0.53	0.44	
IDH-2–MP-5	H,M	214	0.50	0.00	
IDH-2–PEP-2	H,M;C,M;M,Mi	366	0.51	0.27	
IDH-2–PEP-3	H,M	43	0.40	1.88	
IDH-2–PGM	H,M	537	0.50	0.05	
IDH-2–6PGD	H,M;C,M;M,Mi	591	0.49	0.08	
IDH-2–PGK	H,M	86	0.43	1.67	
IDH-2–PGAM-1	H,M	58	0.40	2.48	
IDH-2–PGAM-2	H,M	83	0.46	0.59	
IDH-2–PK-1	H,M	86	0.52	0.19	
IDH-2–PK-2	M,Mi	25	0.68	3.24	
IDH-2–Sd	H,M;C,M	373	0.51	0.32	
IDH-2–Tf	H,M;C,M;M,Mi	464	0.48	0.86	
IDH-2–TPI-1	H,M;C,M;M,Mi	151	0.51	0.06	
IDH-2–UMPK	H,M;M,Mi	93	0.55	0.87	
LDH-1–MPI	H,M	101	0.16	47.14**	II
LDH-1–MP-5	H,M	235	0.53	0.72	
LDH-1–PEP-2	H,M	127	0.50	0.01	
LDH-1–PGM	H,M	235	0.51	0.21	
LDH-1–6PGD	H,M	242	0.48	0.60	
LDH-1–Sd	H,M	116	0.49	0.03	

(*continued*)

Table III (*Continued*)

Locus pair	Species tested in crosses	N	P	χ^2	Linkage group assignment
LDH-1–Tf	H,M	219	0.46	1.32	
LDH-1–TPI-1	H,M	116	0.43	2.21	
MDH-2–MPI	H,M;C,M	220	0.52	0.29	
MDH-2–MP-2	H,M	236	0.48	0.27	
MDH-2–MP-3	C,M	17	0.59	0.53	
MDH-2–MP-4	H,M;C,M	51	0.51	0.02	
MDH-2–PEP-2	H,M;C,M	212	0.51	0.08	
MDH-2–PEP-3	H,M	49	0.39	2.47	
MDH-2–PGM	H,M	240	0.50	0.02	
MDH-2–6PGD	H,M;C,M	265	0.55	2.75	
MDH-2–PGK	H,M	89	0.56	1.36	
MDH-2–PGAM-1	H,M	60	0.53	0.27	
MDH-2–PGAM-2	H,M	84	0.45	0.76	
MDH-2–PK-1	H,M	89	0.46	0.55	
MDH-2–Sd	H,M;C,M	225	0.47	1.00	
MDH-2–Tf	H,M;C,M	196	0.43	3.45	
MDH-2–TPI-1	H,M;C,M	31	0.52	0.03	
MDH-2–UMPK	H,M	70	0.41	2.06	
MPI–MP-1	M	76	0.47	0.21	
MPI–MP-2	H,M;M,Mi	272	0.48	0.53	
MPI–MP-3	C,M	17	0.29	2.88	
MPI–MP-4	H,M;C,M	116	0.53	0.55	
MPI–MP-5	H,M	86	0.47	0.42	
MPI–PEP-2	H,M;C,M;M,Mi	287	0.48	0.42	
MPI–PEP-3	M;H,M	256	0.50	0.02	
MPI–PGM	H,M	332	0.46	1.73	
MPI–6PGD	H,M;C,M;M,Mi	388	0.49	0.26	
MPI–PGK	H,M	75	0.55	0.65	
MPI–PGAM-1	H,M	60	0.47	0.27	
MPI–PGAM-2	H,M	70	0.46	0.51	
MPI–PK-1	H,M	75	0.52	0.12	
MPI–PK-2	M,Mi	25	0.28	4.84*	II ?
MPI–Sd	M;H,M;C,M	466	0.48	0.70	
MPI–SOD	M	100	0.50	0.00	
MPI–Tailspot	M	132	0.47	0.48	
MPI–Tf	M;H,M;C,M;M,Mi	602	0.52	0.66	
MPI–TPI-1	M;H,M;C,M;M,Mi	244	0.45	2.77	
MPI–UMPK	H,M;M,Mi	81	0.58	2.09	
MP-1–PEP-3	M	9	0.44	0.11	
MP-1–6PGD	M	2	0.00	2.00	
MP-1–Sd	M	137	0.55	1.23	
MP-1–Tailspot	M	138	0.54	0.72	
MP-1–Tf	M	94	0.49	0.04	
MP-2–MP-4	H,M	92	0.52	0.17	

Table III (*Continued*)

Locus pair	Species tested in crosses	*N*	*P*	χ^2	Linkage group assignment
MP-2–PEP-2	H,M;M,Mi	230	0.50	0.02	
MP-2–PEP-3	H,M;C,M	74	0.51	0.05	
MP-2–PGM	H,M	293	0.54	1.81	
MP-2–6PGD	H,M;M,Mi	317	0.51	0.08	
MP-2–PGK	H,M	78	0.45	0.82	
MP-2–PGAM-1	H,M	49	0.55	0.51	
MP-2–PGAM-2	H,M	76	0.46	0.47	
MP-2–PK-1	H,M	78	0.50	0.00	
MP-2–PK-2	M,Mi	25	0.36	1.96	
MP-2–Sd	H,M;Co,M	269	0.53	1.07	
MP-2–Tf	H,M;M,Mi	242	0.52	0.60	
MP-2–TPI-1	H,M;M,Mi	45	0.47	0.20	
MP-2–UMPK	H,M;M,Mi	84	0.57	1.71	
MP-3–MP-4	C,M	17	0.59	0.53	
MP-3–PEP-2	C,M	17	0.35	1.47	
MP-3–6PGD	C,M	17	0.41	0.53	
MP-3–Sd	C,M	17	0.41	0.53	
MP-3–Tf	C,M	16	0.31	2.25	
MP-3–TPI-1	C,M	17	0.53	0.06	
MP-4–PEP-2	H,M;C,M	76	0.51	0.05	
MP-4–PEP-3	H,M	36	0.47	0.11	
MP-4–PGM	H,M	134	0.54	1.07	
MP-4–6PGD	H,M;C,M	147	0.50	0.01	
MP-4–Sd	H,M;C,M	69	0.54	0.36	
MP-4–Tf	H,M;C,M	101	0.53	0.49	
MP-4–TPI-1	H,M;C,M	19	0.74	4.26*	
MP-5–PEP-2	H,M	124	0.50	0.00	
MP-5–PGM	H,M	212	0.55	2.28	
MP-5–6PGD	H,M	212	0.52	0.47	
MP-5–Sd	H,M	105	0.50	0.01	
MP-5–Tf	H,M	206	0.49	0.08	
MP-5–TPI-1	H,M	108	0.56	1.81	
PEP-2–PEP-3	H,M	54	0.43	1.19	
PEP-2–PGM	H,M	321	0.52	0.38	
PEP-2–6PGD	H,M;C,M;M,Mi	374	0.48	0.39	
PEP-2–PGK	H,M	89	0.52	0.10	
PEP-2–PGAM-1	H,M	60	0.63	4.27*	
PEP-2–PGAM-2	H,M	84	0.46	0.43	
PEP-2–PK-1	H,M	89	0.42	2.53	
PEP-2–PK-2	M,Mi	25	0.44	0.36	
PEP-2–Sd	H,M;C,M;Co,M	258	0.46	1.55	
PEP-2–Tf	H,M;C,M;M,Mi	343	0.45	3.17	
PEP-2–TPI-1	H,M;C,M;M,Mi	108	0.40	4.48*	
PEP-2–UMPK	H,M;M,Mi	95	0.48	0.09	

(*continued*)

Table III (*Continued*)

Locus pair	Species tested in crosses	N	P	χ^2	Linkage group assignment
PEP-3–PGM	H,M	54	0.50	0.00	
PEP-3–6PGD	M;H,M	57	0.49	0.02	
PEP-3–PGK	H,M	14	0.43	0.29	
PEP-3–PGAM-2	H,M	14	0.36	1.14	
PEP-3–PK-1	H,M	14	0.36	1.14	
PEP-3–Sd	M;H,M;Co,M	129	0.50	0.01	
PEP-3–SOD	M	85	0.46	0.58	
PEP-3–Tailspot	M	56	0.43	1.14	
PEP-3–Tf	M;H,M	238	0.53	1.08	
PEP-3–TPI-1	M;H,M	153	0.50	0.01	
PEP-3–UMPK	H,M	14	0.43	0.29	
PGM–6PGD	H,M	549	0.50	0.02	
PGM–PGK	H,M	84	0.45	0.76	
PGM–PGAM-1	H,M	55	0.46	0.89	
PGM–PGAM-2	H,M	82	0.50	0.00	
PGM–PK-1	H,M	84	0.48	0.19	
PGM–Sd	H,M	375	0.48	0.45	
PGM–Tf	H,M	415	0.45	4.46*	
PGM–TPI-1	H,M	115	0.52	0.22	
PGM–UMPK	H,M	65	0.46	0.38	
6PGD–PGK	H,M	89	0.51	1.01	
6PGD–PGAM-1	H,M	60	0.42	1.67	
6PGD–PGAM-2	H,M	84	0.56	1.19	
6PGD–PK-1	H,M	89	0.49	0.01	
6PGD–PK-2	M,Mi	25	0.52	0.04	
6PGD–Sd	H,M;C,M	395	0.48	0.57	
6PGD–Tf	H,M;C,M;M,Mi	471	0.47	1.33	
6PGD–TPI-1	H,M;C,M;M,Mi	157	0.50	0.01	
6PGD–UMPK	H,M;M,Mi	95	0.49	0.01	
PGK–PGAM-1	H,M	60	0.15	29.40**	Unassigned
PGK–PGAM-2	H,M	84	0.55	0.76	
PGK–PK-1	H,M	89	0.46	0.55	
PGK–Sd	H,M	89	0.46	0.55	
PGK–Tf	H,M	84	0.54	0.43	
PGK–TPI-1	H,M	14	0.36	1.14	
PGK–UMPK	H,M	70	0.51	0.06	
PGAM-1–PGAM-2	H,M	56	0.59	1.79	
PGAM-1–PK-1	H,M	60	0.53	0.27	
PGAM-1–Sd	H,M	60	0.50	0.00	
PGAM-1–Tf	H,M	59	0.53	0.15	
PGAM-1–UMPK	H,M	56	0.52	0.07	
PGAM-2–PK-1	H,M	84	0.44	1.19	
PGAM-2–Sd	H,M	84	0.46	0.43	
PGAM-2–Tf	H,M	79	0.52	0.11	

Table III (*Continued*)

Locus pair	Species tested in crosses	N	P	χ^2	Linkage group assignment
PGAM-2–TPI-1	H,M	14	0.71	2.57	
PGAM-2–UMPK	H,M	66	0.53	0.24	
PK-1–Sd	H,M	89	0.40	3.25	
PK-1–Tf	H,M	84	0.45	0.76	
PK-1–TPI-1	H,M	14	0.43	0.29	
PK-1–UMPK	H,M	70	0.50	0.00	
PK-2–Tf	M,Mi	25	0.52	0.04	
PK-2–TPI-1	M,Mi	25	0.20	9.00**	?
PK-2–UMPK	M,Mi	23	0.48	0.04	
Sd–SOD	M	76	0.51	0.05	
Sd–Tailspot	M	188	0.54	1.04	
Sd–Tf	M;H,M;C,M	550	0.46	3.85*	
Sd–TPI-1	M;H,M;C,M	168	0.52	0.38	
Sd–UMPK	H,M	70	0.41	2.06	
SOD–Tf	M	97	0.44	1.25	
SOD–TPI-1	M	100	0.51	0.04	
Tailspot–TPI-1	M	15	0.53	0.06	
Tf–TPI-1	M;H,M;C,M;M,Mi	302	0.54	1.60	
Tf–UMPK	H,M	68	0.10	42.88**	
	M,Mi	25	0.12	14.44**	
	Total	93	0.11	57.30**	VI
TPI-1–UMPK	H,M;M,Mi	38	0.50	0.00	

should be emphasized that approximately two deviations of this magnitude are expected in a sample of 42 loci by chance sampling error. In the case of *MP-1,* a parvalbumin locus, an additional complication of possible phenotypic misclassification exists; only one allelic polymorphism has thus far been studied, in intraspecific crosses using *X. maculatus* stocks from the Río Jamapa and Río Coatzacoalcos. The protein migrates very rapidly toward the anode, and often the two bands in heterozygotes are difficult to resolve (Fig. 1). Thus, a small proportion of individual backcross and F_2 hybrids assigned to homozygous classes could represent unresolvable heterozygotes, thus causing a spurious homozygote excess. Misclassification is very unlikely in the case of *IDH-2* (Fig. 1), since the electrophoretic patterns are eminently scorable. The consistent excess of heterozygotes in several cross types could indicate a relatively minor heterosis component; larger sample sizes should conclusively answer this question.

Only one case of significant segregation distortion has been documented in *Xiphophorus* crosses, detected by heterogeneity among cross types in linkage group I (Morizot *et al.*, 1977). This deviation results from

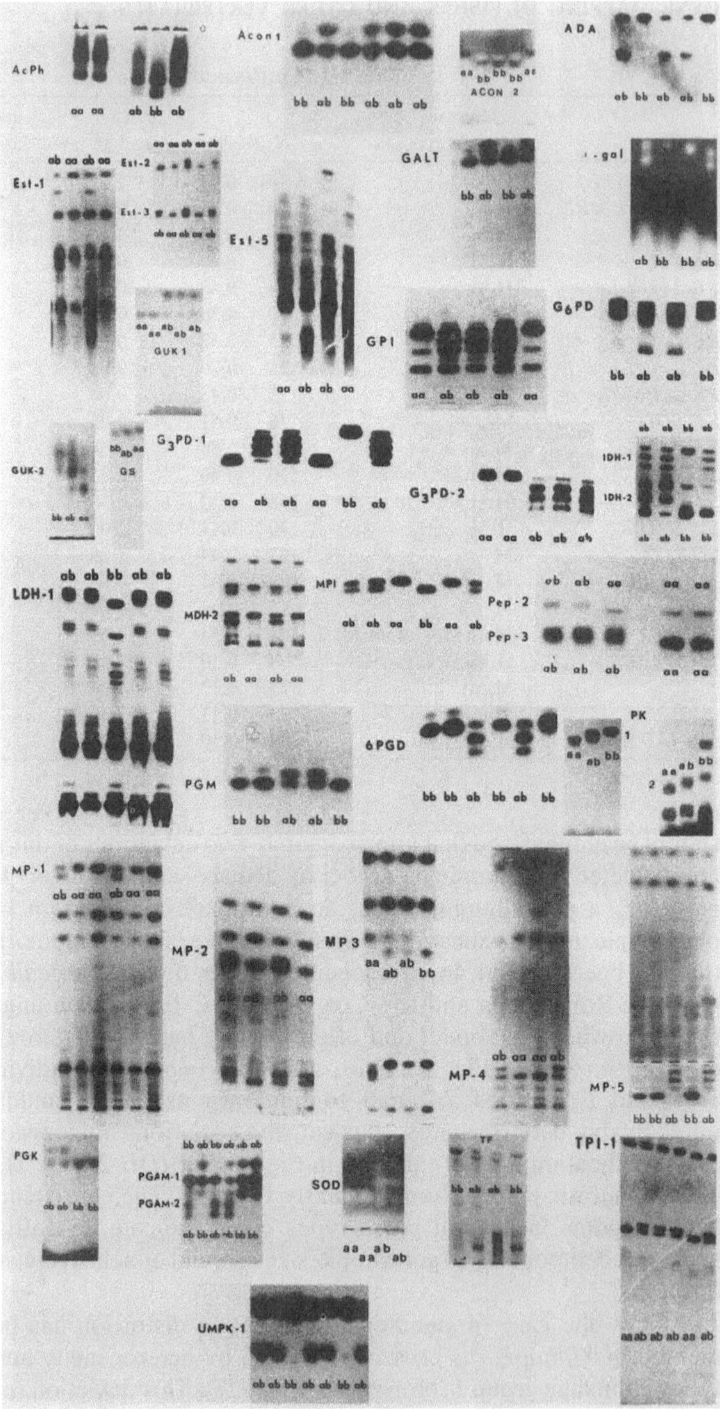

AcPh
Acon 1
ACON 2
ADA
Est-1
Est-2
Est-3
GUK 1
Est-5
GALT
GPI
G6PD
GUK-2
GS
G3PD-1
G3PD-2
IDH-1
IDH-2
LDH-1
MDH-2
MPI
Pep-2
Pep-3
PGM
6PGD
PK
MP-1
MP-2
MP 3
MP-4
MP-5
PGK
PGAM-1
PGAM-2
SOD
TF
TPI-1
UMPK-1

an excess of swordtail alleles at *ADA* and *G6PD* in *helleri* × *maculatus* derived backcrosses; thus an excess of heterozygotes is observed in backcrosses to *maculatus* and an excess of homozygotes in backcrosses to *helleri*. Segregation at *ADA* is normal in intraspecific crosses, and the resulting segregation total (Table II) exhibits no distortion, due to balancing effects of including multiple cross types. No other significant heterogeneities of this type have yet been observed.

The careful reader will no doubt notice that of the 43 segregation totals in Table II, 31 exhibit an excess of heterozygotes, a figure clearly different from 50%. We have begun a detailed analysis of overall segregation in different types of crosses: preliminary results indicate that significant excesses of heterozygotes over all loci occur only in backcrosses to *X. helleri* from *helleri* × *maculatus* F_1 hybrids; the reciprocal backcross to *maculatus* exhibits no such excess. The significance of these results is at present unknown, and the assessment of the contribution of each locus to the overall pattern will necessitate the generation of much larger sample sizes. It remains an intriguing possibility that such a widespread retention of platyfish genes in these backcrosses somehow relates to the regulation of pigment cell tumors in platyfish–swordtail hybrids (Siciliano *et al.*, 1976).

The magnitude of the deviations noted is small enough not to interfere seriously with assessment of linkage relationships among the loci. Obviously, when very large numbers of pairwise comparisons are made, stringent criteria for rejecting the hypothesis of independent assortment must be employed to minimize type I errors. For evidence of linkage we require statistical significance at the 0.01 level in both orthogonal function and contingency table analyses (Mather, 1957; Bailey, 1961) and, where possible, agreement among multiple cross types.

Having established acceptably normal segregation for most biochemical loci in backcross hybrids, the linkage relationships among such loci can be assessed. Pairwise linkage comparisons are presented in Table III. From these data, six independent linkage groups can be described. Each will be considered here with particular reference to the number of *Xiphophorus* species thus far assessed in crosses.

Figure 1. Zymograms illustrating homozygous and heterozygous patterns for isozymes studied for segregation in *Xiphophorus* hybrids. Anodal end of each zymogram is toward the top. With the exception of PK-1, the isozyme studied is indicated either on the zymogram or to the left. The zone of PK-1 activity is indicated to the right of the zymogram. Homozygous (*aa* or *bb*) and heterozygous (*ab*) phenotypes for each sample are indicated in each channel.

2.1.1. Linkage Group I

Three loci have been assigned to LG I, coding for adenosine deaminase, glucose-6-phosphate dehydrogenase, and 6-phosphogluconate dehydrogenase (Morizot *et al.*, 1977). The recombination map from the total data is: *ADA*-13%-*G6PD*-17%-*6PGD*. The linkage group *in toto* has been assessed thus far only utilizing *Xiphophorus maculatus* × *X. helleri* hybrids, though *ADA–6PGD* pairs in backcrosses derived from *X. milleri* × *X. maculatus* and *X. clemenciae* × *X. maculatus* F_1 hybrids are consistent with linkage, albeit with very small sample sizes. Further analysis of LG I is of considerable interest for several reasons. The segregation distortion identified at *ADA* away from *G6PD* and *6PGD* in *X. helleri* × *X. maculatus* derived backcrosses [see above and Morizot *et al.* (1977)] deserves further study. Additionally, both sex- and population-specific recombination differences have been identified in crosses informative for LG I loci (Morizot *et al.*, 1977). To assess the prevalence of these recombination differences in the genus, we have produced intraspecific backcrosses using swordtail strains (*X. helleri guentheri* and *X. h. helleri*) informative for the three LG I loci; these backcross hybrids will be analyzed in the near future, upon attainment of sexual maturity.

2.1.2. Linkage Group II

The original description of *Xiphophorus* linkage group II (Morizot and Siciliano, 1979) included the assignment of four enzyme loci, coding for esterase-2, esterase-3, lactate dehydrogenase-1, and mannosephosphate isomerase. Data accumulated since the original description indicate that two additional loci, coding for esterase-5 and pyruvate kinase-2, should be assigned to LG II. The recombination map based on the total data (Table III) indicates an order of: *ES-2*-16%-*PK2*-8%-*ES-3*-0%-*ES-5*-23%-*LDH-1*-16%-*MPI*. It should be noted that no data assessing recombination between *ES-2* and *PK-2* have been collected as yet; the position of *PK-2* is inferred from recombination with *ES-3* and *MPI*.

Substantial data for LG II locus pairs in both intraspecific and interspecific crosses indicate the conservation of the linkage group in several *Xiphophorus* species (Table III). We should point out, however, the limitations of the analyses and the cases of significant heterogeneity. Except for *maculatus* × *maculatus* and *maculatus* × *helleri* crosses, the evidence for linkage of LG II must be considered provisional, based upon the analysis of only 25 *milleri* × *maculatus* backcross hybrids. In this instance the agreement of *ES-3–ES-5* and *ES-3–MPI* recombination frequency estimates with those observed in other cross types is strong supporting evidence for the presence of similar gene arrangements in *milleri*, *maculatus*, and *helleri*. Recombination between *ES-2* and *ES-3* in *maculatus* × *couchianus* crosses is significantly lower than in other cross types, even though only 26 individuals from a second backcross to *couchianus* have been analyzed. Sex- and species-reciprocal backcrosses between *maculatus* and *couchianus* are needed to assess the gene arrangment of LG II loci in *X. couchianus*.

No recombinants between *ES-3* and *ES-5* have been observed in sampling 100 backcross individuals from *helleri* × *maculatus* and *milleri* × *maculatus* crosses. The lines of evidence for the two isozymes being coded by separate genetic loci thus are necessarily somewhat indirect, but convincing. While *ES-3* is expressed in most tissues, *ES-5* activity has been detected thus far only in liver extracts, anodal to the *ES-3* allozymes. Second, allelic products of the two presumptive loci often require different buffer systems for their resolution. While it is possible that *ES-5* could represent a modified form of *ES-3*, or vice versa, we feel such an explanation is unconservative, due to the necessity of postulating differential regulation as well as modification. The demonstration of a recombinant between *ES-3* and *ES-5* through further sampling would conclusively answer this question.

At present, we view the inconsistent association between *TPI-1* and *PK-2*, *ES-3*, *ES-5*, and possibly *LDH-1* (as indicated by low recombination frequency estimates in Table III), but not with *ES-2* and *MPI*, as probably due to sample error. Variability of the gene position of *TPI-1* among *Xiphophorus* species remains a possibility that at this time cannot be excluded; further crosses are being constructed to analyze these relationships.

2.1.3. Linkage Group III

Guanylate kinase-2 and glyceraldehyde-3-phosphate dehydrogenase-1 were assigned to *Xiphophorus* LG III by Morizot and Siciliano (1982*b*). The recombination map of *GUK-2-8%-GAPD-1* thus far has been examined only in *maculatus* × *helleri* backcross hybrids. Guanylate kinase-2 in *Xiphophorus* is a rather difficult enzyme with respect to resolution. The extensive anodal "satelliting," also observed in mammalian guanylate kinase (Harris and Hopkinson, 1977), renders accurate phenotypic determination possible only in backcrosses to the parent possessing the more anodal allozyme; in such crosses the presence of a less anodal band of activity can be used to distinguish heterozygotes unequivocally. Though independent assortment of *GUK-2* from loci assigned to other linkage groups has been assessed to date with only small sample sizes, the tight linkage of *GUK-2* and *GAPD-1* allows the presumption of independent assortment from the extensive data assembled for *GAPD-1*. We have recently discovered new variants in *X. montezumae* (*sensu lato*) for both GAPD-1 and GUK-2; such variants will be of use in testing the generality of LG III relationships in *Xiphophorus* species.

2.1.4. Linkage Group IV

Xiphophorus LG IV was first described by Morizot and Siciliano (1982*c*), containing loci coding for pyruvate kinase-1, glucosephosphate isomerase-1, and isocitrate dehydrogenase-1. The available data indicate a recombination map of *PK-1*-10%-*GPI-1*-41%-*IDH-1*. Though resolution of PK-1 variants has been accomplished only recently and less than 100 informative backcross hybrids have been analyzed for the enzyme, the independent assortment of LG IV from other *Xiphophorus* linkage groups has been amply demonstrated in crosses informative for *GPI-1* and *IDH-1*, where sample sizes in pairwise comparisons often exceed 400 individuals. Thus far the linkage of loci assigned to LG IV has been demonstrated only in *X. maculatus* × *X. helleri* crosses. The relatively loose linkage of *GPI-1* and *IDH-1* makes the assessment of their linkage relationships in other *Xiphophorus* species difficult. The addition of *PK-1* to LG IV, with its tight linkage to *GPI-1*, provides a useful marker for testing the generality of the group throughout the genus. Several stocks possessing PK-1 allozymes variant from our *X. maculatus* reference strain have been identified.

2.1.5. Linkage Group V

Siciliano *et al.* (1976) demonstrated the linkage of *ES-1* (*esterase*-1) and a factor associated with the severity of melanomas in backcross hybrids between *X. maculatus* and *X. helleri*. Increased sample sizes revealed the loose linkage between *ES-1* and *MDH-2* (malate dehydrogenase-2) and added a liver esterase (*ES-4*); this group has been designated LG V (Morizot and Siciliano, 1983*a*). The report of the association of a highly anodal liver esterase with melanoma phenotype (Ahuja *et al.*, 1980) suggests homology with the *ES-4* locus assigned to LG V. While Ahuja *et al.* assume the liver esterase to be the same as our esterase-1, Leslie and Pontier (1980), in an extensive study of poeciliid esterases, concluded that *Xiphophorus ES-1* was not expressed in liver tissue. As in most fish species, the liver esterase gel patterns of *Xiphophorus* are quite complex. Further studies will be required to resolve the identity of the esterases coded by LG V loci. This example of apparent discrepancies in results obtained in different laboratories clearly illustrates the need for precise characterization of enzymes utilized in linkage studies, particularly when multiple genetic loci are involved.

2.1.6. Linkage Group VI

Xiphophorus LG VI, described by Morizot and Siciliano (1983*b*), contains loci coding for glutamine synthetase, transferrin, and uridine monophosphate kinase. The available data indicate a recombination map of *GS*-0%-*Tf*-11%-*UMPK*. The *Tf–UMPK* linkage has been observed in both *maculatus* × *helleri* and *maculatus* × *milleri* crosses, while the *GS–Tf* linkage has been tested only in *maculatus* × *milleri* hybrids. Electrophoretic variation at all three loci has recently been found in several *Xiphophorus* species; the generality of LG VI linkage relationships in the genus should be easily assessed.

2.1.7. Linkage of *PGK–PGAM-1* and *GALT–PGAM-2*

Two other pairs of loci have been shown to be linked in *Xiphophorus* (Table III): phosphoglycerate kinase with phosphoglycerate mutase-1 (15% recombinants), and galactose-1-phosphate uridyltransferase with phosphoglycerate mutase-2 (2% recombinants). Both these linkages have been studied only in *maculatus* × *helleri* hybrids. Though the available data suggest independent assortment from linkage groups I–VI, these data at present are too few to justify the description of independent linkage groups. In particular, the rather low recombination estimates between *GALT–PGAM-2* and some LG IV loci suggest caution. Premature description of linkage groups can only result in confusion in the literature, necessitating retractions of previously claimed linkages and coalescence of presumed independently assorting groups.

2.1.8. Independent Assortment of Other Enzyme Loci

As described above, linkage groups I–VI presently contain 20 loci of the 41 thus far studied. Of the remaining 21, *GALT–PGAM-2* and *PGK* and *PGAM-1* are known to be linked; thus, the loci presently unassigned could mark as many as 19 chromosomes, which, plus the six chromosomes marked by LG I–VI, sets a maximum of 25 chromosomes marked, one more than the haploid chromosome number of *Xiphophorus*. Both the sample size and the completeness of pairwise testing vary enormously among the unassigned enzyme loci (Table III). For example, *IDH-2* has been tested against 36 of the 40 other loci studied, with sample sizes of fewer than 20 to more than 500 individuals; *ACON-2*, on the other hand, has been tested against only eight other loci, with no sample size exceeding 16 individuals.

2.1.9. Approaches for Expansion of Gene Maps

At this point in our gene mapping studies, then, it is important to evaluate the possible approaches for producing a precise genetic map of *Xiphophorus*. Three types of expansion are apparent, each with its advantages and disadvantages. The first is simply to increase the sample sizes of already initiated pairwise comparisons. A considerable advantage of such an approach is that it requires the use only of previously characterized genetic strains and enzymes. Its major disadvantage is that, while independent assortment may be demonstrated much more convincingly, few new linkages are likely to be detected. A second approach is to design crosses that provide data concerning the approximately 20% of pairwise comparisons not yet initiated involving loci presently under study. This method shares the advantage of often utilizing characterized strains and enzymes. However, many more genetic crosses are required and again, relatively few new linkages are likely to be demonstrated. The third approach, which we feel is usually by far the most desirable, is to introduce additional loci into crosses constructed for linkage analyses. The rather obvious advantage of such an approach is the high likelihood of new linkages, particularly when a large proportion of the genome is marked by loci already tested to some degree for linkage. The major, and usually serious, disadvantage inherent in such an approach is the requirement of large numbers of aquaria to house numerous parental stocks and F_1 hybrids.

While these considerations in planning crosses for linkage analysis may appear obvious, their import upon the availability of data for comparisons of linkage relationships among diverse species is large, as is clearly illustrated in Section 2.2 in our attempts to assess linkage group homology among fish species. For comparative gene mapping studies, it is often much more informative to generate relatively complete pairwise comparison data utilizing few loci than to generate extremely incomplete pairwise data for many loci. For workers with limited space desiring to contribute data useful for comparative gene mapping studies, the judicious choice of genetic crosses for linkage analyses is of paramount importance.

2.1.10. Linkage Group Conservation in *Xiphophorus*

In summary, among the crosses of *Xiphophorus* species analyzed to date, no case of nonhomology between species has been demonstrated.

Two cases of significant recombination differences have been documented, between *X. helleri* swordtail populations in LG I and between species in LG II. Whether such differences reflect differences in chromosomal gene arrangement remains to be investigated. On the other hand, evidence for linkage group conservation comes from three linkage groups and five *Xiphophorus* species. It is clear that many if not most gene arrangements are homologous within the genus; additional crosses and loci will undoubtedly improve the estimation of the rate of linkage group divergence.

With the data from *Xiphophorus* species offering a sizable basis for comparison, we can attempt to assess the degree of linkage group homology in other fishes. Though comparative data are extremely few, some tentative conclusions can be reached.

2.2. Homology of *Xiphophorus* Proteins with Those Studied in Other Fishes

Table IV lists our proposed homologies between protein-coding loci in *Xiphophorus* and those that have been studied for linkage in other fish species. The primary criterion utilized.in presuming homology is a similarity in tissue distribution of the particular isozymes; in many cases these are the only data available for comparison. Ideally, data concerning substrate specificity, kinetic properties, inhibition, subcellular localization, subunit structure, relative electrophoretic mobilities, and cofactor requirements (to mention only some useful properties) should be available for protein homology to be firmly established, but such complete characterizations are rarely available in fishes. Even with these limitations, we feel the proposed homologies not marked with question marks in Table IV have a reasonable probability of being correct.

It is worthwhile to examine some of the more uncertain relationships of Table IV both to appreciate some of the difficulties inherent in comparative mapping attempts and to suggest areas where additional data are needed. With three exceptions, the homologies between *Xiphophorus* and *Poeciliopsis* (both genera of the family Poeciliidae) seem reasonably certain, primarily because of several communications with Dr. J. F. Leslie. Two of the difficulties probably are due to electrophoretic resolution differences: IDH-1 in *Xiphophorus* is a very labile enzyme and we feel that IDH-1 of *Poeciliopsis* is most probably homologous to IDH-2 of *Xiphophorus;* and similarly, GAPD-2 of *Xiphophorus* in muscle is much more active than GAPD-1, perhaps making the former the more likely homologous protein to *Poeciliopsis* GAP-1. In the last case, PEP-2 of *Xiphophorus* is the only peptidase active with glycyl-leucine as substrate in the presence

Table IV
Proposed Homologies of Protein-Coding Loci Studied for Linkage in Fish Species[a]

Xiphophorus	*Poeciliopsis*[b]	*Poecilia*[c]	Centrarchids[d]	Salmonids[e]
ACP				*Mup*?
ADA	*Ada*			*Ada-1*?
ADH	*Adh-2*			*Adh*
DIA				*Dia*??
ES1	*Es-1*		See text	
ES2	*Es-4*	*Es-2*?		
ES3	*Es-5*			
βGAM				*Gam*
GPI-1	*Pgi-1*		*GPI-B*	*Gpi-1, Gpi-2*
GPI-2	*Pgi-2*		*GPI-A*	*Gpi-3*
βGUS				*Gus*
GOT-1	*Aat-1*			*Aat-1, Aat-2*
GOT-2	*Aat-3*			
GOT-3				*Aat-4*?
GPT				*Gpt*
GAPD-2	*Gap-1*?			
αGPD			*αGPDH*?	*Agp-1*??, *Agp-2*??
IDH-1	*Idh-1*?			*Idh-3*?
IDH-2	*Idh-1*?			*Idh-3*?
IDH-3	*Idh-2*			*Idh-1,2*?
LDH-1	*Ldh-1*		*LDH-C*	*Ldh-5*
LDH-2	*Ldh-2*		*LDH-A*	*Ldh-1, Ldh-2*
LDH-3	*Ldh-3*		*LDH-B*	*Ldh-3, Ldh-4*
MDH-1	*Mdh-1*		*MDH-B*	*Mdh-3, Mdh-4*
MDH-2	*Mdh-3*	*MDH-1*?	*MDH-A*	*Mdh-1, Mdh-2*
MDH-3	*Mdh-2*			
ME				*Me-1*?, *Me-2*?
MPI				*Pmi*
MP-4	*Mp-3*			*Cpk-1, Cpk-2*
ODH[f]				*Odh*
PALB[f]				*Palb-(1,2)*
PEP-1				*Pep-(1,2)*
PEP-2	*Pep*?			
PGM	*Pgm*	*PGM*		*Pgm-1*?, *Pgm-2*?
6PGD	*Pgd*		*6PGDH*	
SDH[f]				*Sdh*
SOD	*To*	*TO*	*TO*	*Sod*

[a]See text for criteria and further details.
[b]Data from Leslie and Vrijenhoek (1977, 1978), Leslie and Pontier (1980), Leslie (personal communication).
[c]Data from Shami and Beardmore (1978).
[d]Data from Champion and Whitt (1976), Fisher *et al.* (1980), Wheat *et al.* (1972), Wheat and Whitt, (1973), Whitt *et al.* (1973, 1976).
[e]Data from Allendorf (1975), Allendorf *et al.* (1975, 1977), Allendorf and Utter (1976), Aspinwall (1974), May *et al.* (1979*a,b*), Stoneking *et al.* (1979), May *et al.* (1980), Wright *et al.* (1983).
[f]Loci not yet studied in *Xiphophorus*.

of EDTA; we thus assume *Poeciliopsis* PEP-gl to be homologous, though additional substrate and inhibitor specificity comparisons are desirable.

Homology of ES-2 and MDH-1 of the guppy, *Poecilia reticulata* (Shami and Beardmore, 1978), to *Xiphophorus* enzymes is much more uncertain, for several reasons. First, whole young fish were used in the guppy experiments, precluding the use of tissue specificity criteria to aid in establishing homology; second, with regard to ES-2, a substrate mixture was utilized without presenting data on substrate specificity; and third, the type of gel used for the electrophoretic separations was not specified (though from the description we presume it was starch). We have run samples from several guppy tissues and, notwithstanding the difficulties of comparison, feel that the ES-2 and MDH-2 of Shami and Beardmore are probably homologous to our ES-2 and MDH-2 of *Xiphophorus*, based almost entirely on relative electrophoretic mobility considerations.

The relative ease with which homology between enzymes of centrarchid sunfishes and *Xiphophorus* can be presumed is due to the careful and exhaustive studies of G. S. Whitt and others at the University of Illinois. Only two enzymes present some difficulty, αGPD and an esterase. The problems with αGPD are of some general interest: in sunfishes, two isozymes apparently encoded by different loci exist in liver and muscle (Wheat and Whitt, 1973). In *Xiphophorus* and apparently in *Poeciliopsis* (Leslie, personal communication), the rarity of polymorphic variation makes even the number of independently coded isozymes difficult to determine, though the muscle forms are well resolved. Further study in a variety of fish species obviously is warranted (see also the discussion of αGPD in salmonids below).

Whether homology of any esterases in centrarchids and poeciliids can be established without sequence data is in our opinion a matter of conjecture. The use of relative electrophoretic mobility criteria even in a single tissue (skeletal muscle in the present case) to constitute evidence of homology at the family level seems hazardous; even within a genus or a species after presumably careful study, doubts as to homology of esterases often remain [see Leslie and Pontier (1980) with regard to *Xiphophorus* esterases]. Exhaustive tissue distribution studies resulting in restrictive expression profiles may prove to be the most useful tool in facilitating comparative gene mapping of esterases.

Only the existence of several excellent descriptive papers (in particular, Allendorf, 1975; Allendorf *et al.*, 1977; May *et al.*, 1979*b*, 1980, 1982; Wright *et al.*, 1983) allows the interpretation of homology of salmonid enzymes with those of other fish species to be only difficult rather than impossible. Tissue distribution data have often yielded insights into the homology of a pair of salmonid "reduplicated" isozymes with one member of a "duplicated" pair in diploid fish species. The continuing search for new electrophoretic variants in salmonid species likewise refines the knowledge of how many loci code for a particular enzyme; such studies serve both to ascertain homology with duplicated diploid locus pairs and to assess the extent of gene silencing. The ancestrally tetraploid salmonid duplicated and reduplicated locus pairs have become increasingly well defined; however, a few questions of homology linger, particularly with respect to conservation of linkage relationships. For example, apparently a single locus codes for NADH diaphorase, but activity can be detected either in liver (Allendorf *et al.*, 1977) or in muscle (May *et al.*, 1979*a*) as a single band in homozygotes. This situation contrasts with that in *Xiphophorus,* where we have detected invariant two-banded patterns in liver and no activity in skeletal muscle samples. Can homology be presumed in such cases? It is quite possible that duplicated (or reduplicated) diaphorase loci in salmonids have diverged in chromosomal location and in tissue specificity and that tissue distribution differences in different species have become fixed through gene silencing.

The αGPD (or AGP or G3P in salmonid terminology) story is complicated for different reasons. Allendorf *et al.* (1977) describe the products of three loci, two expressed in muscle and one in liver. May *et al.* (1979*b*) suggest two muscle enzyme loci that are not members of a duplicate pair. These points of view parallel questions raised in diploid poeciliid and centrarchid species as described above; the future study of αGPD in a variety of fish species is of obvious interest.

The duplication tables are turned with respect to IDH in salmonid species. Two liver supernatant IDH loci have been described in *Xiphophorus* (Siciliano and Wright, 1973), plus a mitochondrial form present in most tissues. Two presumably duplicate pairs of IDH loci are observed in salmonids (Allendorf *et al.*, 1977; May *et al.*, 1979*b*) in muscle and in liver and eye, but what of a third pair of loci? In this case subcellular localization data are required. Perhaps, as suggested for *Poeciliopsis,* a second liver system has not been described due to extreme enzyme lability.

The homology of malic enzyme (ME) loci is problematic only because no evidence has been adduced in our studies that would suggest more than one ME locus in poeciliid fishes. Since three loci are assumed to code for ME by Allendorf *et al.* (1977) in salmonids, a two-locus system (with one isozyme always comigrating or unexpressed) should exist in diploid fish species, unless a second duplication has occurred in salmonids. For either case, locus homology is difficult to establish.

Whether only one PGM locus is present in poeciliid fishes is at present questionable. We have from time to time resolved additional bands of PGM activity in *Xiphophorus,* as has Leslie (personal communication) in

Poeciliopsis. The existence of two unlinked PGM loci in *Zoarces* (Hjorth, 1971; Simonsen and Frydenberg, 1972) further suggests the possibility of duplicated loci in diploid fish species. Thus, the homology of the two PGM loci suggested for brown trout by Allendorf *et al.* (1977) must remain uncertain at this time.

We can conclude that, although numerous difficulties arise in attempting to establish homology of protein-coding loci among diverse groups of fish species, a substantial area of reasonable certainty remains after the most severe objections have been aired. Turning to the linkage data thus far amassed for fish species, we can now assess whether evidence for linkage group conservation exists.

2.3. *Poeciliopsis,* Poeciliidae

Of the 17 loci tested for linkage in *Poeciliopsis* (J.F. Leslie, personal communication; Leslie, 1982), 12 are probably homologous to enzyme loci studied for linkage in *Xiphophorus* (Table IV). Most of the proposed homologies between *Poeciliopsis* and *Xiphophorus* are quite certain, due no doubt to the confamilial status of the genera and to joint efforts in enzyme characterization. Unfortunately, only about half the pairwise comparisons have been tested in *Poeciliopsis,* due primarily to a lack of variation among clonally inherited genomes. For example, linkage tests between *ADA* and *6PGD* (linked in *Xiphophorus* LG I) are impossible at present due to lack of variation in *Poeciliopsis* species pairs (Leslie, 1982).

However, at least one striking similarity in gene arrangements is evidenced in the *Poeciliopsis* linkage data. A five-point linkage group has been described in *Poeciliopsis* (Leslie, 1982) containing loci homologous to *Xiphophorus ES-2, IDH-3, LDH-2, ES-3* and *LDH-1*. Three of these loci (*ES-2, ES-3,* and *LDH-1*) are also linked (Table III) in *Xiphophorus.* Furher, the similarity in gene arrangements extends to recombination frequency estimates: between *ES-3* and *LDH-1,* 23% recombinants are observed in *Xiphophorus* crosses, 25% in *Poeciliopsis*. Between *ES-2* and *ES-3,* the discrepancy is greater: 24% averaged over all *Xiphophorus* crosses, 36% in *Poeciliopsis*. It should be noted, however, that recombination in this region of *Xiphophorus* LG II varies from 8% to 40% among crosses. Among the 35 pairs of probably homologous loci tested both in *Xiphophorus* and in *Poeciliopsis* (Table V), no conclusive disagreements are observed. In the only questionable case, *Gpi-1* and *Idh-1* of *Poeciliopsis,* it is quite probable that *Idh-1* of *Poeciliopsis* is homologous to *IDH-2* of *Xiphophorus,* which is unlinked to *GPI-1* in both genera.

The comparison of *Xiphophorus* and *Poeciliopsis* gene maps illustrates the feasibility of extensive comparative mapping among fish species.

Table V
Comparative Linkage Relationships of Probably Homologous Protein-Coding Loci in Fish Species, Derived from Proposed Homologies in Table IV[a]

Locus pair	*Xiphophorus*	*Poeciliopsis*	*Poecilia*	Centrarchidae	Salmonidae[b]
ACP–ADA	U				U (1/1)
ACP–DIA					U (1/1)
ACP–GPI-1	U				U (1/2)
ACP–GOT-3					U (1/1)
ACP–GPT					U (1/1)
ACP–IDH-1 or *2*	U, U				U (1/1)
ACP–LDH-3					U (1/2)
ACP–MDH-2	U				U (1/2)
ACP–ME					U (2/2)
ACP–MP-4	U				U (1/2)
ACP–ODH					U (1/1)
ACP–SDH					U (1/1)
ACP–SOD	U				U (1/1)
ADA–ADH					U (1/1)
ADA–DIA					U (1/1)
ADA–ES-3	U	U			
ADA–F6P					U (1/1)
ADA–GPI-1	U	U			U (2/2)
ADA–GPI-2					U (1/1)
ADA–βGUS					U (1/1)
ADA–GOT-1					U (1/2)
ADA–GOT-3					U (1/1)
ADA–GPT					U (1/1)
ADA–GAPD-2		U			
ADA–αGPD					U, L (2/2)
ADA–IDH-1 or *2*	U, U	U			U (1/1)
ADA–IDH-3					U (1/2)
ADA–LDH-1	U	U			U (1/1)
ADA–LDH-3					U (2/2)
ADA–MDH-1					U (2/2)
ADA–MDH-2	U	U			U (1/2)
ADA–ME					U (2/2)
ADA–MPI	U				U (1/1)
ADA–MP-4	U				U (1/2)
ADA–ODH					U (1/1)
ADA–PGM	U	U			U (1/2)
ADA–PGK	U				U (1/1)
ADA–SDH					U (1/1)
ADA–SOD					U (1/1)

[a]Abbreviations and symbols: A, associated (nonrandomly assorting at $P < 0.05$, linkage phase unknown); L, linked at $P < 0.05$; P, pseudolinked at $P < 0.05$; U, unlinked. Locus symbols are those of *Xiphophorus:* for proposed homologies and data sources, see Table IV.

[b]Numbers in parentheses indicate completeness of pairwise comparisons of duplicate loci.

Table V (*Continued*)

Locus pair	*Xiphophorus*	*Poeciliopsis*	*Poecilia*	Centrarchidae	Salmonidae[b]
ADH–DIA					U (1/1)
ADH–ES-2		U			
ADH–ES-3		U			
ADH–GPI-1		U			U (2/2)
ADH–GOT-1					U (1/2)
ADH–GOT-3					U (1/1)
ADH–GPT					U (1/1)
ADH–αGPD					A, U (2/2)
ADH–IDH-1 or *2*					U (1/2)
ADH–IDH-3		U			
ADH–LDH-1		U			
ADH–LDH-2		U			
ADH–LDH-3					U (2/2)
ADH–MDH-1					U (1/2)
ADH–MDH-2					U (1/2)
ADH–ME					U (2/2)
ADH–MPI					U (1/1)
ADH–MP-4					U (1/2)
ADH–ODH					U (1/1)
ADH–PEP-2		U			
ADH–PGM					U (1/2)
ADH–6PGD		U			
ADH–SDH					U (1/1)
ADH–SOD					U (1/1)
DIA–GPI-2					U (1/1)
DIA–βGUS					U (1/1)
DIA–GOT-1					U (1/2)
DIA–GOT-3					U (1/1)
DIA–GPT					U (1/1)
DIA–αGPD					U (1/1)
DIA–IDH-1 or *2*					U (1/2)
DIA–LDH-3					U (2/2)
DIA–MDH-1					U (2/2)
DIA–MDH-2					U (2/2)
DIA–ME					U (2/2)
DIA–MPI					U (1/1)
DIA–MP-4					L, U (1/2)
DIA–ODH					U (1/1)
DIA–PGM					U (1/2)
DIA–SDH					U (1/1)
DIA–SOD					P (1/1)
ES-2–ES-3	L	L			
ES-2–GPI-1	U	U			
ES-2–GOT-2		U			
ES-2–IDH-1 or *2*	U, U	U			

(continued)

Table V (*Continued*)

Locus pair	*Xiphophorus*	*Poeciliopsis*	*Poecilia*	Centrarchidae	Salmonidae[b]
ES-2–IDH-3		L			
ES-2–LDH-1	L	L			
ES-2–LDH-2		L			
ES-2–MDH-2	U		U		
ES-2–MDH-3		U			
ES-2–MP-4	U	U			
ES-2–PEP-2	U	U			
ES-2–PGM	U		U		
ES-2–6PGD	U	U			
ES-2–SOD	U		U		
ES-3–GPI-1	U	U			
ES-3–GOT-2		U			
ES-3–GAPD-2		U			
ES-3–IDH-1 or *2*	U, U	U			
ES-3–IDH-3		L			
ES-3–LDH-1	L	L			
ES-3–LDH-2		L			
ES-3–MDH-3		U			
ES-3–MP-4	U	U			
ES-3–PEP-2	U	U			
ES-3–PGM	U	U			
ES-3–6PGD	U	U			
ES-3–SOD	U	U			
FDP–αGPD					U (2/2)
FDP–IDH-3					U (1/2)
FDP–IDH-1 or *2*					U (1/1)
FDP–LDH-3					U (2/2)
FDP–LDH-1					U (1/1)
FDP–MDH-1					U (1/2)
FDP–ME					U (1/2)
FDP–PGM					U (1/2)
FDP–PGK					U (1/1)
FDP–SDH					U (1/2)
FDP–SOD					U (1/1)
GPI-1–GPI-2				U	
GPI-1–GOT-1					U (2/4)
GPI-1–GOT-3					U (2/2)
GPI-1–GPT					U (1/2)
GPI-1–αGPD					U (3/4)
GPI-1–IDH-1 or *2*	L (IDH1), U (IDH2)	U			U (2/4)
GPI-1–IDH-3		U			
GPI-1–LDH-1	U	U			
GPI-1–LDH-2		U			
GPI-1–LDH-3					U (2/4)
GPI-1–MDH-1					U (2/4)
GPI-1–MDH-3		U			

Table V (*Continued*)

Locus pair	*Xiphophorus*	*Poeciliopsis*	*Poecilia*	Centrarchidae	Salmonidae[b]
GPI-1–ME					U (3/4)
GPI-1–MPI	U				U (1/2)
GPI-1–ODH					U (2/2)
GPI-1–PEP-2	U	U			
GPI-1–PGM	U	U			U (1/4)
GPI-1–6PGD	U	U			
GPI-1–SDH					L (1/2)
GPI-1–SOD		U			
GPI-2–GOT-1					U (1/2)
GPI-2–GOT-3					U (1/1)
GPI-2–GPT					U (1/1)
GPI-2–αGPD					U (2/2)
GPI-2–IDH-1 or *2*					U (1/1)
GPI-2–LDH-3					U (2/2)
GPI-2–MDH-1					U (1/2)
GPI-2–MDH-2					U (1/2)
GPI-2–ME					U (2/2)
GPI-2–MPI					L (1/1)
GPI-2–MP-4					U (1/2)
GPI-2–ODH					U (1/1)
GPI-2–PGM					U (1/2)
GPI-2–SDH					U (1/1)
GPI-2–SOD					U (1/1)
βGUS–GOT-3					U (1/1)
βGUS–αGPD					U (1/2)
βGUS–IDH-1 or *2*					U (1/1)
βGUS–LDH-3					U (2/2)
βGUS–MDH-1					U (2/2)
βGUS–MDH-2					U (1/2)
βGUS–ME					U (2/2)
βGUS–MP-4					L (1/2)
βGUS–SDH					U (1/2)
βGUS–SOD					U (1/1)
GOT-1–GOT-3					U (1/2)
GOT-1–αGPD					L, U (3/4)
GOT-1–IDH-1 or *2*					A, U (1/4)
GOT-1–LDH-3					U (2/4)
GOT-1–MDH-1					U (1/4)
GOT-1–MDH-2					L (1/4)
GOT-1–ME					U (2/4)
GOT-1–MPI					U (1/2)
GOT-1–MP-4					U (1/4)
GOT-1–ODH					U (1/2)
GOT-1–PGM					U (2/4)
GOT-1–SDH					U (2/4)
GOT-1–SOD					U (1/2)

(continued)

Table V (*Continued*)

Locus pair	*Xiphophorus*	*Poeciliopsis*	*Poecilia*	Centrarchidae	Salmonidae[b]
GOT-2–IDH-3		U			
GOT-2–LDH-1		U			
GOT-2–LDH-2		U			
GOT-2–PEP-2		U			
GOT-2–6PGD		U			
GOT-3–GPT					U (1/1)
GOT-3–αGPD					U (2/2)
GOT-3–IDH-1 or *2*					U (1/1)
GOT-3–IDH-3					U (2/2)
GOT-3–MDH-1					U (2/2)
GOT-3–MDH-2					U (1/2)
GOT-3–ME					U (2/2)
GOT-3–MPI					U (1/1)
GOT-3–MP-4					U (1/2)
GOT-3–ODH					U (1/1)
GOT-3–PGM					U (1/2)
GOT-3–SDH					U (1/2)
GOT-3–SOD					U (1/1)
GPT–αGPD					U (2/2)
GPT–IDH-1 or *2*					U (1/1)
GPT–LDH-3					U (2/2)
GPT–MDH-1					U (1/2)
GPT–MDH-2					U (1/2)
GPT–ME					U (2/2)
GPT–MPI					U (1/1)
GPT–MP-4					U (1/2)
GPT–ODH					U (1/1)
GPT–PGM					U (1/2)
GPT–SDH					U (1/2)
GPT–SOD					U (1/1)
GAPD-2–IDH-3		U			
GAPD-2–LDH-1		U			
GAPD-2–LDH-2		U			
αGPD–IDH-1 or *2*					A, U (2/2)
αGPD–IDH-3					U (2/2)
αGPD–LDH-1					U (1/2)
αGPD–LDH-3					A, U (4/4)
αGPD–MDH-1				U	U (3/4)
αGPD–MDH-2				U	P, U (2/4)
αGPD–ME					A, U (4/4)
αGPD–MPI					U (2/2)
αGPD–MP-4					U (2/4)
αGPD–ODH					U (2/2)
αGPD–PGM					U (2/4)
αGPD–6PGD				L	

Table V (*Continued*)

Locus pair	*Xiphophorus*	*Poeciliopsis*	*Poecilia*	Centrarchidae	Salmonidae[b]
αGPD–PGK					U (2/2)
αGPD–SDH					U (2/4)
αGPD–SOD				U	U (2/2)
IDH-1 or *2–IDH-3*		U			U (1/2)
IDH-1 or *2–LDH-1*	U, U	U			
IDH-1 or *2–LDH-2*		U			
IDH-1 or *2–LDH-3*					U (2/4)
IDH-1 or *2–MDH-1*					U (2/4)
IDH-1 or *2–MDH-2*	U, U				U (1/4)
IDH-1 or *2–MDH-3*		U			
IDH-1 or *2–ME*					L, U (2/2)
IDH-1 or *2–MPI*	U, U				U (1/1)
IDH-1 or *2–MP-4*	U, U				U (1/2)
IDH-1 or *2–ODH*					U (1/1)
IDH-1 or *2–PGM*	U, U				U (1/2)
IDH-1 or *2–PGK*	U				U (1/1)
IDH-1 or *2–SDH*					U (1/2)
IDH-1 or *2–SOD*		U			U (1/1)
IDH-3–LDH-1		L			U (1/2)
IDH-3–LDH-2		L			
IDH-3–LDH-3					U (2/4)
IDH-3–MDH-1					U (1/4)
IDH-3–MDH-3		U			
IDH-3–ME					U (1/4)
IDH-3–MP-4		U			
IDH-3–PEP-2		U			
IDH-3–PGM		U			U (1/4)
IDH-3–6PGD		U			
IDH-3–PGK					U (1/2)
IDH-3–SDH					U (1/4)
IDH-3–SOD		U			U (1/2)
LDH-1–LDH-2		L			
LDH-1–LDH-3					U (2/2)
LDH-1–MDH-3		U			
LDH-1–MP-4		U			
LDH-1–PEP-2	U	U			
LDH-1–PGM	U	U			
LDH-1–6PGD	U	U			
LDH-1–PGK					U (1/1)
LDH-1–SOD		U			U (1/1)
LDH-2–MDH-3		U			
LDH-2–MP-4		U			
LDH-2–PEP-2		U			
LDH-2–PGM		U			
LDH-2–6PGD		U			

(continued)

Table V (*Continued*)

Locus pair	*Xiphophorus*	*Poeciliopsis*	*Poecilia*	Centrarchidae	Salmonidae[b]
LDH-2–SOD		U			
LDH-3–MDH-1					U (3/4)
LDH-3–MDH-2					U (2/4)
LDH-3–ME					U (4/4)
LDH-3–MPI					U (2/2)
LDH-3–MP-4					U (2/4)
LDH-3–ODH					U (2/2)
LDH-3–PGM					U (2/4)
LDH-3–PGK					U (1/2)
LDH-3–SDH					U (2/4)
LDH-3–SOD					U (2/2)
MDH-1–MDH-2				U	U (2/4)
MDH-1–ME					U (4/4)
MDH-1–MPI					U (2/2)
MDH-1–ODH					U (2/2)
MDH-1–PGM					U (2/4)
MDH-1–6PGD				U	
MDH-1–PGK					U (1/2)
MDH-1–SDH					U (2/4)
MDH-1–SOD				U	U (1/2)
MDH-2–ME					U (2/4)
MDH-2–MP-4	U				U (1/4)
MDH-2–ODH					U (1/2)
MDH-2–PGM	U		U		U (1/4)
MDH-2–6PGD	U			U	
MDH-2–SDH					U (1/4)
MDH-2–SOD			U	U	U (1/2)
MDH-3–PGM		U			
MDH-3–SOD		U			
ME–MPI					U (1/2)
ME–MP-4					U (2/4)
ME–ODH					U (2/2)
ME–PGM					U (2/4)
ME–PGK					U (1/2)
ME–SDH					U (2/4)
ME–SOD					U (2/2)
MPI–ODH					L (1/1)
MPI–PGM	U				U (1/2)
MPI–SOD	U				U (1/1)
MP-4–ODH					U (1/2)
MP-4–PEP-2	U	U?			
MP-4–PGM	U				U (1/4)
MP-4–6PGD	U	U			
MP-4–PGK					U (1/2)
MP-4–SDH					U (1/4)

Table V (*Continued*)

Locus pair	*Xiphophorus*	*Poeciliopsis*	*Poecilia*	Centrarchidae	Salmonidae[b]
MP-4–SOD					P, U (1/2)
ODH–PGM					U (1/2)
ODH–SDH					U (1/2)
ODH–SOD					U (1/1)
PEP-2–6PGD	U	U			
PGM–PGK	U				U (1/2)
PGM–SDH					U (1/4)
PGM–SOD		L?	U		U (1/2)
6PGD–SOD				U	
PGK–SDH					U (1/2)
PGK–SOD					U (1/2)
SDH–SOD					U (1/2)

Given the variability of *Xiphophorus* species, further crosses should increase the number of comparisons that can be made between the two genera. Additional linkage studies in other poeciliid species should allow precise estimation of linkage group conservation.

2.4. *Poecilia reticulata*, Poeciliidae

It might be expected that the ubiquitous guppy of aquarium hobbyist fame might be genetically well studied, but such unfortunately is not the case. Only four enzyme loci have been studied for linkage (Shami and Beardmore, 1978) and with very small sample sizes. The probable homologous loci in *Xiphophorus* and *Poeciliopsis* are also unlinked (Tables IV and V), with one possible exception: *Pgm* and *To* of *Poeciliopsis* may be linked ($P < 0.05$), but sample error is quite possibly the correct explanation. It seems probable that the large number of inbred strains and wild populations of guppies should provide a rich source of protein variation. Though linkage data from domestic strains long selected for mutant phenotypes must be interpreted with caution, genetic maps of the guppy would greatly extend our knowledge of gene arrangements in poeciliid fishes.

2.5. Freshwater Sunfishes, Centrarchidae

The first linkage investigations utilizing interspecific sunfish hybrids were undertaken to examine the linkage relationships of presumably duplicated loci. Wheat *et al.* (1972) reported the absence of close linkage between supernatant malate dehydrogenase loci *MDH-A* and *MDH-B*.

Somewhat later, no evidence was found for linkage between *GPI-A* and *GPI-B* (Whitt *et al.*, 1976). Unfortunately, similar tests have not yet been made in other fish species (Table V). A third study assessed the linkage relationships of six enzyme loci and found linkage (15% recombination) between an α-glycerophosphate dehydrogenase locus and the 6-phosphogluconate dehydrogenase locus (Wheat and Whitt, 1973). Since *6PGD* has been assigned to *Xiphophorus* LG I, linked to *ADA* and *G6PD*, the mapping of *αGPD* in poeciliids is of considerable interest, as is the mapping of *ADA* and *G6PD* in centrarchids; neither task has yet been accomplished. The only pairs of rather certain homology tested in centrarchids and poeciliids are *MDH-2–SOD* (*Poecilia* and sunfishes) and *MDH2–6PGD* (*Xiphophorus* and sunfishes) (Table V); both pairs are unlinked in the crosses examined. The assessment of linkage conservation between centrarchids and salmonids will be deferred until the following section.

2.6. Trout and Salmon, Salmonidae

Though the tasty carcass byproducts of salmonids are undoubtedly a great incentive, the interpretation of electrophoretic patterns of the tetraploid salmonids is a difficult and challenging task for biochemical geneticists. Because of their considerable economic importance both in commercial and sport fisheries, hybrids between many salmonid species have been produced, and several laboratories have investigated intensively the biochemical genetics of natural and hatchery stocks and their hybrids. The unique phenomenon of pseudolinkage between loci duplicated in a relatively recent tetraploidization event (Ohno, 1970) has spurred a continuing interest in enzyme gene mapping (Wright *et al.*, 1983; May *et al.*, 1979*b*, 1980, 1982).

Over 30 enzyme loci have been utilized in linkage studies (Table V), but the degree of completion of pairwise comparisons is somewhat low. Several crosses have been constructed to assess the linkage relationships of presumably duplicated loci. Many of these studies suffer from the absence of knowledge of the linkage phase of the alternate alleles of the presumably duplicate loci; thus, parental and recombinant types cannot be identified *per se*. Aspinwall (1974) reported nonrandom segregations in crosses of pink salmon (*Oncorhynchus gorbuscha*) at two muscle malate dehydrogenase loci, which he considered to be duplicated *Mdh-A* loci; May *et al.* (1979*b*) consider these pink salmon variants to be at *Mdh-B* loci (*Mdh-3,4*). Clayton *et al.* (1975) also reported aberrant progeny distributions of *Mdh-3* and *-4* in rainbow trout (*Salmo gairdneri*), while Allendorf (1975) observed independent assortment between *Mdh-3* and *-4* in rainbow trout. In none of these studies was linkage phase known. May *et al.* (1979*b*) found evidence for pseudolinkage (excess of nonparental types) between *Mdh-3* and *Mdh-4* in brook trout (*Salvelinus fontinalis*) × lake trout (*S. namaycush*) hybrids, in a study of linkage between *Mdh-B* loci where linkage phase was known. In contrast to the pseudolinkage observed using splake hybrids, random assortment was observed in pure brook trout crosses (male heterozygous).

Other loci representing probable duplicated pairs also exhibit nonrandom segregations in salmonid crosses. The duplicated muscle *Got* loci (*Got-1,2* or *Aat-1,2*) exhibit deviant joint segregations in cutthroat trout (*Salmo clarki*), linkage phase unknown (Allendorf and Utter, 1976), and brook trout (*Salvelinus fontinalis*), male heterozyous, linkage phase unknown (May *et al.*, 1979*b*). In crosses involving females heterozygous for *Got-1* and *-2* in brook trout (May *et al.*, 1979*b*) and heterozygous males of chum salmon, *Oncorhynchus keta* (May *et al.*, 1975), random segregation is observed. May *et al.* (1979*b*) postulate that, although disomic inheritance certainly is the rule among the salmonid loci thus far tested for segregation, the *Mdh-3,4* and *Aat-1,2* pairs may be examples indicative of "residual tetrasomy." Such a phenomenon, if operative, could prevent the fixation of alternate alleles at the duplicate loci involved.

The most studied example of pseudolinkage in salmonids is that between *Ldh-3* and *-4* (heart and liver lactate dehydrogenases). The two loci were first reported to be linked by Morrison and Wright (1966) in crosses using splake (brook × lake trout) hybrids. Pseudolinkage was first demonstrated by Morrison (1970) in splake crosses of known linkage phase; an excess of recombinants in males and independent assortment in females was observed. Most importantly, a cytogenetic analysis of these hybrids suggested a mechanism of spontaneous, selective Robertsonian centric fusions and fissions (M. J. Davisson *et al.*, 1973). Subsequent studies demonstrated *Ldh-3* and *-4* to be pseudolinked in crosses of pure brook trout (M. J. Davisson *et al.*, 1973) and rainbow trout (J. E. Wright *et al.*, 1975).

Only two other pairs of duplicate loci have been analyzed for linkage in salmonids. Independent assortment was observed for malic enzyme (*Me-1,2*) loci in crosses derived from splake hybrid males (Stoneking *et al.*, 1979). Joint segregation of *Mdh-1* and *-2* has been studied in brown trout (*Salmo trutta*) crosses; independent assortment was demonstrated in crosses derived both from heterozygous males and females (May *et al.*, 1979*a*).

Though pesudolinkage has most often been demonstrated between presumably duplicated loci, the phenomenon has been documented in creatine kinase-diaphorase and creatine kinase–superoxide dismutase pairs (May *et al.*, 1979*a*). The widespread occurrence of pseudolinkage in salmonids [excellently reviewed by J. E. Wright *et al.* (1983)] and the variability of progeny ratios produced by different heterozygous males severely diminishes the usefulness of linkage data from crosses of unknown linkage phase. The "associations" (nonrandom joint segregations) in Table V thus must be viewed only as indications of locus pairs requiring further study to establish whether classical linkage exists between the genes. The nonrandom association of *Agp-1* with at least six other enzyme loci well illustrates how cautiously linkage data must be interpreted in salmonid crosses of unknown linkage phase.

Enzyme locus homologies as assessed primarily by similarities in tissue distribution and subcellular localization are of necessity somewhat more uncertain when comparing the often duplicate loci of salmonids with diploid fish taxa. Thus, the homologies proposed in Table IV are at present only best guesses, though in most cases probably are correct.

Given the difficulties in determining linkage phase and haploid chromosome arm numbers often more than 50 in salmonids, it is not surprising that relatively few classical linkages have been detected (Table V). A cursory examination of locus pairs reveals that only a few probably homologous genes have been mapped in both poeciliids and salmonids. However, two cases of possibly homologous linkages deserve special attention.

As we have described, *αGPD* and *6PGD* are linked in sunfishes and *ADA* and *6PGD* are linked in *Xiphophorus* LG I. The linkage of *ADA-1* with an αGPD locus (*AGP-2* or *G3P-3*) suggests a similar linkage relationship in salmonids. More conclusive evidence will be obtained by mapping *G6PD* and *6PGD* in salmonids, *ADA* and *G6PD* in centrarchids, and *αGPD* in *Xiphophorus*. We have found a rare *αGPD* variant in *X. variatus*, which we will attempt to reacquire from wild populations.

The second salmonid linkage suggesting homology with poeciliids is the classical linkage of *Ldh-1*, an *Ldh-A* locus, with *Idh-1*, presumably a mitochondrial *Idh* locus. We have previously noted a similar linkage in *Poeciliopsis* LG I, homologous with *Xiphophorus* LG II. The association of *Ldh-5* (*Ldh-C*) with *Ldh-1* and *Idh-1* in salmonids (unfortunately, in crosses of unknown linkage phase) perhaps indicates the same linkage relationship as in *Xiphophorus*. On the other hand, *MPI*, a member of *Xiphophorus* LG II, appears to be unlinked to the salmonid genes presumed homologous to *Xiphophorus* LG II (J. E. Wright *et al.*, 1983). Since *MPI* was probably once duplicated and one copy silenced, such a lack of homology must be interpreted with caution.

Table V attempts to summarize pairwise recombination tests in fish species other than *Xiphophorus*, for which only those pairs representing tests of probable homologies are included. Table V has two major uses:

first, to illustrate the great extent of similar independent assortment relationships among probably homologous locus pairs, and second, to provide a summary of locus pairs thus far studied in fishes to workers beginning linkage analyses. For details regarding particular locus pairs, researchers should refer to the primary literature.

Only rarely can independent assortment be compared between homologous locus pairs in salmonids and other fish species. In only a few instances have both members of a pair of duplicated loci been tested against other loci (or pairs of loci) for linkage. This incompleteness of pairwise comparisons makes it impossible at this point to assess meaningfully the evolutionary conservation of linkage groups between salmonids and other fish groups. The discovery of new variants and the performance of linkage tests among them will continue to chip away at the problems that now exist in comparative gene mapping in fishes.

3. Comparisons of Linkage Groups of Fishes with Other Vertebrates

Comparative gene mapping among vertebrate classes is rendered extremely difficult by the lack of critical data assessing homology of proteins, as was seen to be true to a lesser extent even in fishes. In perhaps a majority of cases, a different number of loci code for a given enzyme in mammals than in, e.g., *Xiphophorus*. Mammals seem to have only one locus coding for glyceraldehyde-3-phosphate dehydrogenase; *Xiphophorus* has at least three. Thus, to establish nonhomology of a fish linkage group with a *GAPD*-containing mammalian linkage group requires the mapping of all three *GAPD* loci in the fish. This ubiquitous difficulty reduces drastically the body of usefully complete comparative data.

Homology difficulties notwithstanding, a few suggestive gene arrangments exist that indicate the possibility of stable linkage relationships over many millions of years of evolutionary divergence. For example, an interesting gene mapping parallel involves the *G6PD* (*H6PD*)–*6PGD* linkage. Both *G6PD* and *6PGD* are linked relatively closely in *Xiphophorus* LG I (see above), and appear to be linked in the frog *Rana pipiens* (D. A. Wright, 1975). In mammals there are two forms of G6PD: a sex-linked (Childs *et al.*, 1958), highly specific form and an autosomally coded form with a broad enough substrate specificity to be able to metabolize galactose-6-phosphate (Shaw and Koen, 1968); the latter form has been referred to as hexose-6-phosphate dehydrogenase (H6PD). Chapman (1975) has reported linkage between the locus coding for H6PD (*Gpd-1* in mouse nomenclature) and *6PGD* in mice. The *H6PD* locus has been identified in man (Shaw, 1966; Ohno *et al.*, 1966) and has recently been assigned to human chromosome 1, syntenic with *6PGD* (Hameister *et al.*, 1978). If the "*G6PD*" of *Xiphophorus* and *Rana* is homologous to the *H6PD* of mammals, long-term stability of the linkage to *6PGD* possibly is indicated. The third member of *Xiphophorus* LG I, *ADA* (possibly exhibiting homologous linkages in other fishes as outlined previously), has been mapped to chromosome 20 in man (Tischfield *et al.*, 1974), and thus apparently is not syntenic with *H6PD, G6PD,* or *6PGD*.

Turning to salmonid gene maps, the linkage of two GPI loci with *Pep-1* and *Pep-2* (the two groups being also further associated by pseudolinkage), as reported by J. E. Wright *et al.* (1983), perhaps represents another case of fish–mammal linkage group homology. The *Pep-1* and *Pep-2* in salmonids code for prolidases and thus are probably homologous to mammalian *PEP-D,* which is linked to *GPI* on human chromosome 19, mouse chromosome 17 (Lalley *et al.*, 1978*a*), and in Chinese hamster (Athwal *et al.*, 1979).

Perhaps the most compelling evidence for long-term evolutionary stability of a linkage group comes from a search for possible homologies to the composite linkage group of *Xiphophorus* LG II and *Poeciliopsis* LG I described earlier. Let us first consider the linkage relationships of lactate dehydrogenase loci in vertebrates. In *Xiphophorus* we have demonstrated the linkage of *LDH-1* (= fish *LDH-C*) to *MPI* in LG II. The *LDH-C* in fishes codes for a retinal (or sometimes a liver) isozyme derived from an *LDH-B* duplication and is perhaps homologous to the *LDH-C* gene expressed in testes of birds and mammals (Markert *et al.*, 1975). In *Poeciliopsis LDH-1* and *LDH-A* (or *Ldh-2*) are linked, and in frogs of the genus *Rana LDH-B* and *MPI* are linked (D. A. Wright and Richards, 1980). Loci *LDH-B* and *LDH-C* are tightly linked in pigeons (Zinkham *et al.*, 1969). These linkages suggest the duplication and reduplication of an LDH locus ancestrally linked to *MPI;* in particular, it is perhaps conservative to extrapolate that the *LDH-B–LDH-C* duplication occurred in an ancestor in which the duplicated *LDH-A* and *LDH-B* loci still were linked.

Assessment of loci syntenic with *MPI* on mammalian chromosomes (chromosome 15 in man and chromosome 9 in mouse) provides additional evidence suggesting evolutionary stability. In man *MPI* is syntenic with loci coding for a pyruvate kinase isozyme (M2) and mitochondrial isocitrate dehydrogenase (Shows and McAlpine, 1978). In the mouse *MPI* is syntenic with a locus coding for PK-3, homologous to the human PK-M2 isozyme (Lalley *et al.*, 1978*b*); this synteny has been reported also in the cat (O'Brien and Nash, 1979) and pig (Gellin *et al.*, 1979). The mitochondrial isocitrate dehydrogenase locus (*Id-2*) of the mouse is found to be on chromosome 7 with *LDH-A* (Lalley *et al.*, 1978*a*). *LDH-A* in man is located on chromosome 11 and *LDH-B* on chromosome 12 (Boone *et al.*, 1972; Chen *et al.*, 1973); unfortunately, the testis-specific *LDH-C* locus has not yet been mapped in mammals. *MPI*-containing linkage groups or syntenic associations probably homologous to human chromosome 15 have been identified in the rat and in at least two great ape species (Pearson and Roderick, 1978).

It is difficult for us to attribute to a rather remarkable coincidence that in *Xiphophorus* LG II a muscle pyruvate kinase locus [*PK-2*, probably homologous to the human kidney–muscle pyruvate kinases M1 and M2 (Morizot, 1983)] is linked to *MPI* and *LDH-C*, and that in *Poeciliopsis* LG I and probably in the salmonids (J. E. Wright *et al.*, 1983) the mitochondrial isocitrate dehydrogenase locus (*Idh-2*) is linked to *LDH-C* and *LDH-A*. It seems much more reasonable to suppose that the limited data available in fish, frog, pigeon, rat, mouse, ape and man indicate an ancestral vertebrate gene arrangement and illustrate patterns of divergence during the passage of evolutionary time. It is noteworthy that the probable asyntenies that have been reported are all in species with more than 48 chromosomes; *Ldh-B* unlinked in salmonid fishes (Table V), and *PK* and *MPI* unlinked in sheep and cattle (Saidi *et al.*, 1979; Hors-Cayla *et al.*, 1979).

In conclusion, we should emphasize that we are not suggesting that most, or even many, gene arrangements are stably maintained over millions of years of evolutionary time due to as yet undefined functional or structural constraints. There are far too few data available to estimate meaningfully the rate of evolutionary divergence of genetic linkage relationships. What we do suggest is that the current rapid expansion of gene maps of enzyme loci in fishes, amphibians, and mammals (paralleled, of course, in many nonvertebrate species) promises to make possible the reconstruction of ancestral gene arrangements in the very near future. Instances of extremely conserved linkage relationships may, as in the case of highly conserved amino acid sequences of some proteins, imply functional significance on a structural or regulatory level.

The question of how many certain nonhomologies can be demonstrated between fish and mammal linkage groups perhaps is best addressed by examples. It might reasonably be asked if the members of the X-chromosome group of loci highly conserved among mammalian species (Pearson and Roderick, 1978) could not be compared for linkage or independent assortment with the presumably homologous loci in *Xiphophorus*. Several difficulties arise in attempting such a comparison. First, the X-chromosome inactivation and Y-chromosome partial nullisomy characteristic of mammals are not known to occur in *Xiphophorus* (Kallman, 1975); rather, the X and Y fish chromosomes are completely homologous as far as is known. Second, and a purely fish problem, is that since many electrophoretic variants can be studied only in interspecific crosses where assignment of biological sex of an individual is often equivocal, direct linkage analyses of biochemical markers with sex are frequently difficult. The usual alternative of analyzing linkage relationships between biochemical loci and sex-linked pigment pattern genes is not completely satisfying: with the present sample size ranges it is impossible to conclude that a locus is not loosely linked to a pigment pattern gene with any degree of certainty. Without cytogenetic information about the probable map length of the *Xiphophorus* sex chromosomes, independent assortment data must remain inconclusive until an extremely large number of loci have been analyzed for linkage. The third problem complicating analyses of homology of sex linkage in fishes and mammals is the diversity of methods used to map genes. In man, for example, by using deletion and translocation mapping, the gene order and fairly precise regional localization of *G6PD, PGK, HPRT,* and *αGAL* has been established (Miller *et al.,* 1978). In the mouse, classical mapping methods have been utilized until very recently, and regional localization using chromosome banding techniques has only rarely been achieved (M. T. Davisson and Roderick, 1978). Thus, comparative mapping even between man and mouse is often fraught with difficulties: the mouse X is acrocentric, the human X submetacentric (Ohno, 1967); if banding patterns are not obviously related, comparison of gene-to-band maps is extremely difficult. It should be apparent that the ease with which genes are mapped to the X chromosome in mammals is due in large measure to their hemizygous expression in males. In species such as fish, where such gene dosage phenomena are absent, classical linkage with a sex-linked marker must be established; to date, this has not been accomplished for any enzyme locus. Much more data are required before the conservation of sex-linked groups of enzyme loci throughout vertebrates can be confirmed or contradicted. All that can be said with assurance is that in *Xiphophorus* the sex-linked pigment pattern *Sd, PGK,* and *αGAL* are not closely linked (Table III).

It is hoped that the preceding, rather unsatisfying attempt at sex linkage comparisons illustrates some of the many difficulties in reaching firm conclusions of nonconservation of linkage relationships between groups as evolutionarily divergent as fish and mammals. The only satisfactory basis for analyzing the change in linkage relationships over evolutionary time is the comparison of very large numbers of certainly homologous loci in a large number of vertebrate species. Such a goal is not impossible to achieve. The current rapid expansion of gene maps of enzyme loci in fishes, amphibians, and mammals promises to make much more meaningful comparisons of gene arrangements possible in the very near future.

4. Potential for Expansion of Linkage Maps in Fishes

Gene mapping studies in fishes have yielded respectable bodies of data only in systems utilizing interspecific hybrids—*Xiphophorus, Poeciliopsis, Lepomis, Salmo, Salvelinus,* and *Oncorhynchus*. The relatively high level of electrophoretic variation detectable between fish species assures at least some success in any interspecific hybrid system where reasonable numbers of backcross hybrids can be produced. The question should be asked, then, whether the listed genera represent very unusual instances where fertile F_1 hybrids are produced by crossing congeneric species. The answer is an unqualified no. Schwarz (1972) has compiled a bibliography of over 1800 papers relating to fish hybrids belonging to numerous genera of a large number of different families. The breeding of hybrids has been extremely important both in the aquarium trade and in commercial fisheries research and development. Thus, it appears that the major reason for the paucity of linkage data in fishes is not the lack of producible hybrids, but rather the lack of use of such hybrids for genetic analysis, particularly with respect to mapping of enzyme loci. We are aware of at least one instance where large numbers of different killifish hybrids, including backcrosses, were produced but ended up in vials of formalin with apparently little or no consideration as to their possible genetic value. It is hoped that researchers will become aware of this potential contribution of interspecific fish hybrids to the broad area of comparative gene mapping. Populations of backcross or F_2 hybrids and their parents should be brought to the attention of one of the biochemical genetics laboratories engaged in fish research in every possible instance.

From our research in *Xiphophorus* we can conclude that the observed degree of genetic variation in fishes coupled with the capacity to resolve the products of over 100 different genes should allow the construction of extensive linkage maps in any genus where fertile interspecific hybrids can be produced. Even crosses within species can be expected to yield a substantial body of data, particularly when isolated populations are used as parental stocks. A large data base comparable to the mammalian data could be generated in only a few years by a few laboratories if the needed quantity and variety of biological material were available.

5. Uses of Linkage Maps

Detailed linkage maps in a variety of species promise to allow the identification with considerable precision of the types of chromosomal rearrangements that have accompanied speciation events. The tracing of elements of *Xiphophorus* LG II in other fishes, frogs, birds, and mammals raises the possibility that divergence of many linkage groups may be slow enough to allow extrapolation of the gene arrangements of ancestral vertebrate groups. Whether conservation of linked genes will prove to indicate common or related regulatory mechanisms remains a matter for conjecture. In any event, it seems very probable that comparative gene mapping will yield important insights into the organization of the vertebrate genome.

We are particularly interested in the use of linkage maps to detect the number and location of other genes. In *Xiphophorus* genetic factors responsible for control of hybrid melanoma severity (Siciliano *et al.*, 1976; Ahuja *et al.*, 1980; Morizot and Siciliano, 1983*a*) and carcinogen susceptibility (Schwab *et al.*, 1979) have been detected through linkage to biochemical and pigment pattern markers. Many other such uses can be envisioned. Improved estimates of risk of a child of inheriting a genetically transmitted disease could be made if the genotypes of polymorphic linked loci could be determined in the pedigree. Genes involved in the inheritance of polygenic characters could possibly be identified by linkage to mapped markers. This last prospect may prove to be particularly important in fishes: perhaps the genetic bases for characters often used in systematics can at least be determined. A unified taxonomy could result from such discoveries.

Lastly, linkage maps promise to become increasingly useful in predicting gene arrangements in other species as rates of evolutionary divergence become more accurately estimated. Correct predictions from man to mouse have already been made (Lalley *et al.*, 1978*d*); some authors even believe that predictiveness of linkage in mammals is generally highly correlated with recombination distance (Pearson and Roderick, 1978). Only time will tell whether linkage group divergence reflects a regular evolutionary "clock," but many such fundamental assessments will be able to be made in the near future.

ACKNOWLEDGMENT. This work was supported by U.S. Public Health Service Grant CA-28909.

References

Abbott, U. K., and Yee, G. W., 1975, Avian genetics, in: *Handbook of Genetics,* Vol. 4 (R. C. King, ed.), Plenum Press, New York, pp. 151–200.

Ahuja, M. R., Schwab, M., and Anders, F., 1980, Linkage between a regulatory locus for melanoma cell differentiation and an esterase locus in *Xiphophorus, J. Hered.* **71**:403–407.

Allendorf, F. W., 1975, Genetic variability in a species possessing extensive gene duplication: Genetic interpretation of duplicate loci and examination of genetic variations in populations of rainbow trout, Ph. D. dissertation, University of Washington, Seattle, Washington.

Allendorf, F. W., and Utter, F. M., 1976, Gene duplication in the family Salmonidae. III. Evidence of linkage between two duplicated loci coding for aspartate aminotransferase in the cutthroat trout (*Salmo clarki*), *Hereditas* **82:**19–24.

Allendorf, F. W., Utter, F. M., and May, B. P., 1975, Gene duplication within the family Salmonidae: II. Detection and determination of the genetic control of duplicate loci through inheritance studies, in: *Isozymes IV* (C. L. Markert, ed.), Academic Press, New York, pp. 415–432.

Allendorf, F. W., Mitchell, N., Ryman, N., and Stahl, G., 1977, Isozyme loci in brown trout (*Salmo trutta* L.): Detection and interpretation from population data, *Hereditas* **86:**179–190.

Aspinwall, N., 1974, Genetic analysis of duplicate malate dehydrogenase loci in the pink salmon, *Oncorhynchus gorbuscha, Genetics* **76:**65–72.

Athwal, R., Minna, J. D., and McBride, O. W., 1979, Comparative gene mapping of seven Chinese hamster genes assigned to human chromosomes 1p, 15, 19, and 6 and to mouse chromosomes 4, 7, and 9 using somatic cell hybrids and gene transfer clones, *Cytogenet. Cell Genet.* **25:**132–133.

Bailey, N. T. J., 1961, *Introduction to the Mathematical Theory of Genetic Linkage,* Oxford University Press, London.

Boone, C., Chen, T.-R., and Ruddle, F. H., 1972, Assignment of three human genes to chromosomes (LDH-A to 11, TK to 17, and IDH to 20) and evidence for translocation between human and mouse chromosomes in somatic cell hybrids, *Proc. Natl. Acad. Sci. USA* **69:**510–514.

Bruns, G. A. P., Pierce, P., Regina, V. M., and Gerald, P. S., 1978, Expression of GADPH and TPI in dog–rodent hybrids, *Cytogenet. Cell Genet.* **22:**547–551.

Champion, M. J., and Whitt, G. S., 1976, Synchronous allelic expression at the glucose-phosphate isomerase A and B loci in interspecific sunfish hybrids, *Biochem. Genet.* **14:**723–737.

Chapman, V. M., 1975, 6-Phosphogluconate dehydrogenase (PGD) genetics in the mouse: Linkage with metabolically related enzyme loci, *Biochem. Genet.* **13:**849–856.

Chapman, V. M., and Shows, T. B., 1976, Somatic cell genetic evidence for X-chromosome linkage of three enzymes in the mouse, *Nature* **259:**665–667.

Chen, S.-H., Malcolm, L. A., Yoshida, A., and Giblett, E. R., 1971, Phosphoglycerate kinase: An X-linked polymorphism in man, *Am. J. Hum. Genet.* **23:**87–91.

Chen, T.-R., McMorris, R. A., Creagan, R., Ricciuti, F., Tischfield, J., and Ruddle, F. H., 1973, Assignment of the genes for malate oxidoreductase decarboxylating to chromosome 6 and peptidase B and lactate dehydrogenase B to chromosome 12 in man, *Am. J. Hum. Genet.* **25:**200–207.

Childs, B., Zinkham, W., Browne, E. A., Kimbro, E. L., and Torbet, J. W., 1958, A genetic study of a defect in glutathione metabolism of the erythrocyte, *Bull. Johns Hopkins Hosp.* **102:**21–37.

Clayton, J. W., Tretiak, D. N., Billeck, B. N., and Ihssen, P., 1975, Genetics of multiple supernatant and mitochondrial malate dehydrogenase isozymes in rainbow trout (*Salmo gairdneri*), in: *Isozymes IV: Genetics and Evolution* (C. L. Markert, ed.), Academic Press, New York, pp. 433–488.

Commission on Biological Nomenclature, 1978, *Enzyme Nomenclature,* Academic Press, New York.

Cook, P. R., 1975, Linkage of the loci for glucose-6-phosphate dehydrogenase and for inosinic acid pyrophosphorylase to the X chromosome of the field vole *Microtus agrestis*, *J. Cell Sci.* **17**:95–112.

Davisson, M. J., Wright, J. E., and Atherton, L. A., 1973, Cytogenetic analysis of pseudolinkage of LDH loci in the teleost genus *Salvelinus*, *Genetics* **73**:645–658.

Davisson, M. T., and Roderick, T. H., 1978, Status of the linkage map of the mouse, *Cytogenet. Cell Genet.* **22**:552–557.

Donald, J. A., and Hope, R. M., 1981, Mapping a marsupial X chromosome using kangaroo–mouse somatic cell hybrids, *Cytogenet. Cell Genet.* **29**:127–137.

Donald, L. J., and Hamerton, J. L., 1978, A summary of the human gene map, 1973–1977, *Cytogenet. Cell Genet.* **22**:5–11.

Echard, G., Gellin, J., Benne, F., and Gillois, M., 1981, The gene map of the rabbit (*Oryctolagus cuniculus* L.). I. Synteny between the rabbit gene loci coding for HPRT, PGK, G6PD, and GLA: Their localization on the X chromosome, *Cytogenet. Cell Genet.* **29**:176–183.

Elandt-Johnson, R. C., 1971, *Probability Models and Statistical Methods in Genetics*, Wiley, New York.

Epstein, C. J., 1969, Mammalian oocytes: X chromosome activity, *Science* **163**:1078–1079.

Epstein, C. J., 1972, Expression of the mammalian X chromosome before and after fertilization, *Science* **175**:1467–1468.

Evans, H. J., and Atwood, K. C., 1978, *In situ* hybridization, *Cytogenet. Cell Genet.* **22**:146–149.

Fischer, S. E., Shaklee, J. B., Ferris, S. D., and Whitt, G. S., 1980, Evolution of five multilocus isozyme systems in the chordates, *Genetica* **52/53**:73–85.

Garver, J. J., Pearson, P. L., Estop, A., Dijksman, T. M., Wijnen, L. M. M., Westerveld, A., and Meera Khan, P., 1978, Gene assignments to the presumptive homologs of human chromosomes 1, 6, 11, 12, and X in the Pongidae and Cercopithecoidea, *Cytogenet. Cell Genet.* **22**:564–569.

Gellin, J., Benne, F., Renard, C., Vaiman, M., Hors-Cayla, M. C., and Gillois, M., 1979, Pig gene mapping: Synteny, attempt to assign the histocompatibility complex (SLA), *Cytogenet. Cell Genet.* **25**:159.

Gold, J. R., 1979, Cytogenetics, in: *Fish Physiology*, Vol. 8 (W. S. Hoar, D. J. Randall, and J. R. Brett, eds.), Academic Press, New York, pp. 353–405.

Hameister, H., Rogers, H.-H., and Grzeschik, K. H., 1978, Assignment of the gene for human glucose dehydrogenase (E.C. 1.1.1.47) to chromosome 1 using somatic cell hybrids, *Cytogenet. Cell Genet.* **22**:200–202.

Harris, H., and Hopkinson, D. A., 1977, *Handbook of Enzyme Electrophoresis in Human Genetics*, American Elsevier, New York.

Hjorth, J. P., 1971, Genetics of *Zoarces* populations. I. Three loci determining the phosphoglucomutase isoenzymes in brain tissue, *Hereditas* **69**:233–242.

Hors-Cayla, M. C., Heuertz, S., Van Cong, N., and Benne, F., 1979, Cattle gene mapping by somatic cell hybridization, *Cytogenet. Cell Genet.* **25**:165–166.

Kallman, K. D., 1975, The platyfish, *Xiphophorus maculatus*, in: *Handbook of Genetics*, Vol. 4 (R. C. King, ed.), Plenum Press, New York, pp. 81–132.

Kallman, K. D., and Atz, J. W., 1966, Gene and chromosome homology in fishes of the genus *Xiphophorus*, *Zoologica* **55**:1–16.

Lalley, P. A., Francke, U., and Minna, J. D., 1978*a*, Assignment of the genes coding for pyrophosphatase and hexokinase-1 to mouse chromosome 10: Implications for comparative gene mapping in man and mouse, *Cytogenet. Cell Genet.* **22**:570–572.

Lalley, P. A., Francke, U., and Minna, J. D., 1978*b*, Comparative gene mapping: the linkage relationships of the homologous genes for phosphoglucomutase and peptidase S are conserved in man and mouse, *Cytogenet. Cell Genet.* **22**:573–576.

Lalley, P. A., Francke, U., and Minna, J. D., 1978*c*, Comparative gene mapping in man and mouse: Assignment of the genes for lactate dehydrogenase-A, peptidase-D, and isocitrate dehydrogenase-2 to mouse chromosome 7, *Cytogenet. Cell Genet.* **22:**577–580.

Lalley, P. A., Francke, U., and Minna, J. D., 1978*d,* The genes coding for pyruvate kinase (M2) and mannosephosphate isomerase are linked in man and mouse, *Cytogenet. Cell Genet.* **22:**581–584.

Leslie, J. F., 1982, Linkage analysis of seventeen loci in poeciliid fish (Genus: *Poeciliopsis*), *J. Hered.* **73:**19–23.

Leslie, J. F., and Pontier, P. J., 1980, Linkage conservation of homologous esterase loci in fish (Cyprinodontoidei: Poeciliidae), *Biochem. Genet.* **18:**103–115.

Leslie, J. F., and Vrijenhoek, R. C., 1977, Genetic analysis of natural populations of *Poeciliopsis monacha:* Allozyme inheritance and pattern of mating, *J. Heredity* **68:**301–306.

Leslie, J. F., and Vrijenhoek, R. C., 1978, Genetic dissection of clonally inherited genomes of Poeciliopsis. I. Linkage analysis and preliminary assessment of deleterious gene loads, *Genetics* **90:**801–811.

Markert, C. L., Shaklee, J. B., and Whitt, G. S., 1975, Evolution of a gene, *Science* **189:**102–114.

Mathai, C. K., Ohno, S., and Beutler, E., 1966, Sex-linkage of the glucose-6-phosphate dehydrogenase gene in *Equidae, Nature* **210:**115–116.

Mather, K., 1957, *The Measurement of Linkage in Heredity,* Methuen, London.

May, B., Utter, F. M., and Allendorf, F. W., 1975, Biochemical genetic variation in pink and chum salmon: Inheritance of intraspecies variation and apparent absence of interspecies introgression following massive hybridization of hatchery stocks, *J. Hered.* **66:**227–232.

May, B., Stoneking, M., and Wright, Jr., J. E., 1979*a,* Joint segregation of malate dehydrogenase and diaphorase loci in brown trout (*Salmo trutta*), *Trans. Am. Fish. Soc.* **108:**373–377.

May, B., Wright, J. E., and Stoneking, M., 1979*b,* Joint segregation of biochemical loci in Salmonidae: Results from experiments with *Salvelinus* and review of the literature on other species, *J. Fish. Res. Board Can.* **36:**1114–1128.

May, B., Stoneking, M., and Wright, Jr., J. E., 1980, Joint segregation of biochemical loci in Salmonidae. II. Linkage associations from a hybridized *Salvelinus* genome (*S. namaycush* × *S. fontinalis*), *Genetics* **95:**707–726.

May, B., Wright, Jr., J. E., and Johnson, K. R., 1982, Joint segregation of biochemical loci in Salmonidae. III. Linkage associations in Salmonidae including data from rainbow trout (*Salmo gairdneri*), *Biochem. Genet.* **20:**29–40.

Meera Kahn, P., Vriesendorp, H., Saisson, R., Volkers, W. S., Los, W. R. T., and Doppert, B. A., 1978, Linkage between PGM_3 and the genes determining the major histocompatibility complex (MHC) in *Canis familiaris* (the domestic dog), *Cytogenet. Cell Genet.* **22:**585–587.

Miller, O. J., Sanger, R., and Siniscalco, M., 1978, Report of the committee on the genetic constitution of the X and Y chromosomes, *Cytogenet. Cell Genet.* **22:**124–128.

Morizot, D. C., 1983, Tracing linkage groups from fishes to mammals, *J. Hered.,* in press.

Morizot, D. C., and Siciliano, M. J., 1979, Polymorphisms, linkage and mapping of four enzyme loci in the fish genus *Xiphophorus* (Poeciliidae), *Genetics* **93:**947–960.

Morizot, D. C., and Siciliano, M. J., 1982*a,* Protein variability in fishes of the genus *Xiphophorus* (Poeciliidae) and segregation of electrophoretic alleles in intra- and interspecific hybrids, *Genetics* **102:**539–556.

Morizot, D. C., and Siciliano, M. J., 1982*b,* Linkage of *guanylate kinase-2* and *glyceraldehyde-3-phosphate dehydrogenase-1* in fishes of the genus *Xiphophorus* (Poeciliidae): Designation as linkage group III, *J. Hered.* **73:**163–167.

Morizot, D. C., and Siciliano, M. J., 1982*c,* Linkage group IV of fishes of the genus *Xiph-*

ophorus (Poeciliidae): Assignment of loci coding for pyruvate kinase-1, glucosephosphate isomerase-1, and isocitrate dehydrogenase-1, *Biochem. Genet.* **20:**505–518.

Morizot, D. C., and Siciliano, M. J., 1983*a*, Linkage group V of platyfishes and swordtails of the genus *Xiphophorus* (Poeciliidae): Linkage of loci for malate dehydrogenase-2 and esterase-1 and esterase-4 with a gene controlling the severity of hybrid melanomas, *J. Natl. Cancer Inst.* **71:**809–813.

Morizot, D. C., and Sicilinao, M. J., 1983*b*, Linkage group VI of fishes of the genus *Xiphophourus* (Poeciliidae): Assignment of genes coding for glutamine synthetase, uridine monophosphate kinase, and transferrin, *Biochem. Genet.*, in press.

Morizot, D. C., Wright, D. A., and Siciliano, M. J., 1977, Three linked enzyme loci in fishes: Implications in the evolution of vertebrate chromosomes, *Genetics* **86:**645–656.

Morrison, W. J., 1970, Non-random segregation of two lactate dehydrogenase subunit loci in trout, *Trans. Am. Fish. Soc.* **99:**193–206.

Morrison, W. J., and Wright, J. E., 1966, Genetic analysis of three lactate dehydrogenase isozyme systems in trout: Evidence for linkage of genes coding subunits A and B, *J. Exp. Zool.* **163:**259–270.

Motulsky, A. G., and Yoshida, A., 1969, Methods for the study of red cell glucose-6-phosphate dehydrogenase, in: *Biochemical Methods in Red Cell Genetics* (J. J. Yunis, ed.), Academic Press, New York, pp. 52–93.

Nyhan, W. L., 1968, Clinical features of the Lesch-Nyhan syndrome. Introduction—clinical and genetic features, *Fed. Proc.* **27:**1027–1033.

O'Brien, S. J., and Nash, W. G., 1979, A biochemical genetic map of the domestic cat (*Felis catus*), *Cytogenet. Cell Genet.* **25:**192–193.

Ohno, S., 1967, *Sex Chromosomes and Sex-Linked Genes,* Springer-Verlag, New York.

Ohno, S., 1970, *Evolution by Gene Duplication,* Springer-Verlag, Berlin.

Ohno, S., 1973, Ancient linkage groups and frozen accidents, *Nature* **244:**259–262.

Ohno, S., Poole, J., and Gustavsson, I., 1965, Sex-linkage of erythrocyte glucose-6-phosphate dehydrogenase in two species of wild hares, *Science* **150:**1737–1738.

Ohno, S., Payne, H. W., Morrison, M., and Beutler, E., 1966, Hexose-6-phosphate dehydrogenase polymorphism, *Science* **153:**1015–1016.

Ohno, S., Muramoto, J., Klein, J., and Atkin, N. B., 1969, Diploidtetraploid relationship in clupeoid and salmonoid fish, in: *Chromosomes Today,* Vol. 2 (C. D. Darlington and K. R. Lewis, eds.), Oliver and Boyd, Edinburgh, pp. 139–147.

Opitz, J. M., Stiles, F. C., Wise, D., Race, R. R., Sanger, R., von Gemmingen, G. R., Kierland, R. R., Cross, E. G., and de Groot, W. P., 1965, The genetics of angiokeratoma corporis diffusum (Fabry's disease) and its linkage relations with the Xg locus, *Am. J. Hum. Genet.* **17:**325–342.

Pathak, S., and Stock, A. D., 1974, The X chromosomes of mammals: Karyological homology as revealed by banding techniques, *Genetics* **78:**703–714.

Pearson, P. L., and Roderick, T. H., 1978, Report of the committee on comparative mapping, *Cytogenet. Cell Genet.* **22:**150–162.

Richardson, B. J., Czuppon, A. B., and Sharman, G. B., 1971, Inheritance of glucose-6-phosphate dehydrogenase variation in kangaroos, *Nature* **230:**154–155.

Robinson, R., 1971, *Gene Mapping in Laboratory Mammals,* Part A, Plenum Press, London.

Ruddle, F. H., 1969, Enzyme and genetic linkage studies employing human and mouse somatic cells, in: *Problems in Biology: RNA in Development* (E. W. Hanly, ed.), University of Utah Press, Salt Lake City, Utah, pp. 11–31.

Saidi, N., Hors-Cayla, M. C., Van Cong, N., and Benne, F., 1979, Sheep gene mapping by somatic cell hybridization, *Cytogenet. Cell Genet.* **25:**200.

Schiff, R., and Stormont, C., 1970, The biochemical genetics of rabbit erythrocyte esterases: Two new esterase loci, *Biochem. Genet.* **4:**11–23.

Schroder, J., 1976, *Genetics for Aquarists,* Tropical Fish Hobbyist Publications, Neptune City, New Jersey.

Schwab, M., Kollinger, G., Haas, J., Ahuja, M. R., Abdo, S., Anders, A., and Anders, F., 1979, Genetic basis of susceptibility for neuroblastoma following treatment with *N*-methyl-*N*-nitrosourea and x-rays in *Xiphophorus, Cancer Res.* **39:**519–526.

Schwarz, F. J., 1972, *World Literature to Fish Hybrids, with an Analysis by Family, Species, and Hybrid,* Gulf Coast Research Laboratory Museum, Ocean Springs, Mississippi, Publication No. 3.

Semenoff, R., 1972, Esterase polymorphism in *Microtus ochrogaster:* Interaction and linkage, *Biochem. Genet.* **6:**125–138.

Shami, S. A., and Beardmore, J. A., 1978, Genetic studies of enzyme variation in the guppy, *Poecilia reticulata* (Peters), *Genetica* **48:**67–73.

Shaw, C. R., 1966, Glucose-6-phosphate dehydrogenase: Homologous molecules in deer mouse and man, *Science* **153:**1013–1015.

Shaw, C. R., and Koen, A. L., 1968, Glucose-6-phosphate dehydrogenase and hexose-6-phosphate dehydrogenase of mammalian tissues, *Ann. N.Y. Acad. Sci.* **151:**149–156.

Shimizu, N., Shimizu, Y., Kondo, I., Woods, C., and Wenger, T., 1981, The bovine genes for phosphoglycerate kinase, glucose-6-phosphate dehydrogenase, alpha-galactosidase, and hypoxanthine phosphoribosyl-transferase are linked to the X chromosome in cattle–mouse cell hybrids, *Cytogenet. Cell Genet.* **29:**26–31.

Shows, T. B., and McAlpine, P. J., 1978, The catalog of human genes and chromosome assignments: A report on human genetic nomenclature and genes that have been mapped in man, *Cytogenet. Cell Genet.* **22:**132–145.

Shows, T. B., Brown, J. A., and Chapman, V. M., 1976, Comparative gene mapping of HPRT, G6PD, and PGK in man, mouse and muntjac deer, in: *Human Gene Mapping 3: Third International Workshop on Human Gene Mapping, Birth Defects: Original Article Series,* Vol. 12, No. 7, The National Foundation, New York, pp. 436–439.

Siciliano, M. J., and Wright, D. A., 1973, Evidence for multiple unlinked genetic loci for isocitrate dehydrogenase in fish of the genus *Xiphophorus, Copeia* **1973:**158–161.

Siciliano, M. J., Morizot, D. C., and Wright, D. A., 1976, Factors responsible for platyfish–swordtail hybrid melanoma—many or few?, in: *Melanomas: Basic Properties and Clinical Behavior,* Vol. 2 (V. Riley, ed.), S. Karger, Basel pp. 47–54.

Simonsen, V., and Frydenberg, O., 1972, Genetics of *Zoarces* populations: II. Three loci determining esterase isozymes in eye and brain tissue, *Hereditas* **70:**235–242.

Stoneking, M., May, B., and Wright, Jr., J. E., 1979, Genetic variation, inheritance, and quaternary structure of malic enzyme in brook trout (*Salvelinus fontinalis*), *Biochem. Genet.* **17:**599–619.

Szymura, J. M., and Farana, I., 1978, Inheritance and linkage analysis of five enzyme loci in interspecific hybrids of toadlets, genus *Bombina, Biochem. Genet.* **16:**307–320.

Tischfield, J. A., Creagan, R. P., Nichols, E. A., and Ruddle, F. H., 1974, Assignment of a gene for adenosine deaminase to human chromosome 20, *Hum. Hered.* **24:**1–11.

Trujillo, J. M., Walden, B., O'Neil, P., and Anstall, H. B., 1965, Sex-linkage of glucose-6-phosphate dehydrogenase in the horse and donkey, *Science* **148:**1603–1604.

Warren, D. C., 1949, Linkage relations of autosomal factors in the fowl, *Genetics* **34:**333–350.

Westerveld, A., Visser, R. P. L. S., Freeke, M. A., and Bootsma, D., 1972, Evidence for linkage of 3-phosphoglycerate kinase, hypoxanthine-guanine phosphoribosyltransferase, and glucose-6-phosphate dehydrogenase loci in Chinese hamster cells studied by using a relationship between gene multiplicity and enzyme activity, *Biochem. Genet.* **7:**33–40.

Wheat, T. E., and Whitt, G. S., 1973, Linkage relationships of six enzyme loci in interspecific sunfish hybrids (genus Lepomis), *Genetics* **74:**343–350.

Wheat, T. E., Whitt, G. S., and Childers, W. F., 1972, Linkage relationships between the homologous malate dehydrogenase loci in teleosts, *Genetics* **70**:337–340.

Whitt, G. S., Childers, W. F., Tranquilli, J., and Champion, M., 1973, Extensive heterozygosity at three enzyme loci in hybrid sunfish populations, *Biochem. Genet.* **8**:55–72.

Whitt, G. S., Childers, W. F., Shaklee, J. B., and Matsumoto, J., 1976, Linkage analysis of the multilocus glucosephosphate isomerase system in sunfish (Centrarchidae, Teleostei), *Genetics* **82**:35–42.

Womack, J. E., and Sharp, M., 1976, Comparative autosomal linkage in mammals: Genetics of esterases in *Mus musculus* and *Rattus norvegicus, Genetics* **82**:665–675.

Wright, D. A., 1975, Expression of enzyme phenotypes in hybrid embryos, in: *Isozymes IV: Genetics and Evolution* (C. L. Markert, ed.), Academic Press, New York, pp. 649–664.

Wright, D. A., and Richards, C., 1980, Mapping enzyme loci in the frog, *Isozyme Bull.* **13**:60.

Wright, D. A., Richards, C. M., and Nace, G. W., 1980, Inheritance of enzymes and blood proteins in the leopard frog, *Rana pipiens:* Three linkage groups established, *Biochem. Genet.* **18**:591–616.

Wright, J. E., Heckman, R., and Atherton, L. M., 1975, Genetic and developmental analysis of LDH isozymes in trout, in: *Isozymes III: Developmental Biology* (C. L. Markert, ed.), Academic Press, New York, pp. 375–401.

Wright, J. E., Jr., Johnson, K., Hollister, A., and May, B., 1982, Meiotic models to explain classical linkage, pseudolinkage and chromosome pairing in tetraploid derivative salmonid genomes, in: *Isozymes: Current Topics in Biological and Medical Research,* Vol. 10 (M. C. Ratazzi, J. G. Scandelios, and G. S. Whitt, eds.), A. R. Liss, New York, pp. 239–260.

Yoshida, M. C., 1978, Rat gene mapping by rat–mouse somatic cell hybridization and a comparative Q-banding analysis between rat and mouse chromosomes, *Cytogenet. Cell Genet.* **22**:606–609.

Zinkham, W. H., Isensee, H., and Renwick, I. H., 1969, Linkage of lactate dehydrogenase B and C loci in pigeons, *Science* **164**:185–187.

CHAPTER 5

The Evolutionary Genetics of Xiphophorus

RICHARD BOROWSKY

1. Introduction

Fishes of the genus *Xiphophorus,* the platyfish and swordtails, are common inhabitants of the streams and rivers of Central America. The 15 species described occupy a variety of habitats, ranging widely through the Atlantic drainage from northeastern Mexico, south-east through Guatemala, Honduras and Belize. Like most other members of the family Poeciliidae, *Xiphophorus* species are internally fertilizing and ovoviviparous (Rosen and Bailey, 1963). Females can store sperm and produce broods for several months after a successful mating.

Xiphophorus species are phenomenally variable in genetically determined melanic pigmentation and coloration patterns. Eight of the species are polymorphic for melanic pigmentation patterns, which are controlled by at least six different sex-linked and autosomal loci (Table I). At the tailspot locus in *X. maculatus* alone, eight different alleles segregate in some populations (H. Gordon and Gordon, 1957). Furthermore, several of the species are highly variable for yellow and red (YR) coloration patterns (Kallman, 1975). For example, no fewer than 19 different sex-linked alleles control YR patterns in *X. maculatus* (Borowsky and Kallman, 1976; Kallman, 1965, 1970), and populations in one of the river systems (Belize River) contain at least 15 of these (Kallman, 1975). The known variation in two of the species, *X. maculatus* and *X. variatus,*

RICHARD BOROWSKY • Department of Biology, New York University, New York, New York 10003.

Table I
The Distribution of Melanic Spotting Patterns among the Species of *Xiphophorus*[a]

	Micromelanophore		Macromelanophore	
Species	Tailspot	Caudal blot	Sex-linked	Autosomal
X. gordoni			1	
X. variatus	6		9	
X. xiphidium	3		2	
X. maculatus	8		7	
X. milleri	3		2	
X. nigrensis		1		
X. montezumae		1		3
X. cortezi		1		3
X. helleri				1

[a]The table lists the number of distinct patterns observable. The distinction between micromelanophore and macromelanophore patterns follows M. Gordon (1927). Data compiled from Kallman (1975), Kallman and Atz (1966), Kallman and Borowsky (1972), and the present study.

equals or exceeds that found in "textbook" examples of polymorphic species such as *Cepaea nemoralis, Papilio dardanus,* or *Maniola jurtina* (Ford, 1971). Furthermore, the full extent of the polymorphism is not yet known. This degree of polymorphism is remarkable and demands explanation.

One of the key questions in population biology is how genetic variation is maintained in natural populations. Without variation there can be no evolution, yet adaptive evolution tends to reduce population variation because its mechanism is selection. Evolution proceeds and populations remain variable, however. Thus there must be factors or relationships having considerable effect in natural populations that shield variation from selectional loss and promote polymorphism. *Xiphophorus* is a suitable organism to address this problem because of its extensive polymorphism. While *Xiphophorus* is atypical because the polymorphism in question is conspicuous, there is no reason to believe that the mechanisms fostering the polymorphism are atypical. Furthermore, the conspicuousness of the variation aids its study. The research summarized here indicates that environmental variability, both cyclic-seasonal and spatial variation, plays an important role in this polymorphism.

To discuss all aspects of polymorphism in *Xiphophorus* is a task beyond the scope of this chapter. I have chosen, therefore, to emphasize one aspect of the melanic variation, the tailspot polymorphism (Table I), and one species, *X. variatus*. In addition, this paper documents previously undescribed variation at the bodyspot locus in *X. variatus* and supports the prediction of Kallman and Atz (1966) that *X. variatus* rivals *X. maculatus* in degree of polymorphism. The evidence presented and reviewed in this chapter demonstrates that these two polymorphisms are subject to strong selective pressures. I also present data on allozyme variation, which, in contrast to the melanic variation, appears to be effectively neutral in these fishes. This difference in the relative effects of selection and drift is reflected in differences among the loci in degree of polymorphism and geographic distribution of allelic diversity. Finally, data on two other types of polymorphic systems, tailspot pattern modifiers and size-determining genes, are presented because both of these systems may interact with the tailspot polymorphism.

The tailspot polymorphism is one aspect of the biology of a species group that exists in a highly variable environment. The following account describes the situation in *X. variatus* in the Tamesi/Panuco system, but can safely be applied to *X. variatus* from other areas, and to most populations of *X. maculatus* and *X. xiphidium*. Habitat descriptions for other species have been provided by M. Gordon (1953) and Zander (1967).

Xiphophorus variatus in nature must contend with variation in rainfall and temperature (air temperature and insolation). These factors change seasonally and are predictable to an extent, but their effects on water temperature, flow, and oxygenation vary considerably locally. Rainfall in northeastern Mexico peaks in the summer months (Fig. 1). From May through November, the streams supporting *X. variatus,* partly fed by runoff and partly by springs, continue to flow freely. As the dry season progresses, the streams cease to flow and the stream beds often become divided into isolated, spring-fed pools.

The breeding season coincides with the dry season, starting in December when air temperature is low (Fig. 1), and continues through April or May, when the first rains disperse the fish. Judging by the general health of fish collected during different seasons, the environment becomes more stringent as the dry season progresses (R. Borowsky, unpublished observations). Also, the patchiness of the environment increases. Some pools are in sunlight, while others are in shade. These differ not only in mean water temperature but also in diurnal temperature variability. Some areas have highly oxygenated water, while others are nearly depleted. Some become more crowded than others as they contract. Table II documents this spatial heterogeneity for temperature and dissolved oxygen concentration. While physical differences among areas of a stream may exist early in the season, the free flow of water tends to minimize them and the fish have the opportunity to disperse relatively freely. Later in

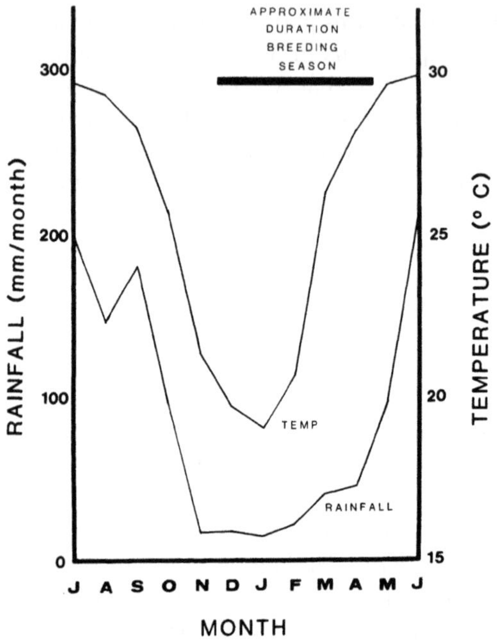

Figure 1. Average rainfall and temperatures recorded at the Río Sabinas near Rt. 85 (Tamaulipas, Mexico) during the period 1966–1980.

the dry season the differences among areas are magnified and the fish are more restricted in their potential for dispersal. While individuals may exercise some choice of the pool in which they get trapped, a strong element of chance must figure in their fate.

In summary, as the dry season progresses the physical environment changes from effectively fine-grained and permissive to coarse-grained and stringent. Any individual. *X. variatus* must contend, therefore, with a predictable seasonal variation in mean temperature and a degeneration of the environment, the nature of which is unpredictable.

2. Materials and Methods

Most of the materials and methods are described where the results are presented, but certain details are clarified here.

2.1. Allozyme Variation

Starch gel electrophoresis (TVB or TC buffer systems) and standard staining techniques were used to measure the electrophoretic mobilities of enzymes and other proteins (Siciliano and Shaw, 1976). Loci were numbered in descending order of anodal mobility and alleles were named

Table II
Some Examples of Variation in Temperature and Oxygen Concentration at Mexican Localities[a]

Locality	Oxygen (ppm)	Temperature, °C	Percent saturation
Sarco (h)	3.2	20.0	35
Encino (g)	3.4	22.5	39
Tributary of Río Comandante (p)	2.8	17.0	29
Río Coy at Rt. 85 (x)	6.0	24.0	71
Pequeseh (y)	8.5	17.0	88

Pool	Temperature, °C	Pool	Temperature, °C
1	21.7	6	17.5
2	23.4	7	20.4
3	23.6	8	17.1
4	23.4	9	22.8
5	23.5		

[a]The temperature measurements in the top part are averages over periods of 5–7 days obtained from recording thermographs placed in the stream beds in January 1981. Mean *RC* is the relative condition of Ct mates collected at the localities. The measurements in the bottom part are averages of four to six measurements per pool made in Arroyo Sarco during the hours of 10 AM–5 PM, January 27–February 25 (1967).

by their mobilities relative to that of a standard. The standard consisted of tissue extracts from the 163 A, Río Jamapa strain of *X. maculatus* and was included on all gels run. The standard strain is highly inbred and isogenic and was provided by K. D. Kallman. Amylase was obtained from extracts of pancreatic tissue. The three loci demonstrated by general protein staining are designated *Pt; Pt-4* is uncharacterized, but the others have been identified as parvalbumins (Borowsky, unpublished). Esterase activity (*ES-2* and *ES-3*) was scored in muscle tissue, as were the loci for malate dehydrogenase (*MDH*). Isocitrate dehydrogenase loci (*IDH*) were scored in muscle (*IDH-3*) and liver tissue (*IDH-1* and *-2*). Mannosephosphate isomerase (*MPI*) was scored in muscle tissue.

2.2. Symbols, Calculations, and Statistics

Allele and pattern symbols are defined at first use. The convention followed in this paper is that the symbol for a phenotype is printed in roman type (e.g., C stands for the crescent pattern), while the symbol for an allele is italicized (*C*). The frequency of an allele is given as "*P* of *X*," e.g., *P*(*C*).

Tailspot allelic frequencies for *X. maculatus* were taken directly from the paper of H. Gordon and Gordon (1957). For *X. variatus* they were calculated from the data in Table III according to the methods detailed by Spiess (1977, pp. 80–81).

Genetic diversity is indexed as *G* which is defined as 2 raised to the power of *H*, where *H* is the Shannon–Wiener diversity index calculated using base 2. *G* is a measure of the effective number of alleles.

Probabilities are reported as one-sided or two-sided (os*P* or ts*P*), depending upon whether or not the observed relationship was predicted prior to testing. Standard methods were used for the manual calculation of statistics, or standard software packages (BMDP and SPSS) were used for computer calculations.

Relative condition and relative fecundity are defined at first use and are symbolized throughout as *RC* and *RF*. Since these measures are derived from regression analyses of raw data, two degrees of freedom are lost in their calculation. Student's *t*-tests to determine whether morph classes differ in *RC* were adjusted to account for this loss.

2.3. Collecting Localities

The locality letters in the following list index the drainage map (Fig. 2) and the table of gene frequencies (Table III). Since the locality map shows no roads, road information has been provided here in many cases. Most of the localities are described, but some were not visited personally. Museum numbers (UMMZ or MCZ) refer to collections at the University of Michigan Museum of Zoology and the Museum of Comparative Zoology, Harvard University. Place names, obtained from natives or from museum jar labels, that were not verifiable from published maps have been enclosed in quotation marks. Mitchell *et al.* (1977) provide a map, useful in detail and accuracy, which covers the region from localities i–bb.

Río Tigre Drainage

- **a.** "Las Yucas, north of Aldama, Tamps." MCZ 41362.

Río Guayalejo Drainage

- **b.** "La Cienega," 0.5 km west of Jaumave, Tamps. Swiftly flowing spring-fed stream with clear water; bottom varies from silt to small stones. (Locality 1, Borowsky, 1978.)
- **c.** Rio Guayalejo at Llera, Tamps. Broad river, poorly flowing, with many shallow pools in winter. Poorly shaded with pebble and rock bottom.
- **d.** Small stream tributary 7.5 km east of Rt. 85 on road to Compuertas, Tamps.

e. Stream ford at road and irrigation ditches near Compuertas, Tamps. Flow often rapid, water turbid, and bottom variable from mud to small stones. (Locality 2, Borowsky, 1978.)

Not on locality map. "Tributary of the Rio Guayalejo, 16 miles north of Xicotencatl." UMMZ 108668. This description is ambiguous and might refer to a station in the Río Sabinas Drainage.

Río Sabinas Drainage

f. Arroyo Encino at the town of Santa Fe, Tamps., 0.3 km east of Rt. 85, and 2 km north of Encino. Shallow, broad stream, clear water over a silt-covered, rocky bottom.

g. Stream tributary of the Arroyo Encino, 0.25 km east of Rt. 85 at Encino, Tamps. Spring-fed pools and shallow riffles, water clear to turbid, bottom silty or covered with decomposing vegetation. (Locality 5 in Borowsky, 1978.)

h. Arroyo Sarco, 1.8 km to the west of Encino, Tamps. Medium-sized stream tributary of Arroyo Encino. (Locality 4 in Borowsky, 1978). This stream supports a second, distinct population of *X. variatus* in a series of pools "5 km north-west of Encino. Slowly flowing clear water, no vegetation, dense tree canopy." This area is referred to as "Sarco—upstream" in Table III and was not visited by the author. A collection made there in 1981 by K. D. Kallman and D. Morizot was made available for examination through their courtesy.

i. Arroyo La Flor. Rocky pools 0.25–4 km above the Nacimiento of the Río Sabinas. Seepage and some flow during the dry season, with torrential runoff during the rainy season. (Locality 3 in Borowsky, 1978.)

j. Río Sabinas below mouth of Arroyo Encino at Rancho Calabazas or "Storm's Ranch" (Darnell, 1963). Clear, flowing water in main channel; bottom gravel to large rocks. Some collections made in main stream, others made in backwaters or stagnant, rocky pools. A second collection, referred to as "stream tributary" in Table III, was made in a small stream feeding the west bank of the river at this point. Bottom mud.

k. Río Sabinas at Rt. 85. Water clear or turbid, flow variable, bottom mud, sand, gravel, or rocks.

l. Stream tributary cutting across field near Adjuntas, Tamps. Slowly flowing water, bottom mud. Series of collections from several pools, shaded and unshaded. (Locality 7 in Borowsky, 1978.)

Table III
Allelic Frequencies at 12 Protein-Coding Loci in 28 Populations of *Xiphophorus*[a]

Locus and allele	*X. variatus*																	*X. montezumae*					*X. cortezi*			*X. nigrensis*		*X. pygmaeus*
	b	e	i	h	g	j	m	p	x	y	hh	cc	pp	qq	rr	ss	tt	i	j	n	OA[b]	RS[c]	y	hh	cc	r	x	cc
MPI																												
0.83													24	2														
0.84																		50	28	12	4	6						
0.92	28	30	34	20	44	16	28	24	43	36	16	22		34	12	16							28	38	36	10	26	20
0.98									1																			
MDH-1																												
0.71	9																											
0.85	19																									5		
0.90							1																					1
1.00		30	34	20	46	16	27	24	44	38	16	28			12	16							28	38	36	5	26	19
1.18													24	30														
1.19																		50	32	16	4	6						
1.29														6														
MDH-2																												
0.69																										10	26	20
0.72																		50	32	16	4	6	30	38	36			
0.82									42	36	16	16	24	36	12	16												
1.00	28	30	34	20	46	16	28	24																				
1.11									2	2		12																
1.18																												

IDH-1																												
1.00																		40	22	8	4	6	24	22	10	6	8	20
1.13	29	20	16	19	42	12	26	20	28	38	16	20		1	10	9	4											
1.20													24	17	2	3												
IDH-2																												
0.46																	1											
0.79	29	28	33	22	35	16	30	24	44	36	16	28	24	26	4		3											
0.82																				10								
0.92				1	9				2	2					8	16												
1.00			1															54	30		4	6	24	36	34	10	25	20
1.15																											1	
IDH-3																												
0.85									1	3																		
1.00	28	30	16	20	46	14	28	23	25	35	18	28	24	36	12	16	4	50	20	14	4	6	24	38	14	10	12	20
1.18								1																				
ES-2																												
0.87																										10	11	2
0.93																		24	20	7	4	6				1		
0.96							1		1					6			1			3								
0.98							1	5						1						1					3[d]			
1.00							2		16	16		17	24	6	11	7	10		2				5[d]	34	3[d]			
1.03	28	30		18	30	14	24	19	5	4		11		5	1	1	1			1			5[d]					
ES-3																												
0.92													1				2											
0.95								1	1				23	10	10	10	2											
0.97								1														1						

(continued)

Table III (*Continued*)

Locus and allele	*X. variatus*																	*X. montezumae*					*X. cortezi*			*X. nigrensis*		*X. pygmaeus*
	b	e	i	h	g	j	m	p	x	y	hh	cc	pp	qq	rr	ss	tt	i	j	n	OA[b]	RS[c]	y	hh	cc	r	x	cc
1.00	28	30		18	30	14	28	20	21	12	21			7	2	6		26	20	12	4	5		8	8			
1.02																										10	12	2
1.06										4	1												14	28				
1.18														1														
Amylase																												
1.15	10	14		12	21	12	25	6	12	16	2	8	24	26	12	16	4											
1.29																								1	4	10	12	5
1.49																		28	25	10	4	6	14	21	3			
PT-1																												
0.74																		26	20	2	4	6	16	12	2	14		
0.79																												20
1.00	10	12	34	20	42	10	14	22	12	22	20	18	24	8	12	8	4											
PT-2																												
1.00																		26	20	2	4	6	20	4	12	2	2	20
1.01	10	12	34	20	42	10	14	22	12	22	20	18	24	8	12	8	4											
PT-4																												
1.00	10	12	34	20	42	10	14	22	12	22	20	18	24	8	12	8	2											
1.10																		26	20	2	4	6	20	4	12	2	14	20

[a]The table lists the numbers of alleles detected in each sample. Locality letters are keyed in Section 2.
[b]"Ojo de Agua," tributary of the Río Tamuin 16 km west of Ciudad Valles.
[c]Tributary of the Río Salto.
[d]All eight individuals from these two populations, which were scored for ES-2, exhibited double banding and may not have been true heterozygotes.

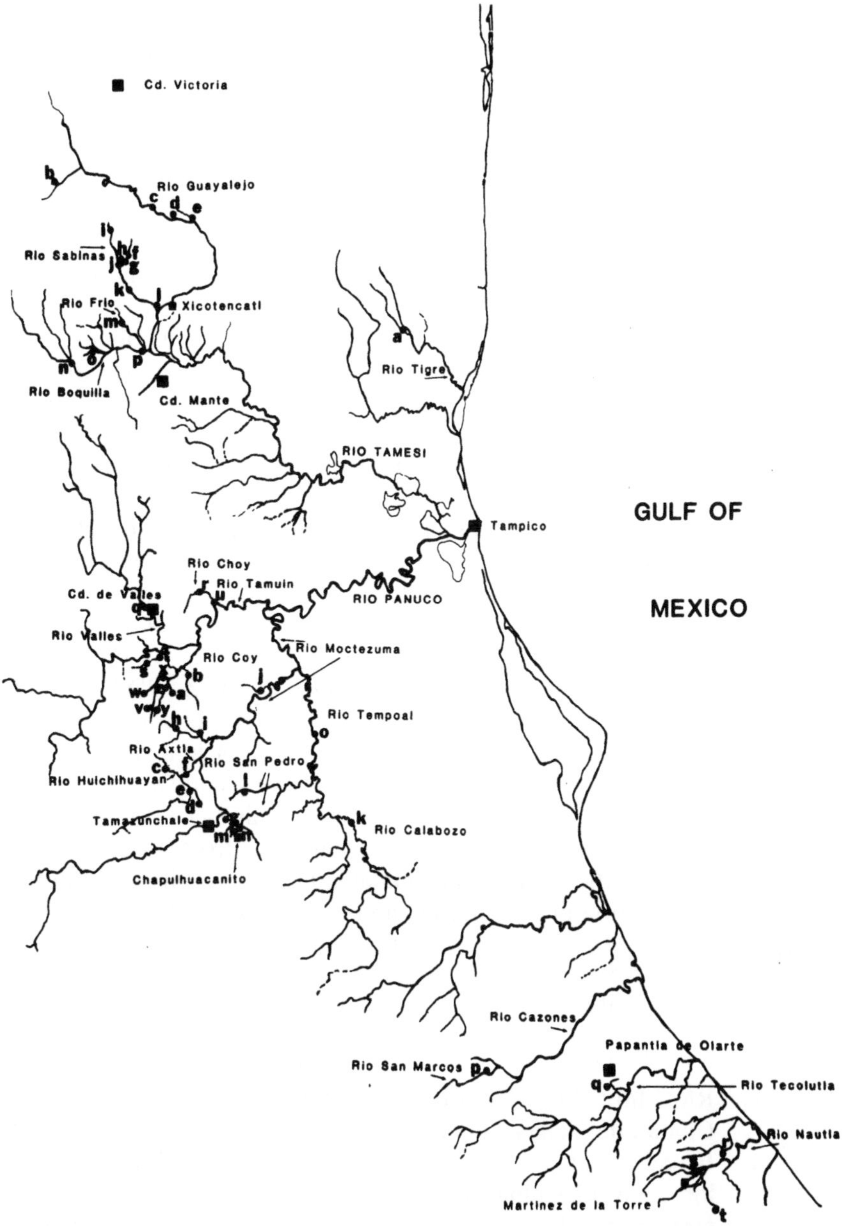

Figure 2. Collecting localities for *X. variatus* in Mexico. Lower case letters are keyed in the locality list in Section 2. The second series of letters running north to south corresponds to the double-lettered series in the list.

Ríos Frio/Comandante/Boquilla Drainage

m. Río Frio mainstream about 0.5 km below Nacimiento del Río Frio [cf. Fig. 5 of Mitchell *et al.* (1977)]. About 20 km WSW of Xicotencatl, Tamps. Deep, swiftly flowing river, water clear with greenish tint. A distinct collecting site at this point is a small stream tributary on the east bank of the river, referred to as "stream tributary" in Table III. (Locality 8 of Borowsky, 1978).

n. Stream crossing road from Chamal to Ocampo, Tamps. at La Muralla, Río Boquilla drainage.

o. Broad, shallow stream tributary of Río Boquilla about 4 km north of Chamal, Tamps., along "Al Mateo" road. Bottom covered with large stones, fish taken from broad pool and backwaters, little flow.

p. Stream tributary of Río Comandante at river, 3 km west of Rt. 85 on road to Chamal, Tamps. Narrow stream with sluggish flow through thick forest skirting cane fields. Water clear or turbid, bottom thick mud and decomposing vegetation. (Locality 9 of Borowsky, 1978.)

Río Mante Drainage

Not shown on map. "Juarez, Tamps." MCZ 39587. (Juarez is the former name of Ciudad Mante and this collection may either have been made in the Río Mante or in a small stream tributary). The Mante drainage has been greatly altered by the construction of dams and irrigation canals since the collection was made in 1939. Although small, and from an ambiguous locality, the collection is valuable because it fills a gap in the distribution of the species, and is no longer replicable.

Río Valles Drainage

q. "Río Valles at Valles, S.L.P." UMMZ 108667.

Río Tamuin Drainage

r. Río Choy about 2 km below Nacimiento; 16 km due east of Ciudad de Valles, S.L.P.

s. Broad stream crossing road near town of San Francisco (22 km SSW of Ciudad de Valles, S.L.P.). Rapid flow in main stream, but fish taken in backwaters and under the banks. Natives provided information that the stream was tributary to the Río Tampaon (Río Tamuin) but, based on the map position of the locality, the stream could be in the Río Coy drainage instead.

- **t.** Arroyo in deep cut under road just north of San Vicente (19 km SSW of Ciudad de Valles, S.L.P.). Tributary of Río Tampaon.
- **u.** Río Tamuin at the town of Tamuin, S.L.P.

Río Coy Drainage

- **v.** Three small streams, 1.0, 1.3, and 1.9 km north of the central square of Aquismon, S.L.P.
- **w.** Small arroyo at Santa Cruz, S.L.P. Poor flow, mud bottom.
- **x.** Río Coy at Rt. 85. Deep, swiftly flowing river with clear, greenish water and steep banks of clay. *Xiphophorus variatus* were collected from quiet backwater, while *X. nigrensis* were collected from main stream. (Locality 10 of Borowsky, 1978.)
- **y.** Small stream at "San Jose Pequeseh" about 3 km north of Ciudad Santos, S.L.P. Reached by dirt road heading east from Rt. 85 at Aquismon turnoff. (Locality 11 of Borowsky, 1978.)
- **z.** Small stream 1.8 km east of Rt. 85 on road ("road A") to Tanlajas, S.L.P.
- **aa.** Small stream crossing dirt road to San Joachin a km NW of "road A" (see z).
- **bb.** Small stream at San Joachin, S.L.P., about 11 km northeast of Tanlajas.

Río Moctezuma Drainage

- **cc.** Río Huichihuayan at Huichihayan, S.L.P. Broad, swiftly flowing spring-fed river with clear waters. *Xiphophorus cortezi* and *X. pygmaeus* taken in the main channel and *X. variatus* collected among the vegetation under the banks. [Locality 13 in Borowsky (1978); incorrectly labeled "Río Axtla" in that publication.]
- **dd.** "Río Matlapa at Matlapa, S.L.P., 13 km north of Tamazunchale." UMMZ 124338. This tributary is referred to now as "Arroyo Matlapa."
- **ee.** "Arroyo Plan de Jalpilla, 29 km north of Tamazunchale." UMMZ 124351. Presumably, this collecting site was the Arroyo at Plan de Jalpilla and the 29 km distance was measured along the road rather than as a true map distance.
- **ff.** "Río Axtla 3 km west of Axtla, S.L.P." UMMZ 108604. The current name for Axtla is Alfredo M. Terrazas.
- **gg.** Two stream localities, 10.4 and 11.0 km northwest of Chapulhuacanito, S.L.P.

hh. Arroyo 10 km ESE of Ciudad Santos, S.L.P., and 3.5 km north of Coxcatlan. (Locality 12 of Borowsky, 1978.)

ii. Arroyo crossing road from Ciudad Santos to Tanquián Escobedo, 1.5 km east of Tampamolón, S.L.P. Stream in deep cut, flow variable, fish taken in quiet backwater.

jj. Stream crossing road to Tamuin, 16.3 km north of Tanquián Escobedo, S.L.P.

Río Tempoal Drainage

kk. "Río Calabozo 14 km southwest of Tantoyuca, Ver." UMMZ 108666.

ll. North fork of the Río San Pedro at San Martin, S.L.P. Habitat variable, swift flow in mainstream but fish collected from quiet backwaters and stagnant ditches.

mm. Two stream tributaries of the south fork of the Río San Pedro, 1.8 and 2.7 km northwest of Chapulhuacanito, S.L.P.

nn. Stream tributary of the south fork of the Río San Pedro, 5.1 km north of Chapulhuacanito, S.L.P.

oo. "Stream tributary of the Río Tempoal 4 km north of Tempoal, S.L.P." UMMZ 108672.

Río Cazones Drainage

pp. Stream crossing road to Poza Rica, Ver., 8.8 km east of Apapantilla, Pue. Tributary of the Río San Marcos.

Río Tecolutla Drainage

qq. Small stream draining orange grove. 1 km south of El Chote on road from Papantla de Olarte, Ver., to Espinal.

Estero Dulce

rr. Small stream tributary draining meadow about 3 km north of El Pital at "Portrero Nuevo." Crosses Rt. 129 between Nautla and Martinez de la Torre, Ver. The stream is a tributary of a small river system (Estero Dulce) draining independently into the Gulf of Mexico just north of Nautla.

Río Nautla Drainage

ss. Stream tributary of the Brinco del Tigre, about 5.5 km east of Martínez de la Torre. Narrow stream deeply dissecting sandy hill with some deep pools.

tt. Small stream tributary of the Río Quila, 7 km south of Independencia on road from Martínez de la Torre to Misantla. Clear water, moderate flow, rock bottom.

3. Five Sets of Polymorphic Loci

This section details the population and Mendelian genetics, phenotypic expression, and geographic distribution of alleles at five different sets of polymorphic loci:

1. The tailspot locus: The tailspot polymorphism is shared by *X. maculatus, X. variatus, X. xiphidium,* and *X. milleri* (Kallman, 1975; Rosen, 1979). The locus is autosomal and has a series of alleles that cause the formation of geometric patterns of melanophores (circles of different sizes, axe-heads, crescents, etc.) at the base of the caudal fin (M. Gordon, 1956). It is characteristic of tailspots that each pattern varies little in morphology among individuals, and that the patterns are present in young as well as older fish. Extensive evidence is presented in this chapter that allelic variation at the tailspot locus is correlated with important aspects of physiology and metabolism.
2. The tailspot modifier loci: Two autosomal alleles that change tailspot patterns have been documented in *X. maculatus* and *X. helleri* (M. Gordon, 1956). A third example, in *X. variatus,* is described below (pp. 259–263). A tailspot modifier allele interacts with a particular allele at the tailspot locus to alter its phenotypic expression, but has no effect on the expression of other tailspot alleles. Evidence presented in this chapter suggests that the modifiers may also have physiological correlates.
3. The sex-linked bodyspot locus: This is a complex gene consisting of several closely linked loci with a large number of alleles causing different numbers and patterns of distribution of melanophores on the flanks or the unpaired fins (Kallman, 1975). Unlike the case in mammals, the two sex chromosomes in *Xiphophorus* (X and Y) carry comparable sets of loci and the bodyspot locus is found on both.

   In the sense that each allele causes spots of a certain size distribution and degree of definition, concentrated on a particular part of the body, these are "pattern alleles." Each pattern is variable among individuals, however. Since the patterns are absent at birth, initially appear as a single spot in juveniles or at sexual maturity, and develop fully over the course of months, juveniles are often difficult to phenotype reliably (Kallman and Atz, 1966). Adults are more easily classified.

4. Allozyme loci: Allelic variation at a number of enzyme-coding loci was studied in populations of five species. One major finding is that polymorphic and polytypic variation in *Xiphophorus* is low, which contrasts with findings for other species and with variation at the tailspot and bodyspot loci. The geographic distribution of allozymes in *Xiphophorus* appears to be determined by genetic drift.
5. The *Pit* locus: This locus is tightly linked to the bodyspot locus and controls the timing of maturity, adult size in males, and fecundity in females (Kallman and Borkoski, 1978). "Pituitary" gene variation was discovered by Kallman and has been documented in *X. maculatus* and *X. nigrensis* (Kallman, 1975; Kallman and Schreibman, 1973) and *X. montezumae* (Kallman, 1983). *Xiphophorus variatus* also shares this polymorphism, as will be documented.

3.1. The Tailspot Locus

The Mendelian genetics of the tailspot polymorphism is uncomplicated. A series of alleles at the locus cause melanic pigmentation patterns at the base of the tailfin. The pattern alleles are codominant with respect to one another and dominant over wild-type (+). Wild-type is the unpatterned condition and individuals lacking tailspots are +/+ in genotype. Individuals having a single pattern are either homozygous for that pattern allele or heterozygous with + (e.g., crescent fish are either *C/C* or *C*/+). Individuals that have two different pattern alleles are "pattern heterozygotes" and exhibit both patterns (e.g., *C/Ct* fish are crescent/cut-crescent in phenotype). In rare cases, one pattern may mask another, so the full set of phenotypes possible may not be observable (H. Gordon and Gordon, 1954). There is no difference between the sexes in expression of the patterns, penetrance is full, and they appear shortly after birth, although they may be difficult to discern without magnification until the animals reach the age of about 1 month.

Fish without tailspot patterns are illustrated in Figs. 3–7. The crescent pattern (C) is illustrated in Figs. 3, 5, 8, 9, 11, and 12. The cut-crescent pattern (Ct) is illustrated in Figs. 3–6, 8–10, and 12. Crescent/cut-crescent heterozygotes are shown in Figs. 5, 10, and 12. The peduncular spot pattern (Ps) is illustrated in Figs. 3, 4, 13, and 14. Upper cut-crescent (Cu) is shown in Figs. 3 and 11. The moon pattern is illustrated in Fig. 15 and the dot pattern (D) in Figs. 11 and 16. Moon and dot differ from penduncular spot only in size.

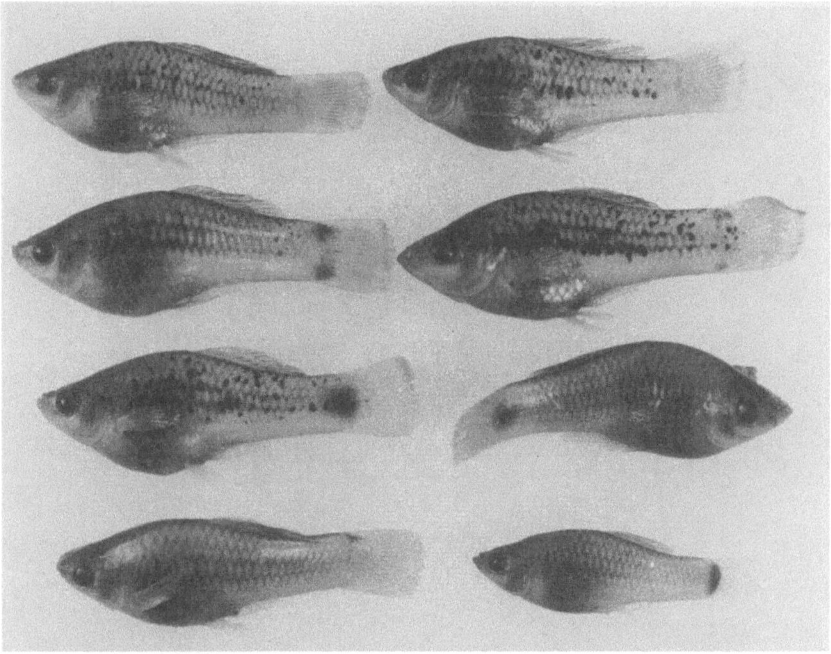

Figures 3–18 are examples of Xiphophorus variatus illustrating the variable melanic pigmentation patterns. Individuals are identified by column (L, M, or R), and numbered from the top. Bodyspot P patterns of uncertain identification are listed as "?".

Figure 3. The specimens were collected 1 km north of Aquismon, 1977. L1 is a mature male with P^1; L2 is a mature male with P^1, Ct; L3 is a mature male with P^1, Ps; L4 is a mature male with Cu; R1 is a mature male with P^1; R2 is a mature male with P^1, Li, Ct; R3 is a juvenile with Ps; R4 is a juvenile with C.

Both moon and dot have been bred in the laboratory and shown to be autosomally inherited (Borowsky, unpublished). Neither, however, has been tested to determine whether it is caused by different alleles than *Ps*, nor have they been tested for allelism with other pattern genes at the tailspot locus. The Ps pattern is exceptional among tailspots because it shows considerable variability in expression and generally gets darker and relatively larger as fish grow. In some lines, it appears to be better expressed in males than in females (Figs. 16 and 17). For these reasons, the designation of three different Ps-like patterns given here is tentative. On the other hand, the patterns shown in Figs. 15 and 17 derived from fish collected at the same locality and the differences in expression between the two lines is so great that at least two distinct alleles must be present. Kallman (1975) has pointed out that a similar size-graded series of Ps-like patterns exists in *X. maculatus*.

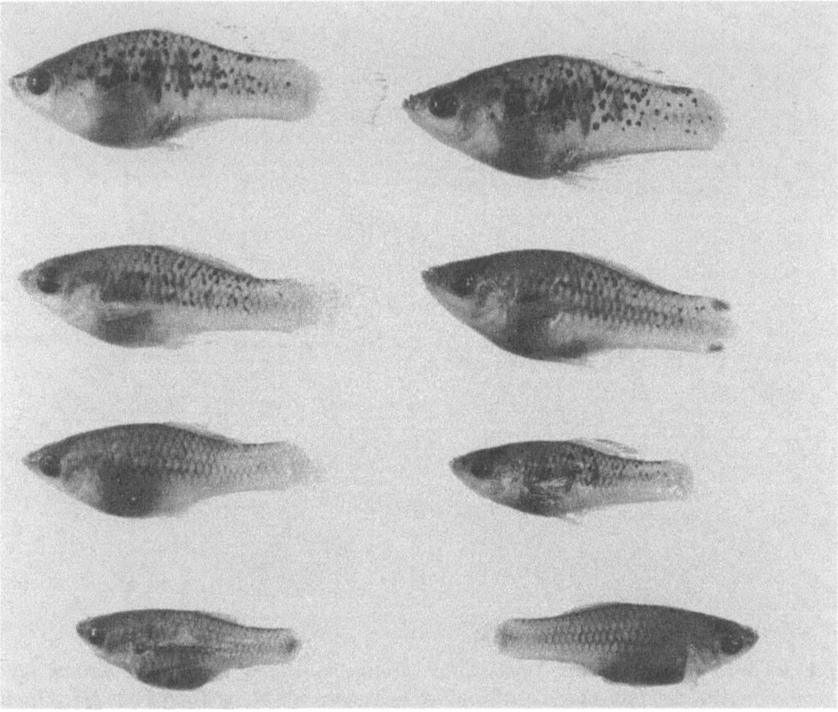

Figure 4. Collected at Río Frio stream tributary in 1977. L1 to L3 and R1 to R3 illustrate P^1; L1, R1 and R2 illustrate Gs; L4 illustrates Ps; R4 has C; R2 has Ct. L4 and R4 are mature females and L2 and L3 are juveniles.

Most of the time, tailspot alleles obey Mendel's first law. In six cases, out of tens of thousands of fish phenotyped, exceptional fish with three patterns were found. Two of these fish were preserved in the field, but four others were successfully bred and it was discovered that two of the patterns were linked on the same autosome (Kallman, 1975). This observation might be interpreted as an instance of gene duplication due to nonreciprocal recombination. On the other hand, if the tailspot series is pseudoallelic rather than allelic, recombination within the locus would produce the same result. This question cannot be resolved with current data, but is of interest because pseudoallelism is one of the attributes of the "supergene" concept (Ford, 1971). The tailspot locus exhibits other supergene attributes also, including the "universal recessive" (+) and correlated or pleiotropic physiological effects. In this context, it is also of interest that some of the more complex tailspot patterns appear to be compounded from simpler ones. For example, in *X. maculatus* the moon-complete pattern appears to be compounded from the moon and twin-spot patterns (Fig. 1 of M. Gordon, 1956) and in *X. variatus,* well-developed Ps-like patterns sometimes have what appears to be an added crescent component (Figs. 15 and 17).

Xiphophorus maculatus is the most variable of the species, having at least eight different tailspot alleles. All eight of the patterns are found together in the populations of the Ríos Papaloapan and Coatzacoalcos (H. Gordon and Gordon, 1957; Kallman, 1975). *Xiphophorus variatus* is the second most variable species and has at least six different tailspot alleles (Table IX). *Xiphophorus milleri* and *X. xiphidium* each have three tailspot patterns. All populations of the polymorphic species contain the + allele, which is the only allele at the locus that is ubiquitous. Presumably, the + allele is also fixed in the nonpolymorphic species. Table 11 in Kallman (1975) lists the patterns in the four polymorphic species and this chapter adds to the list for *X. variatus*.

In *X. cortezi, X. montezumae,* and *X. nigrensis,* there is a different tailspot pattern, caudal blot, that is caused by an allele at another, unlinked

Figure 5. Collected at Arroyo Sarco in 1976. L1 is a mature female with P^1 or P^2 and C/Ct; L2 is a mature male with C/Ct; M1 is a juvenile with "Ct"; M2 is a mature male with P^1, Gn and Ct; M3 is a mature male with P^1, P^2, Gn, and C; M4 is a mature female with P^2, P^3; R1 is a mature male with P^1, Ct; R2 is a mature male with P^2, P^3, Ct; R3 is a mature male with P^1 and Ct; R4 is a mature female with P^1 and C.

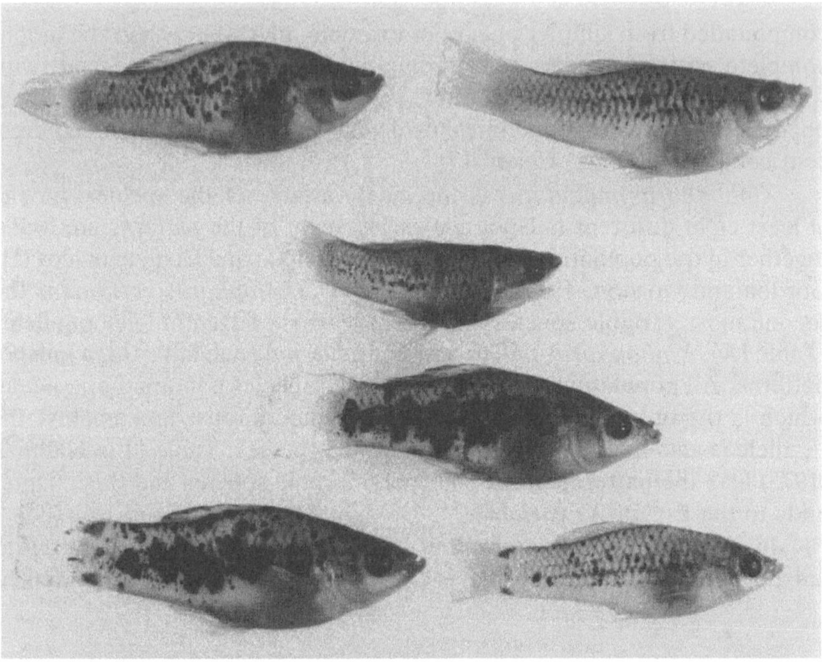

Figure 6. Collected at Arroyo Sarco in 1975. L1 is a mature male with P^1; L2 is a mature male with P^1, P^2, Ct; R1 is a mature male with P^1; M1 is a mature male with P^3; M2 is a mature male with P^2; R2 is a mature male with P^1, P^2, P^3 (?) and Ct.

autosomal locus (Kallman, 1975). This pattern and its genetics are not discussed in this chapter.

3.2. Tailspot Pattern Modifiers

3.2.1. Modifiers in *Xiphophorus maculatus*

The straightforward Mendelian genetics of the tailspot polymorphism is complicated by the existence of one or more other polymorphic loci that modify the morphological effects of tailspot alleles. Kallman (1975) summarized the work of M. Gordon (1946, 1956) and Atz (1962), which established that two modifier genes, extensor (*E*) and Guatemala crescent (*Cg*), are found in natural populations of *Xiphophorus*. The *Cg* gene changes the expression of the twin-spot (*T*) gene in *X. maculatus* into a pattern that superficially resembles a crescent/twin-spot heterozygote. The *E* gene changes the expression of *comet* (*Co*) in *X. maculatus*, causing a general darkening of all of the fins. The visible effects of *Cg* and *E* are dominant. The *Cg* gene has no effect on *Co*, and *E* has no effect on *T*. The two

Figure 7. N329 parental fish, laboratory bred from Río Cazones population. 1 is a wild-type mature male; 2 and 3 are mature females with Sc and P[4].

alleles apparently do not interact with one another, and do not affect the expression of any other tailspot alleles. The *Cg* gene is autosomally inherited, but the linkage of *E* is unknown.

The *Cg* gene is found in natural populations of *X. maculatus* that contain the *T* gene, but it also found in natural populations of the swordtail, *X. helleri,* which are not polymorphic for tailspot patterns at all. The *E* gene has only been verified in *X. helleri* stocks after hybridization with

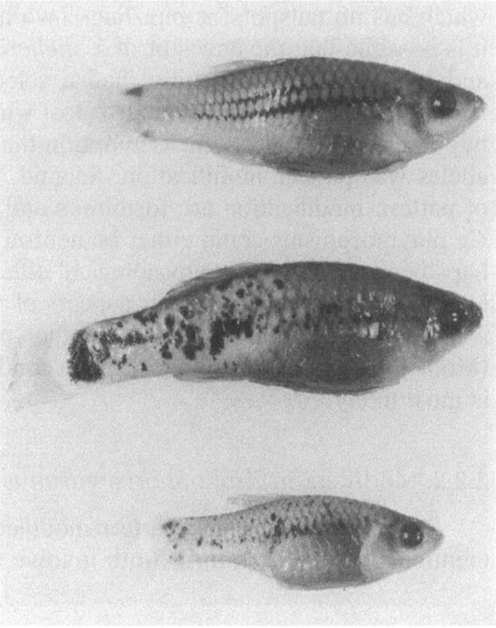

Figure 8. Collected at Aquismon in 1977. 1 is a mature female with Li and Ct; 2 is a mature female with P[6] and C; 3 is a mature female with "P[1]-like" spotting and Ct.

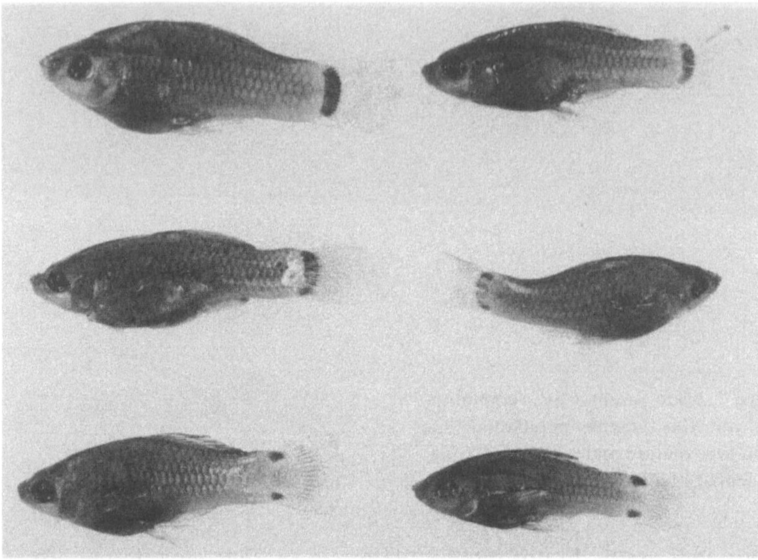

Figure 9. Collected at Arroyo Sarco in 1977. L1 is a mature male with C; L2 is a mature female with P? and "Ct"; R3 is a mature male with Gs and Ct; R1 is a mature male with C; R2 is a mature male with "Ct"; R3 is a mature male with Ct.

X. maculatus, although a similar allele has been detected in *X. xiphidium* (Atz, 1962).

Why the tailspot pattern modifiers *Cg* and *E* should exist in *X. helleri,* which has no tailspots, is puzzling. Two hypothesis come to mind: first, it is possible that the ancestor of *X. helleri* was polymorphic for tailspots and the *Cg* and *E* variability had a selectional basis because of their modification properties, which was lost when the tailspots were lost. This hypothesis is based on the assumption that the original function of these alleles was pattern modification. Second, it could be that the properties of pattern modification are fortuitous and coincidental. If so, the *E* and *Cg* polymorphisms could either be neutral or have a selective basis, perhaps because of cryptic physiological effects. Considering the strong selection demonstrable for other aspects of pigmentation polymorphism in *Xiphophorus* and the effects of another pattern modifier in *X. variatus* (Mod-1, discussed in Section 3.2.2) on condition factor, the last possibility is most likely.

3.2.2. Modifiers in *Xiphophorus variatus*

There are two possible pattern modifiers in *X. variatus* that have been mentioned in the literature. Both involve alterations of the cut-crescent phenotype. The pattern upper cut-crescent (Cu) in *X. variatus* differs from cut-crescent only in the degree of expression of the ventral component of the pattern, it being much lighter in Cu than in Ct fish. Kallman and Atz (1966) have suggested, on the basis of morphological considerations alone, that the two patterns might arise from differences at a modifier locus rather than from two distinct alleles at the tailspot locus. However, I have treated *Ct* and *Cu* as distinct tailspot alleles in the tabulations of gene frequencies because Cu appears to breed true in laboratory crosses. Two lines containing the Cu pattern were independently derived from fish collected at Brinco del Tigre in the Río Nautla drainage (Table IV). In both lines, Cu bred true and Ct did not appear. Although there was some variation in the degree of expression of the ventral component of the pattern, the variation was continuous, rather than bimodal. Since the Brinco del Tigre population contains the Ct phenotype at a somewhat higher frequency than the Cu phenotype (Table IV), the breeding data

Figure 10. Collected in the Río Boquilla drainage in 1977. L1 is a mature male with P^1 and "Ct"; L2 is a mature female with Li and Ct or "Ct"; L3 is a male with Ct or "Ct"; R1 is a mature male with Ct; R2 is a mature female with Li (poorly expressed) and C/Ct; R3 is a juvenile with Li; The fish in the center is a male with Li and "Ct".

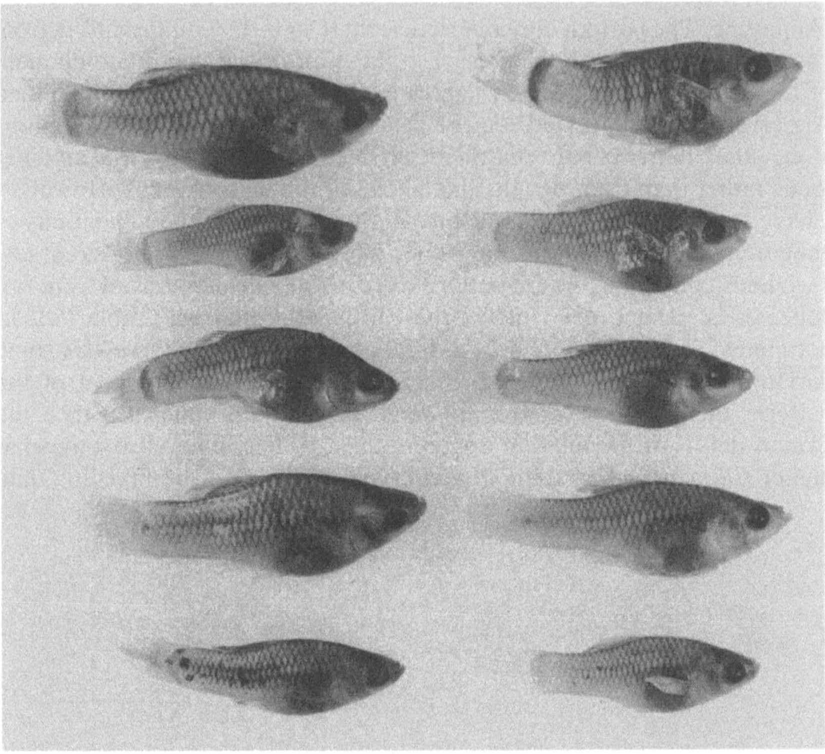

Figure 11. Collected at the Estero Dulce location in 1976. L1 to L3 and R1 illustrate C, which, in fish from this area, extends further up and down and is lighter in expression than the similar pattern in areas to the north (cf. Figures 3, 5, 8 and 9). L4 illustrates D; L5 has Cu; R2 has P?, C or no tailspot; R3 has C/D; R4 has D; R5 has P? and Cu.

suggest that the difference between the two patterns lies at the tailspot locus rather than at a second locus. To attribute the difference to a modifier locus one would have to postulate an unlikely set of coincidences. For example, that the modifier causing Cu was dominant and, although rare in the population, was homozygous in the female progenitor of N256, and accounted for three of the four alleles in the progenitors of N283a. The only data inconsistent with the interpretation that *Cu* is a distinct tailspot allele is that the male parent of N256 was classified as Ct. This individual was not available for reexamination, having been sacrificed for electrophoresis. Since Cu is variable in expression, it might have been Cu in phenotype with a heavily expressed ventral component. Thus, the balance

Figure 12. Collected at the Coxcatlan locality in 1967. L1 is a mature male with Li, Gs and Ct; L2 is a mature male with C/Ct; R1 is a mature female with Li and C; R2 is a mature female with Ct.

of the data suggest that Cu is caused by a distinct allele at the tailspot locus.

There is another pattern in *X. variatus,* however, that appears to be due to modification. Variant expressions of cut-crescent, ranging from a typical Ct pattern with a few extra micromelanophores between the dorsal and ventral spots, to a pattern superficially resembling a C/Ct heterozygote, are common in this fish (Figs. 5, 9, 10, and 18). Originally, I attributed this to a distinct allele at the tailspot locus and designated it pseudo cut-crescent (''*Ct*''; Borowsky, 1969). Atz, however, reported a similar range

Figure 13. Collected at Jaumave in 1967. All four have Ps. R1 is a mature male with P^7.

Figure 14. Laboratory-raised male parent of N336 pedigree illustrating P^5 and Ps.

in expression of the Ct pattern in the hybrid offspring of a cross between a female *X. xiphidium* with cut-crescent and a wild-type male *X. variatus* and attributed it to a pattern modifier in the male parent (Atz, 1962, Fig. 7). This explanation seems more likely, especially since the variable expression of "Ct" surpasses that of Cu and is almost unique among tailspot phenotypes. Furthermore, the modifier hypothesis for "Ct" is supported by some genetic analyses.

Table V gives the frequency of Ct and "Ct" patterns among the offspring of wild-caught females from four localities. These females were .fertilized in the wild and the genotypes of the inseminating males were deduced from the genotypes of the progeny (Borowsky and Kallman, 1976; Borowsky and Khouri, 1976). Only the data on Ct-like progeny (Ct and "Ct") are presented here. At two of these localities (Compuertas and Río Frio) "Ct" is rare or absent. Ninety-seven young from four of these Ct females exhibited Ct, and none exhibited "Ct". Two of the localities (Encino and Arroyo Sarco) have the "Ct" pattern in moderate frequency (0.24–0.30). Two females from these localities were heterozygous *+/Ct* and exhibited the "Ct" pattern. Their offspring exhibited both Ct and "Ct" in equal frequencies (N172 and N188). Four females from these localities exhibited the Ct pattern, and their Ct-like offspring were almost

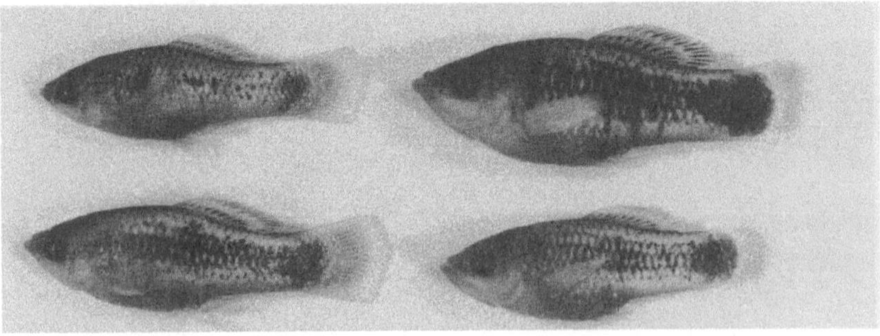

Figure 15. Laboratory-raised N250 mature males descended from fish collected at Brinco del Tigre in 1976. L1 exhibits P? and Ps or M; the others exhibit P^5 and M.

Figure 16. Laboratory-raised N253 individuals descended from fish collected at Estero Dulce in 1976. L1 to L4 are mature females while R1 to R4 are mature males. L1, R1, and R2 are wild-type; L2 and L3 have D; L4, R3, and R4 exhibit D/Cu.

exclusively Ct (pedigrees N182, N184, and N191, with three exceptional "Ct" progeny in N164). Two of the females were heterozygous *C/CT* but appeared to exhibit the "Ct" modification because the crescent portion extended as dark streaks into the tailfin. Their Ct-like progeny (N183 and N199) exhibited both Ct and "Ct" in equal frequencies. These data are consistent with the hypothesis that "Ct" is a modification of Ct caused by an unlinked allele. The modification shows up in another pedigree (N346), which is currently being bred to test the hypothesis. The pedigree N346 was derived from a cross between a wild-type female and homozygous cut-crescent male exhibiting the "Ct" modification (Table IV). The Ct and "Ct" patterns show up in equal frequency in both sexes of N346, which implies that the modifier locus is autosomal.

I have designated this putative modifier as *Mod*-1 and calculated its frequency in various populations (Tables VI and IX) on the assumption that its visible effects are dominant. The frequency of *Mod*-1 was taken as the frequency of "Ct" divided by the summed frequencies of "Ct" and Ct. Because of the variability in expression of the "Ct" pattern, the phenotypic frequencies reported are probably underestimates of the true frequencies.

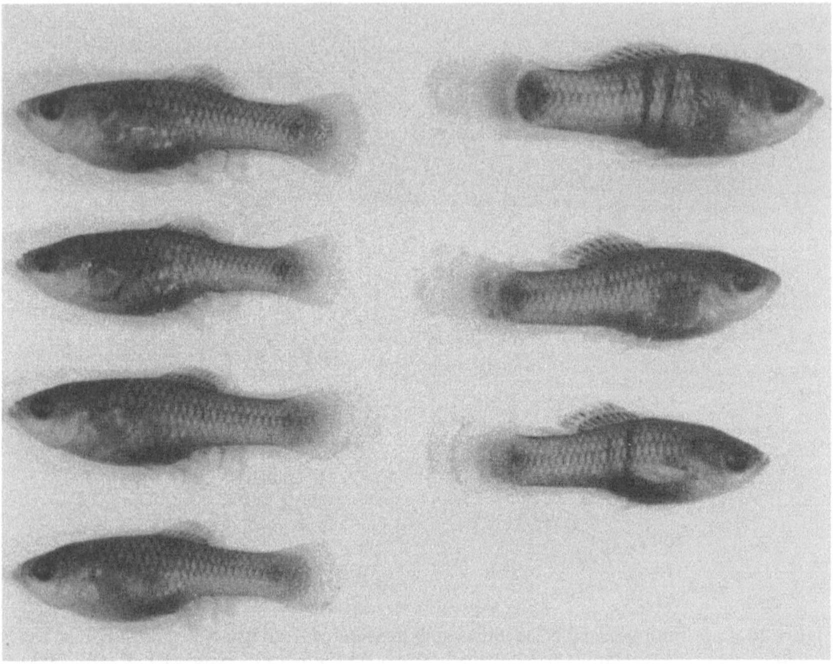

Figure 17. Laboratory-raised N254 individuals descended from fish collected at Brinco del Tigre in 1976. L1 to L4 are mature females while R1 to R3 are mature males. All have Ps, which is better expressed in these males than in the females. R1 illustrates the added crescent component that shows up in some Ps fish when the pattern is well expressed.

Mod-1 appears to have a mosaic distribution. It is common in populations of the Río Sabinas, uncommon in the Río Frio to the south, but common in the two samples from the Ríos Boquilla and Comandante to the west of the Río Frio. Similarly, it is relatively common in populations from the Río Coy, but it was not detected in three small samples from the nearby Río Moctezuma.

Two observations suggest that Mod-1 may be more common in rocky upstream areas than in downstream areas. First, the relative frequency of "Ct" is significantly higher in the Sarco upstream population (47/66) than it is downstream (170/563; $\chi^2 = 42.0$, df = 1, $P << 0.001$). Second, in the downstream Sarco area, there appears to be a cline in relative frequency of "Ct", which increases from 0.26 at the lower and (pool 1 and below) to 0.54 at the upper end (pool 9 and above). The distance between the two extremes is about 0.5 km and the upper pools are more exposed and rockier, with less emergent vegetation than the lower pools. The apparent cline is of borderline statistical significance (Table VI).

Mod-1 is correlated with "relative condition," at least in Ct fish, and the data are presented in Section 6.1. Whether Mod-1 has any physiological effects in fish lacking a *Ct* allele is not known.

3.3. Bodyspot Locus

The frequencies of bodyspot alleles are difficult to estimate reliably because the patterns vary in expression. Therefore, bodyspot phenotyping of collected animals is less reliable than tailspot or allozyme classification. In addition, some of the patterns are incompletely penetrant, expression being related to size or sex. For this reason, samples fixed in the field give unreliable estimates of allelic frequencies, even if the alleles present in a population have already been identified. Furthermore, it is known that the expression of a bodyspot allele is influenced by alleles at other loci and that similar patterns in different river systems can be caused by different alleles (Kallman, 1970). Therefore, a firm statement that the same

Figure 18. Collected at Aldama in 1957. The top fish has Ct. The bottom fish has P^2 and Ct. The others have a diffuse pattern that is interpreted as "Ct".

Table IV
The Inheritance of Cu and "Ct" in *Xiphophorus variatus*[a]

Pedigree number	Parental phenotypes: Female	Male	Progeny phenotypes: Ps	Cu	C/Ps	C/Cu	Ps/Cu		Others[b]
N283a	C	Ps/Cu	7	3	9	13			
N314a	Ps/C	Ps/Cu	20		5	6	18		1 WT,5Cu
N314b	Ps	Ps/Cu	6	8[c]			6		6WT
N256[d]	+	Ps/Ct	22	30					
			Female:	Ct	"Ct"	Male:	Ct	"Ct"	
N346	+	"Ct"		3	6		3	3	

[a]Pedigrees N283a and N283b were derived from pedigrees N266a and N266b. Pedigrees N266 and N256 were the offspring of fish collected from the Río Nautla drainage (Brinco del Tigre). Pedigree N346 was derived from a female collected at the Encino locality and a male collected at Arroyo Sarco (Río Sabinas drainage).

[b]Ps and C were variable in darkness in pedigrees N283 and N314. The unexpected "others" represent nonexpression of these patterns.

[c]The ventral component of Cu was variable in expression in pedigrees N314 and N256. In all cases it was expressed, but in some fish it was darker than in others. Three of the eight Cu in N314b could have been scored as Ct in phenotype.

[d]The male parent of N256 was classified as Ps/Ct, not Ps/Cu, but this classification must be attributed to the variable expression of Cu!

allele is found in two different river systems cannot be based solely on the occurrence of similar phenotypic expressions, but requires verification from reciprocal crosses of lines from the two systems. Kallman's work also implies that different phenotypic expressions from two river systems might be caused by identical alleles. For these reasons, a meaningful study of bodyspot diversity and its geographic variation would require an extensive dedicated breeding program. In spite of these difficulties, the evolutionary implications of the bodyspot polymorphism make it worthy of further study.

Because of the difficulties in establishing gene homology for similar bodyspot patterns from different localities, the listings in Table VII are tentative. For example, males with a pigmented gonopodium ("black gonopodium" Gn), or a linelike (Li) pattern are found in all three of the major geographic areas. Both of these patterns are phenotypically variable, but their ranges of expression in all of the areas are similar. Whether fish from different areas have the same *Gn* or *Li* alleles remains to be tested. None of the other patterns appears to be ubiquitous, and some may be unique to particular areas. For example, Sc in *X. variatus* (Fig. 7) has been found only among fish from the Rio Cazones drainage and presents no problems of homology. Also, the patterns from the Aldama and Jaumave populations are tentatively considered to be unique (Figs. 18 and 13). Since each is represented by only a single preserved fish, however, their homologies could not be determined, even if necessary.

The homologies of the other patterns are even less certain. For example, Table VII lists P^1 for all three geographic areas. While the phenotypes in fish from the Tamesi and Panuco systems are almost indistinguishable (Figs. 3–6 and 10), they differ from the P^1-like pattern present in the Southern rivers. In fish from the Tamesi and Panuco drainages, P^1 spots are round and discrete, while in fish from the Southern rivers, P^1-like spots are more elongate, tend to run together, and are fewer in number. This Southern pattern is tentatively classified as P^1 because the distribution of spots on the body is similar to that exhibited by northern fish, but the homology is uncertain. Similar problems exist with respect to the homology of P^2-like patterns of the Southern rivers and the Río Tamesi and the P^4-like pattern of the Río Panuco and the Southern rivers.

All of the patterns listed in Table VII have been bred in my laboratory,

Table V
Partial Pedigrees Showing the Inheritance of Cut-crescent Phenotype and Its Modification Pseudo Cut-crescent (''Ct'')[a]

Pedigree number	P1 Female phenotype	Progeny phenotypes	
		Ct	''Ct''
N164[b]	Ct	14	3
N168	Ct	30	0
N169	Ct	13	0
N170	Ct	26	0
N171	Ct	28	0
N172	''Ct''	11	10
N182	Ct	29	0
N183[c]	C/Ct	17	8
N184	Ct	26	0
N188[d]	''Ct''	12	16
N191[e]	Ct	16	4
N199[c]	C/Ct	3	4

[a]These pedigrees were derived from wild-caught females fertilized in nature. The genotypes of the fertilizing males were established through analysis of the full progeny (Borowsky and Kallman, 1976; Borowsky and Khouri, 1976). Only the phenotypes of the Ct-type young are listed, however. In those cases where the P_1 female was Ct in phenotype, it was still possible for a pattern modifier to be contributed by a fertilizing male.

[b]The three are unambiguously ''Ct''.

[c]The P_1 females might have been C/''Ct'' in phenotype.

[d]Four of the 16 could be C/Ct in genotype.

[e]The four ''Ct'' express the modification only faintly and might really be Ct.

Table VI
Frequencies of Modified Cut-Crescent Patterns Out of All Cut-Crescents (Mod-1 Ratio) Obtained in Collections from Successive Pools of Arroyo Sarco[a]

Pool number	Mod-1 ratio	Distance
1	28/109 = 0.257	1
3–4	5/21 = 0.238	3.5
5	4/14 = 0.286	5
6	54/174 = 0.310	6
7–9	20/37 = 0.541	8

[a]The data are combined from the 1967 and 1977 collections, which were the most extensive at the site. Pools were numbered in order moving upstream and pool number is used as a measure of distance along the stream.

with the exceptions of P^7 and P^8, and are sex-linked (Borowsky, unpublished data). The patterns listed in Table VII as "previously described" are treated by Kallman and Atz (1965) and Borowsky and Kallman (1972). Photographs of the new patterns are provided in this chapter in lieu of descriptions. The new patterns will be formally described and their genetics detailed elsewhere.

Table VII
The Distribution of Bodyspot Patterns in Populations of *Xiphophorus variatus*[a]

Pattern	River system: Tamesi	Panuco	Southern independent	Figure references
Previously described				
Black gonopodium (Gn)	+	+	+	5
Spotted caudal (Sc)			+	7
P^1-like	+	+	+	3, 4, 5, 6, 10
P^2	+		+	
Line (Li)	+	+	+	8, 10, 12
New				
P^3	+			6
P^4		+	+	7
P^5			+	14, 15
P^6		+		8
P^7 (Jaumave)	+			13
P^8 (Aldama)	[b]			18

[a]"Southern independent" rivers include the Cazones, Tecolutla, and Nautla systems.
[b]Independent river north of Tamesi.

In addition to the patterns listed in Table VII, there are three others that are not listed because little is known of their mode of inheritance or geographic distribution. Two of them appear in males at maturity and the third has only been observed in males. Gravid spot (Gs) is exhibited by some males from all river systems and looks like the gravid or "pregnancy" spot of mature females. It is caused by melanic pigmentation in the peritoneal lining and appears as a large, diffuse darkening of the ventral caudal section of the abdomen (Figs. 4, 10, and 12). Another pattern, abdominal spot (As), is a small (2–3 mm) black spot appearing on the abdomen near the base of the gonopodium in some mature males. This pattern appears in some lines derived from the Río Sabinas, but may be more broadly distributed. The genetics of both As and Gs are currently under investigation. A third, unnamed, pattern has only been observed in a laboratory line derived from the Río Sabinas and is also under study. This pattern consists of a series of melanic speckles along the ventral lateral surface of the fish and is unlike any other I have seen in *Xiphophorus*.

Lewontin (1974) has stressed that persistent or widespread linkage disequilibrium is strong evidence that selection is acting upon a locus. Kallman (1970) has shown that sex-linked melanic pigmentation and YR coloration patterns are restricted to the X and Y chromosomes in certain populations of *X. maculatus* and do not occur on the W chromosome. The W chromosome is homologous to the other sex chromosomes, only has the + allele at the pigmentation loci, and is female-determining. Kallman suggested these coloration patterns confer an advantage on males and a disadvantage on females by signalling "maleness" during social encounters. This hypothesis was supported by later work demonstrating that wild-caught females with such patterns gave birth to fewer young in the laboratory than did females lacking patterns (Borowsky and Kallman, 1976). The young derived from stored sperm from inseminations in nature and the difference in numbers of young suggested that females with the patterns had been inseminated by males less often.

Tests for linkage disequilibrium of sex-linked patterns in *X. variatus* have been carried out as well (Borowsky and Khouri, 1976). The data in Table VIII show that the P^2 allele at Arroyo Sarco (h) is restricted, or nearly so, to the X chromosome. None of the other alleles at Sarco or Encino (g) give evidence of linkage disequilibrium. However, the P^1 pattern in the Frio/Boquilla/Comandante drainage is restricted to males and the allele must be excluded from the X chromosome (R. Borowsky, unpublished data). The few data on linkage in these populations support this belief, but the sample size is too small to test the hypothesis (Table VIII). However, the data from localities m, p, and e strongly suggest that *P* alleles, in general, are biased toward Y linkage in those areas. Therefore,

Table VIII
Linkage of Alleles at the Bodyspot Locus in *Xiphophorus variatus*[a]

Locality	Chromosome	Allele +	P1	P2	P3	*Gn*
Sarco (g)	X	75	7	21	1	9
	Y	38	4			4
Encino (h)	X	68				16
	Y	45				4
Frio/Comandante	X	20				2
(m and p)	Y	6	2	2		3
Compuertas (e)	X	11				
	Y	3	3			

[a]The numbers of chromosomes bearing the particular alleles are listed. Data from Borowsky and Khouri (1976) pooled with new data from pedigrees N143, N144, N148, N150, and N155.

linkage disequilibrium for sex-linked alleles does show up in *X. variatus,* but the pattern of linkage varies from one area to the next.

It is also of interest that some of the sex-linked patterns are partially sex-limited in expression and, further, that the degree of expression is correlated with relative size. In *X. milleri,* for example, males raised together differ in their expression of *Gn*, with larger males having more pigmentation (Kallman and Borowsky, 1972). In *X. variatus,* the number of P^1 spots exhibited by a male is also size-correlated when the males are raised in groups (Borowsky, 1969). Furthermore, the expression of P^1 in this species is better in males than in females (Fig. 5). Also, the expression of YR coloration in *X. variatus* is correlated with male size (Borowsky, 1973*a*). In all three of these examples, relative size within a group of males rather than absolute size is the important determinant of expression. This shows that expression of these patterns is under social control and implies that they serve a signal function during social encounters. If these patterns have an epigamic function, linkage disequilibrium between the sex chromosomes might result.

3.4. Allozyme Loci

In striking contrast to the high degree of polymorphism at the pigmentation and *Pit* loci, there is little allozyme variation in *Xiphophorus.* The most probable explanation for this difference is that allozyme polymorphism in these fishes is neutral, or nearly so, while the pigmentation and *Pit* loci are subject to strong balancing selection. While the ancestral *Xiphophorus* may have exhibited a high degree of enzyme polymorphism, this variation, if neutral, would have been lost through drift during the dispersal of the modern species (founder effects) and during their subsequent isolation (small population effects). Both mechanisms were probably important, but the distributions of alleles at two loci (*MDH-1* and *MDH-2*) suggest that the species may have dispersed completely with much of its original variability intact.

The most isolated population of *X. variatus* I have encountered is the one at locality, b, high in the synclinal Jaumave Valley. The platyfish must have been at Jaumave prior to the birth of the Sierras because the descent of the river is too steep to permit migration upstream. (It is apparent that the Río Guayalejo was in its present position prior to the birth of the Sierra Madre because of the river's entrenched meanders on its course from the valley to the coastal plain.) To my knowledge, the orogenic event has not been dated, but it is clear that the isolation of the Jaumave population has been complete for a long time.

It is of interest, therefore, that the Jaumave population contains two alleles at the *MDH-1* locus (0.71 and 0.85) that were not detected in any of the other 16 populations of *X. variatus* sampled (N = 416 alleles). Although it is possible that these two alleles originated in the Jaumave population, the more likely explanation is that *MDH-1* variation was lost by drift subsequent to the dispersal of the species, rather than by founder effects during the dispersal. The distribution of alleles at the *MDH-2* locus further supports this hypthesis. The only allele detected in the populations of the Río Tamesi drainage is 1.00, yet this allele was not found in any of the populations of the Río Panucó drainage or further south. If 1.00 in the Río Tamesi system was simply the sole founder from a larger pool of alleles present to the south, it had to have been present in populations found to the south. Its loss from those populations must have happened subsequent to the dispersal.

Although none of the five species of *Xiphophorus* screened for allozyme variation exhibits much, *X. variatus* is clearly more polymorphic than any of the others (Table IX). There are several possible reasons for this. The other species are swordtails and are generally found in the headwaters, while *X. variatus*, a platyfish, is found more often in the lowlands than in the headwaters. If headwater forms disperse among rivers by a different mechanism than lowland forms, this could account for the difference in genetic variability, because stream capture among the headwaters may be a more effective filter of genetic variation than the merging of waters in the swampy lowlands. A second possibility is that headwater forms have smaller effective population sizes than lowland forms and lose variability more readily through drift. Yet another possibility is that the ancestor of the swordtails was less variable than the

Table IX.
Tailspot Phenotypic and Allelic Frequencies in Natural Populations of *Xiphophorus variatus*

River system population	Year	N	Tailspot phenotypic frequencies: +	C	Ct	Ps	C/Ct	Ct/Ps	C/Ps	Other	Mod-1 ratio	Allelic frequencies: +	*C*	*Ct*	*Ps*	*Cu*	Other	*G*
Río Tigre																		
a Aldama	1957	9	2		7						5/7	0.447		0.553				1.99
Río Tamesi																		
Río Guayalejo																		
b Jaumave	1967	141	120			21						0.923			0.077			
	1970	28	23			5						0.907			0.093			
	1975	20	14			6						0.839			0.161			
	1977	63	51			12						0.900			0.100			
	Mean	252										0.909			0.091			1.36
c Llera	1930	19	14		4	1						0.862		0.112	0.027			1.60
d East of Llera	1966	7	3		4							0.670		0.330				1.89
e Compuertas	1966	16	10		5	1						0.797		0.171	0.032			
	1967	21	17		4							0.901		0.099				
	1975	16	7		9							0.668		0.332				
	Mean	53									0/5	0.797		0.194	0.010			1.73
"16 M North of Xicotencatl"	1930	80	31	16	28		5					0.625	0.141	0.234				2.48
Río Sabinas																		
f Santa Fe	1966	31	10	7	14							0.610	0.123	0.226				
	1967	45	17	17	10		1					0.638	0.229	0.133				
	Mean	76									6/24	0.627	0.186	0.188				2.51
g Encino	1967	58	35	11	11		1					0.781	0.110	0.110				
	1970	127	66	21	35		5					0.720	0.108	0.172				
	1973	447	221	81	118		27					0.694	0.128	0.177				
	1974	99	59	21	18		1					0.780	0.119	0.102				
	1975	238	141	45	50		2					0.779	0.105	0.117				
	1976	49	26	11	11		1					0.737	0.132	0.132				

	1977	100	67	24	9					0.825	0.129	0.046		
	1981	94	35	21	35		3			0.630	0.139	0.231		
	Mean	1212							57/233	0.732	0.121	0.147		2.15
h Sarco	1966	117	43	20	46		8			0.606	0.128	0.266		
	1967	774	330	165	237		42			0.655	0.144	0.201		
	1970	80	26	27	23		4			0.591	0.220	0.189		
	1973	571	233	134	190		24			0.638	0.151	0.211		
	1974	135	50	28	48		9			0.611	0.148	0.240		
	1975	311	127	60	106		18			0.641	0.135	0.225		
	1976	101	40	11	44		6			0.624	0.088	0.288		
	1977	615	251	120	223		21			0.652	0.123	0.225		
	1981	21	6	4	11					0.578	0.103	0.319		
	Mean	2725							170/563	0.641	0.139	0.220		2.44
h′ Sarco upstream	1981	127	61		66				47/66	0.694		0.306		1.85
i La Flor	1967	45	24	7	14					0.747	0.082	0.172		
	1974	17	13	3	1					0.877	0.093	0.030		
	Mean	62							6/11	0.783	0.085	0.133		1.95
j River at	1975	64	36	8	17					0.741	0.089	0.170		
"Storm's"	1977	54	32	8	14					0.782	0.078	0.140		
	Mean	118							12/16	0.760	0.084	0.156		2.03
j′ Stream tributary	1976	96	58	18	18		2		5/14	0.779	0.110	0.110		1.97
k Route 85	1967	23	10	7	4		2		0/2	0.643	0.218	0.139		2.44
1 Adjuntas	1967	37	21		15			1	3/13	0.741		0.245	0.014	1.87
Río Frio														
m River	1967	172	105	1	50	13	1	2		0.781	0.006	0.168	0.045	
	1970	21	5		2	2				0.771		0.101	0.128	
	1975		7		2	3								
	Mean	193							1/38	0.780	0.005	0.161	0.054	1.96
m′ Stream	1967	22	5		4					0.855		0.145		
tributary	1975		11		2									
	1977	91	63		26	2				0.834		0.156	0.011	
	Mean	113							4/15	0.838		0.154	0.009	1.61

(*continued*)

Table IX (*Continued*)

River system population	Year	N	Tailspot phenotypic frequencies +	C	Ct	Ps	C/Ct	Ct/Ps	C/Ps	Other	Mod-1 ratio	Allelic frequencies +	*C*	*Ct*	*Ps*	*Cu*	Other	*G*
Río Boquilla																		
o North of Chamal	1977	36	16	3	16		1				5/7	0.669	0.057	0.274				2.20
Río Comandante																		
p Tributary	1975	28	11	2	9	5		1				0.647	0.037	0.201	0.115			
	1976	46	27		15	1	2		1			0.742	0.033	0.203	0.022			
	1981	76	43	3	19	3	5	1	2			0.716	0.067	0.177	0.039			
	Mean	150									8/36	0.711	0.051	0.189	0.048			2.35
Río Mante																		
"Juarez"	1939	6	2	1	1	1		1			0/1	0.553	0.181	0.181	0.086			3.18
Río Panuco																		
Río Valles																		
q Valles	1930	117	85	15	11		6					0.832	0.093	0.075				1.77
Río Tamuin																		
s San Francisco	1981	77	39	11	15	8	3			1 = cu	7/16	0.717	0.096	0.125	0.054	0.009		2.52
t San Vicente	1981	40	29	3	2	5	1				0/2	0.846	0.051	0.038	0.064			1.81
u Tamuin	1966	7	5	1	1							0.851	0.074	0.074				1.69
Río Coy																		
v 1 km North of Aquismon	1977	27	18	2	4	2				1 = cu		0.825	0.038	0.077	0.038	0.022		
	1981	75	40	10	19	4				2 = cu		0.747	0.070	0.137	0.027	0.018		
	Mean	102									3/26	0.768	0.062	0.121	0.030	0.019		2.25
v′ 1.3 km North of Aquismon	1981	5	2	1	2						0/3	0.663	0.107	0.230				2.34
v″ 1.9 km North of Aquismon	1981	62	19	22	18	3					1/18	0.607	0.203	0.163	0.025			2.76
w Santa Cruz	1977	3	1		1	1					1/1	0.6	0.2	0.2				2.59
x Route 85	1974–77	36	16	7	11		1	1				0.667	0.118	0.201	0.014			

	1981	20	9	2	9							0.687	0.052	0.261			
	Mean	56									2/10	0.674	0.094	0.222	0.009		2.37
y Pequeseh	1974	35	4	1	4							0.682	0.107	0.211			
	1976		11	6	9												
	1977	66	21	16	29							0.607	0.133	0.268			
	1981	255	90	80	66	4	15					0.607	0.210	0.175	0.008		
	Mean	356									2/108	0.614	0.186	0.194	0.006		2.61
z Near Rt. 85	1977	129	49	33	33	6	3	2	2	1 = Cu	3/33	0.635	0.157	0.162	0.040	0.006	2.81
aa Road to San Joachin	1977	12	5	2	4	1					2/4	0.680	0.089	0.187	0.043		2.53
bb San Joachin	1977	12	7	1	3		1				0/3	0.734	0.086	0.180			2.11
Río Moctezuma																	
cc Huichihuayan	1975–77	15	10	4		1						0.822	0.144		0.034		
	1976	30	22	6		2						0.860	0.106		0.034		
	Mean	45										0.847	0.119		0.034		1.66
dd Matlapa	1939	11	6		2	2				1 = Cu		0.760	0.097	0.097	0.047		2.24
ee Plan de Jalpilla	1939	19	9	4		3				3 = Cu		0.719	0.113	0.084	0.084		2.46
ff Axtla	1932	122	70	19		14			3	1 = C/Cu		0.760	0.095	0.077	0.068		2.25
										14 = Cu							
										1 = Cu/Ps							
gg 10.4 km Northwest of Chapulhuacanito	1981	32	13	2		3				13 = Cu	0/13	0.651	0.032		0.065	0.252	2.50
										1 = Cu/Ps							
										8 = Cu/Ps							
gg′ 11 km Northwest of Chapulhuacanito	981	87	30	12		9				4 = C/Cu	0/24	0.571	0.096		0.102	0.232	3.05
										24 = Cu							

(continued)

Table IX (*Continued*)

River system population	Year	N	+	C	Ct	Ps	C/Ct	Ct/Ps	C/Ps	Other	Mod-1 ratio	+	*C*	*Ct*	*Ps*	*Cu*	Other	*G*
		Tailspot phenotypic frequencies										Allelic frequencies						
h Coxcatlan	1967	23	7	4	3		1					0.795	0.115	0.090				
	1974		8															
	1981	25	7	15	3							0.561	0.375	0.063				
	Mean	48									0/2	0.673	0.250	0.076				2.25
ii Tampamolon	1981	53	34	8		3				1 = C/Cu 6 = Cu	0/6	0.795	0.099		0.038	0.068		2.05
jj 16 km North of Tanquian	1981	8	8									1.000						1.0
Río Tempoal																		
kk **Río Calabozo**	1930	32	28			1				3 = Cu		0.936		0.016		0.048		1.31
Río San Pedro																		
11 San Martin	1981	46	13	9	14			1		6 = Cu 3 = C/Cu	15/20	0.535	0.158	0.197		0.110		3.28
mm 2.7 km Northwest of Chapulhaucanito	1981	9	3	2		1				2 = Cu 1 = C/Cu		0.576	0.183		0.058	0.183		3.02
mm′ 1.8 km Northwest of Chapulhuacanito	1981	53	24	12	1	7				3 = C/Cu 5 = Cu 1 = Cu/Ps	0/6	0.675	0.150	0.018	0.091	0.066		2.77

nn 5.1 km North of Chapulhuacanito	1981	21	6	7		1		1	5 = Cu 1 = C/Cu	0/5	0.548	0.246		0.049	0.156		3.04
oo **Río Tempoal**	1930	166	98	7	33	13	4	2	8 = Cu 1 = C/Cu		0.765	0.027	0.118	0.062	0.027		2.30
pp **Río Cazones**	1976	10	8			1			1 = Cu		0.897				0.051	0.051	1.49
qq **Río Tecolutla**	1976	57	36	12					6 = Cu 1 = C/D 2 = D		0.798	0.122			0.054	0.059 = D	2.00
rr Estero Dulce	1976	150	74	41					17 = Cu 13 = D 1 = M 1 = Cu/D 1 = Cu/C 2 = C/D		0.712	0.160			0.066	0.003 = M 0.059 = D	2.46
Río Nautla																	
ss Brinco del Tigre	1976	34	6	6	2	5			4 = Cu 3 = D 3 = M 1 = Ct/D 1 = Cu/D		0.472	0.130	0.096	0.096	0.079	0.049 = M 0.079 = D	5.03
tt **Río Quila**	1976	5	1	1		1			2 = Cu		0.535	0.112		0.112	0.239		3.21

[a]The Mod-1 ratio is the number of Ct animals exhibiting the "Ct" modification out of the total number of Ct fish phenotyped. When data for more than one collection are presented, the total sample size and mean allelic frequencies are given on the line labeled "Mean." Mean allelic frequencies are weighted averages of all years' collections. The index of allelic diversity G is defined on p. 240 and was calculated from the mean allelic frequencies.

ancestor of the platyfish. That is, the difference predated the emergence and dispersal of the modern species. The present data are inadequate to distinguish among these, and other, possibilities.

Genetic distances among species were computed using a Minkowski metric (Powell *et al.*, 1972), and a dendogram was drawn on the basis of the results. The agreement between this dendogram and a phylogeny Rosen (1979) constructed using cladistic methods, is remarkable, considering the small number of variable loci found in this study (Fig. 19). Rosen's determination of the species' relationships is based on far more characters and is more reliable than mine, but the close agreement of the two trees demonstrates that the allozyme data contain considerable taxonomic information. Future work with more discerning biochemical methods should be of value in the taxonomy of this group.

3.5. The Pituitary Locus

The timing of sexual maturity in males of several species of *Xiphophorus* is strongly determined by genetic factors. Kallman and co-workers (Kallman, 1975; Kallman and Borkoski, 1978; Kallman and Schreibman, 1973) have demonstrated the existence of a sex-linked locus with a multiple allelic series controlling the onset of maturation in *X. maculatus* and

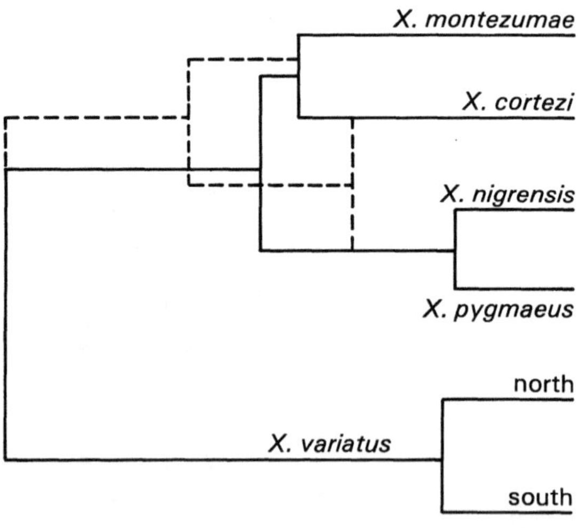

Figure 19. Dendrogram illustrating the relationships among five species of *Xiphophorus* based on allozyme data. Genetic distances among taxa are proportional to the distance from the right edge to the common node. The dotted line alternative in the upper branches gives the result most closely fitting the data, while the solid lines take into account the affinities proposed by Rosen (1979).

X. pygmaeus. Genetic factors for size have also been demonstrated in *X. milleri* (Kallman and Borowsky, 1972). In addition, Kallman (1983) presents evidence that the polymorphism exists in *X. montezumae*. The *Pit* locus determines the timing of development of the gonadotrop layer of the pituitary gland, which triggers the onset of sexual maturity. This occurs earlier and at a smaller size in some *Pit* genotypes than in others. The site of action of the *Pit* locus may not reside in the pituitary gland, a point emphasized by Kallman and Borkoski (1978). Bao (1981) was able to induce early maturation in late-maturing genotypes by the administration of luteinizing hormone releasing factor (LHRF), which suggests that the focus of the *Pit* locus lies elsewhere.

Until recently, I looked in vain for evidence that such a polmorphism exists in *X. variatus*. The search was conducted by mating females that were heterozygous for X-linked pigment markers to males lacking Y-linked pigment markers that would mask, in their male progeny, the patterns inherited from their mates. Matings of this kind produce two classes of male offspring, which can be distinguished on the basis of their X-linked pigmentation markers. Since recombination between the pigmentation marker locus and the *Pit* locus is rare (Kallman and Borkoski, 1978), the two classes of male progeny should differ in size, as well as pigmentation, if their female parent had been a *Pit* heterozygote.

The male progeny of these crosses were measured at maturity, and in all 16 pedigrees (27 broods) no significant size differences between the two male classes were observed (Table X). Although small differences in size between the two classes might not have been detected because of the small sample sizes in some cases, the consistency of the observations suggests that X-linked variation at the *Pit* locus is not of major significance in *X. variatus*.

In contrast, two lines of evidence suggest that Y-linked *Pit* variation is important in this species. First, a comparison between the lengths of the male parent and male progeny in those of the 16 lines where paternal data were available revealed strong resemblances (Fig. 20, $r = 0.91$, $t = 5.06$, os$P = 0.002$). In this regression, environmentally caused variation in offspring length was minimized by selecting a subset of all progeny for analysis. The mean progeny length for each pedigree was taken as that of the single brood in which the males were the largest. The rationale for the selection was that it eliminated from consideration broods in which males were stunted because of suboptimal growing conditions.

The second line of evidence suggesting that male adult size is heritable comes from two other pedigrees, N293 and N295, which exhibit a difference in adult male size persisting over a wide variety of environmental conditions. These two pedigrees are of particular interest because they

Table X
Mean Adult Lengths (mm) of *Xiphophorus variatus* Males[a]

Pedigree and brood	Class I X	Class I SD	Class I (N)	Class II X	Class II SD	Class II (N)	P Male	Source population
2126								
4/7	30.9	± 2.4	(8)	31.4	± 1.9	(11)	—	Encino
5/8	31.8	± 1.1	(5)	30.9	± 2.4	(6)		
10/20	30.6	± 1.7	(6)	33.0	± 1.9	(2)		
12/14	31.6	± 1.3	(10)	32.2	± 0.1	(2)		
2/26	29.8	± 1.6	(10)	30.6	± 2.1	(8)		
2132								
3/11	32.9	± 5.0	(4)	30.5	± 4.1	(7)	—	Sarco
2/8	29.7	± 4.2	(2)	28.5	± 3.0	(7)	—	Río Tamesi
2343								
12/13	31.8	± 0.8	(7)	32.2	± 1.2	(3)	—	Mixed Río Sabinas
N164								
10/17	23.4	± 1.0	(4)	24.2	± 0.4	(4)	—	Sarco
N183								
7/14	23.6	± 0.9	(7)	22.5	± 0.4	(3)	—	Sarco
N184								
7/22	26.1	± 1.9	(4)	25.1	± 0.4	(3)	—	Sarco
N188								
6/4	23.9	± 1.6	(4)	24.2	± 2.4	(5)	—	Sarco
N188								
6/4	23.9	± 1.6	(4)	24.2	± 2.4	(5)	—	Sarco
N251								
5/29*	28.3	± 3.2	(9)	26.8	± 2.7	(15)	29.6	Río Nautla
7/5	24.0	± 2.1	(6)	25.6	± 1.7	(5)		
8/30	25.3	± 2.5	(7)	26.1	± 2.2	(10)		
N254								
6/14*	29.1	± 1.7	(3)	26.8	± 1.2	(3)	30.6	Brinco
7/26	24.4	± 1.9	(7)	24.2	± 2.2	(4)		
9/18	20.8	± 1.2	(3)	21.6	± 2.4	(5)		
N283								
Mixed*	23.4	± 1.0	(6)	23.5	± 2.4	(6)	26.6	Brinco
N285								
3/5	28.5	± 0.9	(4)	29.3	± 0.9	(6)		Brinco
N288								
Mixed*	27.9		(1)	30.2	± 1.4	(2)	30.9	Brinco
N305								
3/6	27.4	± 4.5	(5)	25.1	± 4.0	(3)	32.7	Brinco
5/6*	29.5	± 1.4	(8)	30.5	± 1.3	(7)		
N307								
3/8*	28.5	± 0.8	(2)	29.3	± 0.3	(2)	30.4	Río Cazones
N314								
5/1	24.0		(1)	25.7	± 1.6	(4)	27.0	Brinco
6/20*	27.9	± 1.0	(4)	25.5	± 2.4	(5)		
N315								
Mixed	26.0	± 4.3	(5)	29.2	± 1.0	(5)	—	Río Tecolutla

[a]The two classes of siblings differ for X-chromosome markers. The "Río Tamesi" stock is described by Kallman (1975). The starred broods form the subset used for the regression of offspring length on male parent length (Fig. 20).

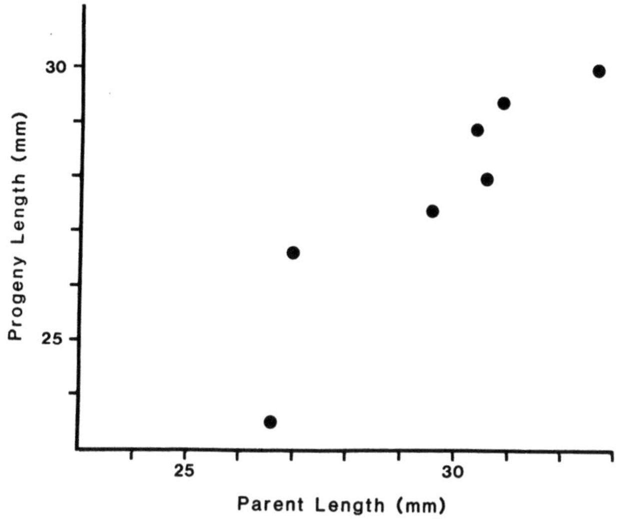

Figure 20. Regression of mean male offspring length on the length of the male parent for seven pedigrees of *X. variatus*.

were derived by reciprocal crosses from the same Río Sabinas stocks. Whether the fish were raised in groups or in isolation, at a variety of temperatures and feeding schedules, N293 males are consistently smaller than N295 males (Table XI). The hypothesis that the difference between N293 and N295 is due to Y-linked variation is currently being tested with reciprocal crosses between the pedigrees and their derivatives. The hypothesis predicts that male progeny of such crosses should resemble their fathers in size, rather than their maternal uncles. Preliminary data from four such matings, two of each type, are now available and support the Y-linkage hypothesis. Under uniform conditions in isolation, male progeny of "small" males start to mature at 20.7 ± 1.3 mm ($N = 5$) and reach maturity at 24.1 ± 1.7 mm ($N = 5$). For male progeny of "large" males the comparable figures are 24.2 ± 2.1 ($N = 14$) and 27.8 ± 7.5 ($N = 11$). The two lines differ significantly in size at both stages (student's $t = 3.46, 2.81$, os$P < 0.01$). Data are still being accumulated and the test will be carried out for one more generation, but Y-linked *Pit* allele-like variation in *X. variatus* is clearly indicated by these results.

Finally, it is worth noting that one male (N226-20) collected at Estero dulce (locus *rr*, Fig. 2) in November 1976 and maintained in the laboratory until its death 31 months later never matured. The animal reached a length of 45 mm. Another male (N313-11), derived from the same population also showed the same failure to develop (47 mm length, and immature at

Table XI

Adult Male Lengths in Four Pedigrees of *Xiphophorus variatus*[a]

Pedigree and brood	Mean length and standard error, mm	Sample size	Comments
N293			
2/10/80	23.2 ± 0.60	6	Isolated until start of maturation, then paired in 9-liter tanks
2/25/80	22.7 ± 0.27	4	Isolated in 9-liter tanks
5/14/80	22.8 ± 0.22	11	Isolated in 9-liter tanks
6/25/80	26.1 ± 0.45	18	Paired in 9-liter tanks
8/10/80	24.4 ± 0.37	11	Mixed: isolated physically but some visual contact with or circulating water from another tank, 9 or 5 liters
N295 (A and B)			
Mixed	31.5 ± 0.92	10	Raised in groups (A and B)
4/17/80	30.3 ± 0.30	10	Same as N293 8/10/80 (N295A)
N325			
9/11/80	24.9 ± 0.79	8	Raised in group
10/24/80	23.3 ± 0.25	12	Isolated or in visual contact with one other male, 9 liters
N326			
10/10/80	27.3 ± 0.60	4	Raised in a group
11/20/80	22.4 ± 0.29	9	Same as N325 10/24/80

[a]All fish were raised at 22°C and fed Tetramin staple food. N325 and N326 were derived from N293.

the age of 39 months). Kallman and Borkoski (1978) have reported that certain homozygous *Pit* genotypes in *X. maculatus* mature very late and at a large size, and Bao (1981) has demonstrated that certain genotypes in hybrid *Xiphophorus* never mature in the laboratory, presumably because their critical size for the onset of maturation exceeds what can be attained under even the most favorable of growth conditions. Alleles causing extreme delay in maturation could not survive in natural populations unless their properties in heterozygous combination overcame this disadvantage. The implications of N226-20 and N313-11 are that the *Pit* polymorphism is present on the X chromosome as well as the Y chromosome, at least in the source population. If all X chromosomes from the source population carried the same allele, the Y-linked allele in these fish would be, in effect, a dominant lethal. Its presence in the population implies the existence of X-linked alleles capable of nullifying this disadvantage. It is clear, however, from the results presented in Table X, that X-linked *Pit* variation is much less important than Y-linked variation in this species.

How is the *Pit* polymorphism maintained and why is it generally more variable in males than in females? It appears that males are subjected to two opposed evolutionary forces that promote, or permit, size polymorphism, while females are not. In poeciliid fish, male size is a good determinant of social dominance (Sohn, 1977), and dominant males generally are more successful than subordinates in fertilizing females (Gandolfi, 1971; McKay, 1971). Late-maturing males attain greater size than early-maturing ones and should, therefore, be favored by sexual selection. On the other hand, early maturation confers the advantage of a short generation time and an attendant increase in fitness via an increased intrinsic rate of increase *r*. In a stable, predictable environment, therefore, a balance between sexual and *r* selection would favor males of intermediate size. Spatial heterogeneity, and perhaps variation among years in the amplitude of seasonal changes, would favor male size polymorphism.

While sexual selection is a major determinant of fitness for *X. variatus* males, its direct effect on the fitness of females is less important. The majority of mature females collected from natural populations have been inseminated, usually more than once (Borowsky and Diffley, 1981; Borowsky and Kallman, 1976; Borowsky and Khouri, 1976). Therefore, in females, the effects of *r* selection will not be opposed by those of sexual selection and the optimal phenotype will be early-maturing. Since females, unlike males, continue to grow after reaching maturity, even the advantage of greater fecundity exhibited by larger females will not select for later maturation. Therefore, alleles for late maturation will be selected against when linked to the X chromosome, but will not be selected against when linked to the Y chromosome. This difference in fitness effect, conditional upon linkage, would lead to linkage disequilibrium at the locus.

4. Geographic Clines in Genetic Diversity

4.1. Tailspot Locus

The overall geographic distribution of tailspot diversity in *Xiphophorus* exhibits a striking pattern. Allelic diversity varies clinally in both *X. maculatus* and *X. variatus*, increasing in both species toward the center of their joint distribution. *Xiphophorus variatus* is least variable in the north (northwest), with an effective number of tailspot alleles of one or two. This increases to three or four in populations to the south (southeast). In *X. maculatus*, the cline runs in the opposite direction. Western (northern) populations are more variable than eastern (southern) populations, with diversity decreasing from four or five tailspot alleles in the Ríos Jamapa and Papaloapan down to three alleles in the Belize and Hondo rivers. Considering the two species as a single taxonomic unit, it is more variable towards the center of its distribution than at either end (Table XII, Fig. 21). This suggests that their common ancestor originated in the Jamapa/Papaloapan region and radiated out in both directions along the coast, speciating while losing tailspot allelic diversity through founder effects during the radiation.

It is of great interest that the clinal change is steeper in *X. variatus* than in *X. maculatus*. This suggests that the potential for gene flow among river systems is lower for the former species than for the latter, which is in accord with the fact that the Gulf lowlands along the Isthmus of Tehuantepec are extensively flooded during the rainy season.

Table XII
Genetic Diversity (2^H) in Two Species of *Xiphophorus* and Distance from the Río Jampa[a]

Species	River system or locality	Allozyme diversity	Tailspot diversity	Allelic diversity	Distance from Río Jamapa, km
X. variatus	Jaumave	1.11	1.00	1.36	640
	Aldama		1.00	1.99	520
	Guayalejo	1.00	1.27	1.73	600
	Sabinas	1.05	1.96	2.33	590
	Frio–Boquilla	1.15	2.05	2.04	570
	Valles–Taumin–Coy	1.28	2.61	2.53	460
	Moctezuma	1.24	3.63	2.61	420
	Tempoal		3.89	2.66	380
	Cazones–Nautla–Tecolutla	1.75	4.19	2.81	170
X. maculatus	Jamapa		3.82	4.77	0
	Papaloapan		5.58	5.34	120
	Coatzacoalcos		3.90	3.84	250
	Tonala		3.30	4.02	360
	Grijalva		4.16	3.16	450
	Usamacinta		3.48	3.25	570
	Hondo		2.76	3.25	870
	Belize		3.06	3.65	910

[a]Genetic diversities (see p. 240) were calculated from data in Tables III and IX and H. Gordon and Gordon (1957). For the allozyme loci, the indices were calculated by averaging the allelic frequencies in the river system and calculating an index of diversity for each locus, then averaging over the eight polymorphic loci. Tailspot diversity was calculated for the pattern alleles after excluding the wild-type allele. "Allelic diversity" includes the wild-type allele as well as the pattern alleles at the tailspot locus. Distances were measured between the centers of the drainages.

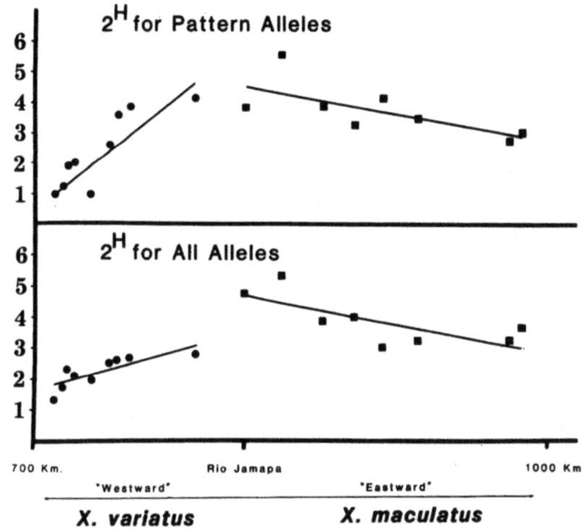

Figure 21. Tailspot allelic variability declines in *X. variatus* (●) and *X. maculatus* (■) with increasing distance from the Río Jamapa. The ordinate is 2^H (see p. 240) calculated from allelic data. Calculations for the lower half were done using the frequencies of all alleles at the tailspot locus. Calculations for the upper half excluded the frequencies of the + allele.

An alternative to the drift hypothesis is that the Jamapa/Papaloapan region provides more varied conditions for *Xiphophorus* than are found at the extremes of the distribution and that the greater genetic diversity there is an adaptive response to the increased environmental variability. While this may be true, my personal observations of *X. variatus* collecting sites do not support this hypothesis.

4.2. Allozyme Loci

The overall distribution of allozyme variability in *X. variatus* is similar to that of tailspot variability (Table XII). Mean genetic diversity (G, p. 240) at the eight polymorphic loci is greatest in the Ríos Cazones, Tecolutla, and Nautla (G = 1.75) and decreases at localities to the north (G = 1.00–1.11). The correlation between genetic diversity and distance from the Río Jamapa is striking [r = 0.97, $t(5)$ = 8.4, ts$P$ = 0.0004; Table XII]. Allelic diversities at the tailspot and allozyme loci are also well correlated (r = 0.84, $t(5)$ = 3.4, ts$P$ = 0.02; Table XII], which suggests that similar factors determine, at least partly, the overall distributions of allozyme and tailspot diversity. An important factor must have been drift and founder effects removing diversity as the species spread north from its point or origin.

A key difference between the tailspot locus and the allozyme loci is that nearly all populations are polymorphic for tailspots, but relatively few exhibit allozyme polymorphism. The contrast is more striking because the most variable enzyme loci were selected for analysis. The contrast probably reflects a difference in the importance of selection in the systems.

Tailspots are subject to strong selective forces but allozyme loci are apparently not. While alleles at both sets of loci would risk loss by drift at transfers from one river to the next (founder effects), once in a population, tailspot alleles would be protected from loss by selection, while allozymes would not. For example, if a tailspot allele was reduced in frequency at a population bottleneck, balancing selection in subsequent generations would adjust its frequency back toward equilibrium. On the other hand, allozymes reduced in frequency at a bottleneck would continue to drift in subsequent generations with an increased risk of eventual loss.

Nothing is known about the distribution of allozyme diversity in *X. maculatus*. However, the hypothesis of drift to account for the observations on tailspot diversity in both species and allozyme diversity in *X. variatus* permits three predictions to be made: First, allozyme diversity in *X. maculatus* will vary clinally, being greatest in the region of the Ríos Jamapa and Papaloapan and decreasing in populations to the east. Second, this cline will be less steep than the one in *X. variatus*. Third, *X. maculatus* will prove to be more polymorphic than *X. variatus* because greater gene flow among areas will increase the probability that lost alleles will be recovered from populations in adjacent river systems.

4.3. Bodyspot Locus

For reasons already discussed, genetic diversity of bodyspot patterns is more difficult to study than allozyme or tailspot diversity. Therefore, whether or not the overall distribution of bodyspot diversity mirrors the patterns exhibited at the other loci is unknown.

It is clear, however, that the number of bodyspot patterns in *X. variatus* is much higher than the four previously reported (Atz, 1961; Kallman and Atz, 1966). Within the lowland populations of the Río Tamesi drainage in the north, there are at least five distinctly different macromelanophore patterns. The same is true of the lowland populations of the Río Panuco. Within the populations of the Ríos Cazones, Tecolutla, and Nautla, at the southernmost range of the species (for this discussion this set of rivers will be labeled "Southern" rivers), there are seven. Furthermore, at the isolated Jaumave and Aldama localities at the northern range of the species there are two patterns, one per locality, that appear unique (Table VII).

Thus, at first glance, it would appear that the geographic distribution of bodyspot patterns does not support the generalization that genetic diversity in *X. variatus* increases from north to south. The Tamesi basin has a total of six patterns, the Panuco a total of five, and the Southern rivers have seven. On the other hand, the populations of the Tamesi drainage are the best known; many field collections have been made in that area and fish from those populations have been bred extensively. In contrast, relatively few lines of *X. variatus* from the Panuco drainage have been bred in the laboratory and few field collections have been made. Finally, the list of seven patterns for the populations of the Southern rivers comes from collections made on only one field trip. Other patterns must have been missed. Furthermore, the collections from the Southern rivers are poor indicators of bodyspot population diversity because they were made early in the breeding season (November 1976), few adults were collected, and immature fish express the patterns poorly (Fig. 11). It is likely, therefore, that further work on bodyspot patterns will support the generalization that genetic diversity is greater in southern populations of *X. variatus*.

5. The Tailspot Hypothesis

The essential elements of the tailspot hypothesis (Borowsky, 1981) are as follows:

1. Phenotypic differences among genotypes depend upon environment and are conditionally expressed.
2. These differences among genotypes reflect underlying physiological differences.
3. Each tailspot morph is a specialist to a different environmental subset, while the wild-type morph is a generalist.
4. The polymorphism is maintained by environmental heterogeneity.

The original observation that suggested that tailspot alleles had important physiological correlates came from extensive field studies of two populations, Sarco and Encino (g and h, Fig. 2), which have only the +, *C*, and *Ct* alleles. Within any collection, the crescent and cut-crescent classes of adult males occasionally differed significantly in mean length. The odd thing about the observation, however, was that in some collections crescents were larger than cut-crescents while in others the reverse was true. This suggested that the differences between the morphs was not dependent upon genotype alone.

Data on these same populations gathered over a 10-year period demonstrated that mean male size varied widely among collections (Borowsky, 1978). Since the collections were repeated samples of the same populations, it was argued that the phenotypic variance among collections had a stronger component of environmental than background genetic variance. That is, unknown environmental factors with significant effects on size and growth differed in level among the collecting areas.

The same population samples also revealed that when males were large in size, crescent males tended to be the largest. When males were small, however, cut-crescents tended to be the largest. The mean length of wild-type males differed little from the collection mean. This observation suggested that the three morphs differ in their physiological responses to environmental factors (Borowsky, 1978), and that the patterned morphs are "specialists" for different subsets of the environment, while wild-type is a generalist morph. Specifically, the data indicated that crescents were selected for by "good conditions" and cut-crescents by "poor conditions" (Borowsky, 1978, 1981). If this proved to be correct, the tailspot polymorphism could be maintained by "Ludwig effects." A sufficiently patchy environment with strong selection for specialist morphs can maintain genetic polymorphism (Ludwig, 1950).

At the time it was proposed, the data supporting this hypothesis included the following:

1. The original observation that the relative sizes of crescent and cut-crescent males was a function of the mean size of all males in the collection (Borowsky, 1978). Since then, six more points have been added to the data set bringing the total to 18 (Fig. 22). The correlation between

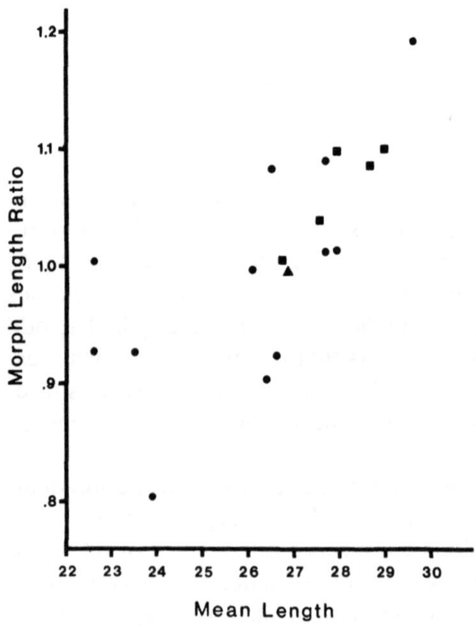

Figure 22. The relative size of crescent and cut-crescent fish is correlated with the mean size of all fish in any collection of *X. variatus*. The "morph length ratio" is the mean length of crescent adult males divided by the mean length of cut-crescent adult males for each collection. The circles denote data previously reported (Borowsky, 1978). The squares and the triangle are data obtained from collections made in 1977 and 1981.

the ratio of the mean sizes of the two morph classes and the mean size of all males in the collection is high and significant ($r = 0.71$, t (16) = 4.03, ts$P$ = 0.001). Since the mean size of all males includes the mean sizes of the two tailspot morph classes, there is some question about the independence of the two variables. This potential problem can be avoided by choosing totally independent variables for the regression analysis. Using as the independent variable only the mean size of wild-type males, rather than the mean size of all males, we demonstrated the same dependence of the relative size of the two morphs on average size of all (Borowsky, 1981).

A different type of analysis on the sister species *X. maculatus* verified that tailspot phenotype was a determinant of male size. Using a single large collection made from the Río Jamapa, it was demonstrated that tailspot pattern heterozygotes had sizes intermediate between the sizes of the appropriate single pattern classes (e.g., the mean size of adult CD males was intermediate between the mean sizes of the C and D classes). The correlation between predicted and observed sizes for pattern heterozygotes was weak but significant (Borowsky, 1981). This analysis not only verifies the relationship between genotype and size, but suggests that pattern heterozygotes exhibit physiological properties of both alleles.

2. The hypothesis that crescent is a "specialist" for environmental conditions producing large fish, and cut-crescent for conditions producing small fish, led to the prediction that the relative frequencies of the two alleles in and population should correlate well with the mean size of males at that site. That is, populations found in localities where good growth conditions prevail should have a higher frequency of *C*, while populations from localities where poor growth conditions prevail should have a higher frequency of *Ct*. This prediction was tested with data from 10 populations in the Tamesi basin and supported (Borowsky, 1978). Data accumulated since then have increased the sample sizes for the 10 original populations and two more have been added. For the 12 test populations, the correlation between $P(C)/[P(C) + P(Ct)]$ and mean size of all adult males collected from the population is 0.75 ($P < 0.001$). Eliminating the two populations with the smallest sample sizes from the analysis decreases the correlation, but it remains significant ($r = 0.64$, $t = 2.35$, ts$P$ = 0.05, Fig. 23). The revised data set permits a more reliable test of the hypothesis than originally made, because of the larger sample sizes, and supports the original interpretation. The populations included in this test are f, g, h, i, j, l, m, m′, p, and y.

3. The hypothesis also prediced that genetic variation in populations should be correlated with the degree of variability in adult male size

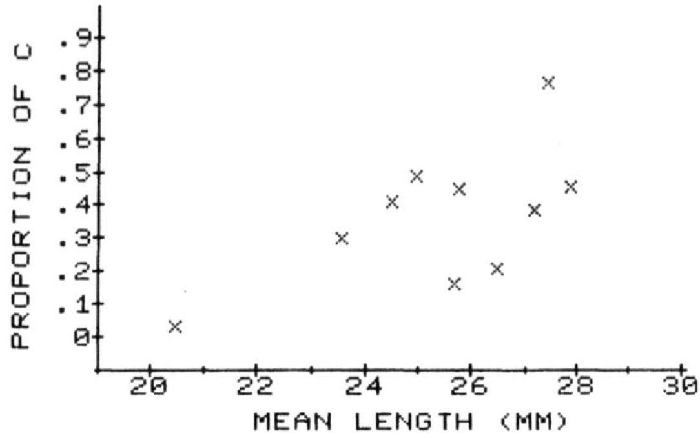

Figure 23. The relative frequencies of the *C* and *Ct* alleles can be predicted from the mean size of adult males collected at the locality. "Proportion of C" was calculated as the ratio of the frequency of the *C* allele to the sum of the frequencies of the *C* and *Ct* alleles. "Mean length" is the average standard length of all adult males collected at the given localities.

exhibited by the populations. The assumption behind this prediction is that size variability is an indirect measure of environmental variability. This prediction was tested and supported by data from nine different populations (Borowsky, 1978). It was retested, however, with the larger data set presently available (populations b, f, g, h, i, j, l, m, m′, p, y, and hh). The larger data set reveals no correlation between size and genetic variability and does not support the prediction.

I believe that the original correlation was spurious and derived from the inappropriate inclusion of the Jaumave locality in the sample. The Jaumave population has low genetic variability because of its isolation and it is only coincidental that it exhibits low size variability as well. The failure to confirm the prediction with this reanalysis does not necessarily mean that the prediction is incorrect. Any test of the prediction should take into account the effects of drift and isolation on genetic diversity and depend upon comparisons made within river systems.

As an alternative means of testing the hypothesis that the tailspot polymorphism is maintained by environmental variation, I have compared the mean genetic diversity within entire river systems to the genetic diversity in the river only. The former measure includes small stream populations, which are subject to greater environmental variation than those of the rivers. According to this hypothesis, the mean river system genetic diversity should be greater than the diversity of the river populations. This is indeed the case (Table XIII) and supports the original hypothesis.

4. The field observation that morph classes often differed significantly in relative condition factor (*RC*; Borowsky, 1981). This observation has been supported by numerous subsequent field and laboratory studies and is discussed at length in Section 6.

5. The observation that the morph classes differed in their reproductive success in natural populations (Borowsky, 1981). The data were obtained by collecting fertilized females from nature and transporting them live to the laboratory to obtain their young. From the genotypes of the female parent and the progeny, the genotypes of the inseminating males were inferred (Borowsky and Kallman, 1976; Borowsky and Khouri, 1976). The frequencies of inseminations by males of the different classes were then compared with their proportional representation in the population to calculate their relative mating success. Wild-type males had a far higher insemination rate than tail-spotted males at the start of the breeding season, while patterned males were far more successful than wild-types at the end of the season. At the end of the season, the frequencies of inseminations by the crescent and cut-crescent male classes were correlated with their relative mean lengths. In collections where crescents were larger than cut-crescents, they had a higher insemination efficiency. In collections where cut-crescents were larger, they were more successful at inseminating females (Borowsky, 1981). The differences in insemination rates were far greater in magnitude than the differences in size and I have termed this phenomenon "fitness amplification." A similar analysis for *X. maculatus* from Belize showed that reproductive success of the different morph classes could also be predicted from size data (Borowsky, 1981).

Table XIII
Comparisons between Tailspot Genetic Diversity (2^H) Averaged Over All Sites within River Systems, and Diversity within Populations Occupying the Main River[a]

River system	Mean diversity	River sites
Río Guayalejo[b]	1.71	1.60 (c)
Río Sabinas	2.31	2.10 (j and k)
Río Frio	1.83	1.96 (m)
Río Tamuin	2.24	1.69 (u)
Río Coy	2.58	2.37 (x)
Río Moctezuma	2.33	1.66 (cc)
		2.25 (ff)

[a]Average diversity (see p. 240) taken as weighted means using data from Table II. Letter designations are identified in the locality list.
[b]Excludes locality b and the collection of uncertain location.

My interpretation of these findings is that different environmental factors select for different tailspot morph classes and that this selection is manifested as differential reproduction. At the beginning of the breeding season, environmental conditions are optimal. There is little crowding, because population sizes are relatively small and water levels are high. Furthermore, temperatures are relatively low and dissolved oxygen levels are probably higher than at the end of the season. Under these conditions, wild-type males are favored over patterned males and have a higher insemination rate. But as the season progresses, the water levels fall, often resulting in isolated pools within the stream bed. Each pool traps its resident fish within a distinct and different set of environmental conditions, which generally are stringent. These pools differ in temperature, dissolved oxygen, degree of crowding, and other factors. As a group, patterned males are favored over wild-type males under these conditions and have higher insemination rates. Yet, some pools are better for Ct males than for C males or vice versa and these differences are reflected in their relative sizes and insemination rates.

Data on relative fecundity of wild-type and patterned females demonstrate exactly the same seasonality and are discussed in Section 6. Data on gene frequency variation within populations also support the interpretation of differential reproduction and are also discussed in Section 6.

6. New Support for the Tailspot Hypothesis

Additional support for the tailspot hypothesis comes from laboratory and field studies that demonstrate that the alleles have physiological correlates affecting growth, reproduction, and the accumulation of energy reserves. These effects permit a variety of environmental factors, including temperature, crowding, food availability, and dissolved oxygen, to exert selection upon the tailspot locus.

The additional support for the hypothesis is outlined in the three following sections, which cover the concept of "relative condition," additional field observations, and the results of pilot laboratory experiments.

6.1. Genetic Correlates of Relative Condition

6.1.1. Definitions

Condition factor in fishes is a length-normalized measure of weight. Heavier fish are said to be in better condition than lighter fish of the same length, and the regression of weight on length has often been used to compare different stocks of fishes. Samples of the same species collected during different seasons or simultaneously in different areas have been shown to differ in condition factor, and such studies aid in understanding how the environment affects growth and health in fishes of commercial importance (Weatherley, 1972).

The concept of condition factor should also be useful in comparisons of different genotypes within populations, because condition is a phenotypic character resulting from the interaction between environment and genotype. In practice, relative condition (*RC*), rather than condition factor (*sensu* Weatherley) is most useful for comparisons of genotypes within populations.

Relative condition is defined here as the observed weight of a fish divided by its expected weight, the latter derived from the regression of log weight on log standard length for all individuals in a set. Usually the set is restricted to a single collection, in order to facilitate a straightforward interpretation of the results. However, more than one collection could enter into the regression, although this would complicate the interpretation of *RC*. The *RC* analyses were only performed on males because the ovarian cycle precludes meaningful analysis of adult females.

6.1.2. Significance of Relative Condition

As will be shown, a significant portion of the residual variation in regressions of log weight on log length is accounted for by tailspot genotype. That is, the tailspot classes often prove to differ significantly in *RC*. The concept of *RC*, therefore, is useful in two ways: first, differences in *RC* among classes demonstrate that the environment distinguishes among tailspot genotypes. Second, *RC* analysis can aid in identifying the relevant environmental factor(s), if comparisons are made among collections or populations. For example, if one genotype consistently has a higher *RC* than the others in warm waters but consistently has a lower *RC* than the others in cold water, it suggests that variation in temperature, or a correlate, affects that genotype differently than the others. It is clear, therefore, that *RC* analysis is a powerful tool for the study of population genetics in these fish.

Before continuing, it should be noted that the concept of *RC* has its limitations because it is a complex phenotype. The relationship between *RC* and general health or other components of fitness is not linear. For example, of two unequally fed fish, the better fed one might have a lower

RC than the other, because fishes store excess energy in the form of lipids [review by Love (1970)], which are less dense than water. Love has pointed out that in fishes with relatively high muscle lipid content (6–15% wet weight), the proportions of lipid and water are inversely correlated. *Xiphophorus variatus* has a moderate muscle lipid content (2–4%; unpublished) and may also exhibit a negative relationship between the two. If so, as lipid content increases in well-fed *X. variatus, RC* may actually decline. Therefore, the relationship between *RC* and nutritional state in *Xiphophorus* is probably curvilinear. Seriously underfed fish will be in poor condition and their *RC* will increase as their nutritional state is improved. At some point, however, as they begin to accumulate significant lipid reserves, *RC* may decline. In consequence, *RC* analysis gives an incomplete picture of nutritional variation within a population and cannot be interpreted safely in terms of fitness differences among genotypes without additional information. For this reason, I am currently assaying preserved fish from populations previously reported on to determine the relationship of *RC* to energy reserves. Reanalyses are possible for fish that have been stored in formalin, and will be most reliable for recent, rather than older, collections.

In any case, the data on the relationships between genotype and *RC* discussed in the following sections clearly show that *RC* is correlated with genotype and also identify collection temperatures as a significant correlate of the relationship between genotype and *RC*.

6.1.3. Tailspots and RC

Previous analyses of field collections (Borowsky, 1981) had revealed that adult males sometimes exhibited differences in condition among tailspot morphs. In two of the six collections analyzed (Sarco pool 1, 1973 and 1975) the differences among morphs were statistically significant, but the direction of these differences could not be fully explained. In the 1973 collection, the ranking of *RC* of the morphs was Ct > C > +, while in 1975 it was Ct > + > C. It was noted that both of those collections were made at times when the water level in pool 1 was low and, at least in 1975, the fish were clearly distressed. On the hypothesis that Ct fish were selected for by "poor environmental conditions," their higher *RC* was to be expected, but it was not obvious why C > + in 1973 and + > C in 1975. Field work done in 1981 has provided a tentative answer to this question.

The relative condition of each individual male collected during the

1981 field trip was calculated. Average *RC* for the three morph classes +, C, and Ct in each collection, along with the temperature and the oxygen concentration of the water at the collecting locality, are listed in Table XIV. The *RC* of wild-type males was strongly and positively correlated with temperature [$r = 0.73$, $T(12) = 3.68$, ts$P = 0.003$, Fig. 24]. This was an unexpected observation, but is highly significant and has considerable explanatory power. The 1973 and 1975 Sarco pool 1 collections were made at different temperatures. Water temperature at the 1975 collection (+ > C) was relatively high (24°C), while the temperature at the 1973 collection (C > +) was moderate (20.6°C). Thus, the temperature difference was consistent with the rankings of *RC*.

The relative condition of cut-crescent males, on the other hand, appeared to be related to oxygen tension in the water at the time of collection. Unfortunately, oxygen concentrations were only determined at five localities and the correlation is only of borderline significance [$r = -0.77$, $t(3) = 2.12$, os$P = 0.062$). The apparent relationship, however, is fully consistent with the prior hypothesis (Borowsky, 1978, 1981) that cut-crescent is a specialist for "poor environmental conditions" and accounts for the observation that cut-crescent was the morph in best relative condition in the 1973 and 1975 Sarco pool 1 collections. Based on these field observations, the experiments on oxygen consumption (Section 6.3.1), and the relationship between oxygen tension, growth, and thyroid histology (Section 6.3.3), it seem reasonable to suggest that "poor environmental conditions" can be defined here as low oxygen concentrations and that the nature of the physiological adaptation of cut-crescent involves decreased metabolic demand. Experiments to test the effects of both oxygen concentration and temperature on relative condition of the different morphs are now in progress.

6.1.4. Tailspot Modifiers and RC

The Sarco upstream population contained "Ct" in such high proportion that it was possible to test whether Ct and "Ct" fish differed in condition factor. I compared the condition of adult males of the two classes and found that Ct individuals had a significantly higher mean *RC* than "Ct" individuals (1.032 ± 0.012 SE, $N = 7$, versus 0.981 ± 0.015, $N = 11$, $t = 2.79$, corrected df = 14, ts$P = 0.014$). This observation suggests that *Mod-1,* like the tailspot alleles, also has physiological correlates that interact with environmental variables.

Table XIV
Relative Condition (RC) of the Tailspot Morphs in Natural Populations of *X. variatus*[a]

Locality	Phenotypes						Temperature, °C	Dissolved O_2, ppm
	+	C	Ct	Ps	Cu	Others		
g	1.008 ± 0.036 (8)	0.969 ± 0.043 (11)	1.045 ± 0.080 (8)				22.5	3.4
h	0.997 ± 0.029 (4)	0.998 (1)	1.013 ± 0.000 (2)				20.0	3.2
p	1.009 ± 0.053 (20)	1.014 ± 0.006 (2)	1.022 ± 0.048 (4)	0.994 (1)		0.953 ± 0.088 (5)	17.0	2.8
s	0.993 ± 0.090 (21)	1.025 ± 0.055 (6)	1.016 ± 0.038 (4)	1.022 ± 0.058 (5)		0.935 (1)	15.0	High
v	1.009 ± 0.063 (22)	1.013 ± 0.041 (6)	0.982 ± 0.043 (9)	0.972 ± 0.011 (2)			18.5	High
v''	0.996 ± 0.017 (6)	1.008 ± 0.024 (4)	1.000 ± 0.029 (6)			1.000 (1)	18.5	High
x	1.031 ± 0.034 (4)	1.032 ± 0.009 (2)	0.965 ± 0.024 (5)				24.0	6.0

y	0.985 ± 0.026 (7)	1.034 ± 0.081 (8)	0.980 ± 0.087 (5)			0.988 ± 0.062 (2)	17.0	8.5
z	0.978 ± 0.067 (10)	1.028 ± 0.025 (9)	0.992 ± 0.067 (8)	0.999 ± 0.031 (2)		1.017 ± 0.037 (5)	15.0	
gg	0.999 ± 0.017 (4)	1.009 (1)		0.993 (1)	1.001 ± 0.047 (8)		19.0	
gg′	1.005 ± 0.052 (5)	0.937 ± 0.058 (3)		0.966 ± 0.029 (2)	1.001 ± 0.043 (10)	1.041 ± 0.079 (5)	18.0	
hh	0.991 ± 0.072 (4)	0.994 ± 0.036 (6)	1.045 ± 0.043 (2)				17.0	
ii	1.011 ± 0.024 (14)	0.959 ± 0.035 (5)			1.019 ± 0.016 (3)	1.002 (1)	17.0	
mm′	1.012 ± 0.052 (15)	0.980 ± 0.035 (9)		1.008 ± 0.094 (4)	0.982 ± 0.030 (4)	1.078 (1)	20.0	

[a]Only data from males were used. The table lists mean *RC* ± SD, sample sizes, collection temperatures, and dissolved oxygen (in ppm). In some streams where dissolved oxygen was not measured, the water was rapidly moving and the value is designated "high." The collections were made in January 1981. Letter designations are referenced in the locality list.

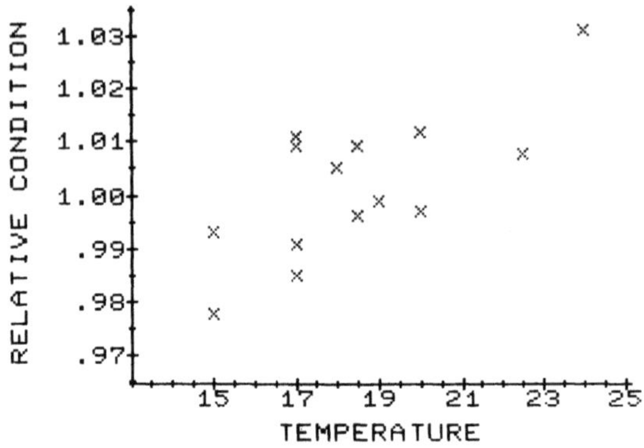

Figure 24. The relative condition (*RC*) of wild-type males is correlated with water temperature (°C) measured at the time of collection. Data from fourteen collections made in January, 1981.

6.2. Additional Field Data

6.2.1. Relative Fecundity and Temperature

Since the relative reproductive success of wild-type males varies seasonally, I decided to check whether the fecundity of females exhibits a similar pattern. Air temperatures provide a major seasonal cue, increase during the breeding season (Fig. 1), and are closely correlated with water temperatures at localities with high surface-to-volume ratios such as Arroyo Sarco. Therefore, I decided to check whether any relationship existed between temperature and the fecundity of wild-type females.

Fecundity is difficult to compare among collections because the relationship between length and egg or embryo number varies widely among collections (Borowsky and Diffley, 1981), presumably reflecting local conditions of crowding and food availability. To permit the comparison of data from different collections, I computed relative fecundities (*RF*) in a way analogous to the calculation of *RC*. The regression of log fecundity on log length for each collection was used to derive the expected fecundity of each female in the collection. Observed fecundity divided by expected fecundity is defined as *RF*.

Data on air temperatures from a nearby meteorological station (Río Sabinas at Rt. 85) were obtained through the courtesy of the Mexican Department of Climatology. Mean monthly air temperatures recorded at the Sabinas station correlate well with water temperatures measured at the Arroyo Sarco locality ($r = 0.93$, $t = 5.99$, df = 6, os$P = 0.0005$;

R. Borowsky unpublished data) and were employed as a measure of water temperatures at Sarco.

To test the relationship, I compared the mean *RF* for wild-type females in each collection that had been made at Arroyo Sarco, with the mean temperature recorded at the Sabinas station during a prior time period. The choice of an appropriate time period was arbitrary, but I chose the preceding 3 months because they corresponded to two full ovarian cycles in this species (Borowsky and Diffley, 1981). The relative fecundity of wild-type females was higher following a period of low temperatures than it was following high temperatures (Fig. 25; r = -0.74, t = 2.66, df = 6, os$P$ = 0.019).

Data from the Encino locality were also tested and a weak, statistically insignificant correlation was obtained. Water temperatures at the Encino locality do not track air temperatures as closely as at the Sarco locality because the stream at Encino has a stronger spring flow (unpublished).

6.2.2. Allelic Frequencies and Temperature

The data also reveal an effect of temperature on the frequency of the + allele [$P(+)$] at the Sarco locality. Allelic frequencies from the nine collections made at Arroyo Sarco were used for the analysis. A strong relationship [$r = -0.75$, $F(1,7) = 9.15$, ts$P = 0.019$] exists between $P(+)$

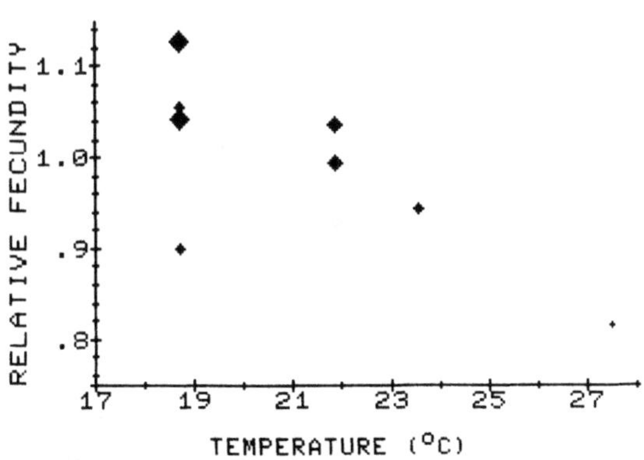

Figure 25. The relative fecundity of wild-type females varied among collections at the Sarco locality. This was correlated with air temperatures measured over the preceding 45 days. Symbol size is a correlate of total sample size (n) and the number of wild-type females within the sample. From smallest to largest, the ranges of "n" for the four symbols were: 7, 11 to 17, 18 to 19, and 38 to 47.

and air temperature averaged over the preceding three months. Allelic frequencies of + are significantly higher in the Sarco population following a cool period than they are following a warm period (Fig. 26).

A similar analysis for the Encino locality revealed only a weak correlation, which was not statistically significant. While it is probable that temperature affects allelic frequencies at the Encino locality the same way as at the Sarco locality, the weaker relationship probably reflects the lower correlation between water temperature at Encino and air temperatures. Nevertheless, the allelic frequencies of the + allele are highly correlated between the two nearby localities ($r = 0.73$, $N = 8$, $t = 2.60$, os$P = 0.02$). Since there is no gene flow between the two populations, this demonstrates that some external factor is affecting both populations similarly, and temperature is the most likely candidate.

6.2.3. Growth Rates in Natural Populations

During the 1981 field trip, growth rates of fish from the Encino population were determined. Females and juvenile males were captured, marked, measured, released, and recaptured 9 days later. Crescents grew the fastest and wild-types the slowest [Table XV; $F(2,20) = 4.25$, ts$P = 0.029$]. A similar study made in 1975 revealed the same pattern of relative growth rates, but the differences among the morph classes were not significant.

6.3. Laboratory Studies

During the last 5 years, we have performed many laboratory experiments on *X. variatus* in order to test whether the tailspot morphs differ

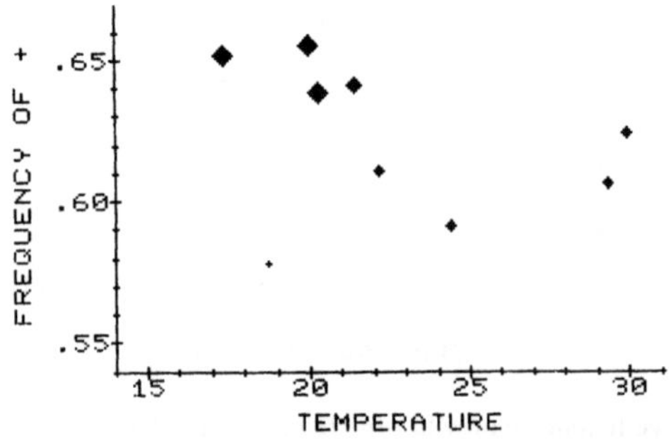

Figure 26. Among collections made at the Sarco locality, the allelic frequency of + is correlated with air temperatures (°C) recorded over the preceding 3 months. Symbol size is a correlate of the size of the sample used to calculate allelic frequencies. From smallest to largest, the sample size ranges for the four symbols were: 21, 80 to 135, 311, and 571 to 774.

Table XV
Growth Rates of *Xiphophorus variatus* at the Encino Locality[a]

Collection year	Phenotypic class		
	Wild-type	Crescent	Cut-crescent
1975			
Mean	−0.104	0.025	−0.090
Standard error	0.117	0.061	0.110
Sample size	7	4	8
1981			
Mean	−0.076	0.100	0.033
Standard error	0.031	0.077	0.029
Sample size	8	5	10

[a]Data from females and immature males only. Growth rates are given as percent change in length per day. Negative growth rates are observed in the laboratory when fish are ill or maintained under stringent conditions (R. Borowsky, unpublished).

from one another in basic attributes such as growth rates, size, weight or age at maturation in males, egg number in females, and relative liver and gonadal sizes. Some of these experiments showed significant differences among classes, while others failed to reveal such differences or revealed significant differences in the opposite direction from those originally seen. We concluded that whatever differences might exist among phenotypes, their manifestations were strongly affected by inadequately controlled environmental factors. We then measured variation among experimental aquaria of simple factors such as temperature, dissolved oxygen, and *p*H and discovered that the ranges were unacceptably large, even among adjacent tanks on the same rack.

Our experiments now employ rigid control of feeding, temperature, *p*H, oxygen tension, and photoperiod. In addition, all fish in our experiments are now maintained in chemical and visual isolation in their own tanks to avoid social effects on physiology (Borowsky, 1973*a,b*). Water for each tank is drawn from a common pool at the start of the experiment and evaporative losses are made up with deionized water. A small piece of sea shell in each tank maintains *p*H between 7.2 and 7.5. All the tanks to be maintained at the same temperature sit in a common circulating water bath. Oxygen tensions are maintained near saturation by aeration, or at lower levels by bubbling the tanks periodically with a mixture of 5% O_2 in N_2 or pure N_2. Feeding is controlled by measuring the daily ration of each fish. Photoperiod of the principal light source is set with timers (although other light sources in the room are not so controlled).

Several experiments have been completed with some or all of the new controls and the results are reported here. The experiments are small in scale and preliminary, although the data from them are far more reliable than earlier laboratory data because of the better control of the environmental factors. Although preliminary, the experiments show that physiological differences among tailspot morphs can be detected in the laboratory, and, taken as a whole, present a unified picture of the nature of the variation that has considerable explanatory power. Similar experiments are now in progress and a fuller report will be made in the future.

6.3.1. Oxygen Consumption

A set of experiments was run by Jorge Flores in order to test whether cut-crescents differed from wild-types in oxygen uptake. Adult male siblings were used for the experiment in order to minimize variation among individuals due to background genetics, sex, or stage of development. The experimental animals were tested in matched flow through respirometers, which were housed in a circulating water bath and shielded from laboratory traffic by a curtain (Flores, 1981). Periodic observations on the fish showed them to be at rest or occasionally swimming slowly. An independent observation series revealed no difference in spontaneous activity levels between the two morphs. We believe that we were measuring routine metabolism. Cut-crescent fish had a consistently lower oxygen consumption than wild-types, averaging 2.00 O_2/g wet weight per hr ($N = 9$). Wild-type males averaged 2.80 mg/g per hr ($N = 12$) [Table XVI; $t(16) = 2.72$, ts$P = 0.015$]. The mean rate of oxygen consumption

Table XVI
Oxygen Consumption (VO_2) of Adult Male *Xiphophorus variatus*[a]

		VO_2, mg/g of fish per hour	
Measurement date	TD, °C	Wild-type ($N = 4$)	Cut-crescent ($N = 3$)
12/12	0.5	3.98 ± 1.27	2.56 ± 1.08
12/18	0.0	2.45 ± 0.86	1.90 ± 0.56
12/22	−0.6	1.98 ± 0.55	1.53 ± 0.31

[a]Water temperature during measurement differed from acclimatization temperature by the factor of TD degree Celsius. Statistical analysis revealed that cut-crescents had a relative oxygen consumption of 74% of the wild-type value. The statistical significance was determined from a multiple regression analysis in which TD and standard length were the independent variables and oxygen consumption was the dependent variable. The observed VO_2 for each fish was divided by the expected VO_2 (determined from the regression) to obtain a relative VO_2 (RV). Mean RV was then compared between morphs using Student's t: RV wild-type = 1.137 ± 0.284, RV cut-crescent = 0.839 ± 0.155; $t = 2.72$, df = 16, ts$P = 0.015$; 0.839/1.137 = 74%.

by cut-crescents was 74% of the wild-type value. Lower metabolism compared to that of wild-type may only be correlated with cut-crescent in the pedigree tested, or might be a general attribute of cut-crescents, or perhaps of all fish with tailspots.

6.3.2. Muscle Lipids

In another experiment, 10 adult male siblings were maintained at 22°C on limited rations (9 mg Tetramin per day, 5 days per week) for 3 weeks prior to measurement and sacrifice. Their muscle tissue was then vacuum-dried and assayed for lipid content by weighing it before and after extraction with chloroform:methanol (2:1). Lipid content differed between genotypes. Crescent/cut-crescent heterozygotes had a mean muscle lipid content 30% higher than wild-types [N = 5, 12.3 ± 0.84 SE, versus N = 5, 9.4 ± 0.75%; $t(8)$ = 2.53, ts$P$ = 0.035] This result is consistent with the observations on oxygen consumption. The feeding ration was skimpy for fish of their size, and the lower percentage of lipid stored in the muscles of wild-types could be attributed to a higher metabolic rate.

6.3.3. Growth Rates and Thyroid Activity

An experiment was performed by Isaac Bao and me in order to determine the relationship between thyroid activity and growth. Eighteen adult sibling males (pedigree N293) were maintained in isolation at 25°C and fed 10 mg of Tetramin five times per week. Nine of them were maintained in air-saturated water, and the others in water partially depleted of oxygen by bubbling with nitrogen gas once daily. During the 4-week treatment period the oxygen levels in the first treatment averaged 7–8 ppm while in the second treatment they averaged 3–4 ppm, rising from a low of about 2 ppm after bubbling to a high of about 5 ppm 24 h later. Six of the fish in each treatment were wild-type and three were cut-crescent. Growth is slow in adult males and declined during the experimental period, so growth rates were calculated with data from the first 2 weeks of the period only. Table XVII lists growth rates for the 18 fish along with a measure of thyroid activity—number of small thyroid follicles.

Fish in the high-oxygen treatment grew more rapidly than those in the low-oxygen treatment [0.53 versus 0.31% per day in weight, $t(16)$ = 2.0, os$P$ = 0.03]. This difference was an expected result of the experiment and its interpretation is that oxygen was a limiting factor for growth in the low-oxygen treatment. An unexpected observation was also made, however. In the high-oxygen treatment there was a strong positive correlation between growth rate and thyroid activity [r = 0.78, $F(1,7)$ = 11.1, ts$P$ = 0.013], while in the low-oxygen treatment there was a strong

Table XVII
Growth Rates and Mean Numbers of Small Thyroid Follicles per Section for Adult Males

Treatment	Genotype	Growth rate[a]	Thyroid follicles
High O_2	+/+	0.954	50.1
		0.702	22.7
		0.668	28.5
		0.603	16.2
		0.319	22.7
		0.201	15.7
	Ct/+	0.806	28.7
		0.262	15.8
		0.227	18.2
Low O_2	+/+	0.606	18.8
		0.449	22.2
		0.359	24.2
		0.330	37.0
		0.264	34.3
		0.073	43.5
	Ct/+	0.340	37.9
		0.309	35.1
		0.039	32.0

[a]Growth rates are expressed as percent change in weight per day.

negative correlation between the two variables [$r = -0.72$, $F(1,7) = 7.5$, ts$P = 0.03$]. The difference between the correlations in the two treatments is highly significant [$F(2,14) = 14.0$, ts$P = 0.0005$] and suggests that low-oxygen treatment penalized individuals with greater thyroid activity more than those with lower activity.

Of greatest interest, however, is the observation that the two genotypes responded differently to the oxygen variable. The correlation between thyroid activity and growth was positive for cut-crescent fish at both oxygen levels ($r = 0.97$, $r = 0.92$), while for wild-type fish it was positive ($r = 0.77$) at high oxygen but negative ($r = -0.92$) at low oxygen levels. For the wild-type fish, the difference between the two correlations is significant (results of ANCOVA: $F = 14.8$, df = 1, 8, ts$P = 0.005$), while for the cut-crescent fish there is no hint of a difference. This observation is also consistent with the data on oxygen consumption, suggesting that 3–4 ppm is more limiting for wild-types than for cut-crescents because of their higher metabolic rate.

This experiment was flawed in design because the two oxygen treatments differed not only in level but also in that one of them fluctuated on a daily basis. The experiment has been repeated with better control of the oxygen variable and slides of the animals have been prepared, but the counts of thyroid follicles have not yet been done.

6.3.4. The Interaction between Temperature and Feeding

An experiment in which wild-type and cut-crescent males were raised to maturity at either of two temperatures and feeding levels was performed by Stephanie Tzall. The results will be presented in greater detail elsewhere. The fish were measured and weighed every week and their feeding ration was adjusted to be either 3% or 10% of their weight, five times per week. Temperature was maintained at either 21.5 or 24.5°C. In general, their weight gain was greater at the higher temperature (2.30% per day) and feeding levels (2.30%) than at the lower temperature (1.99%) and feeding (1.95%) levels.

Of greatest interest, however, is the observation that the two genotypes differed significantly in their response to the feeding variable at the higher temperature but not at the lower temperature. At the high temperature and high feeding level, wild-type exceeded cut-crescent in growth rate (2.99% versus 2.40%), while at the low feeding level, the opposite was true (1.58% versus 2.35%; F-test for significance of interaction between genotype and feeding: $F = 2.74$ with df = 1, 20; ts$P = 0.038$). At the low temperature, no interaction between genotype and feeding was observed. This observation—temperature–dependent differences in response to the feeding variable—is also consistent with the observation that wild-types have a greater oxygen consumption than cut-crescents. The argument runs as follows: at 24.5°C the extra ration was fully utilized for growth by wild-types because of their higher metabolic rate, but was treated as excess by cut-crescents. Therefore, wild-types exceeded cut-crescents in growth rates. On the low ration, however, cut-crescent growth rates exceeded wild-types' because the ration was inadequate for the latter. At 21.5°C, on the other hand, the low ration was adequate for both genotypes and the high ration excessive, so no differences among genotypes were exhibited.

Another finding in this experiment was that the two morph classes differed significantly in RC at sexual maturation. Cut-crescents had a higher RC than wild-types and the difference was consistent over all combinations of temperature and feeding ($RC = 1.04$, $N = 23$, versus $RC = 0.96$, $N = 19$; $F = 19.8$ with df = 1, 34; ts$P = 0.001$). A significant interaction between *temperature* and *feeding* was also observed. At the high temperature the better-fed fish had a higher *RC* (1.02 versus 0.99; $N = 10, 11$), but at the low temperature the better-fed fish had a lower *RC* (0.98 versus 1.03; $N = 13, 8$; *F*-test for interaction: $F = 4.11$ with df $= 1, 34$; ts$P = 0.05$). Similar experiments are now in progress to clarify the meaning of these observations.

The work summarized in this section confirms the hypothesis that tailspot genotype has important physiological correlates and, especially, that wild-type differs from cut-crescent, and perhaps other pattern morphs. While many of these observations need further testing to clarify their implications, the consistent picture that emerges is that wild-type will be favored when food and oxygen are abundant. When either food or oxygen are limiting, however, and especially in warm waters, cut-crescent, and perhaps other pattern morphs, will be favored over wild-type. In cooler waters the differences between the genotypes will be less pronounced.

The experimental observations account for the seasonality of reproduction observed in the field studies. At the start of the breeding season the water is flowing well and presumably is well-oxygenated, and populations are small and dispersed so that competition for resources is relaxed. These environmental conditions will select for wild-type. Near the end of the breeding season, however, the water is warmer and not flowing well and the fish are more abundant and concentrated in distribution because of the drying of the streams. These environmental conditions will select against wild-type.

7. Maintenance of the Tailspot Polymorphism

Genetic polymorphism may be maintained by selection in a variable environment (Levene, 1953; Ludwig, 1950). The data reviewed here show that both seasonal and spatial environmental variation strongly influence the tailspot polymorphism. During the breeding season the environment progresses from effectively fine-grained and permissive to coarse-grained and stringent. It is fine-grained early in the season because the flow of water minimizes differences among areas and permits the fish to move about freely. Population sizes are smaller at the start of the breeding season and the fish are dispersed, so competition for resources is less intense than later. In addition, the cooler temperatures, freely flowing water, and greater surface areas found early in the breeding season promote fuller oxygenation of the water. Later in the season, temperatures are higher and the streams tend to dry down to poorly connected, sluggish pools. Competition for food must increase as the swollen populations are concentrated in the smaller water volumes. Conditions, therefore, are more stringent. But conditions vary among pools and the fish are restricted in their movement, so effective environmental grain size is larger at the end of the breeding season than at its start.

It appears that the tailspot polymorphism is maintained by two sets of balances: the first between wild-type and pattern alleles and the second among the various pattern alleles. Early in the season, wild-types have an advantage over patterned fish because their higher metabolic rate permits a more efficient utilization of plentiful resources under optimal conditions. Later in the season, when conditions are suboptimal, their higher metabolism penalizes wild-types. Patterned fish, having a lower metabolism or other adaptations for survival in a stringent environment, are less affected.

Spatial variation among pools balances the relative frequencies of the different spotting morphs. Each morph is selected by a different pool type or subniche. It is the distribution of subniches among localities that determines local frequencies of the tailspot alleles.

Roughgarden (1979) analyzed the conditions for maintenance of polymorphism in a temporally varying environment at a locus with two alleles, one of which is recessive. It is sufficient that the arithmetic fitness of the recessive phenotype to be greater than 1.0, while its geometric fitness be less than 1.0. These conditions are met if the recessive is usually more fit than average, but is subject to occasional severe setbacks. This model is applicable to the tailspot polymorphism by reducing the multiple-allelic system to the dichotomy between wild-type and grouped pattern alleles. The reproductive advantage enoyed by wild-type early in the breeding season may be sufficient to maintain its arithmetic mean fitness above 1.0, but the burden of the seasonal deterioration of the environment falls more heavily on wild-type and could reduce its geometric mean fitness below 1.0. *The frequency of wild-type in a given population, therefore, would be governed by the severity and duration of conditions at the end of the breeding season. The more stringent the environment, the lower the frequency of the + allele in the population.*

Deakin (1968) analyzed the conditions for polymorphism at a locus with two alleles and no dominance in a population subdivided among distinct subniches. Applicability of this model to the problem depends upon the physiological properties of the pattern heterozygotes. If we assume that they are either codominant or incompletely dominant, Deakin's model may apply. Sufficient conditions for the maintenance of polymorphism are that the relative fitness of each homozygote, in at least one subniche, be less than the product of the relative frequency of that subniche times the probability of staying in the subniche. In this model, therefore, polymorphism is promoted by major differences in relative fitness among subniches and by barriers that inhibit migration among them. Both of these factors are important to *Xiphophorus* populations at the end of the breeding season when the streams are divided into distinct pools with strongly divergent selective regines. *Therefore, pattern morph frequencies in a given population would also be governed by events toward the end of the breeding season, i.e., by the proportions of the different types of environmental patches at the locality.*

It should be noted that the tailspot hypothesis was formulated when there was no evidence that *X. variatus* was polymorphic for sex-linked size-determining genes. Since then, it has become clear that they share the *Pit* polymorphism with other species of *Xiphophorus*. Since the "pituitary locus" has major effects on reproductive fitness of males, the assumption that background genotype will not vary much among repeated samples of the same populations is not justified for this locus. If alleles at the tailspot locus interact with those at the "pituitary locus," the size-dependent differences between the C and Ct classes of males might reflect variation in background genotype rather than in environmental factors. If this is the case, the maintenance of tailspot variation would be indirectly rather than directly dependent upon environmental factors. The two alternative hypotheses remain to be tested.

An unanswered question is why the tailspot alleles have visible effects at all. Given the physiological correlates, the polymorphism would be maintained in the absence of the pigmentation differences, and it is difficult to imagine why the variation is conspicuous. I have suggested that the visible effects of the alleles exist to advertise a fish's genotype to other members of the population and are exploited through a strategy of nonrandom mating having the effect of increasing the heterozygosity of progeny and, therefore, fitness (Borowsky, 1981). This hypothesis remains to be tested experimentally.

8. Summary

The evolutionary aspects of pigmentation, size determination, and allozyme genetics in the poeciliid genus *Xiphophorus* have been discussed, with special reference to the tailspot polymorphism and the variable platyfish, *X. variatus*. A principal theme of this work is that genetic variability in this genus results from the action of selection in a variable environment. Even so, the effects of drift and the founder effect can be seen in the geographic distribution of pigmentation alleles and allozymes.

Xiphophorus variatus is shown to be more polymorphic for tailspot and bodyspot patterns than previously believed. While the tailspot locus is strongly influenced by selection, the distribution of allelic diversity in the two sister species, *X. variatus* and *X. maculatus,* is also clearly affected by drift. These two species are most polymorphic at the center of their joint distribution in the region of the Ríos Jamapa and Papaloapan in eastern Mexico and decline in variability in rivers to the east and west. The identical pattern of distribution is seen for allozyme variation in *X. variatus,* but allozyme variability is lower than tailspot variability because the allozyme loci, if subject to selection at all, are not as strongly selected as is the tailspot locus.

Field and laboratory studies on *X. variatus* show that alleles at the tailspot locus have metabolic correlates that can account for the strong selection observed. In particular, wild-type fish have a higher metabolic rate than cut-crescents and, perhaps, other spotted morphs as well. It is hypothesized that the tailspot morphs are specialists for survival under stringent conditions, while the wild-type is selected for by optimal conditions. The polymorphism seems to be the result of two sets of balances: the first between wild-type and the pattern morphs, and the second among the pattern morphs. The balance between wild-type and patterned fish is strongly influenced by seasonal changes, with wild-type males reproducing better than patterned males at the start of the season. Relative condition of wild-type males and relative fecundity of wild-type females are also influenced by temperature in natural populations. The balance between the different pattern morphs seems to be maintained by spatial variation toward the end of the breeding season when environmental conditions are more severe. Crescent seems to be selected for by "good conditions," while cut-crescent seems to be selected for by "poor conditions." The tailspot hypothesis also permits the prediction of allelic frequencies in natural populations from direct and indirect measure of the environment (temperature and mean size of collected fish).

With reference to variation at the tailspot locus, the concept of relative condition (*RC*) and its employment in population genetic research have been discussed. *RC* analysis can detect allelic variation for physiological variables and also help to identify environmental variables that interact with the alleles.

The existence of a gene designated *Mod-1* has been demonstrated. The *Mod-1* gene alters the phenotypic expression of the cut-crescent allele but has no visible effect in other fish. Like the tailspot alleles, *Mod-1* also appears to have physiological correlates. It is not linked to the tailspot locus.

New macromelanophore (bodyspot) patterns in *X. variatus* have been tabulated and figured. Evidence for linkage disequilibrium at this sex-linked locus has been presented and it has been inferred from this that the locus is subject to selection. The expressions of several patterns at this locus are under social control and the suggestion of Kallman (1970) that some of the patterns serve a signal function during social encounters has been reiterated.

Allozyme frequencies have been reported for populations of *X. variatus* and the swordtails *X. montezumae, X. cortezi, X. pygmaeus,* and *X. nigrensis*. Variation at allozyme loci is low in all five species but is higher in *X. variatus* than in the swordtails. The difference is attributed to the swordtails' residence in headwaters rather than lowland streams, which necessitates dispersal through stream capture rather than confluence in the lowlands. Stream capture would be a more effective filter for allelic variation.

"Pituitary gene" (*Pit*) variation has been documented in *X. variatus*. The alleles having strong affects on male size are Y-linked but it can be inferred that X-linked variation also exists in this species. It is suggested that the *Pit* polymorphism is also maintained by environmental variability and that alleles at this locus may interact with tailspot alleles.

ACKNOWLEDGMENTS. I wish to thank my students and colleagues, who have aided this research with their ideas and efforts in the field and laboratory: Isaac Bao, John Diffley, Jorge Flores, Rene Herrera, Nicholas Hoffman, Jules Markofsky, Elizabeth Russin, and Stephanie Tzall. Klaus Kallman has generously supplied materials, information, and ideas that have aided this research significantly. Betty Borowsky has been a patient source of good ideas and editorial advice.

The field work was done with the permission of the Mexican Department of Fisheries and the data on rainfall and temperature were kindly supplied by their Department of Climatology. The research was funded by NSF grants BMS 75-13464 and DEB 79-12341. This support is gratefully acknowledged.

References

Atz, J. W., 1962, Effects of hybridization on pigmentation in fishes of the genus *Xiphophorus*, *Zoologica* **47**:153–181.

Bao, I., 1981, The genetic and endocrine control of sexual maturation and sterility in hybrids between *Xiphophorus belleri* and *X. maculatus*, Ph. D. dissertation, New York University, New York.

Borowsky, R. L., 1969, Ecological and social factors influencing the polymorphic pigmentation systems of *Xiphophorus v. variatus* (Pisces: Poeciliidae), Ph. D. dissertation, Yale University, New Haven, Connecticut.

Borowsky, R. L., 1973*a*, Relative size and the development of fin coloration in *Xiphophorus variatus*, *Physiol. Zool.* **46**:22–28.
Borowsky, R. L., 1973*b*, Social control of adult size in males of *Xiphophorus variatus*, *Nature* **245**:332–335.
Borowsky, R. L., 1978, The tailspot polymorphism of *Xiphophorus* (Pisces: Poeciliidae), *Evolution* **32**:886–893.
Borowsky, R. L., 1981, Tailspots of *Xiphophorus* and the evolution of conspicuous polymorphism, *Evolution* **35**:345–358.
Borowsky, R. L., and Diffley, J., 1981, Synchronized maturation and breeding in natural populations of *Xiphophorus variatus*, *Environ. Biol. Fish.* **6**:49–58.
Borowsky, R. L., and Kallman, K. D., 1976, Patterns of mating in natural populations of *Xiphophorus* (Pisces: Poeciliidae). I. *X. maculatus* from Belize and Mexico, *Evolution* **30**:693–706.
Borowsky, R. L., and Khouri, J., 1976, Patterns of mating in natural populations of *Xiphophorus*. II *X. variatus* from Tamaulipas, Mexico, *Copeia* **1976**:727–734.
Darnell, R. M., 1962, Fishes of the Rio Tamesi and related coastal lagoons in East-Central Mexico, *Publ. Inst. Mar. Sci. Port Aransas Texas* **8**:299–365.
Deakin, M. A. B., 1968, Genetic polymorphism in a subdivided population, *Aust. J. Biol. Sci.* **21**:165–168.
Flores, J., 1981, Oxygen consumption in two morphs of E1 Charral, *Xiphophorus variatus* (Pisces: Poeciliidae), Master's thesis, New York University, New York.
Ford, E. B., 1971, *Ecological Genetics*, 3d ed., Chapman and Hall, London.
Gandolfi, G., 1971, Sexual selection in relation to the social status of males in *Poecilia reticulata* (Teleostei: Poeciliidae), *Boll. Zool.* **38**:35–48.
Gordon, H., and Gordon, M., 1954, Colour patterns and gene frequencies in natural populations of a platyfish, *Heredity* **4**:61–73.
Gordon H., and Gordon, M., 1957, Maintenance of polymorphism by potentially injurious genes in eight natural populations of the platyfish, *Xiphophorus maculatus*, *J. Genet.* **55**:1–44.
Gordon, M., 1927, The genetics of a viviparous top minnow *Platypoecilus;* the inheritance of two kinds of melanophores, *Genetics* **12**:253–283.
Gordon, M., 1946, Introgressive hybridization in domesticated fishes. 1. The behavior of comet, a *Platypoecilus maculatus* gene in *Xiphophorus helleri*, *Zoologica (N. Y.)* **31**:77–88.
Gordon, M., 1953, The ecological niche of the pygmy swordtail, *Xiphophorus pygmaeus*, in the Rio Axtla, Mexico, *Copeia* **1953**:148–150.
Gordon, M., 1956, An intricate genetic system that controls nine pigment cell patterns in the platyfish, *Zoologica* **41**:153–162.
Kallman, K. D., 1965, Genetics and geography of sex determination in the poeciliid fish, *Xiphophorus maculatus*, *Genetics* **60**:811–828.
Kallman, K. D., 1970, Sex determination and the restriction of sex-linked pigment patterns to the X and Y chromosomes in populations of a poeciliid fish, *Xiphophorus maculatus*, from the Belize and Sibun rivers of British Honduras, *Zoologica (N. Y.)* **55**:1–16.
Kallman, K. D., 1975, The Platyfish, *Xiphophorus maculatus*, in: *Handbook of Genetics*, Vol. 4 (R. C. King, ed.), Plenum Press, New York, pp. 81–132.
Kallman, K. D., 1983, The sex-determination mechanism in the poeciliid *Xiphophorus montezumae*, Jordan and Snyder and the genetic control of the sexual maturation process and adult size, *Copeia* **1983**:755–769.
Kallman, K. D., and Atz, J. W., 1966, Gene and chromosome homology in fishes of the genus *Xiphophorus*, *Zoologica* **51**:107–135.

Kallman, K. D., and Borkoski, V., 1978, A sex-linked gene controlling the onset of sexual maturity in female and male platyfish (*Xiphophorus maculatus*), fecundity in females and adult size in males, *Genetics* **89:**79–119.

Kallman, K. D., and Borowsky, R., 1972, The genetics of gonopodial polymorphism in two species of poeciliid fish, *Heredity* **28:**297–310.

Kallman, K. D., and Schreibman, M. P., 1973, A sex-linked gene controlling gonadotrop differentiation and its significance in determining the age of sexual maturation and size of the platyfish, *Xiphophorus maculatus*, *Gen. Comp. Endocrinol.* **21:**287–304.

Levene, H., 1953, Genetic equilibrium when more than one ecological niche is available, *Am. Nat.* **87:**331–333.

Lewontin, R. C., 1974, *The Genetic Basis of Evolutionary Change,* Columbia University Press, New York.

Love, R. M., 1970, *The Chemical Biology of Fishes,* Academic Press, New York.

Ludwig, W., 1950, Zur Theorie der Konkurrenz: die Annidation (Ennischung) als funfter Evolutionsfaktor, *Neue Ergeb. Probl. Zool. Klatt-Festschrift* **1950:**516–537.

McKay, F., 1971, Behavioral aspects of population dynamics in unisexual-bisexual *Poeciliopsis* (Pisces: Poeciliidae), *Ecology* **52:**778–790.

Mitchell, R. W., Russell, W. H., and Elliot, W. R., 1977, Mexican eyeless characin fishes, genus *Astyanax:* Environment, distribution, and evolution, Museum of Texas Tech University Special Publication No. 12.

Powell, J. R., Levene, H., and Dobzhansky, T., 1972, Chromosomal polymorphism in *Drosophila pseudoobscura* used for diagnosis of geographic origin, *Evolution* **26:**553–559.

Rosen, D. E., 1979, Fishes from the uplands and intermontane basins of Guatemala: Revisionary studies and comparative geography, *Bull. Am. Mus. Nat. Hist.* **162:**267–376.

Rosen, D. E., and Bailey, R. M., 1963, The poeciliid fishes (Cyprinodontiformes), their structure, zoogeography, and systematics, *Bull. Am. Mus. Nat. Hist.* **126:**1–176.

Roughgarden, J., 1979, *Theory of Population Genetics and Evolutionary Ecology: An Introduction,* Macmillan, New York.

Siciliano, M. J., and Shaw, C. R., 1976, Separation and visualization of enzymes on gels, in: (Ivor Smith, ed.), *Chromatographic and Electrophoretic Techniques,* Vol 2 4th ed., Heineman, London, pp. 185–209.

Smatresk, N. J., and Herreid, C. F., 1980, Group metabolism in swordtails, *Xiphophorus helleri,* under controlled oxygen conditions, *Copeia* **1980:**562–564.

Sohn, J. J., 1977, Socially induced inhibition of genetically determined maturation in the platyfish, *Xiphophorus maculatus, Science* **195:**199–201.

Spiess, E. B., 1977, *Genes in Populations,* Wiley, New York.

Weatherley, A. H., 1972, *Growth and Ecology of Fish Populations,* Academic Press, New York.

Zander, C. D., 1967, Okologische und morphologische Beitrage zur Systematik und geographischen Verbreitung der Gattung *Xiphophorus* (Pisces), *Mitt. Hamb. Zool. Mus. Inst.* **64:**87–125.

CHAPTER 6

Apomictic Reproduction in the Amazon Molly, Poecilia formosa, and Its Triploid Hybrids

PAUL J. MONACO, ELLEN M. RASCH, and JOSEPH S. BALSANO

1. Introduction

The Amazon molly, *Poecilia formosa,* initially described in 1859 by Girard, is the first vertebrate in which unisexuality was recognized (Hubbs and Hubbs, 1932). *Poecilia formosa* is intermediate in form between two sexual species, *Poecilia latipinna* (LeSueur) and *Poecilia mexicana* Steindachner and is thought to have arisen in nature as a hybrid of these species (Hubbs and Hubbs, 1932; Hubbs, 1955, 1964; Abramoff *et al.,* 1968; Prehn and Rasch, 1969; Balsano *et al.,* 1972). *Poecilia formosa* is not a true thelytokous parthenogen as defined by White (1973); instead, it reproduces by gynogenesis, a mechanism that is sperm dependent. Males of related sexual species provide sperm, which serve only to activate the egg. Functional syngamy does not occur. Thus, developing embryos are derived from diploid ova that contain only maternal chromosomes and each daughter is a genetic copy of its mother (Meyer, 1938; Hubbs and Hubbs, 1946; Kallman, 1962*a,b,c;* Hubbs, 1964; Rasch *et al.,* 1965; Darnell *et al.,* 1967; Schultz, 1969; Uzzell, 1970; White, 1973). The obligatory dependence on sperm that is a hallmark of gynogenetic reproduction compels *P. formosa*

PAUL J. MONACO and ELLEN M. RASCH • Department of Biophysics, Quillen-Dishner College of Medicine, East Tennessee State University, Johnson City, Tennessee 37614 *JOSEPH S. BALSANO* • Biomedical Research Institute, University of Wisconsin-Parkside, Kenosha, Wisconsin 53141.

to behave in nature as a sexual parasite on closely related, sympatric bisexual species (Hubbs, 1964). Although a variety of poeciliid males can "father" *P. formosa* in the laboratory (Hubbs and Hubbs, 1946), the survival and success in nature of these all-female forms require their coincident distribution with at least one of the bisexual host species.

As yet the mechanism that routinely inactivates or excludes the male genome during gynogenetic reproduction in *P. formosa* is not known. Its occasional failure, however, apparently accounts for reported cases of *de novo* production of triploid progeny from diploid *P. formosa* in the laboratory (Rasch *et al.*, 1965; Schultz and Kallman, 1968; Rasch and Balsano, 1974; Strommen *et al.*, 1975*a*) and for the spontaneous production of triploids in nature (Rasch *et al.*, 1970). Wild-caught triploids are reproductively competent, transmit triploid genomes to their all-female progeny, presumably by gynogenesis, and comprise a significant portion of the total number of females in some naturally occurring populations of *Poecilia* (Rasch and Balsano, 1973, 1974; Strommen *et al.*, 1975*b*).

Two distinctive triploid forms of *Poecilia* have been identified in nature (Rasch *et al.*, 1970, 1978). As yet a formal nomenclature has not been established for these triploid biotypes. The two triploid hybrids might be designated as: *Poecilia 2 mexicana-latipinna* and *Poecilia mexicana-2 latipinna,* respectively. The former occur where *P. formosa* is sympatric with *P. mexicana* in the Soto la Marina drainage of northeastern Mexico (Rasch, 1968; Rasch *et al.*, 1970; Rasch and Balsano, 1974). The latter occur where *P. formosa* is sympatric with *P. latipinna* in the Río Grande drainage of southeastern Texas (Rasch *et al.*, 1970, 1978). Such designations reflect the proposed hybrid origins of these fishes and are patterned after the types of designations used for diploid hybridogens and triploid gynogens of *Poeciliopsis* (Schultz, 1969). Because there is evidence, however, that muscle protein genes from three distinct forms of *Poecilia* are present in contemporary triploid unisexuals (Monaco *et al.*, 1982; Turner *et al.*, 1983), the above designations really are not appropriate for triploid *Poecilia*.

The issue of appropriate nomenclature for contemporary triploid biotypes of *Poecilia* becomes complicated when one considers that specimens of *P. mexicana* that are sympatric with unisexuals in the upper and middle portions of the Soto la Marina drainage are currently recognized at the subspecific level as *Poecilia mexicana limantouri,* and the forms of *P. mexicana* found in drainages much further south are recognized as *Poecilia mexicana mexicana,* whereas those found in the intervening drainages between Tampico and Tuxpan are recognized as morphological intergrades (Menzel and Darnell, 1973). *Poecilia formosa* often occurs with these intergrades in several localities within the Río Tamesi and the Río

Tuxpan. Based upon comparisons of muscle protein phenotypes, allozyme patterns, and plasma albumin phenotypes found in these different populations of *Poecilia,* it is now clear that either an intergrade of *P. mexicana,* or the southern subspecies, along with *P. latipinna,* may well have been a progenitor of *P. formosa* (Balsano and Rasch, 1974; Monaco *et al.*, 1982; Turner *et al.*, 1983).

As yet only a few triploid hybrids have been identified where *P. formosa* is sympatric with *P. latipinna* [four specimens from the Río Grande (Rasch *et al.*, 1970, 1978)] or with *P. m. mexicana* [one specimen from the Río Tuxpan (Monaco, Rasch, and Balsano, unpublished analysis of June 1982 field collections)]. The vast majority of triploid unisexuals identified to date have been taken in upper tributaries of the Río Soto la Marina, where *P. formosa* is sympatric with *P. m. limantouri*. It is thus likely that the unisexual *P. formosa* initially arose in a southern drainage and later migrated north, where it now occupies significant stretches of the Soto la Marina drainage. The origin of triploid unisexuals probably was, and may continue to be, a distinctly different biological event, occurring subsequent to the origin and migration of *P. formosa*. Thus, Soto la Marina triploids might be better identified as *P. formosa-limantouri* (where *P. formosa* actually is *P. m. mexicana* × *P. latipinna* and the triploids are *P. m. mexicana* × *P. latipinna* × *P. m. limantouri*). Following this style of nomenclature, the four Río Grande triploids would be *P. formosa-latipinna* and the one Río Tuxpan triploid would be *P. formosa-mexicana*. We recognize that this style of nomenclature implies the resurrection of a specific status for *P. limantouri* and *P. mexicana* (Rosen and Bailey, 1963). At present, however, such a proposal may be premature. Work is now underway to resolve the taxonomic status of individuals within the *P. mexicana* complex and only when that analysis is completed can the complex problem of the taxonomy of various triploid forms of *Poecilia* be adequately addressed (cf. Rasch *et al.*, 1970).

Unlike some types of fish (Anders *et al.*, 1973, 1981; Kallman, 1975; Yamamoto, 1975), an array of genetic markers is not available in *Poecilia,* making it difficult to identify diploid and tripolid unisexuals or to discern their exact mechanism of reproduction by breeding experiments (cf. Vrijenhoek, this volume, Chapter 8). Plasma albumin phenotypes (Balsano *et al.*, 1972) and muscle protein phenotypes (Monaco *et al.*, 1982) have been useful in identifying triploids from tributaries of the Río Soto la Marina. By far the most reliable way to screen natural populations of *Poecilia* for the occurrence of triploid unisexuals, however, has been a systematic survey of blood smears from all specimens to determine the DNA content of erythrocyte nuclei by DNA-Feulgen cytophotometry (Rasch *et al.*, 1970, 1978; Rasch and Balsano, 1974). The diploid genome size of *P. formosa* and *P. mexicana,* where $2n = 46$ (Figs. 1A and 1B), is approximately 1.6 pg DNA, while their triploid hybrids, where $3n = 69$ (Fig. 1C), contain 2.3–2.4 pg DNA per nucleus.

Recently, multilocus analyses of enzymic proteins have shown that triploid relatives of *P. formosa* are clonally quite diverse and perhaps are the most evolutionarily dynamic members of unisexual–bisexual breeding complexes of *Poecilia* (Turner *et al.,* 1983). Although the role of triploid unisexuals in these breeding complexes remains enigmatic, their frequency and apparent ecological success may be important for evaluating the evolutionary potential of *P. formosa* and other diploid forms that reproduce asexually (Schultz, 1969; Uzzell, 1970; Cuellar, 1977). It is possible, although not yet documented, that the triploids of *Poecilia* can serve as genetic bridges, permitting flow of genetic information into diploid unisexual populations that otherwise would have an essentially fixed gene pool (Rasch *et al.,* 1970). Since a thorough review of the evolutionary genetics of *P. formosa* has recently appeared (Turner, 1982), the present discussion will deal primarily with the cytological basis for gynogenetic reproduction in the unisexuals of *Poecilia.*

2. Cytological Considerations

The cytological mechanisms responsible for unisexual reproduction in *P. formosa* and its related triploids apparently preclude normal meiotic recombination. Not only do they mandate all-female progeny, but these mechanisms also must result in isogenic, matroclinous inheritance (Kallman, 1964). A major obstacle to understanding the origin and persistence of unisexual *Poecilia,* and of many other unisexual vertebrates for that matter, has been the difficulty in obtaining materials suitable for direct observations of meiosis in females. The critical maturation divisions apparently occur over a brief interval of time during the long process of oocyte development. In *Poecilia* such events are probably triggered by a fertilization reaction involving sperm that may be stored for prolonged periods in infoldings of the ovigerous lamella (Monaco *et al.,* 1978), the so-called fertilization valleys of Meyer (1938).

There are fairly well-documented reports for several species of unisexual vertebrates in which individual females produce exclusively female progeny without genetic recombination [for reviews see Maslin (1968), Schultz (1969, 1980), White (1973), and Cuellar (1974, 1977)]. In cases where the oogenic pathways are known, it is assumed that suppression of karyokinesis and cytokinesis occurs in oogonia during the last premeiotic mitosis. This yields a primary oocyte with twice the number of

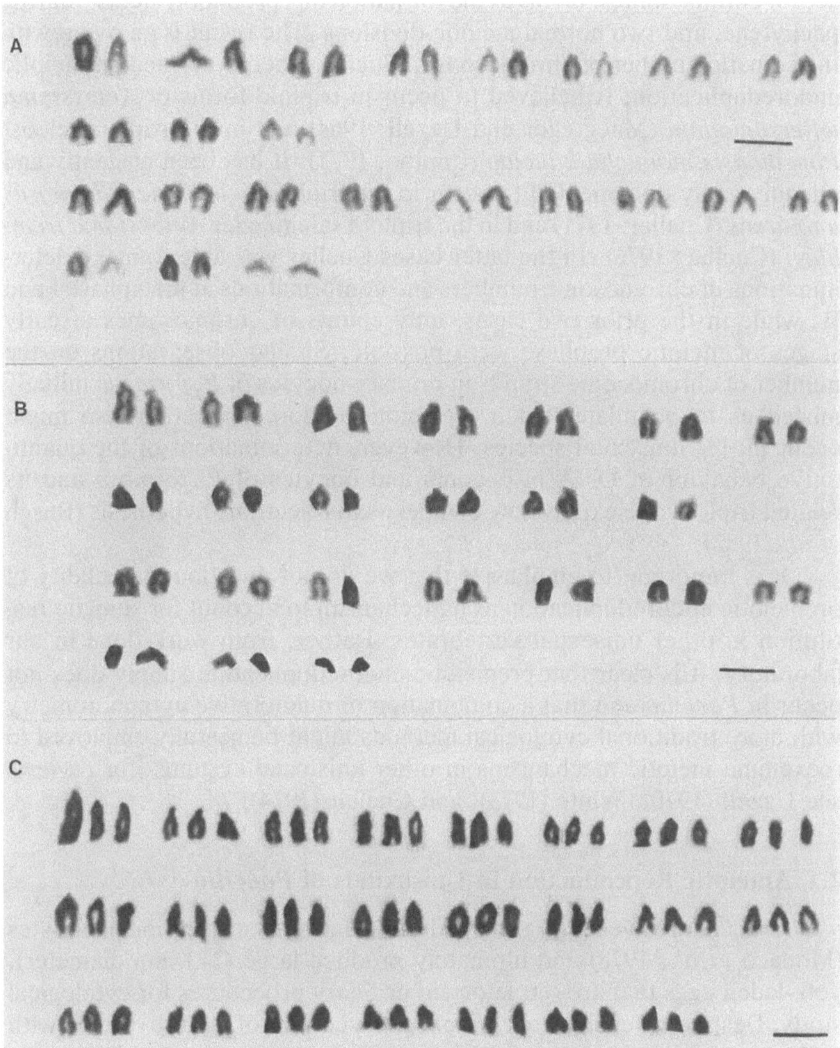

Figure 1. Karyotypes from gill epithelium of (A) an adult *Poecilia mexicana*, (B) *P. formosa*, and (C) a triploid unisexual. For *P. mexicana* and *P. formosa* $2n=46$ and for the triploid $3n=69$. Scale bars are equivalent to 5 μm.

sets of chromosomes, involves the formation of "pseudobivalents" during pachytene, and two normal meiotic divisions. The result is an ovum with the somatic number of chromosomes. Such a process, termed premeiotic endoreduplication, is believed to occur in triploid forms of *Ambystoma jeffersonianum* (Macgregor and Uzzell, 1964) and in the triploid teleost *Poeciliopsis monacha 2-lucida* (Cimino, 1971). It has been elegantly and unequivocally documented to occur in the triploid lizard *Cnemidophorus uniparens* (Cuellar, 1971) and in the triploid salamander *Ambystoma tremblayi* (Cuellar, 1976). In the latter cases Cuellar was able to make determinations of chromosome numbers and conformations at metaphase I and II, while in the prior two cases, only counts of chromosomes at early stages of meiotic prophase were possible. Similar observations on the number of chromosome strands in primary oocytes of *P. formosa* initially misled us to postulate that a premeiotic endoreduplication also might occur in this unisexual species. However, determinations of the quantitative behavior of DNA in oogonia and oocytes of *P. formosa* and its related triploid unisexuals now compel us to reject this hypothesis (Rasch *et al.,* 1982).

It is important to emphasize that we do not question the validity of premeiotic endoreduplication as a mechanism to account for somatic restitution in other unisexual vertebrates. Rather, from work done in our laboratory, it is clear that premeiotic endoreduplication simply does not occur in *Poecilia* and that a combination of quantitative cytophotometry with more traditional cytological methods might be usefully employed to reexamine meiotic mechanisms in other unisexual systems [for reviews see Uzzell (1970), White (1973), and Cuellar (1974)].

2.1. Ameiotic Reproduction in Unisexuals of *Poecilia*

Poecilia are livebearing fish with asynchronous maturation of oocytes (Monaco *et al.,* 1978) and ultimately produce large (2–3 mm diameter), yolk-laden eggs that are refractory to ordinary procedures for cytological study. Despite extensive analysis of serial sections of mature ovaries with yolky eggs from *P. formosa,* we have not yet found suitable meiotic metaphase configurations analogous to those demonstrated in parthenogenetic lizards and salamanders by Cuellar (1971, 1976). For this reason we focused our efforts on analysis of early stages of oogonial proliferation and oogenesis by using absorption cytophotometry, electron microscopy, and autoradiography, as well as conventional cytological techniques. Details of these methods are described elsewhere (Monaco *et al.,* 1978, 1981*a,b;* Monaco, 1982; Rasch *et al.,* 1982). Here we will summarize the salient results.

It appears that *P. formosa* and its related triploids propogate by apomixis, a mechanism of asexual reproduction that (1) involves suppression of the first meiotic division and its replacement by a mitotic division, (2) is asynaptic and is without the formation of bivalents, and (3) does not effect a reduction in chromosome number. Our intrepretation of apomixis is essentially equivalent to gonoid thelytoky as described for soft-scale insects (*Coccus hesperidum*) by Nur (1979). Although to our knowledge this is the first case of apomixis ascribed to a vertebrate species, there is prior biological precedent for its occurrence in a triploid form of unisexual goldfish (Kobayasi, 1976) and in several cases involving abnormal production of triploid human embryos (Jacobs *et al.*, 1978). Apomixis is also known to account for the unisexual reproduction of invertebrates from several major phyla [for a review see White (1973)] and, of course, is known to occur in a variety of plant species (Nygren, 1954; Swanson *et al.*, 1967).

Aside from the number of chromosome strands found in oocyte nuclei at zygotene or pachytene, there are no obvious differences in the general cytological details of oocyte growth and differentiation in unisexual and bisexual females of *Poecilia* (Monaco *et al.*, 1978, 1981*a;* Rasch *et al.*, 1982). As shown in Figs. 2A and 2B, counts of chromosome associations in primary oocyte nuclei reveal 23 bivalent chromosome strands for *P. latipinna,* a bisexual relative and one of the presumptive ancestors of *P. formosa*. There are 46 thin chromosome strands in primary oocytes from the diploid unisexual (Figs. 2C and 2D) and 69 thin strands in pachytene nuclei from triploid unisexuals (Figs. 2E and 2F). Similar illustrations, showing 23 chromosome pairs in primary oocytes from *P. mexicana* (=*P. m. limantouri*), have appeared elsewhere (Monaco *et al.*, 1981*a*). If one considers only the number of chromosome strands in pachytene oocyte nuclei from the unisexuals of *Poecilia,* these observations, at first glance, could be construed as evidence for a doubling of the number of chromosome sets prior to the initiation of meiotic prophase and the subsequent formation of pseudobivalents (cf. Macgregor and Uzzell, 1964; Cimino, 1971). When the diameters and staining intensities of these chromosomes are considered, however, it is apparent that the 46 chromosome strands seen in pachytene oocytes from *P. formosa* (Figs. 2C and 2D) do not represent 46 bivalents, which would imply a total of 92 chromosomes. Rather, there are 46 univalents, each of which contains a pair of sister chromatids. When viewed in this perspective, the chromosome strands seen in pachytene oocytes from triploid hybrids of *Poecilia* (Figs. 2E and 2F) also can be interpreted as univalents, and thus do not necessarily reflect formation of 69 pseudobivalents from the 138 chromosomes that would be produced if there were a premeiotic endoreduplication. Direct

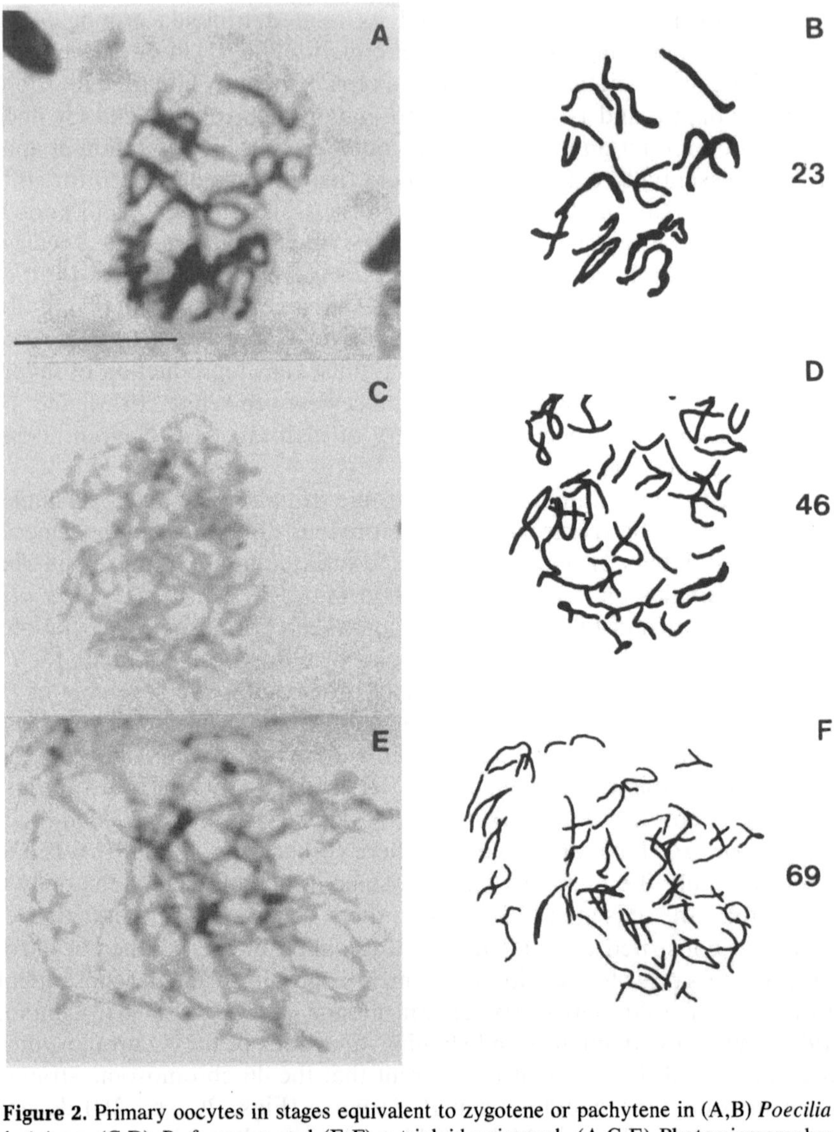

Figure 2. Primary oocytes in stages equivalent to zygotene or pachytene in (A,B) *Poecilia latipinna,* (C,D) *P. formosa,* and (E,F) a triploid unisexual. (A,C,E) Photomicrographs; (B,D,F) *camera lucida* drawings of the respective fields. The number of chromosomes counted from each oocyte is shown in the upper right. Scale bar in (A) is equivalent to 10 μm.

support of these unorthodox inferences comes from quantitative analyses of DNA levels before and during meiotic prophase in these different fishes and from ultrastructural studies of their chromosomes in pachytene oocytes.

Table I summarizes cytophotometric data on the behavior of DNA during oogenesis in *Poecilia*. The haploid or 1*C* amount of DNA from young spermatids of *P. latipinna* is included for reference. Estimates of nuclear DNA content in oogonia and primary oocytes of *P. latipinna* correspond to the 2*C* and 4*C* DNA levels established from measurements of interphase and dividing ovarian follicle cells. The 4*C* amount of DNA in primary oocyte nuclei of these bisexuals, as expected, is twice that of oogonial nuclei and will be reduced to the haploid or 1*C* level in mature ova by conventional meiotic divisions that are stimulated by the fertilization reaction and entry of the sperm nucleus. Essentially similar results have been found for *P. mexicana* (Rasch *et al.*, 1982).

Interphase follicle cell and oogonial nuclei from ovaries of *P. formosa* also fall into the diploid or 2*C* DNA class. Primary oocytes have twice the DNA content of diploid follicle cell nuclei (Table I). No oogonial nuclei or primary oocytes were found with four times the somatic cell DNA content (8*C*), as would be predicted if oogonia undergo a premeiotic endoreduplication.

The general pattern of behavior for DNA in follicle cells, oogonia, and primary oocytes of triploid unisexuals related to *P. formosa* is es-

Table I
Summary of DNA Determinations[a]

Fish species	Cell type	DNA class	DNA per nucleus, pg Mean ± SE	Number of nuclei
P. latipinna	Early spermatid	1*C*	0.85 ± 0.003	180
P. latipinna	Ovarian follicle	2*C*	1.68 ± 0.006	342
P. latipinna	Oogonia	2*C*	1.85 ± 0.025	92
P. latipinna	Primary oocyte	4*C*	3.35 ± 0.060	192
P. formosa	Ovarian follicle	2*C*	1.66 ± 0.006	674
P. formosa	Oogonia	2*C*	1.57 ± 0.009	198
P. formosa	Primary oocyte	4*C*	3.61 ± 0.020	318
Triploid	Ovarian follicle	3*C*	2.43 ± 0.016	165
Triploid	Oogonia	3*C*	2.74 ± 0.023	85
Triploid	Primary oocyte	6*C*	5.31 ± 0.025	137

[a]These data are based on Feulgen cytophotometry from individual nuclei, made at 560 nm with a Vickers M86 scanning and integrating microdensitometer, as described elsewhere (Rasch *et al.*, 1982).

sentially like that observed for the diploid forms, i.e., there is no evidence for a premeiotic doubling of the $3n$ chromosome complement. The observed DNA levels clearly indicate that the 69 chromosomes seen in triploid oocytes do not represent paired sets of daughter chromatids, i.e., meiotic tetrads, and thus cannot be considered pseudobivalents. Nuclei in the triploid forms, as expected, contain 50% more DNA than comparable nuclei from their diploid, unisexual relatives (Table I).

DNA values for primary oocyte nuclei in both diploid and triploid unisexuals are equivalent to the DNA levels found in their respective somatic cell metaphases (Rasch *et al.*, 1982). Thus, in both unisexuals only a single cycle of chromosome replication occurs during the S phase of DNA synthesis prior to the onset of oogenesis. Twice as much DNA in primary oocytes would be predicted if the last oogonial mitosis were restitutional, i.e., if a premeiotic endoreduplication had occurred.

To account for the observation of somatic numbers of chromosome strands in primary oocytes of *P. formosa* and its related triploids we must look to other mechanisms to explain maintenance of a constant chromosome number, fixed heterozygosity, and the isogenic patterns of inheritance so characteristic of these fishes. Two types of thelytoky can be distinguished: automictic (or meiotic) and apomictic (or ameiotic). In the automictic type, chromosomes synapse and cross over, bivalents or pseudobivalents are formed, and reduction occurs. For those organisms that show premeiotic endoreduplication, synapsis and crossing over of the homeologous chromosomes (Huskins, 1932) would have no genetic consequences. The chromosome complement in such a primary oocyte would contain twice as much DNA as a somatic metaphase nucleus, that is, an $8C$ level of DNA for a diploid organism and $12C$ for a triploid form.

In apomictic thelytoky there is no synapsis or bivalent formation and no reduction in chromosome number occurs. Thus, the somatic number of chromosomes is present at pachytene or diplotene and at metaphase I in an ameiotic primary oocyte, not half the somatic number in the form of bivalents, as would be expected for an occyte undergoing conventional meiosis. The maturation division in an apomictic oocyte is mitotic, not meiotic, in character, and results in the separation of sister chromatids, not in the separation of dyads from a pseudobivalent association. The latter distinction is a subtle but important point, since the essential issue in designating an oocyte division as meiotic or ameiotic depends primarily on what cytological criteria are considered most critical. For example, if reduction and segregation of synapsed homologous chromosomes during the first or heterotypic division are considered the essence of meiosis, then divisions during oogenesis that do not show these uniquely meiotic attributes are essentially ameiotic, or mitotic in nature.

If gynogenesis in *P. formosa* and its triploid relatives constitutes an automictic type of thelytoky, the 46 and 69 chromosome strands observed in their primary oocytes would be pseudobivalents and therefore contain twice as much DNA as replicated univalents. On the other hand, replicated univalents should have exactly the same DNA content as mitotic metaphase configurations in adjacent follicle cells. The latter is what we have found.

As noted above, a criterion for distinguishing between automixis and apomixis is the presence or absence of synapsed chromosomes. If the chromosomes are synapsed, gynogenesis is automictic; if they remain univalent, gynogenesis is apomictic. Electron microscopic examinations of primary oocytes of *P. formosa* have consistently failed to show structures that resemble characteristic tripartite configurations of synaptonemal complexes, a feature that is often considered diagnostic of synapsis and bivalent formation [for a review see Moses (1968)]. In the unisexuals of *Poecilia* we see only single elements (Fig. 3), suggesting that these chromosomes do not pair and thus are univalent. Coupled with the cytophotometric and cytological data outlined above, our ultrastructural findings further indicate that there is no synapsis during oogenesis in unisexual *Poecilia* and that gynogenetic reproduction in these forms occurs as a result of apomixis.

2.2. Alternatives to Apomixis

In considering possible oogenic pathways that could account for unisexual reproduction in *Poecilia*, it is important to consider their genetic consequences. Since unisexual *Poecilia* show matroclinous inheritance, have essentially fixed heterozygosity, and produce isogenic offspring, any alternative to apomixis as a mechanism of reproduction must account for these constraints.

Genomic reduplication, of course, allows for isogenecity and fixed heterozygosity, but, as we have shown, premeiotic endoreduplication does not occur in *Poecilia*. Results of radiolabeled thymidine incorporation experiments to monitor total DNA replication (Rasch *et al.*, 1982) and *in situ* hybridization experiments to follow selective DNA amplification (Monaco *et al.*, 1981*a*) clearly show that there is no detectable synthesis of DNA in oocyte nuclei after the onset of leptotene in either unisexual or bisexual mollies. Thus, there is no compensatory doubling of the genome during meiotic prophase, i.e., there is no endopachytene replication in these fishes, like that described for parthenogenetic stick insects (Koch *et al.*, 1972; Pijnacker and Koch, 1975).

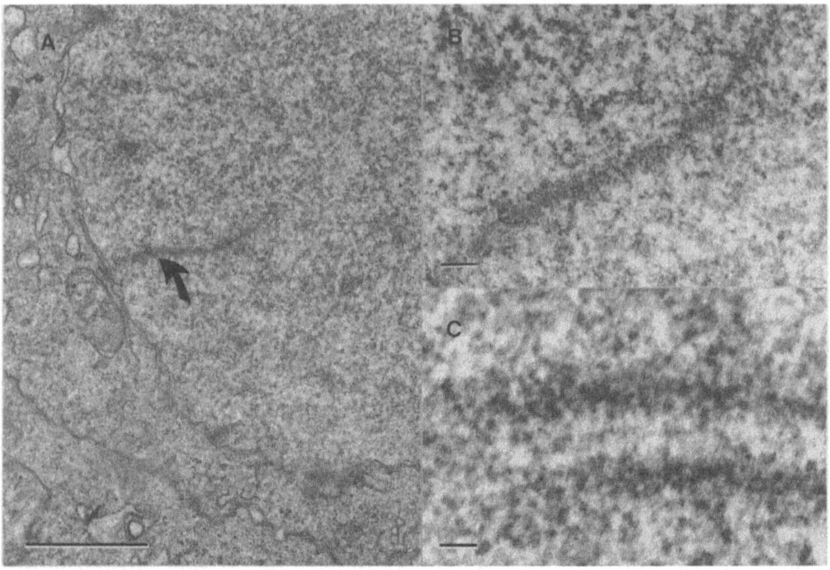

Figure 3. (A) Low-power electron micrograph of a primary oocyte from *Poecilia formosa*. The chromosome associations (arrow) resembling lateral elements of synaptonemal complexes are often contiguous with the nuclear envelope. Scale bar is equivalent to 1 μm. (B) Higher magnification of the area marked by arrow in (A). Although these chromosomal structures suggest synaptonemal complex formation, they do not show the tripartite organization that is typical of synapsed chromosomes. Scale bar is equivalent to 0.1 μm. (C) Electron micrograph of a primary spermatocyte nucleus from *P. mexicana* showing the tripartite character typical of synaptonemal complexes in bivalent chromosomes. Scale bar is equivalent to 0.1 μm.

It is logical to ask why we use terminology uniquely associated with meiosis (leptotene, pachytene, or diplotene) if we claim that unisexuals of *Poecilia* produce oocytes ameiotically. In one sense the issue is semantic, since the primary function of prophase during oogenesis involves not only the reassortment and transmittal of genetic information from one generation to the next, but it also is a necessarily prolonged stage in order to accommodate the biosynthesis of ribonucleoproteins, mRNA, and yolk that will ultimately be used during early embryonic development. The cytological features associated with oogenesis in unisexual *Poecilia,* such as diffuse lampbrush chromosomes and the elaboration of nucleolar ribonucleoproteins (Monaco *et al.*, 1978, 1981*a*), reflect a fundamental program of cellular activity that occurs in oocytes. An ameiotic oocyte would be expected to show the cytological features of a developing egg, although its chromosome mechanics may be quite different.

Additional pathways that could lead to all female offspring have been discussed more fully elsewhere (Rasch *et al.,* 1982). They involve either a suppression of the first or second meiotic division, reentry and fusion with a polar body, a selective maturation involving premeiotic exclusion, or a suppression of the first zygotic division. None of these alternatives can account for the matroclinous inheritance of fixed heterozygous gene loci by *P. formosa*. There are two points worth mentioning, however, both related to perturbations of the normal sequence of events during oogenesis in unisexuals:

1. If synapsis, crossing over, and the first meiotic metaphase are suppressed, the offspring that develop from such unreduced ova will be isogenic and retain the original heterozygosity of their mother. Such a developmental pathway is in fact functionally equivalent to apomixis.
2. Although the spontaneous origin of triploid female offspring from a diploid *P. formosa* and rare cases of clonal heterogeneity in triploid stocks (Balsano, Rasch, and Monaco, unpublished data) superficially may resemble the phenomenon of hybridogenesis in which a male genome is "borrowed" for a generation and is selectively excluded prior to meiosis (Schultz, 1969; Cimino, 1972), all of our findings indicate that triploids consistently produce triploid oogonia and oocytes and yield invariably female, triploid progeny.

The possession of three sets of chromosomes in a triploid unisexual of *Poecilia* should certainly lead to difficulties in synapsis and in the subsequent reductional division of meiosis. There is direct cytological evidence for difficulties of this type during spermatogenesis in the unique triploid hybrid male of *P. formosa* × *P. vittata* (Rasch *et al.,* 1965). Spermatocytes in the rudimentary testis of this infertile specimen showed a high incidence of chromosomal bridging and fragmentation during metaphase I, leading to the production of both "giant" and "micro" spermatid nuclei at the completion of meiosis II. Similar cytological abnormalities have been described by White *et al.* (1977) for triploid grasshoppers. Furthermore, triploids of *Poecilia,* like their diploid, unisexual relatives, do not show inheritance of pigment, muscle protein genes, or allozyme patterns when black molly males are used to sire triploid young in the laboratory. Thus both diploid and triploid forms of *Poecilia* maintain their genetic identities by apomictic reproduction.

3. Concluding Remarks

Although *P. formosa* and its related triploid unisexuals show strictly matroclinous inheritance and exist in natural populations as a series of clones, their gene pools do not remain immutable. As pointed out by Crow and Kimura (1965), for a given amount of genetic variability, the efficiency of selection is greater in an asexually reproducing population than it is in one with free recombination, since the rate of selection is measured by the total genotypic variance. In asexual populations with fixed heterosis, such as naturally occurring clonal stocks of *P. formosa* or its related triploid biotypes, the selection limit is determined by the most fit existing genotype. Apomixis, which bypasses meiotic recombination and genetic segregation, perpetuates fixed heterozygous genomes and thereby should enchance the rate at which favorable gene combinations can be incorporated into these populations. New genetic combinations in these fishes arise primarily by mutations in germ-line tissues that do not affect the contemporary fitness of that particular fish, but would be perpetuated and tested by selection in subsequent generations of her progeny, which, by definition, constitute a new clone.

Anomalous, rare tissue graft rejections among individuals within certain sibships of *P. formosa* and triploid stocks (Balsano, Rasch, and Monaco, in preparation) and unusually high allozyme diversity in some population samples of triploids (Turner *et al.*, 1983) also suggest that somatic mutation and/or somatic crossing over may play a greater than anticipated role in the genetics of these fish. These aspects of the genetics in unisexual vertebrates warrant further intensive research.

There is now a growing body of evidence in *Drosophila* and mouse to suggest that synapsis and crossing over are temporally disjunct from meiosis proper, and take place during the period of DNA synthesis prior to the visible onset of meiotic prophase (Grell, 1973; Grell *et al.*, 1980). If this nontraditional view of the timing of synapsis and genetic recombination also applies to unisexuals of *Poecilia*, then it would provide an additional, and perhaps significant, source of variability in these fishes.

In view of the usual genetic consequences of meiosis and the results described here, it seems that the continued propagation of heterozygous genomes by *P. formosa* and its related triploid unisexuals can stem only from ova with somatic ploidy levels produced in a manner that bypasses the first, reductional division of meiosis and is therefore functionally equivalent to apomixis. The fixed heterosis in such unisexual systems may function to replace the type of allelic variations that normally accompany sexual reproduction and thus may be of critical importance to the long term survival and evolutionary success of these all-female species. By its very nature, apomictic or ameiotic thelytoky serves to maintain a constant chromosome number and confers the highly characteristic, remarkably stable genetic identity of these unisexual fishes.

ACKNOWLEDGMENTS. We thank Dr. Bruce Turner for helpful discussions relating to the preparation of this manuscript. Research was supported in part by grants from the National Geographic Society, the Veterans Administration, and the National Science Foundation (most recently grants DEB 78-07753 to EMR and DEB 80-23277 to JSB). The cooperation of the Mexican Department of Fisheries in granting collecting permits No. 2858 (JSB) and No. 5512 (EMR) is gratefully acknowledged.

References

Abramoff, P., Darnell, R. M., and Balsano, J. S., 1968, Electrophoretic demonstration of the hybrid origin of the gynogenetic teleost *P. formosa, Am. Nat.* **102:**555–558.

Anders, A., Anders, F., and Klinke, K., 1973, Regulation of gene expression in the Gordon–Kosswig melanoma system. II. The arrangement of chromatophore determining loci and regulating elements in the sex chromosomes of Xiphophorin fish, *Platypoecilus maculatus* and *Platypoecilus variatus*, in: *Genetics and Mutagenesis of Fish* (J. H. Schroder, ed.) Springer-Verlag, Berlin, pp. 53–67.

Anders, F., Chatterjee, K., Schwab, M., Scholl, E., and Anders, A., 1981, Tumor gene expression and interphase chromatin appearance in *Xiphophorus, Am. Zool.* **21:**535–548.

Balsano, J. S., and Rasch, E. M., 1974, Biochemical and cytogenetic studies of *Poecilia* from eastern Mexico I. Comparative microelectrophoresis of plasma proteins of seven species, *Rev. Biol. Trop.* **21:**229–257.

Balsano, J. S., Darnell, R. M., and Abramoff, P., 1972, Electrophoretic evidence of triploidy associated with populations of the gynogenetic teleost *Poecilia formosa, Copeia* **1972:**292–297.

Cimino, M. C., 1971, Meiosis in triploid all-female fish (*Poeciliopsis*, Poeciliidae), *Science* **175:**1484–1486.

Cimino, M. C., 1972, Egg production,polyploidization, and evolution in a diploid all female fish of the genus *Poeciliopsis, Evolution* **26:**294–306.

Crow, J. F., and Kimura, M., 1965, Evolution in sexual and asexual populations, *Am. Nat.* **49:**439–450.

Cuellar, O., 1971, Reproduction and the mechanism of meiotic restitution in the parthenogenetic lizard *Cnemidophorus uniparens, J. Morphol.* **133:**139–166.

Cuellar, O., 1974, On the origin of parthenogenesis in vertebrates: The cytological factors, *Am. Nat.* **108:**625–648.

Cuellar, O., 1976, Cytology of meiosis in the triploid gynogenetic salamander *Ambystoma tremblayi, Chromosoma (Berl.)* **58:**355–364.

Cuellar, O., 1977, Animal parthenogenesis, *Science* **197:**837–843.

Darnell, R. M., Lamb, E., and Abramoff, P., 1967, Matroclinous inheritance and clonal structure of a Mexican population of the gynogenetic fish *Poecilia formosa, Evolution* **21:**168–178.

Grell, R. F., 1973, Recombination and DNA replication in the *Drosophilia melanogaster* oocyte, *Genetics* **73**:87–103.

Grell, R. F., Oakberg, E. F., and Generoso, E. E., 1980, Synaptonemal complexes of premeiotic interphase in the mouse spermatocyte, *Proc. Natl. Acad. Sci. USA* **77**:6720–6723.

Hubbs, C. L., 1955, Hybridization between fish species in nature, *Syst. Zool.* **4**:1–20.

Hubbs, C., 1964, Interactions between a bisexual fish species and its gynogenetic sexual parasite, *Bull. Tex. Mem. Mus.* **8**:1–72.

Hubbs, C. L., and Hubbs, L. C., 1932, Apparent parthenogenesis in nature, in a form of fish of hybrid origin, *Science* **76**:628–630.

Hubbs, C. L., and Hubbs, L. C., 1946, Experimental breeding of the amazon molly, *Aquarium J.* **17**:4–6.

Huskins, C. L., 1932, A cytological study of Vilmorin's unfixable dwarf wheat, *J. Genet.* **25**:113–124.

Jacobs, P. A., Angell, R. R., Buchanan, I. M., Hassold, T. J., Matsayama, A. M., and Manuel, B., 1978, The origin of human triploids, *Ann. Hum. Genet.* **42**:49–57.

Kallman, K. D., 1962*a*, Gynogenesis in the teleost *Mollienesis formosa* with a discussion of the detection of parthenogenesis in vertebrates by tissue transplantation, *J. Genet.* **58**:7–21.

Kallman, K. D., 1962*b*, Population genetics of the gynogenetic teleost, *Mollienesia formosa, Evolution* **16**:497–504.

Kallman, K. D., 1962*c*, Population structure of the all female gynogenetic teleost, *Molliensia formosa* (Girard), *Proc. XVI Int. Cong. Zool.* **2**:170, abstract.

Kallman, K. D., 1964, Homozygosity in a gynogenetic fish, *Poecilia formosa, Genetics* **50**:260–262.

Kallman, K. D., 1975, The platyfish, *Xiphophorus maculatus*, in: *Handbook of Genetics*, Vol. 4 (R. C. King, ed.), Plenum Press, New York, pp. 81–132.

Kobaysi, H., 1976, A cytological study on the maturation division in the oogenic process of the triploid ginbuna *Carassius auratus Iangsdorfii, Jpn. J. Ichthyol.* **22**:234–240.

Koch, P., Pijnacker, L. P., and Kreke, J., 1972, DNA reduplication during meiotic prophase in the oocyte of *Carausius morosus* Br. (Insecta, Cheleutoptera), *Chromosoma (Berl.)* **36**:313–321.

Macgregor, H. C., and Uzzell, T. M., 1964, Gynogenesis in salamanders related to *Ambystoma jeffersonianum, Science* **143**:1043–1045.

Maslin, T. P., 1968, Taxonomic problems in parthenogenetic vertebrates, *Syst. Zool.* **17**:219–231.

Menzel, B. W., and Darnell, R. M., 1973, Systematics of *Poecilia mexicana* (Pisces: Poeciliidae) in northern Mexico, *Copeia* **1973**:225–227.

Meyer, H., 1938, Investigations concerning the reproductive behavior of *Molliensia formosa, J. Genet.* **36**:329–366.

Monaco, P. J., 1982, Reproductive biology of *Poecilia formosa* and its related species, Ph.D. dissertation, Marquette University.

Monaco, P. J., Rasch, E. M., and Balsano, J. S., 1978, Cytological evidence for temporal differences during the asynchronous ovarian maturation of bisexual and unisexual fishes of the genus *Poecilia, J. Fish Biol.* **13**:33–44.

Monaco, P. J., Rasch, E. M., and Balsano, J. S., 1981*a*, Nucleoprotein cytochemistry during oogenesis in a unisexual fish, *Poecilia formosa, Histochem. J.* **13**:747–761.

Monaco, P. J., Rasch, E. M., and Balsano, J. S., 1981*b*, Sperm availability in naturally occurring bisexual–unisexual breeding complexes involving *Poecilia mexicana* and the gynogenetic teleost *Poecilia formosa, Environ. Biol. Fish.* **6**:159–166.

Monaco, P. J., Rasch, E. M., Balsano, J. S., and B. J. Turner, 1982, Muscle protein phenotypes and the probable evolutionary origin of a unisexual fish, *Poecilia formosa,* and its triploid derivative, *J. Exp. Zool.* **221**:265–274.

Moses, M. J., 1968, Synaptinemal complex, *Annu. Rev. Genet.* **2**:363–412.

Nur, U., 1979, Gonoid thelytoky in soft scale insects (Coccidae: Homoptera), *Chromosoma (Berl.)* **72**:89–104.

Nygren, P., 1954, Apomixis in angiosperms, *Bot. Rev.* **20**:577–649.

Pijnacker, L. P., and Koch, P., 1975, Complete and incomplete extra DNA reduplication during spermatogenesis of *Carausius morosus* Br. (Insecta, Phasmida), *Chromosoma (Berl.)* **49**:269–278.

Prehn, L. M., and Rasch, E. M., 1969, Cytogenetic studies of *Poecilia* (Pisces) I. Chromosome numbers in naturally occurring poeciliied species and their hybrids from eastern Mexico, *Can. J. Genet. Cytol.* **11**:880–895.

Rasch, E. M., 1968, Use of deoxyribonucleic acid-Feulgen levels as an index of triploidy in naturally occurring interspecific hybrids of poeciliid fishes, *J. Histochem. Cytochem.* **16**:508–509, abstract.

Rasch, E. M., and Balsano, J. S., 1973, Cytogenetic studies of *Poecilia* (Pisces) III. Persistence of triploid genomes in the unisexual progeny of triploid females associated with *Poecilia formosa, Copeia* **1973**:810–813.

Rasch, E. M., and Balsano, J. S., 1974, Biochemical and cytogenetic studies of *Poecilia* from eastern Mexico II. Frequency, prepetuation, and probable origin of triploid genomes in females associated with *Poecilia formosa, Rev. Biol. Trop.* **21**:351–381.

Rasch, E. M., Darnell, R. M., Kallman, K. D., and Abramoff, P., 1965, Cytophotometric evidence for triploidy in hybrids of the gynogenetic fish, *Poecilia formosa, J. Exp. Zool.* **160**:155–170.

Rasch, E. M., Prehn, L. M., and Rasch, R. W., 1970, Cytogenetic studies of *Poecilia* (Pisces) II. Triploidy and DNA levels in naturally occuring populations associated with the gynogenetic teleost) *Poecilia formosa* (Girard), *Chromosoma (Berl.)* **31**:18–40.

Rasch, E. M., Monaco, P. J., and Balsano, J. S., 1978, Identification of a new form of triploid hybrid fish by DNA-Feulgen cytophotometry, *J. Histochem. Cytochem.* **26**:218, abstract.

Rasch, E. M., Monaco, P. J., and Balsano, J. S., 1982, Cytophotometric and autoradiographic evidence for functional apomixis in a unisexual fish *Poecilia formosa* and its related triploid unisexuals, *Histochemistry* **73**:515–533.

Rosen, D. E., and Bailey, R. M., 1963, The poeciliid fishes (Cyprinodontiformes), their structure, zoogeography, and systematics, *Bull. Am. Mus. Nat. Hist.* **126**:1–176.

Schultz, R. J., 1969, Hybridization, unisexuality, and polyploidy in the teleost *Poeciliopsis* (Poeciliidae) and other vertebrates, *Am. Nat.* **103**:605–619.

Schultz, R. J., 1980, Role of polyploidy in the evolution of fishes, in: *Polyploidy: Biological Relevance* (W. H. Lewis, ed.), Plenum Press, New York, pp. 313–340.

Schultz, R. J., and Kallman, K. D., 1968, Triploid hybrids between the all female teleost *Poecilia formosa* and *Poecilia sphenops, Nature* **219**:280.

Strommen, C. A., Rasch, E. M., and Balsano, J. S., 1975*a,* Cytogenetic studies of *Poecilia* (Pisces) IV. Epithelial cell biopses to identify triploid females associated with *Poecilia formosa, Copeia* **1975**:568–572.

Strommen, C. A., Rasch, E. M., and Balsano, J. S., 1975*b,* Cytogenetic studies of *Poecilia* (Pisces) V. Cytophotometric evidence for the production of fertile offspring by triploids related to *Poecilia formosa, J. Fish Biol.* **7**:1–10.

Swanson, C. P., Merz, T., and Young, W. S., 1967, *Cytogenetics,* Prentice-Hall, Englewood Cliffs, New Jersey.

Turner, B. J., 1982, The evolutionary genetics of a unisexual fish, *Poecilia formosa,* in: *Mechanisms of Speciation,* A. R. Liss, New York, pp. 265–305.

Turner, B. J., Balsano, J. S., Monaco, P. J., and Rasch, E. M., 1983, Clonal diversity and evolutionary dynamics in a diploid–tripolid breeding complex of unisexual fishes (*Poecilia*), *Evolution,* **37(4)**:798–809.

Uzzell, T. M., 1970, Meiotic mechanisms of naturally occurring unisexual vertebrates, *Am. Nat.* **104**:433–445.

White, M. J. D., 1973, *Animal Cytology and Evolution,* Cambridge University Press, London.

White, M. J. D., Contreras, Cheney, J., and Webb, G. C., 1977, Cytogenetics of the parthenogenetic grasshopper *Warramaba* (formerly *Moraba*) *virgo* and its bisexual relatives II. Hybridization studies, *Chromosoma (Berl.)* **61**:127–148.

Yamamoto, T., 1975, The medaka, *Oryzias latipes,* and the guppy, *Lebistes reticularis,* in: *Handbook of Genetics,* Vol. 4 (R. C. King, ed.), Plenum Press, New York, pp. 133–149.

CHAPTER 7

Evolutionary Ecology of Unisexual Fishes

WILLIAM S. MOORE

1. Introduction

The evolutionary ecology of unisexual fishes is distinguished from that of closely related sexual species by two peculiar aspects of their genetics: First, they reproduce, in effect, asexually, and second, their genomic constitutions are invariably identical to those of F_1 hybrids. Thus, a major difficulty in analyzing the evolutionary ecology of these species is to determine the extent to which the distinct phenomena of asexuality and hybridity contribute to their successes and failures. That is, where abundance indicates the success of a particular unisexual fish species, is it successful there because it reproduces asexually or because it possesses a hybrid genotype? The ecological consequences of asexuality, in turn, can be more finely divided. Asexually reproducing organisms do not pay the cost of meiosis (Williams, 1975), and, hence, an asexual species has twice the intrinsic rate of increase r of a comparable sexual species (Maynard Smith, 1978). On the other hand, asexual species lack genetic recombination and cannot generate the multiplicity of genotypes possible in a sexual species. This paucity of genotypic variation may limit the effectiveness of selection and, hence, evolutionary potential or it may reduce the aggregate effectiveness of resource utilization because there cannot be a spectrum of "specialized" genotypes in the population.

These are major themes that have dominated thinking and research

WILLIAM S. MOORE • Department of Biological Sciences, Wayne State University, Detroit, Michigan 48202.

concerned with the evolutionary ecology of unisexual fishes during the past two decades. In this review, I will try to identify the major hypotheses concerned with the peculiar evolutionary ecology of parthenogenetic reproduction, trace, to a limited extent, the historical origin and development of each hypothesis, and, finally, evaluate the validity of the hypotheses by comparing each with the data. In some few instances one hypothesis is quite likely true whereas competing hypotheses are clearly false, but more often it is not clear which of two or even several hypotheses is the most plausible explanation of a particular phenomenon. Nonetheless, I have emphasized hypothetical discussions rather than simple exposition of data because I believe that scientific progress occurs only when competing hypotheses are conceived, analyzed, and tested. Although the number of known vertebrate species that are unequivocally or likely to be parthenogenetic exceeds 50, parthenogenesis remains a rare phenomenon among vertebrates and is probably of little consequence in the overall sweep of vertebrate evolution. Thus, the unguided acquisition of data on vertebrate parthenospecies would be unjustifiable, but the peculiar genetics, reproductive biology, and ecology of these species could—and, I believe *are*—providing important insights into a spectrum of evolutionary phenomena, including the origin of duplicate genes, the relationship between specific genotypes and resource utilization, niche specialization, adaptive radiation, hybridization, speciation, and the adaptive value of sexual reproduction. Furthermore, it is my belief that generalization is a virtue in science, and in this sense there exists a commonality in the ecological patterns of *all* naturally occurring parthenogenetic vertebrates that have been studied in any detail. Therefore, I will treat the evolutionary ecology of parthenogenetic amphibians and reptiles in nearly comparable detail. (I know of no naturally occurring parthenogentic species of birds or mammals.)

The biology of unisexual fishes has been exhaustively reviewed (Schultz, 1969, 1971, 1977). Here, I will briefly review the reproductive biology of these fishes for the benefit of the reader unfamiliar with this somewhat esoteric literature.

I use the term parthenogenesis in a broad sense to mean all modes of reproduction in which one or more genomes are inherited clonally. All of the parthenogentic reptilian species are all-female and reproduce by spontaneous parthenogenesis (thelytoky); i.e., egg nuclei have the same chromosome number as somatic cells and embryogenesis occurs without fertilization. All of the parthenogenetic fishes and amphibians, in contrast, require at least the physical stimulus of sperm to induce embryogenesis. Thus, all parthenogenetic fishes and amphibians must mate with males of closely related sexual species to achieve reproduction. This fact has a profound effect on their population biology, since these parthenospecies must coexist with closely related sexual species and must succeed in seducing males of sexual species. When such a species interaction is translated into a mathematical formulation, it has the essential properties of a host–parasite interaction.

Although all are sperm dependent, two very distinct reproductive mechanisms exist emong the parthenogenetic fishes and amphibians: gynogenesis and hybridogenesis. Gynogenesis is the simpler of the two mechanisms. Gynogenetic species produce eggs with chromosome numbers identical to those of the somatic cells; sperms merely stimulate development and male chromatin is never expressed in the progeny. Thus, the only difference between this and spontaneous parthenogenesis as seen in reptiles is the necessity of sperm to stimulate development.

The important genetic consequence of gynogenesis is that daughters are genetic replicates of their mothers. This fact was demonstrated for morphological characters by Hubbs and Hubbs (1932, 1946*a,b*), Clanton (1934), Meyer (1938), and Schultz (1967, 1969), for histocompatibility antigens by Kallman (1962*a*), Darnell *et al.* (1967), and Moore (1977*a*), and for electrophoretic markers by Vrijenhoek (1972). Most of the gynogenetic parthenospecies are also triploid relative to the sexual species from which they were derived. The two gynogenetic salamanders in the complex of Jefferson salamanders, *Ambystoma tremblayi* and *A. platineum* (Uzzell, 1963, 1964; Uzzell and Goldblatt, 1967), are triploid, as are the three gynogenetic species of poeciliid fishes in the genus *Poeciliopsis* (Schultz, 1967, 1969, 1971).

The situation with the Amazon molly (*Poecilia formosa*), however, is more complex. The original forms of *P. formosa* described by Hubbs and Hubbs (1932) were probably diploid, as appears to be the case with most Amazon mollies from the Río Grande drainage (Drewry, 1964; Prehn and Rasch, 1969). More recently, naturally occurring triploid forms of *P. formosa* have been reported from the Río Soto la Marina drainage in northeastern Mexico (Prehn and Rasch, 1969; Rasch *et al.*, 1970; Rasch and Balsano, 1973, 1974). Diploid *P. formosa* have been bred and studied for years in the laboratory. The production of triploid offspring by diploid females does occur rarely in the laboratory (Rasch *et al.*, 1965; Schultz and Kallman, 1968), but the persistence of histocompatible lineages over time indicates that a stable diploid mechanism of reproduction by gynogenesis is the norm in at least some if not most populations of *P. formosa*. Triploid *P. formosa* appear to be gynogenetic also, although there seems to be a level of instability in the meiotic mechanisms of Río Soto la Marina populations not seen elsewhere (Rasch and Balsano, 1974). Although crosses between *Poecilia mexicana* and *P. latipinna* have not resulted in the production of unisexual gynogentic offspring (Turner *et al.*, 1980*a*), much morphological (Hubbs and Hubbs, 1932; Menzel and Darnell, 1973) and electrophoretic (Balsano *et al.*, 1972; Balsano and Rasch, 1974; Rasch and Balsano, 1974; Turner *et al.*, 1980*b*) evidence indicates that both diploid and triploid *P. formosa* are hybrids: the diploids comprise one *mexicana* and one *latipinna* genome, whereas the triploid comprises two *mexicana* and one *latipinna*.

Meiosis has been studied in several of these cases, and, except perhaps in *Poecilia formosa*, the same meiotic mechanism seems to obtain (Macgregor and Uzzell, 1964; Cimino, 1972*a;* Cuellar, 1976*b;* E. M. Rasch, personal communication). Further, this same mechanism is found in the spontaneously parthenogenetic triploid lizard *Cnemidophorus uniparens* (Cuellar, 1971). The mechanism involves a duplication of chromosomes in germ-line cells prior to meiosis without a concomitant cytokinesis. Thus, the typically triploid somatic chromosome number is elevated to a hexaploid state in germ-line cells. The subsequent meiosis is apparently normal except that sister chromosomes preferentially pair and this results in matroclinous inheritance. If homologous but nonsister chromosomes paired at the hexaploid prophase I, a kind of chromosome segregation would result, leading to a loss of heterozygbsity. Suppose, for example, that a triploid gynogen had the genotype AAA' at some locus. If three chromosomes were randomly chosen from the six at the reduction division, then eventually the fixed genotypes AAA or $A'A'A'$ would segregate. This pattern of inheritance has never been observed for electrophoretic (Vrijenhoek, 1972) or antigenic markers (Kallman, 1962*a;* Moore, 1977*a*.)

Hybridogenetic species are invariably diploid and, indeed, triploidy would seem to preclude hybridogenesis. In hybridogens, a reduction division occurs and haploid gametes are produced, but the same genome is always segregated, intact, into functional gametes. Thus, a single genome is inherited clonally. *Poeciliopsis monach-lucida* [earlier termed C_x (Schultz, 1966)] is the classic example and serves to illustrate this mode of inheritance (Schultz, 1961, 1966, 1969; Vrijenhoek, 1972). Schultz (1969) developed the nomenclature systems used here to designate various species of *Poeciliopsis:* Schultz's system has the advantages that it reflects the hybrid origin of the parthenospecies, indicates its genomic composition, and enables one more easily to keep in mind the specific pattern of inheritance. *Peociliopsis monach-lucida* arose via hybridization between two bisexual species *P. monacha* and *P. lucida* (Schultz, 1973). The *monacha* genome is transmitted through the germ line, whereas the *lucida* genome is discarded each generation. *Poeciliopsis monacha-lucida* has been synthesized in the laboratory by crossing female *P. monacha* with male *P. lucida* (Schultz, 1973). The cross results in only female offspring, which then reproduce by hybridogenesis. In nature *P. monacha-lucida* mates with males of *P. lucida;* thus, the genotype of an F_1 hybrid between *monacha* and *lucida* is perpetuated *ad infinitum* where a newly assorted *lucida* genome is combined with the clonally inherited *monacha* genome in each generation. If *P. monacha-lucida* were mated to males of other related species, such as *P. latidens* or *P. occidentalis,* the result would be the production of *P. monacha-latidens* and *P. monacha-occidentalis,* respectively (Schultz, 1961, 1966). Extensive natural populations of both these combinations exist, and the present thinking is that *P. monacha-latidens* did, in fact, arise this way (Schultz, 1977). While *P. monacha-occidentalis* may have arisen this way also (Moore *et al.,* 1970), with the more recent discovery of limited sympatry between *P. monacha* and *P. occidentalis* in the Río Mayo of Sonora, Mexico, it seems more likely that *P. monacha-occidentalis* arose directly by hybridization between the bisexual species rather that via a *P. monacha-lucida* intermediary (Schultz, 1977). *Poeciliopsis monacha-occidentalis* can be synthesized either way in the laboratory (Schultz, 1961; Vrijenhoek, 1979) and so the exact origin of *P. monacha-occidentalis* remains somewhat uncertain.

Evidence for the meiotic mechanism (or, perhaps more descriptively, meiotic aberration) that results in hybridogenesis was reported by Cimino (1972*b*). Apparently, a unipolar spindle is formed in germ-line cells prior to vitellogenesis. A haploid number of chromosomes, presumably the *monacha* genome, is seen to align on the equatorial plate, whereas the remaining half of the chromosomes are left scattered in the cytoplasm, where they eventually disappear. The equatorial chromosomes move to the single pole and the nuclear membrane is reconstituted to envelop this set of chromosomes but not the remainder. Since this mechanism precludes synapsis, inheritance of the maternal genome is strictly clonal. Leakage of paternal (*lucida*) genes into the maternal genome has not been observed in *Poeciliopsis* although some hybridogenetic lines have been maintained in the laboratory for over 33 generations (Schultz, 1977).

An important consequence of hybridogenesis from the perspective of ecological genetics is that an entire genome is clonally—i.e., asexually—inherited. When the essential features of this reproductive mode are abstracted in a mathmatical model, the model is no different from that of a triploid gynogen, namely, a set of genes is transmitted without segregation or assortment, and neither form pays the cost of meiosis. On the other hand, hybridogens should acquire some measure of the adaptedness to the local environment from the host species.

Finally, with regard to hybridogenesis, the European edible frog *Rana esculenta* Linnaeus is hybridogenetic in its essential features, but, unlike hybridogenetic *Poeciliopsis,* this extraordinary species has both males and females and both sexes produce gametes by hybridogenesis (Berger, 1968; Tunner, 1973; Uzzell and Berger, 1975). This long-recognized species is also clearly a hybrid involving the parental species *Rana lessonae,* the pool frog, and *R. ridibunda,* the lake frog (Berger, 1966, 1968, 1973; Blankenhorn *et al.,* 1971; Halfman and Mueller, 1972; Gunther, 1969, 1973; Tunner, 1970, 1972, 1973; Engelmann, 1972, 1973; Uzzell and Berger, 1975). *Rana esculenta* is usually syntopic with either *R. lessonae* (the L-E system) or *R. ridibunda* (the R-E system), although in some places all three are found together (Berger, 1973). Completely allopatric populations are the rule in northern Germany and Poland (Uzzell *et al.,* 1977). Furthermore, two classes of triploid *Rana esculenta* have been identified. Analogous to the situation in *Poeciliopsis,* one class comprises two genomes from *R. lessonae* and one from *R. ridibunda,* whereas the other class is one *R. lessonae* and two *R. ridibunda* (Uzzell and Berger, 1975).

At present, the literature on *Rana esculenta* is more a collection of intriguing facts than a cohesive evolutionary story. Presumably, both male and female *R. esculenta* maintain hybridogenetic relationships with their host species, and there is evidence that the progeny of intraspecific *R. esculenta* crosses are less viable than interspecific crosses within the L-E and R-E systems (Berger, 1973), but it is not clear who is mating with whom in natural populations. It is also unclear whether the triploids are a stable component of natural populations or whether they also are hybridogenetic, gynogenetic, or spontaneously parthenogenetic. It is known that diploids produce some triploid progeny (Uzzell *et al.,* 1975) and that the normal hybridogenetic mechanism "leaks" somewhat (Uzzell *et al.,* 1977). In sum, the picture with *Rana esculenta* is presently too clouded to allow an extensive discussion of the ecological genetics of this species.

2. The Evolutionary Ecology of Parthenogenetic Vertebrates

I have decided to treat the evolutionary ecology of parthenogenetic vertebrates according to the following outline: (1) the adaptive value in being a parthenogenetic vertebrate (i.e., what selection factors control the existence and distribution of the vertebrate parthenospecies), (2) the problem of coexistence, and (3) clonal diversity. This scheme of classification is both somewhat arbitrary and artificial. Ideally, I would like a scheme that represents distinct phenomena associated with parthenogenesis and at the same time reflects the chronology with which each topic was taken up in the literature. Alas, the three phenomenological classes are not mutually exclusive, as several hypotheses are germane to more than one class, and often several ideas were broached, or at least emerged in the literature, more or less simultaneously. On the other hand, in defense of this scheme, the three titles do represent major classes of phenomena taken up in the literature in more or less the sequence indicated. This is not surprising, since there is doubtless a simple historical dynamic involved: for example, it would not have been fruitful to have attacked the third phenomenon without first having given some thought to phenomena one and two. However, again, the chronology is not as sequential as I would like.

3. The Adaptive Value in Being a Parthenogenetic Vertebrate

A presumption in this section is that natural selection plays a role in the determination of where parthenospecies do and do not occur. It could be argued that parthenospecies are rare simply because they have limited opportunities to come into being. This is essentially Williams' (1975) explanation for the infrequent occurrence of asexuality among vertebrates; i.e., vertebrates have a developmental, morphological, and physiological commitment to sexual reproduction deeply embedded in their evolutionary history. Thus, Williams argued, sexuality is the rule among vertebrates not because it is adaptive, but because these species cannot escape the phylogenetic inertia of sexuality. With the parthenogenetic vertebrates that do exist, however, it is apparent that selection does affect their distributions. In the cases of *Poeciliopsis,* there are only a few places where the parthenospecies could conceivably exist that they do not; i.e., few populations of potential host species are free of the sexually parasitic forms. Furthermore, the relative abundances of the parthenogentic and sexual species vary enormously and *consistently* between locales. The most extensive chronological series of collections for a unisexual vertebrate is for the hybridogenetic fish *Poeciliopsis monacha-occidentalis* in Sonora, Mexico. These collections were summarized by Moore *et al.* (1970) and Moore (1976). The collections cover the period 1936–1973 and represent the entire species range. *Poeciliopsis monacha-occidentalis* is always very abundant at the southern end of its range, the Río Mayo, where it comprises from 65% to 97% of mixed unisexual–bisexual populations; whereas at it northern limit in the Río Concepcíon it comprises from 1% to 20% and usually less than 5% of populations. In the Río Mayo at Navajoa, *P. monacha-occidentalis* comprised 73% of the population in March 1940 and 78% in November 1969.

Thibault (1974*a,* 1978) studied short-term variation (January–May) as well as geographic variation in unisexual–bisexual ratios for several complexes of *Poeciliopsis*. He found substantial and statistically significant temporal changes at some locales for the complex involving *Poeciliopsis lucida;* changes involved both the unisexual–bisexual ratio and the relative frequencies of the diploid and triploid unisexual forms. Similar results were reported for the *P. monacha* complex and the *P. latidens* complex, although both of these showed greater seasonal stability, especially the *P. monacha* complex.

Vrijenhoek (1978) reported unisexual–bisexual ratios for the *Poeciliopsis monacha–P. 2 monacha-lucida* complex in headwater tributaries of the Río Fuerte. He found that the unisexual becomes more abundant as populations are sampled from upstream to downstream habitats. I sampled these same populations in April 1973, May 1975, and September 1975. These unpublished data show the identical trend and that the extreme upper reaches of the Platanos are devoid of unisexuals.

Rasch and Balsano (1974) and Balsano *et al.* (1981) reported a similar trend for *Poecilia formosa* in the Río Purificacíon of northeastern Mexico. The bisexual host species is most abundant in headwater habitats, whereas the triploid and diploid unisexuals, respectively, peak in abundance progressing downstream.

In sum, it is easy to accurately predict where the various unisexual fishes can be found in abundance, and this clearly indicates the existence of geographic selection differentials between unisexual and bisexual species.

What factors, then, determine these geographic patterns of abundance? Again, one must be cautious about confounding asexuality with hybridity. That is, the geographic variation in fitness perceived in parthenospecies may be caused by variation in the adaptive value of hybridity and not asexuality—and, in fact, I suspect that to be the case. In either case, there can be little doubt that the population biology of *hybrid parthenogenetic* vertebrates is modulated by selection, and therefore it makes sense to explore the adaptive value of being such an organism.

Papers that report the discovery of new parthenogenetic vertebrate species are invariably concerned with establishing the simple and historically incredible fact that the species in question is parthenogenetic. The harbinger of these reports was that of Hubbs and Hubbs (1932) on *Poecilia formosa*. This was an influential paper not only because it showed in a clearly reasoned way that the Amazon molly was an all-female species that reproduced by gynogenesis, but also because it established the hybrid nature of this extraordinary fish. At about the same time, Clanton (1934) recognized that the all-female salamanders of the *Ambystoma jeffersonianum* complex in southeastern Michigan arose through hybridization of two bisexual species subsequently recognized as *A. jeffersonianum* and *A. laterale* (Uzzell, 1964). Thus, with the spate of parthenogenetic vertebrate species discovered beginning late in the 1950s, the possibility that they might be of hybrid origin was entertained, and borne out in most cases, rather early on. In fact, by the end of the 1960s, a pattern involving hybridization, parthenogenesis, and polyploidy was thought to underlie the evolution of most if not all parthenogenetic vertebrates (Lowe and Wright, 1966; Wright and Lowe, 1968; Schutlz, 1969). The hypothetical relationship among these phenomena was certainly fostered, if not inspired, by the analogy of evolution by allopolyploidy known to be common among angiosperms, although at that time hybridization was thought to be of little consequence in vertebrate evolution. Wright and Lowe (1968) cited Stebbins (1950, 1966) and Johnson *et al.* (1965) in this context, while Schultz (1969) cited Stebbins (1966) and Mayr (1963).

Two papers, both originally presented at a symposium on parthenogenesis organized by the American Society of Zoologists at the AAAS meetings in Chicago, December 1970, one by R. Jack Schultz (Schultz, 1971) and the other by T. Paul Maslin (Maslin, 1971), were the first to systematically formulate and explore a range of alternative hypotheses regarding the adaptive value in being a hybrid parthenogenetic vertebrate. Wright and Lowe (1968) formulated the "weed hypothesis" in 1968. Maslin (1968) also clearly stated the essential postulates of the weed hypothesis, although he did not call it that.

From these foundation papers and others, nine more or less distinct hypothetical adaptations can be discerned for parthenogenetic vertebrates:

1. Production of only female offspring
2. High colonizing and recolonizing ability
3. Preservation of adaptive gene complexes
4. High heterozygosity
5. Heterosis
6. Adaptation to more than one niche
7. Adaptation to "weedy" habitats
8. Ecological intermediacy
9. Relict ecology

Obviously, most of these are simply logical extensions of what would be expected in an organism that is simultaneously asexual and an F_1 hybrid. Thus, it is no surprise that most of the hypotheses seem to have been broached independently more than once. In addition, there was considerable correspondence between individuals or research groups working on the various taxa of parthenogenetic vertebrates during this period. Maslin entitled a section of his 1971 paper, "The Advantages and disadvantages of parthenogenesis." The first two sentences in that section read as follows: "Parthenogens have been known among animals for so long and their peculiar nature discussed so frequently that it is difficult to trace the source of any one idea. None of the available hypotheses is substantiated by experimental work, but most of them seen reasonable." Although some experimental work now has been done and other observational facts have been brought to bear on the various hypotheses, either directly or through mathematical models, this statement would not be wholly inaccurate if written today. It should also be pointed out that the nine hypotheses listed above are not mutually exclusive, and, in fact, all nine could be true (or false). Nevertheless, it is apparent that "camps" have formed where individuals tend to champion some hypotheses while disparaging others.

In the following, I will attempt to explicate the logic of each hypothesis, trace its historical origin, and review merits of each in context of present data. In so doing, I will discuss *heterosis* and *high heterozygosity* together under a single heading, and *adaptation to more than one niche, adaptation to "weedy" habitats, ecological intermediacy,* and *relict ecology* under a single heading, *the competitive interaction hypothesis.*

3.1. Production of Only Female Offspring

Perhaps the most obvious advantage in being parthenogenetic is the production of two females for every one produced by a comparable sexual genetype—if there is no concomitant loss in fecundity. In terms of a more contemporary vocabulary, not producing males is simply the profit when the "cost of meiosis" is not paid (Maynard Smith, 1978, pp. 2–3). Oddly enough, this simple advantage in unisexuality has not been discussed as much as the other hypotheses. Perhaps this is because it is a self-evident adaptation upon which there would be unanimous agreement. However, White (1970), for example, states that, "Probably more significant than the increase in the innate rate of reproduction is the fact that every individual of a parthenogenetic form carried accidentally to a new locality, or left behind in an old locality after local extermination, can regenerate a new colony." Aside from the fact that White's argument would not apply to sperm-dependent parthenogenetic fishes, I believe it is symptomatic of a broader lack of appreciation for the *very high* cost of meiosis.

In discussing the cost of meiosis in his book, *The Evolution of Sex,* Maynard Smith said, "Although absurdly simple, the point [twofold disadvantage of producing males] is so fundamental and so often misunderstood." Maynard Smith continued by pointing out that if parthenogenetic females arose from sexual females in a bisexual species, then the proportion of parthenogens would increase from $n/(2N + n)$ to $n/(N + n)$ in each generation, where N is the number of sexual females and n is the number of parthenogenetic females, and sexuals and parthenogens are identical in all respects (i.e., fecundity, survivorship, etc.). I reiterated Maynard Smith's equation for a population of 10,000 adult sexuals (5000 males, 5000 females) where at generation one a single parthenogenetic female arises by mutation. This result is illustrated in Fig. 1. By generation 28, the parthenogenetic lineage has multiplied to the extent that the sexual species has been driven to extinction.

There is also the vague assertion in the literature that the high intrinsic rate of increase inherent in parthenospecies has adaptive value only when populations are in an exponental growth phase (r) and is of no value when the population is at or near its carrying capacity (K). This is not true. By inspection of the ratio of the Lotka–Volterra competition equations (e.g., Roughgarden, 1979, pp. 412–413) it can be seen that a parthenogenetic species would grow twice as rapidly as a comparable sexual species regardless of how close the overall population was to K. Even at carrying capacity, the parthenogens would "reoccupy space" vacated in the carrying capacity at twice the rate of the sexual species and, again, the asexuals would rapidly exclude the sexuals by competition. The cost of meiosis is paid at all phases of population growth—not just the exponential growth phase.

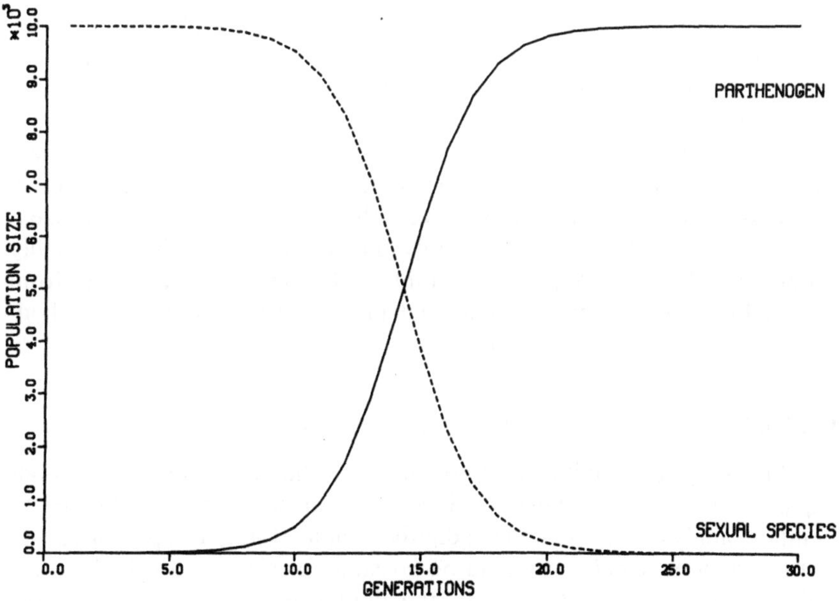

Figure 1. The competitive exclusion of a sexual species by an asexual mutant parthenogen. A single mutant to asexuality arises in generation one.

Beyond theoretical considerations, there is some evidence that the production of twice as many females is, indeed, the *most* important contributor to the adaptive value of unisexual fishes, at least in terms of distributional success. Through a selection model, Moore (1976) analyzed the zoogeography of the hybridogenetic complex involving the unisexual *Poeciliopsis monacha-occidentalis* and its host species *P. occidentalis* in Sonora, Mexico. I chose this complex as a study case because it is the only *Poeciliopsis* complex involving just one unisexual and one bisexual species. All of the other complexes either involve both diploid and triploid parthenospecies or situations where the unisexual form mates with males of more than one species, such as *P. monacha-latidens*. These phenomena could greatly confound the results. I will discuss the logic of this analysis in more detail in Section 4. Suffice it for now that it can be inferred that the unisexual occupies greater than three-fifths of its range solely as a result of not paying the cost of meiosis; in the remaining two-fifths, other components of fitness, for example, ecological intermediacy, heterosis, etc., contribute significantly.

The considerable advantage of all-femaleness would be mitigated if a reduction in fecundity were a concomitant of parthenogenesis. For example, if all-femaleness resulted from male zygotic mortality. The known meiotic mechanisms of parthenogenetic vertebrates, however, should not increase zygotic mortality. To further explore this point, I have tabulated (Table I) the fecundities for those parthenogenetic vertebrate species for which data are available and for putative sexual parental species. In no instance is there a significant reduction in fecundity in the parthenospecies.

At this juncture, I believe that simply producing twice as many female offspring as comparable sexual species is *the major* adaptation among parthenogenetic vertebrates. However, there are disadvantages in asexuality, some of which are only dimly understood, and others probably not perceived at all, that prevent extant parthenogenetic vertebrates from gaining broader distributions and that prevent other would-be parthenospecies from succeeding at all.

3.2. High Colonizing and Recolonizing Ability

An exceptional ability to colonize a new habitat is an obvious consequence of parthenogenesis. This ability can be further subdivided to facilitate analysis of its actual adaptive value. First, in spontaneously parthenogenetic species it would be possible for single individuals to establish populations, and, second, the double intrinsic rate of increase would allow parthenospecies to fill newly opened habitats much faster than comparable sexual species. As noted in the preceding section, White (1970) emphasized the potential value of single individuals establishing whole populations, and Maslin (1971, p. 376) echoed this notion: "The chief advantage of parthenogenesis seems to be the ease of colonization of a new area. Theoretically a single individual could discover a suitable unoccupied habitat and commence to fill it with her offspring." Similarly, one might expect parthenospecies to be at advantage in habitats where

Table I
Fecundities of Naturally Occurring Parthenogenetic Vertebrates and Closely Related Sexual Species

Species[a]	Average brood size[b]	Reference
(U) *Poeciliopsis monacha-lucida + P. monacha-2 lucida*	7.32 ± 4.19 (SD)	Thibault (1974*a*, Table XXXV), averaged over several samples
(U) *P. 2 monacha-lucida*	5.36 ± 1.39 (SD)	Thibault (1974*a*, Table XXXV), averaged over several samples
(B) *P. monacha*	8.16 ± 4.71 (SD)	Thibault (1974*a*, Table XXXV), averaged over several samples
	11.8 ± 0.96 (SE)	Thibault and Schultz (1978, Table 1)
(B) *P. lucida*	9.39 ± 3.67 (SD)	Thibault (1974*a*, Table XXXV), averaged over several samples
	11.0 ± 0.67 (SE)	Thibault and Schultz (1978, Table 1)
(U) *P. monacha-latidens*	11.62 ± 5.15 (SD)	Thibault (1974*a*, Table XXXV), averaged over several samples
(B) *P. latidens*	9.8 ± 4.92 (SD)	Thibault (1974*a*, Table XXXV), averaged over several samples
(U) *Cnemidophorus tesselatus*	3.23 ± 1.1 (SD)	Schall (1978, Table 2)
(U) *C. exsanguis*	2.96 ± 1.1 (SD)	Schall (1978, Table 2)
(B) *C. tigris*	2.02 ± 0.7 (SD)	Schall (1978, Table 2)
(B) *C. gularis*	3.13 ± 1.4 (SD)	Schall (1978, Table 2)
(B) *C. inornatus*	2.37 ± 0.9 (SD)	Schall (1978, Table 2)
(U) *Lacerta armeniaca (valentini × mixta)*	2–5 (3–5)[c]	Uzzell and Darevsky (1975, Table 5)
(B) *L. valentini*	3–8 (5)[c]	Uzzell and Darevsky (1975, Table 5)
(B) *L. mixta*	2–4	Uzzell and Darevsky (1975, Table 5)
(U) *L. dahli (portschinskii × mixta)*	2–5 (4)[c]	Uzzell and Darevsky (1975, Table 5)
(B) *L. portschinskii*	2–5 (4)[c]	Uzzell and Darevsky (1975, Table 5)
(U) *L. rostombekovi (portschinskii × r. raddei)*	2–4	Uzzell and Darevsky (1975, Table 5)
(B) *L. r. raddei*	2–5 (4)[c]	Uzzell and Darevsky (1975, Table 5)
(U) *L. unisexualis (valentini × r. nairensis)*	2–7 (5)[c]	Uzzell and Darevsky (1975, Table 5)
(B) *L. r. nairensis*	2–5 (4)[c]	Uzzell and Darevsky (1975, Table 5)

[a]U, Unisexual; B, bisexual.
[b]SD, Standard deviation; SE, standard error.
[c]More frequent clutch sizes.

occasional local extinctions occur. Thus, areas occasionally denuded by fire or flood might be more readily recolonized by parthenospecies for exactly the same reasons: single colonizers with a high intrinsic rate of increase could found whole populations. This is the major thesis in Cuellar's (1977*b*) hypothesis for the ecology of spontaneously parthenogenetic animals, and Maslin (1971) listed the same possibility among the alternatives he considered.

Recolonization potential has also been connected to the so-called "geographic parthenogenesis hypothesis." This is not so much a hypothesis as it is the repeated observation that parthenospecies often occur in areas of more extreme environmental conditions (cooler, dryer, etc.) than those of the bisexual parental species. In some instances the occurrence of parthenogenesis seems to be associated with Quaternary glaciation (Seiler, 1943, 1946, cited by Darevsky, 1966; Suomalainen, 1953, cited by Darevesky, 1966; Darevsky, 1966). The suggested cause–effect relationship is that parthenospecies can more rapidly colonize the denuded areas vacated by retreating glaciers. Darevsky (1966, p. 147) stated, for example:

> After glacial regression, in the Holocene (=Recent) the ranges of the parthenogenetic forms [*Lacerta*] expanded to their present limits. . . . The rapid radiation of these lizards was promoted by the parthenogenetic mode of reproduction, because even one partheogenetic individual can potentially give rise to a new population. The hypothesis explaining the transition from bisexual to unisexual reproduction in Rock Lizards by means of climatic changes during glacial times finds a parallel in the American parthenogenetic teiid lizards.

[Wright and Lowe (1968) pointed out Darevsky's error in drawing this parallel with parthenogenetic *Cnemidophorus*.]

Although colonizing ability has been significant in the thinking about the adaptive value of parthenogeneis, I doubt that colonizing ability *per se* is a major adaptive component in any parthenogenetic vertebrate species. The value of the high intrinsic rate of increase was discussed in the preceding section, and I simply reiterate that this is an advantage at *all* stages of population growth; i.e., at carrying capacity as well as during phases of exponential growth. The obvious failure of the single-colonizer hypothesis, however, is that it simply is *not* applicable to parthenogenetic fishes and amphibians, all of which are sperm-dependent. Furthermore, there are great similarities between the ecologies of parthenogenetic fishes and lizards, and, since the potential for single individuals to found colonies is not a common property of the two groups, it cannot be the cause of the similarities.

With regard to geographic parthenogenesis, I again caution against confounding the effects of parthenogenesis with those of hybridity. Although, to my knowledge, these cases have not been extensively collected, reviewed, and summarized, several hybrid zones not involving parthenogenesis seem to be associated with ancient glacial fronts or glacial peninsulas that once divided floristic and faunistic assemblages: for example, the red- and yellow-shafted flickers (Short, 1965); the myrtle and Audubon's warblers (Hubbard, 1969; Barrowclough, 1980) and the eastern red cedar and creeping juniper (Fassett, 1945; Palma *et al.*, 1983). Remington (1968) gives apparently the most complete compilation of cases [see also Moore (1977*b*)]. In the syndrome of geographic parthenogenesis, hybridity rather than parthenogenesis may be the cause.

Orlando Cuellar (1977*b*) published an article entitled, "Animal Parthenogenesis: A new evolutionary-ecological model is needed." This article has been subjected to more criticism than any paper concerned with parthenogenetic vertebrates that I know (Vanzolini, 1978; Wright, 1978; Cole, 1978). Although I greatly admire most of Cuellar's contributions to the literature of parthenogenetic *Cnemidophorus* and to the theoretical discussion of parthenogenesis, I also find myself in substantial disagreement with the basic arguments put forth in this paper. I raise these criticisms here because Cuellar's major conclusions are that "the two unique features allowing parthenogenetic species to invade and occupy open habitats faster than bisexuals are (i) a double intrinsic rate of increase and (ii) the ability of one individual to establish a new colony" (Cuellar, 1977*b*, p. 843).

Cuellar argues that the existence of parthenogenetic species of *Cnemidophorus* depends upon the existence of disclimax communities where catastrophes such as flooding, glaciation, or fire eliminate populations, thus allowing parthenospecies to capitalize on their founder potential and high intrinsic rate of increase in recolonizing the denuded area. *A priori*, Cuellar limits the applicability of his hypothesis to exclude gynogenetic (and, implicitly, hybridogenetic) modes of reproduction. Furthermore, he asserts that: "Parthenogenesis can only evolve in areas devoid of the generating bisexual species because such species would prevent newly formed unisexuals from establishing clones due either to hybridization or competition" (Cuellar, 1977*b*, p. 843). Cuellar is among those who think that not paying the cost of meiosis is advantageous only if the population is in an exponential growth phase. According to Cuellar (1977*b*, p. 838):

> [M]ost species have long passed through the exponential growth phase early in their evolution. . . . Hence, most species remain numerically stable from year to year. However, precisely the opposite conditions may be necessary for the origin of parthenogenesis, for, only if a certain degree of instability exists in nature, can new habitats develop in which the high intrinsic rate of increase latent in parthenogenesis be expressed in isolation from bisexuality.

As pointed out in Section 3.1, not paying the cost of meiosis is advantageous during either an r or K growth phase and this fact obviates any theoretical need for parthenospecies to be associated with environments where exponential growth would be the rule. Furthermore, excluding gynogenetic and hybridogenetic forms from the realm of explanation of a hypothesis that purports to explain the evolutionary ecology of parthenospecies is parochial and premature. Logically, it may well be that sperm-dependent parthenospecies are successful for different reasons, but at this juncture there is no compelling reason to believe that they are, and indeed all major groups of parthenogenetic vertebrates share common ecological properties.

With regard to the contention that newly evolved parthenospecies must be protected from hybridization and competition with their generating parental species, the following counterarguments can be raised. It is true that parthenogenetic species of whiptail lizards (*Cnemidophorus*) and rock lizards (*Lacerta*) occasionally hybridize in nature with males of closely related sexual species (Cole, 1979). Such unusual specimens can be triploid or tetraploid, depending upon the ploidy of the parthenospecies; in either case, the hybrid has an additional genome. In fact, it is thought that triploid parthenospecies in the genus *Cnemidophorus* arose by this mechanism (Lowe and Wright, 1966; Parker and Selander, 1976; Cole, 1979). In the case of the Caucasian rock lizards, hybridization may indeed limit the distributions of some parthenospecies. For example, Uzzell and Darevsky (1975, p. 216) state "Mating of males of *L. valentini* with *L. armeniaca* [unisexual] may also be an important factor in limiting the range of *L. armeniaca* since such matings result in the production of sterile triploids. . . . In mixed populations at the ecological margin of *L. valentini's* range about 70% of the lizards are *L. valentini,* about 20% of *L. armeniaca,* and about 10% sterile triploids." This is of course speculative, but it seems likely that a majority of the unisexual females in this population avoid hybridization, because it would seem doubtful that the 20% that is *L. armeniaca* could be immigrants generated in areas free of the dangers of hybridization.

Getting more directly to the point, what is the likelihood that a parthenogenetic female would be mated by a male if the species were sympatric? No data conclusively answer this question; however, an experiment conducted by Charles Cole suggests hybridization would be unlikely. Cole (1979) caged a parthenogenetic *Cnemidophorus sonarae* with, at times, two *C. tigris* males. Only one of the males was seen to copulate with the female, and this was observed on several occasions. The successful male and the female were caged together a total of 15 months. The female produced nine clutches totalling at least 33 eggs. Most of these did not survive for various reasons, mostly as a result of improper environmental conditions. Of seven progeny that reached an age where hybridity could be determined, only one was a hybrid; this one was the sole survivor of a single clutch (Cole, personal communication). Cole (personal communication) also found that when two males are caged with a parthenogenetic female, the first male to mature preferentially mates with the other male rather than with the female. Cole (personal communication) thus believes that natural hybridization between unisexual and bisexual whiptails is rare even where there is sympatry. Thus, Cuellar's premise that hybridization would preclude the establishment of clones in sympatry with bisexual species is not supported by the data.

There are no data to determine whether or not competition with bisexual species would impede clone expansion other than the fact that all gynogenetic and hybridogenetic species hold their own. However, in theory, competition presents no problem since the newly arisen asexual should rapidly outcompete a comparable sexual species solely as a result of its high intrinsic rate of increase. Consider the simplest case first, where a single parthenogen arises in a sexual population in a spatially and temporally homogeneous environment. As pointed out in Section 3.1 and illustrated in Fig. 1, the new clone would rapidly exclude, by competition, the sexual species from which it arose. The sexual species would fare better in heterogeneous environments, but even there, in both spatially (Maynard Smith, 1976) and temporally heterogeneous (Hines and Moore, 1981; Moore and Hines, 1981) environments a newly arisen parthenogenetic genotype would win the competition unless the alternative environments were rather extreme. All else being equal, the twofold advantage of not producing males gives the parthenogen an advantage that is difficult to beat. In terms of competition, the wonder is not whether the parthenogen could establish a clone, but, rather, that parthenospecies are not geographically rampant.

3.3. Preservation of Adaptive Gene Complexes

Zweifel (1965) was apparently the first worker on parthenogenetic vertebrates to point out that it would be adaptive for an organism already well adapted to its environment to congeal that genotype by becoming parthenogenetic because the comparable sexual genotype would continue to segregate less fit genotypes in its progeny. Maslin (1971) and Schultz (1971) both considered this hypothetical adaptation in their papers, but neither was able to determine the extent to which this contributes to the fitness of parthenogenetic taxa. This is no fault of Maslin or Schultz, or any one else working in the field for that matter, because it is difficult to conceive of observational or experimental data that might test the hypothesis. Nonetheless, it is a fact that with some few exceptions parthenogenetic vertebrates produce offspring that are "Xerox copies" of themselves, and this does present theoretical ramifications that can be explored.

Firstly, although this would be difficult to detect after the fact, there is no evidence that a parthenogenetic vertebrate species has ever reverted to sexuality. Therefore a parthenogenetic lineage is stuck with whatever constellation of genes it had at its origin, plus what few adaptive mutants it might sequentially accumulate, and is limited by the adaptive potential of that constellation. A lineage that becomes parthenogenetic might initially increase in number because it is a highly adapted genotype and because it no longer pays the cost of meiosis. Inevitably, however, the environment would change and the then "senescent" parthenospecies would become restricted to relict populations and eventually extinguished. This was Uzzell and Darevsky's (1975) major line of reasoning in their development of the relict ecology hypothesis, which they considered the most plausible explanation for the zoogeography of parthenogenetic rock lizards (*Lacerta*). They reasoned that when parthenogenesis arose by hybridization some 5000–7000 years ago, the parental species were adapted to more xeric conditions than at present. As climatic conditions ameliorated, the parental bisexual species adapted by gene frequency changes but the parthenospecies were trapped by their asexuality and their adaptation to the environments that existed at the time of their origins. This hypothesis accounts for the observations that parthenogenetic *Lacerta* species arose through hybridization but often have disjunct distributions or are no longer in contact with one or both parental species, tend to occur in relatively xeric habitats, and are more tolerant than parentals of aridity during development.

Now, considering short-term advantages of adaptive genotypes congealed by parthenogenesis, it is obvious that in a temporally homogenous environment it would be advantageous for the *most* fit genotype to become parthenogenetic, thus reaping the profit of not paying the cost of meiosis while simultaneously doing away with segregation load. When there is some temporal heterogeneity, however, it might be possible for a sexual species to track the changing environment whereas an asexual species could not. Oddly enough, however, when temporal uncertainty becomes great, "lag load" can become so severe for the sexual population that asexuality is adaptive even if the cost of meiosis is discounted (Moore and Hines, 1981).

There may be additional disadvantages in having a genotype "frozen" by asexuality. What is actually being argued here is the advantage of asexual versus sexual reproduction, and, although there was once considerable enthusiasm for an ecological explanation of sexuality (e.g., environmental uncertainty), now there is skepticism that such an explanation can be reasonable (Maynard Smith, 1971, 1976, 1978; Hamilton *et al.*, 1981; Hines and Moore, 1981; Moore and Hines, 1981). What is left, then, is some sort of physiological explanation, and, if this line of reasoning is correct, then *all* parthenospecies must suffer or begin to accrue the physiological detriments of asexuality, whatever these detriments may be. The most plausible physiological explanation for the malevolence of asexuality is the accumulation of gene load. Muller (1964) pointed out that gene load that arises from dysgenic mutations can *only* stay the same or increase in asexual populations. Consider, first, a primitive haploid asexual species. Since most mutations are dysgenic, a significant number of progeny would carry new detrimental mutations in each generation, and these would be lost to selection eventually, or if per chance such a mutant did become fixed, the resultant population would actually be less fit than the ancestral population. In any case, a lineage that became diploid would be favored by selection because the duplicated genome would mask dysgenic mutations as they arose. This would be a temporary solution, however, because a dysgenic allele would exist eventually at most gene loci and additional mutations would probably result in homozygosity for dysgenic mutants. At this point, two adaptive possibilities are apparent: (1) increase the level of ploidy or (2) become sexual. The first solution, again, only temporarily masks the expression of dysgenic mutations, whereas individuals that become sexual actually have a chance of producing progeny more fit than themselves. Asexuality, then, results in a ratchet-like mechanism wherein gene load can only stay the same or increase; thus, sexuality may be an adaptation to "flush" gene load.

In an elegant experiment, Leslie and Vrijenhoek (1978) showed that a hybridogenetic form of *Poeciliopsis* has in fact accumulated a significant gene load. *Poeciliopsis monacha-lucida* normally mates with males of *P. lucida*. However, Leslie and Vrijenhoek crossed *P. monacha-lucida* with *P. monacha,* and thus, "extracted" the clonally inherited *monacha* genome from the all-female species. The progeny of this cross, all of which were male, were genetically equivalent to pure *P. monacha* males, were fully fertile, and produced gametes by normal Mendelian processes. These males were then backcrossed to the unisexual *P. monacha-lucida* from which they were derived. The progeny of the backcrosses would be expected to be homozygous at 50% of their loci for alleles originally extracted from the unisexual genome. If the unisexual genome, "frozen" by asexuality, had accumulated gene load, then the mortality rate of progeny from the backcross should be higher than that of control crosses. This was precisely the result obtained, and Leslie and Vrijenhoek estimated that one of the unisexual strains carried at least two lethal equivalents (the equivalent in increased embryonic and postparturition mortality of two lethal alleles assorting at independent loci), whereas the second strain carried a minimum of four to six lethal equivalents.

Furthermore, silent alleles have been reported for the *Ldh-1* locus in *Poeciliopsis monacha-lucida* (Leslie and Vrijenhoek, 1978), and for the *Es-5* locus in *P. monacha-occidentalis* (Vrijenhoek *et al.*, 1977). The evolution of histocompatibility clones in *P.* 2 *monacha-lucida* also involves loss of H-gene (histocompatibility) expression (Eisenbrey and Moore, 1981). Thus it is clear that parthenogenetic fishes accumulate deleterious genes.

To summarize, it would be difficult to show by observational or experimental data that preservation of an adaptive genotype congealed by parthenogenesis does in fact contribute to the fitness of parthenospecies. In the long run, parthenospecies may not be able to adapt in a changing "struggle for existence," and their populations would become relict and eventually extinct. Furthermore, the accumulation of mutational load might hasten the inexorable extinction of parthenospecies. In the short run, a congealed genotype would prevent parthenospecies from tracking short-term environmental variation, but this may be advantageous when environmental variation is extreme, because a generalized genotype that does moderately well in all environments may be the best strategy.

3.4. Heterosis

Since most, if not all, parthenogenetic vertebrate species are effectively F_1 hybrids, it seems very plausible that high levels of heterozygosity *per se* or heterosis are important contributors to overall fitness. Schultz (1971) suggested this and White (1970, p. 239) stated, "the majority of cases of thelytoky . . . seem to involve considerable levels of heterozygosity and must be interpreted as genetic systems that exploit the advantages of heterosis and adaptive polymorphism. This conclusion does not seem to have been appreciated until now." It is useful at the onset of this discussion to distinguish the term heterosis as commonly used in population genetics from the term as used in animal breeding and plant cultivation. To a population geneticist, heterosis means that the heterozygote is more fit than either homozygote (in the case of a single gene locus with two alleles); to an agriculturist, heterosis means somatic vigor or luxuriance in growth. In describing the clearly evident heterosis in laboratory reared *Poeciliopsis* Schultz (1971, p. 358) said:

> [A]ll of the unisexual forms are better equipped to survive in this new environment [an aquarium in Connecticut] than any of the 14 bisexual species reared under the same conditions. Such adaptation presumably derives from

> heterosis. . . . Furthermore, the superior qualities exhibited here by *Poeciliopsis* unisexuals (Schultz, 1961; unpublished data) would seem to be of a nature as to provide greater fitness in any environment; forexample, (1) the young are more robust at birth, (2) survival from birth to reproductive age is better, (3) length of reproductive life is longer, (4) they grow larger and consequently most of them produce more young, and (5) they are more disease resistant.

White's usage seems to have the connotation of population genetics, whereas Schultz' implies somatic vigor. Presumably, there is a connection between the two, but at present this is obscure. About all that would be widely agreed upon is that heterozygosity is probably the cause of somatic vigor, but somatic vigor does not necessarily result from heterozygosity. Furthermore, somatic heterosis is observed in habitats created by humans and may be of little or no importance in nature. With regard to parthenogenetic vertebrates, it is an unequivocal fact that they possess extraordinarily high levels of heterozygosity, but there is little evidence that either heterozygosity or somatic heterosis are *general* explanations for the success of parthenospecies in nature.

Turning to the data from which I believe these conclusions can be reasonably drawn, I have summarized the considerable data on heterozygosity in Table II. Average heterzygosities based on electrophoretic data are available for seven species of parthenogenetic fishes and three lizard parthenospecies. For comparison, the percent heterozygosities are tabulated for bisexual species involved in the hybrid origin of the parthenospecies. The heterozygosity averaged across all parthenospecies is 46.7%, whereas the same statistic for progenitor species is 4.2%. The parthenospecies are the most heterozygous species known and are 10 times as heterozygous as comparable bisexual species, an astonishing fact.

But is either heterozygosity or heterosis an important component of fitness in natural populations of parthenogenetic vertebrates? To begin answering this question, one must focus on a subset of the prediction set associated with the heterosis hypothesis that might be tested by the data. The five signs of vitality quoted from Schultz would be expected of heterotic fishes. Similarly, Hubbs (1955), in describing heterosis in F_1 progeny of the green sunfish and bluegill, noted that, relative to the parentals, the progeny grow faster, dominate in social hierarchies, hold their fins more erect, and excel in brightness and intensity of color—but these hybrids all proved to be sterile. In general, then, a prediction of the heterosis hypothesis is that naturally occurring vertebrates should exhibit some subset of these vitality properties.

A second class of testable predictions follows from the fact that all

Table II
Comparison of Heterozygosity between Parthenogenetic Vertebrate and Sexual Ancestral Species

Parthenospecies	Percent loci heterozygous ± SD	Reference	Sexual ancestors	Percent loci heterozygous ± SD	Reference
Poeciliopsis monacha-lucida (2*n*)	42.6 ± 3.9	Vrijenhoek (1979, Table 2)	*P. monacha*	4.7 ± 2.5	Vrijenhoek (1979, Table 2)
P. 2 monacha-lucida (3*n*)	50.7 ± 1.9	Vrijenhoek (1979, Table 2)	*P. lucida*	2.1 ± 1.6	Vrijenhoek (1979, Table 2)
P. monacha-2 lucida (3*n*)	54.4 ± 2.0	Vrijenhoek (1979, Table 2)			
P. monacha-occidentalis (2*n*)	42.5 ± 1.1	Vrijenhoek (1979, Table 2)	*P. monacha*	4.7 ± 2.5	Vrijenhoek (1979, Table 2)
			P. occidentalis	1.8 ± 2.7	
P. monacha-latidens (2*n*)	38.5 ± 0	Vrijenhoek (1979, Table 2)	*P. monacha*	4.7 ± 2.5	Vrijenhoek (1979, Table 2)
			P. latidens	0.6 ± 0.5	
P. monacha-lucida-viriosa (3*n*)	52.0 ± 0	Vrijenhoek (1979, Table 2)	*P. monacha*	4.7 ± 2.5	Vrijenhoek (1979, Table 2)
			P. lucida	2.1 ± 1.6	
			P. viriosa	—	
Poecilia formosa (2*n*)	31–33	B. J. Turner (personal communication)	*P. mexicana*	1.8 – 3.6	B. J. Turner (personal communication)
			P. latipinna	—	
Cnemidophorus tesselatus (2*n*)	56.0	Parker and Selander (1976, Table 6)	*C. tigris*	5.0	Parker and Selander (1976, Table 6)
			C. septemvittatus	5.8	
C. tesselatus (3*n*)	71.4	Parker and Selander (1976, Table 6)	*C. tigris*	5.0	Parker and Selander (1976, Table 6)
			C. septemvittatus	5.8	
			C. sexlineatus	7.0	
C. laredoensis	26.7	McKinney *et al.* (1973) [cited by Parker and Selander (1976)]	*C. gularis*	—	McKinney *et al.* (1973) [cited by Parker and Selander (1976)]
			C. sexlineatus	7.0	
Average percent of loci heterozgous	46.7 ± 3.6			4.2 ± 1.4	

members of a parthenospecies are genetically identical in the sense that they are all F_1 hybrids. If heterosis is an *important* component of fitness, then the fitness (and success) of the parthenospecies should be more or less uniform over its entire range. Thus, for example, if a given unisexual species consistently comprised 80% of mixed populations in one locale and 5% at another, then heterosis must be of little consequence in determining the success of this species relative to its bisexual host species. Examination of the zoogeography of appropriate complexes should determine the validity of this prediction.

Finally, it might be possible to measure some parameters of heterosis, including the vitality parameters just cited, in the laboratory. However, conspicuous and consistent diminution of size and fecundity of virtually all laboratory-reared bisexual species of *Poeciliopsis* relative to wild-caught individuals indicates that bisexuals do not live up to their genetic potential in laboratory environments (Thibault, 1974*a;* personal observation). Inferences drawn from laboratory vitality measures, therefore, must be considered tenuous. For this reason, I believe that most of the symptoms of heterosis observed in the laboratory are artificial; however, I can find no reason to discount the laboratory thermal tolerance studies of *Poeciliopsis* recently reported by Bulger (1978) and Bulger and Schultz (1979).

How do the data measure up to the predictions? With regard to vitalistic properties, only a few data are available from natural populations. Data on survivorship and longevity are nonexistent. Thibault (1974*a*) compared the modes of various size classes determined over a 6-month period for natural populations of *Poeciliopsis* to measure the growth rates of unisexual and bisexual species. In general, there was no trend indicative of higher unisexual growth rate. For example, at Agua Caliente in the Río Fuerte drainage, where all five members of the *P. monacha-lucida* complex are sympatric, Thibault estimated the following monthly growth rates (B = bisexual, U = unisexual): *P. monacha* (B), 3.0 mm; *P. 2 monacha-lucida* (U), 2.0 mm; *P. monacha-lucida* (U) lumped with *P, monacha 2-lucida* (U), 3.5 mm; *P. lucida* (B), 5.0 mm. These estimates suggest that unisexual growth rates do no exceed those of bisexuals as the heterosis hypothesis predicts. The data on disease resistence are even sparser. There would probably be unanimous agreement among *Poeciliopsis* workers that the unisexuals are more disease resistant in the laboratory than bisexuals, but the only set of data for natural populations is a tabulation of the numbers of females with internal mesentary nematodes in *Poeciliopsis* (Table XXIX, Thibault, 1974*a*). For *P. lucida* and its associated unisexuals (combined) the frequencies of infestation were 0.06 and 0.03 respectively (P = 0.02, χ^2 test); for *P. latidens* and its unisexual the frequencies of infestation were identical (0.03); and for *P. monacha* and its unisexual the frequencies of investation were 0.01 and 0.03, respectively ($P = 0.3$, χ^2 test). These data indicate that it is likely that the unisexuals associated with *P. lucida* are more resistant to mesentary nematodes, but otherwise do not corroborate heterosis in natural populations.

Although increased fecundity is not necessarily expected as a concomitant of heterosis, given that the parthenogenetic vertebrates are fertile, it is tempting to compare the fecundities of closely related unisexual and bisexual species for evidence of increased vigor manifest in brood size. The temptation is fostered by the fact that relatively many data are available, as summarized in Table I. In *Poeciliopsis,* broods are similar in all species, with the bisexuals *P. monacha* and *P. lucida* having slightly larger broods in nature than their unisexual derivatives. If there is a hint of heterosis in fecundity data, it is in *P. monacha-latidens,* where clutches average approximately 1.4 times those of *P. monacha* and 1.2 times those *P. latidens* (Thibault, 1974*a*). In the whiptail lizards (*Cnemidophorus*) the pattern is inconsistent and Schall (1978) noted that parthenospecies may resemble their parental species with regard to clutch size. Similarly, the pattern in Caucasian rock lizards (*Lacerta*) is indeterminate and Uzzell and Darevsky (1975, p. 219) noted that "in the two instances [parthenospecies] in which the parental species differ noticeably in clutch size (*L. armenica* and *L. unisexualis*), the parthenogenetic species clearly has intermediate sized clutches." In sum, the data pertinent to vitalistic properties are too few to be conclusive, but the trend does *not* corroborate heterosis.

Examination of the zoogeography of the sperm-dependent unisexual–bisexual species complexes could indirectly test the heterosis hypothesis. If heterosis is an *important* contributor to unisexual fitness, then unisexuals should be uniformly successful relative to bisexuals over the entire range of a complex, because the unisexuals are everywhere F_1 hybrids. [Strictly speaking, the prediction is that primary fitness would be geographically uniform if heterosis were an important component of fitness, where primary fitness is fitness after all-femaleness is discounted (Moore, 1976).] Contrary to this prediction, the frequency distributions of *Poeciliopsis monacha-occidentalis* and its bisexual host (illustrated in Fig. 7) show a stable pattern of geographic variation over the range of the complex in Sonora, Mexico. Similar trends of predictable geographic variation are evident in three other unisexual–bisexual complexes of *Poeciliopsis* and in the gynogenetic fish *Poecilia formosa*. In the *P. monacha* complex, the bisexual, *P. monacha,* comprises 100% of the extreme headwater populations in the Arroyo de los Platanos, but its unisexual associate, *P. 2 monacha-lucida,* increases steadily moving downstream until the entire complex is superseded by the *P. lucida* complex. At the downstream limit, the unisexual comprises as much as 75% of the females for this complex (Thibault, 1974*a*, Table VIII; Vrijenhoek, 1978, Fig. 2). The pattern with *P. lucida* and its unisexual associates is more complex, but again there is substantial and predictable variation. Thibault (1974*a*) noted that generally the unisexuals of *P. lucida* tend to decrease in relative abundance as stream size increases. Similarly, *P. monacha-latidens,* the hybridogen associated with *P. latidens,* becomes relatively abundant in the upstream habitats similar to those favored by *P. monacha.* Balsano *et al.* (1981) reported that *Poecilia mexicana,* the bisexual host of *P. formosa* (the Amazon molly), is most abundant in headwater habitats of the Río Purificacíon in northeastern Mexico, but then progressing downstream, the triploid unisexual becomes most abundant, followed by the diploid unisexual. Thus, the zoogeography of all unisexual–bisexual species complexes that have been studied in detail exhibit stable patterns of substantial geographic variation in unisexual–bisexual ratio. Collectively, these data suggest that heterosis is not *generally* an important determinant of fitness in naturally occurring parthenogenetic vertebrates.

Recently, Bulger (1978) and Bulger and Schultz (1979) reported an extensive analysis of thermal tolerances for two unisexual–bisexual complexes of *Poeciliopsis.* Although these experiments were conducted in the laboratory, it is not apparent that a laboratory environment would affect these kinds of experiments, nor is there reason to believe these data would not be indicative of thermal tolerances in natural environments. Three measures of thermal tolerance were made in these studies: acute cold stress, acute heat stress, and critical thermal minima. Comparisons were made between the bisexual *Poeciliopsis occidentalis* and its hybridogenetic associate *P. monacha-occidentalis,* which together form the northernmost complex of *Poeciliopsis,* and between *P. lucida* and its two unisexual associates *P. monacha-lucida* and *P. monacha-2 lucida,* which collectively form a more southerly complex. (Interclonal comparisons were also made within each of these three unisexual species.) The acute cold and acute heat stresses were measured as percent survivorships for fish immersed in an ice-water bath for 8 min at 1°C or a warm water bath at 40°C for 20 or 30 min (depending upon the species), respectively. Critical thermal minima were measured as the temperatures at which fish lost equilibria when the water temperature was decreased 5°C per day. Controls were incorporated into the experimental design such that inbreeding and acclimation could be eliminated as causes of different thermal tolerances. Furthermore, consistent differences in test results between inbred strains, between unisexual clones, and between species clearly established that thermal responses are heritable.

The major results from Bulger (1978) and Bulger and Schultz (1979) are summarized in Table III. The unisexual *P. monacha-occidentalis* is not heterotic, generally, compared to parental species, although tolerance

Table III

Thermal Tolerances in Unisexual and Bisexual *Poeciliopsis*

		Young female			Adult female			
		Percent survivorship acute cold	Percent survivorship acute heat; 40°C		Percent survivorship acute cold	Percent survivorship acute heat; 40°C		Mode of critical thermal
Species	Type of reproduction	(1°C; 8 min)	20 min	30 min	(1°C; 8 min)	20 min	30 min	minimum, °C
P. occidentalis[a]	Bisexual	48[c]	56[c]	—	94[c]	44[c]	—	4
P. monacha-occidentalis[a]	Unisexual	21[c]	81[c]	—	88[c]	67[c]	—	5
P. monacha[a]	Bisexual	13	80[d]	46	10	56[d]	23	7
P. monacha-lucida[b]	Unisexual	75	—	89	92	—	98	6
P. monacha-2 lucida[e]								
Clone 1961	Unisexual	—	—	—	100	—	4	6
Clone 1963/1974	Unisexual	—	—	—	0	—	61	7
P. lucida[b]	Bisexual	35	—	511	39	—	35	7

[a]Based on Bulger (1978, Tables V–VII and X).
[b]Based on Bulger (1978, Table I).
[c]Based on totals summed over all drainages.
[d]Based on Bulger (1978, Fig. 8).
[e]Based on Bulger and Schultz (1979, Table 6).

to acute heat stress did seem to be enhanced in adult unisexual females. More generally, the unisexual seemed intermediate between the parental species. In contrast, *P. monacha-lucida* is clearly heterotic for all types of thermal stress in both young and adults. This is the clearest evidence that heterosis is an important component of fitness in parthenogenetic vertebrates. Of the two clones of the triploid *P. monacha-2 lucida* tested by Bulger and Schultz (1979), one seemed to be cold-adapted whereas the other was warm-adapted. More detailed anaysis of specific clones of both *P. monacha-occidentalis* and *P. monacha-lucida* revealed interclonal variation in thermal tolerances within both of these species as well.

In one additional critical experiment, Bulger and Schultz (1979) compared the thermal tolerances of clones of *P. monacha-lucida* recently synthesized in the laboratory with those of naturally occurring clones to determine whether heterosis was an immediate consequence of a hybrid origin or has evolved subsequently. The results were mixed: adults of newly synthesized clones were as heterotic as natural clones, but the young were *less* tolerant of thermal stress than both young of natural clones and young of the parental species. Thus, the thermal tolerance manifest in young of naturally occurring *P. monacha-lucida* clones is apparently evolved, either within clones or as a result of interclonal selection.

Although this relates only indirectly to heterosis, there also is evidence that the thermal tolerances seen in *P. monacha-occidentalis* are not a direct result of hybridization but have evolved subsequent to clonal origin. All of the populations of *P. monacha-occidentalis* found north of the Río Mayo in Sonora, Mexico, appear to have derived from a single clone that arose in the Río Mayo and dispersed northerly (Vrijenhoek *et al.*, 1977; Angus, 1980), and yet populations in the Río Matape are more heat-sensitive than are related unisexuals in adjacent drainages. The only point to be made here is that the variations in thermal tolerances seen in *Poeciliopsis* may be, to some extent, adaptations to local environments rather than a geographically uniform expression of heterosis.

With regard to thermal heterosis, then, Bulger and Schultz (1979) provided clear evidence that *Poeciliopsis monacha-lucida* is heterotic for thermal tolerances, which must be important components of fitness in nature. However, comparable evidence in *P. monacha-occidentalis* is generally negative and there is evidence that thermal tolerances have evolved to some extent subsequent to the hybrid origins of the major unisexual taxa studied.

To summarize the evidence pertaining to the hypothesis that high heterozygosity and heterosis are major components of fitness in unisexual fishes: Parthenogenetic vertebrate species, including unisexual fishes, have the highest levels of heterozygosity ever reported for vertebrates, but it is not certain that heterosis is a direct consequence of heterozygosity. Parthenogenetic vertebrates do not express exceptional somatic vigor in nature, although parthenogenetic fishes do in laboratory environments. If heterosis were an *important* component of fitness in natural environments, then the abundance of unisexual fishes should be more or less uniform relative to the bisexual host species over the range of the complex, since the unisexuals are F_1 hybrids everywhere. The fact that there are stable patterns of substantial variation in unisexual–bisexual ratio is, therefore, at odds with this hypothesis. Broad thermal tolerances in *Poeciliopsis monacha-lucida* are the most favorable evidence for heterosis, but heterosis for these characters is not evident in either *P. monacha-2 lucida* or *P. monacha-occidentalis*.

3.5. The Competitive Interaction Hypothesis (Parthenospecies as "Weeds"; Ecological Intermediacy; Multiple Niches)

Four seemingly disparate hypotheses explore the genetic peculiarities of parthenogenetic vertebrates in the context of contemporary niche theory. I believe three of these (the "weed" hypothesis, ecological intermediacy, and adaptation to more than one niche) are special cases of a single phenomenon; therefore, I will treat them collectively. The fourth is a *clonal specialization hypothesis* formulated and articulated by Vrijenhoek [Vrijenhoek (1978) and, especially, Vrijenhoek (1979)]. In simple terms, the latter hypothesis states that individual clones are ecological specialists each adapted to a narrow range of environmental conditions, and therefore clonal diversity begets numerical success in a parthenospecies. Since the first three hypotheses are essentially the same, I will discuss the subsuming hypothesis, which might be termed the *competitive interaction hypothesis*. The essential features of this hypothesis can be understood by verbal description and reference to Fig. 2. In the upper panel of Fig. 2, the fitnesses of two hypothetical bisexual species involved in the hybrid origin of the unisexual species are plotted over the range of some geographic variable representing the ranges of the two species. Each bisexual species is adapted to a distinct composite of physical and biotic conditions; therefore, each species has its peak fitness at a distinct point along the geographic range. The competitive interaction hypothesis postulates that a parthenospecies that arose via hybridization of the two bisexual species would have a comparatively broad, flat fitness profile as suggested in the sketch. The distribution of species that would result from this model if the unisexual were *spontaneously* parthenogenetic is illustrated below the fitness graph. The two parental species are competitively excluded by the parthenospecies from the center area of the range where

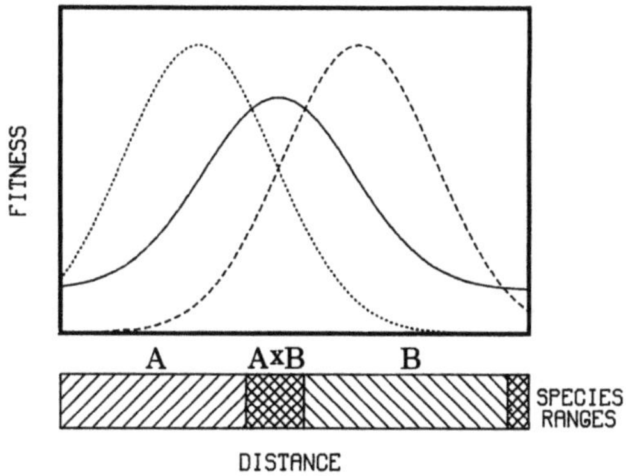

Figure 2. The competitive interaction hypothesis. The fitnesses of two bisexual species (A and B) and a parthenospecies (A × B) of hybrid origin are plotted in the upper panel as functions of some geographic variable (distance) (bisexual species A, dotted line; bisexual species B, dashed line; parthenospecies, solid line). The lower panel illustrates the ranges of the three species that would result if the parthenospecies were spontaneously parthenogenetic.

neither parental species is particularly well adapted. Ignoring for a moment the right-hand extreme of the diagram, species A would occupy the left-hand portion of the range, species B the right, and the parthenospecies a narrow portion of the intermediate range. All three species, as a result of competitive exclusion, would be allopatric (or at least not syntopic). If the parthenospecies were sperm dependent, that is, either gynogenetic or hybridogenetic, then the zoogeography becomes more complex. Suppose, for example, that parthenospecies A × B arose via hybridization between bisexual species A and B but becomes a sexual parasite on species A. In this set of circumstances, the parthenospecies would be sympatric everywhere with species A [provided its primary fitness was at least approximately one-half that of species A; see Moore (1976) and the discussion of the problem of coexistence in this chapter, Section 4], but the relative frequencies of the species would vary sharply over the range: the unisexual would be most abundant in the narrow intermediate belt but would not competitively exclude species A from that area. In the left portion of the range, unisexual frequency would decline precipitously, but nonetheless the unisexual would be present at low frequency over the remainder of the composite range provided it had sufficient primary fitness. In this supposition, parthenospecies A × B is not a sexual parasite of species B, and therefore both species would be competitively excluded from the right-hand portion of the range by species B. It is interesting that in this scheme the range of species A could actually expand as a result of its association with the sperm-dependent parthenospecies.

Furthermore, the competitive interaction hypothesis could explain at least some of the peculiar distributional patterns in parthenogenetic species of rock lizards that Uzzell and Darevsky (1975) attributed to relict distributions. If the parthenospecies has a modicum of heterosis, for example, it is possible that the fitness of the parthenospecies is actually greater than that of the bisexual species in one or both tails of the fitness curves. This is the situation drawn at the right-hand extreme in Fig. 2. As illustrated in the lower panel, this circumstance would result in an outlying population of the parthenospecies peripheral to the range of either sexual species involved in its hybrid origin. By supposing variously complex fitness curves, one can use the competitive interaction hypothesis to explain almost any distributional pattern, and indeed this is both a virtue and a weakness of the hypothesis: the virtue is the explanatory power, the weakness is that the hypothesis may be unfalsifiable by zoogeographic data.

Historically, the "weed" hypothesis was proposed more or less simultaneously by Wright and Lowe (1968) and Maslin (1968), but it was Wright and Lowe who first applied the term weed. In generalizing about ecological conditions that typify parthenogenetic species of whiptail lizards, Wright and Lowe (1968, p. 137) said:

> There is a significant similarity in all of the habitats that are occupied by the parthenogenetic species of *Cnemidophorus*. This similarity is apparent in the habitat descriptions for the species, with the use of such terms as disclimax, ecotone, marginal, edge, transient, extreme, and perpetually disturbed, to characterize the habitats. We conclude that each of the known all-female species of *Cnemidophorus* is a "weed" in the botanical sense, both geographically and ecologically. Moreover, as usual for weeds, the parthenogenetic species of *Cnemidophorus* are successfully established and widely dispersed within habitats that appear to be little or not at all available for bisexual species in the genus.

The essence of the "weed" hypothesis is implicit in the last sentence; namely, parthenospecies are successful only where competition from the generating bisexual species is low. Maslin (1968, p. 227) stated the essence of the "weed' hypothesis clearly:

> Another consequence of the origin of parthenogens from hybrid stock is that they will not be adapted to any one habitat or niche. . . . They behave as opportunists, propagating rapidly in areas where competition is weak or absent. But where a species is well entrenched and an integral part of a community that has not been disrupted by man, the hybrid parthenogen is at a disadvantage and is excluded.

The application of the "weed" hypothesis to parthenogenetic lizards has been widely discussed (Wright and Lowe, 1968; Maslin, 1968, 1971; Uzzell and Darevsky, 1975; Cuellar, 1977*b;* Schall, 1978), while its application to unisexual fishes has received little attention. However, I believe the "weed" hypothesis, as a subsumption of the more general competitive interaction hypothesis, has not been given just consideration as an explanation of the zoogeography of these fishes. Thibault (1974*a*) concluded that the "weed" hypothesis was not consistent with his analysis of distribution patterns of the unisexual species associated with *Poeciliopsis monacha* and *P. lucida*. Rather, he argued that the unisexual forms were ecologically intermediate and, in fact, could be ordinated according to relative abundance along ecological gradients passing from depauperate, bedrock, headwater pools inhabited by *P. monacha* through transitional (ecotonal) habitats dominated by unisexuals to broader downstream habitats where *P. lucida* abounds. This distribution, however, is *not* at odds with the "weed" hypothesis; it is simply a realization of another special case of the subsuming hypothesis. The common denominator of the "weed" and ecological intermediacy hypotheses is that competition determines unisexual success, and therefore unisexuals are successful in *any* environment where bisexual parental species are poorly competitive. This includes ecotones or ecologically transitional areas, and, in fact, ecotones are one of the specific "weed" habitats cited by Wright and Lowe (1968).

Similarly, the unisexual *Poeciliopsis monacha-occidentalis* achieves its greatest abundance (as high as 97% of females) at the interface of the ranges of its parental species, *P. monacha* and *P. occidentalis*. Theoretically, the geographic distribution of fitness for this complex should conform rather closely to the hypothetical fitness distributions sketched in Fig. 2: *P. occidentalis* can be equated to species A, *P. monacha* to species B, and the unisexual, *P. monacha-occidentalis,* to the parthenospecies, A × B. The central area where neither parental species is particularly well adapted would correspond to the Río Mayo, which is both where the ranges of *P. monacha* and *P. occidentalis* come together and where *P. monacha-occidentalis* is very abundant.

More generally, the ecological intermediacy hypothesis predicts that the relative abundance of unisexual and bisexual forms would be determined by genomic dosage. In the case of *Poeciliopsis monacha, P. lucida,* and their associated unisexuals, the model would predict an ordination of abundances with *P. monacha* most abundant in the headwaters, followed by *P. 2 monacha-lucida, P. monacha-lucida, P. monacha-2 lucida,* and, finally, *P. lucida* in a downstream progression. Although the data are complex because these five species comprise two distinct unisexual–bisexual complexes, which themselves must have a higher level competitive interaction, the prediction is clearly corroborated by the data gleaned from Thibault (1974*a*, Tables VII and VIII) and from Vrijenhoek (1978, Fig. 2). I have summarized the data from Thibault (1974*a*) in Fig. 3. The data are limited in that *P. monacha-lucida* and *P. monacha-2 lucida* are a combined category in Thibault's Table VIII and data from more downstream habitats are not available for the Arroyo Jaguari. However,

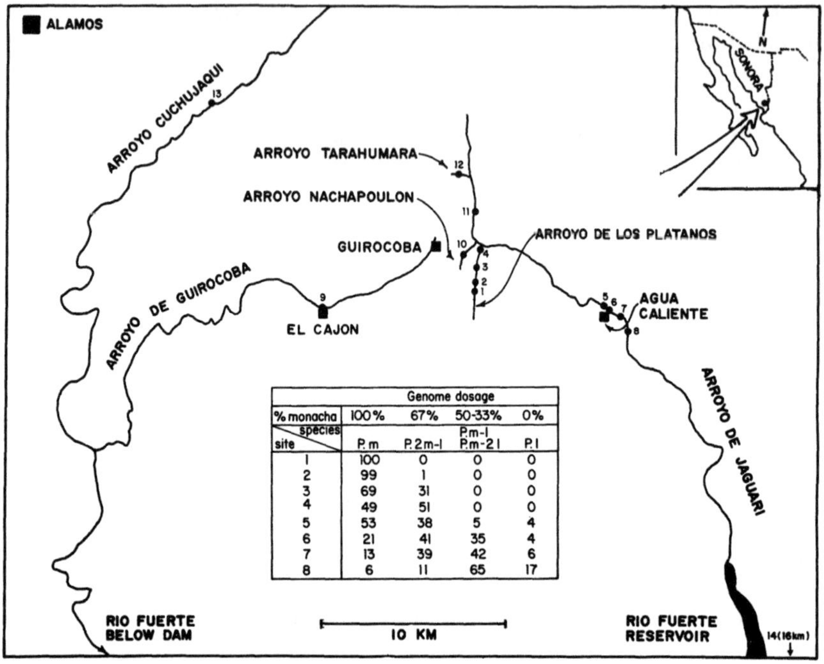

	Genome dosage			
% monacha	100%	67%	50-33%	0%
site \ species	P.m	P.2m-1	P.m-1 P.m-21	P.1
1	100	0	0	0
2	99	1	0	0
3	69	31	0	0
4	49	51	0	0
5	53	38	5	4
6	21	41	35	4
7	13	39	42	6
8	6	11	65	17

Figure 3. The ordination of *Poeciliopsis* according to genomic dosage and position along an upstream–downstream gradient. The body of the table is percent of the female population: P.m, *Poeciliopsis monacha;* P.2m-1, *P.2 monacha-lucida;* P.m-1, *P. monacha-lucida;* P.m-2 1, *P. monacha-2 lucida;* P.1, *P. lucida*. Data derived from Thibault (1974*a*, Tables VII and VIII).

it is my observation that the trend continues downstream, where *P. lucida* becomes still more abundant and the *P. monacha* complex disappears altogether. The data clearly corroborate the prediction that species abundances can be ordinated by genomic dosage.

Furthermore, the predicted trend also obtains in the Amazon molly complex. Balsano *et al.* (1981) reported an extensive series of molly collections from the Río Soto la Marina drainage in northeastern Mexico. *Peocilia formosa* (the Amazon molly) is a diploid gynogen that is a hybrid between the southerly and inland *P. mexicana* and the more coastal (even brackish) *P. latipinna* (Hubbs and Hubbs, 1932; Abramoff *et al.*, 1968; Turner *et al.*, 1980*a*). In addition to *P. formosa* there is a triploid gynogen comprising two *mexicana* genomes and one *latipinna* genome (Prehn and Rasch, 1969; Balsano *et al.*, 1972; Menzel and Darnell, 1973; Rasch and Balsano, 1974); both gynogens are sexual parasites of *P. mexicana* in the

Río Soto la Marina. Since *P. mexicana* is the more headwater species, the prediction that follows from the competitive interaction hypothesis is that *P. mexicana* would be most abundant in the headwaters, followed by the triploid and then *P. formosa* progressing downstream. A subset of the collections reported by Balsano *et al.* (1981) afford a test of this prediction: Five collections were taken along the length of the Río Purificacíon from the upstream locale of Hidalgo downstream approximately 60 km to just above Lago Guerrero. Over the gradient represented by the five collections, bisexual females decline from 99% to 47% of all females. Among the unisexuals, the triploid becomes most abundant first, followed by the diploid. This is as expected because the former is two-thirds *mexicana* whereas the latter is only one-half.

Although the notion that the ecology of parthenogenetic vertebrates is determined by dosage summations of paternal genomes might seem naively oversimplified, I know of no data inconsistent with this notion and there are considerable supporting data. Equally striking is the fact that there is a correlation between genomic dosage and cannibalism in the various species of *Poeciliopsis* constructed from *monacha* and *lucida* genomes, with *P. monacha* the most cannibalistic, *P. lucida* the least, and the three hybrids falling at the predicted intermediate positions (Thibault, 1974*b*). Morphological intermediacy among parthenogenetic fishes according to genomic dosage is well known (Hubbs and Hubbs, 1932; Schultz, 1961, 1966, 1969; Menzel and Darnell, 1973).

As a final note with regard to the competitive interaction hypothesis, Uzzell and Darevsky (1975) pointed out that parthenospecies of Caucasian rock lizards in some places occupy more extreme environments than either of their parental species. However, with regard to some environmental or geographic parameters they seem to be intermediate: *Lacerta armeniaca* is intermediate to its parental species with regard to altitude, but with regard to humidity it resembles neither. Similarly, *L. unisexualis* occurs at intermediate elevations. *Lacerta rostembekovi* is not altitudinally intermediate, but its geographic range is between the ranges of its parental species, and *L. dahli* is intermediate between its parental species with regard to humidity and vegetation (shrubby versus woodland). Uzzell and Darevsky (1975) proposed the relict ecology hypothesis as an explanation for those occurrences where the parthenospecies have environmental tolerances or preferences distinct from their parental species. The essence of the relict ecology hypothesis is that the unisexual species are "stuck" with the ecologies permitted by their genomic constitutions at the time of their origins at least 5000 years ago. The environment has ameliorated since then and the parental species, by virtue of sexuality, have tracked the amelioration, whereas the parthenospecies were locked into their original adaptations by asexuality and are thus left in isolated extreme environments comparable to those that prevailed at the time of their origins. As Uzzell and Darevsky (1975) pointed out, however:

> Nothing in this hypothesis suggests that the original hybridization may not have occurred in intermediate habitats. It explains the occupancy of more extreme habitats not as an escape from competition with the parental species, but as a relic from times when those habitats were occupied by the parental species which are now excluded from them in part by evolution of new ecologies, but in large part by competition from the unisexual species.

Thus, Uzzell and Darevsky's (1975) interpretation of the zoogeography of *Lacerta* is not inconsistent with the competitive interaction hypothesis, although *they* feel that it is not only distinct from but a preferred alternative to the "weed" hypothesis. With regard to unisexual fishes, the relict ecology hypothesis is fruitless. No unisexual fish exhibits a relict distribution nor outstanding success in an environment more extreme than those inhabitated by the aggregate of parental species. The obvious counterexample is *Poeciliopsis monacha-occidentalis,* where a single clone and perhaps one or two derived clones have greatly expanded the range of the species from its locus of origin.

Now, to reflect on the larger picture of the evolutionary ecology of parthenogenetic vertebrates, I believe that the reason parthenospecies persist at all is primarily a result of the simple fact that they produce twice as many female progeny as do comparable sexual species, but most if not all of their varied patterns of zoogeography are the result of the forces postulated in the competitive interaction hypothesis. (Although Vrijenhoek's clonal diversity hypothesis is a plausible alternative.) Reflecting back to the question posed in the first paragraph of the introduction, to the extent that these tentative conclusions are true, it would be possible to disentangle asexuality from hybridity in understanding the adaptive successes and failures of parthenogenetic vertebrates: they owe their existence to asexuality, but the varied and interesting nuances of their zoogeography are a consequence of hybridity. In the latter sense regions where parthenospecies achieve great abundance may be comparable to vertebrate hybrid zones (Moore, 1977*b*).

4. The Problem of Coexistence

The most striking problem in the population ecology of unisexual fishes and other gynogenetic and hybridogenetic species is: How do the unisexual forms coexist with their bisexual "host" species? This problem was anticipated by Clanton (1934) and was a major focus of McKay (1971),

Moore and McKay (1971), Moore (1975), and Vrijenhoek (1978). Briefly stated, the problem is that the unisexual population has twice the growth potential of the bisexual. If the two species compete for a common limiting resource, the unisexual should competitively exclude the bisexual and, in so doing, bring about its own demise. Moreover, since unisexual fishes have polyclonal population structures, the question must be extended to ask: How do clones coexist?

Two rather distinct explanations have been hypothesized with regard to unisexual–bisexual coexistence. The two hypotheses are not mutually exclusive, but it seems unlikely that both hypotheses are true, although some data support each and no data clearly falsify either. The first hypothesis is that there is a behavioral mechanism inherent in the social and courtship behavior of fishes involved in unisexual–bisexual species complexes that regulates sperm availability to the unisexuals such that they are not likely to competitively exclude their bisexual "host" species (McKay, 1971; Moore and McKay, 1971; Moore, 1975, 1976). The second hypothesis is that the unisexual and bisexual forms are sufficiently distinct that they do not compete for a common limiting resource, and hence they coexist by resource partitioning (Vrijenhoek, 1978). Reviewing the two hypotheses and the data now available is worthwhile if, for no other reason, it serves to suggest the more critical experiments that might be done.

The behavioral regulation hypothesis is based on the observation that the males of bisexual *Poeciliopsis* have a complex of behaviors that would result in coexistence of unisexual and bisexual species even if they occupied identical niches. Males exhibit a well-developed mate preference for conspecific females (McKay, 1971), and, in fact, this discriminatory behavior is so strong that one might wonder how unisexuals *ever* become impregnated. The answer seems to be at least twofold. First, males align themselves in dominance hierarchies in which dominant males, to a great extent, are able to deny subordinate males access to the preferred conspecific females. Consequently, subordinate males, presumably with a sex drive heightened by denial, mate with whatever females are available in hasty forays when there is neither the time nor inclination to be "fastidious." Constanz (1975) described a class of males in *Poeciliopsis occidentalis,* called "sneak males," that matured at a small size, never gained dominance in social hierarchies, never assumed the jet black color of dominant males, and always mated in hasty encounters afforded by chance. Second, although mate preference is clearly hereditary, it is modulated by learning, and immature males may make mistakes during the maturation and learning process (McKay, 1971).

When these behaviors are incorporated into a population growth and competition model, the result is a mechanism that regulates the frequencies of unisexual and bisexual components of the population. When bisexuals are relatively abundant, large numbers of subordinate and naive males inseminate unisexuals at a high rate. Because unisexuals produce twice as many females, their numbers increase at the expense of bisexuals. As bisexuals decrease, unisexual inseminations decrease and, subsequently, the unisexual population declines. In terms of a regulatory system, these factors combine to form a negative feedback loop, which, for parameter values (fecundities, survivorship, insemination rates, etc.) measured thus far, should regulate the unisexual–bisexual ratio at a steady state. The two behavioral components, hierarchy formation and naive males or "sneak males," could both contribute to the regulation of unisexual–bisexual ratio, but either one alone would suffice. What is required is that some fraction of the bisexual males make "mistakes." The fraction can be small and constant (naive and "sneak" males) or proportional to the frequency of bisexuals (subordinate males in dominance hierarchies).

This system was modeled first by computer simulation wherein many detailed complexities (age-specific birth and death rates, etc.) were incorporated into the models at the expense of analytical tractability (Moore and McKay, 1971; Moore, 1975). A much simpler model of the same system, however, provides greater insight and yields extraordinarily similar results (Moore, 1976). That is, the trends are the same and the precision gained by the more complex model is negligible. Thus, I will briefly review the simpler model to explicate the essential properties of behavioral regulation. In so doing, I will try to avoid the tedious mathematical details [these can be found in Moore (1976)] and go directly from the description of the model to the results. A summary of the symbols and equations is given in Table IV. The dependent variable is the frequency of unisexual females Q or, equivalently, the frequency of bisexuals P ($P + Q = 1$). The model is very much the same as any discrete generation selection model, but here the fitnesses W_P and W_Q are decomposed into three multiplicative components: (1) the probability that a female will be pregnant, (2) the probability that a given progeny will be female, and (3) primary fitness. If it is assumed that unisexuals produce only female offspring and that the sex ratio of bisexual progeny is 1:1, the probabilities of progeny being female, then, are 1 and ½ for the unisexual and bisexual species respectively. The probabilities that each of the two female types will become pregnant is the key to regulation. In trying to determine a reasonable impregnation submodel, I have worked most extensively with the *Poeciliopsis monacha–P. monacha-occidentalis* complex. This is the simplest *Poeciliopsis* complex in that it involves a single unisexual and a single bisexual species. I fear that other complexes, where a diploid and

Table IV
Summary of the Mathematical Model for Unisexual–Bisexual Coexistence Based on Sperm Allocation[a]

Phenotype	Bisexual female	Unisexual female
Frequency	P	Q
Fitness		
Primary fitness	W'_P ($W'_P/W'_P = 1$)	W'_Q ($W'_Q/W'_P = R$)
Probability of being pregnant	I_P	I_Q
Probability of a female offspring	S ($\frac{1}{2}$, assuming a 1:1 sex ratio)	1
Frequency after selection	$PI_PS/\bar{W}$	$QI_QR/\bar{W}$

$$P + Q = 1, \qquad \bar{W} = PI_PS + QI_QR$$

[a]See text for a full explanation. From Moore (1976).

triploid unisexual comingle with a single bisexual species or where unisexuals can mate with more than one species of male, might obscure the functional relationship sought. For the unisexual *P. monacha-occidentalis*, the impregnation probability is a function of the relative abundances of the two types of females: as the frequency of bisexuals increases, the probability of unisexual insemination increases (Fig. 4). Bisexual females, in contrast, are almost always pregnant (Moore, 1976). Primary fitness W' includes all remaining components normally considered in fitness measures of any species or hybrid, i.e., heterosis, thermal tolerance, fecundity, etc. Finally, for the sake of mathematical simplicity, it is useful to work with the ratio of primary fitnesses $W'_Q/W'_P = R$ rather than W'_Q and W'_P.

The essential properties of the population ecology of a unisexual–bisexual complex regulated in this way can be seen in Figs. 5 and 6. Figure 5 represents a set of phase-plane curves determined from the model for various primary fitness ratios. The abscissa represents unisexual frequency Q and the ordinate represents the amount of change in unisexual frequency ΔQ per generation. Consider, for example, $R = 1$ (unisexual and bisexual equally fit): when $Q = 0.5$, the $R = 1$ curve indicates that the unisexuals would increase in frequency approximately 0.18 or 18% in one generation. If $Q = 0.9$, on the other hand, ΔQ is approximately -0.08, that is, unisexuals would decrease approximately 8% in one generation. Thus, the phase-plane curves indicate that the unisexual frequency will move toward internal equilibrium ($\Delta Q = 0$) as long as the primary fitness of unisexuals is at least one-half that of the bisexuals [more precisely, $R = 0.4738$ (Moore, 1976)].

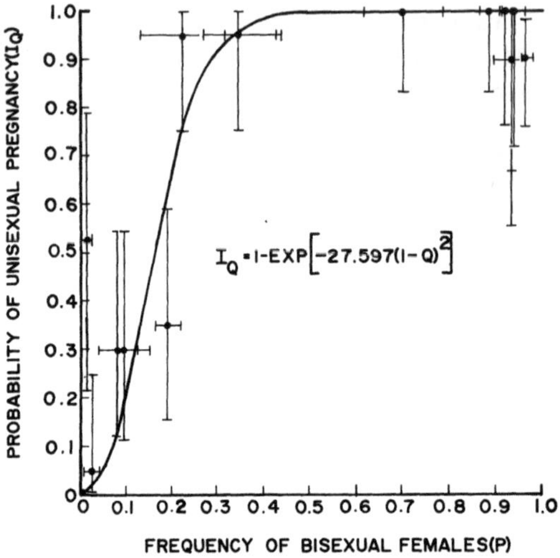

Figure 4. The probability I_Q of a unisexual being pregnant as a function of the frequency of bisexual females P. The curve is the least-squares estimate for the *P. monacha-occidentalis* complex. Ninety-five percent confidence intervals are illustrated for both I_Q and P. For clarity, some confidence intervals in the cluster around $P = 0.95$ have been deleted. From Moore (1976).

The behavioral regulation model predicts coexistence, but it also predicts the possibility of spatial and temporal variation in unisexual—bisexual ratio. The latter result stems from the fact that primary fitness ratio R determines the equilibrium unisexual–bisexual ratio, and primary fitness, in turn, would be expected to vary through time and space. This is evident in either Fig. 5 or Fig. 6. In Fig. 5, the point where each phase-plane curve crosses the $\Delta Q = 0$ line is the equilibrium for that particular primary fitness ratio. Figure 6 is a plot of equilibrium unisexual frequency as a function of primary fitness ratio. An interesting result apparent in this figure is that a linear change in fitness ratio produces a nonlinear response in equilibrium frequency, and, for fitness ratios between 0.4783 and 1.0, a small change in R could produce a large change in equilibrium frequency. A more subtle regulatory property can be seen in Fig. 5; namely, the rate at which the unisexual–bisexual ratio responds to changes in primary fitness ratio is high when the fitness ratio is high and low when the fitness ratio is low.

Moore (1976) used this model to interpret the distribution of *Poeciliopsis monacha-occidentalis* and its bisexual host species *P. occidentalis* in Sonora, Mexico. A distributional map is reproduced in Fig. 7. If this distribution is to be explained by the behavioral regulation model, one must conclude that the primary fitness ratio R of the two species ranges, on average, from somewhere less than 0.4738 to approximately one and that the sharp decline in unisexual frequency between the Río Mayo and Río Yaqui corresponds to the steep portion of the unisexual frequency

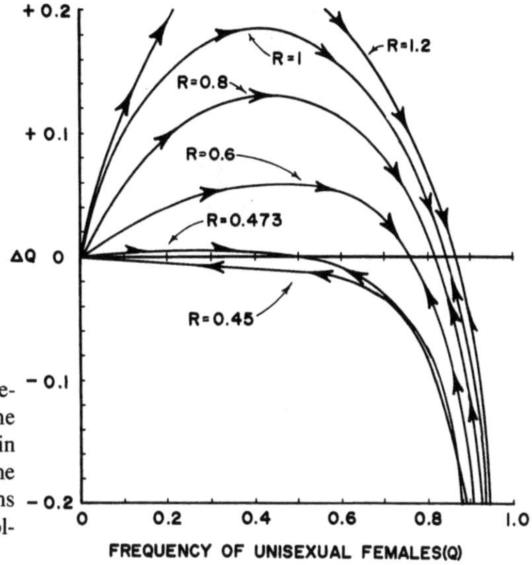

Figure 5. Phase-plane curves for selected primary fitness ratios R. The symbol (ΔQ) represents the change in unisexual frequencies/generation. The arrows on the curves indicate the paths that unisexual frequency (Q) will follow. From Moore (1976).

curve. The variances in unisexual frequency are similar in the five drainages inhabited by the complex, and one must account for this by supposing that in the south (Río Mayo) a given amount of change in R causes a small change in equilibrium, which is quickly tracked. In the north (Río de la Concepcíon), an equal change in R produces a larger change in Q, but this is tracked only very slowly, and the trend reverses before the population fully responds, thus limiting the perceived variance in unisexual frequency.

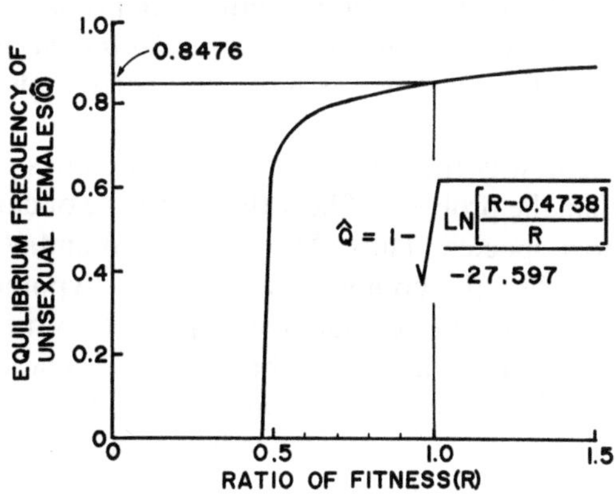

Figure 6. The equilibrium frequency $\hat{Q}$ of unisexual females as a function of the ratio of primary unisexual fitness (R) to primary bisexual fitness; $R = W'_Q/W'_P$. From Moore (1976).

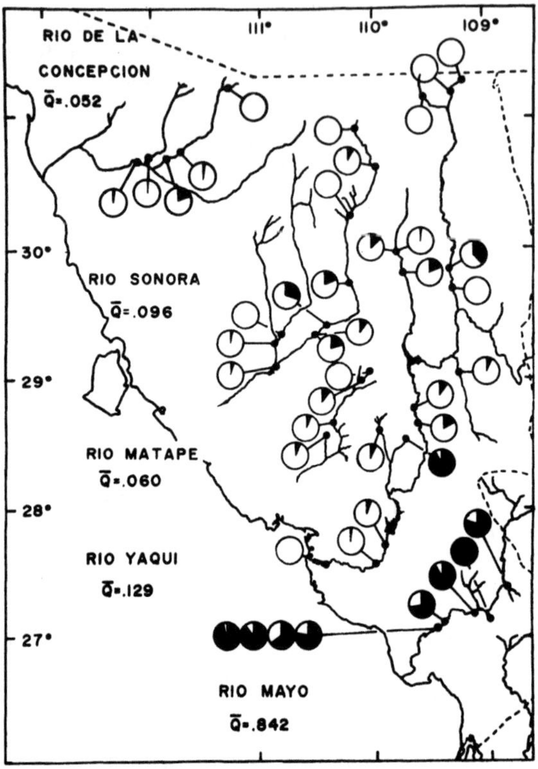

Figure 7. Drainage map of Sonora, Mexico. The circles indicate relative frequencies of unisexual and bisexual females for the *P. monacha-occidentalis–P. occidentalis* complex: black represents the unisexual; white the bisexual. From Moore (1976).

Vrijenhoek (1978) studied the problem of coexistence in the complex involving *Poeciliopsis 2 monacha-lucida* and its bisexual host *P. monacha*. This complex is more southerly in distribution (the Río Mayo is its northern limit) and much more restricted. In addition, Vrijenhoek's major focus was the coexistence of two clones of *P. 2 monacha-lucida* rather than the problem of how the aggregate of clones coexist with their bisexual host species. Thus, Moore's (1976) and Vrijenhoek's (1978) studies are not directly comparable, but the hypothetical explanation of unisexual–besexual coexistence implicit in Vrijenhoek's study would obviate any need to invoke the behavioral regulation hypothesis, because, if two closely related clones can coexist, then clearly there is no difficulty in understanding coexistence between the less closely related unisexual and bisexual species.

In retrospect, it is clear that both my work and Vrijenhoek's work regarding coexistence warrant criticism. With regard to my work, it was a mistake to assume, *a priori,* that coexistence is a problem; i.e., some sort of ecological release experiment needed to be done—and still does! If such an experiment, well conceived and conducted in natural habitats, failed to show competition, then behavioral regulation would be irrelevant, and the question would indeed become: How do such closely related species avoid competition? Resource partitioning would then become the most plausible explanation.

In this context Vrijenhoek (1978) concluded that the two major electrophoretic clones (E-clones I and II) of *Poeciliopsis 2 monacha-lucida* that occur in the headwaters of the Río Fuerte apparently coexist with one another and with their host species by resource partitioning resulting from feeding specializations. In other words, the taxa behave ecologically as distinct species each occupying a separate niche. Vrijenhoek (1978) based this inference on three lines of evidence: (1) differences in dentition, (2) stomach content analysis, and (3) observations of feeding behavior in laboratory-held fishes. Dentition differences between E-clones I and II are unequivocal. However, the stomach contents of E-clones I and II collected from natural populations were *not* significantly different, but the two clones collectively were different from the host species *P. monacha.* Furthermore, Eisenbrey *et al.* (1981) attempted without success to replicate the observation reported by Vrijenhoek that E-clones I and II forage differently in the laboratory. Vrijenhoek (1978 p. 550) said that, "clone 1 [I] individuals primarily engage in scraping algae from rocks .. . . Clone 2 [II] individuals browse within the floating filamentous algae and in the detritus that accumulate in aquaria." The experiment of Eisenbrey *et al.* was more rigorously designed. Five individuals of each clone were acclimated in aquaria set up such that feeding behavior could occur in five recognizable subhabitats: (1) the surface, (2) open water, including algae and other small particles, (3) plants and mats of filamentous algae, (4) rocks or other hard surfaces, and (5) the gravel bottom and detritus. As a control, the two parental species, *P. monacha* and *P. lucida,* were tested in the identical design. Although Eisenbrey *et al.* (1981) found a marked and significant difference between the parentals, they found no consistent or significant differences between clones. The two clones, collectively, showed a weak preference for open water, gravel bottoms and detritus, and plants, whereas *P. monacha* preferred open water and *P. lucida* preferred the surface, gravel bottom and detritus, and rocks. *Poecilliopsis 2 monacha-lucida* tended to use available substrates more evenly than parentals and with regard to feeding event frequency were intermediate but more like *monacha* than *lucida* (which is expected, assuming genomic dosage influences this behavior). In sum, the data indicate that there is really very little if any difference between the foods taken in nature by the two major E-clones of *P. 2 monacha-lucida,* and their foraging behaviors do not differ.

Although the competition question cannot be answered now, a number of additional experiments and observations are pertinent. Wilbur (1971) conducted a field experiment designed to test for competition between larvae of the gynogenetic salamander *Ambystoma tremblayi* and its bisexual host *A. laterale*. The experiment involved caging pure and mixed groups of larvae in cages partially submerged in the native ponds of the two species. Survivorship to metamorphosis, length of larval period, and body size of metamorphosed individuals were assumed to be measures of competitive success. The experimental design was complex, as is interpretation of the results. It is evident, however, that the unisexual and bisexual larval forms did compete, although response differences also indicated that the two species were not ecologically identical.

Cuellar (1979) conducted a rather different type of experiment to determine the competitive relationship between the unisexual whiptail lizard species *Cnemidophorus uniparens* and one of its bisexual parental species, *C. tigris*. Working in natural habitats, Cuellar removed adult *C. uniparens* from a weedy clearing not occupied by *C. tigris,* but the clearing was adjacent to a mesquite-creosote community supporting a *C. tigris* population. If competition determines the ranges of the two species, *C. tigris* should fill the "vacuum." Removals were done in 1975, 1976, and 1978. In 1975 and 1976 only *C. uniparens* were seen in the clearing and these were removed. In 1978, three *C. tigris* were seen in the clearing where it gives way to the mesquite-creosote association, and a change toward smaller *C. uniparens* was evident. Based on these data Cuellar tentatively concluded that habitat rather than competition preference was the crucial factor determining their distributions. In a note added in proof, however, Cuellar reported 14 *C. tigris* scattered throughout the clearing in 1979 and he concluded that "perhaps direct competition is a critical factor after all." The experiment was continued into 1980, when 12 *C. tigris* were found in the clearing (Cuellar, personal communication). Cuellar's present interpretation of the experiment (personal communication) is that habitat preference and competition interact to determine the distributions of the unisexual and bisexual species: *C. uniparens* does exclude *C. tigris* by competition, but the clearing is not good *C. tigris* habitat. In the absence of *C. uniparens, C. tigris* would colonize the clearing, but they would not achieve the densities found in the adjacent mesquite-creosote association. A final *caveat* with regard to this experiment is that the clearing was also adjacent to a gallery forest of cottonwoods inhabited by *C. uniparens*. The logistics of the experiment did not prevent *C. uniparens* from recolonizing the clearing from the gallery forest. Thus, the *Cnemidophorus* composition in the clearing following denudation may have been determined by a recolonization race, in which case the all-femaleness of *C. uniparens* may have been decisive early in the experiment.

It is noteworthy that the conclusions drawn by both Wilbur and Cuellar are identical; namely, that the unisexual and bisexual species differ ecologically, but interspecific competition is significant.

An additional shred of evidence suggests that, in the absence of sperm dependence, unisexual and bisexual species would competitively exclude one another. That is, if the unisexual species were to become spontaneously parthenogenetic, the range of each species would become identical to the geographic areas where they had the greatest fitnesses (fitness would be the product of primary fitness and the probability that a given offspring is female). The shred of evidence is that this appears to be the case with all of the unisexual lizards, all of which are spontaneously parthenogenetic; i.e., the unisexual species are not generally syntopic with their parental species, and usually they are not even sympatric (Wright and Lowe, 1968; Cuellar, 1979; Uzzell and Darevsky, 1975). All of these experimental and observational results must be interpreted with caution because they represent very disparate taxa relative to unisexual fishes. Nonetheless, unisexual lizards and amphibians bear the same relatedness to their parental species as do unisexual fishes.

Another point regarding the argument over the cause of coexistence is that all of the elements necessary to make the behavioral regulatory mechanism work seem to exist, and therefore the behavioral regulation hypothesis is a sufficient if not necessary explanation. The crucial element is mate preference and the resultant frequency-dependent allocation of inseminations to the unisexual component of the population. The insemination model Moore (1976) used is purely empirical: it is a curve fitted to the data for the *Poeciliopsis monacha-occidentalis* complex. With regard to mate preference, there is no doubt that it is well developed in *P. lucida* and *P. monacha* males (McKay, 1971). Furthermore, the Amazon molly, *Poecilia formosa,* is generally less successful in achieving insemination than are females of the host species, *P. latipinna,* in Texas (Hubbs, 1964). With regard to gynogenetic salamanders, Uzzell (1964, 1969) reported strong discrimination by courtship males against gynogenetic females, and eggs of the latter go largely unfertilized in natural populations (Wilbur, 1971). Thus, the elements necessary for behavioral regulation of the unisexual–bisexual ratio seem to exist in most, perhaps all, gynogenetic and hybridogenetic vertebrates.

The *fact* that argues most strongly against the behavioral regulation hypothesis is that multiple clones of unisexual fishes *do* coexist. Vrijenhoek (1978) argued that they accomplish this by partitioning food resources, but regardless of how they coexist, the fact is that they do, and this could not conceivably be explained by regulation via sperm allocation. The crux of the argument is that if two genetically very similar clones can coexist, then coexistence of the more dissimilar unisexual and bisexual species presents no problem. It is possible to explain clonal "coexistence" in terms of a stepping stone model wherein the relative fitnesses of the clones differ among pools and there is some dispersal among these pools (see Section 5) and still cling to the behavioral explanation of unisexual–bisexual coexistence, but the lack of parsimony in such an explanation is not very appealing.

To summarize this section and draw some conclusions, the most important population ecological question concerning unisexual fishes is: How do the unisexual species coexist with the bisexual species upon which they depend for sperm? A mating preference of males for conspecific females could regulate the allocation of sperm to the unisexuals and bring about coexistence regardless of whether or not the unisexual and bisexual species compete. Alternatively, the unisexual and bisexual species may have sufficient niche separation to avoid competitive exclusion. Neither hypothesis can be falsified by the present data. There is some evidence that parthenogenetic salamanders and lizards compete with their host species, but no one has bothered to find out whether or not unisexual fishes really compete with their host species! The population ecology of unisexual fishes will not progress until the competition question is answered, and the critical experiments, obviously, remain to be done. The only conclusive test of competition is an experiment based on ecological release. That is, if the unisexual component of a population were removed, and the bisexuals gained numerically, or *vice versa,* then competition would be evident. Alternatively, a mixed population could be "driven" by addition of one species or the other. For example, if addition of unisexuals decreased the density of bisexuals, that would be evidence for competition. The latter experiment is appealing because large numbers of genotypically identical individuals could be reared in the laboratory to add to natural populations.

Such experiments are logistically difficult and probably would be impossible with *Poeciliopsis* and Mexican populations of *Poecilia formosa*. However, the *Poecilia formosa–P. latipinna* populations around Brownsville, Texas, particularly the feral populations in the ponds and drainage canals at the now abandoned Lula Sams State Fish Hatchery (Kallman, 1962*b;* Hubbs, 1964), might afford an experimental opportunity. Ecological release experiments would only be meaningful if conducted in natural habitats, as the fitnesses of the bisexual species are obviously very much less in the laboratory than in nature, relative to unisexual fitness. If there is no competition, it is meaningless to consider behavioral regulation, but if there is competition, it is as meaningless to look for resource specialization as an explanation of coexistence. The search for resource partitioning approaches a kind of limiting absurdity; i.e., by looking ever more intensely one is bound to find some ecological difference between any two genotypes, and if one argues that such a difference explains coexistence, then the conclusion that two species coexist by niche partitioning is inevitable.

Subsequent to writing this section on coexistence, an important and germane paper on the reduction of competition between *Poecilia formosa* and *P. mexicana* by Balsano *et al.* (1981) was published. After studying their paper in detail, I find that I would not change by summary statement at all. Their results, however, are not only too important to ignore, but are seemingly disparate with results reported by Hubbs (1964) on Texas populations of *Poecilia formosa* and with results reported by McKay (1971), Moore and McKay (1971), and Moore (1976) on *Poeciliopsis*. Therefore, I attach this brief addendum to summarize their results pertinent to the question of coexistence and briefly discuss the apparent disparity. In so doing, I apologize for not giving the paper the extended treatment it deserves.

Although Balsano *et al.* (1981) were very cautious in interpreting their results, they seemed to favor the hypothesis that niche separation allows the three species (*P. mexicana,* diploid and triploid *P. formosa*) to coexist. Most surprisingly, they found that *P. mexicana* males apparently do *not* discriminate against unisexual females in courtship and that both types of unisexuals were as successful in becoming impregnated in nature as were *P. mexicana* females. These two results stand in sharp contrast to those reported by Hubbs (1964) for Texas populations of *P. formosa* where *P. latipinna* serves as the host species [see Hubbs (1964), Fig. 11 and discussion on p. 32]. Not only were unisexuals considerably less likely to be pregnant, but impregnation rates decreased as unisexual abundance increased. This is comparable to the trend reported for *Poeciliopsis monacha-occidentalis* (Moore, 1976), and in fact it was this trend that originally suggested that sperm allocation might permit coexistence. Furthermore, in mate choice tests, sympatric *P. latipinna* directed 76.6% of their gonopodial thrusts at conspecific females and only 24.4% at unisexuals (Hubbs, 1964). Mate preference is even more acute in two species of *Poeciliopsis* (McKay, 1971).

How can the study of Balsano *et al.* (1981) be reconciled with the earlier studies, particularly with that of Hubbs (1964)? First, it is possible that coexistence results from different causes in Mexican and Texas populations—after all, the host species are distinct. Such a possibility is an anathema to a pathological generalizer such as myself, but nonetheless a logical possibility. An alternative reconciliation was articulated by Balsano *et al.* (1981) themselves. They pointed out that the Mexican populations they studied tend to have only low to moderate unisexual frequencies (20–40%) relative to bisexual females. In fact unisexual frequencies this low (high bisexual frequencies) result in 100% impregnation in *Poeciliopsis monacha-occidentalis* as well. Only when unisexuals begin to exceed ~ 70% of total females does sperm limitation become apparent (Fig. 3). In the Texas populations, the unisexual frequencies were 81% and 90% for the two sites. If the situation in *Poeciliopsis* is at all paradigmatic, it is conceivable that sperm allocation does regulate unisexual–bisexual ratio in the populations studied by Balsano *et al.* (1981), but that these populations are operating on a portion of the insemination curve where it cannot be seen. That sperm is limiting and yet 100% of unisexuals are pregnant is one of a number of unintuitive (at least to me) results that came out of that model, and yet that is the nature of the hypothesis. Indeed, the discovery of unintuitive relationships is a major reason for exploring mathematical models. However, this does not seem an adequate explanation for the lack of mate discrimination on the part of *Poecilia mexicana* males in laboratory trials.

5. Clonal Diversity

Pioneering studies on clonal diversity were reported by Kallman (1962*b*) and Darnell *et al.* (1967) on *Poecilia formosa* and by Maslin (1967) on *Cnemidophorus tesselatus*. These studies utilized tissue grafting as a technique for identifying genetically dissimilar parthenogens. Although limited in scope, these three early studies suggested that clonal diversity may be high in unisexual fishes but relatively low in unisexual lizards. During the decade 1970–80, descriptive studies reporting clonal diversity in additional parthenospecies proliferated. In addition to the continued application of tissue grafting (Cuellar, 1976*a*, 1977*a*; Cuellar and McKinney, 1976; Moore, 1977*a*; Angus and Schultz, 1979; Moore and Eisenbrey, 1979; Angus, 1980), several studies were based on protein electrophoresis (McKinney *et al.*, 1973; Parker and Selander, 1976; Vrijenhoek *et al.*, 1977, 1978; Vrijenhoek and Leslie, 1977) and one study was based on immunological cell typing techniques (Eisenbrey and Moore, 1981). Major conclusions that can be drawn at the end of this decade of great activity *vis-à-vis* clonal diversity are as follows: The contrast between parthenogenetic lizards (at least in the genus *Cnemidophorus*) and parthenogenetic fishes may or may not be a general trend—the considerable data are inconsistent. However, the amount of diversity detected by histocompatibility techniques substantially exceeds that detected by protein electrophoresis. Finally, recurring hybridization appears to be a major cause of clonal diversity, but new clones that arise by hybridization can evolve, and even radiate, by "point" mutations, most of which seem to be loss mutations, i.e., loss of genetic information by gene silencing or possibly by recombinational events where heterozygous loci become homozygous.

In the following I have attempted to abstract, summarize, and compare clonal diversity data for parthenogenetic vertebrates. More is known about clonal diversity in *Poeciliopsis* than in other groups; in fact, only in this genus have the same species been assayed consistently by both electrophoresis and tissue grafting. In the hybridogenetic species of *Poeciliopsis*, a single genome is inherited matroclinously. Vrijenhoek *et al.* (1977) termed these clonal genomes *hemiclones* and the genotypes of specific hemoclones *haplotypes*. The clonal structure of *P. monacha-occidentalis* has been studied in depth electrophoretically (Vrijenhoek *et al.*, 1977; Angus, 1980) and by tissue grafting (Angus, 1980). The geographic distribution of hemiclones based on these two studies is summarized in Fig. 8. Three of the four E-clones (electrophoretic clones) are composites of two or more H-clones (histocompatibility clones). In the extreme case, E-clone II is a composite of at least 16 H-clones. The clonal structures of *P. monacha-occidentalis* and two other *Poeciliopsis* parthenospecies are summarized in Table V.

The clonal structure of hybridogenetic *Poeciliopsis monacha-lucida* was studied by Vrijenhoek *et al.* (1978) and Angus and Schultz (1979). Again, E-clones are usually composites of H-clones (Table V). The geographic distribution of *P. monacha-lucida* hemiclones is illustrated in Fig. 9.

The gynogenetic triploid *Poeciliopsis 2 monacha-lucida* comprises three E-clones (refer to Table V): clone III is found only in the Río Mayo, whereas clones I and II occur in some headwaters of the Río Fuerte drainage (Fig. 10). The E-clone III has not been studied by tissue grafting. The E-clone I seems to correspond to a single H-clone, whereas E-clone II is a composite of five or six H-clones (Moore and Vrijenhoek, 1977; Moore and Eisenbrey, 1979).

Comparative clonal diversity data are summarized in Table VI for the vertebrate parthenospecies for which data are sufficient to calculate a diversity index. [Parker (1979) tabulated diversity indices for several parthenogenetic invertebrate taxa as well.] As a diversity index *D*, I followed Parker's advice and used the complement of Simpson's index *SI*.

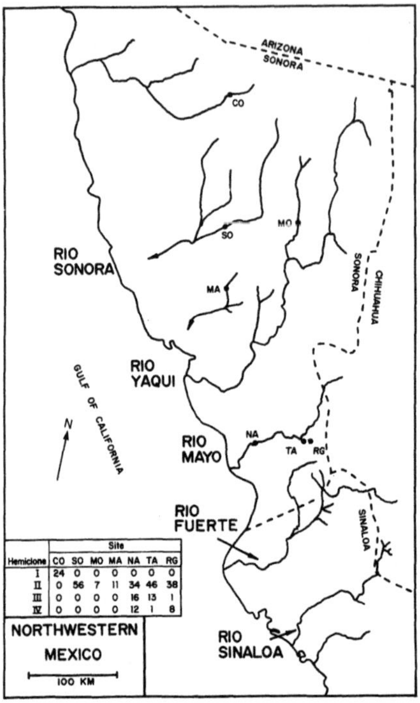

Hemiclone	Site						
	CO	SO	MO	MA	NA	TA	RG
I	24	0	0	0	0	0	0
II	0	56	7	11	34	46	38
III	0	0	0	0	16	13	1
IV	0	0	0	0	12	1	8

Figure 8. The geographic distribution of electrophoretic hemiclones (E-clones) in *Poeciliopsis monacha-occidentalis*. There are four distinguishable hemiclones (I–IV). The body of the table is the number of genetic strains from each locale assigned to each hemiclone. The data were synthesized from Angus and Schultz (1979) and Angus (1980). Abbreviations: (CO) Río Concepcíon; (MA) Río Matape; (SO) Río Sonora; (MO) Río Moctezuma; (NA) Navajoa; (TA) Tabelo; (RG) Rancho Guamuchil.

Thus, $D = 1 - SI$, and D can be interpreted as the probability that two individuals drawn at random from the population represent *different* clones. Furthermore, my calculations are commensurable with those of Parker (1979, Table 2). The formulas used are from Fager (1972, Table 1) for D and Simpson (1949) for the standard error of the estimate (SE).

Data sufficient to calculate diversity indices (Table VI) are available for four species of parthenogenetic fishes and four species of whiptail lizards (*Cnemidophorus*). Adequate data are lacking for other groups. The data are most complete for *Poeciliopsis,* where a concerted effort has been made to assay diversity in the same populations by both electrophoretic and tissue grafting methodologies. Comparable data are available for only one population (Pueblo, Colorado) of triploid *Cnemidophorus tesselatus*. Comparative data are also available for a single population of the Amazon molly (*Poecilia formosa*). Both E-clone and H-clone diversity indices can be calculated for a total of 10 populations representing the four species of fishes. In the two northernmost populations of *Poeciliopsis monacha-occidentalis,* the Río Concepcíon and Río Sonora, the populations are apparently monoclonal, and therefore all diversity indices are

Table V
Clonal Structure of *Poeciliopsis* Parthenospecies

Species	E-Clone	Number of components H-clones (sample size)	Definitive electrophoretic alleles[a] (number of loci assayed)				
			Ldh-2	*Mp-3*	*Es-5* (21)		
Poeciliopsis monacha-occidentalis	I	1 (12)	*b*	0	*c*		
(Vrijenhoek *et al.*, 1977; Angus 1980)	II	16 (61)	*b*	*f*	*a*		
	III	2 (4)	*a*	*f*	*a*		
	IV	3 (3)	*b*	*d*	*a*		
			Ldh-1	*Pgd*	*Aat-3*	*Mp-3*	*Es-5* (22)
Poeciliopsis monacha-lucida	I	6 (10)	*b*	*a*	*b*	*a*	*f*
(Vrijenhoek *et al.*, 1978;	II	2 (6)	*b*	*a*	*b*	*a*	*c*
Angus and Schultz, 1979)	III	2 (2)	*b*	*a*	*a*	*a*	*c*
	IV	4 (6)	*b*	*a*	*b*	*a*	*d*
	V	1 (2)	*b*	*a*	*a*	*a*	*f*
	VI	1 (1)	*b*	*a*	*a*	*a*	*d*
	VII	1 (2)	*a*	*a*	*b*	*b*	*e*
	VIII	1 (1)	*a*	*c*	*b*	*a*	*d*
			Ldh-1	*Aat-3*	*Idh-2*	*Pgm* (23)	
Poeciliopsis 2 monacha-lucida	I	1 (5)	*b/b/c*	*b/b/c*	*a/a/a*	*d/d/d*	
(Vrijenhoek and Leslie, 1977;	II	5–6 (19)	*b/b/b*	*b/b/b*	*a/a/c*	*d/d/e*	
Moore and Vrijenhoek, 1977;	III	? (0)	*b/b/b*	*b/b/b*	*a/a/c*	*d/d/d*	
Moore and Eisenbrey, 1979)							

[a]The definitive alleles combine to form an electromorph profile for each clone; for *P.* 2 *monacha-lucida*, each locus is characterized by a triploid genotype.

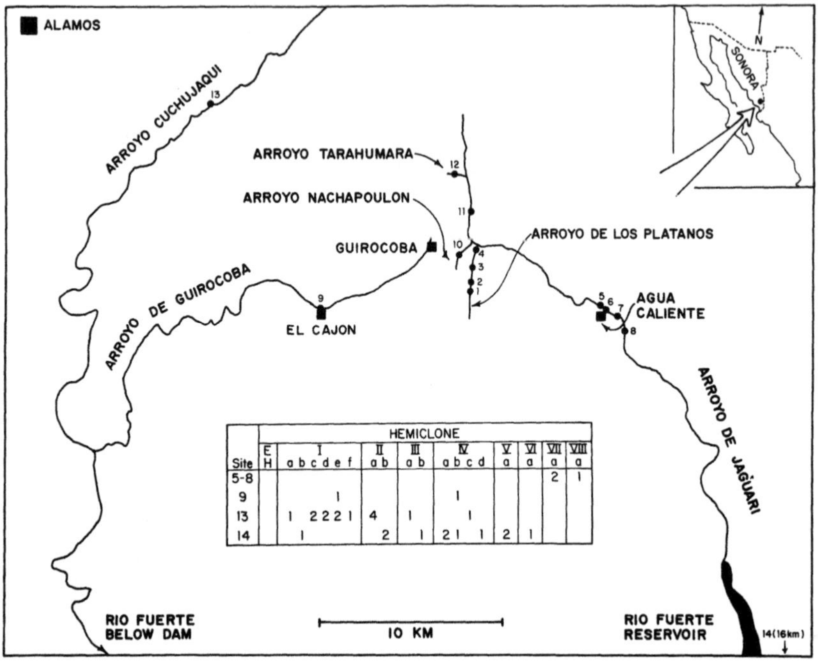

Site	E/H	I a	I b	I c	I d	I e	I f	II a	II b	III a	III b	IV a	IV b	IV c	IV d	V a	VI a	VII a	VIII a
5-8																		2	1
9						1							1						
13		1		2	2	2	1	4		1				1					
14			1						2		1	2	1		1	2	1		

Figure 9. The geographic distribution of hemiclones in *Poeciliopsis monacha-lucida*. There are eight distinguishable E-clones (I–VIII), several of which are composites of H-clones as indicated (lower case letters). The body of the table is the number of genetic strains from each locale assigned to each hemiclone. The data were derived from Angus and Schultz (1979). Site 14 is located south of the mapped area (see lower right-hand corner).

zero. In all other cases, histocompatibility diversity exceeds electrophoretic diversity. In four comparisons (marked with asterisks) approximate 95% confidence intervals ($\pm$ 2 SE) for E-clone diversity and H-clone diversity do *not* overlap; it is conservative to say, at the 0.05 significance level, that, in these four instances, histocompatibility diversity exceeds electrophoretic diversity. Furthermore, a pairwise comparison of E- and H-clone diversities made over *all* parthenogenetic fishes shows that histocompatibility diversity is greater than electrophoretic diversity at the 0.005 significance level [Wilcoxon signed ranks test (Zar, 1974)]. Thus, among parthenogenetic fishes at least one trend is evident: parthenospecies tend to be richer in histocompatibility clones than electrophoretic clones.

The data are too sparse to allow a comparable inference regarding parthenogenetic *Cnemidophorus*. Triploid populations of *C. tesselatus* from around Pueblo, Colorado, were assayed coincidentally by both tissue grafting (Maslin, 1967) and electrophoresis (Parker and Selander, 1976), but only a single clone was detected by either technique. It would be most worthwhile to do an extensive histocompatibility study of those diploid populations of *C. tesselatus* in New Mexico for which Parker (1979) reported high electrophoretic diversity: It would be surprising if these populations were not even richer in H-clones, not only because the trend observed in parthenogenetic fishes is expected, but also because there are grounds for believing that the subset of the genome detected by tissue grafting is both larger and inherently more polymorphic than that detected by protein electrophoresis [see Moore and Eisenbrey (1977) for more discussion].

In the early papers, Kallman (1962*b*) found at least four H-clones in a single population of *Poecilia formosa* [and subsequently found eight additional clones (Kallman, personal communication)], whereas Maslin's

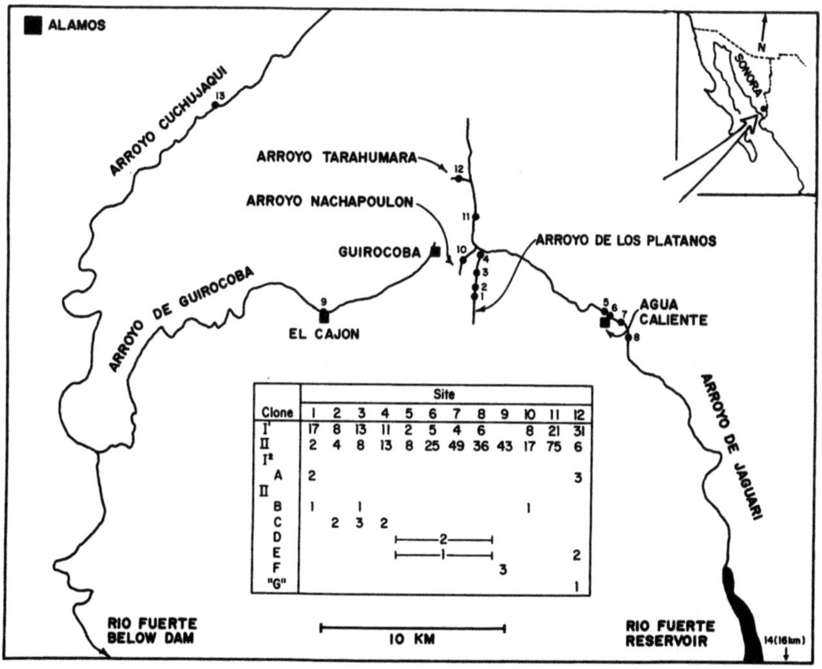

Clone	Site 1	2	3	4	5	6	7	8	9	10	11	12
I[1]	17	8	13	11	2	5	4	6		8	21	31
II	2	4	8	13	8	25	49	36	43	17	75	6
I[2]												
A	2											3
II												
B	1		1							1		
C		2	3	2								
D					├—	—	2	—┤				
E					├—	—	1	—┤				2
F									3			
"G"												1

Figure 10. The geographic distribution of clones of *Poeciliopsis 2 monacha-lucida*. There are two distinguishable E-clones, I and II. The E-clone II is a composite of five or six H-clones, B–F and probably "G," whereas E-clone I corresponds to H-clone A. The body of the table is the number of genetic strains from each locale assigned to each clone. The data were synthesized from (1) Vrijenhoek (1978) and (2) Moore and Eisenbrey (1979).

Table VI
Clonal Diversity in Parthenogenetic Vertebrate Populations

Species and population	Electrophoretic clones: Number of clones (sample size)	Electrophoretic clones: Diversity $D \pm 2$ SE	Histocompatibility clones: Number of clones (sample size)	Histocompatibility clones: Diversity $D \pm 2$ SE
Poeciliopsis monacha-occidentalis				
Río Concepcíon	1 (24)[a,b]	0	1 (12)[b]	0
Río Sonora	1 (56)[a,b]	0	1 (17)[b]	0
Río Matape	1 (11)[b]	0	2 (11)[b]	0.33 ± 0.31
Río Moctezuma	1 (7)[b]	0	1 (7)[b]	0
Río Mayo at:				
Navajoa	3 (62)[a,b]	0.61 ± 0.08	7 (10)[b]	0.93 ± 0.12[n]
Tabelo	3 (60)[a,b]	0.37 ± 0.13	9 (17)[b]	0.91 ± 0.08[n]
Rancho Guamuchil	3 (47)[a,b]	0.32 ± 0.15	4 (6)[b]	0.87 ± 0.26[n]
P. monacha-lucida				
Arroyo Cuchujaqui	4 (18)[c]	0.69 ± 0.17	8 (14)[d]	0.90 ± 0.12
San Pedro	6 (12)[c]	0.85 ± 0.15	8 (11)[d]	0.95 ± 0.11
P. 2 monacha-lucida				
Arroyo Jaguari (collectively)	2 (369)[k]	0.45 ± 0.03	6 (21)[e]	0.82 ± 0.10[n]
El Cajon	1 (48)[k]	0	1 (3)[e]	—
Poecilia formosa				
Olmito, Texas	2 (63)[j]	0.49 ± 0.06	≥4 (47)[f]	0.64 ± 0.11

Cnemidophorus tesselatus (2*n*)				
Espanola, New Mexico	—	—	1 (5)[i]	0
"Conchas"	6 (27)[g]	0.78 ± 0.09[g]	—	—
"West Conchas"	4 (24)[g]	0.57 ± 0.19[g]	—	—
"West Engle"	3 (31)[g]	0.54 ± 0.08[g]	—	—
"Engle"	4 (16)[g]	0.64 ± 0.20[g]	—	—
"Presidio"	2 (18)[g]	0.37 ± 0.21[g]	—	—
"El Pueblito"	2 (15)[g]	0.24 ± 0.22[g]	—	—
"Ruidosa"	2 (13)[h]	0.15 ± 0.25[g]	—	—
"Elephant Butte Lake"	2 (9)[h]	0.39 ± 0.33	—	—
"Eleven monoclonal populations" (various locales)	0 (N ≥ 5)[h]	0	—	—
C. tesselatus (3*n*)				
Pueblo, Colorado	1 (57)[h]	0	1 (19)[i]	0
Higbee, Colorado	1 (10)[h]	0	—	—
C. uniparens				
Bosque del Apache, New Mexico	—	—	1 (9)[l]	0
Frau Cristobol Mt., New Mexico	—	—	1 (10)[l]	0
C. laredoensis				
Laredo, Texas	1 (72)[m]	0	—	—

[a]Vrijenhoek *et al.* (1977); [b]Angus (1980); [c]Vrijenhoek *et al.* (1978); [d]Angus and Schultz (1979); [e]Moore and Eisenbrey (1979); [f]Kallman (1962*b*); [g]Parker (1979); [h]Parker and Selander (1976); [i]Maslin (1967); [j]B. J. Turner (personal communication); [k]Vrijenhoek (1978); [l]Cuellar (1976*a*); [m]McKinney *et al.* (1973); [n]No overlap between 95% confidence intervals of H-clone and E-clone diversity index (CI = ±2 SE).

(1967) study indicated that at least some populations of *Cnemidophorus tesselatus* comprised a single clone spread over a broad area. This in turn suggested that parthenogenetic fishes might, as a rule, be richer in clonal diversity that parthenogenetic lizards. The more extensive data garnered in subsequent studies make this inference somewhat uncertain; nonetheless, there is at least a hint that parthenogenetic fishes are usually more clonally diverse than parthenogenetic lizards. Although the clonal diversity data collected in Table VI are too scattered to allow a complete statistical comparison, it is possible to classify populations as diverse (more than one clone detected either by electrophoresis or tissue grafting) or not diverse (monoclonal) and then to test the null hypothesis that fishes and lizards are equally likely to be diverse. I have tabulated in Table VI all populations I could find for which a sample of five or more was reported and for which it was possible to categorize the population as diverse or not. The tallies are: diverse-fish, eight; monoclonal fish, four; diverse-lizards, eight; monoclonal lizards, 17. Formulating the tallies into a 2 × 2 contingency table and testing the null hypothesis by Fisher's exact probability test (MIDAS, Statistical Research Laboratory, The University of Michigan, 1976), the attained probability is 0.0508, which is marginally significant.

Now, turning to those populations that are diverse, is the mean clonal diversity for fishes greater than that for lizards? Commensurable data are available only from electrophoretic studies and based on these data the answer is yes, *but not significantly* (t-test, $P \geq 0.45$, 13 df, MIDAS). Taken at face value, then, the data suggest that parthenogenetic fish populations are more likely to be clonally diverse, but given that a population is diverse, the level of diversity is as high in lizards as fishes. These conclusions must be considered very tentative, however, because of the sparsity of data and incompleteness in the geographic pattern of sampling. As mentioned above, histocompatibility studies of the electrophoretically diverse *Cnemidophorus tesselatus* populations would be very worthwhile. Cuellar (1977*a*) reported additional tissue grafting studies that support the notion that lizard parthenospecies have relatively low clonal diversity. Although his experimental design does not lend itself to compilation of the data in the format of Table VI, it is clear that the two study species, *Cnemidophorus neomexicanus* and *C. velox,* both have simple clonal structures. A total of 132 inter- and intrapopulational grafts were exchanged between 56 *C. neomexicanus*. All grafts were accepted except one where unidirectional histocompatibility was evident within a pair, suggesting that one clone was derived from the other by some sort of simple mutation. The *C. neomexicanus* populations were separated by as much as 277 km. *Cnemidophorus velox* comprises at least three clones, but again histocompatibility is the rule, as 38 of 50 inter- and intrapopulational grafts were accepted, including grafts between members separated by as much as 625 km. Unidirectional histocompatibility was observed within one pair of *C. velox* collected at Espanola, New Mexico. Parker (1979) cited an unpublished electrophoretic survey of *C. neomexicanus:* there are apparently only two E-clones over the entire species range and no one population is polyclonal. Thus, these additional studies on *C. neomexicanus* and *C. velox* are consistent with the inference that parthenogenetic lizard species tend to be less clonally diverse than parthenogenetic fishes.

A serious problem regarding the adequacy of available data is that the populations of at least some parthenospecies are geographically structured: it is possible that the apparent contrast in diversities is attributable to the happenstance sampling of *Cnemidophorus* from atypically depauperate areas. Geographic structuring is most evident in *Poeciliopsis monacha-occidentalis* and *Cnemidophorus tesselatus,* where in both cases some populations are highly diverse while others are monoclonal. At some future time this geographic structuring will prove important in testing hypotheses regarding the historical biogeography of parthenospecies and the forces that determine clonal diversity. For example, if clonal diversity accumulates through time, then the northern populations of *Poeciliopsis monacha-occidentalis* must be of recent origin, whereas electromorph clone II of *P. 2 monacha-lucida* must be ancient because it is rich in H-clones.

A decade ago it was hardly known whether parthenogenetic vertebrate species were clonally diverse or not. Fortunately, within the past decade not only has a moderately complete picture of clonal diversity emerged, but it is equally clear how new clones arise. Angus and Schultz (1979) pointed out that there are two potential sources of clonal diversity for unisexual populations: (1) multiple hybrid origins, and (2) genetic divergence within clones subsequent to hybrid origins. It is now apparent that both of these events have in fact contributed to clonal diversity. Furthermore, genetic divergence is a composite in that divergence could result from simple mutational events, recombinational events, or even "leakage" of genes from the host species into the unisexual genome. It is also apparent that each of these mechanisms has resulted in the origin of new clones.

The compelling evidence for multiple hybrid origins is that although major E-clones of the same parthenospecies differ at as many as four out of 22 or 23 loci, all of the allelic variants can be found in the gene pools of parental species. It is much more plausible that these clones simply represent different "draws" from the parental gene pools than the alternative that these several allelic forms arose by mutation within the parthenopsecies but happen to have electrophoretic mobilities identical to those of preexisting alleles (Vrijenhoek *et al.*, 1977, 1978). Eight clones of *Poeciliopsis monacha-lucida* (Vrijenhoek *et al.*, 1978; Angus and Schultz, 1979) apparently have arisen in distinct hybridizations, at least three of *P. monacha-occidentalis* (Vrijenhoek *et al.*, 1977; Angus, 1980), and at least two of *P. 2 monacha-lucida* (Vrijenhoek and Leslie, 1977). Parker and Selander (1976) thought four of 12 clones of diploid *Cnemidophorus tesselatus* arose via distinct hybridizations.

Evidence of the opposite sort suggests that new "minor" clones have derived by simple mutational events within "major" clones. That is, occurrence of clones whose electromorph profiles differ from other clones by one or two alleles that are not found in the parental species. Often these "mutant" clones are restricted to single locales. Parker and Selander (1976) suggested that three of 12 clones of *Cnemidophorus tesselatus* arose in this manner. The clearest case of clonal divergence by point mutation was developed by Vrijenhoek *et al.* (1977) and Angus (1980) for a northern population of *Poeciliopsis monacha-occidentalis*. As can be seen in Fig. 8, four clones comprise *P. monacha-occidentalis*. Clone II as well as clones III and IV, seem to have arisen in the Río Mayo, where the two parental species are sympatric, and then colonized more northerly drainages in a stepwise manner. Thus, clone II is found in the Río Yaqui, Río Matapi (MA), and the Río Sonora. It is likely that clone II reached the Río Concepcíon as well, where it gave rise to and then was competitively excluded by clone I. Clone I differs from clone II by two mutant alleles of 25 presumptive loci assayed. One allele ($Es\text{–}5^0$) is a null allele, whereas the other codes for a detectable muscle protein ($Mp\text{–}3^c$); neither allele is known from any other species of *Poeciliopsis* (Vrijenhoek *et al.*, 1977). The clinching evidence, however, is that clone I has a close histocompatibility affinity to clone II; in fact, tissue grafted from I to II is accepted, whereas tissue grafted from II to I is rejected (Angus, 1980). In other words, clone II must have *all* of the histocompatibility antigens clone I does (although clone II must have one or more additional antigens missing from clone I). To complete the scenario, then, primitive clone II colonized the Río Concepcíon from the Río Sonora. Following colonization it gave rise to clone I by two mutational events involving the loss of an esterase allele (or enzymatic function) and the mutation of the "primitive" muscle protein to a new electromorph. Concurrently, histocompatibility genes (perhaps only one) were lost from the Río Concepcíon population. Other hypothetical scenarios are possible but very much less probable.

The unique prediction for clones arising via recombinational events would be the occurrence of a locus that was homozygous when it should be heterozygous. For example, one local clone of *Cnemidophorus tesselatus* differed from the most common widely spread clone by being homozygous at an *Me* (malic enzyme) locus, whereas the common clone was heterozygrous for the allelic forms found in the putative parental species. Parker and Selander (1976) attributed the origin of five of 12 clones of *C. tesselatus* to recombinational events. Recombinational events appear to be rare in *Poeciliopsis,* as such a case has yet to be reported, although the phenomenon of unidirectional histocompatibility prevalent in unisexual *Poeciliopsis* could be explained by this model (Moore, 1977*a*). However, breeding experiments in *P. 2 monacha-lucida* designed to detect recombinational events showed that recombination is a rare or nonexistent phenomenon in this fish (Moore, 1977*a*).

"Leakage" of host species genes into sperm-dependent parthenospecies could be a copious source of novel clones and does seem to occur in some parthenospecies but not others. Occasional recombinational events in the normally hybridogenetic European edible frog result in "leakage" and, potentially, the production of new clones (Uzzell *et al.,* 1977). Rasch and Balsano (1974) reported the irregular but not infrequent incorporation of a *Poecilia mexicana* paternal genome into the zygote of diploid gynogenetic *P. formosa*. The result is the production of triploid progeny. The rare production of triploid progeny by diploid *P. formosa* has been known for years to occur in the laboratory (Rasch *et al.,* 1965; Schultz and Kallman, 1968). Although the unisexual species of *Cnemidophorus* are spontaneously parthenogenetic, occasional matings result in the production of triploid or tetraploid progeny in the cases of diploid and triploid parthenospecies, respectively (Lowe *et al.,* 1970; Neaves, 1969; Cuellar and McKinney, 1976; Cole, 1979). The several triploid species of *Cnemidophorus* seem to have arisen this way (Parker and Selander, 1976; Cole, 1979). In contrast, cases of host genome "leakage" into matroclinous *Poeciliopsis* genomes have not been reported despite the fact that these species have been more extensively bred and assayed than other parthenospecies. Again, breeding experiments were specifically designed to detect such events in *P. 2 monacha-lucida* but failed to do so (Moore, 1977*a*). However, the predominantly *monacha* matroclinous genome of *Poeciliopsis* in the Río Mocorito (Sinaloa, Mexico) has some admixture of *P. viriosa* genes. *Poeciliopsis viriosa* is a sexual species related to *P. monacha* (Vrijenhoek and Schultz, 1974).

The most completely studied case of clonal origin and radiation in a parthenospecies is that of gynogenetic *Poeciliopsis 2 monacha-lucida*. Considering just the populations occurring in the Río Fuerte drainage (see Fig. 10), it is likely that the two major electromorph clones arose in distinct hybridization events possibly separated by considerable time. The E-clone

I corresponds to a single H-clone, whereas E-clone II has radiated into a composite of at least five H-clones. The most plausible explanation for the lack of histocompatibility diversity in E-clone I is that it is of relatively recent origin and simply has not had time to acquire histocompatibility mutations. Other explanations are less plausible because E-clones I and II are sympatric and E-clone I would seem to "need" any hypothetical advantage gained by H-clone diversity as much as E-clone II.

Within E-clone II, two pairs of minor histocompatibility clones are characterized by unidirectional histocompatibility: tissues grafted from H-clone C to B or from E to D take, whereas the reciprocal grafts are acutely rejected (Moore and Eisenbrey, 1979). This means, in the case of the C–B pair, for example, that H-clone B possesses all the antigens found in C plus some additional antigen or a set of antigens lacking in C. Graft reactions are determined by so many gene loci, and H-gene loci are so polymorphic, that it would be extremely unlikely that the antigens of one clone would be a precise subset of those of another clone unless one clone was derived from the other by either a gain or loss mutation (Eisenbrey and Moore, 1981). Furthermore, the two unidirectional histocompatibility pairs themselves seem to be related: Tissues grafted from H-clone C (C–B pair) to either H-clones D or E are sometimes accepted but sometimes chronically rejected; however, reciprocal grafts are always acutely rejected.

What is the direction of divergence in *Poeciliopsis 2 monacha-lucida?* That is, have new H-clones arisen by gaining new antigen specificities or by losing old ones, or both? To test the suite of hypotheses implicit in these questions, Eisenbrey and Moore (1981) used a microcytotoxicity test developed by Eisenbrey (1980) to determine the occurrences of specific antigens in several species of *Poeciliopsis* involved in unisexuality. Briefly, antisera against specific clones of *P. 2 monacha-lucida* were produced in rabbits. Whole antisera were then absorbed on cells of various clones, leaving antisera specific for various antigen subsets. Thus, for example, whole antiserum prepared against H-clone B and absorbed on cells from H-clone C (B/C) would be left with only the antibodies specific for the subset of antigens that distinguish clones B and C. These "essence" antigens could be single-gene products but are not necessarily so. In this manner a cell-typing panel was constructed, A absorbed on B (A/B), A on C (A/C), etc. A large sample of fish representing various unisexual and bisexual species of *Poeciliopsis* was then assayed by the typing panel to determine the occurrence of various antigen subsets. The critical evidence is that *all* of the specific subsets were found in at least one species of *Poeciliopsis* in addition to *P. 2 monacha-lucida,* and both of the "essence" antigens, B/C and D/E, were found in both *P. monacha* and *P. lucida*. Therefore, it can be inferred that H-clones B and D are primitive and that H-clones C and E were derived by some sort of loss mutation. Furthermore, by a process of elimination of unlikely alternatives, Eisenbrey and Moore (1981) reasoned that these loss mutations involve simple gene silencings. To the extent that the case of *Poeciliopsis 2 monacha-lucida* is paradigmatic, then, major electrophoretic clones arise in repeated hybridizations and simple loss mutations are an important cause of the proliferation of component H-clones. No doubt many of these loss mutations are shielded from natural selection by the absence of meiotic segregation inherent in the parthenogenetic life style, but it is plausible that even this seemingly "degenerate" diversification of histocompatibility genotypes is an adaptive defense against pathogens.

Over the past decade, then, a fairly clear description of clonal diversity in parthenogenetic vertebrates has emerged, and, to a considerable extent, it is evident how clones originate. The outstanding question for present and future research is: How is this clonal diversity maintained? This is a challenging question because there are severe theoretical limitations on the number of clones than can be maintained in asexual populations, and an important question because it is possible, even likely, that insight into the ecological aspects of such general evolutionary phenomena as specialization and adaptive radiation could be gained if the nature of clonal coexistence were understood. The theory of genetic polymorphism in spatially and temporally heterogenous environments is so complex and permutable that a comprehensive summary of possibilities is impossible here [Felsenstein (1976) provides a good review with references]. Nonetheless, two points are basic even to begin to understand the conceptual challenge created by the existence of both mono- and polyclonal populations of parthenogenetic vertebrate species. First, an asexual population will become monoclonal in a homogeneous environment, in contrast to a sexual population, where simple overdominance suffices to maintain polymorphism. Even random temporal variation does not protect polymorphisms in asexual populations, where, again, a sexual population will remain polymorphic provided there is geometric mean overdominance averaged over generations. Spatial heterogeneity does permit polyclonal populations, but, and this is the second point, many variables interact in complex ways to determine the clonal diversity expected in this circumstance. These include geographic structuring of populations (island models, stepping stone models, etc.), dispersal rates, local, geographic, and temporal selection patterns, population sizes, and mutation rates (to include all sources of new clones, recurring hybridization, recombination, etc.). Thus, a general expectation is that parthenospecies should manifest many fewer genotypes than comparable sexual species, but the derivation of a precise set of expectations regarding clonal diversity would require a very detailed set of population data for the parthenospecies under investigation.

Since the answers to questions regarding maintenance of clonal diversity seem so uncertain, I will not extensively compare alternative hypotheses against the meager data. However, I will examine briefly three alternatives because the data that need to be collected presumably would be collected more judiciously and efficiently if collecting efforts were guided by theory. The first possibility is that the number of clones in a population may represent a dynamic state where new clones arise by the mechanisms described above and move through the population in random walks until they inevitably become extinct. The levels of clonal diversity apparent in the several parthenospecies may or may not represent equilibria under such a model. When a parthenospecies first arises, it would comprise a single clone or a small number of "sibling" clones, but clonal diversity should increase through time as "mutations" accrue. As clonal diversity increases, the average numbers of individuals per clone would decrease, leading to an increase in the probability of individual clonal extinction. Thus, this model predicts that the number of clones is balanced by rates of clonal origin and extinction.

A second explanation was proposed by Vrijenhoek [Vrijenhoek (1978) and, especially, Vrijenhoek (1979)], who hypothesized that clonal diversity depends upon the opportunity for recurring hybridization. This leads to interclonal selection, which results in clonal specialization; i.e., utilization of multiple niches, and this in turn results in an overall increase in unisexuals relative to bisexuals:

> Where polyphyletic origins [multiple clones] are possible, the asexual population size can increase through exploitation of the between-phenotype component of niche width contained in the ancestral sexual populations. Clonal reproduction could freeze adaptive complexes of genes which decrease niche overlap among clones and also provide high efficiency within specific subniches. The fundamental niches of the different clones could be completely included within those of the sexual ancestors, as long as the clones are more efficient within specific subniches. (Vrijenhoek, 1979, p. 792).

This hypothesis is worth serious consideration not only because it is plausible, but because it is a powerful hypothesis that bridges all levels of the evolutionary ecology of parthenospecies: it solves the problems of clonal and unisexual–bisexual coexistence (niche specialization), and explains geographic patterns of clonal diversity and unisexual–bisexual ratios (clonal diversity is high where parental species are sympatric and unisexuals are most abundant where clonal diversity is highest). Furthermore, for two hybridogenetic species of *Poeciliopsis*, *P. monacha-*

lucida and *P. monacha-occidentalis,* unisexual frequency is, indeed, correlated with clonal diversity (Vrijenhoek, 1979), and clonal diversity in turn seems to be high in areas where recurring hybridization is a possibility (Vrijenhoek *et al.*, 1977, 1978; Vrijenhoek, 1979; Angus and Schultz, 1979; Angus, 1980). Thus, the hypothesis is corroborated to this extent, although the latter point is somewhat equivocal. Rather than belabor the data, however, I would point out that the samc data corroborate, to an equal extent, the hybrid intermediacy hypothesis discussed above and herein we have a salient illustration of a classic problem in the philosophy of science; namely two fundamentally different hypotheses usually have very broad overlap in their prediction sets, and most data will corroborate both while falsifying neither. When judging mutually exclusive hypotheses, progress can be made only if predictions that distinguish tbe theories are sought, found, and tested. To my knowledge, no such data presently exist in this case.

To compare and contrast the two, the hybrid intermediacy hypothesis states that hybrid parthenospecies would be most abundant in ecologies intermediate between those best suited for the parental species and these presumably would be the areas where the ranges of the parental species overlap (see Fig. 2). Thus, although both hypotheses predict unisexual abundance in areas of parental sympatry, the causal explanations are fundamentally different: Vrijenhoek's hypothesis says that abundance is achieved by a proliferation of narrowly adapted clones, whereas the ecological intermediacy hypothesis postulates the existence of rather broadly adapted clones. Clonal diversity is not even an essential part of the latter hypothesis; the fact that there is clonal diversity in ecologically intermediate areas is a coincidental byproduct of sympatry. The fundamental distinction between the two, then, is niche breadth, and this could be the premise for a critical test of the alternatives. The problem is: How does one measure niche breadth? An alternative approach would be through competition. Vrijenhoek's hypothesis predicts the existence of narrowly adapted clones that avoid competition via specialization. In contrast, competition should be significant between the broadly adapted generalist clones implicit in the hybrid intermediacy hypothesis. Thus, ecological release experiments might be fruitful where specific clones were systematically introduced or removed from natural populations. Such an experiment might be equivocal if it turned out, for example, that driving a population by adding clone A caused a reduction in the absolute density of B; it could be argued still that the excess clone A simply compressed the *realized* niche of B and that the two had distinct *fundamental* niches [in the sense of Miller (1967)]. Thus, such a result would cast doubt on

Vrijenhoek's hypothesis but would not falsify it. The opposite result, no effect of clone A on B, would clearly falsify the ecological intermediacy hypothesis, however.

If this brief discussion serves no greater purpose, I would hope that it brings to focus the ease with which hypotheses are corroborated and the extreme difficulty in formulating critical tests of alternative hypotheses. In the case of testing explanatory hypotheses of clonal diversity and the relative successes of vertebrate parthenospecies, considerable thought must antecede data gathering, otherwise the data gathered are likely to be superfluous.

In terms of elegance and parsimony, Vrijenhoek's niche specialization hypothesis is the more outstanding of the two. As pointed out above, this is a theory that bridges all levels of unisexual ecology. The ecological intermediacy hypothesis says absolutely nothing about clonal diversity, and therefore a completely distinct hypothesis is required to explain the observed patterns of clonal diversity. This is not a formidable problem, however, because appropriate hypotheses are well known from the theoretical literature. In theory, clonal diversity (in equilibrium) is an impossibility in a homogeneous environment, and temporal heterogeneity would be of little help. However, many patterns of spatial variation, in conjunction with genetic drift, migration, and selection, are conducive to clonal polymorphism. The geographic structuring of *Poeciliopsis 2 monacha-lucida* corresponds to one such pattern. In the headwaters of the Río Fuerte, this species occurs in a network of small, semiisolated "island" populations between which a small amount of dispersal occurs. In the absence of interdeme dispersal, small populations such as these should become randomly fixed for clones. Addition of even a modicum of geographically variable selection makes it likely that locally adapted clones become fixed at specific locales, but in either case it is likely that *inter*deme polymorphism would persist provided clonal variation was there to begin with. Add interdeme dispersal, and virtually any level of *intra*deme as well as *inter*deme polymorphism could be explained, depending upon population size, geographic selection differentials, and dispersal rate.

In the case of *Poeciliopsis 2 monacha-lucida,* a factual basis for the maintenance of clonal polymorphism clearly exists. The overall network of populations in the Arroyo Jaguari was estimated to total just over 3000 adults divided into bedrock pools and scattered in the interconnecting streamlets in several small headwater tributaries; individual pool populations averaged 133 adults, but the variance is substantial (Moore and Eisenbrey, 1979). Dispersal between pools within a tributary was measured directly. Twenty-one of 716 fish tagged prior to the rainy season were recovered at the end of the rainy season; six of these had moved to other pools within the same tributary. Intertributary dispersal was not detected directly, but the occurrence of histocompatible fish in distinct tributaries is unequivocal evidence that this event occurs, presumably at a relatively low level (Moore and Eisenbrey, 1979). Finally, the existence of geographic clines in unisexual–bisexual ratio in the Arroyo Jaguari complex and in the frequency of electrophoretic clone II suggests geographic selection differentials.

In sum, then, theoretical arguments based on substantiated premises would make it surprising if clonal diversity were not maintained in *Poeciliopsis 2 monacha-lucida* by the interplay of selection, drift, and dispersal among "island" populations. Furthermore, it seems to me that the biogeography of this parthenospecies does not conform well to Vrijenhoek's hypothesis in that only two clones have arisen via hybridization, the species is nonetheless locally abundant, and a *single* electrophoretic clone varies in abundance geographically. In Vrijenhoek's defense, however, the single E-clone that is most geographically variable in abundance is the E-clone that is a composite of histocompatibility genotypes. It is possible that these H-clones, too, represent distinct adaptive peaks, but I doubt that the apparently simple genetic differences that distinguish these clones could result in the amount of resource partitioning required.

As a final note on clonal diversity, histocompatibility and electrophoretic diversity may be maintained by distinct causal agents. It is plausible, for example, that major electrophoretic clones are distinct adaptive complexes *vis-à-vis* resource partitioning, whereas the histocompatibility clones that have proliferated within electrophoretic clones are either adaptively neutral or responses to frequency-dependent selection exerted by pathogens. Eisenbrey (1980) found that the 22 subsets of antigens he could detect with his cell typing panel occurred with rather uniform frequencies not only within natural populations of *Poeciliopsis 2 monacha-lucida* but within populations of closely related species as well. This is consistent with the trend reported for other vertebrates; namely, numerous alleles segregate at the major histocompatibility loci and allelic frequencies are rather uniform within populations (Bodmer, 1975). Varying explanations have been hypothesized for the extraordinarily high degree of polymorphism seen at histocompatibility loci, but it is believed generally that frequency-dependent selection, wherein more common genotypes are more susceptible to pathogens, is the major cause.

ACKNOWLEDGMENTS. I wish to thank the following individuals for the very significant contribution they made to the development of my thoughts on the evolutionary ecology of parthenogenetic vertebrates: R. Jack Schultz, Michael Cimino, Bradley Eisenbrey, Gerard Joswiak, Francis McKay, Roger Thibault, and Robert Vrijenhoek. I thank M. Patricia Hensler, Eileen D. Koglin, and Donna M. Laciak for the expert typing of this manuscript. I thank Paul Vogel for his help in preparation of Fig. 1 and 2. This research was supported by grant DEB 77-15352 from the National Science Foundation.

References

Abramoff, P., Darnell, R. M., and Balsano, J. S., 1968, Electrophoretic demonstration of the hybrid origin of the gynogenetic teleost *Poecilia formosa, Am. Nat.* **102:**555–558.

Angus, R. A., 1980, Geographic dispersal and clonal diversity in unisexual fish populations, *Am. Nat.* **115:**531–550.

Angus, R. A., and Schultz, R. J., 1979, Clonal diversity in the unisexual fish *Poeciliopsis monacha-lucida:* A tissue graft analysis, *Evolution* **33:**27–40.

Balsano, J. S., and Rasch, E. M., 1974, Microspectrophotometric and enzymatic analyses of fish plasma proteins electrophoretically separated in thin polyacrylamide gels, *J. Fish Biol.* **6:**51–59.

Balsano, J. S., Darnell, R. M., and Abramoff, P., 1972, Electrophoretic evidence of triploidy associated with populations of the gynogenetic teleost *Poecilia formosa, Copeia* **1972:**292–297.

Balsano, J. S., Kurcharski, K., Randle, E. J., Rasch, E. M., and Monaco, P. J., 1981, Reduction of competition between bisexual and unisexual females of *Poecilia* in northeastern Mexico, *Environ. Biol. Fish.* **6:**39–48.

Barrowclough, G., 1980, Genetic and phenotypic differentiation in a wood warbler (genus *Dendroica*) hybrid zone, *Auk* **97:**655–668.

Berger, L. S., 1966, Biometrical studies on the population of green frogs from the environs of Poznan, *Ann. Zool.* **23:**303–324.

Berger, L. S., 1968, Morphology of the F1 generation of various crosses within *Rana esculenta*-complex, *Acta Zool. Carcov.* **13:**301–324.

Berger, L. S., 1973, Systematics and hybridization in European green frogs of *Rana esculenta* complex, *J. Herpetol.* **7:**1–10.

Blankenhorn, H. J., Heusser, H., and Vogel, P., 1971, Drei Phaenotypen von Gruenfroeschen aus dem *Rana esculenta*-Komplex in der Schweiz, *Rev. Suisse Zool.* **78:**1243–1247.

Bodmer, W. F., 1975, Evolution of HL-A and other major histocompatibility systems, *Genetics* **79:**s293–304.

Bulger, A. J., 1978, Heterosis and thermal adaptations in desert fishes, Ph. D. dissertation, University of Connecticut.

Bulger, A. J., and Schultz, R. J., 1979, Heterosis and clonal variation in thermal tolerance in unisexual fishes, *Evolution* **33:**848–859.

Cimino, M. C., 1972*a*, Meiosis in triploid all-female fish (*Poeciliopsis,* Poeciliidae), *Science* **175:**1484–1486.

Cimino, M. C., 1972*b*, Egg-production, polyploidization and evolution in a diploid all-female fish of the genus *Poeciliopsis, Evolution* **26:**294–306.

Cimino, M. C., and Schultz, R. J., 1970, Production of a diploid offspring by a gynogenetic triploid fish of the genus *Poeciliopsis, Copeia* **1970:**760–763.

Clanton, W., 1934, An unusual situation in the salamander *Ambystoma jeffersonianum* (Green), *Occ. Pap. Mus. Zool.* (University of Michigan) **290:**1–15.
Cole, C. J., 1975, Evolution of parthenogenetic species of reptiles, in: *Intersexuality in the Animal Kingdom* (R. Reinboth, ed.), Springer-Verlag, Berlin, pp. 340–355.
Cole, C. J., 1978, Parthenogenetic lizards, *Science* **201:**1153–1155.
Cole, C. J., 1979, Chromosome inheritance in parthenogenetic lizards and evolution of allopolyploidy in reptiles, *J. Hered.* **70:**95–102.
Constanz, G. D., 1975, Behavioral ecology of mating in the male gila topminnow, *Poeciliopsis occidentalis* (Cyprinodontiformes: Poeciliidae), *Ecology* **56:**966–973.
Cuellar, O., 1971, Reproduction and the mechanism of meiotic restitution in the parthenogenetic lizard *Cnemidophorus uniparens, J. Morphol.* **133:**139–166.
Cuellar, O., 1976*a,* Intraclonal histocompatibility in a parthenogenetic lizard: Evidence of genetic homogeneity, *Science* **193:**150–153.
Cuellar, O., 1976*b,* Cytology of meiosis in the triploid gynogenetic salamander *Ambystoma tremblayi, Chromosoma* **58:**355–364.
Cuellar, O., 1977*a,* Genetic homogeneity and speciation in the parthenogenetic lizards *Cnemidophorus velox* and *C. neomexicanus,* evidence from interspecific histocompatibility, *Evolution* **31:**24–31.
Cuellar, O., 1977*b,* Animal parthenogenesis, *Science* **197:**837–843.
Cuellar, O., 1978, Parthenogenetic lizards, *Science* **201:**1155.
Cuellar, O., 1979, On the ecology of coexistence in parthenogenetic and bisexual lizards of the genus *Cnemidophorus, Am. Zool.* **19:**773–786.
Cuellar, O., and McKinney, C. O., 1976, Natural hybridization between parthenogenetic and bisexual lizards: Detection of uniparental source by skin grafting, *J. Exp. Zool.* **196:**341–350.
Darevsky, I. S., 1966, Natural parthenogenesis in a polymorphic group of Caucasian rock lizards related to *Lacerta saxicola eversman, Ohio Herpetol. Soc.* **5:**115–152.
Darnell, R. M., Lamb, E., and Abramoff, P., 1967, Matroclinous inheritance and clonal structure of a Mexican population of the gynogenetic fish, *Poecilia formosa, Evolution* **21:**168–173.
Drewry, G. E., 1964, Appendix I—Chromosome number, in: C. Hubbs, Interactions between a bisexual fish species and its gynogenetic sexual parasite, *Tex. Mem. Mus. Bull.* **8:**67–68.
Eisenbrey, A. B., 1980, Evolution of histocompatability diversity in an asexual species, *Poeciliopsis 2 monacha-lucida* (Pisces: Poeciliidae), Ph. D. dissertation, Wayne State University.
Eisenbrey, A. B., and Moore, W. S., 1981, Evolution of histocompatibility diversity in an asexual vertebrate, *Poeciliopsis 2 monacha-lucida* (Pisces: Poeciliidae), *Evolution* **35:**1180–1191.
Eisenbrey, A. B., Buchanan, D. B., and Joswiak, G. R., 1981, Laboratory analysis of feeding behavior of two genetic clones of the asexual Poeciliid, *Poeciliopsis 2 monacha-lucida, The Biologist* **63:**92–96.
Engelmann, W. E., 1972, Disk-elektrophorise der serumproteine von Wasserfroschen, ein Beitrag zur Diskussion uber den hybrischarakter von *Rana esculenta* L, *Acta Biol. Med. German.* **29:**431–435.
Engelmann, W. E., 1973, Zur frage der verwandtschaftlichen Bezienhungen europaischen Grunfrosche (Guttung *Rana*), *Zool. Jahrb. Abt. Syst. Oekol. Geogra. Tiere* **100:**183–196.
Fager, E. W., 1972, Diversity: A sampling study, *Am. Nat.* **106:**293–310.
Fasset, N. C., 1945, *Juniperus virginiana, J. horizontalis* and *J. scopulorum*—IV. Hybrid swarms of *J. virginiana* and *J. horizontalis, Bull. Torrey Bot. Club* **72:**379–384.

Felsenstein, J., 1976, The theoretical population genetics of variable selection, *Annu. Rev. Genet.* **10**:253–280.

Gunther, R., 1969, Untersuchungen zum Artproblem an europaischen anuren der Gattung *Rana* (Amphibia), Dissertation, Humboldt-Universitat, Berlin.

Gunther, R., 1973, Uber die verwandtschaftlichen beziehunger zwischen den europaischen Grunfroschen und den Bastardcharakter von *Rana esculenta* L. (Anura), *Zool. Anz.* **190**:250–285.

Halfman, H., and Mueller, P., 1972, Populations untersuchungen an Gruenfroeschen in Saar-Mosel-Raum, *Salamandra* **8**:112–116.

Hamilton, W. D., Henderson, P. A., and Moran, N. A., 1980, Fluctuation of environment and coevolved antagonist polymorphism as factors in the maintenance of sex, in: *Natural Selection and Social Behavior:* Recent Research and Theory (R. D. Alexander and D. W. Tinkle, eds.), Chiron Press, New York. pp. 363–381.

Hines, W. G. S., and Moore, W. S., 1981, An analysis of sex in random environments, I. *Adv. Appl. Prob.* **13**:453–463.

Hubbard, J. P., 1969, The relationships and evolution of the *Dendroica coronata* complex, *Auk* **86**:393–432.

Hubbs, C., 1964, Interactions between a bisexual fish species and its gynogenetic sexual parasite, *Tex. Mem. Mus. Bull.* **8**:1–72.

Hubbs, C. L., 1955, Hybridization between fish species in nature, *Syst. Zool.* **4**:1–20.

Hubbs, C. L., and Hubbs, L. C., 1932, Apparent parthenogenesis in nature, in a form of fish of hybrid origin, *Science* **76**:628–630.

Hubbs, C. L., and Hubbs, L. C., 1946*a,* Breeding experiments with the invariably female, strictly matroclinous fish, *Mollienesia formosa, Genetics* **31**:218.

Hubbs, C. L., and Hubbs, L. C., 1946*b,* Experimental breeding of the Amazon molly, *Aquarium J.* **17**:4–6.

Johnson, A. W., Parker, J. G., and Reese, G., 1965, Polyploidy, distribution and environment, in: *The Quaternary of the United States* (H. E. Wright and D. G. Frey, eds.), Princeton University Press, Princeton, New Jersey, pp. 497–507.

Kallman, K. D., 1962*a,* Gynogenesis in the teleost, *M. formosa* with discussion of the detection of parthenogenesis in vertebrates by tissue transplantation, *J. Genet.* **58**:7–21.

Kallman, K. D., 1962*b,* Population genetics of gynogenetic teleost, *M. formosa, Evolution* **16**:497–504.

Leslie, J. F., and Vrijenhoek, R. C., 1978, Genetic dissection of clonally inherited genomes of *Poeciliopsis.* I. Linkage analysis and preliminary assessment of deleterious gene loads, *Genetics* **90**:801–811.

Lowe, C. H., and Wright, J. W., 1966, Evolution of parthenogenetic species of *Cnemidophorus* (whiptail lizards) in western North America, *J. Ariz. Acad. Sci.* **4**:81–87.

Lowe, C. H., Wright, J. W., Cole, C. J., and Bezy, R. L., 1970, Natural hybridization between teiid lizards *Cnemidophorus sonorae* (parthenogenetic) and *Cnemidophorus tigris* (bisexual), *Syst. Zool.* **19**:114–127.

Macgregor, H. C., and Uzzell, T. M., 1964, Gynogenesis in Salamanders related to *Ambystoma jeffesonianum, Science* **143**:1043–1045.

Maslin, T. P., 1967, Skin grafting in bisexual Teiid lizard *Cnemidophorus sexlineatus* and in unisexual *C. tesselatus, J. Exp. Zool.* **166**:137–150.

Maslin, T. P., 1968, Taxonomic problems in parthenogenetic vertebrates, *Syst. Zool.* **17**:219–231.

Maslin, T. P., 1971, Parthenogenesis in reptiles, *Am. Zool.* **11**:361–380.

Maynard Smith, J., 1971, What use is sex?, *J. Theor. Biol.* **30**:319–335.

Maynard Smith, J., 1976, A short term advantage for sex and recombination through self-competition, *J. Theor. Biol.* **63**:245–258.

Maynard Smith, J., 1978, *The Evolution of Sex,* Cambridge University Press.
Mayr, E., 1963, *Animal Species and Evolution,* Belknap Press, Cambridge, Massachusetts.
McKay, F. E., 1971, Behavioral aspects of population dynamics in unisexual–bisexual *Poeciliopsis* (Pisces: Poeciliidae), *Ecology* **52**:778–790.
McKinney, C. O., Kay, F. R., and Anderson, R. A., 1973, A new all-female species of the genus *Cnemidophorus, Herpetologica* **29**:361–366.
Menzel, B. W., and Darnell R. M., 1973, Morphology of naturally occurring triploid fish related to *Poecilia formosa, Copeia* **1973**:350–352.
Meyer, H., 1938, Investigations concerning the reproductive behavior of *Mollienesia formosa, J. Genet.* **36**:329–366.
Miller, R. S., 1967, Pattern and process in competition, *Adv. Ecol. Res.* **4**:1–74.
Moore, W. S., 1975, Stability of small unisexual–bisexual populations of *Poeciliopsis* (Pisces: Poeciliidae), *Ecology* **56**:791–808.
Moore, W. S., 1976, Components of fitness in the unisexual fish *Poeciliopsis monacha-occidentalis, Evolution* **30**:564–578.
Moore, W. S., 1977*a,* A histocompatibility analysis of inheritance in the unisexual fish *Poeciliopsis 2 monacha-lucida, Copeia* **1977**(2):213–223.
Moore, W. S., 1977*b,* An evaluation of narrow hybrid zones in vertebrates, *Q. Rev. Biol.* **52**:263–277.
Moore, W. S., and Eisenbrey, A. B., 1979, The population structure of an asexual vertebrate, *Poeciliopsis 2 monacha-lucida* (Pisces: Poeciliidae), *Evolution* **33(2)**:563–578.
Moore, W. S., and Hines, W. G. S., 1981, Sex in random environments, *J. Theor. Biol.* **92**:301–316.
Moore, W. S., and McKay, F. E., 1971, Coexistence in unisexual–bisexual species complexes of *Poeciliopsis* (Pisces: Poeciliidae), *Ecology* **52**:791–799.
Moore, W. S., and Vrijenhoek, R. C., 1977, The population structure of an asexual vertebrate: *Poeciliopsis 2 monacha-lucida,* American Society of Ichthyol. Herpetol. Annual Meeting Program, Gainesville, Florida.
Moore, W. S., Miller R. R., and Schultz, R. J., 1970, Distribution, adaptation and probable origin of an all-female form of *Poeciliopsis* (Pisces: Poeciliidae) in northwestern Mexico, *Evolution* **24**:789–795.
Muller, H. J., 1964, The relation of recombination to mutational advance, *Mutat. Res.* **1**:2–9.
Neaves, W. B. 1969. Gene dosage at the lactate dehydrogenase b locus in triploid and diploid lizards (Cnemidophorus), *Science* **160**:557–559.
Palma-Otal, M., W. S. Moore, and R. P. Adams, 1983, Morphological Chemical and Biogeographical analyses of a hybrid zone involving *Juniverus virginiana* (L.) and *J. horizontalis* (Moench) in Wisconsin, *Can. J. Botany* **61**:(in press).
Parker, E. D., 1979, Ecological implications of clonal diversity in parthenogenetic morphospecies, *Am. Zool.* **19**:753–762.
Parker, E. D., and Selander, R. K., 1976, The organization of diversity in the parthenogenetic lizard *Cnemidophorus tesselatus, Genetics* **83**:791–805.
Prehn, L. M., and Rasch, E. M., 1969, Cytogenetic studies of *Poecilia* (Pisces). I. Chromosome numbers of naturally occurring poecliid species and their hybrids from eastern Mexico, *Can. J. Genet. Cytol.* **11**:888–895.
Rasch, E. M. and Balsano, J. S., 1973, Cytogenetic studies of Poecilia (Pisces). III. Persistence of triploid genomes in the unisexual progeny of triploid females associated with Poecilia, *Copeia* **1973**:810–813.
Rasch, E. M., and Balsano, J. S., 1974, Biochemical and cytogenetic studies of *Poecilia* from eastern Mexico. II. Frequency, perpetuation, and probable origin of triploid genomes in females associated with *P. formosa, Rev. Biol. Trop.* **21**:351–381.
Rasch, E. M., Darnell, R. M., Kallman K. D., and Abramoff, P., 1965, Cytophotometric

evidence for triploidy in hybrids of the gynogenetic fish, *Poecilia formosa, J. Exp. Zool.* **160:**155–170.

Rasch, E. M., Prehn, L. M., and Rasch, R. W., 1970. Cytogenetic studies of *Poecilia* (Pisces). 11. Triploidy and DNA levels in naturally occurring populations associated with the gynogenetic teleost *Poecilia formosa* (Girard), *Chromosoma (Berl.)* **31:**18–40.

Remington, C. L., 1968, Suture zones of hybrid interaction between recently joined biotas, in: *Evolutionary Biology,* Vol. 2 (T. Dobzhansky, M. K. Hecht, and W. C. Steere, eds.), Plenum Press, New York, pp. 321–428.

Roughgarden, J., 1979, *Theory of Population Genetics and Evolutionary Ecology: An Introduction,* Macmillan, New York.

Schall, J. J., 1978, Reproductive strategies in sympatric whiptail lizards (*Cnemidophorus*), two parthenogenetic and three bisexual species, *Copeia* **1977:**108–116.

Schultz, R. J., 1961, Reproductive mechanism of unisexual and bisexual strains of the viviparous fish *Poeciliopsis, Evolution* **25:**302–325.

Schultz, R. J., 1966, Hybridization experiments with an all-female fish of the genus *Poeciliopsis, Biol. Bull.* **130:**415–429.

Schultz, R. J., 1967, Gynogenesis and triploidy in the viviparous fish *Poeciliopsis, Science* **157:**1564–1567.

Schultz, R. J., 1969, Hybridization, unisexuality, and polyploidy in the teleost *Poeciliopsis* (Poeciliidae) and other vertebrates, *Am. Nat.* **103:**605–619.

Schultz, R. J., 1971, Special adaptive problems associated with unisexual fishes, *Am. Zool.* **11:**351–360.

Schultz, R. J., 1973, Unisexual fish: Laboratory synthesis of a species, *Science* **179:**180–181.

Schultz, R. J., 1977, Evolution and ecology of unisexual fishes, in: *Evolutionary Biology,* Vol. 10 (M. K. Hecht, W. C. Steere, and B. Wallace, eds.), Plenum Press, New York, pp. 277–331.

Schultz, R. J., Kallman, K. D., 1968, Triploid hybrids between the all-female teleost *Poecilia formosa* and *Poecilia sphenops, Nature* **219:**280–282.

Seiler, J., 1943, Über den ursprung der Parthenogenese und Polyploidie bei Schmetterlingen, *Arch. Klaus-Stift. Vererb.-Forsch.* **18:**691–699.

Seiler, J., 1946, Die Verbreitungsgebiete der verschiedenen Rassen von *Solenobia triquetrella* (Psychidae) in der Schweiz, *Rev. Suisse Zool.* **53:**529–533.

Short, L. L., 1965, Hybridization in the flickers (*Colaptes*) of North America, *Bull. Am. Mus. Nat. Hist.* **129:**307–428.

Simpson, E. H., 1949, Measurement of diversity, *Nature* **163:**688.

Stebbins, G. L., 1950, *Variation and Evolution in plants,* Columbia University Press, New York.

Stebbins, G. L., 1966, Chromosomal variation and evolution, *Science* **152:**1463–1469.

Suomalainen, E., 1953, Die Polyploidie bei den parthenogenetischen Rüsselkäfern, *Zool. Anz. Suppl.* **17:**280–289.

Thibault, R. E., 1974*a,* The ecology of unisexual and bisexual fishes of the genus *Poeciliopsis:* A study in niche relationships, Ph. D. dissertation, University of Connecticut.

Thibault, R. E., 1974*b,* Genetics of cannibalism in a viviparous fish and its relationship to population density, *Nature* **251:**138–140.

Thibault, R. E., 1978, Ecological and evolutionary relationships among diploid and triploid unisexual fishes associated with the bisexual species, *Poeciliopsis lucida* (Cyprinodontiformes: Poeciliidae), *Evolution* **32:**613–623.

Thibault, R. E. and R. J. Schultz, 1978, Reproductive adaptations among viviparous fishes (Cyprinodontiformes: Poeciliidae), *Evolution* **32:**320–333.

Tunner, H. G., 1970, Das serumeiweisbild einheimischer wasserfrosche und der hybridcharakter von *Rana esculenta, Verh. Dsch. Zool. Ges.* **64:**352–358.

Tunner, H. G., 1972, Serologische and morphologische untersuchungen zur frage der artabgrenzung bei wasserfroschen aus der umgebung von Mainz (Rhein-Maingebert), *Z. Zool. Syst. Evolutionsforsch.* **10**:127–132.

Tunner, H. G., 1973, Das albumin und andere bluteiweisse bei *Rana ridibunda* Pallas, *Rana lessonae* Camerano, *Rana esculenta* Linne und deren hybriden, *Z. Zool. Syst. Evolutions forsch.* **11**:219–235.

Turner, B. J., Brett, B. H., and Miller, R. R., 1980*a*, Interspecific hybridization and the evolutionary origin of a gynogenetic fish, *Poecilia formosa, Evolution* **34**(5):917–922.

Turner, B. J., Brett, B. H., Rasch, E. M., and Balsano, J. S., 1980*b*, Evolutionary genetics of a gynogenetic fish, *Poecilia formosa,* the Amazon Molly, *Evolution* **34**(2):246–258.

Uzzell, T. M., 1963, Natural triploidy in salamanders related to *Ambystoma jefferisonianum, Science* **39**:113–115.

Uzzell, T. M., 1964, Relations of the diploid and triploid species of the *Ambystoma jeffersonianum* complex (Amphibia, Caudata), *Copeia* **1964**:257–300.

Uzzell, T. M., 1969, Notes on spermatophore production by salamanders of the *Ambystoma jeffersonianum* complex, *Copeia* **1969**:602–612.

Uzzell, T. M., and Berger, L., 1975, Electrophoretic phenotypes of *Rana ridibunda, Rana lessonae,* and their hybridogenetic associate *Rana esculenta, Proc. Acad. Nat. Sci. Phila.* **127**:13–24.

Uzzell, T. M., and Darevsky, I. S., 1975, Biochemical evidence for the hybrid origin of the parthenogenetic species of the *Lacerta saxicola* complex (Sauria: Lacertidae), with a discussion of some ecological and evolutionary implications, *Copeia* **1975**:204–222.

Uzzell, T. M., and Goldblatt, S. M., 1967, Serum proteins of salamanders of the *Ambystoma jeffersonianum* complex, and the origin of the triploid species of this group, *Evolution* **21**:345–354.

Uzzell, T. M., Berger, L., and Gunther, R., 1975, Diploid and triploid progeny from a diploid female of *Rana esculenta* (Amphibia Salientia), *Proc. Acad. Nat. Sci. Phila.* **127**:81–89.

Uzzell, T. M., Gunther, R., and Berger, S., 1977, *Rana ridibunda* and *R. esculenta:* A leaky hybridogenetic system (Amphibia Salientia), *Proc. Acad. Nat. Sci. Phil.* **128**:147–171.

Vanzolini, P. E., 1978, Parthenogenetic lizards, *Science* **201**:1152.

Vrijenhoek, R. C., 1972, Genetic relationships of unisexual-hybrid fishes to their progenitors using lactate dehydrogenase isozymes as gene markers (*Poeciliopsis,* Poeciliidae), *Am. Nat.* **106**:754–766.

Vrijenhoek, R. C., 1978, Coexistence of clones in a heterogeneous environment, *Science* **199**:549–552.

Vrijenhoek, R. C., 1979, Factors affecting clonal diversity and coëxistence, *Am. Zool.* **19**:787–797.

Vrijenhoek, R. C., and Schultz, R. J., 1974, Evolution of a trihybrid unisexual fish (*Poeciliopsis, Poeciliidae), Evolution* **28**:306–319.

Vrijenhoek, R. C., Angus, R. A., and Schultz, R. J., 1977, Variation and heterozygosity in sexually vs. clonally reproducing populations of *Poeciliopsis, Evolution* **31**:767–781.

Vrijenhoek, R. C., Angus, R. A., and Schultz, R. J., 1978, Variation and clonal structure in a unisexual fish, *Am. Nat.* **112**:41–55.

White, M. J. D., 1970, Heterozygosity and genetic polymorphism in parthenogenetic animals, in: *Essays in Evolution and Genetics in Honor of T. Dobzhansky* (M. K. Hecht and W. C. Steere, eds.), pp. 237–262.

Wilbur, H. M., 1971, The ecological relationship of the salamander *Ambystoma laterale* to its all-female, gynogenetic associate, *Evolution* **25**:168–179.

Williams, G. C., 1975, *Sex and Evolution,* Princeton University Press, Princeton, New Jersey.

Wright, J. W., 1978, Parthenogenetic lizards, *Science* **201**:1152–1154.

Wright, J. W., and Lowe, C. H., 1968, Weeds, polyploids, parthenogenesis and the geographical and ecological distribution of all-female species of *Cnemidophorus*, *Copeia* **1968:**128–138.

Zar, J. H., 1974, *Biostatistical Analysis,* Prentice-Hall, Englewood Cliffs, New Jersey.

Zweifel, R. G., 1965, Variation in and distribution of the unisexual lizard, *Cnemidophorus tesselatus, Am. Mus. Novit.* **2235:**1–49.

CHAPTER 8

The Evolution of Clonal Diversity in Poeciliopsis

ROBERT C. VRIJENHOEK

> I am sure that many readers have already concluded that I really do not understand the role of sex in . . . evolution. At least I can claim, on the basis of the conflicting views in the recent literature, the consolation of abundant company. Clearly the contest of ideas on these fundamental problems has only just begun. History has afforded a rare opportunity to ardent participants and alert spectators in the years ahead (Williams, 1975, p. 169).

1. Introduction

Despite the elegant body of theoretical literature on sexuality, recently reviewed by Williams (1975) and Maynard Smith (1978), very little empirical data exist to assess this "contest of ideas." I will not attempt in this chapter to address the question, "Why is there sex?" Instead, I will examine a common assumption in most of these theoretical studies—that asexual populations lack genetic variation. It is commonly argued that asexual reproduction is an evolutionary dead end, that the absence of recombinational variability results in genetic inflexibility and sure extinction in a changing environment. The clonal genomes of asexual organisms are looked upon as rigid structures that can only change through mutation, and most mutations are deleterious (Muller, 1964). Despite the occasional success of some asexual populations, the long-term prospects appear grim (White, 1978). Yet, recent genetic studies have reported abundant clonal

ROBERT C. VRIJENHOEK • Department of Biological Sciences and Bureau of Biological Research, New Brunswick, New Jersey 08903.

diversity in a number of asexually reproducing organisms [see reviews by Parker (1979*a*), Vepsalainen and Jarvinen (1979), and Vrijenhoek (1979*a*)]. What then is the source of this clonal diversity and how might it contribute to the ecological and evolutionary success of asexual populations?

The existence of multiclonal asexual populations is in itself problematic. It seems inescapable that asexual populations should consist, at any one time, of only the single "best-adapted" clone, since coexistence of many genetically similar clones should be prevented by competitive exclusion (Crow and Kimura, 1965). If the differences among clones are adaptively neutral, random processes in finite populations should lead to clonal drift and a concomitant loss of variation (Maynard Smith, 1978). Even with different clones adapted to different patches in a heterogeneous environment, clonal diversity should decay gradually in finite populations (Maynard Smith, 1978). How then is clonal diversity maintained?

Students of unisexual fish have been afforded one of these "rare opportunities" to examine sexual and asexual reproductive processes in both experimental and natural settings. This chapter will address the questions raised above by reviewing the current status of our ongoing research into the roles that polyphyletic hybrid origins, mutation, and recombination play in generating clonal diversity in all-female populations of fish in the genus *Poeciliopsis* (Poeciliidae). The potential contributions of these phenomena to the evolution of niche-width variation in asexual populations will be discussed.

Sexually reproducing species of *Poeciliopsis* are dioecious, with males generally smaller than mature females. Formal genetic studies reveal that they employ conventional forms of recombination: Mendelian segregation and assortment, crossing over, and syngamy (Leslie, 1982). Several sexual species, living in northwestern Mexico, hybridized with *Poeciliopsis monacha* and gave rise to a series of diploid and triploid, all-female, hybrid biotypes (Table I). The hyphenated names of each of the biotypes reflects

Table I
The Unisexual Forms of *Poeciliopsis:* Their Sexual Hosts and Reproductive Modes

Unisexual biotype	Ploidy	Sexual host	Reproductive mode
P. monacha-lucida	2*n*	*P. lucida*	Hybridogenetic
P. monacha-occidentalis	2*n*	*P. occidentalis*	Hybridogenetic
P. monacha-latidens	2*n*	*P. latidens*	Hybridogenetic
P. 2 monacha-lucida	3*n*	*P. monacha*	Gynogenetic
P. monacha-2 lucida	3*n*	*P. lucida*	Gynogenetic
P. monacha-viriosa-lucida	3*n*	*P. viriosa*	Gynogenetic

its hybrid composition and genomic dosages (Schultz, 1969). For a detailed description of their hybrid origins, breeding systems, biogeography, and ecology see the review by Schultz (1977). Assignment of names to the hybrid biotypes does not imply that we consider them valid species. Furthermore, several hybrid biotypes arose polyphyletically (Schultz, 1973; Vrijenhoek *et al.*, 1977, 1978; Angus, 1980). The biological species concept (Mayr, 1963) is not applicable to asexual biotypes.

The three triploid biotypes are gynogenetic (Fig. 1), a strictly asexual form of reproduction. Fertilization of the eggs by sperm from a coexisting sexual species is required to activate embryogenesis, but the sperm contribute nothing genotypically or phenotypically to the offspring (Schultz, 1967). In these females, the chromosomes are doubled during an endomitotic division prior to meiosis. Pairing of replicated sister chromosomes preserves the maternal genotype (Cimino, 1972*a*). Their clonal pattern of inheritance was confirmed using biochemical genetic and tissue grafting studies (Vrijenhoek, 1972; Moore, 1977*a*).

The three diploid all-female biotypes are hybridogenetic, a semisexual reproductive mechanism (Schultz, 1969). During hybridogenetic oogenesis, only the haploid chromosome set derived from *P. monacha* (designated M in Fig. 1) is transmitted to the ova; the paternal set (designated L) is expelled from the oogonium, thereby precluding synapsis and potential recombination (Cimino, 1972*b*). The haploid *monacha* ova produced by these hybrids are fertilized by sperm from a coexisting host species (*P. occidentalis, P. lucida,* or *P. latidens*), reestablishing diploid hybrids that express morphological and electrophoretic traits encoded by both

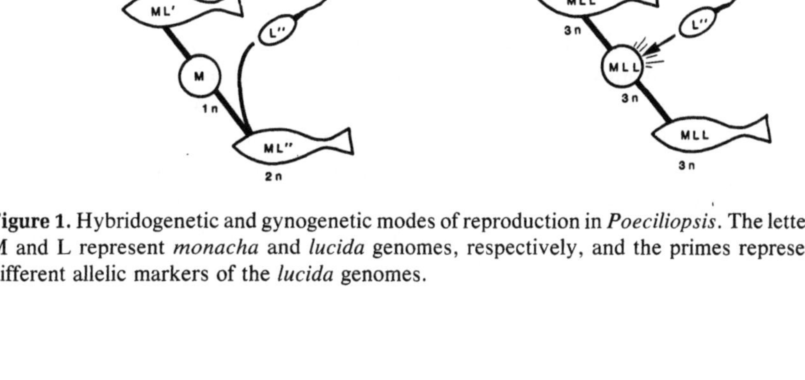

Figure 1. Hybridogenetic and gynogenetic modes of reproduction in *Poeciliopsis*. The letters M and L represent *monacha* and *lucida* genomes, respectively, and the primes represent different allelic markers of the *lucida* genomes.

parental genomes (Schultz, 1966; Vrijenhoek, 1972). Hybridogenetic strains are not clones in the strict sense because their paternal genomes are substituted in each generation; thus natural populations of hybridogens have access to all the allelic variation in the gene pool of the sexual host (Vrijenhoek *et al.*, 1977). Nevertheless, the *monacha* genome is inherited clonally; therefore discussions of clonal variation in the hybridogens refer only to the genotypic differences among the haploid *monacha* genomes, which we have previously referred to as "hemiclones" (Vrijenhoek *et al.*, 1977).

Unlike truly parthenogenetic species, gynogenetic and hybridogenetic *Poeciliopsis* can never escape from their sexual hosts to invade new habitats, nor can they competively exclude their hosts, for in doing so they lose their sperm source and ensure their own extinction. Sperm dependence undoubtedly plays an important role in the evolution of unisexual populations since they are forced to coexist and compete with sexual species with which they share considerable genetic, and presumably ecological, overlap. In attempts to model the population dynamics of the all-female sperm parasites and their sexual hosts, the two forms were treated as if they were simply a pair of alternative phenotypes with identical niche requirements (Moore and McKay, 1971; Moore, 1976). The model elucidated the sufficient conditions leading to a dynamic coexistence between unisexual and sexual biotypes: the twofold advantage of all-female reproduction is counterbalanced with low unisexual mating success, due to a preference for conspecific mates by males of the sexual host (see Moore, this volume, Chapter 7). Unfortunately, the model was developed prior to our knowledge of widespread clonal diversity. It cannot explain how numerous genetically distinct clones can coexist with one another and with their sexual host. Only intensive studies of phenotypic variation, niche relationships, and life histories of the participating sexual and clonal genotypes are likely to answer this question.

2. The Frozen Niche Variation Hypothesis

According to the reasoning of Roughgarden (1972), the "niche width" of a species should be composed of two components: (1) the within-phenotype component, due to the variety of resources used by each phenotype; and (2) the between-phenotype component, due to differences among phenotypes. As long as a population contains genetic variation for the between-phenotype component, cloning should lead to niche diversification more rapidly than sexual reproduction. Recombination of sexual genotypes can significantly retard the rate of adaptive evolution at gene loci involved in epistatic interactions (Eshel and Feldman, 1970). For *Poeciliopsis*, each hybridization between the sexual species *P. monacha* and *P. lucida*, or between *P. monacha* and *P. occidentalis*, isolates another *monacha* genome from the sexual gene pool. If the new hybrids prove fertile and hybridogenetic, they "freeze" whatever combinations of morphological, physiological, behavioral, and ecological characteristics the *monacha* genomes encode. Polyphyletic hybrid origins from a genetically variable sexual ancestor should give rise to an array of new clonal genomes, each with its unique temporal, spatial, and trophic niche requirements. This process may be advanced even further in the origins of new triploid clones from diploid ancestors.

The adaptive success of a unisexual *Poeciliopsis* population can be expressed as a function of its frequency relative to that of females of the sexual host species (Moore, 1976). Based on a study of these proportions, I proposed that the numerical success of unisexual populations depends largely on the opportunity for polyphyletic hybrid origins (Vrijenhoek, 1979*a*). In river systems where recurrent hybridizations can give rise to multiclonal unisexual populations, the unisexuals generally comprise from 30% to 80% of the *Poeciliopsis* females (Table II). Rivers lacking this

Table II

The Proportion of Unisexual Females Relative to the Total Number of Females in Unisexual/Bisexual Complexes of *Poeciliopsis monacha-occidentalis* and *Poeciliopsis monacha-lucida*

River systems[a]	Number of localities	Percentage unisexuals[b]	Range
Monoclonal localities			
Río de la Concepción	4	1.5 ± 1.1	0–3.0
Río Sonora	10	9.6 ± 9.9	0–30.0
Río Matape	3	6.0 ± 4.6	0–11.0
Río Yaqui	16	7.5 ± 10.0	0–38.0
Mean		7.2 ± 9.3	
Multiclonal localities			
Río Mayo	7	82.1 ± 10.6	65.0–97.0
Río Fuerte	4	63.8 ± 20.6	33.6–91.0
Río Sinaloa	2	45.2 ± 5.2	40.0–50.4
Río Mocorito	2	36.9 ± 5.3	31.6–42.2
Mean		66.3 ± 21.60	

[a]Rivers are listed in order from north to south. The first five contain *P. monacha-occidentalis* and the last four contain *P. monacha-lucida*.

[b]The percentage of unisexual females ± its standard deviation.

opportunity for generating clonal diversity have monoclonal unisexual populations that comprise less than 10% of the *Poeciliopsis* population. A multiclonal population that efficiently exploits the between-phenotype component of niche width should achieve significantly higher population densities than a monoclonal population. Furthermore, clonal variation will lead to a continuous evolutionary contest among clones. Interclonal selection should result in a highly structured unisexual population, as long as selection is not overridden by a high rate of clonal synthesis.

To examine the "frozen niche variation" hypothesis, we must question each of its assumptions: (1) What is the evidence for polyphyletic origins of clones? (2) Does ecologically relevant variation exist among clones? (3) Do the sexual ancestors contain sufficient genetic variation to drive this process?

2.1. Polyphyletic Hybrid Origins

Electrophoretic studies provided strong evidence for polyphyletic hybrid origins in natural populations of the hybridogenetic biotypes (Vrijenhoek *et al.,* 1977, 1978). The genetic basis for the allozyme variation used in these studies was established through the excellent formal genetic studies by Leslie (1982). Nearly all of the allozymes that distinguished among the numerous electromorph hemiclones (i.e., electrophoretically distinguishable clonal genomes) were also found in nearby populations of *P. monacha*. The most parsimonious way that so many alleles at numerous loci could repeatedly occur in various clonal combinations is through multiple hybrid events (Table III). Alternatively, one would have to postulate the occurrence of numerous electrophoretically parallel mutations in separate lineages derived from a single ancestral hemiclone, an unlikely prospect, indeed. The existence of several endemic hemiclones carrying local *monacha* alleles (VII, VIII, X, XII, XIII, and XIV) provides the strongest evidence for polyphyletic hybrid origins.

Corroboration of the polyphyletic origins hypothesis was provided in the elegant tissue grafting experiments by Angus and Schultz (Angus and Schultz, 1979; Angus, 1980). Because histocompatibility systems in vertebrates involve many alleles at many gene loci, multiple hybrid events should freeze greatly dissimilar histocompatibility genotypes in various hemiclones. Tissue grafts between polyphyletic clones should result in vigorous immune responses, due to "major" differences in the histocompatibility genotypes. If hemiclonal differences arose through small, progressive mutational changes within ancestral hemiclones, one would predict the kind of chronic graft rejections associated with "minor" histocompatibility differences. With only a single exception, grafts be-

Table III
Allelic Variation among Hemiclonal *monacha* Genomes of *Poeciliopsis monacha-lucida*

	Diagnostic loci[a]								
Hemiclone	*Es-4*[b]	*Es-5*[b]	*Ldh-1*[b]	*Aat-3*[b]	*Pgm*	*Pgd*	*Mp-3*	*Mdh-2*	*Mdh-1*
I	*b*	*f*	*b*	*d*	*d*	*a*	*a*	*a*	*a*
II	*b*	*c*	*b*	*b*	*d*	*a*	*a*	*a*	*a*
III	*b*	*c*	*b*	*a*	*d*	*a*	*a*	*a*	*a*
IV	*b*	*d*	*b*	*b*	*d*	*a*	*a*	*a*	*a*
V	*b*	*f*	*b*	*a*	*d*	*a*	*a*	*a*	*a*
VI	*b*	*d*	*b*	*a*	*d*	*a*	*a*	*a*	*a*
VII	*b*	*e*	*a*	*b*	*d*	*a*	*b*	*a*	*a*
VIII	*b*	*d*	*a*	*b*	*d*	*c*	*a*	*a*	*a*
IX	*b*	*e*	*b*	*b*	*d*	*a*	*a*	?	*a*
X	*b*	*g*	*b*	*b*	*d*	*a*	*a*	*a*	*a*
XI	*b*	*e*	0	*b*	*d*	*a*	*a*	*a*	*a*
XII	*c*	*d*	*b*	*b*	*d*	*a*	*a*	*a*	*b*
XIII	*c*	*f*	*b*	*b*	*d*	*a*	*a*	*a*	*a*
XIV	*b*	*c*	*b*	*b*	*e*	*a*	*a*	*a*	*a*

[a]Twenty-five loci were examined; nine were polymorphic among hemiclones.
[b]The first four loci carry common *monacha* alleles that occur in two or more hemiclones.

tween distinct electromorph hemiclones were vigorously rejected. These studies provided another important lesson for students of asexual populations. Electrophoresis seriously underestimated the levels of clonal diversity in localities where polyphyletic origins are possible. For example, in the headwaters of the Río Mayo, *P. monacha* segregates for two of the 25 electromorph markers we monitored (Vrijenhoek, 1979*b*). With two alleles at each locus, four *monacha* haplotypes are possible. Only three haplotypes have been found in these unisexuals, but they comprise at least 17 histocompatibility hemiclones (Angus, 1980). In river systems where *P. monacha* is absent, and polyphyletic origins of hybridogens presently are not possible, clonal diversity is low (Vrijenhoek, 1979*a*). Such localities contained a single electromorph clone and tissue grafting failed to reveal additional differentiation (Angus, 1980).

2.2. Ecological Studies of Clones

The first evidence for ecologically relevant differences among co-occurring clones was found in a study of the triploid gynogenetic biotype *P. 2 monacha-lucida*. Clones I and II differ electrophoretically and immunologically (Vrijenhoek, 1978; Moore and Eisenbrey, 1979). A morphological study revealed no significant differences among them, except for trophic characters, which suggested that coexistence among these two clones and their sexual host, *P. monacha,* might be aided by resource partitioning (Vrijenhoek, 1978). The three forms inhabit the small headwater streams of the Río del Fuerte, wherein they segregate on a productivity gradient. *Poeciliopsis monacha* dominates shaded, rocky habitats of low productivity and clone II reaches its highest frequencies in open, sunny areas with high productivity. The inverse numerical relationship between clone II and *P. monacha* suggested a negative competitive interaction between the two forms. The frequency of clone I was relatively low in all the habitats and independent of the other two.

Since the 1978 study, we have begun detailed analyses of the microhabitat distributions, natural food habits, and foraging behaviors of the unisexual and sexual forms inhabiting these streams. Much of this work is currently in progress as portions of various thesis projects undertaken by students in my laboratory. I will briefly summarize the salient points that have emerged. Russell Schenck is examining samples of fish taken over a 12-month period (March 1980–March 1981) from a series of pools in the Arroyo de los Platanos of the Río del Fuerte (Vrijenhoek, 1978, Fig. 1). Each pool was sampled four times at approximately 3-month intervals. We chose pools that offered a diversity of feeding sites. Some fish were observed "drift-feeding," that is, they oriented toward the incoming current and darted after food that washed by. Other fish were feeding in the deeper portions of the pools, on the detritus-covered bottom and among the macrophytes. We will refer to them as "pool-feeders" in contrast to the "drift-feeders." Samples of fish occupying each feeding mode were taken separately, frozen in the field, and sorted electrophoretically in the laboratory.

We originally expected that spatial and dietary overlap would be low near the end of the dry season, when fish are densely crowded into small residual pools. Such competitive displacement should be relaxed during the rainy season, when more space, and presumably other resources, would be available. Zaret and Rand (1971) described such a pattern of niche shifts in small, tropical freshwater fish in Costa Rica. The samples of *Poeciliopsis* examined by Schenck do not show this pattern. Instead, a remarkably consistent pattern of habitat preferences occurred over the 12-month period (Table IV). The two triploid clones are significantly more frequent in the drift-feeding mode than in the pool-feeding mode. Contrastingly, females of the sexual host species, *P. monacha,* are more frequent in the pool-feeding mode. It is possible that the *lucida* genome contained in the *P. 2 monacha-lucida* triploids imparts an inflexible preference for feeding in slight currents rather than pools. The *P. lucida* tend

Table IV
Proportions (in percent) of Sexual and Asexual Forms Taken in the Pool and Drift Feeding Modes in the Arroyo de los Platanos[a]

	Monacha females	Clone I	Clone II	*Monacha* males	Sample size	χ^2
June 1980						
Drift	59.3	20.4	6.5	13.9	108	15.23
Pool	69.7	7.9	1.3	21.1	152	[b]
Difference[c]	−10.4	+12.9	+ 5.2	− 7.2		
October 1980						
Drift	41.3	36.4	15.2	7.1	184	17.39
Pool	59.3	27.0	6.2	7.5	226	[b]
Difference[c]	−18.0	+ 9.4	+ 9.0	− 0.4		
January 1981						
Drift	57.4	26.1	5.9	10.6	357	34.27
Pool	75.9	12.8	3.9	7.4	486	[b]
Difference[c]	−18.5	+13.3	+ 2.0	+ 3.2		
March 1981						
Drift	59.3	29.9	3.7	7.1	351	51.61
Pool	80.3	12.5	1.5	5.7	543	[b]
Difference[c]	−21.0	+17.4	+ 2.2	+ 1.4		

[a] χ^2 contingency tests were performed on the raw numbers to test the hypothesis that feeding mode is independent of biotype.
[b] All χ^2 values are highly significant.
[c] The difference is equal to the frequency of a particular form in the drift sample minus its frequency in the pool sample.

to occupy sandy-bottomed streams rather than the headwater pools preferred by *P. monacha*. Direct observations of individually marked fish in natural habitats revealed that aggressive interactions and interference play a significant role in establishing this pattern of habitat separation (R. Schenck, unpublished).

The analysis of gut contents is incomplete, but some patterns are beginning to emerge. Schenck finds that *Poeciliopsis* are generally omnivorous, but in any particular habitat *P. monacha* always takes more insects and *P. lucida* eats more detritus. For the three members of the *monacha* complex inhabiting the Arroyo de los Platanos, *P. monacha*, clone I, and clone II have indistinguishable gut contents when taken together in the drift-feeding mode. This result is expected if drift-feeders do not choose food items but take food in whatever proportions wash by. However, when the three forms are taken together in the pool-feeding mode, the triploids obtain significantly fewer insects than *P. monacha*. The triploids' increased success in capturing insects in the drift-feeding mode suggests one possible explanation for their higher frequencies in areas having a current. Near the end of the dry season, in some residual pools that offer no current, both *P. monacha* and clone II tend to be insectivorous. Dietary overlap is often higher between these two forms than between either one of them and clone I (Horn's index, based on volumetric estimates of invertebrate taxa and algal genera). This finding is concordant with the distributional data in which *P. monacha* and clone II tend to displace one another whereas the frequency of clone I is independent of the other two. However, patterns of dietary overlap can change over short distances and time spans. They are tied to the diversity of resources available in a pool at any point in time.

The food habits in the field are matched by predatory behaviors in the laboratory. Kurt Spindler examined handling times and capture rates with small prey (*Artemia salina*) and large prey (chironimid larvae). Again, *P. monacha* and clone II were nearly indistinguishable, regardless of the size of the prey. Clone I individuals were more variable. Smaller fish have very poor handling times, but larger ones are roughly equivalent to *P. monacha* and clone II. We also observed foraging behaviors in fish that had been denied prey items and commercial fish foods for at least 3 weeks prior to the experiment. Clone I directs approximately 30% of its feeding pecks to scraping solid surfaces (rocks and aquarium walls), while clone II directs less than 5% of its pecks to scraping (based on four cumulative hours of observations of six fish of each clone, these differences are significant at the 0.05 level). Clone II more commonly foraged near the water's surface.

Schenck, Spindler, and Vrijenhoek are also studying *P. lucida* and two hybridogentic clones, VII and VIII, with which it coexists in the Arroyo de Jaguari of the Río del Fuerte. A remarkably similar picture is emerging. Both *P. lucida* and clone VII are predominantly detritivorous; their prey handling times are very poor, and they tend to partially displace one another within these streams. Frequencies of clone VIII tend to be independent of the other two. It is largely insectivorous and has a prey handling time at least as fast as that of *P. monacha*.

An interesting and unanticipated result occurred in the foraging behavior experiments. The observations took place over a 30-day period. Thus, these fish had been denied commercial fish food and prey items for at least 50 days. They were forced to survive on whatever plant materials, microinvertebrates, and detritus accumulated in the experimental aquaria. After resumption of regular feedings, the six clone I individuals regained a healthy condition and began to reproduce. The six clone II individuals did not. Four died within 2 weeks and the two survivors never reproduced. Similar results occurred in my earlier foraging behavior experiments (Vrijenhoek, 1978). It is possible that clone II requires a higher quality, insectivorous diet to survive and reproduce. Such a requirement might explain their increased frequency in highly productive, sunny habitats and their scarcity in shady, unproductive arroyos (Vrijenhoek, 1978). We are currently undertaking laboratory experiments aimed at quantifying the differential growth rates, reproduction, and survivorships of these two clones and *P. monacha* when subjected to differing dietary regimes.

Bulger and Schultz (1979) reported differential survival for a pair of *P. monacha-2 lucida* clones inhabiting a different tributary of the Río del Fuerte. When subjected to thermal stresses, clone I was better able to withstand heat shocks and clone II was better able to survive cold shocks. Differential survival in the summer and winter might contribute to the coexistence of these two clones. Life history studies of these clones are underway in R. J. Schultz's laboratory. To date, all the detailed ecological studies of *Poeciliopsis* clones have revealed differences that might contribute to the spatial and temporal partitioning of resources and to the differential survival and reproduction in these heterogeneous environments.

2.3. Variation in the Sexual Ancestors

The "frozen niche variation" hypothesis assumes that the sexual ancestors of the all-female populations contain sufficient between-phenotype genetic variation to drive clonal differentiation. This capacity must reside in *P. monacha,* the common ancestor for all the unisexual biotypes (Table I). It appears to be a habitat specialist and food generalist. It is generally restricted to bedrock outcroppings in a few small streams and permanent springs in the headwater tributaries of the Ríos Mayo, Fuerte, and Sinaloa (Miller, 1960), where it is often the only sexually reproducing species of fish. Yet, within these restricted habitats, it has an omnivorous diet, ranging from snails and insect larvae to algae and detritus (Vrijenhoek, 1978; Schenck, unpublished).

Poeciliopsis monacha exhibits substantial phenotypic variation that could contribute to clonal diversity. Despite its limited distribution, it is one of the most polymorphic sexual species of *Poeciliopsis* in electrophoretic studies (Vrijenhoek, 1979*a,b*). Yet enzymatic variability might not be predictive of variability at other loci (Lewontin, 1974). Morphological variation was examined for 10 characters in 30 adult females of *P. monacha* and *P. lucida* and compared to four asexual forms: *P. monacha-lucida* hemiclones VII and VIII, and *P. 2 monacha-lucida* clones I and II (Table V). All were collected at the same time from the same series of pools in the Arroyo de Jaguari of the Río del Fuerte. The morphological subsample comprised adults in the 25–35 mm range; there were no differences in the means or variances of standard length between taxa within this subset. A comparison of the coefficients of variation of these char-

Table V
Coefficients of Variation for 11 Meristic Traits in Sexual and Clonal Forms of *Poeciliopsis*

			monacha-lucida		*2 monacha-lucida*	
Trait	*monacha*	*lucida*	VII	VIII	I	II
Scales						
Lateral-line	2.6	2.5	1.8	1.7	1.9	2.3
Dorsal rows[a]	33.0	12.5	10.2	9.9	9.1	9.0
Fin rays						
Caudal	3.3	5.2	2.8	2.7	1.8	1.9
Dorsal	0.2	4.1	0	0	0	0
Anal	2.8	1.5	0	0	0	0
Pectoral	3.9	4.4	0	0	0	0
Pelvic	1.2	0	0	0	0	0
Vertebrae						
Abdominal	2.0	3.4	3.3	3.2	2.5	2.9
Caudal	0.3	2.0	0	0	0	0
Dentary teeth						
Inner	48.1	47.9	28.4	33.0	16.4	36.1
Outer	14.2	13.5	9.7	10.3	2.3	7.2
Rank sum[b]	57	58	33	31	23	29

[a]Dorsolateral scale row count (Vrijenhoek and Schultz, 1974).
[b]The CVs in each row were ranked and summed over the columns. H_0: There are no differences among the six forms; Freidman's rank $\chi^2 = 29.6$ (df $= 5$); $P < 0.01$; reject. H_1: There are no differences between the mean CVs for the sexuals versus the unisexuals. Wilcoxan's signed rank test, $P < 0.001$; reject. H_2: There are no differences between the mean CVs for diploid versus triploid unisexuals. Wilcoxan's signed rank test; not significant.

acters reveals that the sexual species, *P. monacha* and *P. lucida,* are substantially more variable than any of the unisexual forms. The discovery of greater phenotypic variation in sexual populations is not surprising given the opportunity for additive genetic variation, which is lacking in true clones (cf. Atchley, 1977; Parker, 1979*b*). The hybridogens, however, are hemiclonal, and the substitutable paternal genomes might contribute some of the variability that resides in the *lucida* population. The gynogen, electromorph clone I, is a true clone; tissue grafting studies by Moore and Eisenbrey (1979) revealed that it contains no hidden histocompatibility variation. Electromorph clone II contains some minor histocompatibility variation apparently due to mutational differences that have accumulated since the origin of this clone (Eisenbrey and Moore, 1981). Whether this mutational variation contributes to phenotypic or ecological diversity in clone II is not known.

A parallel study of the *P. occidentalis* and *P. monacha-occidentalis* also revealed significantly greater phenotypic variation in the sexual species in comparisons over five river systems. The multivariate coefficient of variation over 28 size-corrected mensural characters was consistently higher in the sexual populations (Michael Douglas, personal communication).

2.4. Stability of the Phenotype

A potential error exists in concluding that the increased variation in the sexual forms is due to additive genetic sources alone, because the sexual and unisexual forms might not respond to environmental sources of variation in the same way. Following the logic of Lerner (1954), it is possible that the reduced variation in the unisexual hybrids simply reflects greater developmental homeostasis, a byproduct of their hybridity. To test this hypothesis, we undertook comparisons of fluctuating asymmetry in *P. monacha* and *P. 2 monacha-lucida* clone I (Vrijenhoek and Lerman, 1982). Fluctuating asymmetry reflects random accidents of development and as such provides a measure of the developmental stability (Van Valen, 1962). On average, for 10 independently varying bilateral characters, the highly heterozygous clonal form is no more homeostatic than a typical outcrossed population of *P. monacha*. However, a nearby population of *P. monacha* with very low heterozygosity due to a recent founder event had a highly unstable phenotype. Developmental stability probably reaches a plateau in relatively large outcrossing sexual populations and additional heterozygosity resulting from interpopulational or interspecies hybrids is unlikely to produce further gains.

The high phenotypic variability of the sexual forms is due primarily to additive genetic variation rather than random developmental noise. Formal quantitative genetic studies of phenotypic variation in *P. monacha* would be desirable, but a more direct test of the "frozen niche variation" hypothesis is described in Section 2.5.

2.5. Synthetic Clones

The potential for polyphyletic hybrid origins in creating ecologically diverse clones can be tested in laboratory experiments. Schultz (1973) synthesized several new hybridogenetic strains through crosses of *P. monacha* females with *P. lucida* males. Significant differences in growth rates exist between two of the synthetic clones (Schultz, 1982). More recently, we have synthesized over 120 F_1 hybrids of *P. monacha* × *P. occidentalis,* using outbred *P. monacha* stocks from a variety of localities in the Ríos Mayo, Fuerte, and Sinaloa. All the hybrids have been females. The new hybrid strains will be used to directly examine the potential for "freezing" morphological, behavioral, and life-history variation in synthetic hemiclones. Before such comparisons can be made, the substitutable paternal genomes of these new hybridogens must be standardized. We are currently in the process of combining all the synthesized hemiclonal genomes with a standard paternal genome from an isogenic *lucida* strain (M61-31). Most of these attempts have proven unsuccessful. In many cases the F_1 *monacha-occidentalis* hybrids were sterile, often having no functional gonads or associated ducts. Apparently most *monacha* genomes are sufficiently compatible with *occidentalis* genomes to produce viable hybrids, but significantly fewer are capable of producing fertile hybrids. This variation in compatibilities suggest a first and most intense level of selection acting on potentially new hemiclones. Similar patterns of clonal selection were reported for *Drosophila mecatorum* (Templeton, 1979). Nevertheless, genetic analysis demonstrated that all fertile *monacha-occidentalis* hybrids have proven to be hybridogenetic. Unique *occidentalis* alleles at four enzyme loci and a locus determining a dorsal fin spot were all replaced by *lucida* alleles in the F_2 generation. Hybridogenesis is not an evolved mechanism. It is a spontaneous byproduct of the failure of conventional oogenetic processes. Detailed cytochemical studies of this meiotic failure are needed, for they may shed light on chromosome mechanics under normal control. We know even less about the origin of triploid gynogenetic clones (Cimino, 1972*b*).

3. Mutations and Muller's Ratchet Mechanism

Mutations also provide a source of clonal variation, yet one wonders how much they contribute to adaptive evolution in asexual populations. Much of the theoretical debate over the rates of evolution in sexual versus asexual lineages has been concerned with the relative significance of recurrent and novel mutations as sources of adaptive variation (Crow and Kimura, 1965, 1969; Maynard Smith, 1968, 1971; Felsenstein, 1974). But harmful mutations occur far more frequently than beneficial ones. Muller (1964) proposed the "ratchet mechanism" hypothesis to explain why genetic recombination was beneficial. Accordingly, asexual lineages progressively degrade through an accumulation of mutations that cannot be purged by the recombinational processes characteristic of sexual lineages. The following development of this hypothesis derives from Maynard Smith's (1978, pp. 33–36) exploration of the requirements for Muller's ratchet.

The total rate at which new mutations accumulate is a product of the per gene mutation rate u and the number of gene loci l. In *Drosophila melanogaster*, the total mutation rate $U = ul$ for polygenes with small viability effects might be as high as 35% of the gametes (Mukai *et al.*, 1972). The Poisson probability of producing an optimal gamete (having no new deleterious mutations) is 0.705 (Fig. 2). A bacterium having a genomic mutation rate of $U = 0.002$ (Strickberger, 1976) is much more likely ($P = 0.998$) to produce optimal offspring. However, vertebrates such as humans and *Poeciliopsis* have about 25 times as much DNA as *Drosophila* (Bachmann, 1972; Sparrow *et al.*, 1972). If only 10% of this DNA affects fitness, the total mutation rate probably exceeds one new mutation per gamete. Given $U = 1.0$, only 36.8% of the gametes are expected to have no new deleterious mutations.

Both sexual and asexual lineages initially free of deleterious genes will accumulate mutations at the same rate until an equilibrium between mutational gain and selective loss is reached (Muller, 1964). Should the selection intensity increase, a sexual lineage can purge all, or at least some, of its mutational load, since recombination is likely to produce some offspring that are relatively load-free. However, an asexual lineage

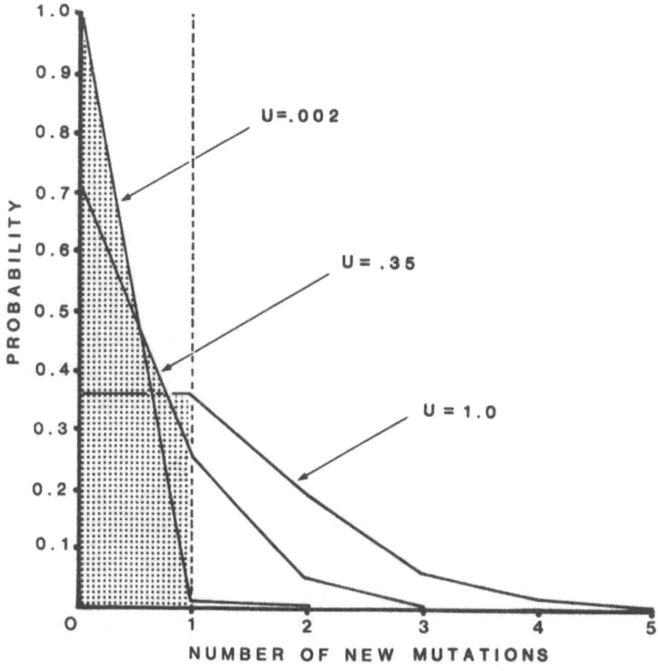

Figure 2. Poisson probabilities of acquiring one or more deleterious new mutations per gamete given various genomic mutation rates U.

at the mutation/selection equilibrium is subject to Muller's ratchet; its load of mutations cannot decrease below that already present in its "least loaded" clone. But it can increase if the "least loaded" clone is lost by chance. Then the ratchet has turned one notch and it cannot reverse, excepting back mutations, which are generally rare (Muller, 1964). The advance of the ratchet depends upon the total mutation rate U, the intensity of selection against such mutation(s), and the effective number of gametes N that establish the new generation. If the effective gamete number N is small, then the number of gametes falling in the optimal class will be even smaller, $Ne^{-U/s}$. This class of gametes might fail to establish progeny by chance alone. In higher organisms with large genomes and relatively small population sizes, Muller's ratchet will inevitably lead to genetic deterioration of clones (Maynard Smith, 1978). However, in prokaryotes, with few gene loci, low per gene mutation rates, and large effective population sizes, Muller's ratchet is probably unimportant. Muller originally proposed his ratchet mechanism for haploid asexual organisms, but it also applies to diploid organisms, as long as the mutations are expressed to some degree in the heterozygous condition. In general, most new mutations are not completely recessive (Muller, 1950; Crow and Temin, 1964).

3.1. Silent Mutations of Enzymes

Direct evidence for mutations in hemiclonal genomes was provided by the discovery of silent alleles at several loci encoding enzymes: *Es-5*, *Ldh-1*, and *Adh-2*. For each locus the unisexual hybrids express only the paternally derived *lucida* or *occidentalis* allozyme, rather than the typical codominant pattern found in nonmutant hybrid strains. Extensive population surveys of *P. monacha* failed to reveal the presence of the silent alleles in the sexual ancestor (Vrijenhoek, 1979*b*). The *Es-5*^{0} mutant has been studied most extensively. This locus encodes a carboxylesterase expressed predominantly in liver tissue (Leslie and Pontier, 1980). It resides within a linkage group (LG I) that contains much of the intra- and interspecific variation found in *Poeciliopsis* (Leslie, 1982).

In most *P. monacha-occidentalis* strains, the *Es-5* locus expresses a two-banded phenotype upon starch-gel electrophoresis. A fast-migrating allozyme is contributed by the *occidentalis* genome, while the slow-migrating allozyme is contributed by the *monacha* genome. Strain Ia(0) of *P. monacha-occidentalis* (from the Arroyo Altar of the Río de la Concepcíon) expresses only a single-banded phenotype corresponding to the fast *occidentalis* allozyme. The hemiclonal genome contributes nothing to the esterase phenotypes of these hybrids; it carries a silent allele (Vrijenhoek *et al.*, 1977). Using anti-esterase-5 antiserum, Spinella and Vrijenhoek (1982) were able to detect an enzymatically inactive protein product of the *Es-5*^{0} allele.

The silent *Es-5*° allele apparently arose by mutation from the *Es-5*^{f} allele carried by a hemiclone IIa(0)-type ancestor, which occurs in the next river to the South, the Río Sonora. Angus (1980) found that tissue grafts from "standard-bred" strains of Ia(0) to IIa(0) are accepted, but the reciprocal grafts are slowly rejected. These closely related strains have major histocompatibility differences with *P. monacha-occidentalis* strains from rivers to the south. The protein produced by the *Es-5*^{0} allele has undergone only slight mutational change, perhaps as little as a single amino acid substitution affecting its active site. Its mobility is not demonstrably different from that of the functional *Es-5*^{f}-coded allozyme of hemiclone IIa(0). The mutation also had little effect on the antigenicity of the silent protein, since it cross-reacts identically with antiserum directed against wild-type *Es-5*-coded allozymes. Similar investigations are underway with the *Ldh-1* and *Adh-2* silent alleles. An anti-LDH-1 has been produced and we are currently using it to probe for cross-reacting material in hemiclone XI of *P. monacha-lucida*.

3.2. Dominant and Recessive Lethals

Genetic dissection of hemiclonal *monacha* genomes from hybridogenetic *Poeciliopsis* also provided evidence for an accumulation of potentially deleterious genes (Leslie and Vrijenhoek, 1978, 1980). The hybridogenetic mechanism persists as long as the unisexuals mate with males of their natural sexual host: *P. lucida, P. occidentalis,* or *P. latidens*. However, with the appropriate laboratory crosses, the hemiclonal genomes can be forced to recombine (Fig. 3). When the hemiclonal genomes are extracted from the hybrids and placed in a homospecific *monacha* genetic background, they exhibit Mendelian segregation and assortment plus crossing over. Conventional meiotic processes are reestablished. The F_1 progeny from such matings were used to assess the genetic composition of the hemiclonal *monacha* genomes.

To date, 14 natural hybridogenetic strains have been mated with *monacha* males in an attempt to extract hemiclonal genomes into a homospecific background (Leslie and Vrijenhoek, 1980 Table 1). Seven of the 14 strains contain hemiclonal genomes that are no longer compatible with a wild-type *monacha* genome. We were unable to produce viable hemiclonal/*monacha* embryos. Crosses with seven other strains did result in F_1 hybrid offspring. High juvenile mortality and stillbirths characterized the hemiclonal genomes of several of these strains. Morphological abnormalities and asynchronous somatic growth were evident in the F_1 progeny of other strains. Only three unisexual strains [Id, Va, and IIIa(0)] did not show exceptionally high mortality in the F_1 generation.

The complete failure of so many of these crosses and the high mortality and developmental abnormalities in others suggest that substantial genetic incompatibilities have accumulated between many of the hemiclonal *monacha* genomes and the wild-type *monacha* genomes. Similar incompatibilities are not found in hybrids between geographic strains of *P. monacha,* or between *P. monacha* and a related species, *P. viriosa* (Leslie, 1982). Thus, these incompatibilities did not result from divergence among the ancestral *P. monacha* populations prior to the origins of these hemiclonal genomes. Furthermore, the alleles causing these incompatibilities are unlikely to exist at anything but extremely low frequencies in sexual populations. Although they apparently are recessive in the *monacha-lucida* or *monacha-occidentalis* genetic backgrounds, they behave as partial dominants in the *monacha* genetic background. Since partially dominant deleterious alleles would be maintained at low (mutation/selection equilibrium) frequencies, it is inconceivable that so many could be "frozen" into new hemiclonal genomes at the moment of their hybrid origins.

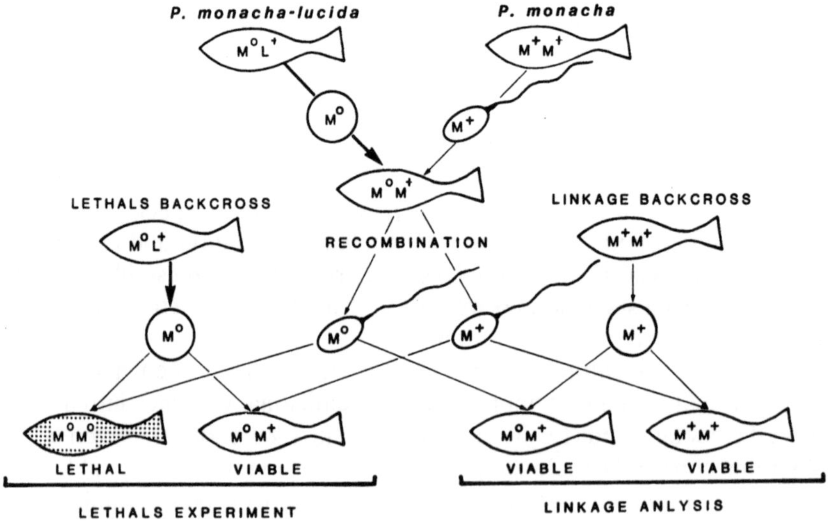

Figure 3. Crossing experiments used to detect the mutational load and linkage arrangements of hemiclonal *monacha* (M) genomes. The superscripts (+) and (0) represent a wild-type allele and a recessive lethal mutation respectively.

Attempts were made to better quantify the lethal gene loads of various hemiclonal genomes (Leslie and Vrijenhoek, 1978). Backcrosses with the same unisexual strain produce progeny expected to be homozygous at 50% of their gene loci. These offspring were used to detect deleterious recessive alleles. For example, a single lethal allele in a hemiclonal genome (designated by the superscript 0 in Fig. 3) has a 50% chance of being transmitted to a gamete by the F_1 parent and a 100% chance of being transmitted by the hybridogenetic parent. Thus, zygotes should exhibit 50% mortality due to homozygosity of a single lethal allele. Similarly, 75% mortality is expected for two lethals, etc. Crude estimates of lethal equivalent gene loads were made by comparing embryonic and juvenile survivorships from these matings with those of parallel outcrosses.

Of the seven unisexual strains that produced F_1 progeny, only IIIb and Va produced males that proved useful in backcrosses with the parental unisexual strain. We ascertained that the Va genome contains approximately two lethal equivalent alleles and the IIIb genome contains at least four. We could not determine from the present data whether these lethal equivalents represent a few discrete lethal recessives or many, diffusely spread semilethals and subvitals. Such an analysis is further complicated by the potential for variable expressivity and incomplete penetrance of deleterious alleles. For the present, these values must be considered conservative estimates of the deleterious gene loads of these two genomes.

Although I argued previously that partially dominant lethals are unlikely to be "frozen" into hemiclonal genomes at the moment of their origins, completely recessive lethals might be maintained at significantly higher frequencies in the ancestral sexual populations. Also, it is conceivable that placement of a *monacha* genome into the hybridogenetic background might produce "hybrid dysgenesis," resulting in significantly higher mutation rates, as observed in some interstrain hybrids of *Drosophila melanogaster* (Thompson and Woodruff, 1980). Tests of these hypotheses are currently underway. It is necessary to examine laboratory-synthesized unisexual strains for lethal equivalents in the same manner as the natural hemiclones. To date, only two of the synthetic clones have been subjected to genetic dissection (strains A and D). In both cases, crosses with *monacha* males encountered no difficulties. All the matings produced offspring and juvenile mortality was not significantly higher than expected. Spending two or more generations in the hemiclonal mode of existence had not altered the compatibilities of these "frozen" *monacha* genomes with "wild-type" ancestors, lending no support to the "hybrid dysgenesis" hypothesis. Only the progeny of strain A included males that could be used for the lethal equivalents backcross. Although only a small number of backcross offspring were born, juvenile mortality is not higher than expected. We plan to proceed with these difficult crossing experiments as more synthetic clones become available for testing.

In view of the inevitability of Muller's ratchet, how can unisexual *Poeciliopsis* persist as clones? For completely recessive mutations, potentially deleterious effects are masked by diploidy or triploidy. In the diploid hybridogenetic unisexuals, recessive mutations in the *monacha* genomes are permanently sheltered in the heterozygous condition by the paternally derived (*occidentalis, lucida,* or *latidens*) alleles that are substituted in every generation. The hybridogens benefit from the selective purging of deleterious recessives that occurs in the sexual host population. In addition, they escape the costs of genetic load paid by their hosts. Although lacking hybridogenetic substitution, the triploid gynogenetic forms also are buffered against deleterious mutations through gene duplication. The probability of obtaining a recessive, triploid homozygote through mutation alone is extremely low (Lokki, 1976). This conclusion appears correct as long as recombination between tbe heterospecific genomes never occurs in the gynogens (see Section 4). Thus Muller's ratchet can only affect these fish to the degree that mutations have partially dominant effects. In general, most new mutations are not completely recessive (Muller, 1950; Crow and Temin, 1964). For example, many of the accumulated mutations may resemble the enzymatically "silent" *Es-5*^{0} allele. Perhaps the single dose of this enzyme contributed by the paternal allele is adequate to perform its metabolic function, but an accumultion of many such mutations might be costly in terms of the synthesis of unnecessary proteins, wasted energy that might be more economically budgeted.

3.3. Mutations and Sexual Mimicry

Not all silencing mutations need be deleterious. A situation described by Vrijenhoek and Schultz (1974) reveals how some clonal mutants could contribute to unisexual mating success. Males of the sexual species, such as *P. lucida,* prefer conspecific females as mates over the unisexual hybrids (McKay, 1971). Apparently, visual, tactile, and olfactory cues are important in this discrimination in poeciliid fish. Sexual selection by the males would favor unisexual clones that most closely resemble the conspecific sexual females for these traits.

As a result of their hybrid constitution, *P. monacha-lucida* females are already similar in form and habit to *P. lucida*. Discrimination by *lucida* males is directed against mate selection characteristics expressed by the hemiclonal *monacha* genome. Laboratory-synthesized *monacha-lucida* hybrids are intermediate in their genital pigmentation (Fig. 4). They have reduced expressions of the pre-anal bar (*lucida* character) and dark black pigment in and around the genital pit (*monacha* character). This pattern is also typical of many wild hybridogens, such as hemiclones VII and VIII of the Río Fuerte. However, hybridogens from the Río Mocorito are very good mimics of the local *P. lucida* females. They lack the dark genital pit pigmentation altogether and the pre-anal bar is nearly complete (Fig. 4). One Mocorito strain (Ie) was subjected to the genetic dissection experiment in order to reveal the contents of its hemiclonal genome (Vrijenhoek and Schultz, 1974). In combination with a "wild-type" *monacha* (M) genome, the Ie/M hybrid expressed a reduced genital pigment pattern, suggestive of a single dose for these pigment genes. Silencing of the *monacha* pigment genes allows full expression of the *lucida* pattern, i.e., *lucida* pseudodominance. Such hemiclones should have significantly higher mating success than those expressing the intermediate pattern; however, this prediction remains to be tested in controlled experiments.

4. Recombination

Genetic recombination can be dissected into its major components: (1) Mendelian segregation and assortment, or recombination among chromosomes; (2) crossing over, or recombination within chromosomes; and

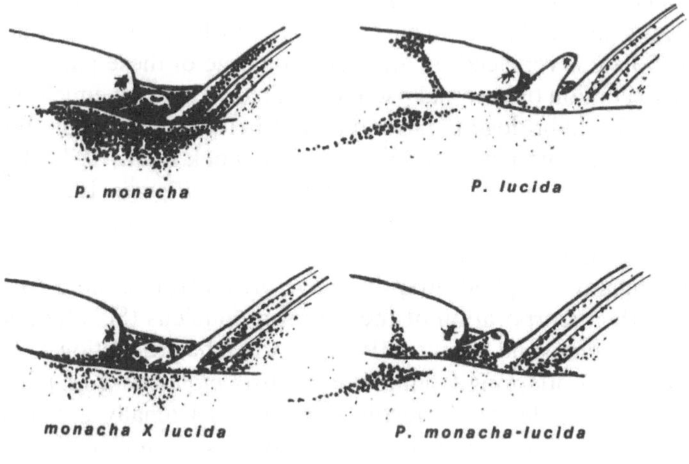

Figure 4. A ventral view of genital pigmentation in *Poeciliopsis*. Not all wild *P. monacha-lucida* strains show this degree of *lucida* mimicrv.

(3) syngamy, the fusion of gametes. The first two processes generate a tremendous diversity of gametes, which through syngamy generate an even higher diversity of zygotes. Although none of these processes occur regularly in gynogenetic *Poeciliopsis*, syngamy is a regular feature of hybridogenetic reproduction. Hybridogens have access to all the allelic variation that exists in the gene pool of their sexual hosts. *Poeciliopsis monacha-occidentalis* populations were found to express *occidentalis* alleles in the same frequencies that they occurred in the local host populations (Vrijenhoek *et al.*, 1977). Nevertheless, the paternal variation in hybridogens is not inheritable; it is used for one generation and discarded. As long as paternal alleles cannot leak into *monacha* genomes, they cannot contribute to adaptive evolution of the hemiclones, but they can contribute to the immediate fitness of the hybrids. Upon migration to a new locality or river system, the hybridogens acquire locally adapted genes through mating with the indigenous host species. Bulger and Schultz (1982) reported that the cold-adapted *occidentalis* genome confers cold tolerance to *P. monacha-occidentalis* hybrids and thus to their success in the colder northern rivers. Perhaps this ability to adopt local adaptations has permitted hybridogens to spread from a few centers of origin in the Ríos Mayo, Fuerte, and Sinaloa to rivers far to the north and to the south.

Despite their ability to express paternal traits, no evidence exists for the leakage of paternal genes into the hemiclonal *monacha* genomes of hybridogenetic *Poeciliopsis*. Fourteen electromorph hemiclones have been identified in *P. monacha-lucida* (Table III), four in *P. monacha-occidentalis* (Vrijenhoek *et al.*, 1977), and two in *P. monacha-latidens* (Vrijenhoek, unpublished). Surveys of 25 loci in several thousand specimens of these hybridogens have failed to reveal a single case where a species-diagnostic *monacha* allele is replaced by an allele from one of these paternal hosts. With the exception of paternal pseudodominance due to hemiclonal silent mutations, the hybridogens have retained heterozygosity for those loci that characterize the parental species (Vrijenhoek *et al.*, 1977, 1978).

Syngamy does not occur in gynogenetic *Poeciliopsis*. It is interesting in this regard that their distributions are far more restricted than those of the hybridogenetic biotypes. Gynogenetic *Poecilia formosa* occasionally include syngamy. Triploid offspring have arisen from diploid parents as a result of the incorporation of the sperm pronucleus (Schultz and Kallman, 1968). Cimino and Schultz (1970) reported the production of a diploid *monacha-lucida* offspring from a *Poeciliopsis monacha-2 lucida* parent. It was a sterile male, but if fertile diploids occasionally are produced, genome replacements could occur. This process is not unlike hybridogenesis except that it would be more sporadic in its occurrence.

4.1. Trihybrid Unisexuals

A roundabout pathway for recombination apparently has occurred in hybridogens inhabiting the Río Mocorito (Vrijenhoek and Schultz, 1974). *P. monacha* does not occur in this river, but a closely related species, *P. viriosa,* does. Schultz (1971) postulated that the hybridogen occurring there was a *viriosa-lucida* hybrid, with the *viriosa* genome playing the hemiclonal role. But attempts to synthesize this form through crosses of *P. viriosa* × *P. lucida* (and the reciprocal cross) failed to produce viable embryos. Crossing experiments similar to the genetic dissections discussed in the previous section were employed to reveal the contents of the Mocorito hemiclonal genome. An examination of morphological characters indicated that it transmitted mostly *monacha* traits, but some unique *viriosa* traits were also expressed, such as the "brassy pigment" display and scalation patterns. Vrijenhoek and Schultz (1974) postulated that *P. monacha-lucida* migrants established themselves in the Río Mocorito. Occasional hybridizations with males of *P. viriosa* would produce *monacha* × *viriosa* hybrids, which would produce fertile recombinant gametes. Fertilization of these recombinant *monacha/virosa* (1*n*) gametes by *lucida* sperm would create a new diploid hybridogen with a mixed hemiclonal genome, i.e., (*monacha/viriosa*)-*lucida*. We could not prove this scenario, but we could demonstrate its feasibility by duplicating the steps in the laboratory. *Poeciliopsis monacha-lucida* were crossed with *P. viriosa*. The *monacha* × *viriosa* F_1's had conventional meiosis, producing recombinant gametes. This latter point was corroborated by Leslie (1982), who examined electrophoretic markers in *monacha* × *viriosa* hybrids. The recombinant *monacha/viriosa* (1*n*) eggs were fertilized with *lucida* sperm and new hybridogenetic hemiclones were synthesized. Recent electrophoretic studies of the Mocorito hybridogens corroborated the fact that they are *monacha-lucida* hybrids. None of the four species-diagnostic *viriosa* allozymes have been found. Nevertheless, the "brassy display" and scalation traits do indicate that some *viriosa* genes have penetrated these *monacha* genomes. The Río Mocorito also contains a novel triploid biotype, *P. monacha-viriosa-lucida*. It is a trihybrid form that expresses triallelic allozyme phenotypes at several species-diagnostic gene loci. It is likely that this form arose by the addition of a *viriosa* genome to an unreduced (2*n*) egg from *monacha-lucida*.

Matings between the hybridogens and males of *P. monacha* or *P. viriosa* could result in a flow of genes from the hemiclones to the local *monacha* or *viriosa* population. If unique hemiclonal mutations or combinations of genes can persist for any length of time in the sexual population, they might become "frozen" again in new combinations in newly synthesized hybridogens. The huge hybridogenetic populations might also serve as a reservoir of genetic variation for the isolated headwater populations of *P. monacha* and *P. viriosa* (Vrijenhoek, 1979*b*). If this phenomenon occurs at all, it must be rare. The common release of hemiclonal genomes into the local sexual population would give rise to noticeable linkage disequilibria for hemiclonal gene combinations; these have not been found in *P. monacha* (Vrijenhoek, 1979*b*). It needs to be explored in *P. viriosa,* where hemiclonal *monacha* alleles could be recognized more easily.

4.2. The Triploids

Moore (1977*a*) performed an experiment that demonstrated the absence of recombination in gynogenetic triploid *P. 2 monacha-lucida.* Given the following genotype for a hypothetical locus in these triploids, M′ML, recombination between the heterospecific (M or M′ versus L) or homospecific (M versus M′) genomes would result in tissue graft rejections between siblings and between parents and offspring. Since no rejections occurred within any particular strain, Moore concluded that recombinations had not occurred. The persistence of heterozygous genotypes in these triploids for all those loci that distinguish *P. monacha* and *P. lucida* confirm his conclusion, at least for the heterospecific genomes. Cimino (1972*a*) proposed that an endomitotic division elevates oogonia to the hexaploid level and that synapsis occurs between replicated homologs; only M′M′, MM, and LL bivalents are possible. If rare mispairings of homospecific chromosomes occur (e.g., M′M), the triploids eventually would give rise to diverse clones that are completely homozygous for homospecific alleles [e.g., M′M′L and MML; see Asher and Nace (1971) and Eisenbrey and Moore (1981) for a more complete treatment of this phenomenon]. Thus, if the strains examined by Moore (1977*a*) were already homozygous for M alleles, he could not have observed recombinational effects even if mispairings had occurred. But his observations cannot exclude the possibility that these events had occurred in the history of his laboratory strains, or that they might have contributed to the generation of clonal diversity in gynogenetic triploids. Through such a process, a silent mutation in one M genome could be spread to both M genomes. The result could be *lucida* pseudodominance for a specific trait. A single mutation in an L gene could lead to *monacha* pseudodominance for that trait without recombination.

4.3. Linkage Arrangements

Because synapsis of maternal and paternal chromosomes does not occur in hybridogenetic unisexuals, and because replicated homologs pair in gynogens, there would appear to be no mechanical barriers to the accumulation of chromosomal mutations in unisexual *Poeciliopsis*. The small size of the *Poeciliopsis* chromosomes (Cimino, 1972*b*) has prevented any attempts to look for karyotypic differences among diploid and triploid clones. However, Leslie and Vrijenhoek (1980) examined this question through genetic dissections of hemiclonal genomes from the hybridogenetic forms. Linkage group I in *Poeciliopsis* carries five enzyme markers. We examined three of these markers in six hemiclonal genomes that had been removed from the hybridogenetic background (Table VI). We found no evidence for statistically significant differences in these linkage relationships among the six genomes. Apparently they retain the ancestral *monacha* condition.

5. Summary and Conclusions

Unisexual populations of *Poeciliopsis* do not appear to be destined for extinction due to a lack of genetic flexibility. Abundant clonal diversity exists in river systems, where interspecific hybridizations between sexual species continue to give rise to numerous genetically distinct clones. Ecological studies suggest that polyphyletic hybrid origins are primarily responsible for the ecological diversity and numerical success of unisexual populations. Diverse multiclonal assemblages often comprise 30–80% of the local *Poeciliopsis* population. Each hybrid event has the potential for "freezing" a new clone with its unique genetic, physiological, and eco-

Table VI
Recombination Maps for Linkage Group I in Six Hemiclonal Genomes and *monacha/viriosa* Hybrids[a]

Hemiclone Id	*Ldh-1*....17.1.....*Es-5*.....................20.0.....................*Idh-2*
Hemiclone Ie	*Ldh-1*....24.6.....*Es-5*.....................19.7.....................*Idh-2*
Hemiclone IIIb	*Ldh-1*....11.1.....*Es-5*.....................11.1.....................*Idh-2*
Hemiclone Va	*Ldh-1*....16.1.....*Es-5*.....................10.5.....................*Idh-2*
Hemiclone VIIa	*Ldh-1*....20.0.....*Es-5*.....................19.5.....................*Idh-2*
Hemiclone IIIa(0)	*Ldh-1*....24.2.....*Es-5*.....6.4.....*Ldh-2*....10.0.....*Idh-2*
monacha/viriosa	*Ldh-1*....27.4.....*Es-5*.....................12.8.....................*Idh-2*

[a]The lowercase letter designates the histocompatibility type.

logical attributes. The stable assemblages of a few clones in some streams, such as the Arroyo de Jaguari of the Río del Fuerte, suggests the actions of long-term interclonal selection fixing an assemblage of unisexual forms that can coexist ecologically with one another and with their sexual progenitors (Vrijenhoek, 1978, 1979*a*). Other localities, such as the Río Mayo, have such a high diversity of clones that local ecotypic adaptation might be swamped by a continuous supply of new clonal combinations (Angus, 1980). Still other localities contain only a single clone. Only a few monoclonal populations have been found where the unisexuals comprise more than 10% of the *Poeciliopsis* population, although the unisexuals often achieve higher numbers in specific microhabitats (Vrijenhoek, 1984).

While it has been demonstrated that mutations occur in clonal genomes, their role in adaptive evolution is not clear. Genetic studies of unisexual *Poeciliopsis* have confirmed Muller's (1964) prediction that asexual genomes should accumulate potentially deleterious mutations (Leslie and Vrijenhoek, 1978, 1980; Spinella and Vrijenhoek, 1982). Yet most of these deleterious genes are not expressed in the hybrid fish. Although some mutants, such as the silent enzyme genes, might exert a cumulative energetic cost in wasteful protein synthesis, it is unlikely that mutations will contribute significantly to the demise of unisexual *Poeciliopsis*. In most river systems, polyphyletic origins allow for the replacement of debilitated clones. Even in rivers lacking the potential for recurrent hybrid origins, migration would probably contribute new clones before Muller's ratchet could cause complete extinction of a unisexual population. Other silencing mutations apparently have had beneficial effects through increasing unisexual mimicry of genital pigments of the sexual host (Vrijenhoek and Schultz, 1974). It is possible that similar mutations have affected ecologically relevant triats, causing shifts in the dominance and expressivity of the clonal genes contained in various unisexual forms. However, this potential form of clonal diversification might be superfluous if variance in dominance relationships can be "frozen" from the gene pools of the sexual ancestors. Our studies of the synthetic clones should settle this question.

Other than the peculiar situation in the Río Mocorito, we have no evidence that gametic recombination contributes to unisexual diversity. However, syngamy might contribute to the adaptive plasticity of hybridogenetic unisexuals, since migrants acquire local adaptations by incorporating a genome from the local sexual host in their offspring. We still need to test the degree to which variable paternal genomes contribute to phenotypic variation in morphological, physiological, and important life history traits in hybridogenetic *Poeciliopsis* (Bulger and Schultz, 1982). These experiments might also provide much needed information on the levels of additive variation for ecologically relevant traits in the sexual populations (see Stearns, 1977).

The main thesis of this chapter has been the importance of polyphyletic hybrid origins in the adaptive evolution of unisexual populations. However, two competing hypotheses exist. Schultz (1977) and Bulger and Schultz (1979) argued that heterosis plays an important role in unisexual adaptation. They view the unisexuals as capable of expressing the ecological tolerances of both parental species, and thus occupying a broader niche. Although this may be true for thermal tolerance in *P. monacha-lucida,* it is not true for the triploid biotypes or for *P. monacha-occidentalis* (Bulger and Schultz, 1982). Vrijenhoek and Lerman (1982) did not find evidence for overdominance in developmental stability in unisexual *Poeciliopsis*. Furthermore, the argument for a broad unisexual niche is not supported by our studies of food habits and spatial distributions.

Thibault (1974) and Moore (1976, 1977*b*) favor the hypothesis that unisexual *Poeciliopsis* are most successful in habitats that are intermediate to those of the parental species. Accordingly, the hybrids would achieve their highest numbers in ecotones, due to weaker competition from their sexual progenitors. An examination of Table II would appear to support their contention. *Poeciliopsis monacha-occidentalis* achieves its highest frequency in the Río Mayo, which is at the northern end of the *monacha* range and the southern end of the *occidentalis* range. But other than citing geographic position, Thibault and Moore provide no ecological information about these "intermediate" habitats. In fact, the Río Mayo is a highly heterogeneous drainage that includes rocky mountain arroyos, deep canyon pools, and sandy-bottomed, sun-drenched streams. Within any locality, the fish might occupy areas of current or pool, sun or shade, rocks or sandy bottoms. Yet the multiclonal unisexual population in this river comprises approximately 80% of the fish in all these habitats from the headwater arroyos to the coastal mainstream. To which "intermediate" habitats do Thibault and Moore refer?

Neither the the "heterosis" hypothesis nor the "intermediate niche" hypothesis is adequate to explain the persistence of multiclonal unisexual populations. Both hypotheses portray the unisexual population as comprising one phenotype: either a broad-niched heterotic generalist or a narrow-niched ecotonal specialist. In either case, competitive exclusion or random clonal drift should rapidly lead to monoclonal populations (Maynard Smith, 1978). This has not occurred in rivers that sustain polyphyletic origins of unisexuals. Furthermore, the persistence of stable assemblages of clones in many of these localities would indicate that a simple equilibrium between clonal synthesis and extinction is not involved (Vrijenhoek *et al.*, 1977; Angus, 1980). The studies of phenotypic variation, life histories, and food habits reported here all point to the diversity among natural clones. Ultimately, the niche occupied by a population is a multidimensional structure (Hutchinson, 1957). Different clones may turn out to be narrow specialists for one niche parameter and broad generalists for others. Attempts to attribute the adaptive success of unisexual populations to single factors, such as "heterosis" or "intermediate niches," overlook the complexity of interactions we have come to know in the biological world.

If our current experiments with clonal synthesis support the hypothesis that polyphyletic hybrid origins are capable of generating the trophic and life history variation observed in unisexual *Poeciliopsis,* we must come to a rather ironic conclusion: sexual processes are also responsible for adaptation in asexual populations. The opportunity for recruitment of new clonal genomes from a sexual gene pool contributes to the adaptive success of the unisexual population as a whole. Furthermore, if the unisexual reproductive processes occasionally break down, they might serve as a reservoir of variation for the very sexual ancestors that spawned them. Thus, unique clonal mutations can be recycled and incorporated into newly arisen clones. Given the potential for this recycling over the long term, it is probably erroneous to view distinct clones as separately evolving units, or "species." The ultimate fate of each clone is extinction. Instead, we must consider a diverse assemblage of *monacha* genotypes, some occurring in the diploid sexual phase, some occurring in the hemiclonal hybridogenetic phase, and some occurring as either single or double doses in the various triploid gynogenetic forms (Table I). To what extent do these genomes become irreversibly adapted to the unisexual mode of existence (Leslie and Vrijenhoek, 1980)? What unique properties does the *monacha* genome contain that allow it to function in such different ways when paired with another genome?

Because of the diversity of breeding modes, the ease with which they can be cultured and manipulated in laboratory experiments, and the relative ease with which we can synthesize new clones, *Poeciliopsis* provide an almost unique opportunity in the animal world for addressing important questions about ecology and evolution of genetic systems. We have only begun to scratch the surface of the treasure trove that resides within these fascinating little fish.

ACKNOWLEDGMENTS. The success of our studies of unisexual *Poeciliopsis* must be attributed to Dr. R. Jack Schultz, who elucidated their unusual reproductive mechanisms. The time and effort he devoted to nurturing curiosity in his graduate students is greatly appreciated. In turn, I would like to thank my present and past graduate students and Henry Rutgers Scholars for their contributions to the work reported here: J. F. Leslie,

G. Meffe, D. Spinella, J. Wetherington, Susan Morris, and S. Lerman. I especially thank R. Schenck and K. Spindler for allowing me to use their unpublished data. R. J. Schultz, M. Douglas, N. Buroker, R. Schenck, and S. Morris provided helpful criticisms of the manuscript. These studies were made possible through grants from from the National Science Foundation, the most recent of which are DEB79-22854 and DEB82-12150. This work could not have been accomplished without the help and co-operation of the Departmento de Pesca of Mexico (permit number 13).

References

Angus, R. A., 1980, Geographical dispersal and clonal diversity in unisexual fish populations, *Am. Nat.* **115**:531–550.

Angus, R. A., and Schultz, R. J., 1979, Clonal diversity in the unisexual fish *Poeciliopsis monacha-lucida:* A tissue graft analysis, *Evolution* **33**:27–40.

Asher, J. H., Jr., and Nace, G. W., 1971, The genetic structure and evolutionary fate of parthenogenetic amphibian populations as determined by Markovian analysis, *Am. Zool.* **11**:381–398.

Atchley, W. R., 1977, Evolutionary consequences of parthenogenesis: Evidence from the *Warramaba virgo* complex, *Proc. Natl. Acad. Sci. USA* **74**:130–1134.

Bachmann, K., 1972, Genome size in mammals, *Chromosoma* **37**:85–93.

Bulger, A. J., and Schultz, R. J., 1979, Heterosis and interclonal variation in thermal tolerance in unisexual fish, *Evolution* **33**:848–859.

Bulger, A. J., and Schultz, R. J., 1982, Origins of thermal adaptation in northern vs. southern populations of a unisexual hybrid fish, *Evolution* **36**:1041–1050.

Cimino, M. C., 1972*a*, Meiosis in triploid all-female fish (*Poeciliopsis, Poeciliidae*), *Science* **175**:1484–1486.

Cimino, M. C., 1972*b*, Egg production, polyploidization and evolution in a diploid all-female fish of the genus *Poeciliopsis, Evolution* **26**:294–306.

Cimino, M. C., and Schultz, R. J., 1970, Production of a diploid male offspring by a gynogenetic triploid fish of the genus *Poeciliopsis, Copeia* **1970**:760–763.

Crow, J. F., and Kimura, M., 1965, Evolution in sexual and asexual populations, *Am. Nat.* **99**:439–450.

Crow, J. F., and Kimura, M., 1969, Evolution in sexual and asexual populations: A reply, *Am. Nat.* **103**:89–90.

Crow, J. F., and Temin, R. G., 1964, Evidence for the partial dominance of recessive lethal genes in *Prosphilia, Am. Nat.* **98**:21–33.

Eisenbrey, A. B., and Moore, W. S., 1981, Evolution of histocompatibility diversity in an asexual vertebrate, *Poeciliopsis 2 monacha-lucida* (Pisces: Poeciliidae), *Evolution* **35**:1180–1191.

Eshel, I., and Feldman, M., 1979, On the evolutionary effect of recombination, *Theor. Popul. Biol.* **1**:88–100.

Felsenstein, J., 1974, The evolutionary advantage of recombinations, *Genetics* **78**:737–756.

Hutchinson, G. E., 1957, Concluding remarks, *Cold Spring Harbor Symp. Quant. Biol.* **22**:415–427.

Lerner, I. M., 1954, *Genetic Homeostasis,* Oliver and Boyd, Edinburgh.

Leslie, J. F., 1982, Linkage analysis of seventeen loci in poeciliid fish (genus *Poeciliopsis), J. Hered.* **73**:19–23.

Leslie, J. F., and Pontier, P. J., 1980, Linkage conservation of homologous esterase loci in fish (Cyprinodontoidei: Poeciliidae), *Biochem. Genet.* **18**:103–115.

Leslie, J. F., and Vrijenhoek, R. C., 1978, Genetic dissection of clonally inherited genomes of *Poeciliopsis:* I. Linkage analysis and preliminar assessment of deleterious gene loads, *Genetics* **90:**801–811.

Leslie, J. F., and Vrijenhoek, R. C., 1980, Consideration of Muller's ratchet mechanism through studies of genetic linkage and genomic compatibilities in clonally reproducing *Poeciliopsis, Evolution* **34:**1105–1115.

Lewontin, R. C., 1974, *The Genetic Basis of Evolutionary Change,* Columbia University Press, New York.

Lokki, J., 1976, Genetic polymorphism and evolution in parthenogenetic animals, VIII. Heterozygosity in relation to polyploidy, *Hereditas* **83:**65–72.

Maynard Smith, J., 1968, Evolution in sexual and asexual populations, *Am. Nat.* **102:**469–473.

Maynard Smith, J., 1971, The origin and maintenance of sex, in: *Group Selection* (G. C. Williams, ed.), Aldine-Atherton, New York, pp. 163–176.

Maynard Smith, J., 1978, *The Evolution of Sex,* Cambridge University Press, London.

Mayr, E., 1963, *Animal Species and Evolution,* Harvard University Press, Cambridge, Massachusetts.

McKay, F. E., 1971, Behavioral aspects of population dynamics in unisexual–bisexual *Poeciliopsis* (Pisces: Poeciliidae), *Ecology* **52:**778–790.

Miller, R. R., 1960, Four new species of viviparous fishes, genus *Poeciliopsis,* from northwestern Mexico, *Occ. Pap. Mus. Zool. Univ. Mich.* **433:**1–9.

Moore, W. S., 1976, Components of fitness in the unisexual fish *Poeciliopsis monacha-occidentalis, Evolution* **30:**564–578.

Moore, W. S., 1977*a,* A histocompatibility analysis of inheritance in the unisexual fish *Poeciliopsis 2 monacha-lucida, Copeia* **1977:**213–223.

Moore, W. S., 1977*b,* An evaluation of narrow hybrid zones in vertebrates, *Q. Rev. Biol.* **52:**263–277.

Moore, W. S., and Eisenbrey, A. B., 1979, The population structure of an asexual vertebrate, *Poeciliopsis 2 monacha-lucida* (Pisces: Poeciliidae), *Evolution* **33:**563–578.

Moore, W. S., and McKay, F. E., 1971, Coexistence in unisexual–bisexual complexes of *Poeciliopsis* (Pisces: Poeciliidae), *Ecology* **52:**791–799.

Mukai, T., Chigusa, S. T., Mettler, L. E., and Crow, J. F., 1972, Mutation rate and dominance of genes affecting viability in *Drosophila melanogaster, Genetics* **72:**335–355.

Muller, H. J., 1950, Our load of mutations, *Am. J. Hum. Genet.* **2:**111–176.

Muller, H. J., 1964. The relation of recombination to mutational advance, *Mutat. Res.* **1:**2–9.

Parker, E. D., Jr., 1979*a,* Ecological implications of clonal diversity in parthenogenetic morphospecies, *Am. Zool.* **19:**753–762.

Parker, E. D., Jr., 1979*b,* Phenotypic consequences of parthenogenesis in *Cnemidophorus* lizards: I. Variability in parthenogenetic and sexual populations, *Evolution* **33:**1150–1166.

Roughgarden, J., 1972, Evolution of niche width, *Am. Nat.* **106:**683–718.

Schultz, R. J., 1966, Hybridization experiments with an all-female fish of the genus *Poeciliopsis, Biol. Bull.* **130:**415–429.

Schultz, R. J., 1967, Gynogenesis and triploidy in the viviparous fish *Poeciliopsis, Science* **157:**1564–1567.

Schultz, R. J., 1969, Hybridization, unisexuality, and polyploidy in the teleost *Poeciliopsis* (Poeciliidae) and other vertebrates, *Am. Nat.* **103:**605–619.

Schultz, R. J., 1971, Special adaptive problems associated unisexual fish, *Am. Zool.* **11:**351–360.

Schultz, R. J., 1973, Unisexual fish: Laboratory synthesis of a "species," *Science* **179:**180–181.

Schultz, R. J., 1977, Evolution and ecology of unisexual fishes, in: *Evolutionary Biology,* Vol. 10 (M. K. Hecht, W. C. Steere, and B. Wallace, eds.), Plenum Press, New York, pp. 277–331.

Schultz, R. J., 1982, Competition and adaptation among diploid and polyploid clones of unisexual fish, in: *Evolution and Genetics of Life Histories* (H. Dingle and J. P. Hegmann, eds.), Springer-Verlag, New York, pp. 103–119.

Schultz, R. J., and Kallman, K. D., 1968, Triploid hybrids between the all-female teleost *Poecilia formosa* and *Poecilia sphenops, Nature* **219:**280.

Sparrow, A. H., Price, J. H., and Underbrink, A. G., 1972, A survey of DNA content per cell and per chromosome of prokaryotic and eukaryotic organisms: Some evolutionary considerations, in: *Evolution of Genetic Systems* (H. H. Smith *et al.*, eds.), *Brookhaven Symp. Biol.* **23:**451–495.

Spinella, D. G., and Vrijenhoek, R. C., 1982, Genetic dissection of clonally inherited genomes of *Poeciliopsis*: II. Investigation of a silent carboxylesterase allele, *Genetics.* **100:**279–286.

Stearns, S. C., 1977, The evolution of life history traits: A critique of the theory and a review of the data, *Annu. Rev. Ecol. Syst.* **8:**145–171.

Strickberger, M. W., 1976, *Genetics,* 2nd ed., Macmillan, New York.

Templeton, A. R., 1979, The unit of selection in *Drosophila mercatorum,* II. Genetic revolution and the origin of coadapted genomes in parthenogenetic strains, *Genetics* **92:**1265–1282.

Thibault, R. E., 1974, The ecology of unisexual and bisexual fishes of the genus *Poeciliopsis:* A study in niche relationships. Ph. D. dissertation, University of Connecticut, Storrs, Connecticut.

Thompson J. N., Jr., and Woodruff, R. C., 1980, Increased mutation in crosses between geographiclly separated strains of *Drosophila melanogaster, Proc. Natl. Acad. Sci. USA* **77:**1056–1062.

Van Valen, L., 1962, A study of fluctuating asymmetry, *Evolution* **16:**125–142.

Vepsalainen, K., and Jarvinen, O., 1979, Apomictic parthenogenesis and pattern of the environment, *Am. Zool.* **19:**739–752.

Vrijenhoek, R. C., 1972, Genetic relationships of unisexual hybrid fishes to their progenitors using lactate dehydrogenase isozymes as gene markers (*Poeciliopsis,* Poeciliidae), *Am. Nat.* **106:**754–766.

Vrijenhoek, R. C., 1978, Coexistence of clones in a heterogeneous environment, *Science* **199:**549–552.

Vrijenhoek, R. C., 1979*a,* Factors affecting clonal diversity and coexistence, *Am. Zool.* **19:**787–797.

Vrijenhoek, R. C., 1979*b,* Genetics of a sexually reproducing fish in a highly fluctuating environment, *Am. Nat.* **113:**17–29.

Vrijenhoek, R. C., 1984, Ecological differentiation among clones: the frozen niche variation model in: *Population Biology and Evolution* (K. Wohrmann and V. Loschke, eds.), Springer-Verlag, Heidelberg, Federal Republic of Germany (in press).

Vrijenhoek, R. C., and Lerman, S., 1982, Heterozygosity and developmental stability under sexual and asexual breeding systems, *Evolution* **36:**786–776.

Vrijenhoek, R. C., and Schultz, R. J., 1974, Evolution of a trihybrid unisexual fish (*Poeciliopsis;* Poeciliidae), *Evolution* **28:**306–319.

Vrijenhoek, R. C., Angus, R. A., and Schultz, R. J., 1977, Variation and heterozygosity in sexually vs. clonally reproducing populations of *Poeciliopsis, Evolution* **31:**767–781.

Vrijenhoek, R. C., Angus, R. A., and Schultz, R. J., 1978, Variation and clonal structure in a unisexual fish, *Am. Nat.* **112:**41–55.

White, M. J. D., 1978, *Modes of Speciation,* Freeman, San. Francisco.

Williams, G. C., 1975, *Sex and Evolution,* Princeton University Press, Princeton, New Jersey.

Zaret, T. M., and Rand, A. S., 1971, Competition in tropical stream fishes: Support for the competitive exclusion principle, *Ecology* **52:**336–342.

CHAPTER 9

Evolutionary Phenetics and Genetics

The Threespine Stickleback, Gasterosteus aculeatus, and Related Species

MICHAEL A. BELL

1. Introduction

The advent of analysis of allozyme variation within natural populations has touched off a veritable revolution of our knowledge of evolutionary genetics. Many organisms that could not be bred in the laboratory or that exhibit no obvious polymorphism have yielded important insights into the genetic structure of natural populations. Although allozyme variation has allowed investigation of evolutionary phenomena at a level very close to the genome in diverse organisms, it has left very important problems unresolved and engendered a series of problems that are preculiar to allozymes (e.g., Lewontin, 1974; Koehn and Eanes, 1977, 1978).

Paramount among these problems is lack of demonstration of the evolutionary significance of allozyme to other aspects of phenotypic variation. Indeed, there is a growing realization that although there is a correlation between allozyme divergence and taxonomic difference based on morphology (e.g., Dobzhansky, 1972; Avise, 1974), there is no general causal relationship between them (e.g., B. J. Turner, 1974; J. R. G. Turner *et al.*, 1979). There is still a great need to study the evolutionary genetics of morphological and other phenotypic traits that are more readily interpretable from an adaptive point of view. However, the use of more tra-

MICHAEL A. BELL • Department of Ecology and Evolution, State University of New York at Stony Brook, Stony Brook, New York 11794.

ditional phenotypic features restricts study to a relatively small set of organisms that are amenable to this approach.

The threespine stickleback, *Gasterosteus aculeatus,* is well suited for study of evolution of diverse phenotypic features. While it has been possible to study the evolutionary genetics of color polymorphisms in such organisms as land snails (e.g., Clarke *et al.,* 1978), moths (e.g., Kettlewell, 1973), and guppies (Haskins *et al.,* 1961), these traits represent a limited aspect of the phenotype. In *G. aculeatus* the evolutionary genetics of color polymorphisms, trophic structures, locomotory structures, diverse armor structures, and, to a limited extent, behavior and physiology have been studied. This body of knowledge has provided valuable insights into diverse evolutionary problems and forms a solid foundation for progress in the future.

In this review I attempt to draw together diverse information on the evolutionary genetics of *G. aculeatus.* I focus my attention on western North American populations, with which I am most familiar. I also incorporate data concerning threespine sticklebacks from other regions when this information provides unique insights or contrast, or forms part of a global pattern of variation. I consider evolution of other sticklebacks when parallel phenomena in them and *G. aculeatus* exist. I incorporate unpublished data to the extent possible, but do not feel compelled to provide a comprehensive review of phenotypic variation in *G. aculeatus,* since Wootton (1976) has cited most of this literature up to 1975 and Coad (1981) compiled a bibliography of the stickleback literature current through 1979.

1.1. Biology of *Gasterosteus aculeatus*

The threespine stickleback is a small fusiform fish, ranging in adult female size from less than 26 mm standard length (SL; i.e., distance from the tip of premaxilla to the end of the last vertebra) (McPhail, 1977) to nearly 100 mm SL (Moodie and Reimchen, 1976*a*); males tend to be smaller than females. The most impressive morphological features (Fig. 1) include the three dorsal spines. The first two spines are large, denticulated, locking spines (Hoogland, 1951) that are separated from the dorsal fin, and the third spine is small, undenticulated, and immediately precedes the dorsal fin. The pelvic girdle is a complex and advanced structure compared to that of other teleosts (Nelson, 1971*a*). Each bilateral counterpart consists of a basal portion and unique ascending branch, and a prominent spine projects from it laterally. Finally, a single row of bony lateral plates, which may range to 36, occurs on the side of the body. Ventrally the lateral plates interdigitate with the ascending branches of the pelvic girdle and dorsally with the radial bones that support the dorsal spines. Thus, the abdominal region is encased in bone and studded with spines. This complex of structures affords some defense against predatory vertebrates (Hoogland *et al.*, 1957; Moodie *et al.*, 1973; Gross, 1978*a*; Reimchen, 1983).

Gasterosteus aculeatus possibly is best known among biologists for its territorial and reproductive behavior. Wootton (1976) has drawn to-

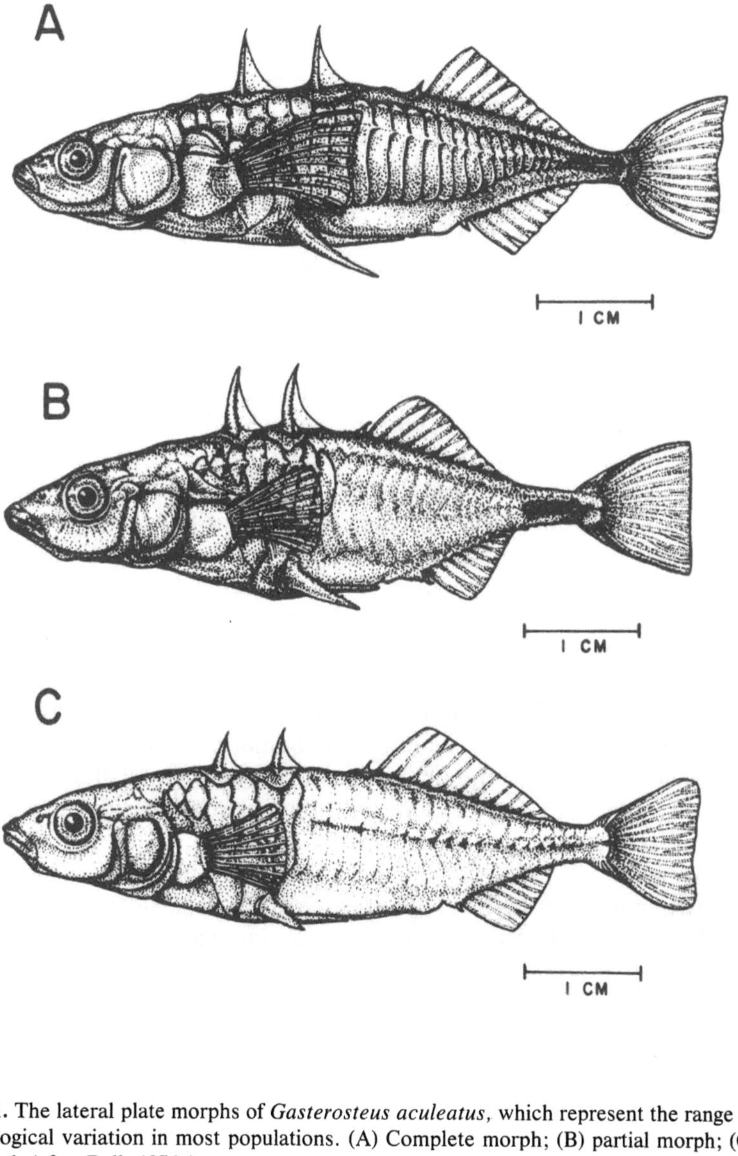

Figure 1. The lateral plate morphs of *Gasterosteus aculeatus*, which represent the range of morphological variation in most populations. (A) Complete morph; (B) partial morph; (C) low morph (after Bell, 1976*a*).

gether the extensive literature on stickleback behavior. Sticklebacks become sexually mature between 1 and 3 years of age. A male stickleback establishes a territory in shallow water, preferably near vegetation (Jenni, 1972; Kynard, 1979*a*), and defends it from conspecific males. The nest is constructed of decaying vegetation fibers, which are held together by glue released from the male's vent. "Creeping through" the nest indicates the male's readiness to court females. Typically in reproductive males, the pupils and much of the body are blue, and the ventrolateral surfaces of the head, shoulder, and other parts of the body are red (Hardy, 1978), but the extent of red nuptial coloration variation is inadequately recognized (Bell, unpublished data). An intricate set of behavioral cues, for which red male coloration is important, is exchanged between the male and a receptive female. After exchanging courtship signals with the female, the male leads her to the nest, which she enters to deposit her eggs. The male enters the nest after she has left and fertilizes the eggs, and the female leaves or is driven from the territory by the male. This courtship sequence may be repeated with several females before the male enters a parental phase in which he guards the territory against predators, including cannibalistic conspecifics (e.g., Kynard, 1978), and fans the nest to produce a flow of water through it. Typically, eggs hatch in about 6 days (at 20° C), and the male guards the fry for a few days until they become mobile enough to leave the territory. The male may repeat this paternal cycle a number of times during the breeding season. The females generally spend the periods between spawning in open habitats adjacent to the breeding grounds, and return to spawn every several days (Kynard, 1978). Thus, in *Gastrosteus* parental investment (*sensu* Trivers, 1972) is divided between the sexes, allowing an intense effort by both sexes and the production of large numbers of relatively well-developed fry. The breeding season varies according to latitude, but usually is restricted to a few months in the spring and summer. In southern California breeding individuals may be found in some populations throughout the entire year, although most breeding activity is in the summer (Baskin, unpublished data; Soltz, personal communication).

Although all threespine sticklebacks apparently enter shallow water to spawn, it is possible to recognize three major life history modes. Marine populations complete the life cycle in marine water, anadromous populations breed in streams but spend the nonbreeding period in marine water, and freshwater populations spend the entire life cycle in freshwater. Freshwater populations occur in diverse habitats, but generally are absent from those that they could not have colonized from the sea without traversing high gradient reaches of streams. Thus, native populations usually are absent from high elevations (Hardy, 1978). However, an introduced population in June Lake, California (Miller and Hubbs, 1969), and an apparently native population in southern California (Bell, 1982*a*) occur at high elevations. Populations from large lakes may resemble marine populations in that some occupy open water except during the breeding season (Moodie, 1972*a*, McPhail, personal communication).

The worldwide distribution of *G. aculeatus* has been reviewed in detail by Wootton (1976) and will only be repeated briefly here (Fig. 2). Marine and anadromous populations occur throughout most of the range, but are absent in the southern perirphery of the distribution. Freshwater populations occur alone in some southerly areas and are present in most regions in which marine and anadromous sticklebacks occur (e.g., Miller and Hubbs, 1969; Münzing, 1963). Bell (1976*a*) postulated that the southern freshwater populations were established during cooler periods when marine populations may have occurred further south.

According to Münzing (1963), threespine sticklebacks occur in Europe from Novaya Zemlya westward through the White Sea, throughout Scandanavia and western Europe and Iceland. They also occur in freshwater of southern Europe tributary to the Mediterranean Sea. Eastern Europe between the Baltic Sea and the Black Sea is the only major inland region occupied by *G. aculeatus,* and it is present in the Black Sea. It is absent from the Arctic Ocean and its tributary fresh waters in Asia, but reappears in fresh and marine waters at the Bering Strait. It extends southward to the southwest coast of Korea and most of Japan (Wootton, 1976) and in western North America to northern Baja California (Miller and Hubbs, 1969). It also is absent from the Arctic coast of North America, but reappears in Hudson Bay and is present along southeastern coasts of Baffin Island, Greenland, and along the northern coast of mainland North America eastward to Newfoundland and thence to Chesapeake Bay (Hardy, 1978; Lee *et al.,* 1980). Eastern North America is unusual because freshwater populations generally are absent to the south of Maine (Lee *et al.,* 1980). It penetrates inland along the St. Lawrence River drainage upstream to Lake Ontario. *Gasterosteus aculeatus* is a widespread circumboreal and north temperate species.

Throughout its range, *G. aculeatus* also is present on numerous islands to which it could not have dispersed over land (e.g., McPhail and Lindsey, 1970; Moodie and Reimchen, 1976*b*; Campbell and Williamson, 1979). Clearly, the distribution of *G. aculeatus* reflects dispersal through the sea, from which freshwater populations have been founded. Thus, it may be considered to be a peripheral freshwater fish (Darlington, 1957).

Threespine sticklebacks provide a number of advantages for evolutionary studies. Their behavior, ecology, parasitology, physiology, geographic variation, and evolution have been studied intensively (Wootton,

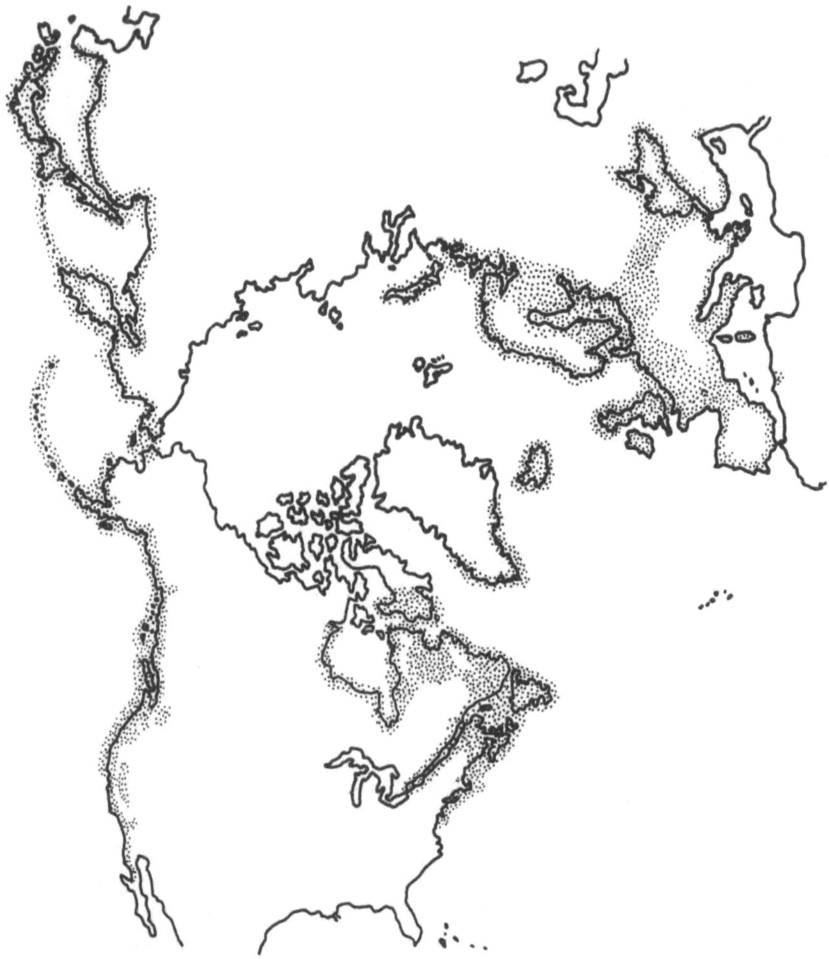

Figure 2. Distribution of the *Gasterosteus aculeatus* species complex shown in a north polar projection. The distribution between marine and freshwater fish is uncertain in eastern Asia. Based on information in Münzing (1963), Miller and Hubbs (1969), McPhail and Lindsey (1970), Wootton (1976), Lee *et al.* (1980), and other sources.

1976). This store of information provides an exceptional background for evolutionary studies, in which subtle phenomena may have important effects. Access to this vast body of literature is facilitated greatly by Wootton's (1976) review of the biology of sticklebacks and Coad's (1981) bibliography.

Its natural history lends this species well to evolutionary studies. It occurs in regions where many evolutionary biologists live, making year-round sampling and collection of live material convenient and inexpensive. *Gasterosteus aculeatus* occurs in diverse habitats and thus is exposed to a variety of selection regimes within a restricted region [e.g., the Queen Charlotte Islands, Moodie and Reimchen (1976*a*)] or even within a single drainage (e.g., Hagen, 1967; McPhail, 1969; Moodie, 1972*a*; Bell and Richkind, 1981). Threespine sticklebacks often are very abundant and are small enough to allow collection of several hundred mature specimens within a restricted area without creating difficulties in the transport and storage of the collections or of significantly affecting the evolutionary phenomena being studied [but see Hagen (1967)]. Many populations attain reproductive condition within 1 year (Wootton, 1976), and it may be possible to reduce generation time to less than 1 year in the laboratory by manipulation of photoperiod and temperature (Baggerman, 1957). Finally, easily recognizable phenotypic variation is very common in natural populations.

From the standpoint of experimental design and analysis, *G. aculeatus* provides numerous advantages. It may be crossed artificially, allowing uniform rearing conditions and a variety of crossing schemes, and all traits of interest usually can be scored within 2 or 3 months of crossing (Hagen and Gilbertson, 1973*a*). Generations appear to be discrete (e.g., Hagen and Gilbertson, 1973*b*; Gilbertson, 1980), making modeling and between-year comparisons relatively simple. Within streams, *G. aculeatus* often occurs in different habitat types, in response to which divergence has occurred (e.g., Bell and Richkind, 1981; Bell, 1982*b*). These systems are ideal for studying small-scale differentiation and the effects of gene flow between divergent demes because the dispersal route is defined by the stream course and one-dimensional distributions within streams are easy to describe and analyze. This stickleback also occurs in many thousands of separate drainages, which may be treated as replicates. It may be collected using minnow seines, beach seines, trawls, and minnow traps, and Hagen and Gilbertson (1973*b*) showed that seined samples are random with respect to some interesting morphological traits. A large number of morphological features vary in *G. aculeatus*, most can be scored rapidly and with minimal training, and almost all of them can be studied in conventional formalin-fixed specimens. Similar morphological variation also occurs in different stickleback species, allowing comparisons of traits between species (e.g., Bell, 1974). This set of attributes allows the use of *G. aculeatus* to approach evolutionary problems from many aspects and the application of appropriate tests of hypotheses.

Disadvantages of using *Gasterosteus* in evolutionary studies are minor and not unusual for fishes. Certain life history events may be difficult to observe in the field. For example, the biology of anadromous and marine sticklebacks while at sea is poorly understood, and it is difficult to observe reproductive behavior of subtidal breeding aggregations. Threespine sticklebacks are readily maintained and bred in the laboratory, but a large number of aquaria are needed to house stock and broods, and good water quality is necessary. Some of the most interesting morphological traits vary ontogenetically and are not stable until a frequently unknown size is attained. Failure to use only specimens in which traits being scored are ontogenetically stable may lead to serious error (Hagen and McPhail, 1970). While ecotypic variation over short distances [often just hundreds of meters, (Hagen, 1967; Hagen and Gilbertson, 1972; Reimchen, 1980; Bell, 1982*b*)] is an important advantage of sticklebacks, it imposes a burden on the investigator of spatial variation to determine whether variation over short distances is confounded with variation at a larger scale. Differential habitat use by the sexes during the breeding season may make estimates of sex ratio difficult under some circumstances (Bell, 1979; Kynard, 1978), and complicates assessment of variation of sexually dimorphic features. On balance, however, problems encountered in the use of *G. aculeatus* are minor compared to its advantages.

1.2. Phylogenetic Relationships

A brief review of the phylogenetic relationships of the sticklebacks is important to allow inference of character state polarity and to provide a context for comparison of variation of *G. aculeatus* to that of other gasterosteids. Nelson (1976) placed the family Gasterosteidae in the order Gasterosteiformes, a member of the superorder Acanthopterygii and the division Euteleostei. Nelson (1976) included 15 orders in the Acanthopterygii, and their interrelationships generally are unclear.

Pietsch (1978) tentatively suggested that the order Gasterosteiformes consists of two suborders, the Gasterosteoidei and the Syngnathoidei. The Gasterosteoidei includes three families, the Aulorhynichidae (tubesnouts) and Hypoptychidae (sand eels), which together form the sister group to the family Gasterosteidae (sticklebacks). Although Nelson (1976) did not regard the hypoptychids as even a distant relative of the Gasterosteiformes, he earlier had considered interrelationships of the gasterosteids and aulorhynchids (Nelson, 1971*a*). Outgroup comparison generally is regarded as the most reliable method of recognizing primitive character states (Waltrous and Wheeler, 1981; Stevens, 1980). Sticklebacks and aulorhynchids share a number of characters that have undergone striking radiation within *Gasterosteus* and some other sticklebacks.

The body is elongate and the mouth tends to be tubular. There is a series of isolated dorsal spines preceding the dorsal fin, and these are relatively numerous (24–26) in the aulorhynchids. In both families a single row of lateral bony scutes, referred to as lateral plates in *Gasterosteus,* replace conventional scales. The pelvic fin consists of a spine and relatively few soft fin rays. These character states are widespread in the gasterosteids, and may be considered to be the primitive states. However, the degree of development of dorsal spines, lateral plates, and the pelvis varies ecotypically in the three stickleback genera, *Gasterosteus, Pungitius* (ninespine), and *Culaea* (brook), and except for lateral plates in *Culaea,* reduction of these structures is the derived state in each stickleback genus.

Intergeneric relationships within the Gasterosteidae have been studied recently by Nelson (1971*a*), Mural (1973), and Chen and Reisman (1970), and Wootton (1976) reviewed their work and incorporated information from the unpublished work of Banister (1967) and Hall (1956). Most characters have a mosaic distribution among the stickleback genera. However, all recent workers seem to agree on the following points: The Gasterosteidae is a monophyletic group, and *Spinachia* (fifteen-spined or sea stickleback) most closely resembles the aulorhynchids. *Gasterosteus, Pungitius,* and *Culaea* represent a monophyletic group, each genus is distinctive, and the interrelationships of the three genera are usually considered uncertain. However, on the basis of their karyotypes Chen and Reisman (1970) considered *Gasterosteus* and *Pungitius* to be more closely related to each other than either is to *Culaea. Gasteroteus* and *Pungitius* have a fossil record (e.g., Bell, 1977; Rawlinson and Bell, 1982), but it is completely uninformative concerning phylogeny. *Apeltes* (fourspine stickleback) is not considered to be closely related to any of the other genera. *Gasterosteus* includes two well-defined species, *G. aculeatus* and *G. wheatlandi,* and *Pungitius* generally is considered to include two species, *P. platygaster* and the widespread *P. pungitius,* but the other genera appear to be monotypic.

2. Variable Features

The threespine stickleback is remarkably variable for a wide array of features. This variability has led to considerable taxonomic confusion (Hubbs, 1929; Penczak, 1966; Wootton, 1976), which is yet to be resolved (Hagen and McPhail, 1970). However, the same variability provides excellent material for evolutionary studies. The extent of our knowledge of the evolutionary significance of variation of different features differs greatly among features. Usually, studies of geographic variation have provided the initial stimulus for studies of genetics and natural selection. Thus, geographic variation often is all that is known about some features, but there has been great progress on the genetics and natural selection of a few features. Knowledge of the evolution of these features rivals that of some of the classic systems in ecological genetics.

In general, the other sticklebacks are less thoroughly studied, and I have been more selective in discussion of species other than *G. aculeatus*. For some features it is worthwhile to consider other sticklebacks for contrast with *G. aculeatus*, but in a number of cases, variation of a feature is better understood in another stickleback. The latter cases are worthy of consideration in their own right, but they also may be the best indication of mechanisms in *G. aculeatus*.

2.1. Lateral Plate Phenotypes

The lateral plates are bony, metameric structures that occur as a single, bilateral row, usually one plate per myomere. Polymorphism and geographic variation of lateral plate phenotypes are extensive, easily quantifiable, and obvious, making them attractive material for evolutionary studies. Variability and attractiveness of lateral plates as characters led to the description between 1829 and 1910 of at least 41 nominal species that now are referred to *Gasterosteus aculeatus* Linnaeus (Penczak, 1966; Wootton, 1976). Bertin (1925) maintained that lateral plate variation in coastal European populations is the product of environmental differences acting upon a uniform genetic background and recognized only one species of *Gasterosteus*. Since Bertin (1925), it has been accepted practice to treat all threespine sticklebacks [excluding *G. wheatlandi* (Hubbs, 1929)] as a single species (Penczak, 1966; Miller and Hubbs, 1969), although Bertin's (1925) assertion of genetic homogeneity of the nominal species was disproven by a series of studies by Heuts and Münzing (Wootton, 1976). With the recognition that lateral plate variation represents underlying genetic variation, explanations invoking divergence in Pleistocene refugia followed by secondary intergradation resulting from range shifts in response to climatic change emerged (Münzing, 1963; Miller and Hubbs, 1969). However, although Münzing (1963) minimized the importance of natural selection in the formation of spatial variation of lateral plate phenotypes in European *G. aculeatus*, the role of selection in such variation seems undeniable from existence of polytopic evolution of the partial morph (Münzing, 1962, 1963; Hagen and Moodie, 1982). It has become apparent from studies on North American populations during the past 10 years that natural selection plays a critical role in geographic variation of lateral plate phenotypes (Hagen and McPhail, 1970; Bell, 1976*a*). The current paradigm that natural selection is the principal agent in lateral plate variation has led to articulation of specific hypotheses that have been tested using ecological genetics. These studies will be the focus of this section.

Lateral plate variation actually involves at least two traits, the lateral plate morph system (*sensu* Hagen and Gilbertson, 1972) and the number of lateral plates within each morph. It is not clear how determination of plate number interacts with morph determination, for plate number genetics has been studied only in one morph (Hagen, 1973). In addition, there is no evidence of a connection between selection on lateral plate morph and lateral plate number. Thus, plate morph and lateral plate number variations are discussed separately.

2.1.1. Lateral Plate Morph Variation in *Gasterosteus aculeatus*

Almost all mature individuals can be assigned to one of three lateral plate morphs (Fig. 1). Hagen and Gilbertson (1972) coined the terms "complete," "partial," and "low" for the three morphs. The complete morph (Fig. 1A) has a series of plates running the entire length of the body, occasionally with a gap of one plate (Hagen and Gilbertson, 1973*a*). Generally, the first plate is immediately behind the expanded dorsal end of the cleithrum (shoulder girdle), but frequently abuts or overlaps the cleithrum and rarely may precede it. The plate row continues posteriad, usually one plate per myomere, terminating just anterior to the caudal fin. The posterior plates on the caudal peduncle form a keel that projects laterally. There is a gradation from broad, flat plates on the abdomen to narrow, projecting plates that form the keel on the caudal peduncle. Lateral plates in complete morphs number between about 30 and 36 on a side, including plates that form the keel.

The partial morph (Fig. 1B) has a row of plates on the abdomen followed by an unplated region of variable width and finally by a caudal row of plates, all or most of which form a keel. The number of lateral plates in partial-morph specimens seems to be very variable, and there is a positive correlation between the number of abdominal and caudal plates (Bell, 1981). Unfortunately, many studies do not include the number of plates in the caudal plate row.

Low morphs (Fig. 1C) possess only the abdominal plate row, and the posterior two-thirds of the body is unplated. The number of lateral plates usually ranges between zero and eight (rarely nine or 10) per side. Zero-plated specimens presumably represent low morphs, because they occur in populations that contain only low morphs (Moodie and Reimchen, 1976*a*; Miller and Hubbs, 1969).

Hay (1974) suggested an alternative method of defining major classes of lateral plate phenotypes based on the presence of specific lateral plates, which are recognized by positions with respect to the pelvic girdle and the dorsal pterygiophores (radial bones). Hay's (1974) method has not been adopted by other workers, but Reimchen (1983) also stressed the importance of plate position.

Lateral plate morph terminology has created some confusion. In the European literature the three lateral plate morphs have been referred to as *trachurus, semiarmatus,* and *leiurus*. Unfortunately, they have developed other connotations over the years (Hagen and Moodie, 1982). *Trachurus* has been used to denote marine and anadromous populations because such populations are dominated by complete morphs or are monomorphic complete. *Leiurus* has been used to refer to freshwater populations because many freshwater populations are monomorphically low-plated. The term *semiarmatus* frequently has been used in reference to putative hybrids between *leiurus* and *trachurus*. Münzing (1962) referred to a monomorphic partial morph population as "*semiarmatus-Typ*" in recognition of the impossibility of a hybrid origin for this phenotype. Thus, the terms *trachurus, semiarmatus,* and *leiurus* have acquired undesirable ecological and evolutionary connotations, and I shall refrain from using them.

Unlike almost all other genetic polymorphisms, the lateral plate morph system has a fossil record. Specimens that can be assigned unequivocally to the complete (Bell, 1977) and low morphs (e.g., Bell, 1973, 1974) have been collected from the late Miocene of western North America. Interestingly, the Miocene complete-morph specimen came from a marine deposit (Bell, 1977) and all low-morph specimens come from lake deposits, where those morphs dominate today. A complete morph specimen from the freshwater Pleistocene of eastern Canada (McAllister *et al.*, 1981) comes from a region where the complete morph dominates modern freshwater populations (Hagen and Moodie, 1982). Thus, the lateral plate morph trimorphism dates at least to the late Miocene, and the current relationship between lateral plate morph and habitat apparently has existed since then.

2.1.1.1. Lateral Plate Morph Genetics. There have been three studies of lateral plate morph genetics. Hagen and Gilbertson's (1973*a*) crosses utilized specimens from populations with a variety of lateral plate morph compositions; they performed a large number of crosses and developed a realistic and general model. Münzing (1959) and Avise (1976) did fewer crosses, and the latter study utilized an unusual low–complete dimorphic population and was flawed by use of very small sibships.

The earliest plate morph genetics study by Münzing (1959) employed

18 crosses including all combinations of morphs. Although he conceded that the results were not clear-cut, Münzing suggested a one-locus, two-allele model with substantial influence from polygenic modifiers and the following mode of plate morph inheritance: low (*leiurus*), *tt*; partial (*semiarmatus*), *Tt;* and complete (*trachurus*), *TT*. Münzing (1959) presented only plate count data, but in another paper (Münzing, 1963) equated the plate count phenotypes to morphs. I have transformed his data (Table I) and find that eight crosses include substantial proportions of unexpected phenotypes and four others produced unexpected phenotype frequencies (χ^2, $P < 0.05$). In all, 12 of 18 crosses are inconsistent with his one-locus model.

Hagen and Gilbertson (1973*a*) performed 77 crosses including every combination of morphs. Progeny mortality was low, family size was large, and morphs were scored after plate development had ended. Crosses

Table I
Results of Lateral Plate Morph Crosses by Münzing (1959)[a]

	Parental phenotypes		Progeny phenotypes			
Cross	♂	♀	L	P	C	*N*
A	C(6)	C(6)	0	1	78	79
B	L(9)	L(11)	199	0	0	199
C*	C(6)	L(11)	0	21	28	49
D*	L(9)	C(6)	0	125	100	225
E*	L(11)	C(6)	0	89	87	176
F*	L(9)	C(6)	0	175	25	200
G*	C(6)	L(6)	0	5	19	24
H*	P(S)	C(C)	6	18	17	41
I*	P(E)	C(E)	5	14	6	25
K**	P(F)	P(F)	31	39	33	103
L**	P(D)	P(D)	15	33	3	51
M**	P(6)	P(6)	40	173	97	310
N	L(9)	P(S)	83	60	0	143
O	L(9)	P(S)	81	68	0	149
P	L(11)	P(D)	139	146	0	285
Q	P(S)	L(11)	46	39	0	85
R*	C(C)	L(11)	49	0	45	94
S**	P(6)	L(11)	7	65	0	72

[a]Crosses are designated by Münzing's (1959) original letter code; lateral plate count data were transformed to lateral plate morphs based on Münzing (1963) and are abbreviated as follows: C, complete (29–35 plates per side); P, partial (9–28 plates per side); L, low (4–8 plates per side); the source of parents is indicated by Münzing's (1959) locality numbers in West Germany (6, Cranz; 9, Oldesloe; 11, Seeve) and crosses from which they were derived (C, D, E, F, S). Crosses that do not fit a one-locus model are marked with asterisks; one asterisk, unexpected phenotypes; two asterisks, unexpected phenotype frequencies.

within monomorphic populations produced monomorphic progeny. Crosses between fish from monomorphic complete and low populations always produced only partial morphs, and their F_2 progeny always have low–partial–complete ratios of 1:2:1. These crosses produced no evidence of sex linkage and are consistent with Münzing's (1959) model. However, crosses between fish from a trimorphic population in Lake Wapato, Washington (Hagen and Gilbertson, 1972), contradicted the one-locus model. For example, using a one-locus model, one expects that all partial × low morph crosses should produce half low and half partial morphs. Instead, the 23 low × partial crosses produced one monomorphic low, 17 trimorphic, and five low–partial dimorphic broods. All other combinations of parents produced morph frequencies in the progeny that are inconsistent with Münzing's (1959) one-locus model.

The simplest model (Fig. 3) that was consistent with the results of these crosses was one with two interacting loci each with two alleles and partial dominance (Hagen and Gilbertson, 1973*a*). Genotypes with three or four alleles for the complete or low morph express that morph, and those with two alleles for each morph are partial morphs, regardless of the locus at which the alleles occur. A maximum likelihood test indicated that the model is adequate to explain the results. Hagen and Gilbertson (1973*a*) performed four F_2 crosses, the results of which are consistent with a two-locus model but do not fit a one-locus model. The only cross that was inconsistent with their model was one between a complete and low morph, which produced all complete progeny, but with reduced plate

LOCUS A

LOCUS B	AA	Aa	aa
BB	Complete 4:0	Complete 3:1	Partial 2:2
Bb	Complete 3:1	Partial 2:2	Low 1:3
bb	Partial 2:2	Low 1:3	Low 0:4

Figure 3. Hagen and Gilbertson's (1973*a*) model for lateral plate morph inheritance in *Gasterosteus aculeatus*. Two major gene loci *A* and *B*, each with two alleles, a "complete allele" (*A* and *B*) and a "low allele" (*a* or *b*), interact to produce the three lateral plate morphs, complete (C), partial (P), and low (L) according to the ratio ("complete alleles" to "low alleles") of alleles at both loci.

size. This model will account for lateral plate morph variation in all but two known populations, and one must assume only complete morph dominance to explain these two low–complete dimorphic populations.

Avise (1976) studied lateral plate morph genetics in a dimorphic low-complete population, but also used fish from a monomorphic low and a monomorphic complete morph population. He performed 49 successful crosses, but family size was small. He suggested a one-locus, two-allele model with complete morph dominance for the dimorphic population. However, small family size prevented estimation of segregation ratios within broods, and virtual absence of partial morphs precluded recognition of heterozygotes. As Avise (1976) acknowledged, these results are ambiguous.

Hagen and Gilbertson's (1973*a*) model for lateral plate morph variation seems to be the most general and realistic model for this character. Obviously, differences at modifier loci are necessary to account for differences between the results of some crosses. In particular, Avise's (1976) interpopulation crosses indicate the evolution of dominance in one population; Hagen and Gilbertson's (1973*a*) observation of a "minute keel" variant of the partial morph and occasional anomalous results of all three studies suggest that there is marked differentiation of the genetic mechanism for lateral plate morph determination. While it seems reasonable to use Hagen and Gilbertson's (1973*a*) model as a general model for trimorphic populations, population studies that depend strongly on a specific genetic mechanism of lateral plate morph determination to explain phenotype frequencies should include crosses of fish from the population being studied.

Bell (1981) suggested that the low and partial morphs are paedomorphs (*sensu* Alberch *et al.*, 1979) of the complete morph, which he took to be the primitive condition for *G. aculeatus*. During development complete morphs pass through stages that resemble the low morph first and later the partial morph (Roth, 1920; Igarashi, 1964). Once plates are present on the abdomen and caudal peduncle, the plate rows grow together by sequential addition of plates to the posterior end of the abdominal row and to the anterior end of the caudal plate row. Thus, there is a correlation between the number of anterior and posterior plates in developing complete morphs. Bell (1981) detected a similar correlation in partial morphs and showed that the difference in number of plates between low and complete morphs is due mostly to differences in the rate of plate addition with respect to body size. Thus, plate morph loci may control the rate of plate addition. This simple mechanism could account for the probable polyphyletic origin of the allele that encodes the low and partial morphs.

2.1.1.2. Geographic Variation of Lateral Plate Morphs. Geographic variation of lateral plate morph frequency was studied first in Europe and was reviewed by Münzing (1963). Although lateral plate number phenotypes were surveyed in western North America as early as 1896 by Rutter (1896), it was not until very recently that morph frequency *per se* was investigated (Hagen, 1967; Hagen and Gilbertson, 1972). Hagen and Moodie (1982) recently performed an extensive survey of eastern Canadian *G. aculeatus,* and they provide a brief overview of plate morph variation elsewhere.

According to Münzing (1963), monomorphic completely plated populations occupy northen European marine waters from at least the White Sea south to the Atlantic and North Sea coasts of Sweden and central Britain (about 55° N, Fig. 2). Monomorphic completely plated marine populations also occur along the coast of Iceland, and in the southern Baltic and Black Seas. Trimorphic marine populations occupy the remaining southern portion of the North Sea to the western half of the English Channel and the northern Baltic Sea. In marine populations along the coast of western Europe, the frequency of complete morphs declines toward the southwest and low morph frequency increases. Trimorphic populations have been reported from inland waters of Poland (Penczak, 1965), Germany (Paepke, 1970), the Netherlands (Heuts, 1947*a*), and Sweden, Finland, and adjacent USSR (Münzing, 1963). Streams in a large area trending southeast from the southern Baltic to the northern Black Sea are inhabited by monomorphic completely plated sticklebacks. However, Münzing (1962) reported a monomorphic partial-morph population from Turkey, and Băcescu and Mayer (1956) reported a complete–partial dimorphic population from Romania. The *G. aculeatus* populations in countries boarding the Mediterranean generally are monomorphic low-plated (Münzing, 1963), but Bianco (1979) reported partial morphs from brackfish water in Italy.

A similar distribution of lateral plate morphs occurs in western North America, although *G. aculeatus* does not penetrate into inland areas as it does in Europe. The distribution of lateral plate morphs in marine and anadromous populations is inadequately documented to allow generalization, but some examples may be cited. McPhail and Lindsey (1970) reported completely plated sticklebacks from St. Lawrence Island and adjacent Seward Peninsula, Alaska, at the northern extreme of the range. Narver (1969) reported on an apparently anadromous population from the Alaska Peninsula, and he implied that it is monomorphically completely plated. Black (1977) sampled populations schooling during the winter near wharves in southern British Columbia, one of which included low frequencies of partial (4.9%) and low (1.6%) morphs. Black (1977) found significantly lower frequencies of low and partial morphs in breeding populations. However, these findings indicate the need to use large samples to allow estimation of the frequency of rare morphs. Hagen's (1967) extensive study of an anadromous population from the Little Campbell River on the British Columbia–Washington border established that it was monomorphic completely plated, but it seems possible that some of the sticklebacks that he demonstrated to be hybrids between the anadromous population and a monomorphic low-plated, freshwater population could run to sea, accounting for Black's (1977) observation of low and partial morphs in marine water. The most southerly report of *G. aculeatus* in marine waters is a single, completely plated specimen taken off of San Francisco Bay, California (Miller and Hubbs,1969), but Howe (1973) suggested that *G. aculeatus* is rare in the marine waters of northern California. As in western Europe, complete morphs dominate in marine and anadromous populations, and low and partial morphs occur at least at low frequencies in some populations, but we are ignorant of possible geographic trends in morph ratio in marine and anadromous populations.

Freshwater populations apparently are absent north of the Bristol Bay region of Alaska (McPhail and Lindsey, 1970). However, within this region all three morphs occur: Narver (1969) reported a trimorphic population from the Chignik River system, Gilbertson (1980) studied a partial–low dimorphic population from Lake Aleknagik, and Greenbank and Nelson (1959) studied monomorphic low-plated populations on Kodiak Island, just to the south. Unfortunately, there are no lateral plate morph data from the region of the Gulf of Alaska between Anchorage and Juneau or from most of the Alexander Archipelago (Lee *et al.*, 1980). Hagen and Gilbertson (1972) presented lateral plate morph data from 50 freshwater samples from the Pacific Northwest and Alaska. Except in Alaska, where partial morphs were abundant and are monomorphic in one lake (Kynard and Curry, 1976), populations tend to be trimorphic or monomorphic for the low or complete morph. Usually the low morph predominates and is most often monomorphic, but the complete morph also occurs monomorphically in an appreciable number of populations. Generally, dimorphic samples reported by Hagen and Gilbertson (1972) comprised low and partial morphs, with the low morph usually predominating.

Miller and Hubbs (1969) reported on lateral plate number variation with the expressed purpose of demonstrating the importance of intergradation and introgression in lateral plate variation. They did not present lateral plate morph data, but in their Table 1 they presented lateral plate number data from which the presence of lateral plate morphs may be inferred. Bell (unpublished data) sampled 40 California drainages, many of which were sampled by Miller and Hubbs (1969). Generally, trimorphic populations occur south to San Luis Obispo Creek. In contrast to the implication of Miller and Hubbs (1969), drainages that yielded trimorphic samples generally also produced monomorphic or dimorphic (low–partial or partial–complete) samples as well. From the Big Sur River northward, monomorphic and dimorphic populations were more likely to contain complete morphs, and to the south low morphs always predominated. From San Luis Obispo Creek south to northern Baja California, Mexico, all but one sample, which included about 1% partial morphs, was monomorphic low-plated. Monomorphic low-plated populations extend south to central Baja California, Mexico, in fresh water.

Lateral plate morph variation in Iceland is similar to that in western Europe and western North America. All three plate morphs occur; marine populations apparently are monomorphic completely plated (Münzing, 1963), but in freshwater all three morphs occur and some populations are trimorphic (Penczak, 1964; Bell, 1981; Bell and Williams, unpublished data).

In eastern North America *G. aculeatus* extends south to Chesapeake Bay in marine water, but in fresh water only one small population from Boston, Massachusetts (Hartel, personal communication; Bell and Baumgartner, 1984) is known south of Maine (Lee *et al.,* 1980). Hagen and Moodie (1980) reported lateral plate morph frequencies from 177 populations of *G. aculeatus* from fresh and marine waters of eastern Canada. Low morphs are lacking in their samples, although Coad and Power (1974) reported a low frequency (2.2%) of low morphs from the Matamek River, Quebec, and the sample from Boston is nearly complete–low dimorphic, with a substantial (8.7%) frequency of low morphs. Hagen and Moodie (1982) found that in marine water 90% of their samples were monomorphic completely plated, compared to 48% in fresh water. It was not uncommon for Hagen and Moodie's (1982) dimorphic freshwater samples to have substantial (>10%) frequencies of partial morphs, up to 100% partial; the marine samples, however, exceeded 2% partial morph frequency only once. Thus, the low morph is all but absent in eastern North America and the complete morph predominates, especially in marine water.

In Asia the situation closely resembles that in eastern North America (Hagen and Moodie, 1982). Low morphs generally are absent in Asia except for the warmer south of Japan, to the north of which populations are either monomorphic complete or partial–complete dimorphic.

Geographic variation in *G. aculeatus* lateral plate morphs was interpreted by Münzing (1963) and Miller and Hubbs (1969) to be a consequence of dispersal from refugia and secondary contact following the most recent holarctic glaciation. Earlier interpretations (Wootton, 1976) are of historical interest only. Münzing (1963) believed that in Europe the distribution of lateral plate morphs reflects division of *G. aculeatus* into two populations during the latest glacial period: a monomorphic completely plated one from the northern North Sea to the Arctic Ocean, and one containing low morphs (and possibly others) in the Mediterranean. Glacial and postglacial climatic changes, transgression of the North Sea through the Straits of Dover, and slucing of faunas by ice lakes were given as causes of movement out of the refugia. However, Münzing (1963) was unable to account for isolated partial morph and complete–partial dimorphic populations near the Black Sea or for monomorphic low-plated populations in freshwaters of Iceland, Greenland, or the Kola Peninsula without invoking selection of lateral plate morphs. He doubted derivation of low morphs from monomorphic complete-morph populations by mutation in the northern freshwater populations, favoring presence of "mixed" (i.e., trimorphic) populations from which monomorphic low populations were derived during some past temperature maximum. However, he invoked mutuation for the origin of partial morphs in Lake Techirghiol, Rumania, which became isolated from the Black Sea in the past 100 years. While it is possible that the present distribution of plate morphs reflects some vestige of lateral plate morph variation among *G. aculeatus* in different glacial refugia, Münzing's (1963) arguments for evidence of dispersal out of glacial refugia are unconvincing, and it is obvious that the origin of derived morphs (i.e., low and partial) *in situ* followed by selection to increase their frequency or even to fix one of the derived morphs is a sufficient and plausible explanation. Plausibility of this explanation is increased by Bell's (1981) demonstration that the partial and presumably the low morph are paedomorphs of the complete morph.

Miller and Hubbs (1969) provided a similar explanation of lateral plate morph variation in western North America. They envisioned the existence of three subspecies of *G. aculeatus* with different lateral plate phenotypes, which came into contact as a result of Pleistocene climatic change. Hagen and McPhail (1970) argued against Miller and Hubb's (1969) proposal that lateral plate morph variation reflected gene flow among monomorphic populations, and subsequent research has supported their position. Hagen and McPhail's (1970) arguments are equally applicable to Münzing's (1963) model for differentiation in Europe. They noted that Miller and Hubb's (1969) lateral plate data may be unreliable because they used small specimens in which counting errors are likely and plate development may be incomplete. They also noted that even Miller and Hubbs (1969) conceded the possibility of "independent loss" (i.e., parallelism) as one of two possible explanations for reduction in size of lateral plates. They stressed that populations falling outside of the range of variation of Miller and

Hubb's (1969) three subspecies are abundant in the Pacific Northwest and that there is clear evidence that many characters used to diagnose the three nominal subspecies of *G. aculeatus* in western North America are demonstrably subject to natural selection. Hagen and Gilbertson (1972) explicitly tested the hypothesis that complete morphs enter freshwater populations by means of introgression, by regressing frequency of the complete morph upon distance of the sample from the sea and obtained no relationship. The existence of monomorphic partial-morph populations (Münzing, 1962; Kynard and Curry, 1976; Hagen and Moodie, 1982) seems to be inconsistent with simple intergradation between complete and low-morph populations. Another flaw in Miller and Hubb's (1969) analysis is the use of single samples to characterize plate morph variation within a stream. Considering the abundant evidence that spatial variation of plate morph frequencies occurs within drainages (Heuts, 1947*a*; Penczak, 1965; Hagen, 1967; Bell,1982*b*; see Section 2.1.1.3), the use of single samples by Miller and Hubbs (1969) to characterize lateral plate morph variation within a drainage seems unjustified. Subsequent sampling by Bell (unpublished data) has demonstrated ubiquity of spatial variation of lateral plate morph ratio within some of the streams that Miller and Hubbs (1969) cited as evidence of unrestrained intergradation between stickleback subspecies when they come into contact in California. Finally, Miller and Hubb's (1969) demonstration in two cases that different morphs in a single population are not F_1 hybrids does not warrant their conclusion that the populations originated from secondary contact. These cases of polymorphism could as easily have resulted from natural selection acting upon an initially monomorphic population. Lateral plate morphs cannot be treated as neutral markers by which the geographic and genealogical origin of a population may be inferred.

If large-scale patterns of lateral plate morph variation cannot be explained by divergence of isolated populations and secondary contact, what has caused the current pattern of geographic variation of lateral plate morphs in *G. aculeatus*? Hagen and Moodie (1982) suggested that climatic differences are directly *or* indirectly responsible for geographic variation of lateral plate morphs. Thus, in western Europe and western North America, where the climate is mild and the difference between winter minimum and summer maximum temperatures is small, the low morph occurs and is monomorphic in the mildest southern regions. In eastern Europe, northwest of the Black Sea, in eastern North America, and possibly Asia, where the climate is more severe, with greater annual temperature variation, the low morph is virtually absent and the complete morph predominates. Some exceptions to the general absence of low morphs in severe climates have been discovered by Hagen and Moodie

(1982), but in two cases where ecological data are available, the populations occupy thermally stable habitats. Thus, the exceptions seem to confirm the rule that low morphs are absent from cold or thermally variable habitats. Many environmental factors vary with climate, any one or several of which could be the selective agent responsible for lateral plate morph distribution.

2.1.1.3. Ecotypic Variation of Lateral Plate Morphs. All three lateral plate morphs occur in western Europe north of the Straits of Dover, in western North America north of southern California, in Iceland, and in southern Japan. In eastern North America and probably elsewhere, complete-partial dimorphic populations occur. Morph ratio varies among populations in all of these areas, often over short distances. This section is concerned with the mechanisms that influence lateral plate morph frequencies locally.

The earliest insight into this problem was Heuts' (1947*a*) observation that breeding populations from Belgian streams generally were monomorphic complete near the mouth, trimorphic upstream, and monomorphic low-plated farther upstream. Heuts (1947*a*) felt that the monomorphic complete and low-plated populations were largely isolated from each other, but it was not until Hagen's (1967) thorough study of the ecological genetics of *G. aculeatus* in the Little Campbell River, British Columbia, that local factors that might affect lateral plate morph polymorphism began to emerge. Hagen (1967) showed that anadromous complete-morph and freshwater low-morph sticklebacks have strikingly different habitat preferences, differ in a number of morphological features, and represent separate biological species in the Little Campbell River and probably in some other streams nearby. Although a trimorphic hybrid zone occurs between the monomorphic populations and laboratory crosses produced no evidence of postzygotic isolating mechanisms, Hagen (1967) concluded that differences in the breeding season and ecological isolation prevented gene flow between the monomorphic populations. Hay and McPhail (1975) added strong evidence for behavioral isolating mechanisms in both populations. The two populations were strongly divergent for scores of a hybrid index, and the most divergent lateral plate number phenotypes in each population occurred adjacent to the hybrid zone. The monomorphic complete- and low-morph populations in the Little Campbell River clearly represent separate biological species, which are free from introgression, despite formation of a hybrid swarm in the area of parapatry.

Although Hagen's (1967) study may have confounded interspecific differences with differences between lateral plate morphs, it suggested that stream gradient or one or more of its correlates could be a selective mechanism for lateral plate morph frequency. Bell (1979, 1982*b*; Baumgartner and Bell, 1984) sampled stickleback populations in a variety of northern California freshwater habitats. Although only preliminary data presently are available, they form a pattern that is consistent with Hagen's (1967) results. Generally, samples from relatively high-gradient reaches of stream are monomorphic or nearly monomorphic for the complete morph. The low morph is the most abundant morph in trimorphic samples, which typically are found in low-gradient streams and in lagoons or estuaries at stream mouths. The low morph is nearly monomorphic in lentic habitats without major tributary streams. Unlike populations from regions with cold winters (Münzing, 1962; Hagen and Gilbertson, 1972; Coad and Power, 1974; Kynard and Curry, 1976; Hagen and Moodie, 1982), the partial morph rarely is the modal phenotype in California (Baumgartner and Bell, 1984). This difference suggests some interaction between selection related to climate and to stream gradient. More detailed knowledge of plate morph variation is needed to adequately address this problem.

Bell (1982*b*) studied a small drainage consisting of a lagoon and one high- and low-gradient tributary. Over a distance of about 3 km from the low- to the high-gradient tributary, the frequency of the low morph declines clinally from 74 to 1% and complete-morph frequency increases from 6 to 94%. Three samples from the lagoon and low-gradient stream contain substantial frequencies of all three morphs, but the two samples from the high-gradient tributary consist of more than 94% complete morphs. A pond near this drainage contained a nearly monomorphic low-plated population. Bell (1982*b*) interpreted the difference in morph ratio in the pond and the lagoon of the stream system to indicate that gene flow from the high-gradient stream contributed to polymorphism within the lagoon, supporting the model of Levene (1953).

The essentially monomorphic complete and trimorphic samples in the stream system also were divergent for the number of gill rakers and lateral plates in complete morphs. If the essentially monomorphic complete-morph samples represent an anadromous population, they should have higher gill raker counts than the trimorphic samples (Hagen, 1967; Hagen and Gilbertson, 1972). The reverse was observed, although all samples had high gill raker counts compared to stream populations reported by Hagen and Gilbertson (1972). Thus, the same relationship between lateral plate morph and stream gradient that Hagen (1967) observed occurs in this stream, though the specific status of the divergent populations is uncertain.

Additional fine-scale sampling to determine the distribution of lateral plate morphs with respect to habitat type is needed. A number of outstanding questions remain: Is the apparent relationship between stream gradient and plate morph ratio a general pattern? Is there an interaction between climatic based selection and that based on stream gradient? Does gene flow between populations in contrasting habitat patches contribute to lateral plate morph diversity within natural populations? What form do lateral plate morph ratio clines between populations with divergent morph frequencies take? Do allele frequencies at other gene loci covary with lateral plate morph ratio? Answers to these questions will contribute to our knowledge of the evolutionary genetics of *G. aculeatus* and to knowledge of the genetic structure of natural populations.

2.1.1.4. Evidence of Selection on Lateral Plate Morphs. MacLean (1980) and Gilbertson (1980) both studied polymorphic lake populations of *G. aculeatus* and obtained complementary results that indicate that lateral plate morphs are subject to natural selection. In both studies lateral plate morph frequencies were heterogeneous both in time and space, and the spatial heterogeneity sometimes was related to habitat differences. MacLean (1980) undertook a 3-year study of two adjacent trimorphic populations with 1-year generation times. He observed a decline in the frequency of partial morphs and low and complete morphs increased over the period of the study. Plate morph ratio changed both within and between cohorts. He attributed temporal decline of the partial morph to parasitism by pleurocercoids of *Schistocephalus solidus* (MacLean, 1974) and to difference in ability to secure resources. He observed a higher incidence of *S. solidus* in partial morphs in one population and lower incidence in the other. Partial-morph frequency declined in the population in which it was more likely to be parasitized and increased when it was less heavily parasitized. MacLean (1980) examined spatial heterogeneity and movements of lateral plate morphs carefully. He found plate morph frequency at a single site to be nonheterogeneous in females with respect to water depth in May, but in June complete morphs occurred at the greatest depth, partial morphs in a narrow band at intermediate depths, and low morphs nearest the surface. There also was morph frequency heterogeneity among sites in the lakes, but no obvious pattern with respect to any morph or in relation to any resource was demonstrated. However, examination of movement indicated that the populations consisted of sedentary territorial individuals and nonterritorial "floaters," that territorial males and females were more likely than floaters to be in reproductive condition, and that sometimes a male's probability of holding a territory was related to its morph. In one of his two study areas, low and complete morphs tended to be more likely to be territorial. The probability of being in reproductive condition also varied among morphs for females, but the favored morphs varied spatially and temporally. MacLean (1980) did not examine pairing between morphs, but unless there is positive assortative mating, his frequent finding of low- and complete-morph advantage over partials will tend to retard the loss of any morph by selection.

Gilbertson (1980) followed morph frequencies for 11 years (1961–1971) in a very large low–partial dimorphic population. He found that the frequency of the partial morph declined from greater than 20% to about 12% during the period of study. He also observed spatial heterogeneity of lateral plate morph ratio, and differential parasitism by *Schistocephalus solidus*. However, in contrast to MacLean's (1980) study, he found that partial morphs had a lower frequency of infection. In this population all breeding occurs in the third year, and all changes in plate morph frequency occurred between third-year adults and their 1-year-old offspring in the following year. MacLean's (1980) observation of differential reproduction certainly would provide a good basis for Gilbertson's (1980) conclusion that differential reproduction was responsible for temporal variation of morph frequency, although Gilbertson did not address this problem directly. He discounted the possibility of differential survival during the first year of life as the cause of this shift in morph ratio between parents and their yearling offspring because he failed to detect selective predation by northern pike (*Esox lucius*) or arctic char (*Salvelinus arcticus*) on older specimens, and he did not observe changes in morph ratio between age I and II or age II and III within any cohort. However, Hay (1974) found that cutthroat trout (*Salmo clarki*) and water scorpion (*Ranatra fusca*) selectively capture smaller individuals and that low morphs initially grow more slowly than partials in his study population. If the growth rate difference between low and partial morphs or size selectivity of predators were reversed in Gilbertson's (1980) system, size selection could account for his observation of reduction of partial-morph frequency between parents and 1-year-old offspring. Further study would be necessary to resolve this uncertainty, but differential reproduction seems to be the most likely explanation at present.

Gilbertson (1980) also noted a negative relationship between the amount of change between generations in partial-morph frequency and the frequency of partial morphs. The negative regression of the change in phenotype frequency Δp upon phenotype frequency p has been taken as an indication of frequency-dependent selection (Whittam, 1981), and frequency-dependent selection has been suggested as an important general mechanism for maintenance of genetic polymorphism (Clarke, 1979). Thus, this finding is of potential general importance. However, Whittam (1981) recently argued that this relationship may arise from random choice of Δp's. Thus, although the negative regression of Δp on p for the partial morph indicates that it is subject to selection, and it is consistent with a hypothesis of frequency-dependent selection, other explanations also are possible (Ginzburg, personal communication).

2.1.1.5. Conclusions. Lateral plate morph variation is extensive and occurs on a global and local scale. Although lateral plate morph variation in *G. aculeatus* has not been emphasized, there has been progress in understanding its significance. It clearly represents heritable variation and most certainly is subject to natural selection. It also is easy to score and the existence of only three common pheontypes allows use of relatively small samples to estimate frequencies. Although determination of the adaptive significance of lateral plate polymorphism is an intrinsically interesting problem, the polymorphism is potentially a useful tool for a variety of evolutionary problems. Two instances in which lateral plate morph polymorphism was applied to the problem of maintenance of genetic polymorphism, for example, were cited above. Lateral plate morph genetics, the roles of selection and gene flow in maintaining local spatial variation, and the factors that mediate the relationship between climate and plate morph variation are unresolved problems. In particular, studies of single populations that examine genetics and selection of lateral plate phenotypes throughout the life cycle are needed to draw firm conclusions about mechanisms for intrapopulation variation of lateral plate morphs and of divergence among populations.

2.1.2. Lateral Plate Number Variation in *Gasterosteus aculeatus*

The number of lateral plates varies within and among populations for each of the three lateral plate morphs. Low-morph lateral plate number has received the most attention to the nearly complete exclusion of the other morphs. First, I consider lateral plate number variation in the complete and partial morphs and then take up the much more extensive information on variation in the low morph.

2.1.2.1. Plate Number Variation in the Complete Morph. Variation of complete morph plate number is at a descriptive stage. Nothing is known about the genetics of their plate counts, and there is virtually no information on associations between complete morph plate number and environmental parameters or with other morphological features. Heuts (1947*a*) indicated in a figure, without otherwise commenting, that his and Bertin's (1925) data show a decrease in the number of complete morph plates from north to south in Europe. Lindsey (1962), however, has shown that in *G. aculeatus,* as in many other fishes, the number of elements in meristic series is affected by the temperature at which fry develop, and this effect could account for the relationship indicated by Heuts (1947*a*). Complete morph plate counts from western North America were reported by Hagen (1967), Hagen and Gilbertson (1972), Miller and Hubbs (1969) and Bell (1982*b*), but the number of populations reported is inadequate to discern any large-scale geographic trends. Plate counts in Europe (Münzing, 1963; Paepke, 1970; Heuts, 1947*a*; Erkinario, 1974), western and eastern North America (Coad and Power, 1973; Thomson *et al.*, 1978), and Japan (Igarashi, 1970) appear to be comparable and usually range between about 30 and 35 on a side. More information on geographic variation of complete morph lateral plate counts is needed.

Variation of complete morph plate counts within drainages has received scant attention. Hagen (1967) observed that among the anadromous sticklebacks in the Little Campbell River there were early migrants that preceded the principal breeding migration by up to 1 month. These individuals constituted a small fraction of the population, and they migrated farther upstream than most anadromous sticklebacks. Their earlier arrival and greater penetration upstream brought them into the hybrid zone, where they encountered freshwater sticklebacks, many of which still were breeding. Thus, the early migrants were most likely to engage in hybridization with freshwater sticklebacks. These early migrants also were characterized by significantly higher lateral plate counts ($\bar{x} = 34.55$ per side compared to the remainder of the population ($\bar{x} = 32.7$), and genes within the hybrid swarm never introgressed either parental population. Consequently, hybridization resulted in selective loss from the anadromous population of genes encoding higher plate counts.

Bell (1982*b*) compared complete morph lateral plate counts among five samples from the Brush Creek drainage, northern California. Samples in which the majority of specimens were low and partial morphs had nonheterogeneous complete-morph plate counts, as did those in which the complete morph was monomorphic or nearly monomorphic. However, the essentially monomorphic complete-morph samples had significantly higher plate counts than the trimorphic samples. The significance of this plate count difference is not known, but possibly suggests absence of gene flow among the divergent populations, in contrast to the plate morph ratio cline, which suggests the presence of gene flow (see Section 2.1.1.3).

2.1.2.2. Plate Number Variation in the Partial Morph. Partial morph plate number variation also is poorly documented. Münzing (1959) performed three crosses between partial morphs, which are consistent with the view that their plate counts have a genetic basis, but more crosses are needed. Hagen and Gilbertson (1972) found a strong mode of seven abdominal plates per side in the pooled counts for low and partial morphs in habitats with abundant predatory fishes. Their sample included sticklebacks from four lakes with more than 80% partial morphs; three of the lakes had abundant predatory fishes, and in the fourth, predatory fishes were present but uncommon. All four sites had seven plate count frequencies, comparable to sites in which the low morph dominates numerically and predatory fishes are present. Gilbertson's (1980) more detailed study of one of these populations indicates that partial morphs with seven abdominal plates are at an advantage over specimens with other abdominal plate counts with respect to fish predation, although they do not have maximal fitness. Kynard and Curry's (1976) lateral plate counts from a monomorphic partial-morph population that is sympatric with predatory fishes contrasts with these results. The minimum abdominal plate count in this population is seven plates per side, and 8- to 10-plated fish are the most abundant phenotypes, together constituting 64% of the population. They pointed out that this exception to the rule of association between abundant predatory fishes and a seven-plate mode (Hagen and Gilbertson, 1972) could result from the presence of preferable alternative prey or unavailability of the sticklebacks to the predators. The bulk of information indicates that predation by fishes on partial morphs favors increase in frequency of sticklebacks with seven abdominal plates on a side.

Hagen and Gilbertson (1972) examined lateral plate number variance in partial morphs and observed that abdominal plate number variance is much higher in populations with low frequencies (<6%) of partial morphs than those in which the partial morph is at a high (>23%) frequency. Coad and Power (1974) and Kynard and Curry (1976) also obtained low variances for partial-morph abdominal plate counts in monomorphic or mostly partial-morph populations. Hagen and Gilbertson (1972) believed that the inverse relationship between variance and frequency of the partial morph "suggested something like the evolution of dominance in a polymorphic population." Thus, at low frequency, plate number in partial morphs may be subject to weaker selection.

2.1.3. Plate Number Variation in the Low Morph of *Gasterosteus aculeatus*

The literature on low-morph lateral plate number is too voluminous to attempt a comprehensive review here, and Wootton (1976) and Coad (1981) should be consulted for additional papers. In this section I consider evidence on genetics, large-scale patterns of spatial variation, correlations with environmental variables, and direct evidence of natural selection on low-morph plate number.

2.1.3.1. Low-Morph Lateral Plate Number Genetics. Hagen (1973) performed 95 artificial intrapopulation crosses, 77 of which involved specimens from Lake Wapato and 18 of which used stock from five other freshwater sites in Washington state. Progeny were reared at constant temperatures of 21 ± 0.8°C or 16 ± 0.5°C until they were 28 mm SL, beyond the size at which lateral plate number becomes developmentally stable (Hagen and McPhail, 1970). Temperature control is important because Lindsey (1962) showed that temperature affects lateral plate number. Lateral plate number heritability h^2 was determined by regression of the mean plate count of progeny upon that of the parents (i.e., the "midparent" plate count). Heritabilities of lateral plate numbers in crosses involving Lake Wapato stock were 0.50 at 21°C and 0.83 at 16°C, and at 21°C was 0.84 based on results of intrapopulation crosses of stock from the other five sites. Heritability of lateral plate number generally did not differ betweeen the sexes and there was no evidence of maternal effects on heritability. Hagen (1973) also detected evidence that lateral plate number asymmetry has a genetic basis and estimated its heritability at about 0.63. He noted that the parents used in his crosses developed in the field under possibly diverse conditions, and thus the heritability values that he obtained should be considered to be minimal estimates. These results indicate that there is a large genetic component to lateral plate number variation in low morphs.

Lindsey (1962) performed 15 successful intrapopulation crosses, only eight of which produced five or more offspring large enough for plate counts. Hagen (1973) noted that five is the minimum family size for reasonably precise estimates of heritability. Based on all 15 crosses, Lindsey (1962) obtained a significant relationship between maternal plate count and the mean for her offspring, but not with the paternal or midparent counts. However, based on the eight crosses with five or more offspring, none of the regressions are significant, though the regression on female plate number explains far more variation than the other two. The difference between Hagen's (1973) and Lindsey's (1962) results may be spurious, resulting from small family size, experimental temperature variation, or a small number of crosses; or they may reflect actual genetic differences between Lindsey's (1962) study population and those studied by Hagen (1973). Two other studies (Münzing, 1959; Ross, 1973) provide limited evidence that supports the existence of a genetic basis for lateral plate numbers, but neither is very informative.

Hagen's (1973) study was the only one designed specifically to measure lateral plate number heritability in the low morph and these results are the most convincing. The other studies suggest that there may be variation among populations for the mode of inheritance of lateral plates.

More intrapopulation crosses for low-morph lateral plate number must be performed before generalization about the mode of inheritance and heritability of plate number is possible. In particular, F_2 crosses that should eliminate most environmental variance are needed. Serious studies of ecological genetics of lateral plate number phenotypes should include crosses to determine the genetics of the phenotypes because it presently is not possible to claim that results of crosses in one population are representative of others. However, it seems very likely at present that lateral plate number has a substantial genetic component and that it is polygenic.

2.1.3.2. Geographic Variation of Low-Morph Lateral Plate Number Phenotypes. Variation of low-morph lateral plate number has been studied most intensively in northwestern Europe and western North America. Heuts (1947*a*) computed the frequency distribution of lateral plate counts in trimorphic European populations. The plot is difficult to evaluate because frequencies apparently were weighted by sample size. However, two modes were present, one peaking within the range of low-morph plate counts at seven plates on a side and a second for complete morphs. The frequency distribution of lateral plate count modes from western North American populations reported in the literature (Miller and Hubbs, 1969; Hagen and Gilbertson, 1972; Moodie and Reimchen, 1976*a*) is shown in Fig. 4. Marked modes occur at zero and seven plates on a side, with a weaker mode at five plates. Although Heuts' (1947*a*) figure did not depict southern European populations, one geographic trend that is common to both western North America and western Europe is a high frequency of low plate counts and presence of zero-plated specimens in Mediterranean climates. Gross (1977) found that in western Europe, low-morph plate counts tended to be highest at intermediate latitudes and to decline to the north and south; at any latitude, they tended to increase to the east. Although Gross (1977) generally did not directly assess the intensity of predation or abundance of predatory fishes, he attributed the observed pattern of geographic variation to selective predation by fishes for higher plate counts. The paucity of predatory fishes and abundance of stickleback populations with low plate counts in the Outer Hebrides, Scotland (Campbell, 1979), is consistent with Gross' (1977) argument. The same argument could be applied to the western North American distribution of low-morph lateral plate number phenotypes; the lowest plate counts and highest frequencies of zero-plated specimens come from populations that are not sympatric with predatory fishes in the Queen Charlotte Islands (Moodie and Reimchen, 1976*a*) and California (Miller and Hubbs, 1969; Miller, 1960; Bell, 1976*b*, 1978, unpublished data). The distribution of very low plate counts among low-morph sticklebacks appears to be associated with

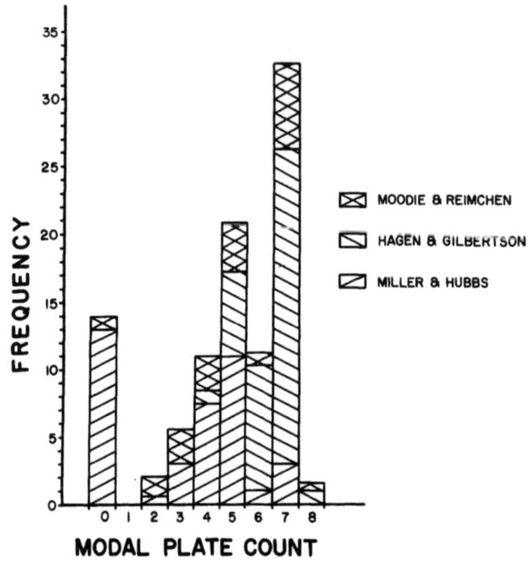

Figure 4. Frequency distribution of modal low-morph lateral plate counts in samples ($N \geq 25$) from western North America reported by Miller and Hubbs (1969) (mostly from California, 39 samples), Hagen and Gilbertson (1972) (mostly from the Pacific Northwest, 49 samples), and Moodie and Reimchen (1976) (from the Queen Charlotte Islands, British Columbia, 19 samples).

the absence of predatory fishes, and this association imparts large-scale trends to the geographic distribution of low plate counts in low-morph populations. Predatory fishes tend to be absent from broad areas in the north and south because of glaciation, other physiographic changes, or inappropriate climate.

2.1.3.3. Natural Selection: Differential Mortality of Low-Morph Lateral Plate Number Phenotypes. The relationship between predatory fishes and a modal plate count of seven abdominal plates per side was first discovered by Hagen and Gilbertson (1972) in the first of a series of studies by them and their colleagues on selection and distribution of lateral plate number phenotypes. Hagen and Gilbertson (1972) demonstrated that there is a strong association between the presence of abundant predatory fishes and a modal plate count in both low- and partial-morph fishes of seven plates per side. Moodie and Reimchen (1976*a*) extended this observation to low-morph stickleback populations on the Queen Charlotte Islands, and Bell and Richkind (1981) gave evidence for its applicability to southern California sticklebacks. The relationship between presence of abundant predatory fishes and a modal plate count of seven plates per fish may be observed over very short distances. The most striking example was reported by Hagen and Gilbertson (1972) for two parapatric populations sampled at points only 200 m apart and having frequencies of seven plates per side of 2 and 73%. Association between predatory fishes and lateral plate number in low-morph *G. aculeatus* is widely distributed in western North America.

Gross (1977) claimed that a modal plate count of seven plates per side in sympatry with predatory fishes did not occur in European populations of *G. aculeatus*. However, he did not report the raw data, his method of assessing the presence of predators was imprecise, and, as noted above, the modal plate count in Heuts' (1947*a*) data set for low morphs was at seven plates per side. In 12 trimorphic German samples reported by Paepke (1970), the modal plate count among low morphs was seven per side in six to eight samples, depending on whether counts from the left or right side of the body are used. A careful study of geographic variation in relationship to ecological variables is critically needed to settle the question of whether there is an association between a modal plate count of seven per side with abundant predatory fishes in Europe.

Low morphs are common in Iceland (Penczak, 1964; Bell and Williams, unpublished data), and the effects of fish predation should be studied there as well. Presumably, low morphs have evolved independently in Europe, western North America, and Iceland (Bell, 1976*a*, 1981), and it would be very interesting to know whether their evolutionary responses to predation by fishes are similar.

Hagen and Gilbertson's (1972) initial study provided the hypothesis that selective predation by fishes on *G. aculeatus* favors survival of individuals with seven plates per side. This hypothesis has been subjected to a series of tests, which generally have proven to be consistent with the hypothesis. Moodie (1972*b*) compared the lateral plate counts of sticklebacks from the stomachs of cutthroat trout (*Salmo clarki*) to those in the prey population collected simultaneously by seining in Mayer Lake, British Columbia. He did not find that fish with seven plates per side or any other plate number phenotype were significantly underrepresented in the predators' stomachs compared to their frequncies in the population, only that eight-plated fish were overrepresented. Hagen and Gilbertson (1973*a*) performed a more extensive comparison of lateral plate number phenotypes of combined low (96%) and partial morphs (4%) in stomachs of rainbow trout (*Salmo gairdneri*) to their frequencies in a random sample (Hage and Gilbertson, 1972) from the prey population in Lake Wapato, Washington. They made four samples from trout stomachs, three in the springs of 1968–1970 and one in the winter of 1968. They observed, as other studies had, that sticklebacks are a more important prey item for piscivorous fishes in the winter than in the spring. There consistently were fewer sticklebacks with seven plates per side in the trout stomachs than expected from their frequency in the prey population. Other plate number phenotypes usually were overrepresented or their representation in predator stomachs fluctuated among samples. During the period that they observed underrepresentation of seven-plated fish in trouts stomachs, the frequency of seven-plated specimens rose in the population. This directional selection resulted from recent introduction of the Lake Wapato sticklebacks from Lake Chelan, where the frequency of seven-plated specimens is lower. Recently, Gilbertson (1980) reported essentially identical results from 3 years of sampling stomachs of northern pike (*Esox lucius*) and arctic char (*Salvelinus arcticus*) in Lake Aleknagik, Alaska. Thus, selective predation by four fish species on lateral plate number phenotypes has been documented in three distant areas, and sticklebacks with seven plates on a side generally enjoyed a survival advantage.

Moodie *et al.* (1973) independently performed laboratory experiments concerning selective predation on low-morph lateral plate number phenotypes. Cutthroat trout (*Salmo clarki*) or northern squawfish (*Ptychocheilus oregonensis*) were used as predators and prey *G. aculeatus* were acclimated to winter (i.e., cold water, short photoperiod) or summer (i.e., warm water, long photoperiod) conditions. Results of experiments run under winter conditions indicate a selective advantage for fish with seven plates on a side, but surprisingly, this phenotype was at a disadvantage under summer conditions. Moodie *et al.* (1973) observed behavioral differences among plate number phenotypes and suggested that general timidity was responsible for differential survival. The rate of survival of seven-plated sticklebacks was independent of the modal plate count in the population from which the prey came, and they did not believe that seven plates *per se* was the reason for excessive survival. They were surprised that seven-plated fish were not at an advantage under summer conditions, but suggested that behavioral changes associated with reproduction, which was prohibited in the laboratory, could solve the discrepancy between laboratory and field results obtained in the summer. This hypothesis has not been tested.

Studies of other potential selective predators have been conducted. Bell and Haglund (1978) noted frequent predation by garter snakes (*Thamnophis couchi hammondi*) on a population of *G. aculeatus* and performed a laboratory study to determine whether predation was selective. They found that predation was selective for lateral plate number among specimens from two monomorphic low-plated populations and that fish with 10 plates (almost always five per side) were virtually immune to garter snake predation. They used populations with modal plate counts of 10 and 14 per fish, and found that success of 10-plated fish was independent of the modal plate count in the source population. Eleven-plated fish, however, have one plate more than 10-plated specimens on only one side of the body, but they had relatively poor survival. This result supports the conclusion by Moodie *et al.* (1973) that plate number itself is not responsible for differential predator avoidance of low-morph lateral plate number phenotypes. However, if a generalized behavioral response such as timidity, as suggested by Moodie *et al.* (1973), were responsible, it should be equally effective against garter snakes. The difference of the optimum plate number for predator avoidance between selective predation by fishes and garter snakes is puzzling, but seems to be inconsistent with an explanation involving a generalized predator avoidance response.

Many other animal species are known to prey on *G. aculeatus* (Wootton, 1976), but the effects of such predation on plate composition of stickleback populations is unknown.

Reimchen (1983) showed that when pressure is applied to the first two dorsal spines of *G. aculeatus,* the force is transmitted to the lateral plates and the dorsal spines are prevented from tilting to the side. Dorsal spine stability should be most important where vertebrates prey on sticklebacks. Dorsal spine stability is conferred by the four or five plates in contact with the pterygiophores of the dorsal spines (Fig. 7), and so predation by vertebrates should favor retention of these plates. Their presence is less variable than that of other plates. However, this mechanism does not explain why plates anterior to this series should be retained in populations subject to predation by fishes, nor does it explain why specimens with plates posterior to the second dorsal radial should be at disadvantage with respect to vertebrate predation (Reimchen, 1983). Thus, while Reimchen's (1983) results account for presence of a minimum number of plates (i.e., those that contact the pterygiophores of the dorsal spines) in populations in which predation by vertebrates is a significant source of mortality, they do not account for the modal plate count of seven per side associated with fish predation (Hagen and Gilbertson, 1972) or for higher plate count phenotypes.

Gasterosteus aculeatus is known to be the host for a large array of parasites, and it has been studied intensively by parasitologists (Wootton, 1976). Only two studies (Gilbertson, 1980; Bell and Baumgartner, unpublished data) have looked for parasitism that is selective with respect to low-morph lateral plate number. Infection by pleurocercoids of the cestode *Schistocephalus solidus,* the digene trematode *Cryptocotyle concavum,* and the copepod *Lernaea cyprinacea* have been examined, and none were selective with respect to lateral plate number.

There is evidence that mortality due to physical stress could be selective for lateral plate number. In intrapopulation comparisons using a monomorphic low-plated population, Heuts (1947*b*) showed that specimens with higher plate counts survived longer when transferred from fresh to sea water (35%). When transferred in fresh water from cold (0°C) to warm (25–28°) water, those with lower plate count survived longer. However, Gross (1977) failed to confirm this result, and Heuts (1947*b*) also failed to obtain this result using a population that was polymorphic for plate morph. Lindsey (1962) found that within a monomorphic low-plated population, females with lower plate counts had higher optimal temperatures for egg survival, confirming Heuts' (1956) earlier results. Bell and Baumgartner (1983) failed to detect nonrandom morality of plate number phenotypes in a pool shrinking due to drought. The high mortality rate observed in this pool probably was due to hypoxia. Giles (1983) suggested that a low concentration of calcium ion in water may favor evolutionary loss of lateral plates. There is some evidence that stress induced by physical factors could contribute to lateral plate number selection in the low morph. Lower plate count phenotypes seem to be at an advantage in warm water, and this observation is consistent with the presence of relatively low plate counts in populations at the southern end of the range of sticklebacks in western Europe and western North America. However, predatory fishes also are absent in this part of the range, and this could account for low plate counts. The relative importance of these two potential mechanisms for low plate counts in the southern part of the range of *G. aculeatus* remains to be determined.

2.1.3.4. Natural Selection: Differential Reproductice Success of Low-Morph Lateral Plate Number Phenotpes. Differential reproductive success is the second major determinant of natural selection. Both male and female reproductive success must be measured separately and then combined for an overall measure of success for each plate number phenotype. Ideally, female reproductive success should measure lifetime fecundity and the efficiency of conversion of eggs to fry. Likewise, male reproductive success should take into account the ability of males to fertilize and successfully brood eggs. Generally, actual reproductive success is impossible to measure in the field, but a number of measures of varying quality and difficulty may be substituted (Howard, 1979).

A reasonably good measure of female reproductive success may be obtained, although lifetime fecundity cannot be measured in natural populations. In many stickleback populations the vast majority of individuals breed only during one year (e.g., Hagen and Gilbertson, 1973*b*; Gilbertson, 1980). In such populations the number of eggs produced by a female in a breeding season still is difficult to determine because females produce several broods of eggs per breeding season (Wootton, 1976). The best approach would be to compare the number of eggs per ripe female and the number of broods produced by females with different lateral plate numbers. The number of broods produced per season is very difficult to measure, but its correlate, the probability of a female being gravid, is not. However, the probability of being gravid is a very noisy measure, as a female may not be gravid either because she is producing few broods or because she recently had spawned. Thus, very large samples must be used to detect differences in the probability of lateral plate number phenotypes being gravid. In addition, the probability of being gravid should be measured periodically, as Kynard (1972) detected differences in the timing of breeding in male plate number phenotypes, and this phenomenon could occur in females. Haglund (1981) attempted to measure female reproductive success as a function of plate number, but his analysis was flawed by inadequate sample size and invalid assumptions underlying his statistical analysis. Kynard (1972) compared the number of eggs of ripe females with seven plates on each side to that of females with eight on each side and found that seven-plated females have more eggs. Additional study is needed to determine whether differential fecundity is important.

Male reproductive success has not been adequately measured either. Kynard (1972) measured a number of variables related to reproductive success of male *G. aculeatus* from Lake Wapato [where Hagen and Gilbertson (1972*b*) investigated selective predation], but it is impossible to obtain an overall measure of reproductive success from his data. He demonstrated that all lateral plate number phenotypes prefer nesting in concealment and in water greater than 45 cm deep (Kynard, 1979*a*). Kynard (1972, in press) concluded tentatively that early-breeding males, among which fish with six plates on each side (i.e., 6–6 fish) are disproportionately represented, obtained more eggs. He also showed that 7–7 males were superior to 6–6 or 8–8 males in acquisition of preferred nesting sites and suffered relatively little loss of eggs during nesting. However, although 7–7 males were more aggressive than 6–7 and 8–8 males, 7–8 and 8–9 males were the most aggressive and produced the most late-stage embryos. [Kynard (1972, in press) was careful to control for variation in the level of aggression associated with age of the eggs.] Unfortunately, Kynard's (1972, in press) data cannot be collapsed into a comprehensive measure of male reproductive success, but his results demonstrate differences in factors that contribute to reproductive success among males with different numbers of lateral plates. Kynard (1972) noted that those phenotypes that are favored between juvenile and adult stages appear to be at a reproductive disadvantage.

Haglund (1981) studied the same population that Bell and Haglund (1978) had used in most of their garter snake predation trials and classified phenotypes according to the total number of plates per fish. In this population lateral plate counts vary between zero and 14 plates per fish.

Haglund (1981) inferred from the rapid reoccupation of territories from which males had been removed that there was competition for territories. He compared the frequency of lateral plate number phenotypes among nesting males and males greater than 28 mm SL in the population. Seven- to 10-plated specimens were underrepresented and 11- to 14-plated fish were overrepresented among nesting males. Haglund (1981) also examined nests containing eggs too old for the male still to be courting but too young for any of them to have hatched. On the average, 11- to 14-plated fish had about 18.5% more eggs than 7- to 10-plated ones. Haglund (1981) assessed male aggression as Kynard (1972) had and concluded that more successful phenotypes were more aggressive. Huntingford (1981) also observed heterogeneity of aggression of male and female low-morph lateral plate number phenotypes toward conspecifics, but did not assess reproductive success. Haglund's (1981) results paralleled Kynard's (1972) in that both found that reproductive success appeared to be lowest among phenotypes that were best able to avoid common predators in their habitats. Huntingford (1976) demonstrated that nonreproductive male stickleback boldness toward a predator (pike, apparently *Esox lucius*) is correlated with its aggressiveness toward territorial intruders during its paternal cycle. Thus, selection for territorial aggressiveness, which produces greater reproductive success, could result in higher rates of predation by fishes or garter snakes. This suggestion is supported by Moodie's (1972*a*) observation that 7–7 males take longer to return to their nests following disturbance than other plate phenotypes in Mayer Lake.

2.1.3.5. Ecotypic Variation of Low-Morph Lateral Plate Number Phenotypes within Drainages. Variation of lateral plate number within drainages is considered for two reasons. Miller and Hubbs (1969) claimed that gene flow between populations is a major determinant of lateral plate number variation, and with knowledge of selective agents it should be possible in some cases to interpret variation within drainages in terms of the distribution of selective agents.

As noted above, predatory fishes are a selective agent that produces a modal plate count of seven plates on a side. Bell and Richkind (1981) studied sticklebacks in the Ventura River, southern California, in which the headwaters are populated by abundant rainbow trout (*Salmo gairdneri*), a known stickleback predator (Hagen and Gilbertson, 1972), but the lower reaches lack abundant predatory fishes. In samples taken from the headwater trout habitat, the frequency of sticklebacks with a total of 14 plates (usually seven per side) ranges between 27 and 48%, but downstream in the absence of trout the frequency of 14-plated fish are rare, and their frequency declines with distance away from the trout habitat.

Bell and Richkind (1981) noted no divergence at 12 protein loci among stickleback samples throughout the drainage and that a dry reach of stream probably severely restricts dispersal (especially upstream) between the trout and trout-free habitats. They interpreted these results to indicate that this stream was invaded only once by *G. aculeatus* and that spatial differentiation had occurred *in situ* [i.e., primary intergradation, *sensu* Mayr (1963)]. Trout predation favors 14-plated fish in the headwaters, but the first sample downstream of the trout habitat sustains a high frequency of 14-plated fish in the absence of trout because of gene flow from the trout-selected population. It also seems possible that this population could have been founded by immigrants from upstream following earlier extinction of sticklebacks in the site. The cline for 14-plate fish and the low frequency of this phenotype downstream of the dispersal barrier were attributed to a generally low level of gene flow from above and a reduction of the effects of gene flow with distance from the trout habitat. If this interpretation is correct, then there must be selection against 14-plate fish in the absence of predatory fishes. Individuals with seven plates on a side must generally be at a disadvantage in the absence of predatory fishes because they appear always to be rare (<10%) or absent where predatory fishes are lacking (Hagen and Gilbertson, 1972; Moodie and Reimchen, 1976*a*).

Bell (1976*b*) performed a similar study in the Santa Clara River drainage nearby. *Gasterosteus aculeatus* was abundant in the river and in two tributaries. Average plate counts of 0.06 and 11.21 per fish occurred in the headwaters of the two tributaries, and these were the extreme mean counts for all samples throughout the drainage. [The lowest plate counts came from the type locality of the nominal subspecies, *G. aculeatus williamsoni* (Miller, 1960)]. The divergent headwater populations were connected by a cline for lateral plate number that stepped at an apparent barrier to dispersal, and maximim diversity [Shannon–Weaver formulation (Peet, 1974)] of lateral plate number phenotypes occurred at the confluence of the two tributaries. Hybridization is necessary to account for the composition of populations and simple mixing of the headwater populations will not account for the cline. Nor does it appear that lateral plate count diversity reflects increased habitat diversity resulting from increased habitat size downstream, because a very small tributary to the Santa Clara River in the area of high plate count diversity also contains a population with high lateral plate number diversity. Subsequent investigations (Bell, unpublished data) show that sticklebacks readily disperse downstream and that samples taken downstream of the confluence of the tributaries also are intermediate to the headwater populations for lengths of the pelvis and pelvic spine, the only other divergent characters measured. After a segment of the stream containing a cline for lateral plate number went dry, it was repopulated by fish with plate counts similar to those of samples collected upstream (Bell, unpublished data). The selective agents responsible for divergence in the headwaters of the tributaries are unknown. However, lateral plate number variation appears to reflect an interaction between divergent selection and gene flow with a downstream bias. Both this study and the study by Bell and Richkind (1981) indicate that gene flow within drainages contributes significantly to lateral plate number variation within drainages.

There are few other studies of lateral plate number variation in low-morph populations within drainages. Hagen's (1967) data on lateral plate number variation within the monomorphic low-morph population in the Little Campbell River indicated an increasing cline for mean lateral plate number toward the headwaters. Hagen and Gilbertson (1972) also noted sharp lateral plate number differentiation between two adjacent sites in Mayer Lake on the Queen Charlotte Islands in association with differences in trout abundance, but Moodie (1972*a*) showed that the populations in these sites represent separate biological species. Hagen and Gilbertson (1973*b*) and Gilbertson (1980) failed to detect lateral plate number variation in different parts of lakes.

Lateral plate number differentiation can occur over short distances. Spatial variation of selective agents favors such differentiation, and it is accentuated by barriers to gene flow. However, within drainages with divergent populations, gene flow appears to be a diversifying factor for lateral plate number within the divergent populations. With this more or less descriptive information in hand, we have numerous hypotheses to test by use of laboratory and field experiments.

2.1.3.6. Conclusions. The number of plates of low-morph *G. aculeatus* appears to have a strong genetic basis, and plate number phenotypes have measurable differences in rates of survival and reproduction. *Gasterosteus aculeatus* is subject to predation by a great variety of predators, including several fish species, garter snakes, many bird species, aquatic insects, and others (Wootton, 1976). Modal plate counts of zero, five, and seven plates per side are common in natural populations, and inasmuch as selective fish predation has caused formation of the seven-plate mode, it seems reasonable to test the hypothesis that predation by other organisms is involved in evolution of other common modal plate counts. Bell and Hagulund (1978) stressed that garter snake predation could not account for all cases of modal plate counts of five per side. Modes at zero plates per stickleback are associated with absence of predatory fishes, but the mechanism that mediates this association is unknown. Reimchen

(1980) showed that predatory insects select against presence of fin spines because they present a grasping surface, and perhaps lateral plates also have this effect. In the absence of other predators, insect predation could possibly be an effective selective mechanism against presence of lateral plates. Many (but not all) populations with zero-plate modes occur in xeric environments, and high temperature could favor lower plate counts. The effects of differential reproductive success are unpredictable from population to population based on present data, but could be an important factor as well. Reimchen (personal communication) noted that 16–18 species of predators in Drizzle Lake, Queen Charlotte Islands, prey on *G. aculeatus* and stressed that simply establishing presence or absence of predatory fishes is inadequate to interpret variation of lateral plates or other defensive structures. More research is needed on the operation of these potential selection mechanisms. The most useful studies will be those in which many of these potential mechanisms are examined in a single population and in which fitness is measured by comparison of phenotype frequencies in successive generations. The genetics of lateral plate number also should be measured in conjunction with studies of selection. Studies on the Lake Wapato population come closest to achieving this goal.

More fundamental problems with which we have not even begun to come to grips are the relationships among different components of fitness. It is not clear, for example, why fish with seven plates on a side should be most likely to avoid predatory fishes and less likely than some other phenotypes to produce offspring. Even more basic is the question of why there are fitness differentials between sticklebacks that differ by as few as one plate on one side of the body! Research to resolve these problems will help add pieces to the puzzle of stickleback lateral plate evolution, but also will contribute to our knowledge of how selection on one trait can influence evolution of others.

Despite remaining uncertainty concerning stickleback lateral plate evolution, some rather important conclusions can be drawn. Clearly they are subject to natural selection. Just as clearly, the same phenotype will have different fitnesses under different conditions. It is possible, however, that the effects of one selection mechanism (i.e., fish predation) can obscure the effects of others. Research on plate number in *G. aculeatus* also demonstrates clearly that one cannot assess fitness on the basis of one selection mechanism or of relative fitness during a restricted part of the life cycle. Further research can build on the extensive foundation already developed and should continue to contribute to knowledge of the operation of such important evolutionary mechanisms as natural selection and gene flow.

2.1.4. Lateral Plate Variation in Other Gasterosteids

Lateral plates are present in all of the gasterosteids except for the fourspine stickleback, *Apeltes quadracus* (Nelson, 1971*a*). In none of the other gasterosteids has lateral plate variation received such close scrutiny as it has in *Gasterosteus aculeatus*. However, some interesting parallels occur between lateral plate variation in *G. aculeatus* and other sticklebacks, and in the light of recent developments in knowledge of lateral plate variation in *G. aculeatus,* we are in a better position to frame useful hypothesis concerning lateral plate variation in other sticklebacks.

2.1.4.1. Lateral Plate Variation in *Gasterosteus wheatlandi.* The blackspotted stickleback, *Gasterosteus wheatlandi,* is the sister taxon to the *G. aculeatus* species complex. It is endemic to eastern North America between northern Newfoundland and western Long Island (Lee *et al.,* 1980) and with few exceptions is restricted to marine and brackish waters (Wootton, 1976). Its biology generally resembles that of marine *G. aculeatus,* with which it always is sympatric and often is confused. However, it is considerably smaller at sexual maturity than *G. aculeatus,* from which it can be distinguished by a number of other characters (e.g., Nelson, 1971*a*; Liem and Scott, 1966).

Wootton (1976), Nelson (1971*a*), and Leim and Scott (1966) characterized *G. wheatlandi* as having only plate counts comparable to those of low-morph *G. aculeatus,* although Perlmutter (1963) has described lateral plate phenotypes corresponding to all three lateral plate morphs in samples from Long Island Sound. Sargent *et al.* (1984) are studying geographic variation of *G. wheatlandi* in New York and New England waters. They confirmed Perlmutter's (1963) observation of three "lateral plate morphs" and found that "complete morphs" are most abundant along northern Long Island, where "low morphs" appear to be virtually absent, and "complete morphs" are absent north of Cape Cod, where samples are nearly monomorphic for the "low morph." In contrast to *G. aculeatus* in western Europe and western North America (Hagen and Moodie, 1982), "complete morphs" are present in the south and absent in the north, where low morphs may be monomorphic (Coad and Power, 1973; Nelson, 1971*a*). The genetic basis of presumptive lateral plate morphs of *G. wheatlandi* is not known, and there is no evidence bearing on their natural selection.

Lateral plate counts in "complete morphs" range between 25 and 30 per side, and "partial morphs" are exceedingly variable for lateral plate number. Partial-plate counts range from 21 abdominal plates per side and an unplated gap just a few plates wide to specimens with eight abdominal plates per side and a single minute ossicle on one side of the caudal peduncle (Sargent *et al.*, 1984). The number of lateral plates in "low morphs" ranges from 17 to five per side, and in none of the samples available to Sargent *et al.* (1984) is there a mode at seven plates per side. Leim and Scott (1966) placed the maximum plate count at six per side, but a sample reported from Quebec by Coad and Power (1973) had a mode at seven plates per side (36%) and plate counts ranged between five and 10.

Because of its similarity and close relationship to *G. aculeatus, G. wheatlandi* provides an interesting opportunity to study evolutionary mechanism in a closely related species. The genetic basis of its presumptive lateral plate morphs must be determined and systematic surveys of lateral plate variation are needed. Unfortunately, *G. wheatlandi* spends most of its life cycle off shore (Krueger, personal communication), and adults can be sampled in inshore waters for only a few months during the breeding season. Thus, although comparisons between *G. wheatlandi* and *G. aculeatus* would be interesting, progress may be slow.

2.1.4.2. Lateral Plate Variation in *Pungitius*. The range of the ninespine stickleback, *Pungitius pungitius,* is circumpolar and extends south along the Asian and eastern North American coasts (Nelson, 1971*b*). The ninespine stickleback is replaced by the nominal species *Pungitius platygaster* (Ukrainian stickleback) in freshwaters between the Black Sea and Balkhash Lake (Münzing, 1969). The lateral plates in *Pungitius* generally are smaller than in *Gasterosteus* and usually form a strong caudal keel that extends well forward from the caudal peduncle. Although not previously recognized as such, phenotypes resembling the lateral plate morphs of *G. aculeatus* occur in *P. pungitius* (Baumgartner, personal communication). Specimens resembling the complete morph have plates running from behind the shoulder girdle onto the caudal peduncle, and a phenotype corresponding to the partial morph has anterior and posterior plate rows separated by an unplated region. There is no phenotype resembling the low morph with only abdominal plates, but a third phenotypic class lacking abdominal plates has only caudal keel plates. Baumgartner (personal communication) found that the number of anterior and posterior plates in adult "partial morphs" are correlated, as they are in *G. aculeatus* (Bell, 1981), and suggested that if paedomorphosis is the cause of this correlation, in "complete morphs" of *Pungitius* the first plates to appear during ontogeny should be those on the caudle peduncle. Thus, it appears that *P. pungitius* also may have a lateral plate trimorphism like that of *G. aculeatus* and the advanced morphs may have evolved by means of paedomorphosis.

There is no direct evidence concerning natural selection of lateral plate phenotypes in *Pungitius*. The species is preyed upon by a variety of predatory fishes (Wootton, 1976) and presumably other predators, and it should be possible to assess the importance of selective predation as a cause of lateral plate phenotype variation, using, for example, the methods of Hagen and Gilbertson (1972, 1973*b*). The genetics of lateral plate phenotypes has not been studied. Thus, much remains to be done to attempt to explain lateral plate variation in this genus.

McPhail (1963) studied variation in the number of abdominal lateral plates (i.e., not including plates in the caudal keel) in North American populations of *P. pungitius*. He found that populations from tidal waters have modal counts of three to five abdominal plates per side, while inland populations have modes between zero and two. McPhail (1963) observed more or less concordant patterns for other traits; inland populations contrast with tidal populations by possession of relatively short pelvic spines, low dorsal spine counts, and high gill-raker counts. He likened the distribution of *P. pungitius* phenotypes to that of *G. aculeautus* in western Europe and suggested that the phenotypic distribution that he observed reflected dispersal from Pleistocene glacial refugia.

Münzing (1969) studied lateral plate variation in Eurasian *Pungitius,* and interpreted the distribution much as McPhail (1963) had for North American *P. pungitius* and as he had earlier for European *G. aculeatus* (Münzing, 1963) to reflect dispersal from Pleistocene glacial refugia. He recognized two subspecies of *P. pungitius, P. p. pungitius* (''partial morph'') and *P. p. laevis* (caudal keel only),as well as *P. platygaster* (''complete morph'') in Europe. The *P. p. laevis* is restricted to warmer inland regions and *P. p. pungitius* and *P. platygaster* occur in colder parts of Europe. *Pungitius p. pungitius* occurs along the Arctic coast to the north Pacific and is joined by two other nominal subspecies in eastern Asia, forming a confusing combination of three nominal subspecies.

McPhail's (1963) and Münzing's (1969) interpretations of lateral plate variation in *Pungitius* as a vestige of divergence in Pleistocene refugia was criticized by Gross (1979). Gross did not study lateral plate variation but analyzed geographic variation of other morphological features in western European *P. pungitius*. He attributed variation in some features to selection based on local environmental conditions and suggested that this mechanism is important for a number of features, including lateral plate variation, throughout the range of *Pungitius*. Another possible mechanism for lateral plate phenotype selection is suggested by Münzing's (1969) data. Southern European and British populations have the lowest plate counts, higher plate counts occur in northwestern *P. pungitius,* and the highest counts occur among *P. platygaster*. This distribution of lateral plate phenotypes bears some resemblence to that of *G. aculeautus,* and thus it may be that Hagen and Moodie's (1982) explanation involving a correlation of the distribution of lateral plate morphs in *G. aculeatus* with climatic severity also applies to *Pungitius*. Regardless of the actual causes of geographic variation of lateral plate phenotypes in *Pungitius,* past experience with *G. aculeatus* causes us to be skeptical of explanations involving divergence in glacial refugia. As Gross (1979) concluded, much work is needed before one can begin to apportion variation between historical and adaptive causes.

2.1.4.3. Lateral Plate Variation in Other Sticklebacks. Nelson (1971*a*) discovered that the brook stickleback, *Culaea inconstans,* possess a single row of at least 30 minute, circular, nonimbricated lateral plates. Moodie (1977) reported mean plate counts of 35.2–39.9 per side in Manitoba populations, but found no association between plate counts and any environmental features.

The fifteen-spined or sea stickleback, *Spinachia spinachia,* has a continuous row of about 41 narrow, strongly keeled lateral plates (Nelson, 1971*a*). Apparently, no one has studied lateral plate variation in this species (Wootton, 1976; Gross,1978*b*).

2.1.5. Conclusions

Lateral plates occur in all sticklebacks except *Apeltes quadracus*. Lateral plate variation has been studied in all species that possess plates, except for *Spinachia spinachia,* but *Gasterosteus aculeatus* has received the most attention by far. Variation of lateral plate phenotypes is poorly understood with respect to genetics and adaptive significance in species other than *G. aculeatus,* and only in the latter species and *Pungitius pungitius* do we have a reasonable understanding of geographic variation.

Lateral plate morphology is controlled by at least two gene systems in *G. aculeatus;* a Mendelian system with at least two interacting loci controls the lateral plate morph, and an apparently polygenic system determines the number of lateral plates in low-morph phenotypes. It is not known whether the same gene system affects the number of lateral plates in the other two morphs (i.e., partial and complete). While Hagen and Gilbertson (1973*a*) have provided a reasonably general model for plate morph inheritance, two other studies either were misinterpreted (Münzing, 1959) or were uninformative (Avise, 1976). Similarly, Hagen (1973) provided strong evidence of low-morph plate number heritability, but one less extensive data set (Lindsey, 1962) seems to be inconsistent with his conclusions. Additional studies of lateral plate genetics in *G. aculeatus* are needed to determine whether or not genetic mechanisms for determination of plate phenotypes vary among populations. Similarly, there is a clear need to study lateral plate genetics in *P. pungitius* and *Gasterosteus wheatlandi,* both of which exhibit marked variation.

The adaptive significance of lateral plate phenotypes has been examined only in *G. aculeatus*. Selection on lateral plate morphs appears to involve a selection mechanism that varies on a large scale and may be related to climate, but local selection differences must be invoked to account for divergence between demes in different parts of the same drainage and in adjacent drainages with contrasting habitats. Available data on selection of lateral plate morphs is limited and inconclusive. Selection of low and, to a limited extent, partial morphs is understood much more thoroughly, principally through the efforts of Hagen and his colleagues. However, it is obvious that the situation is complicated, and detailed studies of natural populations will be needed to tease apart the interacting and sometimes opposing effects of different selection mechanisms.

Lateral plate variation in *G. aculeatus* has been used with good success as a tool to investigate a number of interesting evolutionary phenomena. For example, Hagen (1967) could distinguish between parapatric species of the *G. aculeatus* species complex on the basis of plate morph alone, and Moodie (1972*a*) studied two parapatric species of low-morph sticklebacks that were sharply divergent for lateral plate number, reflecting strong selection differences in their contrasting, adjacent habitats. Hagen and Gilbertson (1972) provided suggestive evidence that low-morph lateral plate number tracks environmental variation in the intensity of predation by fishes, and subsequent studies (see Section 2.1.3.3) have confirmed this evidence. Despite general concordance between spatial variation of selective regime and lateral plate number phenotypes, Bell and Richkind (1981) produced evidence that gene flow within a stream contributes to phenotypic variation and the generation of clines. Although all predatory fishes studied selectively underutilize low-morph fish with seven plates per side, a reasonably similar predator, the garter snake, is virtually unable to capture specimens with five plates on a side (Bell and Haglund, 1978). The effects of numerous other predators (Wootton, 1976) of *G. aculeatus* on the genetic composition of natural populations is unknown. However, the limited data available clearly point out the danger of facile inferences of the effects of predation by many species based on observation of a few of the predator species. Similarity of lateral plate variation in western Europe and western North America indicates that parallel geographic arrays may evolve independently under the direction of natural selection. Although lateral plate variation already has provided interesting insights and important examples of evolutionary mechanisms, the most important lesson has been that phenomena that contribute to evolutionary divergence are numerous and their interactions can be complex. Much more remains to be learned about lateral plate variation in *G. aculeatus,* and as this knowledge grows, important insights of general applicability may be expected.

2.2. Variation in Coloration

Gasterosteus aculeatus occupies diverse habitats among which lighting conditions, the amount of cover, and species and abundance of predators vary greatly. They show the usual associations between the photic environment and body coloration encountered in fishes and thus there is great variation among populations (Hardy, 1978). Specimens collected from open water may be very silvery with some countershading (i.e., darker above). Specimens from shaded habitats or from habitats with abundant vegetation tend to be dark above and lighter below with a series of bars projecting downward along the sides of the body. The dark coloration ranges from brown to green and the light ventral regions usually are various shades of beige to silver. In deeply stained water extreme melanism may occur (Moodie, 1972*a*; McPhail, personal communication). The range of variation of body coloration in *G. aculeatus* is not unusual for a teleost.

Nuptial coloration is much more interesting and has received more attention. Reproductive females show the usual coloration, but as the brood of eggs matures, the abdomen swells, and the swollen region is white. Male nuptial coloration is characterized by red, which may occur inside the bucal cavity, on the ventrolateral surfaces of the head and shoulder, and pelvic girdles, on the pelvic fin membrane, along the lateral surface of the body between the pectoral fin base and the origin of the anal fin, and on the body along the base of the anal fin. The remainder of the body ranges from pale blue to iridescent green, usually superimposed on the normal countershading and barring.

Although the essential features of the reddish male nutpial coloration is expressed to some degree in most populations, the precise distribution of red seems to vary greatly among (Bell, unpublished data) and within populations (Semler, 1971; Moodie,1972*a*; Gilbertson, 1980), and various shades of red are reported (Hardy, 1978). For example, in some freshwater populations from California the red coloration is restricted primarily to the ventral surfaces of the head and shoulder girdle, while marine sticklebacks from Long Island Sound have the maximal distribution of red color described above (Fig. 5). Geographic variation in the distribution of red coloration has not been studied.

Gilbertson (1980) studied a large population in which he recognized four nuptial coloration phenotypes. All breeding males have red inside the mouth, and some males have red on ventrolateral portions of the head and are otherwise metallic green above and pale blue below. A gold phenotype is canary yellow over the entire body, a green one has various shades of green, and the "drab" phenotype exhibits characteristic cryptic

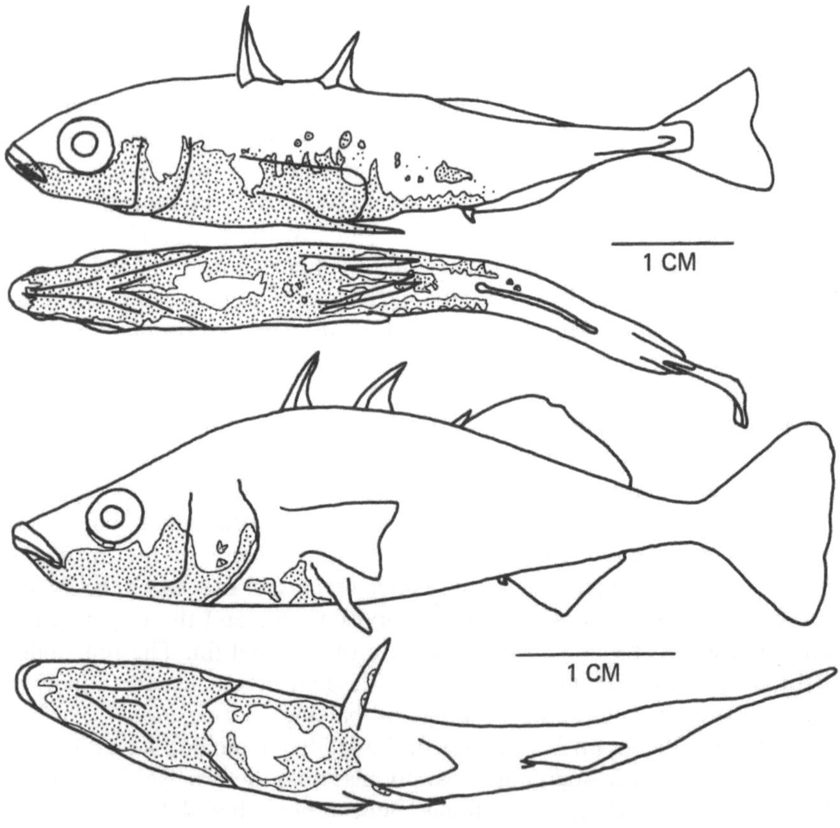

Figure 5. Variation in the distribution of red nuptial coloration (stippled) in male sticklebacks. A specimen (top) from Long Island Sound, eastern North America, showing extensive red coloration characteristic of marine, anadromous, and some freshwater populations, and a specimen (bottom) from Corcoran Lagoon, a salt marsh in California (Bell, 1979), showing limited distribution of red coloration characteristic of freshwater populations in California.

coloration. Gilbertson (1980) did not study the genetics of this variation, but the spatial distribution of color phenotypes is heterogeneous. Red males are more common in bays than the open lake coast, and within bays red and drab phenotypes occur more often on mud to sand substrates, but gold males are equally frequent over mud–sand and gravel–pebble bottom. There were no obvious differences among nuptial phenotypes for nesting behavior and ecology, but some associations with other morphology was detected. Gilbertson (1980) speculated that red males may enjoy a sexual advantage but suffer greater predation by fishes.

Melanism has been reported from a few populations, primarily from the Pacific Northwest, but melanism in this region is more common than indicated in the literature (McPhail, personal communication; Reimchen, personal communication). The two most thoroughly studied cases are a melanistic population from Mayer Lake in the Queen Charlotte Islands, British Columbia (Moodie, 1972*a,b*) and another from the Chehalis River system and some adjacent drainages (McPhail, 1969; Hagen and Moodie, 1979; Hagen *et al.*, 1980). Semler (1971) and Bell (1982*a*) also reported melanism in *G. aculeatus*, and the significance of the latter is that it is the only report of melanism in *G. aculeatus* outside of the Pacific Northwest.

Genetics of nuptial coloration has been studied only in populations from the Chehalis River and adjacent drainages. In these populations sexually mature males become entirely black, and red coloration is lacking. McPhail (1969) showed that allopatric populations with red and black coloration breed true and that the F_1 hybrids are intermediate. Hagen and Moodie (1979) extended these results using F_1 and backcrosses. Their F_1 hybrids resembled McPhail's (1969) with interspersed red and black spots on the throat plus dark brown or green dorsally, and reciprocal crosses yielded equivalent results. The F_2 and backcrosses produced appropriate ratios for a one-locus, two-allele mechanism without dominance. They suggested that polygenic modifier loci influence the expression of this locus, because of variable amounts of red and black pigment in the hybrids and of greater resemblance in backcross progeny to the population to which backcrossing was performed.

There is ample evidence of selection on nuptial coloration. Using nonred males that were artificially colored red or sham-marked with a clear lipstick or nail polish, Semler (1971) demonstrated female preference for red males. McPhail (1969) also demonstrated preference for typical red males over all-black ones by females from the red-male population or by allopatric females from the black-male population, but lack of preference by parapatric females from black-male population appears to represent reproductive character displacement (Bell, 1976*c*). Using male fish artificially colored with red or clear lipstick, but otherwise similar phenotypically, Moodie (1972*b*) showed that predatory trout are more likely to attack red males. McPhail (1969) and Hagen and Moodie (1979) demonstrated a very precise association between the distribution of black-male sticklebacks and the endemic mudminnow, *Novumbra hubbsi,* in drainages on the Olympic Peninsula. McPhail (1969) concluded from experiments in which mudminnows oriented preferentially toward red males over all-black males that *Novumbra* must be more likely to prey on the fry being tended by red males. This conclusion was supported by existence of an avoidance response by fry of the black-male population to orientation by the mudminnow and absence in typical populations, but McPhail (1969) had misgivings about this conclusion because he never found stickleback fry in stomachs of wild *Novumbra*. Hagen *et al.* (1980) also failed to find sticklebacks in *Novumbra* stomachs and demonstrated that it orients away from black males, not toward red ones. They also found that black males have greater nesting success than reds in the laboratory because of greater territorial intrusion by territorial male *Novumbra* into territories of red males. However, in the absence of mudminnows, red males acquire larger territories than black males. They explained the distribution of black male populations as convergence of stickleback nuptial coloration on that of *Novumbra* (which is black) which facilitates territorial defense.

Evolution of coloration is poorly understood compared to some other traits of *G. aculeatus*. Subtle variation in the extent of red male nuptial coloration in otherwise unexceptional populations has not been studied, although it could play a critical role in mate choice and territoriality (Wootton, 1976). A number of melanistic populations have been studied, and though they tend to occur in darkly stained water in sympatry with predators, the critical work to demonstrate an advantage with respect to fish predation has not been done. Red males generally enjoy a reproductive advantage over nonred ones and apparently are at a disadvantage with respect to predation by fishes. Except for the black-male populations that are sympatric with mudminnows, however, our knowledge of the evolution of melanism, loss of red nuptial coloration, and variation among populations for the distribution of red coloration and the shade of background coloration on reproductive males is seriously deficient.

2.3. Gill Raker Number Variation

Gill rakers are bony projections that are joined by connective tissue to each gill arch and run in anterior and posterior rows along the medial surface of the arches. The gill rakers of fishes prevent particles from leaving the buccopharyngeal cavity via the gill slits. The anterior rakers on the first gill arch are larger than those in other rows, and it is the number of gill rakers in this row that is reported. The spaces between the gill rakers determine the maximum size of particles that can escape from the buccopharyngeal cavity. Among closely related populations, the number of gill rakers on the first gill arch is strongly correlated with the spaces between rakers, and the characteristic size of items in the diet (Larson, 1976). Thus, the number of gill rakers may be a good indicator of trophic differences among populations. Gill raker counts are reported as taxonomic characters, and counts appear in numerous publications. However, interpretation of the structure requires knowledge of the ecology of the population that was sampled, and relatively few studies of gill raker counts provide sufficient information.

The genetics, geographic variation, and function of gill rakers have been studied in western North American populations. Hagen (1973) reported that heritability of the number of gill rakers in one population is high ($h^2 = 0.58$), and presumably this measure of heritability is representative of other populations. Hagen and Gilbertson (1972) compared gill raker counts of *G. aculeatus* populations from small streams and large oligotrophic lakes. Although they detected no large geographic trends, they found that mean gill raker counts from populations in large lakes tend to be high (19–21.7) and those from streams to be lower (14.6–17). Stream samples made near the confluence with large rivers tended to have higher counts, but closely adjacent stream and lake populations may differ significantly. Hagen and Gilbertson (1972) reported one stream population with an unaccountably high mean gill raker count, and Bell (1982*b*) also sampled a stream population in which the gill raker counts were in the range of lake populations. However, care must be taken in comparing stream and lake populations, because Hagen (1967) showed that anadromous *G. aculeatus* breeding in a stream have gill raker counts ($\bar{X}$ = 21.8) comparable to those of lake populations. Moodie and Reimchen (1976*a*) detected a possible increasing trend for raker number with increasing lake size, although they pointed out problems with their data. Larson (1976) studied the behavior and morphology of two sympatric species of *Gasterosteus* from a lake. One species is benthic, has low gill raker counts, and eats large food items, which it handles more efficiently than does the second species. The second species is liminetic, has higher gill raker counts, and eats smaller zooplankton, which it handles more efficiently than does the benthic species. Thus, the number of gill rakers is a heritable trait that influences feeding efficiency, and population differentiation according to habitat type may precisely reflect differences between even closely adjacent populations in the size of their food.

2.4. Body Size Variation

Except for McPhail's (1977) study of minimum body size of female *G. aculeatus* at first reproduction, there apparently is no systematic treatment of geographic variation and genetics of body size in any of the sticklebacks. Myers (1930) noted in passing that in moving southward along the west coast of North America one observes that body size in *G. aculeatus* tends to decrease. Indecd populations to the north may contain very large individuals (Moodie, 1972*a*) and populations in southern California, near the southern end of the distribution, tend to be small (Bell, unpublished data). However, because of the difficulty of characterizing body size in an organism with indeterminate growth, geographic variation of body size in *G. aculeatus* is poorly documented. On Vancouver Island alone, minimum female body size at first reproduction varies among populations by 100%! Thus, it is premature to make any generalizations.

McPhail (1977) found that there is a significant relationship between minimum size at first reproduction in field-caught and laboratory-reared *G. aculeatus,* indicating a genetic basis for interpopulation variation. Based on 11 populations (Fig. 6), he suggested that populations subject to the most intense predation either have relatively small or large body size. He tentatively suggested that if the predatory fish are small, large body size might be selected and that presence of large predators should select for small body size, because it might be impossible for a stickleback to become large enough to be unavailable to the predator. However, McPhail (1977) noted that lakes in which predation by fish is not intense also have populations spanning much of the range of body sizes, so unidentified factors must influence evolution of body size in *G. aculeatus*. While it is true that a population with very large body size (Moodie, 1972*a,b*) is subject to intense fish predation, McPhail's (1977) data only are suggestive of a relationship between the intensity of predation and evolution of body size. Finally, McPhail (1977) pointed out that evolution of body size probably occurs rapidly *in situ* because the lakes are reasonably isolated habitats and Vancouver Island was covered with ice until 13,000 years ago.

The only other evidence concerning evolution of body size in *G. aculeatus* is from a study of differential mortality within a population in a drying pothole in a xeric stream in southern California (Bell and Baumgartner, unpublished data). During approximately 3 months of observation dead fish found in the pool were significantly longer than live specimens collected simultaneously, and over the period of observation mean size in the population declined significantly. These results suggest that smaller body size of freshwater sticklebacks toward the southern end of the range

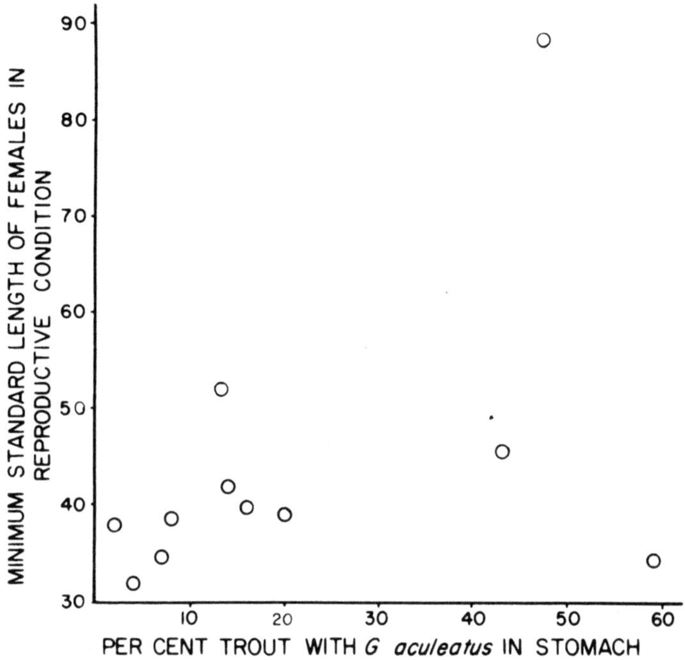

Figure 6. The relationship between minimum size at first reproduction in populations of *Gasterosteus aculeatus* and the intensity of predation by cutthroat trout (*Salmo clarki*) reported by McPhail (1977). The largest minimum size shown was estimated by subtracting two standard errors from the mean reported by Moodie (1972*a*).

could be favored by differential mortality during episodic periods of stress during drought.

Very little is known about evolution of body size in *G. aculeatus*, and nothing is known about it in other sticklebacks. Considering the well-known relationship of fecundity to body size (Wootton, 1973) and its potential as a determinant of male reproductive success (Rowland, 1982), we should expect selection generally to favor increase until other selective factors come to bear on it.

McPhail's (1977) study established a genetic basis for body size and demonstrated the utility of minimum size of reproductive females as a criterion for body size. However, at present there is no compelling evidence concerning selection on body size. Using McPhail's (1977) criterion for body size, it should be relatively simple to determine the relationship between body size and predation and other environmental variables.

2.5. Dorsal Spine Number Variation

The genera of sticklebacks often are characterized by the number of dorsal spines: the fifteen-spined stickleback, *Spinachia,* has 15 (rarely 14 or 16), the ninespine stickleback, *Pungitius,* has eight to 10 (rarely seven to 12), the brook stickleback, *Culaea,* has four to six (rarely one or seven), the fourspine stickleback, *Apeltes,* has four or five (rarely three or six) and the threespine and black-spotted sticklebacks, *Gasterosteus,* have three (rarely 0 to five) (Nelson, 1977; other references cited below). In all of the sticklebacks the dorsal spines are followed by a membrane, and none of them are attached to the dorsal fin. In *Gasterosteus* and *Culaea* the dorsal spines are in the mid sagittal plane, but in *Pungitius, Apeltes,* and *Spinachia* successive spines diverge alternately to either side. Geographic variation of the number of dorsal spines frequently is reported for some sticklebacks, but genetics and selection of spine number generally are not understood.

2.5.1. Dorsal Spine Number Variation in *Gasterosteus aculeatus*

As its name implies, *G. aculeatus* typically has three dorsal spines. Only a few studies have reported dorsal spine number frequenceis, and the genetics and adaptive significance of dorsal spine number variation are poorly understood. Penczak (1963) crossed three-spined and four-spined males to four-spined females, and all offspring had three dorsal spines, providing no evidence for a genetic basis for dorsal spine number. Lindsey (1972) crossed three-spined parents and obtained progeny with reduced numbers of spines in about 5.9% of the offspring and a fourth dorsal spine in 1.8%. His experiment was not designed to determine the genetic basis of spine number, but he observed that one family of 42 progeny included nine with only two spines, while another family of 13 progeny reared under identical conditions produced all three-spined fish. Gross (1977a) performed 26 crosses using parents with unspecified spine number, and 10 produced progeny with other than three spines, two of which included 35% two- and four-spined specimens, respectively, Lindsey's (1962) data also indicate that high rearing temperature may induce four-spined phenotypes. Thus, Lindsey's (1962) results indicate that dorsal spine number has a genetic basis, but is liable to the influence of environmental conditions. Campbell (1979) performed three interpopulation crosses between sticklebacks from a population with reduced dorsal spine number and pelvic structure (to which he referred as "anomalous") and individuals from populations with characteristic pelvic structure and modal dorsal spine counts of three. One cross produced all anomalous, another all normal, and the third anomalous, normal, and intermediate offspring. These results indicate a genetic basis for differences between populations for dorsal spine number. The data are by no means satisfactory, but are consistent with the notion that there is a genetic influence on dorsal spine number that may interact with environmental variables.

Penczak (1965) summarized published dorsal spine number data from Europe and presented original data from Polish populations. Gross (1978*a*) supplemented these data with additional dorsal spine counts from European populations. Both data sets indicate that, on the average, more than 97.6% of specimens have three dorsal spines and the highest intrapopulation frequencies of specimens with other than three spines were 11.1% for two-spined and 17.2% for four-spined specimens. Generally, the frequencies of four-spined specimens tend to be higher than those of two-spined specimens in natural populations, and the frequency of samples with one or more four-spined specimens is greater than that with at least one two-spined specimen. Gross's (1978*a*) data also indicate that low-morph, freshwater populations are much more likely to include spine counts other than three than are marine, completely plated populations; only 33% (n = 12) of the completely plated population samples included specimens with two or four spines, but 73% (n = 53) of the samples from low-morph populations include two- or four-spined specimens. Penczak (1965) noted that two-spined specimens were most likely to occur in southern Europe and Africa, but this conclusion is based on only two samples from this part of the range compared to samples pooled into nine groups from elsewhere in Europe. However, Hoogland *et al.* (1957) showed that dorsal spines deter predatory fishes, and Gross (1977) argued that reduced predation by fishes in southern Europe was responsible for lower plate counts there. If Gross (1977) was correct, apparently higher frequencies of two-spined specimens in southern Europe also may be a manifestation of lower predation intensity by fishes. More work is needed in Europe and elsewhere to determine geographic variation of dorsal spine number.

There certainly appears to be ample dorsal spine number variation in natural populations, and this trait appears likely to have a genetic basis. Under appropriate selection regimes populations that are divergent for dorsal spine number are expected. Despite the ubiquity of four-spined individuals in many populations, only two cases of significantly elevated frequencies of four-spined specimens have been reported (Gross, 1978*a*). Bell and Baumgartner (1984) are studying another population from a small freshwater pond in which 18.7% (N = 230) of the specimens have four spines. This population is isolated and genetic drift could account for abundance of four-spined fish. Elevated frequencies of four-spined *G. aculeatus* in natural populations are rare and never exceed about 20% of the population.

Populations in which spine number reduction occurs at a high frequency also are rare but regularly occur. Such populations have been reported from lakes on islands off of British Columbia (Moodie and Reimchen, 1973, 1976*a*; Bell, 1974) and from islands off of Scotland (Campbell, 1979), and Bell (1974) and Bell and Haglund (1982) reported them from temporally heterogeneous samples of *Gasterosteus doryssus* (a member of the *G. aculeatus* complex) from a Miocene lake deposit in Nevada. Moodie and Reimchen (1976*a*) demonstrated significant dorsal spine deficiency in three lake populations from the Queen Charlotte Islands, and in their Boulton Lake sample ($N = 221$) 46.6% of the specimens had fewer than three dorsal spines (43.4% two-spined). In a sample from Texada Island, British Columbia, 1.5% were one-spined, 38.3% were two-spined, and 60.2% were three-spined ($N = 201$) (Bell, 1974). The fossil assemblage reported by Bell (1974) consisted of 5.6% zero-spined, 68.5% one-spined, and 13.9% two-spined specimens. Subsequent sampling by Bell and Haglund (1982) at six levels over a 150,000-year period revealed that 96.4% of the specimens had zero or one spine, but that the frequency of zero- and one-spined specimens is temporally heterogeneous. Two samples ($N = 129$) within the earliest 39,000 years of the deposit had 75.2% zero-spined specimens, while the remaining four samples ($N = 235$) include 84.7% one-spined specimens. In both the Texada Island and fossil material, two-spined specimens lack the first dorsal spine and one-spined specimens lack the first and second one; in Boulton Lake, Queen Charlotte Islands, it is almost invariably the second spine that is missing in two-spined fish (Reimchen, 1980). Dorsal spine deficiency also was reported from four populations from North Uist, off of Scotland, but little more is known about these populations (Campbell, 1979, personal communication). In all of the populations noted, the pelvic girdle also tends to be reduced to a vestige that lacks pelvic spines. Lateral plate counts in these populations also tend to be low, and in the Texada Island population, specimens with vestigial pelvic girdles had fewer dorsal spines and lateral plates than those with normal pelvic structures (Bell, 1974). Dorsal spines clearly function in defense against predatory fishes (Hoogland *et al.*, 1957; Reimchen, 1983) and there appears to be an association between absence of (Moodie and Reimchen, 1976*a*; Reimchen, 1980) or refuge from (Larson, 1976; Campbell, 1979) predatory fishes and spine deficiency. However, Giles (1983) has suggested that spine deficiency is a response in lakes with low calcium ion concentration to selection for low skeletal calcium content, and populations in which pelvic structures are severely reduced without commitant dorsal spine loss recently have been observed (Bell, Francis and Havens, unpublished data). Thus, the causation of dorsal spine loss requires further study.

The only detailed study of selection on a *G. aculeatus* population with a high frequency of spine-deficient individuals was undertaken by Reimchen (1980) in Boulton Lake, Queen Charlotte Islands. He characterized phenotypes primarily by the presence or absence of the second dorsal spine and the bilateral pelvic spines and believed that predation by birds and macroinvertebrates (primarily dragonfly nymphs) are the major selective mechanisms on spine number. He found that females tend to be more spiny than males and are more frequent in open water, where they are more liable to bird predation. Fully-spined phenotypes were most common in open water and least-spined (two dorsals and no pelvic spines) phenotypes near shore in shallow water. Reimchen (1980) believed that the absence of predatory salmonid fishes allowed piscivorous insects to develop unusually dense populations, which were able to expand into the habitats frequented by fry. He interpreted spatial variation of spine number phenotypes to reflect predation pressure by birds in open water and by insects in the shallows. Unfortunately, experimental attempts to test for selective predation by dragonfly nymphs failed to detect selection, but Reimchen (1980) pointed out the necessity for exceedingly large samples to detect selection. Size dependence of insect selection detected by Hay (1974) is another factor that might have confounded Reimchen's (1980) data. Irrespective of his experimental results, Reimchen's (1980) demonstration of concordant spatial variation of spine number phenotypes and predators, combined with data on *Culaea* and *Pungitius* (see Section 2.7.2) indicates that an interaction between predation by vertebrates, which select for presence of spines, and insects, which favor absence of spines, strongly influences selection for dorsal spine number. The mechanism suggested by Reimchen (1980) could be a general explanation for spine number variation in sticklebacks. Similar interactions have been described by Kerfoot (1975) for cladocerans subject to seasonally fluctuating predation by copepods and sticklebacks. Such interactions may be common features of adaptation to predation (Zaret, 1980).

2.5.2. Dorsal Spine Number Variation in *Apeltes*

Blouw and Hagen (1981) have initiated a major study of dorsal spine number variation in *Apeltes quadracus* and provide an overview of variation of this trait. They reported that spine number varies between one and seven, and rank in abundance as follows: 4,5 >3,6 > 1,2,7. Spatial variation of dorsal spine number is the rule and even adjacent populations may differ. However, within a site, differences between size (age) classes and sexes are negligible. Samples collected at one site at different times of day, stages of the tide cycle, months of the year, or over periods of

1–6 years do not differ significantly. Even samples taken at six localities more than 50 years apart have stable spine number composition. Dorsal spine number phenotypes also did not vary significantly among microhabitats within one of their study sites.

Blouw and Hagen (personal communication) are studying geographical variation, genetics and differential predation on dorsal spine number phenotypes of *Apeltes*. Hagen and Blouw (1983) reported that dorsal spine number is highly heritable (average $h^2 = 0.61$) but other details of this work are not yet available.

A major survey of dorsal spine number variation in *A. quadracus* was conducted by Krueger (1961). His samples came from all but the most northern part of the range, but no geographic trends or correlations with other characters were detected. Most of his large samples were variable for dorsal spine number, and he suggested that dorsal spine counts vary inversely with salinity. However, Hagen (personal communication) has been unable to confirm this relationship.

2.5.3. Dorsal Spine Number Variation in Other Sticklebacks

There is not much information on dorsal spine number variation in other sticklebacks, and most available information is on geographic variation. Gross (1978*b*) failed to detect much variation among populations of *Spinachia spinachia* from northwestern Europe, sample means not deviating much from 15 spines. Similarly, all but two of 433 *Gasterosteus wheatlandi* collected between Maine and Long Island Sound had three spines (Sargent *et al.*, 1984). *Culaea inconstans* exhibits great dorsal spine number variability, but Moodie (1977) did not detect any association between spine number variation and environmental variables. The only spatial pattern that he observed was higher frequencies of six-spined specimens in two more heavily wooded regions. Reist (1980*a*) found that predatory beetles (*Dutiscus* sp.) are a selective predator on *C. insonstans* with and without pelvic spines only when they had five but not six dorsal spines.

McPhail (1963) presented extensive data on dorsal spine number variation in *Pungitius pungitius* in North America. He rejected earlier claims of a north–south cline for dorsal spine number. Generally, coastal populations appeared to have higher dorsal spine counts than inland populations, but there was no relationship to salinity. McPhail (1963) suggested that frequent abrupt transition between high- and low-spined populations could possibly be a strictly phenotypic response to the conditions during development. He attributed the major pattern of spatial variation to postglacial dispersal from Bering and Mississippi refugia, an explanation that has been discredited for lateral plate variation in *G. aculeatus* (Hagen and

McPhail, 1970; Bell, 1976*a*; Hagen and Moodie, 1982). Nelson (1971*b*) observed that dorsal spine number is relatively low in regions of Canada in which *P. pungitius* tended to have reduced pelvic structures, but this region is contained within the inland area where spine number tends to be low. Nelson (1971*b*) failed to observe an association between spine number phenotypes and the degree of development of the pelvis within his largest sample. Dorsal spine number variation in *P. pungitius,* as in the other sticklebacks, is poorly understood.

2.6. Spine Length Variation

Except for *G. aculeatus,* variation in spine length has received little attention, and in the threespine stickleback such study has been restricted mostly to geographic variation. Nothing is known about the genetics of spine length, and except for Moodie's (1972*b*) study, evidence concerning selection on spine length is inferential.

The largest spines in *Gasterosteus* are the first two dorsal spines and the pelvic spine on either side; the spines preceding the dorsal and anal fin are much shorter. Hoogland (1951) described the locking mechanism for the large spines, and Hoogland *et al.* (1957) showed that the spines of *G. aculeatus* are effective deterrents against predation by fishes by piercing the buccal tissues. Given a choice, the predators selectively removed nonspiny prey and became conditioned to avoid threespine sticklebacks on sight, but removal of the spines demonstrated that sticklebacks are palatable. Hagen and Gilbertson (1972) observed associations between greater pelvic spine length (with which dorsal spine length is correlated) and the presence of predatory fishes and a modal plate count of seven lateral plates per side. Similarly, Moodie and Reimchen (1976*a*) sampled freshwater populations in the Queen Charlotte Islands and Gross (1978*a*) sampled marine and freshwater populations in Europe and found that freshwater populations subject to predation by fishes tend to have longer spines. Gross (1978*a*) also found that marine sticklebacks have longer spines than freshwater ones, and he also found that populations with the longest spines have the most strongly denticulated spines. The only direct evidence that there is selection operating on spine length is Moodie's (1972*b*) observation that both male and female sticklebacks taken from trout (*Salmo clarki*) stomachs have relatively short pelvic spines compared to specimens seined from the same population, but the difference was significant only in females. There is no evidence on selection for shorter spines; possibly predation by insects, which seems to favor loss of spines (Section 2.4), also selects for shorter spines. Although the evidence is limited, it appears that predation by fishes selects for greater spine length in *G. aculeatus*.

Nelson (1969) examined lengths of the first dorsal spine and pelvic spine in the brook stickleback, *Culaea inconstans,* which is endemic to North America. Spine lengths were not sexually dimorphic, but varied greatly among populations. Dorsal and pelvic spine lengths generally were greatest in the Great Lakes region and decreased to the northwest, but adjacent populations sometimes differed greatly. Correlations between dorsal and pelvic spine lengths were greater between than within populations, suggesting that allometry is not the cause of the correlation. Nelson (1969) considered a number of hypotheses to explain the trend for shorter spines toward the northwest, but the best-developed hypothesis suggested that the absence of alternate prey species in the northwest of the range made it impossible for predators to switch to less spiny prey, rendering spines an ineffective defense mechanism. However, in the absence of less spiny alternative prey species, there still should be predator-mediated selection for longer spines.

McPhail (1963) measured pelvic spine length in North American populations of *Pungitius pungitius*. Tidal populations tend to have longer spines those from fresh water. He suggested that tidal populations may exhibit a cline for spine length decreasing to the north, but observed no spatial trend in fresh water populations. McPhail (1963) did not consider the possible role of natural selection in spine length variation, and it did not fit into his model of postglacial secondary contact. He speculated that at high salinity, temperature differences might influence pelvic spine length. In light of subsequent work on *Gasterosteus,* the role of fish predation in spine length variation in *Pungitius* must be considered carefully.

2.7. Pelvic Structure Variation

In three of the five gasterosteid genera, *Gasterosteus, Pungitius,* and *Culaea,* the structure of the pelvis is strongly divergent from the primitive condition in teleosts (Nelson, 1971*a*). In each of these three genera, populations in which the pelvis is reduced to a vestige or is absent have been reported. In this section I discuss variation within populations of *G. aculeatus* with typical pelvic structure and then consider those with extreme reduction. Finally I consider extreme pelvic reduction in *Pungitius* and *Culaea,* because similar selection mechanisms for pelvic reduction must occur in all three genera, and pelvic reduction in these genera has been studied in more detail than in *G. aculeatus*.

The pelvis in *Gasterosteus, Pungitius,* and *Culaea* is a bilateral structure consisting on each side of an element of the pelvic girdle, one spine, and one or two fin rays (Fig. 7). Nelson (1971*a*) described the gasterosteid pelvic girdle in detail and gave a synonomy of terms for pelvic structures; I have adopted this terminology. The bilateral elements of the pelvic girdle are joined medially by a deeply emarginate suture. Each element consists of an anterior and a posterior process in the ventral plane, anterior and posterior to the pelvic spine, respectively. The ascending branch is dorsal to the pelvic spine, emanating from the posterolateral margin of the anterior process, and it lies in the lateral plane. The pelvic girdle lies just under the skin. During ontogeny, the pelvic girdle of *Gasterosteus* appears as an anterior ovoid mass of bone and a posterior V-shaped ossification, which correspond to the anterior and posterior processes, respectively; the ascending branch develops as an appendage of the anterior process (Bell, unpublished data). The pelvic spine and ray appear as paired struts of bone that subsequently fuse at their tips, and fusion progresses proximally (Bell, unpublished data).

In these three stickleback genera the pelvic girdle encloses the ventral

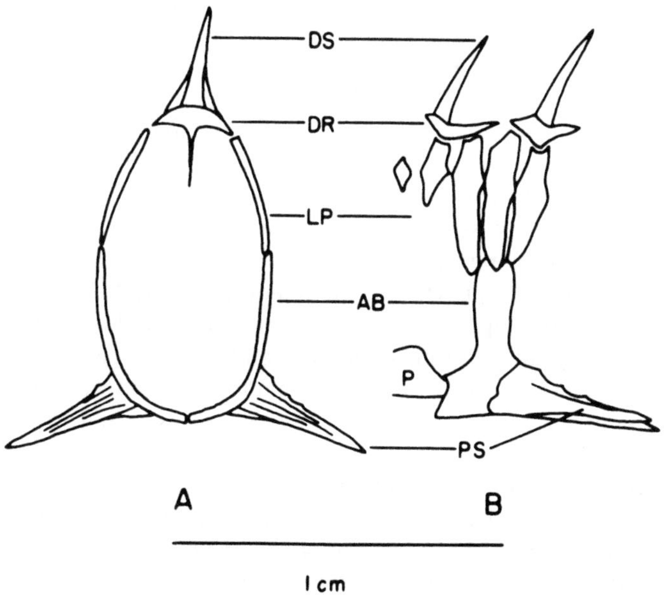

Figure 7. Diagram of the structure of the pelvic skeleton of *Gasterosteus aculeatus*, showing its relationship with the lateral plates and dorsal spines; (A) cross section, (B) lateral view. Abbreviations: (AB) ascending branch of the pelvis; (DR) radial bone of the dorsal spine; (DS) dorsal spine; (LP) lateral plate; (P) pectoral girdle; (PS) pelvic (ventral) spine.

and lateral surfaces of the abdomen, and the pelvic spines provide further protection. In *G. aculeatus* the pelvic girdle abuts or is interdigitated with the lateral plates, which in turn contact the pterygiophores that support the dorsal spines. Thus in *Gasterosteus,* the entire abdomen is encased in a chain-mail-like structure from which spines project at approximately 120° angles. In *Pungitius* and *Culaea* the lateral plates are more weakly developed or are absent and the spines tend to be shorter.

2.7.1. Pelvic Variation in *Gasterosteus aculeatus*

In populations of *G. aculeatus* with normally developed pelvic structures, the shape of the anterior and posterior processes vary and may be sexually dimorphic (Penczak, 1965). The number of forks at the dorsal end of the ascending branches appears to be the most variable portion and may be bilaterally asymmetric, but it is not sexually dimorphic (Penczak, 1965). Penczak (1965) indicated that marine populations tend to have fewer forks to the ascending branch. Gross (1978*a*) examined pelvic girdle structure in relation to the intensity of fish predation and found that ascending branches tend to be longer in populations subject to predation by fishes and in monomorphic complete-morph populations in the northern part of their range. He also found that the symphysis joining the pelvic elements is shorter in populations not subject to predation by fishes. Myers (1930) suggested that the degree of indentation of the anterior margin of the pelvic girdle, which may be inversely related to the length of the pelvic symphysis, increases toward the south in western North American populations of *G. aculeatus*. He provided no data, however, and no one has examined this feature subsequently. Reimchen's (1983) observation that dorsal spine pterygiophores and lateral plates overlap more when dorsal spines are longer may provide an explanation for variation of pelvic girdle structure. This overlap tends to buttress the spines, holding them in place, and the pelvic girdle well may similarly stabilize the pelvic spines when they are grasped by a predator. Additional research on variation of pelvic structures in relationship to environmental variation and to variation in other morphological features should prove interesting.

Gasterosteus aculeatus populations in which the pelvic girdle tends to lack the pelvic spine and to be reduced to a vestige or to be entirely absent have been reported from a number of disjunct sites. The full range of pelvic structures from complete absence to normal development occurs within some populations, and bilateral asymmetry is common, but most specimens with reduced structures have an ovoid vestige or no pelvic girdle at all (Bell, 1974; Reimchen, 1980). The first cases were reported from the Queen Charlotte Islands by Moodie and Reimchen (1973), who also noted existence of similar populations on Texada Island, British Columbia, and in Ireland. Bell (1974) subsequently described reduced pelvic structures from the Texada Island population and in *G. doryssus* from the late Miocene Truckee formation, Nevada. Most recently, Campbell (1979) noted the presence of such populations in islands off of Scotland. Thus, populations of *G. aculeatus* with greatly reduced pelvic structures occur in at least four disjunct regions between which they could not have dispersed; reduction of the pelvis is polytopic in *G. aculeatus* (Bell, 1974).

The genetics of reduced pelvic structures is inadequately studied. Campbell (1979), preformed interpopulation crosses which provided evidence that there is a genetic basis for differences in the structure of the pelvic girdle among populations of *Gasterosteus*. Giles (1983) found that progeny of *G. aculeatus* with vestigial pelvic structures that were bred in the laboratory developed the parental phenotype. The pelvic vestige often resembles in detail the center of ossification that forms the anterior process of the pelvic girdle in normal populations of *G. aculeatus* (Bell, unpublished data). Thus, the pelvic vestige appears to be homologous with the anterior process of the normal pelvis and could be paedomorphic (*sensu* Alberch *et al.*, 1979).

Most populations with reduced pelvic structures also have reduced numbers of dorsal spines (see Section 2.6). Reimchen's (1980) study implicates insect predation on young in the absence of predatory fishes as the selection mechanism for loss of the dorsal and pelvic spines (see Section 2.5.1). He felt that the spines could increase the probability of capture by insects that have grasping appendages, and perhaps this mechanism would account for frequent reduction of the pelvic girdle to a vestige. An apparent association between absence of predatory fishes and reduction of pelvic structure (Bell, 1974; Moodie and Reimchen, 1973, 1976*a*) may reflect a shift from dominance of selective mortality from vertebrate predators to dominance by insects. The Texada Island population (Larson, 1976) and those in Scotland (Campbell, 1979) are sympatric with predatory fishes, but at least the former may be protected from predatory fish (*Salmo clarki*) by its benthic habit and abundance of aquatic vegetation. However, Giles (1983) observed ten Scottish populations with reduced pelvic structures and dorsal spines, that are sympatric with native brown trout (*Salmo trutta*), a reported predator on *Gasterosteus* (Wootton, 1976). These stickleback populations all come from lakes with low calcium ion concentrations, while adjacent stickleback populations with normally developed pelvic structures occupy lakes with a variety of calcium ion concentrations. These observations led Giles (1983) to suggest that reduction of bone to reduce calcium demand is the mechanism for evolution of pelvic reduction. The situation is complicated further by the recent observation in Alaska of populations with reduced pelvic structures but with the normal complement of three dorsal spines (M. A. Bell, R. E. Francis, and A. Havens, unpublished data). Taking into account Reimchen's (1980) and Giles's (1983) results, it appears that pelvic loss may evolve in response to both low environmental calcium availability and elevated insect predation in the absence of fish predation. These hypotheses will have to be evaluated within the same populations before their relative importance can be appreciated.

2.7.2. Pelvic Variation in *Culaea inconstans*

The brook stickleback, *Culaea incostans,* is endemic to fresh water of North America. Its biology was reviewed briefly by Wootton (1976), and geographic variation in *Culaea* has been studied by Nelson (1969) and Moodie (1977). In most populations of *C. inconstans* the pelvic girdle is fully developed with anterior and posterior processes, an ascending branch, and a prominent pelvic spine (Nelson, 1969, 1971*a*). However, some Canadian populations have substantial frequencies of specimens with reduced or absent pelvic structures. It is these unusual populations on which I focus in this section.

Nelson (1977) demonstrated that reduction of the pelvic girdle has a genetic basis. Although pelvic reduction forms a continuous series, for the purpose of his crosses, he recognized three phenotypic classes: A, pelvic girdle completely absent on both sides; I, pelvic structure reduced but ranging from presence of an irregularly shaped vestige on only one side to absence of one pelvic spine only on one side; and C, complete development. Nelson (1977) performed 19 artificial intrapopulation crosses comprising five A × A, one A × I, seven A × C, and six C × C, and he set up four phenotypically homogeneous populations by releasing 30–60 specimens of one phenotypic class (one A, one I, two C) into aritifical ponds. All specimens came from Astotin Lake, Alberta, where the full range of pelvic structure occurs. Artificial laboratory crosses and mass crosses in the ponds produced compatible results. A × A and C × C crosses produce 75–80% A and C phenotypes respectively in the artificial crosses and about 60% in the mass crosses. The C × A cross produced about 50% A and 30% C offspring, and the A × I cross produced mostly A offspring, but more C offspring than the A × A crosses. The I × I mass cross produced a relatively even distribution of phenotypes, but the frequency of A > I > C. Nelson (1977) concluded that there is a genetic basis for these phenotypes, but that more work would be required to elucidate the genetic mechanism for pelvic girdle phenotype determination.

The frequency of incompletely developed pelvic skeletons varies from about 2.7% in apparently normal populations (Moodie, 1977) to 100%. The form of reduced pelvic structures is exceedingly heterogeneous within populations and includes a wide variety of shapes and sizes (Nelson and Atton, 1971). Nelson and Atton (1971) discounted paedomorphosis (*sensu* Alberch *et al.*, 1979) as the developmental mechanism for reduced pelvic structure, because the pelvic spines develop first during ontogeny but sometimes are the only missing structure in reduced pelvic girdles. In addition, the ascending branch, the last structure to appear during ontogeny in *Gasterosteus* (Bell, unpublished data), sometimes is present when the posterior process on the same side is missing or the entire girdle on the other side of the body is absent.

Occurrence of reduced pelvic structures at frequencies greater than the apparent background rate of less than 2.7% is restricted mostly to Alberta, Canada, with two populations known from Ontario and four from Saskatchewan (Nelson and Atton, 1971; Nelson, 1977). In Alberta, at least 39 sites have been reported to harbor populations with at least 20% reduced pelvic phenotypes and about a dozen other sites have populations with low or undetermined frequencies of reduced pelvic structures (Nelson and Atton, 1971; Nelson, 1977; Reist, 1981). Nelson (1977) noted that virtually all localities sampled in different years had stable phenotype frequencies. He also observed that the highest frequencies of pelvic reduction occur in the north and west of Alberta, but that pelvic reduction seems to be absent farther to the west and north in British Columbia and the Northwest Territories. Within Alberta, populations of *C. inconstans* with and without reduced pelvic structures are interspersed and often in close proximity.

Nelson (1969) found that interpopulation variation in dorsal and pelvic spine length and body depth are associated and tend to decrease toward the north. Nelson (1977) noted that populations with reduced pelvic structures occur at the northern terminus of this cline. Greater spine length and body depth combine to produce a greater cross-sectional area, which is more difficult for a predatory vertebrate to ingest. Noting the finding of Hoogland *et al.* (1957) that predatory fishes will eat all soft-rayed fishes before eating sticklebacks, Nelson (1969) suggested that lack of alternate soft-rayed prey fishes in the north might render long spines an untenable defensive strategy. Absence of predatory fishes from many habitats of *C. inconstans* with reduced pelvic girdles also seemed consistent with the view that fish predation selects for robust pelvic structure (Nelson, 1977). Despite geographic trends for spine length, possibly associated with reduced selective predation by fishes, populations with reduced pelvic phenotypes do not tend to have shorter or fewer dorsal spines than those with normal pelvic structures.

In a series of papers on selective mortality and small-scale geographic variation in *C. inconstans,* Reist (1980*a,b,* 1981) attempted to test hypotheses for the selection of reduced pelvic structure. He found that within the Redwater River system, Alberta, streams have a significantly higher frequency of *C. inconstans* with normal structures than lakes, although the difference in frequency only amounts to a few percent. He also noted a significant increase over winter in the frequency of specimens with normal pelvic girdles at the expense of those with reduced structures in Lake Wakomao, Alberta, and these frequencies tended to return to their original values by the following summer. Despite seasonal periodicity of pelvic phenotype frequencies in this lake, they were stable over a 7-year period. It appears that different fitnesses are associated with different pelvic phenotypes during different life history stages, but their frequencies may be stable over several years.

Reist (1980*a,b,* 1981) also examined three possible mechanisms for differential mortality. He observed that apparent hypoxic death was not selective for pelvic phenotype (Reist, 1981). Predation by large (>18 cm SL) northern pike (*Esox lucius*) on a natural population of brook sticklebacks included an overrepresentation of specimens with normal pelvic structures, a strong deficit of those lacking the pelvic girdle, and the expected frequency of those with pelvic vestiges. No selectivity for dorsal spine number was observed, but the largest size class was overrepresented in pike stomachs. In the laboratory, small pike (11–18 cm SL) generally consumed pelvic phenotypes at random, but when cover was provided, specimens lacking the pelvic girdle were removed more rapidly than those with normal pelvic structures in a significant excess of replicate experiments. When given a choice between prey lacking the pelvic girdle and those with normal pelvic structures from which the pelvic spines have been removed, the altered fish were eaten selectively. Absence of tooth marks on survivors of experiments indicated that escape was not important in survival of *C. inconstans* exposed to pike. The effects of pike predation seem to be ambiguous, but may be explained as follows (Reist, 1980*b*): Specimens without the pelvic girdle compensate for the lack of pelvic spines by being more timid. Large pike preferentially removed specimens with normal pelvic structures in the wild, but in the presence of shelter, the large, pelvicless phenotypes are less able to avoid predation by small pike.

Reist (1980*a*) also examined predation by three groups of aquatic insects, adult and nymphal giant water bugs (*Lethocerus americanus*), dragonfly larvae (*Aeschna* sp.), and larval predatory water beetles (*Dytiscus* sp.). *Aeschna* sp. and *Lethocerus* did not appear to be selective predators on pelvic phenotypes. *Dytiscus* sp. selectively removed specimens with normal pelvic structures only when they had five dorsal spines, but not when they had six. Reist (1980*a*) believed that differential predation on pelvic phenotypes resulted from closer approach to the predator. The results of these experiments and those with predatory pike indicate that the cosmopolitan distribution of predatory aquatic insects provides a constant background of selection in favor of pelvic loss that is manifested as reduction of the pelvis only in the absence of countervailing selection by small predatory fishes. However, as Reimchen (1980) noted, predatory fishes also remove insects, thereby reducing their effect on sticklebacks.

2.7.3. Pelvic Variation in *Pungitius pungitius*

The ninespine stickleback, *Pungitius pungitius,* has a broad circumpolar distribution and extends southward along the coast of Asia. The closely related *Pungtius platygaster* is present in a band between the Black Sea and Balkhash Lake. Populations recognized as *P. pungitius tymensis* along the coast of Asia have relatively small pelvic girdles, no soft ray, and short spines, but extreme pelvic reduction and complete loss at high frequencies have been detected only in the Salt River drainage in the Wood Buffalo National Park region of Canada, the Shannon River system, Ireland (Nelson, 1971*b*), and in the Sperchios River system, Greece (Stephanidis, 1971, cited in Gross, 1979). Nelson (1971*b*) figured specimens from Ireland and Canada, and reduced pelvic structures range from absence of one spine and a portion of the girdle to presence of only one or two minute ossicles. Only three of the Irish samples were large enough ($n = 29$) to allow inference of the approximate phenotype frequencies, and in these samples an average of about 88% of the specimens completely lacked the pelvis. In one of these samples 75% lacked the pelvis, 16% had vestigial structures, and 13% had normal pelvic complexes. All 142 specimens in one of the Canadian samples lacked the pelvis; in the other two, reduced phenotypes accounted for about one-third of the specimens; and in one of the latter samples, phenotype frequencies were unchanged over successive years. The Irish samples were inadequate to infer the geographical distribution of the reduced pelvic structures, but in the Canadian samples, populations with reduced pelvic structures occurred in one drainage but either were isolated or separated from each other by streams containing populations of *P. pungitius* composed mostly or entirely of specimens with normal pelvic structures. Coad (1973) examined samples from eastern Canada and the Northwest Territories and found that the frequency of specimens with reduced pelvics averaged 2.08% (0.17–4.22%). Thus, it appears that pelvic reduction may be ubiquitous at a low frequency and only occurs in a substantial fraction of the population when conditions favor its presence (Coad, 1973).

Nelson (1971*b*) noted some other structures that are correlated with pelvic structure. Pelvic spine length and, to a lesser degree, dorsal spine length are shorter in samples with reduced pelvic structures than in those with only normal pelvics. Coad (1973), however, claims to have failed to detect an interpopulation correlation between the frequency (always less than 4.22%) of reduced pelvic structures and mean pelvic spine length. Nelson (1971*b*) also found that the distance between the pelvic spine bases is smaller in samples with reduced pelvics. This difference and the difference between spine lengths combine to produce a smaller cross-sectional area of the body. There also appeared to be a possible association of reduced dorsal spine number with reduced pelvic structure, but the difference between dorsal spine counts was small, and Nelson (1971*b*) did not subject the data to statistical analysis.

Gasterosteus aculeatus sympatric with Irish *P. pungitius* with reduced pelvic structures had normal pelvic girdles, but one of the Canadian populations with reduced pelvics is sympatric with a *Culaea inconstans* population in which all specimens lack the pelvis, and the pelvic reduction is relatively common in *C. inconstans* from this region (Nelson, 1971*b*). The presence of reduced pelvic structures, a rare condition, in both brook and ninespine sticklebacks from the same area suggests a common cause of the phenomenon in both species. The tendency for an association of reduced cross-sectional area and possibly dorsal spine number in *P. pungitius* as well as evidence from *C. inconstans* and *G. aculeatus* implicate differences in the intensity of predation by various predators in selection of pelvic structure in *P. pungitius*.

2.7.4. Conclusions

Extreme reduction of the pelvic girdle in gasterosteids provides an exciting opportunity to study the effects of predation as a selective mechanism on a defensive structure. This phenomenon occurs in the three widespread stickleback taxa, *Gasterosteus aculeatus* (species group), *Pungitius pungitius,* and *Culaea inconstans,* in which the pelvic girdle has an ascending branch. The geographic distribution of the phenomenon and the antiquity of the taxa and of the phenomenon itself compels one to accept the view that reduction and loss of the pelvic girdle has evolved independently in each of the taxa (Bell, 1974). Similarly, the geographic and geological distribution of pelvic reduction in *Gasterosteus* (Bell, 1974) and its geographic distribution within *P. pungitius* (Nelson, 1971*b*) are inconsistent with a monophyletic origin of the character in either taxon. Thus, a genetically based structure has regressed independently in three stickleback genera and natural selection is its likely cause.

A beginning has been made on elucidation of the selective mechanism for pelvic reduction. Gross patterns of geographic variation in relation to environmental variables suggest that absence of predatory fishes, presence of a refuge from predatory fishes, or absence of more palatable alternate prey fish species may remove the selective force that favors strong pelvic structures (Nelson, 1969). Absence of predatory fishes may produce a secondary selective effect favoring loss of the pelvic girdle; in the absence of predatory fishes, aquatic insects may be more abundant, and there is evidence that selective predation by insects may fall most heavily upon spined individuals (Reimchen, 1980; Reist, 1980*a*).

The developmental mechanism for loss of the pelvic girdle also presents an interesting facet of this phenomenon. Nelson and Atton (1971) presented compelling evidence that paedomorphosis (*sensu* Alberch *et al.*, 1979) is not responsible for loss of the pelvic girdle in *C. inconstans*, and Nelson's (1971*b*) figures of reduced pelvic structures in *P. pungitius* appear to rule out this mechanism in this species. In the fossil *Gasterosteus doryssus*, however, pelvic vestiges are very stereotyped (Bell, 1974) and resemble in detail early ontogenetic stages of the anterior process of the pelvic girdle in *G. aculeatus* (Bell, unpublished data). Reduction of the pelvic girdle in *Gasterosteus* may provide novel insights into the role of paedomorphosis in evolution.

Another important difference between genera is that in *Gasterosteus* there is a strong association between pelvic reduction and reduction of other defensive structures; such an association may exist in *P. pungitius* (Nelson, 1971*b*), but it is absent in *C. inconstans*. The reasons for this difference are obscure.

Much remains to be done with the problem of pelvic reduction in sticklebacks. The genetics of pelvic reduction are barely understood, and it is unknown whether the same developmental genetic mechanisms are responsible for pelvic reduction in the different populations and species. The selection mechanisms for pelvic loss apparently involve some balance among selective predators, but Reist (1980*a,b*) demonstrated that there may be subtle effects of predator size and of other prey traits on selection of pelvic phenotypes. Experimental approaches to the measurement of possibly weakly selective predation (Reimchen, 1980) must be devised. The possibility of differential reproductive success of pelvic phenotypes appears to be a real one, as the spines are erected during territorial encounters between males and during courtship in all three species that exhibit pelvic reduction (Wootton, 1976). Pelvic reduction in sticklebacks appears to have a genetic basis and something is known about selection on this genetic variation, but exactly what the genetic basis is and how various selective mechanisms interact to produce observed geographic patterns of phenotype frequencies are not adequately understood.

2.8. Protein Polymorphism

Considering the great interest that *G. aculeatus* and other sticklebacks have held for evolutionary biologists and the prominence of the use of protein polymorphism in recent evolutionary studies, it is remarkable how little this technique has been applied to sticklebacks. A few studies have used data on protein polymorphism to examine population structure. Hagen (1967) showed that parapatric anadromous, monomorphic, completely plated and freshwater, monomorphic, low-plated populations also are fixed for alternate muscle protein alleles. Thus, there is no evidence of gene flow between them, despite presence of a lateral plate trimorphic hybrid population that has both muscle protein alleles. Avise (1976) studied a population that is virtually dimorphic for the low- and complete-plate morphs and found no evidence that the two morphs were divergent at any of 15 loci. Avise (1976) also analyzed protein variation in two other populations and concluded that they are similar, but five loci were polymorphic in either of the populations and not the other. Bell and Richkind (1981) examined variation at 12 protein loci to determine whether divergence for lateral plate number between headwater and downstream populations within a drainage is facilitated by genetic isolation. They found that populations in both areas share monomorphic alleles at 11 loci and the single dimorphic locus was not divergent in frequency among populations. Although protein variation has not been employed extensively to determine the genetic structure of *G. aculeatus* populations, it has proven to be a useful tool.

Raunich *et al.* (1972) found that populations of *G. aculeatus* from brackish habitats around the northwestern Adriatic Sea are dimorphic for hemoglobin, but freshwater populations from scattered sites in Italy and Germany are monomorphic. The generality of this pattern is not known, and Raunich *et al.* (1972) did not speculate on the physiological significance of the dimorphism.

Based on the distribution and differentiation of freshwater populations, I generated a model of the derivation of freshwater populations from anadromous *G. aculeatus,* from which six hypotheses concerning protein variation were suggested (Bell, 1976*a*). Withler (1980) has examined some of these hypotheses, using the products of eight-enzyme loci in samples from 73 marine and freshwater populations on Vancouver Island and adjacent British Columbia. She found that populations from small or isolated freshwater habitats are less genetically diverse than those from large freshwater or marine habitats. Generally, variation among marine populations is far lower than among freshwater populations, but she detected regional differentiation of marine populations. Most interestingly, she found that the genetic distance between pairs of freshwater populations is approximately equal to that between pairs of freshwater and marine populations, as predicted (Bell, 1976*a*). This finding indicates that freshwater populations are just as likely to have been derived independently from marine populations as from other freshwater populations. She also discovered significant frequency differences at specific loci between marine and freshwater populations, among freshwater habitats, and between two areas for marine populations. Some of these frequency differences formed clines that cut across drainages, suggesting that natural selection is responsible for allozyme variation in these populations.

The use of protein variation to examine evolutionary problems in *G. aculeatus* has been limited either to a small number of populations or to few gene loci. Regardless, limited objectives of inferring population structure have been successful, and Withler's (1980) survey of a large number of populations produced interesting insights into the evolution of freshwater populations, despite the use of only six polymorphic loci. The application of this method to test specific evolutionary hypothesis will produce rewarding results.

3. Divergent *Gasterosteus* Populations

In this section I consider a few of the best documented cases among the numerous highly divergent freshwater populations belonging to the *G. aculeatus* species group. These populations often are most instructive in our understanding of evolutionary mechanisms because they exhibit one or more phenotypic traits in an extreme state. Often the extreme phenotypic state reflects an extreme in selection, allowing isolation and analysis of the causal factor. All of the populations examined in this section have been discussed in the sections that consider the specific traits for which they are divergent. However, while it is useful to consider the traits one at a time, in reality, the traits do not evolve in isolation. Whole phenotypes, encompassing suites of coadapted traits, are subject to selection, and the response to selection upon one trait may influence the selection regime to which another is subjected. Thus, it often is revealing to consider the ensemble of traits that distinguish highly divergent populations.

Before considering highly divergent or unusual populations, it is worthwhile to sketch what I regard as typical. In reality, there probably are few populations of *G. aculeatus* that are typical with respect to all characters; most populations seem to be distinctive in some respect or for some combination of characters. However, two phenotypic modes, which European workers usually have referred to as "trachurus" and "leiurus" (e.g., Münzing, 1963), are the forms usually encountered. The "trachurus" form generally is a relatively large marine or anadromous stickleback. It usually is monomorphic completely plated and the lateral plates, pelvic girdle, and spines are robust. The body is fusiform and the gill rakers are relatively numerous, reflecting residence during most of its life in open water, where it must swim efficiently and feed on relatively small zooplankton. "Leiurus" populations usually are smaller and occur in fresh water. They may be monomorphic, dimorphic, or trimorphic, but most often are monomorphic low and complete or dimorphic with the partial and one other morph present. The lateral plates and pelvic girdle are weaker structures and the spines tend to be shorter than in marine or anadromous populations. Body form and gill raker numbers vary considerably, depending on the habitat, but many populations lead a benthic existence and are less streamlined than marine and anadromous populations and have relatively low gill raker counts. This phenotype reflects activity among obstructions and feeding upon relatively large benthic prey. Populations considered to be highly divergent may be typical in many respects, but are notable for their high degree of divergence for one or more characters. Some examples of such populations are considered next.

3.1. Populations with Reduced Armor

As indicated above, freshwater populations tend to have fewer and smaller lateral plates and weaker dorsal and pelvic spines than marine populations. At the southern extremes of the range of *G. aculeatus* in both western Europe and western North America, populations in which many individuals lack lateral plates occur (Regan, 1909; Miller and Hubbs, 1969; Ross, 1973; Bianco, 1980). The "unplated" populations also tend to have relatively short fin spines (Ross, 1973; Gross, 1978*a*; Bianco, 1980), and thus have generally reduced armor. The unplated populations in southern California usually are not sympatric with predatory fishes (Bell, 1978, unpublished data). Unplated populations also have been reported from the Queen Charlotte Islands, British Columbia, where they also seem to be associated with reduced predation by fishes (Moodie and Reimchen, 1976*a*) and from Outer Hebrides, Scotland, where they are sympatric with predatory fishes (Campbell, 1979). However, some populations have independently experienced even greater armor loss. Many specimens in these populations have fewer than three dorsal spines and the pelvis may be reduced or absent. Such populations have been described from the Queen Charlotte Islands (Moodie and Reimchen, 1973,

1976*a*; Reimchen, 1980), from Texada Island in the Strait of Georgia between Vancouver Island and mainland British Columbia (Bell, 1974; Larson, 1976), from a late Miocene lake deposit in Nevada (Bell, 1981), and from the Outer Hebrides, Scotland (Campbell, 1979).

One of four populations (Moodie and Reimchen, 1976*a*) with reduced numbers of spines in the Queen Charlotte Islands was studied over a 10-year period by Reimchen (1980). This population occurs in Boulton Lake, and is the only fish species present. He classified phenotypes according to the presence or absence of the enlarged first two dorsal spines and the pelvic spines. He found that 80% of the sticklebacks lack the second dorsal spine, but that the other dorsal spines rarely are absent. The third dorsal spine is significantly more likely to be absent in specimens lacking the second dorsal spine than in those possessing it. In specimens lacking the second dorsal spine, the first spine is displaced backward and the pterygiophore that normally bears the second spine is reduced in size. The pelvic girdle is represented by an irregular vestige on both sides of the body (65%), on one side only (15%), or is fully developed on both sides (20%), and pelvic spines are absent or greatly reduced unless the pelvic girdle is fully developed. In specimens with unilaterally vestigial pelvic girdles, the left side of the structure is significantly more likely (77%) to be complete. There is no association between absence of the second dorsal spine and of pelvic spines. The second dorsal and pelvic spines are absent in about half of the specimens, but all spines are present in only 4.3% of individuals. Females tend to have more spines than males. Reimchen (1980) did not comment on lateral plate number phenotypes, but the Boulton Lake population is monomorphically low-plated with mean and modal plate counts in both sexes of about three plates per side (Moodie and Reimchen, 1976*a*). Thus, armor structures are poorly developed in this population.

Reimchen (1980) examined the habitat distribution of stickleback phenotypes and known stickleback predators to infer selection. He found that spinier individuals are more abundant in open water than near shore or near the bottom in deep water. This distribution reflected both greater abundance of spinier adult females and of spinier phenotypes among both sexes of subadults in open water, and presence near shore of juveniles too small for development of spines. He also observed that predatory birds feed principally in open water and that dragonfly nayads, the most abundant predatory insect, occur mostly near shore. He reasoned that spiny phenotypes predominate in open water because the spines afford protection from vertebrate predators (i.e., birds) and least spiny phenotypes are abundant near shore because the spines might make it easier for the dragonfly nayads to hold and resist escape by the fish. Reimchen's

(1980) attempt to experimentally demonstrate an escape advantage of less spiny specimens failed, but, as he pointed out, it might require exceedingly large samples to detect differential escape. However, Reist (1980*a*) obtained results for insect predation on *Culaea inconstans* that indicate selectivity with respect to the number of pelvic spines. Reimchen (1980) also emphasized that absence of predatory fishes not only eliminates an important potential cause of selective mortality of sticklebacks that would favor presence of spinier phenotypes, but also allows expansion of the habitat of larval dragonflies into the habitat of juvenile sticklebacks and increase in dragonfly abundance. Thus, in this population, complete absence of predatory fishes appears to shift net selection from favoring presence of spines to favoring their loss, but the uneven distribution of avian predators and predatory insects produces association of spine phenotypes with habitat type. It is not clear in this system whether spine number phenotype frequency is stable or represents a transient polymorphism.

The population with reduced numbers of spines on Texada Island occurs in Paxton Lake and contrasts with the Boulton Lake population morphologically and ecologically. A major contrast is that cutthroat trout (*Salmo clarki*), a stickleback predator, is native to Paxton Lake (Larson, 1976). Paxton Lake also supports two biological species (McPhail, unpublished data) of the *G. aculeatus* group. One species is limnetic and adapted to feed in open water on zooplankton; it is not exceptional morphologically, and is subject to trout predation. The other species is benthic and adapted to feed on larger benthic invertebrates. It lives in association with the bottom and apparently is protected from trout predation by benthic vegetation; this population tends to have reduced armor structures. Bell (1974) investigated variation in armor structures in this population, and there are differences between this population and the Boulton Lake population. Bell (1974) recognized three major pelvic phenotypes, developed (bilaterally fully developed), vestigial (one or two vestiges and no pelvic spines), and absent (no pelvic structures). Most specimens have two (38%) or three (60%) dorsal spines, but two-spined specimens invariably lack the first spine and rare one-spined specimens lack the first two spines. This pattern contrasts with Boulton Lake sticklebacks, which generally lack the second spine, if any, and have only the first spine when two are missing. Another contrast between these populations is that Paxton Lake sticklebacks frequently (57%) lack lateral plates. Although there was no association between dorsal spine number and pelvic structure in the Boulton Lake population (Reimchen, 1980), in the Paxton Lake population, specimens with normal pelvic morphology are more likely to have the full complement of dorsal spines and to have more lateral plates than those with pelvic vestiges or lacking the pelvis. Thus, while both Paxton and Boulton Lake sticklebacks have experienced loss of armor, different structures may be involved and correlations between states of characters differ between populations. Although the genetic basis of armor reduction has not been studied in these populations, there must be differences in the genetics of the advanced phenotypes in the two lakes.

The Outer Hebrides populations merely have been reported (Campbell, 1979), but the fossil material adds a little to the picture. As in the Paxton Lake population, reduced spine counts result from sequential loss of the dorsal spines from first to third (Bell, 1974). Individuals also tend to be either three-spined with a normal pelvis or less than three-spined (usually one or zero) with reduced pelvic structures. However, lateral plates always are absent. Over a 150,000-year sequence, the frequency of spine number and pelvic girdle phenotypes was heterogeneous, but more than one dorsal spine number and pelvic condition phenotype occur throughout the section (Bell and Haglund, 1982). Predatory fishes are very rare in this fossil deposit, but their importance in the evolution of the fossil stickleback cannot be inferred from available information. The fossil material exhibits some morphological similarity with the Paxton Lake population, but most importantly shows that more than one spine number and pelvic structure phenotype can coexist within a single basin (probably within a single population) for tens of thousands of years.

It is worth noting at this point that *Pungitius pungitius* populations with reduced pelvic structures tend to have lower dorsal spine counts, but no intrapopulation correlations were detected (Nelson, 1971*b*). However, Nelson and Atton (1971) failed to detect correlations between dorsal spine number and pelvic condition within *Culaea inconstans* populations and they found that, with the exception of one population, those with reduced pelvic structure do not tend to have lower dorsal spine counts than populations with only normally developed pelvises.

3.2. The Giant Black Stickleback of Mayer Lake

Moodie (1972*a*,*b*) studied a giant melanistic stickleback in Mayer Lake in the Queen Charlotte Islands. The lake water is deeply stained and there is moderately dense vegetation near shore, but three tributary streams are densely vegetated. These ecological differences result in easy access of abundant cutthroat trout (*Salmo clarki*) to sticklebacks in the lake, but complete exclusion from some stream habitats. Thus, the streams are occupied by ordinary freshwater sticklebacks, but the lake population exhibits a number of traits that reflect life in the open lake with intense predation by trout. Moodie (1972*a*) concluded from mate preference tests, sharp boundaries between black and ordinary sticklebacks, scarcity of presumptive hybrids between them, and a 39-year period of phenotypic stability of the black stickleback phenotype that it satisfies the criteria for recognition as a distinct biological species.

The black sticklebacks tend to be more streamlined than stream populations and have higher gill raker counts, both adaptations for life in open water feeding on zooplankton. However, they have a suite of characters that adapt them for resistance to predation by trout. Sexually mature females average about 90 mm standard length and males average about 80 mm. Presumably, larger fish are less liable to predation (McPhail, 1977), though Moodie (1972*b*) was unable to demonstrate this. Members of this population have relatively long pelvic spines; the ratio of spine length to body length ranked fifth among 20 populations sampled by Moodie and Reimchen (1976*a*). Moodie (1972*b*) found that females from trout stomachs have relatively short pelvic spines compared to seined females. Males from trout stomachs also have shorter spines than seined males, but the difference was not significant. Specimens with seven lateral plates per side, the phenotype most able to avoid fish predation (see Section 2.1.3), constituted 73% of the population (Hagen and Gilbertson, 1972). This frequency of seven-plated fish ranked second among 50 populations reported by Hagen and Gilbertson (1972). Even the red male nuptial coloration is suppressed; it is expressed in less than 18% of males and in those is weakly developed (Moodie, 1972*a*). However, Reimchen (personal communication) noted that many *G. aculeatus* populations in the Queen Charlotte Islands lack red male nuptial coloration and are not subject to predation by trout. Thus other selective mechanisms that may be operating in Mayer Lake must favor loss of red nuptial coloration. The black stickleback is markedly divergent by virtue of high gill raker counts and streamlined body, adaptations primarily to feeding in the open lake, and by their large size, long pelvic spines, strong mode at seven plates per side, and suppression of red male nuptial coloration, all of which could be adaptations to resist predation.

The giant black sticklebacks of Mayer Lake are the only reported case of a large suite of characters that adapt a population to avoid predation. Maximum body size is large in some other populations in the Queen Charlotte Islands, but their modal plate counts are not always seven plates per side (Moodie and Reimchen, 1976*a*). Hagen and Gilbertson (1972) reported two other populations with seven-plate-count frequencies in excess of 70%, but while these populations were reported to occur with abundant predatory fishes, other traits were not described. Hagen and Gilbertson (1972) did, however, report that dorsal and pelvic spine lengths tend to be correlated and that populations sympatric with abundant predatory fishes tend to have modal plate counts of seven per side and long pelvic spines. Semler (1971) and Gilbertson (1980) both reported on populations subject to fish predation and with modal plate counts of seven per side and males that frequently lacked red nuptial coloration. Thus, the Mayer Lake population represents an extreme in widespread adaptation to fish predation, but individual elements in its suite of antipredator adaptations commonly occur in other populations. No other studies on the ecological genetics of comparable populations have been reported. Clearly, predation by fishes selects in concert for a variety of traits that decrease liability to predation in the Mayer Lake stickleback, but these traits need not be selected in the same manner in all populations.

The Mayer Lake population provides another important lesson. While it possesses locomotory and trophic adaptations that allow it to exploit zooplankton as a food resource, this resource would be unavailable without adaptations to predatory fish, to which it is vulnerable in the limnetic habitat of the zooplankton. Thus, adaptation for use of this food resource has altered the selection regime for a number of features involved in reducing predation.

3.3. Populations with Male Nuptial Melanism

McPhail (1969) published an extensive study of a series of adjacent freshwater populations of the *G. aculeatus* group from northwest Washington state in which the characteristic red and blue male nuptial coloration (Section 2.2) is replaced by uniform, deep black coloration. He documented the distribution of male nuptial melanism, showed that melanic populations (blacks) are partially reproductively isolated from parapatric typical populations (reds), and suggested a selection mechanism for nuptial melanism and other derived traits of the populations. This study was augmented in two subsequent studies that extended the known range of the black populations, determined the genetics of the melanism, and demonstrated an alternative to McPhail's (1969) hypothesis for the selection of male nuptial melanism (Hagen and Moodie, 1979; Hagen *et al.*, 1980). The only other case of nuptial melanism was reported by Bell (1982*a*); other cases of melanism apparently are not associated with reproduction (Semler, 1971; Moodie, 1972*a,b*; Reimchen, personal communication).

The black stickleback discovered by McPhail (1969) is distributed widely in the Chehalis River system, where it often is parapatric with red sticklebacks, and in three other adjacent drainages (Hagen and Moodie, 1979). In each of these drainages, the distribution of the black stickleback coincides perfectly with that of the western mudminnow, *Novumbra hubbsi,* and in the Chehalis River system sharp clines for nuptial coloration connect divergent stickleback populations that are sympatric (black) and allopatric (red) with *Novumbra*. Concordance of the distribution of black sticklebacks and *Novumbra* suggested to McPhail (1969) that *Novumbra* is involved in selection for nuptial coloration.

As adult sticklebacks are too large for *Novumbra* to eat, McPhail (1969) reasoned that red males may attract *Novumbra* to their fry, but black males should be less conspicuous. He found that *Novumbra* spent more time orienting toward nesting red males than toward nesting black males in a choice situation and that fry from the black population were more evasive of *Novumbra* in the laboratory. However, neither he nor Hagen *et al.* (1980) ever found *G. aculeatus* fry in the stomachs of several hundred *Novumbra*. Hagen *et al.* (1980) suggested an alternative hypothesis for evolution of black nuptial coloration: black nuptial coloration is an interspecific threat display to *Novumbra,* which breeds at the same time as sympatric *Gasterosteus* and which itself has black nuptial coloration. They found that *Novumbra* is as likely to orient away from a brooding black male and toward the unoccupied end of the apparatus as it is to orient toward a brooding red male. They also found that red males introduced into an aquarium occupied by a brooding male *Novumbra* are less likely to hold a territory, more likely to suffer territorial intrusion by *Novumbra,* and experience less hatching success than black males. These data strongly favor the interspecific threat display hypothesis of Hagen *et al.* (1980) over the predation hypothesis of McPhail (1969).

McPhail (1969) had demonstrated that differences between allopatric red and black populations for male nuptial coloration, gill raker number, position of the first two dorsal spines, and salinity tolerance of adults and eggs are inherited. Hagen and Moodie (1979) determined that the difference between red and black nuptial coloration is controlled by a major autosomal locus with two codominant alleles and sex limited expression. They also postulated the presence of polygenic modifier loci (see Section 2.2).

Within the Chehalis River system the ranges of black and red stickleback populations are parapatric at a number of points and the monomorphic populations are connected by steep clines for nuptial coloration (Hagen and Moodie, 1979). McPhail (1969) showed that a cline in another, small drainage (Connor Creek) involved both changes in the frequency of presumptive red and black homozygotes and possible hybrids. It is clear that red male nuptial phenotypes would suffer a disadvantage within the range of the black population because of territorial intrusion by *Novumbra,* but failure of the black stickleback to displace the red populations remains to be explained. In the lower reaches of streams salinity increases, and McPhail (1969) showed that black sticklebacks are less tolerant of salinity than red ones. However, blacks have not displaced red sticklebacks from the headwaters of the Chehalis River system. McPhail (1969) showed that red and even allopatric black females prefer red males. Thus, in the absence of *Novumbra,* male black sticklebacks are at a reproductive disadvantage. Interestingly, McPhail (1969) demonstrated that parapatric black females are significantly more likely to select a black male in laboratory mating experiments than are allopatric black females. This difference suggests that female preference has not been selected in concert with male nuptial coloration; preference has evolved only where opportunities for incorrect mate choice occurs. This phenomenon represents a case of reproductive character displacement (Bell, 1976*c*), and it should sharpen the boundaries between red and black populations.

Bell (1982*a*) reported another population in which black male nuptial coloration occurs. This population is from southern California, showing that nuptial melanism clearly has evolved more than once. However, it differs from the black populations discovered by McPhail (1969) in that it appears to be a rare phenotype coexisting with nonmelanistic males, and the red throat is retained in black males. This population is not well enough known to allow of any further conclusions.

3.4. Other Notable Populations

A number of populations have not been noted above but are interesting either because they are so thoroughly studied or because they are morphologically divergent but are not well enough studied to allow one to say much at present. The most thoroughly studied stickleback population is from Lake Wapato, Washington. The genetics of lateral plate morphs (Hagen and Gilbertson, 1973*b*) and of low-morph lateral plate and gill raker numbers (Hagen, 1973), selective predation by trout (*Salmo gairdheri*) on low- and partial-morph plate number phenotypes (Hagen and Gilbertson,1973*b*), variation in nuptial coloration (Semler, 1971), and differential reproductive success and breeding behavior (Kynard, 1972, 1978, in press) all have been studied in this population. Unfortunately, the Lake Wapato population suffered a severe decline apparently as a result of introduction of exotic fishes (Kynard, 1979*b*).

A number of other interesting cases merely have been reported or currently are being studied. Moodie and Reimchen (1973) reported a number of highly divergent populations from the Queen Charlotte Islands, Canada, and others not reported by them have been discovered. Reimchen (personal communication) is intensively studying a number of these divergent populations that are endemic to lakes. McPhail (personal communication) is studying sympatric biological species that are divergent for trophic and other structures. Campbell (1979) also reported some highly divergent populations from the Outer Hebrides, Scotland, which he is continuing to study (Campbell, personal communication). Hagen (personal communication) is studying an endemic *Gastrosteus* in eastern Canada in which the nuptial coloration is white. Although one can recognize typical morphology for *G. aculeatus,* highly divergent populations are widely distributed and many of them are the subject of current research. This research promises to yield exciting insights into the evolution of phenotypic diversity within the genus, but, more importantly, it will have general significance for evolutionary theory.

4. Parallelism

Simpson (1961) defined parallelism as "the development of similar characteristics separately in two or more lineages of common ancestry and on the basis of, or channeled by, characteristics of that ancestry." He stated that this channelization could result from genetic or other characteristics of the ancestry. The cause of genetic channelization was explained in detail by Throckmorton (1965), who distinguished carefully between genotypic and phenotypic homology. It is possible that all those genes necessary for the formation of an advanced phenotype (character state) could be encoded in an "unassembled" state in the genome of an ancestral population without the advanced phenotype being expressed. The unassembled state might involve heterozygosity or epistatic interactions that prevent expression of the advanced phenotype in the ancestor, but which could form the substrate for natural selection resulting in assembly of the phenotype independently in several descendant lineages. The unassembled state might even involve the presence of an allele that by a single point mutation contributes to the formation of a different phenotype, which then might appear independently in several descendant lineages. Thus, while the appearance of the advanced phenotype in the descendant lineages represents parallelism, a kind of analogy, the genotypic information upon which the analogous phenotypes is based is homologous. A consequence of persistant genetic homology is the recurrence of similar phenotypes over widely separated intervals of time and space where the lineages encounter similar selection regimes. Simpson (1961) also alluded to nongenetic consequences of common ancestry that favor parallelism in the descendant lineages. Related organisms are likely to share features of their natural history that expose them to similar selection pressures. Thus, not only may related forms carry with them the unassembled genetic information that predisposes them to evolution of certain phenotypes, but they also are likely to be exposed to selection regimes that will assemble the same phenotypes repeatedly.

Detection of parallelism should be simple in groups in which divergent evolution is the mode. Character states evolved by means of parallelism should have a chaotic distribution at some taxonomic level over such a phylogeny. However, if the parallelism encompasses a substantial fraction of the phenotype due either to genetic causes or to multifarious consquences of life history similarity, parallelism may be different to distinguish from similarity due to phenotypic homology. Parallelism in the gasterosteids is sometimes specific to a few structures and therefore is obvious, but also may be pervasive and consequently difficult to detect. However, dispersal between freshwater stickleback populations of different drainages is a rare event, and some phenotypes tend to be restricted to freshwater populations. Thus, some phenotypic similarity between freshwater populations must reflect parallelism, and even pervasive phenotypic similarity at least in widely separated populations may be assigned to parallelism as the cause.

The most obvious case of parellelism among gasterosteids is reduction and loss of the pelvic girdle (Bell, 1974). *Gasterosteus, Culaea,* and *Pungitius* include numerous populations in which the usually elaborate pelvic girdle has been reduced to a vestige or is eliminated (see Section 2.7). Populations with reduced pelvic structures in the three genera clearly do not form a monomorhpic group, but rather appear to reflect similar selection upon a largely homologous unassembled genotype latent in each lineage; this is an unequivocal case of parallelism. Although Nelson (1977) makes a convincing argument for parallelism within *C. inconstans* for reduction of pelvis, this species is endemic to North America, and the reduced phenotypes occur only in a restricted region of Canada. Parallelism for pelvic reduction within *Gasterosteus* and *Pungitius,* however, is unassailable. Pelvic reduction in *Gasterosteus* occurs on two recently deglaciated islands off British Columbia (Bell, 1974; Moodie and Reimchen, 1973, 1976*b*; Reimchen, 1980), within a late Miocene fossil assemblage in Nevada referred to *Gasterosteus doryssus* (Bell, 1974), and in the Outer Hebrides, Scotland (Campbell, 1979). Similarly, *Pungitius* with reduced pelvic structures occur in Canada, Ireland (Nelson, 1971*b*; Coad, 1973) and Greece (Stephanidis, 1971, cited in Gross, 1979). The distributions of *Gasterosteus* and *Pungitius* populations with markedly reduced pelvic structures hardly appear to be relictual!

Similarly, it could be argued that melanism has evolved by means of parallelism in *G. aculeatus*. Melanistic populations have been reported from the Queen Charlotte Islands (Moodie, 1972*a*), Washington state (McPhail, 1969; Hagen and Moodie, 1979; Hagen *et al.*, 1980; Semler, 1971), and southern California (Bell, 1982*a*). Many stickleback populations include dusky or darkly barred specimens, but only a few have jet black phenotypes.

The most impressive case of parallelism in *G. aculeatus* involves the lateral plate variation in freshwater populations. Both lateral plate morph and low-morph lateral plate number phenotypes have experienced parallelism. When similar plate phenotypes predominate in western European and western North American freshwater populations, for example, it is reasonably certain that the phenotypes, or at least their high frequencies, have evolved independently. However, even within restricted areas one can make a case for parallelism of lateral plate phenotypes. This case depends on the proposition that populations in different drainages generally have been derived independently from marine populations. This view is supported by the occurrence of morphologically typical freshwater populations on islands that never have been connected to the continent or on islands that were glacially defaunated during the Pleistocene (Penczak, 1964; McPhail and Lindsey, 1970; Moodie and Reimchen, 1976*b*; Bell, 1976*a*). Freshwater populations on each island must have been established at least once independently of continental populations. Bell (1976*a*) hypothesized that if populations in adjacent drainages were derived independently from marine ancestors, the genetic similarity based on allozyme variation (Nei, 1972) should not exceed that expected by chance. Withler (1980) (see Section 2.8) examined allozyme variation of *G. aculeatus* in a number of separate drainages on Vancouver Island and adjacent British Columbia and found that average genetic distance in pairwise comparisons of freshwater populations did not differ from the average genetic distance between freshwater and marine populations. It is reasonable to conclude that, in general, freshwater populations in a significant fraction of drainages are derived directly from marine populations rather than from freshwater populations in other drainages.

Despite independent origins of many freshwater populations, they commonly are distinguishable from marine populations on the basis of lateral plate morphology. Marine populations are predominantly completely plated and usually are monomorphic; freshwater populations often are mostly low plated and low-morph monomorphism is common, particularly in southern Europe and southern California (see Section 2.1.1). Monomorphic partial-morph populations have been reported from Alaska (Kynard and Curry, 1976), eastern Canada (Hagen and Moodie, 1982), and Turkey (Münzing, 1962). At the very least, parallelism of morph frequency occurs, but in some cases it is reasonable to argue that the ancestral populations lacked the morph that predominates in a descendant population. For example, partial morphs in a Romanian population must have evolved recently from monomorphic Black Sea sticklebacks (Băcescu and Mayer, 1956; cited in Münzing, 1963). Münzing (1963) had difficulty accounting for low morphs in freshwater Icelandic populations by his dispersalist model, and low morphs have been reported only from two freshwater sites in eastern North America (Coad and Power, 1974; Bell and Baumgartner, 1984), where they must have evolved in place. Bell (1981) even argued that separate mutational events could account for the same plate morphs in different places because the genetics of lateral plate morphs are simple, and advanced phenotypes may evolve by means of paedomorphosis.

Lateral plate number variation also very clearly represents parallelism. If freshwater populations in different drainages are derived independently from marine populations, the mode of seven plates on a side, so commonly associated in freshwater populations with predation by fishes, must represent parallelism. Similarly, zero-plated sticklebacks occur in Italy (Regan, 1909; Bianco, 1980), southern California (Miller and Hubbs, 1969), and the Queen Charlotte Islands (Moodie and Reimchen, 1976*a*), where this phenotype must have appeared independently.

A similar argument could be constructed for gill raker variation (see Section 2.3). The number of gill rakers in fishes generally is inversely related to the space between rakers and to the size of food particles in the diet (Magnuson and Heitz, 1971). Larson (1976) demonstrated that this relationship holds in two sympatric species of the *G. aculeatus* complex. Typically marine populations of *G. aculeatus* have high gill raker counts and stream populations have low counts; many populations from large lakes probably have been independently derived from stream populations but have high gill raker counts (Hagen, 1967; Hagen and Gilbertson, 1972; Moodie and Reimchen, 1976*a*). Thus, there is ample evidence for parallelism for gill raker number among stream populations and among lake populations.

Parallelism is obvious in *G. aculeatus* because its wide range and the insular distribution of freshwater populations (i.e., restriction to drainages) precludes phenotypic homology among all similar populations. Bell (1974) discussed parallelism in a number of freshwater and other organisms. Some particularly interesting cases have appeared subsequently (e.g., Pierce *et al.,* 1981) and it appears that paedomorphosis may be an important general mechanism for parallelism (Alberch, 1980; Bell, 1981). However, parallelism probably is a much more common phenomenon than is recognized because in most groups morphology is the prime criterion for establishing phylogenetic relationships. This phenomenon may be an important cause of error in systematics and geographic variation. For example, some explanations of geographic variation in *Gasterosteus* (Münzing, 1963; Miller and Hubbs, 1969) and *Pungitius* (McPhail, 1963; Münzing, 1969) incorporate the implicit assumption that each phenotype has evolved only once, and Miller and Hubbs (1969) suggested that the present distribution of *G. aculeatus* phenotypes in western North America resulted from postglacial contact of three subspecies. If one accepts the conclusion that parallelism is ubiquitous in *G. aculeatus*, there is no reason to suppose that three subspecies ever existed. When independent criteria for phylogenetic inference, such as distribution and dispersal ability or allozyme and other biochemical similarities, can be utilized, recognition of parallelism may become more common.

5. Evolutionary Rates

Recent years have witnessed a burst of interest in evolutionary rates. This interest has grown out of the observation by paleontologists that even in groups experiencing explosive adaptive radiation, the transitions between morphologically defined species are not often seen in the fossil record; evolutionary stasis is most common (Eldredge and Gould, 1972; Gould and Eldridge, 1977; Stanley, 1979). This observation has led to dispute over whether extrapolation from phenomena in the realm of population genetics can account for observed adaptive radiation when evidence from the fossil record indicates that evolutionary stasis characterizes almost all of the history of lineages (Gould, 1980; Stebbins and Ayala, 1981). While this problem of evolutionary rates is a central issue in evolutionary theory, its solution will be hindered by a number of sampling problems that introduce uncertainty to analysis of virtually any fossil material (Bell and Haglund, 1982). Even under ideal circumstances, the time interval between samples will profoundly influence inferred rates (Bookstein *et al.*, 1978; Ginzburg, 1981), and in most published studies of evolutionary rates the time between samples is known poorly or not at all. The *Gasterosteus aculeatus* species group has experienced extensive adaptive radiation, and lineages within this group have been studied long enough or occur under circumstances that allow us to infer minimum rates of evolution. Considering the problems of quantifying evolutionary rates, however, I restrict myself to a qualitative treatment.

5.1. Evidence from the Fossil Record

By the end of the Miocene both *Gasterosteus* and *Pungitius* appear in the fossil record (Bell, 1977; Rawlinson and Bell, 1982). Fossils from both genera are indistinguishable from extant representatives, and in *Gasterosteus* one of the most advanced phenotypes and probably extensive geographic variation already existed late in the Miocene (Bell, 1974, 1977). The lacustrine, late Miocene Truckee formation in Nevada contains *Gasterosteus doryssus,* a member of the *G. aculeatus* species group, which varies in the number of dorsal spines (zero to three) and the structure of the pelvis (absent, vestigial, and fully developed with spines) (Bell, 1974). Bell and Haglund (1982) studied a single, apparently continuous exposure spanning 150,000 years, and during the approximately 36,500-year period separating two (of six) samples, the frequency of specimens with one dorsal spine declined from 69 to 21% and zero-spined specimens increased from 29 to 79%. This represents an increase in the frequency of zero-spined specimens of 0.137×10^{-3} per year. However, Bell and Haglund (1982) had no samples within this 36,500-year interval, so this rate is minimum. In fact, during other intervals within this 150,000-year record, significant fluctuations in spine number phenotype frequencies occurred over periods of a few hundred years. For example, during the first half of one 800-year period, the average number of dorsal spines was 0.16 ($N = 37$) and in the second half was 0.67 ($N = 21$), a fourfold increase in mean! Preliminary data from collections of *G. doryssus* made at approximately 5000-year intervals during the summer of 1982 reveal a pattern of almost continuous oscillation of mean dorsal spine number. Thus, Bell and Haglund's (1982) within-pit results did indicate rapid temporal change that was not adequately resolved by between-pit comparisons. As I shall show, there is ample precedent for much higher rates of change in phenotype frequencies. It also must be recognized that both spatial variation and evolution can contribute to temporal change in the fossil record (Gould and Eldridge, 1977; Bell and Haglund, 1982), and Reimchen (1980) demonstrated marked intracustrine spatial variation of spine number in Boulton Lake sticklebacks.

5.2. Divergence in Recently Deglaciated Areas

The presence of *G. aculeatus* in areas that were glaciated during the latest glacial advance provides us with an upper age for these populations (Denton and Hughes, 1981). In many cases it can be argued that freshwater populations in separate drainages have been derived independently from marine sticklebacks since deglaciation of their habitat (see Section 2.8). For such cases it is reasonable to assume that the freshwater populations represent the termini of lineages that had essentially the morphology of extant marine populations when their habitats were deglaciated. Moodie and Reimchen (1976*b*) argued that the freshwater *G. aculeatus* populations of the Queen Charlotte Islands, British Columbia, were derived from marine populations after deglaciation. However, Warner *et al.* (1982) showed that some regions of the Queen Charlotte Islands with habitats that could have supported *G. aculeatus* were ice-free during and since the late Wisconsin glacial maximum in western coastal Canada. Thus, some of these populations could be older than Moodie and Reimchen (1976*b*) believed or could have been derived from other freshwater populations since local deglaciation. However, the frequency with which highly advanced, endemic *Gasterosteus* are encountered in the Queen Charlotte Islands and in other regions that experienced extensive late Wisconsin glaciation indicates that extensive postglacial divergence is a common phenomenon in *Gasterosteus*. Absence of primary freshwater fishes from the Outer Hebrides, Scotland, indicates autochthnous origin (Campbell and Williamson, 1979) of advanced phenotypes there since deglaciation (Campbell, 1979). Low-plate-morph sticklebacks in Iceland (Penczak, 1964) also must have evolved in place since glacial retreat. Obviously, much of the observed geographic variation in *G. aculeatus* in the northern part of its range may have evolved since the last glacial retreat.

5.3. Rates over Historical Periods

Finally, at the scale of ecological or historical time we observe less impressive, but remarkably rapid changes in phenotype frequencies. Gilbertson (1980) observed a decline in the frequency of partial morphs in Lake Aleknagik, Alaska, from about 20 to 12% over an 11-year period. The frequency of partial morphs decreased by 0.73% per year. This rate of change is 531 times greater than the rate of increase of one-spined sticklebacks reported by Bell and Haglund (1982). Similarly, the frequency of fish with seven plates on a side in Lake Wapato, Washington, increased among adult low morphs from 56% in 1968 to 65% 1 year later, but declined to 62% in 1970. Băcescu and Mayer (1956; cited in Münzing, 1963) observed that a population in Lake Techirghiol, Rumania, had been derived from monomorphic completely plated ancestors within the past 100 years but that 85% were partial morphs. Thus, partial morphs had increased in frequency at about the same average rate as they had declined in Lake Aleknagik. In contrast to these observations, Moodie (1972*a*) compared samples from Mayer Lake collected in 1929 and 1968 for seven phenotypic characters, and the means of none differed significantly between samples. Over very short periods of time, phenotype frequencies can experience marked change in natural populations of *G. aculeatus;* on the other hand, they may not change at all.

5.4. Conclusions

Despite the difficulties of comparing evolutionary rates, observations at time scales ranging from a single year to thousands of years indicate that *G. aculeatus* populations are capable of extraordinary rates of evolutionary divergence. The characters that have undergone such extraordinary divergence generally are known to have a genetic basis and to be subject to natural selection. Populations will diverge or experience stasis depending on whether selection is directional, disruptive, or stabilizing. Evolutionary lability of *G. aculeatus* has led to remarkable geographic variation, but while the divergent populations often are of postglacial origin, the process of diversification has been in existence since at least the late Miocene (Bell, 1974). If the rates of divergence observed over a few years could be sustained for thousands of years without reversal. much greater divergence than occurs in the Queen Charlotte Islands sticklebacks, for example, would be observed. If divergence such as that in the Queen Charlotte Islands sticklebacks could be sustained for the millions of years since the origin of *G. aculeatus* without reversal or extinction, we probably would have taxa that would be unrecognizable as gasterosteids. Clearly, orthogenesis has not characterized the history of stickleback evolution! Evolutionary rates must vary greatly among populations and from time to time within lineages. Without the complete history of a lineage, the rate of evolution cannot be measured accurately (Bookstein *et al.*, 1978).

6. Conclusions

I have had two objectives in the preparation of this review. The first objective was to present the major contributions that study of the sticklebacks, and particularly *Gasterosteus aculeatus*, have made to evolutionary theory. Second, I have tried to make the system more accessible to persons who are interested in evolutionary problems to which study of sticklebacks might contribute. Although Wootton's (1976) book provides a very useful general background on sticklebacks, it has a much broader scope than this review.

The threespine stickleback appears to be a particularly useful tool for evolutionary studies both because it is relatively well studied and because it apparently responds very rapidly to natural selection by adapting and speciating. Bell (1976*a*) suggested a general model for the derivation of freshwater from marine populations and subsequent evolution of traits characteristic of freshwater populations (Fig. 8). It may be that

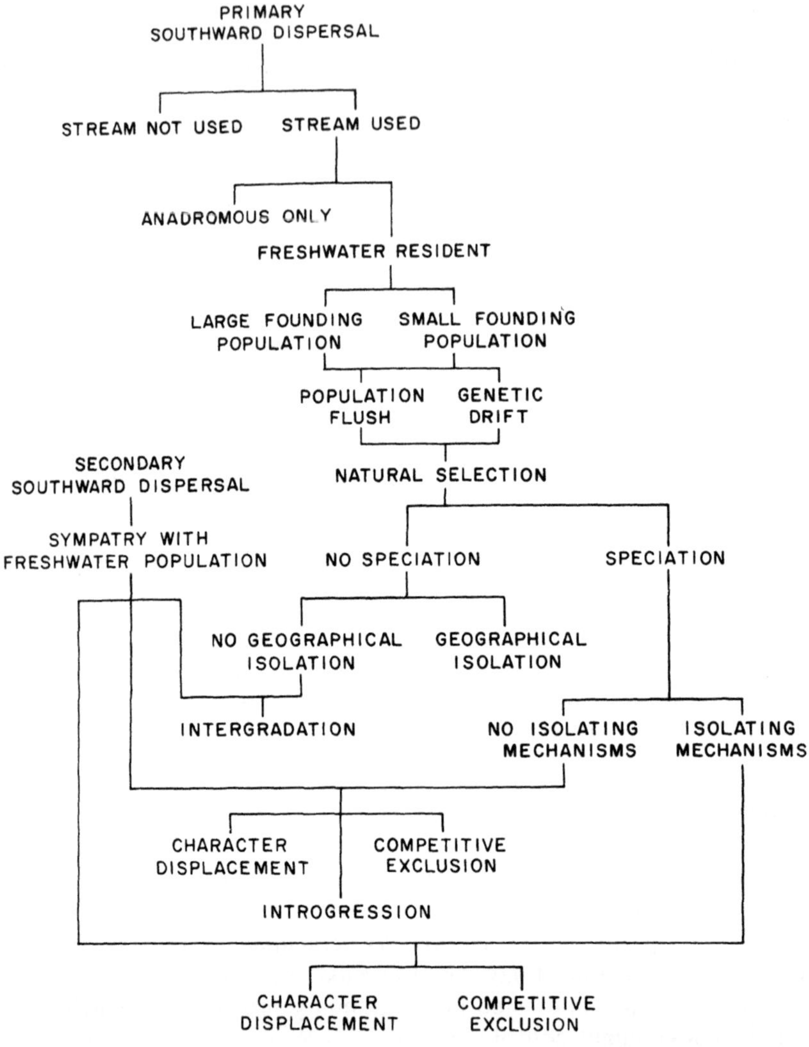

Figure 8. The mechanisms that contribute to evolutionary diversification of *Gasterosteus aculeatus*. After Bell (1976*a*).

the process of isolation within numerous, ecologically diverse habitats is particularly common in *G. aculeatus* and is conducive to evolutionary diversification. The ability of *G. aculeatus* to rapidly invade habitats from which most potential competitors and some predators are excluded probably is an important factor in their evolution. Distribution in numerous isolated freshwater environments also provides the advantage of having a large number of independently evolving populations to study. Thus, it appears that the mode of colonization in *G. aculeatus* and its occupancy of numerous isolated habitats have contributed to its diversification and to our understanding of the process.

The utility of a system for evolutionary genetics depends on a number of factors. Certainly, there must be ample variation to study, and *G. aculeatus* clearly satisfies this criterion. As information about a system accumulates, our ability to pose interesting hypotheses and to marshal the appropriate background information to adequately test the hypothesis grows rapidly. In particular, we must know the genetic basis of traits in which we are interested, and we must understand the function of the trait sufficiently to know what kinds of ecological or life history interactions to investigate. Our knowledge of genetics and function of geographically varying and polymorphic structures in *G. aculeatus* compares reasonably well with what is known about other groups of organisms that have played a central role in the development of ecological genetics.

Regardless of this progress, there are fundamental problems unsolved. Threespine sticklebacks are so widely distributed that some major regions have not been sampled, and some ubiquitous polymorphisms and other forms of intrapopulation variation have not been explained. An important deficiency in our understanding of stickleback evolution can be traced to failure to use protein variation in the study of divergence in sticklebacks. While there are impressive morphological similarities among populations in different parts of the world, no objective comparison of a sample of populations taken throughout the range of the *G. aculeatus* species complex has been performed. Our perceptions of the evolutionary biology of threespine sticklebacks has changed only slightly but has been enriched greatly by research performed since the system was reviewed most recently (Bell, 1976*a*; Wootton, 1976). Considering the continuing activity of workers in North America and the rewakening of interest in Europe, we should learn a great deal about stickleback evolution and from it about broader issues in evolutionary biology in coming years.

Appendix A. Measurement of Morphological Features in *Gasterosteus aculeatus*

Methods for measurement of morphological features in *Gasterosteus aculeatus* and other sticklebacks generally follow standard ichthyological procedure (Hubbs and Lagler, 1947, and subsequent editions). However, some structures and problems are peculiar to sticklebacks, and they are briefly reviewed below.

A1. Sampling

The area over which samples are collected should be minimized, as the scale of spatial variation may be very small and phenotype frequencies may vary among closely adjacent habitats (Hagen, 1967; Reimchen, 1980; Bell, 1982*b*). It follows that characterizing populations on the basis of samples from one site is invalid. Hagen and Gilbertson (1972) showed that a minnow seine provides a random sample with respect to lateral plate morph, plate number, and gill raker number, and other studies indicate that various sampling methods produce similar samples (Reimchen, 1980; Reist, 1981). Thus, sampling bias does not appear to be a serious problem.

A2. Fixation, Staining, and Preservation

It has become general practice to stain specimens for lateral plate analysis with Alizarin Red S, a bone stain, as small plates may be missed even in large unstained specimens (Hagen and McPhail, 1970; Hagen and Gilbertson, 1972). Specimens should be fixed live in 10% formalin without abdominal slitting and stored in 50% isopropyl or 70% ethyl alcohol. Specimens are stained in a ≤2% (wt/vol) solution of KOH in which enough Alizarin Red S has been dissolved to make the solution opaque (Bell, 1979). The best way to tell when staining is complete is to wash adhering stain off of a specimen and examine the fin rays. Staining is complete when the rays are deeply stained purple. Stained specimens should be washed once in 2% KOH, then soaked in the KOH for 1–2 hr and washed several times in water before soaking in water for about 24 h. It is important to minimize the concentration of KOH and the time during which specimens are exposed to it, and to completely wash it out of specimens prior to return to alcohol, as KOH tends to clear the specimens. After thorough rinsing, the stained specimens can be returned to alcohol. They should not be mixed with unstained material, because the stain will be exchanged with unstained specimens.

A3. Scoring Morphological Features

The lateral plate morphs are described in Section 2.1, and in most mature specimens they can be distinguished unequivocally (Hagen and Gilbertson, 1972, 1973*a*; Bell, 1981). In rare mature complete morphs there is a gap in the plate row corresponding to one absent plate, but these specimens segregate as complete morphs (Hagen and Gilbertson, 1973*a*). Specimens with a gap of more than one plate and those with one or more plates isolated on the caudal peduncle should be scored as partials.

Care must be taken in scoring lateral plate morph and plate number phenotypes in specimens less than about 30 mm standard length, because plate development may be incomplete; the size at which plate development terminates apparently varies among populations [compare Hagen and McPhail (1970) and Bell (1981)].

Counting lateral plates and classifying the plate number phenotypes present some problems. When the anterior plate is fused to or overlaps the cleithrum some workers have not counted it and others have. There sometimes are isolated plates between the contiguous abdominal plates and caudal keel plates of partial morphs, and one must be explicit on the procedure used if any isolated plates are pooled with abdominal or caudal keel plate counts. Plate counts are reported from the left side only, from each side separately, or as the sum of both sides. The papers to which new data are to be compared should be consulted for plate counting methodology.

Sex determination is relatively simple because the ovaries are sparsely pigmented with melanophores, are transluscent, and contain ova, and the testes are profusely pigmented and are filled with amorphous, opaque white material.

Dorsal and pelvic spines generally are measured from the distal tip to a depression in the pelvis at the ventral base of the pelvic spine and in the dorsal radial at the posterior base of the dorsal spine.

Gill raker counts are facilitated by alizarin staining and by cutting the soft tissue that constricts the gill slit.

All dorsal and anal fin rays originate separately and are counted separately, in contrast to some fishes in which the last two fin rays are scored as a single one.

In general, great care must be taken to ensure comparability of data collected by different workers.

Appendix B. Crossing and Rearing *Gasterosteus aculeatus*

Knowledge of the genetic basis of a trait and the effect of important environmental variables on its development provide an important element for evolutionary studies. A great number of crosses of *Gasterosteus aculeatus* have been performed, and these are mentioned in the body of this review. The most useful information on performance of crosses is provided by Lindsey (1962), Hagen (1967, 1973), McPhail (1969), Hagen and Gilbertson (1973*a*), and Avise (1976).

Stock can be crossed immediately after collection or may be brought into reproductive condition in the laboratory. Attainment of reproductive condition requires summer conditions (i.e., long photoperiods and warm temperature), which will vary according to locality (Baggerman, 1957). The number of eggs produced per female varies according to body length, and the precise relationship varies among populations. However, a 50-mm (total length) specimen should produce about 130–160 eggs (Wootton, 1976). Reproductive condition in females is recognized by the severely swollen abdomen and in the males by the presence of nuptial coloration or nest-building behavior. Female reproductive maturity also may be verified by allowing a physically isolated female to view a courting male; response with a "heads up" signal (tilting the head up at a 45° angle while facing the male) indicates readiness to spawn.

By means of gentle pressure on the abdomen, ova may be stripped from the female into a petri dish. The testis is removed from a male and minced with a forceps and probe in the same or in a separate dish. Some workers stipulate addition of a few drops of aged tap or well water to the eggs and minced testis of freshwater sticklebacks. Finally, the two preparations are mixed together and allowed to stand for 5–20 min in water for fertilization to occur. Fertilization can be verified by formation of the perivitelline space between the egg and the vitelline membrane. Mucous and the testis preparation may be removed with a pipette to retard fungal growth, and the fertilized eggs are transferred to an incubation chamber.

The incubation chamber is usually a small vessel with numerous minute holes or with a fine nylon mesh bottom, and the chamber is suspended in water. Water of the appropriate salinity and temperature, based on ambient conditions in the source habitat, should be used for rearing. Lindsey (1962) obtained the best survival between about 18 and 24°C, but hatching success varied according to temperature and salinity and varied among lateral plate count phenotypes. The water in the incubating chamber must be vigorously aerated. If fungus begins to grow on the eggs they may be treated in a 1:200,000 dilution of Malachite Green solution for 1 hr (Hagen, 1967). Dead eggs increase in opacity and should be discarded. Hardy (1978) reviewed the embryology of *G. aculeatus;* hatching time was temperature dependent and took about 7 days at 19°C.

One day after hatching, the fry can be transferred to 10–20-gal aquaria. There should be a thin layer of sand or fine gravel at the bottom of the aquarium and the water should have been aged. The water should be filtered and aerated vigorously, but precautions must be taken to prevent the fry from being sucked into the filter. A fraction of the water should be changed periodically and care should be taken to hold salinity constant. Hagen and Gilbertson (1973) never reared more than 30 fish in a 61-liter aquarium, to avoid crowding and resulting elevated mortality.

The newly hatched fry must be fed "infusoria," unless a culture has developed in the aquarium substrate, and after a few days they must be fed brine shrimp nauplii or nematodes. When they become large enough, they may be fed brine shrimp, *Daphnia,* crumbled flaked fish food, or chopped tubifex, meat, fish, or shrimp of appropriate size.

As temperature influences the number of elements in meristic series (Lindsey, 1962), it should be controlled as carefully as possible until the structures being studied are completely developed. Care should be taken to ensure that development is complete when the progeny are sacrificed.

ACKNOWLEDGMENTS. I wish to express my gratitude to many colleagues, principally J. D. McPhail, D. W. Hagen, G. E. E. Moodie, and T. E. Reimchen, who have freely shared their ideas and unpublished manuscripts with me. I also am indebted to J. V. Baumgartner, R. N. Campbell, B. W. Coad, D. E. Hay, T. R. Haglund, B. E. Kynard, and J. S. Nelson. The manuscript benefited from the comments of J. V. Baumgartner, G. E. E. Moodie, and T. E. Reimchen. Some of the original illustrations were drafted by D. Naylor. Unpublished results of my research to which I refer were obtained with the support from the Fishes Section of the Natural History Museum of Los Angeles County, the Theodore Roosevelt Memorial Fund (American Museum of Natural History), the Raney Fund (American Society of Ichthyologists and Herpetologists), and the Henry/Marsh Fund (National Academy of Sciences). My research was supported by National Science Foundation Grant DEB-8111013 during the preparation of this chapter. This chapter is Contribution No. 413 from the Department of Ecology and Evolution at the State University of New York at Stony Brook.

References

Alberch, P., 1980, Ontogenesis and morphological diversification, *Am. Zool.* **20**:653–667.

Alberch, P., Gould, S. J., Oster, G. F., and Wake, D. B., 1979, Size and shape in ontogeny and phylogeny, *Paleobiology* **5**:296–317.

Avise, J. C., 1974, Systematic value of electrophoretic data, *Syst. Zool.* **23**:465–489.

Avise, J. C., 1976, Genetics of plate morphology in an unusual population of threespine sticklebacks (*Gasterosteus aculeatus*), *Genet. Res.* **27**:33–46.

Băcescu, M., and Mayer, R., 1956, Cercetari Aspura Ghidrinilor (*Gasterosteus aculeatus* L.) Din Apele Roministi, *Bull. Inst. Cercet. Piscole Roman.* **15**:19–36.

Baggerman, B., 1957, An experimental study of the timing of breeding and migrations in the three-spined stickleback, *Gasterosteus aculeatus, Arch. Neerl. Zool.* **12**:105–318.

Baumgartner, J. V. and Bell, M. A., Lateral plate morph variation in California populations of the threespine stickleback, *Gasterosteus aculeatus, Evolution* (in press).

Banister, K. E., 1967, The anatomy and classification of the order Gasterosteiformes, Ph. D. dissertation, University of Newcastle upon Tyne.

Bell, M. A., 1973, The Pliocene stickleback, *Pungitius haynesi,* a junior synonym of *Gasterosteus aculeatus, Copeia* **1973**:588–590.

Bell, M. A., 1974, Reduction and loss of the pelvic girdle in *Gasterosteus* (Pisces): A case of parallel evolution, *Nat. Hist. Mus. Los Ang. Cty. Contrib. Sci.* **257**:1–36.

Bell, M. A., 1976*a*, Evolution of phenotypic diversity in *Gasterosteus aculeatus* superspecies on the Pacific coast of North America, *Syst. Zool.* **25**:211–227.

Bell, M. A., 1976*b*, The evolution of phenotypic diversity in threespine sticklebacks (*Gasterosteus aculeatus*), Ph. D. dissertation, University of California, Los Angeles.

Bell, M. A., 1976*c*, Reproductive character displacement in threespine sticklebacks, *Evolution* **30**:847–848.

Bell, M. A., 1977, A late Miocene marine threespine stickleback, *Gasterosteus aculeatus aculeatus*, and its zoogeographic and evolutionary significance, *Copeia* **1977**:277–282.

Bell, M. A., 1978, Fishes of the Santa Clara River system, southern California, *Nat. Hist. Mus. Los Ang. Cty. Contrib. Sci.* **295**:1–20.

Bell, M. A., 1979, Low-plate morph of the threespine stickleback breeding in salt water, *Copeia* **1979**:529–533.

Bell, M. A., 1981, Lateral plate polymorphism and ontogeny of the complete plate morph of threespine sticklebacks (*Gasterosteus aculeatus*), *Evolution* **35**:67–74.

Bell, M. A., 1982*a*, Melanism in a high elevation population of *Gasterosteus aculeatus*, *Copeia* **1982**:829–835.

Bell, M. A., 1982*b*, Differentiation of adjacent stream populations of threespine sticklebacks, *Evolution* **36**:189–199.

Bell, M. A., and Baumgartner, J. V., 1984, An unusual population of *Gasterosteus aculeatus* from Boston, Massachusetts, *Copeia* **1984**:258–262.

Bell, M. A., and Haglund, T. R., 1978, Selective predation of threespine sticklebacks (*Gasterosteus aculeatus*) by garter snakes, *Evolution* **32**:304–319.

Bell, M. A., and Haglund, T. R., 1982, Fine-scale temporal variation of the Miocene stickleback *Gasterosteus doryssus*, *Paleobiology* **8**:282–292.

Bell, M. A., and Richkind, K. E., 1981, Clinal variation of lateral plates in threespine stickleback fish, *Am. Nat.* **117**:113–132.

Bertin, L., 1925, Recherches bionomiques, biométriques et systématiques sur les epinoches (Gastérostéids), *Ann. Inst. Oceanogr. Monaco* **2**:1–204.

Bianco, P. G., 1979, Solla presenza di *Gasterosteus aculeatus* nella pensicola salentina (Pisces, Gasterosteidae), *Natura* **70**:76–80.

Bianco, P. G., 1980, Areale italico, rinvenimento in Calabria e origini delle populaioni mediterranee de *Gasterosteus aculeatus* L. (Pisces, Gasterosteidae), *Boll. Mus. Civ. Stor. Nat. Verona* **7**:197–216.

Black, E. A., 1977, Population differentiation of the threespine stickleback (*Gasterosteus aculeatus*), M. A. thesis, University of British Columbia, Vancouver.

Blouw, D. M., and Hagen, D. W., 1981, Ecology of the fourspine stickleback, *Apeltes quadracus*, with respect to a polymorphism for dorsal spine number, *Can. J. Zool.* **59**:1677–1692.

Bookstein, F. L., Gingerich, P. D., and Kluge, A. G., 1978, Hierarchical linear modeling of the tempo and mode of evolution, *Paleobiology* **4**:120–134.

Campbell, R. N., 1979, Sticklebacks (*Gasterosteus aculeatus* (L.) and *Pungitius pungitius* (L.)) in the Outer Hebrides, Scotland, *Hebridean Nat.* **3**:8–15.

Campbell, R. N., and Williamson, R. B., 1979, The fishes of inland waters in the Outer Hebrides, *Proc. R. Soc. Ed.* **77B**:377–393.

Chen, T. R., and Reisman, H. M., 1970, A comparative chromosome study of the North American species of sticklebacks (Teleostei: Gasterosteidae), *Cytogenetics* **9**:321–332.

Clarke, B. C., 1979, The evolution of genetic diversity, *Proc. R. Soc. Lond. B* **205**:453–474.

Clarke, B., Arthur, W., Horsley, D. T., and Parkin, D. T., 1978, Genetic variation and natural selection in pulmonate molluscs, in: *Pulmonates* (V. Fretter and J. Peake, eds.), Academic Press, New York, pp. 219–220.

Coad, B. W., 1973, Modifications of the pelvic complex in ninespine sticklebacks, *Pungitius pungitius* (L.), of eastern Canada and the Northwest Territories, *Nat. Can.* **100:**315–316.

Coad, B. W., 1981, A biobliography of the sticklebacks (Gasterosteidae: Osteichthyes), *Syllogeus* **35:**1–142.

Coad, B. W., and Power, G., 1973, Observations on the ecology and phenotypic variation of the threespine stickleback, *Gasterosteus aculeatus* L., 1758, and the blackspotted stickleback, *G. wheatlandi* Putnam, 1867 (Osteichthys: Gasterosteidae) in Armory Cove, Quebec, *Can. Field-Nat.* **87:**113–122.

Coad, B. W., and Power, G., 1974, Meristic variation in the threespine stickleback, *Gasterosetus aculeatus,* in the Matarmek River, Quebec, *J. Fis. Res. Board Can.* **31:**1155–1157.

Darlington, P. J., Jr., 1957, *Zoography: The Geographic Distribution of Animals,* Wiley, New York.

Denton, G. H., and Hughes, T. J. (eds.), 1981, *The Last Great Ice Sheets,* Wiley-Interscience, New York.

Dobzhansky, T., 1972, Species of *Drosophila, Science* **177:**664–669.

Eldredge, N., and Gould, S. J., 1972, Punctuated equilibria; An alternative to phyletic gradualism, in: *Models in Paleobiology* (T. J. M. Schopf, ed.), Freeman, Cooper and Co., San Francisco, pp. 82–115.

Erkinario, E., 1974, Variation in some meristic characteristics of the three-spined stickleback, *Gasterosteus aculeatus* L. in northern Finland, *Acta Univ. Oulu A. 42* **3:**121–126.

Gilbertson, L. G., 1980, Variation and natural selection in an Alaskan population of the threespine stickleback (*Gasterosteus aculeatus* L.), Ph. D. dissertation, University of Washington, Seattle.

Giles, N., 1983, The possible role of environmental calcium levels during the evolution of phenotypic diversity in outer Hebridean populations of the three-spined stickleback, *Gasterosteus aculeatus, J. Zool. Lond.* **199:**535–544.

Ginzburg, L. R., 1981, Bimodality of evolutionary rates, *Paleobiology* **7:**426–429.

Gould, S. J., 1980, Is a new and general theory of evolution emerging?, *Paleobiology* **6:**119–130.

Gould, S. J., and Eldredge, N., 1977, Punctuated equilibria: The tempo and mode of evolution reconsidered, *Paleobiology* **3:**115–151.

Greenbank, J., and Nelson, P. R., 1959, Life history of the three-spine stickleback *Gasterosteus aculeatus* Linnaeus in Karluk and Bare Lake, Kodiak Island, Alaska, *U. S. Fish Wildl. Serv. Fish. Bull.* **59:**537–559.

Gross, H. P., 1977, Adaptive trends of environmentally sensitive traits in the three-spined stickleback, *Gasterosteus aculeatus* L., *Z. Zool. Syst. Evolutionsforsch.* **15:**252–278.

Gross, H. P., 1978*a,* Natural selection by predators on the defensive apparatus of the three-spined stickleback, *Gasterosteus aculeatus* L., *Can. J. Zool.* **56:**398–413.

Gross, H. P., 1978*b,* Observations on the geographic variation of a marine coastal fish, *Spinachia spinachia* L., *Mar. Biol.* **47:**297–302.

Gross, H. P., 1979, Geographic variation in European ninespine sticklebacks, *Pungitius pungitius, Copeia* **1979:**405–412.

Hagen, D. W., 1967, Isolating mechanisms in threespine sticklebacks (*Gasterosteus*), *J. Fish. Res. Board Can.* **24:**1637–1692.

Hagen, D. W., 1973, Inheritance of numbers of lateral plates and gill rakers in *Gasterosteus aculeatus, Heredity* **30:**303–312.

Hagen, D. W., and Blouw, D. M., 1983, Heritability of dorsal spines in the fourspine stickleback (*Apeltes quadracus*), *Heredity* **50:**275–281.

Hagen, D. W., and Gilbertson, L. G., 1972, Geographic variation and environmental selection in *Gasterosteus aculeatus* in the Pacific Northwest, America, *Evolution* **26:**32–51.

Hagen, D. W., and Gilbertson, L. G., 1973*a*, The genetics of plate morphs in freshwater threespine sticklebacks, *Heredity* **31**:75–84.

Hagen, D. W., and Gilbertson, L. G., 1973*b*, Selective predation and the intensity of selection acting upon the lateral plates of threespine sticklebacks, *Heredity* **30**:273–287.

Hagen, D. W., and McPhail, J. D., 1970, The species problem within *Gasterosteus aculeatus* on the Pacific coast of North America, *J. Fish. Res. Board Can.* **27**:147–155.

Hagen, D. W., and Moodie, G. E. E., 1979, Polymorphism for breeding colors in *Gasterosteus aculeatus*. I. Their genetics and geographic distribution, *Evolution* **33**:641–648.

Hagen, D. W., and Moodie, G. E. E., 1982, Polymorphism for plate morphs in *Gasterosteus aculeatus* on the east coast of Canada and an hypothesis for their global distribution, *Can. J. Zool.* **60**:1032–1042.

Hagen, D. W., Moodie, G. E. E., and Moodie, P. F., 1980, Polymorphism for breeding colors in *Gasterosteus aculeatus*. II. Reproductive success as a result of convergence for threat display, *Evolution* **34**:1050–1059.

Haglund, T. R., 1981, Differential reproduction among lateral plate phenotypes of *Gasterosteus aculeatus,* the threespine stickleback, Ph. D. dissertation, University of California, Los Angeles.

Hall, M. F., 1956, A comparative study of the reproductive behaviour of the sticklebacks (Gasterosteidae), Ph. D. dissertation, St. Hugh's College, Oxford University, Oxford, England.

Hardy, J. D., Jr., 1978, *Development of fishes of the Mid-Atlantic Bight*, Vol. II, *Anguillidae through Synghathidae,* Fish and Wildlife Service, U. S. Department of the Interior, Washington, D.C.

Haskins, C. P., Haskins, E. F., McLaughlin, J. J. A., and Hewitt, R. E., 1961, Polymorphism and population structure in *Lebistes reticulatus,* an ecological study, in: *Vertebrate Speciation* (W. F. Blair, ed.), University of Texas Press, Austin, Texas, pp. 320–395.

Hay, D. E., 1974, Ecological genetics of threespine sticklebacks (*Gasterosteus*), Ph. D. dissertation, University of British Columbia, Vancouver, British Columbia.

Hay, D. E., and McPhail, J. D., 1975, Mate selection in threespine sticklebacks (*Gasterosteus*), *Can. J. Zool.* **53**:441–450.

Heuts, M. J., 1947*a,* The phenotypic variability of *Gasterosteus aculeatus* L. populations in Belgium, *Verh. K. Vlaam. Acad. Wet. Lett. Schone Kunsten Belg.* **9**:1–63.

Heuts, M. J., 1947*b,* Experimental studies on adapative evolution in *Gasterosteus aculeatus* L., *Evolution* **1**:89–102.

Heuts, M. J., 1956, Temperature adaptation in *Gasterosteus aculeatus* L., *Publ. Staz. Zool. Napoli* **28**:44–61.

Hoogland, R. D., 1951, On the fixing-mechanism in the spines of *Gasterosteus aculeatus* L., *Ned. Akad. Wet. Ser. C* **54**:171–180.

Hoogland, R. D., Morris, D., and Tinbergen, N., 1957, The spines of sticklebacks (*Gasterosteus* and *Pygosteus*) as means of defense against predators (*Perca* and *Esox*), *Behaviour* **10**:205–236.

Howard, R. D., 1979, Estimating reproductive success in natural populations, *Am. Nat.* **114**:221–231.

Howe, K. M., 1973, Systematics of the *Gasterosteus* complex (Pisces: Gasterosteidae) in northern California, M. A. thesis, California State College, Sonoma.

Hubbs, C. L., 1929, The Atlantic American species of the fish genus *Gasterosteus, Occ. Pap. Mus. Zool. Univ. Mich.* **200**:1–9.

Hubbs, C. L., and Lagler, K. F., 1947, Fishes of the Great Lakes Region, Cranbrook Inst. Sci. Bulletin No. 26, Bloomfield Hills, Michigan.

Huntingford, F. E., 1976, The relationship between anti-predator behaviour and aggression among conspecifics in the three-spined stickleback, *Gasterosteus aculeatus, Anim. Behav.* **24**:245–260.

Huntingford, F. E., 1981, Further evidence for an association between lateral scute number and aggressiveness in the threespine stickleback, *Gastrosteus aculeatus, Copeia* **1981**:717–720.

Igarashi, K., 1964, Observations on the development of the scutes in landlocked form of three-spined stickleback, *Gasterosteus aculeatus aculeatus* Linnaeus, *Bull. Jpn. Soc. Sci. Fish.* **30**:95–103.

Igarashi, K., 1970, On the variation of the scutes in the three-spined stickleback, *Gasterosteus aculeatus aculeatus* (Linnaeus) from Nasu area, Tochigi-ken, *Annot. Zool. Jpn.* **43**:43–49.

Jenni, D. A., 1972, Effects of conspecifics and vegetation on nest site selection in *Gasterosteus aculeatus* L., *Behaviour* **42**:97–118.

Kerfoot, W. C., 1975, The divergence of adjacent populations, *Ecology* **56**:1298–1313.

Kettlewell, H. B. D., 1973, *The Evolution of Melanism,* Clarendon Press, Oxford, England.

Koehn, R. K., and Eanes, W. F., 1977, Subunit size and genetic variation of enzymes in natural populations of *Drosophila, Theor. Popul. Biol.* **11**:330–341.

Koehn, R. K., and Eanes, W. F., 1978, Molecular structure and protein variation within and among populations, *Evol. Biol.* **11**:39–100.

Krueger, W. H., 1961, Meristic variation in the fourspine stickleback, *Apeltes quadracus, Copeia* **1961**:442–450.

Kynard, B. E., 1972, Male breeding behavior and lateral plate phenotypes in the threespine stickleback (*Gasterosteus aculeatus* L.), Ph. D. dissertation, University of Washington, Seattle.

Kynard, B. E., 1978, Breeding behavior of a lacustrine population of threespine sticklebacks (*Gasterosteus aculeatus* L.), *Behaviour* **67**:178–202.

Kynard, B. E., 1979*a,* Nest habitat preference of low plate number morphs in threespine sticklebacks (*Gasterosteus aculeatus*), *Copeia* **1979**:525–528.

Kynard, B. E., 1979*b,* Population decline and change in frequencies of lateral plates in threespine sticklebacks (*Gasterosteus aculeatus*), *Copeia* **1979**:635–638.

Kynard, B. E., 1984, Breeding behavior and reproductive success of male low plate number, phenotypes of threespine sticklebacks (*Gasterosteus aculeatus*), *Copeia,* in press.

Kynard, B., and Curry, K., 1976, Meristic variation in the threespine stickleback, *Gasterosteus aculeatus,* from Auke Lake, *Copeia* **1976**:811–813.

Larson, G. L., 1976, Social behavior and feeding ability of two phenotypes of *Gasterosteus aculeatus* in relation to their spatial and tropic segregation in a temperate lake, *Can. J. Zool.* **54**:107–121.

Lee, D. S., Gilbert, C. R., Hocutt, C. H., Jenkins, R. E., McAllister, D. E., and Stauffer, J. R., Jr., 1980, *Atlas of North American Freshwater Fishes,* North Carolina State Museum of Natural History and U.S. Fish and Wildlife Serve (Publication #1980-12 of the North Carolina Biological Survey), Raleigh, North Carolina.

Leim, A. H., and Scott, W. B., 1966, Fishes of the Atlantic coast of Canada, *Bull. Fish Res. Board Can.* **155**:1–485.

Levene, H., 1953, Genetic equilibrium when more than one ecological niche is available, *Am. Nat.* **87**:331–333.

Lewontin, R. C., 1974, *The Genetic Basis of Evolutionary Change,* Columbia University Press, New York.

Lindsey, C. C., 1962, Experimental study of meristic variation in a population of threespine sticklebacks, *Gasterosteus aculeatus, Can. J. Zool.* **40**:271–312.

MacLean, J. A., 1974, Variation and natural selection in a population of sticklebacks (*Gasterosteus*), Ph. D. dissertation, University of British Columbia, Vancouver, British Columbia.

MacLean, J. A., 1980, Ecological genetics of threespine sticklebacks in Heisholt Lake, *Can. J. Zool.* **58**:2026–2039.

Magnuson, J. J., and Heitz, J. G., 1971, Gill raker apparatus and food selectivity among mackerals, tunas and dolphins, *Fish. Bull.* **69:**361–369.

Mayr, E., 1963, *Populations, Species and Evolution,* Belknap Press, Cambridge, Massachusetts.

McAllister, D. E., Cumbaa, S. L., and Harrington, C. R., 1981, Pleistocene fishes (*Coregonus, Osmerus, Microgadus, Gasterosteus*) from Green Creek, Ontario, Canada, *Can. J. Earth Sci.* **18:**1356–1364.

McPhail, J. D., 1963, Geographic variation in North American ninespine sticklebacks, *Pungitius pungitius, J. Fish. Res. Board Can.* **20:**27–44.

McPhail, J. D., 1969, Predation and the evolution of a stickleback (*Gasterosteus*), *J. Fish. Res. Board Can.* **26:**3183–3208.

McPhail, J. D., 1977, Inherited interpopulation differences in size at first reproduction in threespine stickleback, *Gasterosteus aculeatus* L., *Heredity* **38:**53–60.

McPhail, J. D., and Lindsey, C. C., 1970, Freshwater fishes of northwestern Canada and Alaska, *Bull. Fish. Res. Board Can.* **173:**1–381.

Miller, R. R., 1960, The type locality of *Gasterosteus aculeatus williamsoni* and its significance in the taxonomy of California sticklebacks, *Copeia* **1960:**348–350.

Miller, R. R., and Hubbs, C. L., 1969, Systematics of *Gasterosteus aculeatus,* with particular reference to intergradation and introgression along the Pacific Coast of North America: A commentary on a recent contribution, *Copeia* **1969:**52–69.

Moodie, G. E. E., 1972*a,* Morphology, life history and ecology of an unusual stickleback (*Gasterosteus aculeatus*) in the Queen Charlotte Islands, Canada, *Can. J. Zool.* **50:**721–732.

Moodie, G. E. E., 1972*b,* Predation, natural selection and adaptation in an unusual threespine stickleback, *Heredity* **28:**155–167.

Moodie, G. E. E., 1977, Meristic variation, asymmetry, and aspects of the habitat of *Culaea inconstans* (Kirtland), the brook stickleback, in Manitoba, *Can. J. Zool.* **55:**398–404.

Moodie, G. E. E., and Reimchen, T. E., 1973, Endemism and conservation of sticklebacks in the Queen Charlotte Islands, *Can. Field-Nat.* **87:**173–175.

Moodie, G. E. E., and Reimchen, T. E., 1976*a,* Phenetic variation and habitat differences in *Gasterosteus* populations of the Queen Charlotte Islands, *Syst. Zool.* **25:**49–61.

Moodie, G. E. E., and Reimchen, T. E., 1976*b,* Glacial refugia, endemism and stickleback populations of the Queen Charlotte Islands, British Columbia, *Can. Field-Nat.* **90:**471–474.

Moodie, G. E. E., McPhail, J. D., and Hagen, D. W., 1973, Experimental demonstration of selective predation in *Gasterosteus aculeatus, Behaviour* **47:**95–105.

Münzing, J., 1959, Biologie, Variabilität und Genetik von *Gasterosteus aculeatus* L. (Pisces). Untersuchungen im Elbegebiet, *Int. Rev. Gesamten Hydrobiol.* **44:**317–382.

Münzing, J., 1962, Ein neuer *semiarmatus*-Typ von *Gasterosteus aculeatus* L. (Pisces) aus dem Izniksee, *Mitt. Hamb. Zool. Mus. Inst.* **60:**181–194.

Münzing, J., 1963, The evolution of variation and distributional patterns in European populations of the three-spined sticklebacks, *Gasterosteus aculeatus, Evolution* **17:**320–332.

Münzing, J., 1969, Variabilitat, Verbreitung und Systematik der Arten und Unterarten in der Gattung *Pungitius* Coste, 1848 (Pisces, Gasterosteidae), *Z. Zool. Syst. Evolutionsforsch.* **7:**208–233.

Mural, R. J., 1973, The Pliocene sticklebacks of Nevada with a partial osteology of the Gasterosteidae, *Copeia* **1973:**721–735.

Myers, G. S., 1930, The killifish of San Ignacio and the stickleback of San Ramon, Lower California, *Proc. Calif. Acad. Sci.* **19:**95–104.

Narver, D. W., 1969, Phenotypic variation in threespine sticklebacks (*Gasterosteus aculeatus*) of the Chignik River System, Alaska, *J. Fish. Res. Board Can.* **26:**405–412.

Nei, M., 1972, Genetic distance between populations, *Am. Nat.* **106:**283–292.

Nelson, J. S., 1969, Geographic variation in the brook stickleback, *Culaea inconstans,* and notes on nomenclature and distribution, *J. Fish. Res. Board Can.* **26:**2431–2447.

Nelson, J. S., 1971*a,* Comparison of the pectoral and pelvic skeletons and of some other bones and their phylogenetic implications in the Aulorhynchidae and Gasterosteidae (Pisces), *J. Fish. Res. Board Can.* **28:**427–442.

Nelson, J. S., 1971*b,* Absence of the pelvic complex in ninespine sticklebacks, *Pungiitus pungitius,* collected in Ireland and Wood Buffalo National Park Region, Canada, with notes on meristic variation, *Copeia* **1971:**707–717.

Nelson, J. S., 1976, *Fishes of the World,* Wiley-Interscience, New York.

Nelson, J. S., 1977, Evidence of a genetic basis for absence of the pelvic skeleton in brook stickleback, *Culaea inconstans,* and notes on the geographical distribution and origin of the loss, *J. Fish Res. Board Can.* **34:**1314–1320.

Nelson, J., and Atton, F. M., 1971, Geographic and morphological variation in the presence and absence of the pelvic skeleton in the brook stickleback, *Culaea inconstans* (Kirtland), in Alberta and Saskatchewan, *Can. J. Zool.* **49:**343–352.

Paepke, H. J., 1970, Studien zur Ökologie, Variabilität und Populationstrukter des Dreistachligen und Neunstachligen Stichlings. II. Die Variabilität der Lateralbeschildung von *Gasterosteus aculeatus* in eirer brandenburgischen Intergradationszone und ihre zoogeographisch-historischen Hintergründe, *Veroeff. Bezirksheimatatmus. Potsdam* **21:**5–48.

Peet, R. K., 1974, The measurement of species diversity, *Annu. Rev. Ecol. Syst.* **5:**285–307.

Penczak, T., 1963, A report on the crosses between *G. aculeatus* L. with various number of dorsal spines, *Naturwissenchaften* **50:**49.

Penczak, T., 1964, Threespined stickleback from Iceland *Gasterosteus aculeatus icelandicus* Sauvage, *Ann. Zool.* **22:**441–448.

Penczak, T., 1965, Morphological variation of the stickleback (*Gasterosteus aculeatus* L.) in Poland, *Zool. Pol.* **15:**3–49.

Penczak, T., 1966, Comments on the taxonomy of the threespined stickleback, *Gasterosteus aculeatus* Linnaeus, *Ohio J. Sci.* **66:**81–87.

Perlmutter, A., 1963, Observations on fishes of the genus *Gasterosteus* in the waters of Long Island, New York, *Copeia* **1963:**168–173.

Pierce, B. A., Mitton, J. D., and Rose, F. L., 1981, Allozyme variation among large, small and cannibal morphs of the tiger salamander inhabitating the Llano Estacado of west Texas, *Copeia* **1981:**590–595.

Pietsch, T. W., 1978, Evolutionary relationships of the sea moths (Teleosteii Pegasidae) with a classification of gasterosteiform families, *Copeia* **1978:**517–529.

Raunich, L., Callegarini, C., and Cucchi, C., 1972, Ecological aspects of hemoglobin polymorphism in *Gasterosteus aculeatus,* in: *Fifth European Marine Biology Symposium* (B. Battaglia, ed.), Piccini, pp. 153–162.

Rawlinson, S. E., and Bell, M. A., 1982, A stickleback fish (*Pungitius*) from the Neogene Sterling formation, Kenai Peninsula, Alaska, *J. Paleontol.* **56:**583–588.

Regan, C. T., 1909, The species of threespined sticklebacks (*Gasterosteus*), *Ann. Mag. Nat. Hist.* **4:**435–437.

Reimchen, T. E., 1980, Spine deficiency and polymorphism in a population of *Gasterosteus aculeatus* an adaptation to predators?, *Can. J. Zool.* **58:**1232–1244.

Reimchen, T. E., 1983, Structural relationships between spines and lateral plates in *Gasterosteus, Evolution,* **37:**931–946.

Reist, J. D., 1980*a,* Predation upon pelvic phenotypes of brook stickleback. *Culea inconstans,* by selected invertebrates, *Can. J. Zool.* **58:**1253–1258.

Reist, J. D., 1980*b,* Selective predation upon pelvic phenotypes of brook stickleback, *Culea inconstans,* by northern pike, *Esox lucius, Can. J. Zool.* **58:**1245–1252.

Reist, J. D., 1981, Variation in frequencies of pelvic phenotypes of the brook stickleback, *Culaea inconstans* in Redwater drainage, Alberta, *Can. Field-Nat.* **95:**178–182.

Ross, S. T., 1973, The systematics of *Gasterosteus aculeatus* (Pisces: Gasteosteidae) in central and southern California, *Nat. Hist. Mus. Los Ang. Cty. Contrib. Sci.* **243:**1–20.

Roth, F., 1920, Über den Bau und die Entwicklung des Hautpanzers von *Gasterosteus aculeatus, Anat. Anz.* **52:**513–534.

Rowland, W. J., 1982, Mate choice by male sticklebacks, *Gasterosteus aculeatus, Anim. Behav.* **30:**1093–1098.

Rutter, C., 1896, Notes on the freshwater fishes of the Pacific slope of North America. I. On the sticklebacks of California with special reference to individual variations, *Proc. Calif. Acad. Sci. Ser. 2* **6:**245–254.

Sargent, R. C., Bell, M. A., Krueger, W. H., and Baumgartner, J. V., 1984, A lateral plate cline, sexual dimorphism, and phenotypic variation in the blackspotted stickleback, *Gasterosteus wheatlandi, Canadian J. Zool.,* in press.

Semler, D. E., 1971, Some aspects of adaptation in a polymorphism for breeding colors in the threespine stickleback (*Gasterosteus aculeatus*), *J. Zool. Lond.* **165:**291–302.

Simpson, G. G., 1961, *Principles of Animal Taxonomy,* Columbia University Press, New York.

Stanley, S. M., 1979, *Macroevolution, Pattern and Process,* Freeman, San Francisco.

Stebbins, G. L., and Ayala, F. J., 1981, Is a new evolutionary synthesis necessary?, *Science* **213:**967–971.

Stephanidis, A., 1971, About some fishes of the freshwater of Greece, *Biol. Gallo-Hell.* **3:**213–241.

Stevens, P. F., 1980, Evolutionary polarity of character states, *Annu. Rev. Ecol. Syst.* **11:**333–358.

Thomson, K. S., Weed, W. H., III, Tarushi, A. G., and Simanek, D. E., 1978, Saltwater fishes of Connecticut, 2nd ed., Bulletin 105, State Geological and Natural Historical Survey of Connecticut.

Throckmorton, L. H., 1965, Similarity versus relationship in *Drosophila, Syst. Zool.* **14:**221–236.

Trivers, R. L., 1972, Parental investment and sexual selection, in: *Sexual Selection and the Descent of Man* (B. Campbell, ed.), Aldine, Chicago, pp. 136–179.

Turner, B. J., 1974, Genetic divergence of Death Valley pupfish species: Biochemical versus morphological evidence, *Evolution* **28:**281–294.

Turner, J. R. G., Johnson, M. S., and Eanes, W. F., 1979, Contrasted modes of evolution in the same genome: Allozymes and adaptive change in *Heliconius, Proc. Natl. Acad. Sci. USA* **76:**1924–1928.

Waltrous, L. E., and Wheeler, Q. D., 1981, The out-group method of character analysis, *Syst. Zool.* **30:**1–11.

Warner, B. G., Matthews, R. W., and Clague, J. J., 1982, Ice-free conditions on the Queen Charlotte Islands, British Columbia, at the height of late Wisconsin glaciation, *Science* **218:**675–677.

Whittam, T. S., 1981, Is a negative regression of Δp on p evidence of a stable polymorphism?, *Evolution* **35:**595–596.

Withler, R. E., 1980, Genetic relationships among threespine sticklebacks *Gasterosteus aculeatus,* M.Sc. thesis, University of British Columbia, Vancouver, British Columbia.

Wootton, R. J., 1973, The effect of size of food ration on egg production in the female threespined stickleback, *Gasterosteus aculeatus* L., *J. Fish. Biol.* **5:**89–96.

Wootton, R. J., 1976, *The Biology of the Sticklebacks,* Academic Press, New York.

Zaret, T. M., 1980, *Predation and Freshwater Communities,* Yale University Press, New Haven, Connecticut.

CHAPTER 10

Population Genetics of North Atlantic Catadromous Eels (Anguilla)

GEORGE C. WILLIAMS and RICHARD K. KOEHN

1. Introduction

The genetics of catadromous eels of the North Atlantic basin is of special interest because of uncertainties in the taxonomic status of the stocks found on opposite sides of the ocean, because their extraordinary life cycle makes them a source of information otherwise difficult to obtain on evolutionary processes, and because of uncertainties in genotype–environment interactions in the determination of sex.

It is customary to name the juvenile eels of the eastern side of the Atlantic *Anguilla anguilla* and those to the west *A. rostrata,* although the two forms are closely similar in morphology and ecology. Only the vertebral count serves to assign nearly all specimens to one or the other form. Both inhabit coastal and inland waters only as juveniles. Maturing adults migrate to spawning areas in the western tropical Atlantic (see Sections 1.2 and 1.3).

Tucker (1959) proposed that the difference in vertebral count is entirely of environmental origin, and results from slight differences in temperature experienced during embryonic development in the eastern and western parts of the supposed spawning region. He also believed it unlikely that maturing specimens from Europe, with their regressed digestive systems, could migrate all the way to a spawning area near the West

GEORGE C. WILLIAMS and RICHARD K. KOEHN • Department of Ecology and Evolution, State University of New York, Stony Brook, New York 11794.

Indies. European specimens would all be expatriates from the American population.

We reject Tucker's proposal, mostly for reasons adduced by Brunn (1963) (see also Deelder and Tucker, 1960; Jones and Tucker, 1959), and regard the American and European forms as at least partly separate breeding populations, perhaps as a result of largely allopatric spawning. This need not support recognition of separate species, a taxonomic conclusion that would require other evidence. Ideally it might be shown that the two forms have intrinsic reproductive isolation and near absence of gene exchange, despite partly sympatric spawning. In the absence of this kind of information it might be shown that the magnitude of evolutionary divergence is comparable to that found between closely related species in other groups. We believe that the limited evidence in both respects is against the recognition of distinct species. We here review indications of gene exchange between the two forms, and of genetic distance measures well below those commonly found between species. Unless the weight of evidence is reversed by future work, we favor the name *Anguilla anguilla* for all North Atlantic members of this genus.

1.1. Life Cycle of *Anguilla*

The life cycle is indeed remarkable. The eggs are buoyant and planktonic, as established by experimental induction of sexual maturity by hormone treatment of European (I. Boetius and Boetius, 1980; Kokhnenko *et al.*, 1978), American (Edel, 1975), and Japanese (Yamamoto and Yamauchi, 1974) material. A million or more eggs per female would be common. They hatch into leptocephalus larvae, once considered fishes ordinally distinct from eels. These larvae disperse, partly or largely through passive transport by ocean currents, for about 1 year to American coasts, 2 or more years to European and North African coasts. Metamorphosis into the earliest juvenile stage (elver) takes place mostly over the continental shelves, and elvers take up residence in coastal, estuarine, and freshwater environments. They live for a number of years in diverse habitats, from the tropics to the arctic, and the European form covers the enormous longitudinal range from Iceland and the Azores to Black Sea drainages. Females may grow to 1 m in length, the males less. At the approach of maturity they change into the silver-eel phase (described in Section 1.2) and start an autumnal migration back to the western tropical Atlantic to spawn and die.

This life cycle gives genetic information a special importance. Each year's spawning migration, from either side of the Atlantic, must include many millions of individuals, all the successful survivors of a great range of environmental conditions. Yet it is entirely possible that spawning is essentially panmictic. There is no reason to suppose that, for example, an adult migrant from Newfoundland will favor mating with another Newfoundland migrant. One from the Caribbean might be perfectly acceptable. Then, even if mating is monogamous, young of a single pair may scatter to widely different juvenile habitats. Evidence is presented here in favor of this near panmixia, or at least against any self-maintaining local populations. The seeming panmixia of such an enormous population was a particular difficulty for Wynne-Edwards (1962) theory of group selection.

If *Anguilla* reproduction is panmictic, it means that collections of juveniles from any locality are all samples of the same breeding population. Any demonstrable difference between such samples must have arisen as a result of different selection pressures operating between zygote and the stage sampled. For instance, if elvers arriving this year in New York and in Florida are found to differ in the frequency of some genetically determined allozyme, the difference could be attributed to different forces of natural selection in pelagic waters in transit to the sampled localities. Likewise, if those arriving this year differ from those of last year, selection pressures must have differed in the two years. Genetic differences between widely separated localities for most species would be the result of differing evolutionary forces of unknown strength acting for an unknown number of generations. Such differences in either American or European *Anguilla* should represent what natural selection can accomplish within a single generation.

1.2. Classic Evidence of Taxonomy and Life History

Detailed taxonomic studies of the genus *Anguilla,* including the 10 or more species from the Indo-Pacific region, have been carried out with morphometric comparisons (Ege, 1939; Schmidt, 1914; Tesch, 1977, Chapter 2). The North Atlantic members constitute a group phylogenetically separate from the others, and two species are recognized, *A. anguilla* from European and North African waters, and *A. rostrata* from American. These two can be distinguished only by vertebral count, *anguilla* having 110–119 (mean 114.7), and *rostrata* having 103–110 (mean 107.1). Until recently only a minute proportion of either species was reported to have 110 or 111 vertebrae. Other longitudinal meristic differences, for instance dorsal or anal fin-ray counts, are less marked and are so closely correlated with vertebral count as to provide little additional information. The two forms also differ by a fraction of a standard deviation in pectoral fin-ray count, but this character is not yet completed at the elver stage and may be less reliably indicative of genotype or conditions of embryonic development. Earlier workers had recognized differences in color and proportions between the two forms, and even proposed as many as 15 species within Europe, but careful work showed that none of these characters had diagnostic value. The only reliable character seemed to be the vertebral count. Doubts on the validity of distinctions between Indian Ocean species were recently raised by Balon (1975).

The peculiarities of its life history once made *Anguilla* a good example of spontaneous generation. No specimen, however large, ever showed evidence of sexual maturation, and no spawning was ever observed. Aristotle said that the myriads of elvers that appeared along shore every spring arose from the "Entrails of the sea" (Jordan and Barton, 1902). The first progress toward resolution of the mystery was not immediately recognized as such. In the 19th century the use of large plankton nets in oceanic waters provided specimens of some remarkably compressed and transparent fishes, which were attributed to *Leptocephalus, Tilurus, Oxyurus,* and other genera of a taxonomically isolated family Leptocephalidae. Near the end of the century it was discovered that one species, *Leptocephalus brevirostris,* was really the larval form of *Anguilla anguilla,* and that most of the larval biomass was lost in metamorphosis.

Patient investigations by Danish workers led by Johannes Schmidt [summarized by Schmidt (1925)] in the first two decades of this century showed that the smallest and youngest (<10 mm) larvae with the European vertebral count are found in an elongate east–west area of perhaps 100,000 km^2 northeast of the Lesser Antilles. Schmidt concluded that the one and only spawning area of the European eel must be within this region. Small (<15 mm) larvae with the American count were found mainly to the west, in a one million km^2 area between the Greater Antilles and Bermuda.

No really early American larvae nor eggs of either species were identified by Schmidt or subsequent workers [a possible exception is Yevseyenko (1974)]. So the evidence did not closely specify a geographic location or habitat of spawning, but did permit an outline account of the life cycle. Seasonal and geographic distribution of specimens of different sizes showed that spawning occurs mainly in early spring and that larvae are carried away by the clockwise circulation of North Atlantic surface waters. American larvae reach the east coast of North America within 1 year. European larvae need perhaps $2\frac{1}{2}$ years to reach the edge of the European continental shelf. Metamorphosis takes place in shelf waters during the winter. So North American elvers should be about 1 year old, European elvers 3 years old.

Juvenile growth in coastal and continental waters may typically last about 10 years for females in temperate regions, less for males (Gray and Andrews, 1971; Moriarti, 1972; Ogden, 1970; Sinha and Jones, 1967). They are immensely variable in this respect, and captive eels have lived for many decades (Bertin, 1957, Chapter 3; Tesch, 1977, p. 193).

The juvenile period ends with transformation into the silver-eel stage. Color then changes from various muddy shades to dark above and silvery below; the eye enlarges; maximum pressure of buoyancy maintenance increases from about 7 to 16 atm (Kleckner, 1980). These may be considered adaptations for migration through a mesopelagic environment. The gonads start to mature, the digestive system atrophies, and the fish begins its final migration, downstream at night through the river systems, and then out to sea. European migrants have been tracked by sonic tagging (Tesch, 1978) into the open ocean, where they travel at depths of 50–400 m or more, with the greater depths preferred during daylight.

More detailed accounts of the evidence and interpretation are available (Bertin, 1957; Harden-Jones, 1968; Tesch, 1973, 1977). The only serious challenge to Schmidt's interpretations has been Tucker's, mentioned in the introduction. While we reject his conclusions, we believe that some of his criticisms of orthodox views need more attention than they have received.

1.3. More Recent Discussions of the Life History

A number of workers have recognized difficulties with Schmidt's conclusions on the spawning area and subsequent dispersal, especially for the American form. Juveniles are most abundant along the Atlantic coast from the southeastern states to the Maritime provinces, but they are also common in the eastern Gulf states and occur sporadically throughout the Gulf and Caribbean to Trinidad and Surinam. Vladykov (1964) commented on the unlikelihood of larval dispersal to these regions from anywhere in the spawning area proposed by Schmidt. He suggested that some spawning must take place further south, so that the North Equatorial Current could take larvae through the Lesser Antillies into the Gulf and Caribbean. This suggestion is supported by Eldred's (1968, 1971) discovery of larvae in the Gulf of Mexico and Yucatan Straits.

Smith (1968) also commented on the enigma of *Anguilla rostrata* in the Gulf and Caribbean and pointed out another in its presence in Greenland. Only a few specimens have been found there, all identified by vertebral count as American (Jensen, 1937). The only ocean current that reaches Greenland from warmer parts of the Atlantic comes from the east after passing south of Iceland, where the European form is common and the American as yet unreported. The hydrographic pattern would lead us to expect the European eel in both Iceland and Greenland.

A similar problem might be recognized for the presence of the European form in the Azores and its absence from the Caribbean. The Azores are small targets for a planktonic organism riding south in the eastern part of the North Atlantic gyre. For every one that finds shallow water for metamorphosis in the Azores there must be many that pass by into the North Equatorial current that would reach northern South America or the Caribbean a few months later. Hydrographically it would make more sense to find the European form rather than the American in that region. The occurrence of elvers in large numbers in thc eastern Mediterranean (Ezzat and El-Serafy, 1977) certainly strains any model of entirely passive larval dispersal.

These difficulties suggest that dispersal must have some nonrandom or active aspects that would bear on the likelihood of panmixia. Perhaps American larvae as they approach 1 year in age acquire the motivation and sensory capability for swimming west. European larvae might be similarly endowed for swimming east at about 2 years of age. Even this simple a difference might suffice to explain most of the facts of juvenile distribution that would not follow from purely passive transport from the supposed spawning area.

2. Panmixia or Self-Maintaining Local Populations?

2.1. Aspects of the Life History

If the possibility of larval behavioral mechanisms that direct larvae to one or the other side of the Atlantic is to be seriously considered, why stop at merely two such mechanisms? Why not endow some larvae with behavior that assures their arrival in North Carolina, similarly equip others for finding South Carolina, and so on? This possibility was suggested by Wynne-Edwards (1962) as a way of providing *Anguilla* with the self-maintaining local populations that he thought essential to the long-term survival of any species. It would also be necessary that spawning migrants from a given locality mate mostly among themselves.

Such homing by a larva to a special locality to which it has never been would obviously require a more complicated mechanism than one that would merely put it on the proper side of the Atlantic, and may seem intuitively less likely. Such intuitions have proved misleading in the past on questions of animal navigational capabilities, and we feel that the possibility of larval homing to the juvenile habitats of their parents needs careful examination. If real, it would support the concept of self-maintaining local populations, invalidate our assumption of panmixia, and force new interpretations of published data. In this section we deal with several lines of indirect evidence that favor panmixia over local populations:

hydrographic considerations, larval size–frequency distributions, records of juvenile movement, juvenile sex ratios, and geographic distribution of vertebral counts. Later we will present more conclusive evidence from genetic polymorphisms.

Ocean currents in the North Atlantic would make larval navigation to any small and remote target extremely difficult, regardless of the point of origin. Even full-size larvae are weaker swimmers than elvers (Schmidt, 1916), which are unable to maintain a speed of 50 cm/sec even briefly (McCleave, 1980; Creutzberg, 1959). Ocean currents over much of the North Atlantic exceed even the elver speeds, and the main axis of the Gulf Stream may reach 3 m/sec (Stommel, 1965; Hachey, 1961). We can safely assume that larvae, especially the smaller sizes, are normally in water moving at several times their maximum swimming speeds. We might imagine them capable of navigational movements that would keep them on one or the other side of the Gulf Stream, but not counteract the direction of transport.

Ocean currents are highly variable in velocity and position (Neumann, 1968; Stommel, 1965). The main mass of the Gulf Stream at a given latitude can show manifold variations in velocity within a few weeks, and north of the 35th parallel it may show left–right shifts of hundreds of km. The pattern of movement can be extremely complex, with masses of Gulf Stream water left behind in swings from side to side (Hachey, 1961; Kerr, 1977; Watts and Olson, 1978). These detached water masses (termed *rings*) may be hundreds of km across and remain recognizable chemically and biotically as Gulf Stream water for several months. They usually move parallel to the Gulf Stream but in the opposite direction. Rings detached to the left may carry Gulf Stream organisms toward the American coast and mix with waters of the continental slope and shelf. Those detached to the right may move southwest through the Sargasso Sea and rejoin the Gulf Stream off the southeastern states near a point passed many months before. Many other patterns of ring behavior are possible, and water movements are highly variable on both a short-term and annual basis. It seems most unlikely that weakly swimming planktonic larvae could maneuver so as to adjust for such water movements and reach a prescribed locality on the American coast. Similar arguments could be made for any more distant goal on the European coast. Water movements in the northeastern Atlantic are complex and changeable, and vary from a few cm/sec to over 1 m/sec (Lee, 1974).

Larval size distributions in collections from the central North Atlantic suggest a largely random scatter by ocean currents. For instance, one of Schmidt's (1925) collections taken in June 1920 from well to the north of the spawning area contained a great abundance of 17–32 mm larvae and a single specimen of 43 mm. He assumed that the smaller specimens came from that spring's spawning and that the larger one was more than 1 year old. A sample from farther east in September included 63 specimens of 36–44 mm and 13 of 47–50 mm, and he interpreted these two groups as about 6 months and 18 months of age. Co-occurrence of larvae from two years' spawning suggests extensive turbulent mixing of specimens that must have followed entirely different trajectories from their origins to the point of sampling. A wide scatter of specimens from a single pair or group of spawners would result from the same mixing process.

Most of the dispersal between zygote and the end of the juvenile period must take place during the larval phase, but substantial movements by juveniles also occur, especially in fresh water. Large juveniles, many years beyond metamorphosis, are found in salt marshes of the Mississippi Delta and up the Mississippi and its tributaries beyond 45° N in Minnesota (Eddy and Underhill, 1974). Thus the elvers colonizing the coast of Louisiana may subsequently occupy markedly different and geographically widely separate habitats. They clearly show that the great adaptability of this species to different environments is not dependent on any long-term ecotypic divergence of local populations.

Considerable dispersal can occur in brackish water after metamorphosis. Eels of the Great Lakes and their tributaries must have passed through several hundred km of brackish habitats in the St. Lawrence system. They grow to at least 20 cm before reaching Lake Ontario (Hurley, 1972). The same can be said of eels in Baltic drainages. Eels of the Black Sea and its tributaries are the best example. The net flow of water through the Bosphorus is always toward the sea of Marmora and Mediterranean, so that there is no possibility of larval transport into the Black Sea. Understandably, elvers seem to be missing from that region. The juveniles of the Danube and other Black Sea drainages were always found to be at least 15 cm long (Slastenenko, 1936; Zineviei, 1967), prior to recent stocking with elvers.

If even a small fraction of the persistently saltwater eels ever swim great distances it would provide an easy explanation for some of the anomalies in geographic distribution mentioned above. Smith (1968) suggested that specimens found in Greenland might have gotten there by swimming more than 1000 km from Labrador, but thought this unlikely. South American specimens could conceivably have swum there from the Greater Antilles, but some larvae must reach the southern Caribbean, because Breder (1925) found elvers of 54–63 mm in Panama. It is important that future workers record, not only the coasts where *Anguilla* may be found, but also the sizes. Colonization by larval dispersal can be assumed only for areas seasonally inhabited by newly transformed elvers.

There is little information that bears directly on the possibility of significant saltwater dispersal by juveniles. On a short-term basis (weeks or months) American eels can be highly sedentary (Gunning and Shoop, 1962), and marked specimens may return to points of original capture after release as much as 200 km away (Tesch, 1977, and citations therein) in the North Sea region. The most impressive examples of homing were to freshwater localities, but some were wholly or partly through salt water. On the other hand, Aker and Koops (1973) reported considerable random movement and marked seasonal shifts of eels in coastal marine habitats of the North Sea. Coastal waters were often abandoned for the winter, with some specimens entering rivers and others moving to deeper water. Some specimens marked in German waters were recovered from Dutch and Danish localities.

Seasonal shifts between freshwater and marine habitats are also documented for Canadian waters (Medcof, 1969; Smith and Saunders, 1955), mostly an autumn migration into fresh water and a spring return to the sea. There is no evidence on whether specimens leaving a river in the spring will return to the same river in the fall.

It is often reported that eels from certain bodies of water or from major geographic regions are almost entirely of the same sex. Reliable sex identification for even large juveniles is difficult, and many of the older reports may be unreliable (Sinha and Jones, 1966). Vladykov (1966) reported a north–south gradient in sex ratio for eastern North America, with 94% (698/742) of Canadian eels being female. Gray and Andrews (1970) and Dolan and Power (1977) confirmed the scarcity of males in Canadian waters. The observations suggest that much of the fertilization of eggs of females from Canada would be by males from elsewhere. European eels may be mainly female in the Atlantic region and mainly male in the Mediterranean (Kuhlmann, 1975), although there is great local variation in sex ratio in Atlantic localities.

Perhaps the clearest of the classic evidence against local populations was recognized by Schmidt (1915). Despite considerable vertebral-count variation within his samples, there were no significant differences among samples from widely separate geographic regions. In general, widespread species always show geographic variation in means of variable characters, and such variation has been found to have both genetic and environmental components (Barlow, 1961). Geographic uniformity of mean vertebral count indicates a geographic uniformity of genetic factors affecting this trait and a uniformity of environment during embryogeny. Exceptions recently reported on the geographic uniformity of European vertebral counts are discussed in Section 2.2 in relation to the possibility of American–European gene flow. The geographic uniformity of American vertebral counts from the Virgin Islands to the Canadian Maritime provinces has been confirmed by Wenner (1972) by counts on 1746 specimens (original data plus literature review).

2.2. Genetic and Geographic Variation in North Atlantic *Anguilla* Populations

Electrophoretic studies of the genus *Anguilla* in the North Atlantic Ocean have generally been undertaken for two different reasons. First, the study of enzyme polymorphism affords discrete phenotypic variation that can be statistically compared with particular genetic models and thereby form a basis from which to make genetic statements about the structure of eel populations. When these data are interpreted within the context of the accepted life cycle, the pattern of genetic variation (or its absence) allows specific inferences to be made about both the genetic structure of eel populations and the direction and magnitude of evolutionary forces that shape variations in genetic composition. For example, unlike arguments presented in the preceding section, specific tests of panmixia in North American *Anguilla* can be made. We initiated work on the eel because we suspected that natural selection, especially in such prolific organisms as *Anguilla,* is often much stronger than commonly assumed and capable of producing noteworthy effects within a single generation. If our suppositions were true, widely separate stocks of juveniles may have experienced widely different environmental conditions during larval and subsequent development, leading to significant geographic variation as a consequence of natural selection. On the other hand, within the currently accepted life history and implied universal panmixia of the American eel, the absence of genetic differentiation would cause us to reject our initial supposition concerning the strenth of natural selection. It would also support the assumption of panmixia, independently of arguments developed above.

Genetic studies of North Atlantic *Anguilla* have also been directed toward attempting to settle once and for all the validity of the specific designations of *A. rostrata* and *A. anguilla*. Within this context, electrophoretic studies can be important for estimating the genetic uniqueness of eel samples from either side of the North Atlantic Ocean. If samples of the American and European eels could be shown to be qualitatively distinct, the specific status of the two groups would be supported. On the other hand, if samples of *Anguilla* taken throughout the North Atlantic differ no more from one another than local variants of widespread species, this genetic information would support no more than a subspecific distinction between the American and European eels.

Some of the predictions that follow from assumptions of panmixia and strong selection pressures have already been mentioned. Our complete list is as follows:

1. Elvers sampled at widely separate localities should be consistently different in some allozyme frequencies as a result of any consistent differences in selection pressures acting on larvae en route to the widely separate localities.
2. Strongly different environmental conditions (e.g., boreal versus tropical) at different localities should result in subsequent changes in gene frequencies, observable as allozyme frequency differences between elvers and older individuals at the same locality.
3. Allele frequency differences conforming to the first two predictions should do so consistently, with no significant variation from one year to the next.
4. Allele frequencies subject to strong selection that is not consistently directional should vary from year to year (or even from time to time within a year) at the same locality, since the selection that produces these differences must occur prior to recruited elvers taking up residence at a specific locality.
5. For changes produced by selection pressures characteristic of a given site, such changes should be observable between arriving elvers and juveniles of longer residence at this site.
6. In general, differences produced by natural selection should result in different patterns of variation among different characters since selection should not reasonably act in the same way on all characters.

We can also enunciate specific predictions concerning the degree and pattern of genetic differentiation between American and European populations of *Anguilla,* if these populations actually represent noninterbreeding, taxonomically distinct species:

1. Samples compared from the two continents should be genetically distinct to approximately the same degree as other congeneric species.
2. When European and American samples are compared at a reasonably large number of gene loci, a significant proportion of the loci should differ qualitatively, since evolutionarily differentiated species are characteristically fixed for different electrophoretic alleles (Avise, 1976).
3. If "morphologically intermediate" individuals are found, such individuals should be sterile F_1 hybrids, intermediate in most or all diagnostic characters (otherwise gene flow would be indicated).

4. Genetic differences among populations from a single continent should not accurately estimate genetic differences between continental populations, because the evolutionary history and contemporary habitats of the two species should be substantially different from one another, producing differing patterns of geographic differentiation within each.

We must emphasize that the main significance of electrophoretic information for the investigation of population and systematic questions in the genus *Anguilla* derives from our ability to draw genetic inferences. When detected variation cannot be attributed to a specific genetic model, such variation does not allow genetic inferences and provides little information of evolutionary and systematic import. Some of the earlier electrophoretic studies on the genus *Anguilla* (Drilhon *et al.*, 1956; Pantelouris and Payne, 1968; Pantelouris *et al.*, 1970, 1971) described electrophoretic variation that significantly differed from that expected by any known genetic model (Koehn, 1972) and therefore is not discussed in the following paragraphs. Except for our studies on spatial and temporal variation in North American eels (Williams *et al.*, 1973; Koehn and Williams, 1978), the genetic composition of samples of various geographic areas can be illustrated as a characteristic set of allele frequencies at a large number of gene loci (Table I). These data together with our own studies on the North American eel provide the basis for a clear and uncomplicated summary of the pattern and magnitude of genetic differentiation in North Atlantic *Anguilla*.

2.3. Spatial Genetic Variation in North American *Anguilla*

We have found significant geographic variation (Williams *et al.*, 1973) at the alcohol dehydrogenase (*Adh*), sorbitol dehydrogenase (*Sdh*) and glucosephosphate isomerase (*Phi-2*) loci in North American eels (Table II). While the magnitudes of variation at these loci were approximately the same, each exhibited variation in a distinctly different way (Fig. 1). The differences are attributable to selection acting at different times in the life cycle of North American *Anguilla*. For example, clinal variations of frequencies at the *Sdh* locus were the same in elvers and long-term residents sampled between Vero Beach, Florida, and St. Johns, Newfoundland, and the same pattern of geographic variation was observed during several years (Koehn and Williams, 1978). Although clinal variation was also observed at the *Phi-2* locus over a similar latitudinal range in North America, the clines were observed in residents only. There was also some apparent geographic variation at the *Adh* locus, but the spatial

Table I
Summary of Genetic Variation in North Atlantic *Anguilla*[a]

Enzyme	North America[b]	Iceland[c]	Europe
1. Alcohol dehydrogenase (1.1.1.1; *Adh*) (Koehn and Williams, 1978; Koehn and Williams, unpublished)			
Adh^{a}	0.114	0.005	0.030
Adh^{b}	0.659	0.495	0.385
Adh^{c}	0.227	0.500	0.585
2. Sorbitol dehydrogenase (1.1.1.14; *Sdh*) (Koehn and Williams, 1978; Koehn and Williams unpublished; Comparini *et al.*, 1975)			
Sdh^{a}	0.000[d]	0.009	0.000
Sdh^{b}	0.900	0.564	0.506
Sdh^{c}	0.100	0.427	0.483
Sdh^{d}	0.000	0.000	0.011
3. Lactate dehydrogenase (1.1.1.27; *Ldh-2*)[e] (Koehn and Williams, unpublished; Comparini and Rodino, 1980)			
Ldh-2^{a}	1.000	0.941[f]	0.915
Ldh-2^{b}	0.000	0.050	0.085
4. Malate dehydrogenase (1.1.1.37; *Mdh-2*)[g] (Williams *et al.*, 1973; Comparini and Rodino, 1980; de Ligny and Pantelouris, 1973; Rodino and Comparini, 1978; Comparini *et al.*, 1975)			
Mdh-2^{a}	0.958	0.129	0.100
Mdh-2^{b}	0.042	0.819[h]	0.900[i]
5. Isocitrate dehydrogenase (1.1.1.42; *Idh*) (Koehn and Williams, unpublished)			
Idh^{a}	0.048	0.072	0.031
Idh^{b}	0.952	0.918	0.957
Idh^{c}	0.000	0.012[j]	0.012
6. Phosphogluconate dehydrogenase (1.1.1.44; *6-Pgdh*) (Koehn and Williams, unpublished; Rodino and Comparini, 1978)			
6-Pgdh^{a}	0.018	0.150	0.263[k]
6-Pgdh^{b}	0.223	0.130	0.418
6-Pgdh^{c}	0.759	0.720	0.317
7. Superoxide dismutase (1.15.1.1; *To*) (Koehn and Williams, unpublished)			
To^{a}	0.000	0.018	0.000
To^{b}	0.994	0.800	0.990
To^{c}	0.006	0.191	0.010

(*continued*)

Table I (*Continued*)

Enzyme	North America[b]	Iceland[c]	Europe
8. Glutamate oxaloacetate transaminase-1 (2.6.1.1; *Got-1*) (Koehn and Williams, unpublished; Pantelouris, 1976)			
Got-1^{a}	0.020	0.036	0.004
Got-1^{b}	0.770	0.905	0.990
Got-1^{c}	0.210	0.055	0.006
Got-1^{d}	0.000	0.004	0.000
9. Glutamate oxaloacetate transaminase-2 (2.6.1.1; *Got-2*) (Koehn and Williams, unpublished; Pantelouris, 1976)			
Got-2^{a}	0.000	0.005	0.000
Got-2^{b}	0.946	0.986	0.993
Got-2^{c}	0.054	0.009	0.006
10. Glutamate oxaloacetate transaminase-3 (2.6.1.1; *Got-3*)[l] (Koehn and Williams, unpublished)			
Got-3^{a}	1.000	1.000	1.000
11. Adenylate kinase (2.7.4.3; *Ak*) (Koehn and Williams, unpublished)			
Ak^{a}	1.000	1.000	1.000
12. Phosphoglucomutase (2.7.5.1; *Pgm-2*)[m] (Koehn and Williams, unpublished)			
Pgm^{a}	0.000	0.014	0.000
Pgm^{b}	1.000	0.977	1.000
Pgm^{c}	0.000	0.009	0.000
13. Glucosephosphate isomerase-1 (5.3.1.9; *Phi-1*) (Koehn and Williams, unpublished; Comparini *et al.*, 1975; Rodino and Comparini, 1978)			
Phi-1^{a}	0.029	0.091	0.075
Phi-1^{b}	0.941	0.777	0.831
Phi-1^{c}	0.029	0.127	0.094

14. Glucosephosphate isomerase-2 (5.3.1.9; *Phi-2*) (Williams *et al.*, 1973; Comparini *et al.*, 1975, 1977; Rodino and Comparini, 1978; Comparini and Rodino, 1980; Koehn and Williams, unpublished)			
Phi-2[a]	0.000	0.023	0.000
Phi-2[b]	0.853	0.941	0.993
Phi-2[c]	0.147	0.036	0.007
15. Hemoglobin (*Hb*) (Sick *et al.*, 1962, 1967)			
Hb[a]	0.033	[n]	0.000
Hb[b]	0.967	[n]	1.000

[a]Allele frequencies at each locus are representative of each geographic region and there is complete agreement on these values among the several published sources. Some rare alleles are not shown. Data are for adults (residents) only.

[b]Long Island is given as a representative site.

[c]Sample of 110 individuals taken in 1975.

[d]*Sdh*[a] observed in elvers only.

[e]*Ldh-2* locus. *Ldh-1* is monomorphic such that *rostrata* = *anguilla* ≠ *japonica* (Taniguchi and Morita, 1979).

[f]An allele electrophoretically faster than *Ldh*[a] was observed at frequency 0.009.

[g]See Table V for nomenclature. Only two common alleles are shown here. *Mdh-1* is monomorphic such that *rostrata* = *anguilla* ≠ *japonica* (Taniguchi and Morita, 1979).

[h]See Discussion of Table V.

[i]Frequency of *Mdh*[b] is slightly lower in the Mediterranean Sea.

[j]An allele slower than *Idh*[c] observed at 0.005.

[k]The three alleles correspond to *6-Pgd*[b], *6-Pgd*[d], and *6-Pgd*[f], respectively, of Rodino and Comparini (1978).

[l]*A. japonica* is monomorphic for a slower allele, based on a small sample (Koehn, unpublished).

[m]*Pgm-1* is monomorphic and identical at all sites.

[n]No data available.

Table II
Results of χ^2 Homogeneity Tests of Adults and Elvers among Localities and between Adults and Elvers within Localities in North America at Three Loci[a]

Locus	Interlocality comparison of adults	Interlocality comparison of elvers	Interlocality comparison of adults and elvers
Sdh	N.S.[b]	$P < 0.015$	N.S.
	N.S.	$P < 0.025$	N.S.
Adh	N.S.	N.S.	N.S.
	N.S.	N.S.	N.S.
Phi-2	N.S.	N.S.	$P < 0.05$[c]
	$P < 0.05$	N.S.	$P < 0.02$[c]

[a]Localities: Vero Beach, Florida; Long Island, New York; St. John's, Newfoundland. Upper value is a comparison of zygotic frequencies and lower value is a comparison of allele frequencies. Modified from Williams *et al.* (1973).
[b]N.S., not significant.
[c]Difference observed in St. John's, Newfoundland, sample only.

differentiation was of marginal statistical significance. The test of the dependence of *Adh* allele frequency on latitude was significant ($P < 0.05$), but the contingency χ^2 test of *Adh* variation among North American localities was nonsignificant. Spatial variation at the *Sdh* and *Phi-2* loci proved to be temporally stable (i.e., observations are repeatable from year to year), but significant year-to-year variation occurs in the frequency of *Adh* alleles in elvers arriving in North American localities (Table III).

Although statistically significant variation in gene frequency can thus be detected among North American sites, its magnitude is remarkably small, considering the vast geographic range of the American eel and the diversity of environments sampled. In other coastal marine species, genetic differences of a magnitude comparable to those in *Anguilla* can be observed over a few kilometers (Koehn *et al.*, 1976; Powers and Place, 1978) and sometimes over a few meters (Koehn *et al.*, 1973). We will discuss the implication of these data for the population structure of *Anguilla* more fully in Section 3, but the paucity of geographic variation in *Anguilla* stands in contrast to the abundant genetic differentiation of nearly all other widespread marine animals. It was the comparable dearth of geographic variation in morphology that led Schmidt (1925) to postulate a single spawning population for each of the two recognized species in the North Atlantic.

2.4. Spatial Genetic Variation in European *Anguilla*

No significant geographic variation in genetic composition of the European eel has been observed. Comparini *et al.* (1977) compared allele

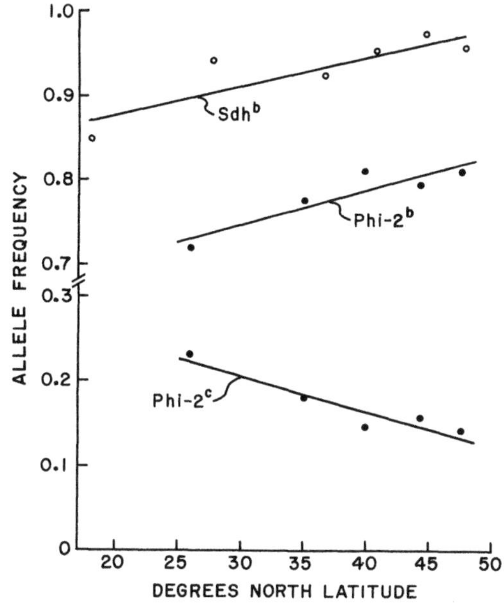

Figure 1. Geographic variation in North America. (○) Regression of Sdh^b variant allele frequency as a function of latitude in combined residents and recruits ($F_{1,4} = 14.89$; $P < 0.025$; $\hat{Y} = 0.822 + 0.00310X$). (●) Regression of $Phi\text{-}2^b$ frequency as a function of latitude in residents only ($F_{1,3} = 16.38$; $P < 0.05$; $\hat{Y} = 0.650 + 0.00466X$) and $Phi\text{-}2^c$ frequency as a function of latitude in residents only ($F_{1,3} = 24.18$; $P < 0.025$; $\hat{Y} = 0.3395 - 0.00428X$).

frequencies of *Pgi-1* and *Pgi-2* between samples from the Atlantic coast of Wales and the Mediterranean (Pisa). Allele frequencies from these two widely separated geographic sites were virtually identical at both *Pgi* loci. The same result was reported by Rodino and Comparini (1978) in a more extensive study of Mediterranean and Atlantic coast samples. In all but one test of homogeneity, involving comparisons at six loci among five Mediterranean and one Atlantic coast locality (Table IV), the frequency of Mdh^b varied only between 0.86 and 0.93 among samples taken from Poland, Holland, Spain, Greece, and the Azores (de Ligny and Pantelouris, 1973). Using serological techniques, no differences could be detected between two samples of *Anguilla* from within the Baltic Sea (Timonina, 1976). Recently published observations on possible differentiation in vertebral counts of northern Europe and unpublished work on associated *Mdh* frequencies are discussed in Section 3.2.

In summary, there is no respectable evidence for geographic variation in allozymes of the European eel, and there is very little differentiation among North American samples of *Anguilla*. Virtually no information is available on genetic differentiation in *Anguilla* from areas outside the North Atlantic, but Taniguchi and Numachi (1978) detected no statistical differences at the *6-Pgdh, Idh,* and *Got-3* loci in samples of *Anguilla japonica* from Kochi Prefecture, Japan, and Fukien, China. The two localities are approximately 2000 km apart.

Table III
χ^2 Comparisons of Among-Year and Among-Locality Variation in Alcohol Dehydrogenase Variant Frequencies[a]

	Florida	New York	Nova Scotia	χ^2 (localities)
Allele *Adh*[a]				
1971	0.0577	0.1215	—	N.S.[b]
1973	0.0340	{0.0671 0.0482	0.0650	N.S.
1974	0.1250	0.1918	0.1583	N.S.
χ^2 (years)	7.95**	20.39***	7.26**	
Allele *Adh*[b]				
1971	0.5096	0.5187	—	N.S.
1973	0.6944	{0.6280 0.6446	0.5950	N.S.
1974	0.5263	0.4658	0.5417	N.S.
χ^2 (years)	10.75***	14.64***	N.S.	
Allele *Adh*[e]				
1971	0.4327	0.3598	—	N.S.
1973	0.2708	{0.3049 0.3072	0.3400	N.S.
1974	0.3487	0.3425	0.3000	N.S.
χ^2 (years)	7.07*	N.S.	N.S.	

[a]From Koehn and Williams (1978). $*P < 0.05$, $**P < 0.025$, $***P < 0.005$.
[b]N.S., not significant.

3. Panmixia with Strong Selection versus Local Populations

Regardless of the systematic status of *Anguilla anguilla* and *A. rostrata,* present information fails to support genetic structuring within either form. The failure of several attempts to find geographic variation in allozyme frequencies in the European eel supports Schmidt's assumption of a single breeding population, to which all specimens belong, no matter how far they may be from each other or how ecologically different their habitats. The slight differentiation that we have found in North America would make it conceivable that marked departures from panmixia characterize the American eel, but the data do not support this view. Rather, the detected differences at the *Sdh, Pgi-2,* and *Adh* loci can only be explained by the action of natural selection operating on the offspring of single panmictic populations, because in each case the action of natural selection is different. For example, identical clinal variation of *Sdh* variant frequencies occur in both recruits and residents (Williams *et al.,* 1973), suggesting that the forces that generate this differentiation must occur

Table IV
The Results of Among-Locality Comparisons by Contingency χ^2 Tests for Six Loci in *Anguilla anguilla*[a]

	A	B	C	D
Mdh-2	6.24	15.58	2.95	5.56
Pgi-1	2.21	6.53	4.42	1.39
Pgi-2	1.12	3.72	0.00	0.61
Sdh	2.62	9.16	0.75	0.05
Ldh-2	2.60	6.20	2.17	4.13*
6-Pgdh	—	—	3.99	—

[a]A, Among Pisa samples; B, among Rome, Sardinia, Marseille, and Goro; C, between A samples pooled and B samples pooled; D, between A–C samples pooled and a sample from Swansea, U.K. Modified from Rodino and Comparini (1978). $^*P < 0.05$.

prior to the time that recruits take up residence in continental localitiies. If a number of gene loci exhibited the same pattern of systematic geographic variation, it would be difficult to reject a hypothesis of subpopulational differentiation in the breeding grounds. However, the genetic composition of recruits arriving throughout North America estimated from the *Phi-2* locus indicates homogeneity of the arriving cohort, with subsequent differentiation occurring among continental sites after the recruits have taken up residences in continental habitats. The temporal stability of this variation (Koehn and Williams, 1973) further suggests that the differences arise anew in each age cohort. It appears that latitudinal differentiation at this locus occurs as a result of differential viability during the growth of elvers to large juveniles.

Recruits to North American localities are homogeneous within a cohort at the *Adh* locus, but there are significant year-to-year differences (Table III). Temporal variation of this type could reflect annual variations in the contributions of various continental sources to a panmictic breeding unit in the tropical Atlantic, if genetic differentiation among the potential source populations could be demonstrated. There is no evidence for such differences. The pattern of variation at the *Adh* and *Pgi-2* loci strongly argues against substructuring of the breeding populations, since each arriving elver cohort is genetically homogeneous. The temporally varying, but continentally stable, allele frequencies of *Adh* must result from the similar action of natural selection upon the breeding population. If substructuring were to exist, subpopulations would be expected to differ from one another in allele frequency and these differences would have to be somehow erased prior to our measure of homogeneity in the offspring of these subpopulations. We know of no mechanism by which this could occur. The evidence is consistent with panmixia within each of the breeding populations from which offspring are dispersed to the North American and European coastal waters.

It is worth emphasizing that the selection we have postulated which would be capable of, for example, changing the frequency of the Adh^2 allele from perhaps 0.10 to something like 0.15 or 0.05 in a single developing cohort, must be intense. Acceptance of our interpretation implies that one-locus selection coefficients in *Anguilla,* and by implication in other prolific organisms, often differ by considerably more than 10%.

3.1. Intercontinental Genetic Differentiation in North Atlantic *Anguilla*

As noted above, it is customary to name eels from the Eastern North Atlantic *Anguilla anguilla* and those from the West *Anguilla rostrata,* and only the number of vertebrae serves to assign nearly all specimens to one or the other form. It is on the basis of this single character that the two forms are recognized as distinct species (Ege, 1939; Tesch, 1977). We believe that this taxonomic distinction, implying intrinsic reproductive isolation, cannot be supported. We have presented evidence for panmixia within each of the American and European forms. In the following paragraphs, we discuss indications of substantial gene flow between the two forms and show that the overall pattern and magnitude of genetic differentiation are comparable to those observed among conspecific populations, not distinct species, in other groups.

The only genetic evidence for two distinct forms of *Anguilla* in the North Atlantic derives from a series of three papers that describe allozyme or muscle protein differences between the two forms. Much of this work has focused on the *Mdh-2* locus (Fig. 2; nomenclatural summary in Table V), for reasons that will become obvious. De Ligny and Pantelouris (1973) were the first to describe the large gene frequency differences that could

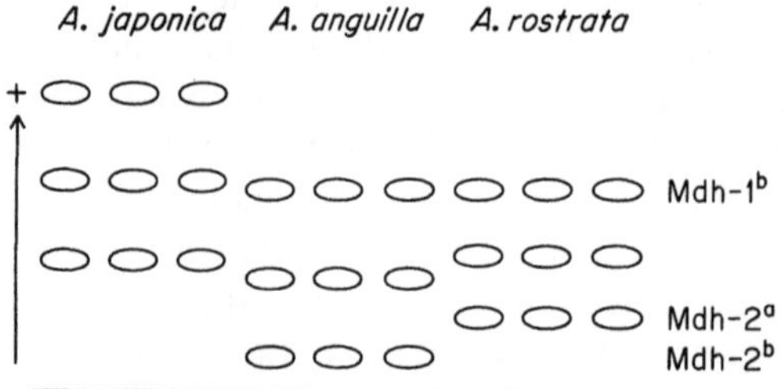

Figure 2. Malate dehydrogenase has been important as a diagnostic system in North Atlantic eels. The schematic representation of the *Mdh* loci illustrates the most common allele only, at each locus, in three studied species (see Tables I and V).

Table V
A Comparison of the Nomenclature for the Malate Dehydrogenase Locus and Its Common Alleles among Individual Studies

		Common alleles in:	
Source	Locus	*A. rostrata*	*A. anguilla*
Williams *et al.* (1973)	*Mdh*	Mdh^a	Mdh^b
Comparini *et al.* (1975), Comparini and Rodino (1980)	*Mdh-2*	Mdh^a	Mdh^c
de Ligny and Pantelouris (1973)	*Mdh*	*H*	*M*
Tanaguchi and Morita (1979)	$Mdh\text{-}A_2$	"Normal"	"Normal"

be detected between American and European forms at the *Mdh-2* locus (Table I). Allele frequencies at this locus seemed to be homogeneous within each of the American and European continental areas. These authors conclude that, "it seems highly unlikely that the difference in *Mdh* frequencies between American and European eels could be due to differential mortality caused by environmental factors." We do not dispute this conclusion, which implies that the two forms do not derive from a single panmictic population. This, however, does not mean that they represent biological species.

This work was significantly extended by Comparini and Rodino (1980), who demonstrated that the frequencies reported by de Ligny and Pantelouris could be observed in samples of *Anguilla* larvae captured near the spawning area in the Sargasso Sea. There was an extraordinarily large deficiency of heterozygotes at the *Mdh-2* locus, suggesting that their samples were composed of a mixture of populations, each characterized by the high frequency of alternative *Mdh-2* alleles (Table I). These data would seem to demonstrate beyond a reasonable doubt that two forms of the genus *Anguilla* can be sampled in the earliest part of the life cycle and that these forms differ in the same characters that distinguish the continental stocks. Again, the existence of two species does not necessarily follow from the existence of two consistently recognizable forms. The most critical information, the degree to which the two forms spawn in sympatry, was not established by Comparini and Rodino (1980), only that leptochephali "were captured in different stations between 22–29° N and 55–70° W." This represents nearly 70,000 km^2, an ample area for the existence of separate spawning areas and the maintenance of separate populations by simple allopatry, rather than any intrinsic isolating mechanisms.

Differences between the North Atlantic forms of *Anguilla* at the *Mdh-2* locus are large *quantitative* differences, but it is *qualitative* genetic differentiation that would constitute the most compelling evidence for genetic isolation. Such a difference was reported for a muscle protein (Jamieson and Turner, 1980) between a sample from Lowestoft, U.K., and one from Halifax, Nova Scotia. Isoelectric focusing patterns of the sarcoplasmic proteins from the two samples differed in the isoelectric points of a single protein constituent. This finding could possibly be interpreted as a qualitative genetic difference, though it is important to note that this is a two-sample study and no attempt has yet been made to assess geographic variation in these phenotypes. Another difficulty with drawing this conclusion is that we do not know of a genetic basis for the phenotypic difference described. One electrophoretic band does not necessarily represent a genetic difference, particularly when we are ignorant of the specific nature of the studied protein. Such a difference could likely be due to environmental or developmental influences. The data of Jamieson and Turner (1980) constitute necessary but not sufficient evidence of genetic isolation between the North Atlantic forms of *Anguilla*.

There is direct evidence that the overall genetic differences between the two forms of *Anguilla* are not as large as we can reasonably expect for two separate species. In a great diversity of organisms, encompassing both invertebrates and vertebrates (including fishes), the average level of genetic similarity among local populations ranges between 0.89 and 1.00 (Avise, 1976), whereas the genetic similarity between species is much lower, mostly in the range of 0.47–0.78. In pairwise comparisons between North America, Iceland, and European samples of *Anguilla,* the range of genetic similarity is 0.896 (Europe versus North America) and 0.983 (Europe versus Iceland) (Table VI and Fig. 3). Hence, the genetic similarity among North Atlantic *Anguilla* is in the usual range of differentiation observed among local populations *in all other studied organisms*. When local populations have been compared, the distribution of genetic similarities among loci is characteristically of an inverse L shape, such as that illustrated from *Drosophila* in Fig. 4. The number of loci studied in *An-*

Table VI
Average Genetic Similarity (above) and Genetic Distance (below) between *Anguilla* Samples from Major Geographic Areas

	North America	Iceland	Europe
North America	—	0.9208	0.8960
Iceland	0.0825	—	0.9832
Europe	0.1100	0.0170	—

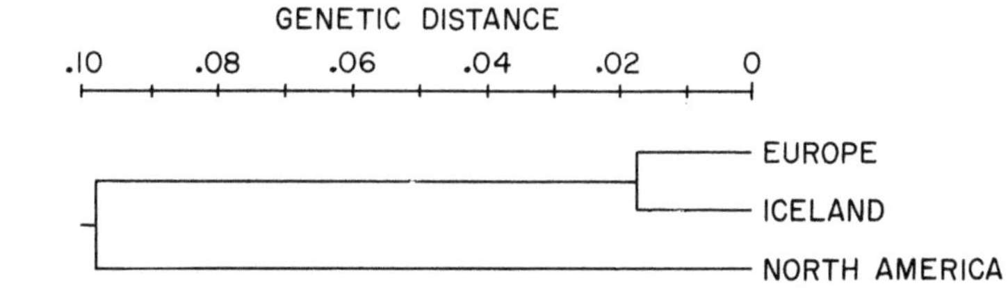

Figure 3. A phenogram of the average genetic distance between North Atlantic *Anguilla* from major geographic regions. The phenogram was constructed by the unweighted pair-group method from data in Table I.

guilla is only about one-half the number studied in *Drosophila,* but the distribution of genetic identities among loci in the two groups is very similar (Fig. 4).

The genetic identities among individual loci in Fig. 4 illustrate another important point: among local populations of *Drosophila,* a very small number of loci are diagnostic for some local populations (i.e., $I = 0.00$). The existence of one or a few loci fixed for different alleles in different populations of *Drosophila* has not led taxonomists to classify these populations as separate species, and the same is true of studies of all other major groups reviewed by Avise (1976). Hence, the existence of a largely diagnostic locus, such as *Mdh-2* in *Anguilla,* does not by itself constitute evidence for the specific status of the two forms. This gene provides evidence of partial, not complete genetic isolation.

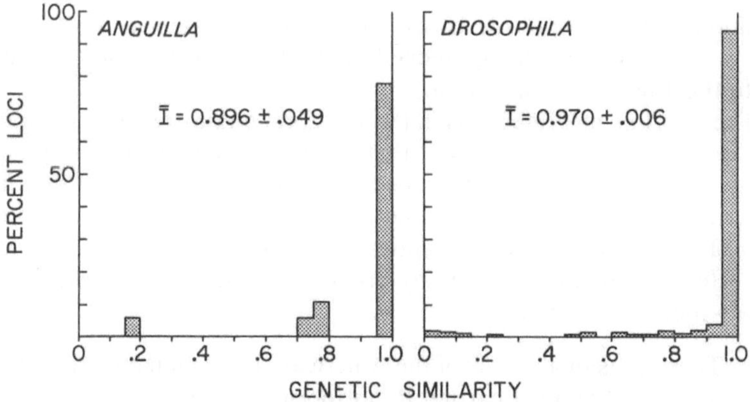

Figure 4. The distribution of genetic identities among loci between local populations of *Drosophila* (Ayala *et al.*, 1974) and *Anguilla* (from Table I). $\bar{I}$ represents the genetic similarity coefficient.

We believe it is not unimportant to note that of the four loci (*Adh, 6-Pgdh, Got-1,* and *Pgi-2;* in addition to *Mdh-2*), at which allele frequencies differ between North America and Europe, three vary significantly within North America as a consequence of the action of natural selection (Williams *et al.,* 1973; Koehn and Williams, 1978). The fourth locus (*6-Pgdh*) has not been studied among sites in North America. This observation would suggest that the differentiation observed between Europe and North America is only an extension of the differentiation that occurs within the North American continent. The situation with the *Phi-2* locus is particularly illustrative. The frequency of *Phi-2*^{b} is homogeneous in elvers over the North American continent at a frequency of approximately 0.77. In more northern continental North American sites, the frequency of this allele increases to approximately 0.85, shortly after North American residence begins (Koehn and Williams, 1978). The frequency of the *Phi-2*^{b} allele is 0.94 in Iceland and 0.99 in Europe. In other words, the forces that increase the frequency of the *Phi-2*^{b} allele in each North American elver cohort, as it takes up residence, could likely produce the even greater increase in the frequency of this allele after more extended dispersal to Iceland and Europe.

3.2. Intercontinental Morphological Differentiation

To recognize a species distinction between two forms as similar morphologically and genetically as the American and European eels could only be justified by convincing evidence that, despite the close similarity, there is intrinsic reproductive isolation between them. In fact there are indications that the opposite is true. Two lines of evidence suggest that there is intergradation and appreciable gene flow at least from the American to the European population.

The first line of evidence is in the vertebral counts recently compiled by J. Boetius (1980). He showed that the mean number of vertebrae is homogeneous over most of the range of the European eel, but that it is lower, i.e., shifted toward the American mean, in Scotish and Scandinavian (Icelandic to Swedish) localities, i.e., those closest to the American coast. He explained this finding by proposing processes consistent with the two-species tradition:

1. That occasional larvae of the American population miss the American coast and get carried to northern Europe
2. That occasional hybridization takes place, but with no suggestion that such events are of any consequence for either parent population

3. That for the northern European stock, "in some years the differentiation of vertebrae number seems to stop a little earlier than usual"

This last effect would presumably depend on environmental fluctuations that affect only specimens destined for northern Europe.

Boetius thus proposed categories of specimens found in northern Europe that are missing or much less frequent elsewhere in Europe and in North Africa. These extra specimens would all have vertebral number reduced for one or another reason, so that they would affect the frequencies of low counts rather than high. The data support, or at least permit, this interpretation (Table VII). The proportions of specimens with counts of 113 or fewer vertebrae are consistently higher for northern Europe. The two distributions are not significantly different over most of the range of counts, from 111 vertebrae up, but they do differ significantly with counts of 110 included ($\chi^2 = 18.7, N = 10, P = 0.044$). The difference becomes highly significant with the inclusion of still lower counts. This analysis shows that some special explanation is needed for the augmented lower tail of the distribution for northern Europe.

We suggest that this explanation ought, in the absence of compelling contrary evidence, to follow conventional procedures with morphologically intermediate specimens from regions geographically intermediate between closely similar organisms. The intermediates should be regarded as intergrades and as evidence that the forms thus connected are members of the same species. There is no evidence against Boetius' three processes occurring in exactly the frequencies required to explain the observations, but their recognition is gratuitous and contrary to normal treatment of geographic variation.

Ideally intermediate specimens (110 and 111 vertebrae) are scarce even where they are most frequent, and the rate of gene exchange between American and European populations is evidently low. This must result, to some extent, from at least partly allopatric spawning, a point established long ago by Schmidt. Boetius' data (Table VII) show that the spawning must be at least partly sympatric, whether his explanation or ours is accepted. We see no reason to doubt Schmidt's assumption that European eels normally spawn where currents will facilitate dispersal in 2 or 3 years to the eastern side of the Atlantic, and American where currents would most likely put larvae close to American coasts within 1 year. There could still be some overlap in spawning ground at least in some years, either as a result of unpredictability of the ideal spawning site or of imperfections in mechanisms of spawning-site selection. A resultant region of sympatry is likely to lie between the main centers of spawning by the two popu-

Table VII

Vertebral Counts from Scottish and Scandinavian Localities Compared with Those from More Southern and Eastern Parts of the Range of the European Eel[a]

	104	105	106	107	108	109	110	111	112	113	114	115	116	117	118	119	120	Σ
Northern Europe																		
N	0	1	3	5	6	9	23	109	431	1349	2423	2750	1668	641	142	18	2	9580
P	0.000	0.000	0.000	0.001	0.001	0.001	0.002	0.001	0.045	0.141	0.253	0.287	0.174	0.067	0.015	0.002	0.000	1.000
Other European sites																		
N	1	0	0	0	1	0	4	64	261	872	1670	1861	1076	382	69	12	1	6274
P	0.000	0.000	0.000	0.000	0.000	0.000	0.001	0.010	0.042	0.139	0.266	0.300	0.172	0.061	0.011	0.002	0.000	1.004

[a]From J. Boetius (1980, pp. 104–110). The American mean is 107.1.

lations. Specimens of mixed parentage from this intermediate region would perhaps have an intermediate sort of larval dispersal that would put them in the geographically intermediate region of northern Europe. As returning adults they might have intermediate migratory mechanisms that would put them back in the intermediate spawning region where they might mate with each other or backcross to one or the other parental form. Our assumption of intergradation between American and European populations requires that individuals of mixed parentage be viable and fertile and that they have some opportunity to mediate gene flow.

Thus in our view the north European spectrum of variation between vertebral counts of 105 and 115 represents a continuum of individuals from those of largely American to largely European ancestry. If so, these same individuals ought to show a parallel spectrum of *Mdh* gene frequencies. Those with low counts ought to show a markedly elevated incidence of the American allele, and those with high counts the normally low European frequency. We were recently able to test this idea with specimens from several localities in Iceland (Williams *et al.*, 1984). The results (Table VIII) thoroughly confirm our expectation and show beyond any reasonable doubt that the morphologically intermediate specimens are also genetically intermediate between the American and European populations. Recognition of the species *Anguilla rostrata* is supported by longstanding tradition, and perhaps desired from some sort of continental pride, but has no biological justification.

Table VIII

The Frequency of *Mdh-2* Alleles in Iceland *Anguilla* with Respect to Numbers of Vertebrae

	N	*Mdh-2*[a]	*Mdh-2*[b]	Number of vertebrae
North America[a]	>1000	0.958	0.042	$\bar{x}$ = 107.1 (103–110)
Iceland	2	0.250	0.750	108
	1	0.500	0.500	109
	5	0.400	0.600	110
	8	0.313	0.687	111
	31	0.226	0.774	112
	28	0.143	0.857	113
	57	0.044	0.956	114
	55	0.055	0.945	115
	41	0.037	0.963	116
	19	0.026	0.974	117
	3	0.000	1.000	117
Europe[a]	>1000	0.100	0.900	$\bar{x}$ = 114.7 (110–119)

[a]From Table I.

4. Sex Determination

There is a large literature on this topic, dating from the 19th century. According to the work of Dolan and Power (1977) and Tesch's (1977) summary, much of it may be based on faulty gonad identification. According to Sola *et al.* (1980) the sex chromosomes have also been misidentified. Only a brief review of some recent work on sex determination seems justified.

As indicated earlier, a number of workers have found that different habitats, or different geographic regions, have either mostly female or mostly male eels. This could mean that there is strict genetic sex determination and that males and females differ in habitat selection. Or it could mean that sex is determined by environmental factors at the elver stage or later.

Clear evidence of environmental influence was found by Parsons *et al.* (1977). Every year from 1965 to 1974 they transplanted elvers from a source on the Irish coast to a lake (Lough Neagh) that had previously contained few eels. They also monitored the sex ratio of maturing seaward migrants through the outlet from this lake. There was a steady shift in the proportion of males from 9% in 1965 to 86% in 1974. This paralleled a steadily increasing density of juvenile eels in the lake. The investigators proposed that nutritional or other effects of crowding and competition induced development into males, which mature at a much smaller size than females.

Other work shows that genetic factors may be of some importance. Passakas and Tesch (1980) examined 61 specimens from the North Sea isle of Helgoland and found every one to be female, but 13 of these had what were interpreted as male karyotypes. It may be that a genetically female eel is more likely than a male to take up residence at Helgoland, and conditions there are such that even the genetically male become phenotypically female. A site on the Elbe River yielded 42 with both identifiable gonads and sex chromosomes, and all of these were male. Yet 16 of these freshwater males had female karyotypes. Perhaps conditions here favor selection by genetically male individuals, and subsequently induce phenotypic maleness in the genetically female.

Despite doubts on the identification or even the existence of sex chromosomes in *Anguilla* (Sola *et al.*, 1980), it seems likely from this work of Passakas and Tesch that sex and karyotype have some relationship. The rearing of elvers from diverse localities in diverse experimental environments to the stage of histological gonad differentiation (30 cm) might be richly informative.

References

Aker, E., and Koops, H., 1973, Untersuchungen über Aalbertände in der Deutschen Bucht, *Arch. Fischereiwiss.* **24:**19–39.

Avise, J. C., 1976, Genetic differentiation during speciation, in: *Molecular Evolution* (F. J. Ayala, ed.), Sinnauer, Sunderland, Massachusetts, pp. 106–122.

Ayala, F. J., Tracey, M. L., Barr, L. G., McDonald, J. F., and Perez-Salas, S., 1974, Genetic variation in natural populations of five *Drosophila* species and the hypothesis of the selective neutrality of protein polymorphisms, *Genetics* **77:**343–384.

Balon, E. K., 1975, The eels of Lake Kariba: Distribution, taxonomic status, age, growth and density, *J. Fish Biol.* **7:**797–815.

Barlow, G. W., 1961, Causes and significance of morphological variation in fishes, *Syst. Zool.* **10:**105–117.

Bertin, L., 1957, *Eels: A Biological Study,* Philosophical Library, New York.

Boetius, I., and Boetius, J., 1980, Experimental maturation of female silver eels, *Anguilla anguilla.* Estimates of fecundity and energy reserves for migration and spawning, *Dana* **1:**1–28.

Boetius, J., 1980, Atlantic *Anguilla.* A presentation of old and new data of total numbers of vertebrae with special reference to the occurrence of *Anguilla rostrata* in Europe, *Dana* **1:**93–112.

Breder, C. M., Jr., 1925, Notes on fishes from three Panama localities: Gatun Spillway, Rio Tapia and Caledonia Bay, *Zoologica* **4**(4)**:**137.

Brunn, A. F., 1963, The breeding of the North Atlantic fresh-water eels, *Adv. Mar. Biol.* **1:**137–169.

Comparini, A., and Rodino, E., 1980, Electrophoretic evidence for two species of *Anguilla* leptocephali in the Sargasso Sea, *Nature* **287:**435–437.

Comparini, A., Rizzotti, M., Nardella, M., and Rodino, E., 1975, Ricerche elettroforetiche sulla variabilita' genetica di *Anguilla anguilla, Boll. Zool.* **42:**283–288.

Comparini, A., Rizzotti, M., and Rodino, E., 1977, Genetic control and variability of phosphoglucose isomerase (PGI) in eels from the Atlantic Ocean and the Mediterranean Sea, *Mar. Biol.* **43:**109–116.

Creutzberg, F., 1959, Discrimination between ebb and flood tide by migrating elvers (*Anguilla vulgaris* Turt.) by means of olfactory perception, *Nature* **184:**1961–1962.

Deelder, C. L., and Tucker, D. W., 1960, The atlantic eel problem, *Nature* **185:**589–592.

De Ligny, W., and Pantelouris, E. M., 1973, Origin of the European eel, *Nature* **246:**518–519.

Dolan, J. A., and Power, G., 1977, Sex ratio of American eels, *Anguilla rostrata,* from the Matamek River system, Quebec, with remarks on problems in sexual identification, *J. Fish Res. Board Can.* **34:**294–299.

Drilhon, A., Fine, J., Uriel, J., and LeBourdelles, F., 1956, Etude electrophoretique des constituants du serum de l'Anguille, *C. R. Acad. Sci.* **243:**1802–1805.

Eddy, S., and Underhill, J. C., 1974, *Northern Fishes,* University of Minnesota Press, Minneapolis.

Edel, R. K., 1975, The induction of maturation of female American eels through hormone injections, *Helgol. Wiss. Meeresunters.* **27:**131–138.

Ege, V., 1939, A revision of the genus *Anguilla* Shaw, *Dana Rep.* **16:**1–255.

Eldred, B., 1968, Larvae and glass eels of the American freshwater *Anguilla rostrata* (LeSueur, 1817), in Florida waters, *Leafl. Ser. Mar. Lab. St. Petersburg, Fla.* **4:**1–4.

Eldred, B., 1971, First records of *Anguilla rostrata* larvae in the Gulf of Mexico and Yucatan Straits, *Leafl. Ser. Mar. Lab. St. Petersburg, Fla.* **4:**1–3.

Ezzat, A., and El-Serafy, S., 1977, The migration of elvers of *Anguilla anguilla* L. in the Mex canal, Alexandria, Egypt, *J. Fish Biol.* **11:**249–256.

Gray, R. W., and Andrews, C. W., 1970, Sex ratio of the American eel (*Anguilla rostrata* LeSueur) in Newfoundland waters, *Can. J. Zool.* **48:**483–487.

Gray, R. W., and Andrews, C. W., 1971, Age and growth of the American eel (*Anguilla rostrata* LeSueur) in Newfoundland waters, *Can. J. Zool.* **49:**121–128.

Gunning, G. E., and Shoop, C. R., 1962, Restricted movements of the American eel, *Anguilla rostrata* (LeSueur), in freshwater streams, with comments on growth rate, *Tulane Stud. Zool.* **9:**265–272.

Hachey, H. B., 1961, Oceanography and Canadian Atlantic waters, *Bull. Fish Res. Board Can.* **134:**1–120.

Harden-Jones, F. R., 1968, *Fish Migration,* Edward Arnold, London.

Hurley, D. A., 1972, The American eel (*Anguilla rostrata*) in Eastern Lake Ontario, *J. Fish Res. Board Can.* **29:**539–543.

Jamieson, A., and Turner, R. J., 1980, Muscle protein differences in two eels *Anguilla anguilla* (Linnaeus) and *Anguilla rostrata* (LeSueur), *Biol. J. Linn. Soc.* **13:**41–45.

Jensen, A. S., 1937, Remarks on the Greenland eel, its occurrence and reference to *Anguilla rostrata, Medd. Gronl.* **118:**1–8.

Jones, J. N., and Tucker, D. W., 1959, Eel migration, *Nature* **184:**1281–1283.

Jordan, D. S., and Barton, W. E., 1902, *American Food and Game Fish,* William Briggs, Toronto.

Kerr, R. A., 1977, Oceanography: A closer look at Gulf Stream rings, *Science* **198:**387–389.

Kleckner, R. C., 1980, Swim bladder volume maintenance related to initial oceanic migratory depth in sliver-phase *Anguilla rostrata, Science* **208:**1481–1482.

Koehn, R. K., 1972, Genetic variation in the eel: A critique, *Mar. Biol.* **14:**179–181.

Koehn, R. K., and Williams, G. C., 1978, Genetic differentiation without isolation in the American eel, *Anguilla rostrata.* II. Temporal stability of geographic patterns, *Evolution* **32:**624–637.

Koehn, R. K., Turano, F. J., and Mitton, J. B., 1973, Population genetics of marine pelecypods. II. Genetic difference in microhabitates of *Modiolus demissus, Evolution* **27:**100–105.

Koehn, R. K., Milkman, R. M., and Mitton, J. B., 1976, Population genetics of marine pelecypods. IV. Selection, migration and genetic differentiation in the blue mussel, *Mytilus edulis, Evolution* **30:**2–32.

Kokhnenko, S. V., Bezdenezhnykh, V. A., and Gorovaya, S. L., 1978, Maturation of the European eel (*Anguilla anguilla*) when artificially reared, *J. Ichthyol.* **17:**878–883.

Kuhlmann, H., 1975, Der Einfluss von temperatur, futter, Gröss und herkunft auf die sexuelle differenzierunz vox glasaalen (*Anguilla anguilla*), *Helgol. Wiss. Meeresunters.* **27:**139–155.

Lee, A. J., 1974, Oceanic circulation of the North Atlantic region, in: *Sea Fisheries Research* (F. R. Harden-Jones, ed.), Wiley, New York, pp. 1–30.

McCleave, J. D., 1980, Swimming performance of European eel (*Anguilla anguilla* (L.)) elvers, *J. Fish Biol.* **16:**445–452.

Medcof, J. C., 1969, Fishermen's reports of freshwater and saltwater migrations of Nova Scotia eels (*Anguilla rostrata*), *Can. Field Nat.* **83:**132–138.

Moriarti, C., 1972, Studies of the eel *Anguilla anguilla* in Ireland I. In the Lakes of the Corrib system, *Ir. Fish. Invest. Ser. A* (*Freshwater*) **10:**3–39.

Neumann, G., 1968, *Ocean Currents,* Elsevier.

Ogden, J. C., 1970, Relative abundance, food habits, and age of the American eel, *Anguilla rostrata* (LeSueur), in certain New Jersey streams, *Trans. Am. Fish. Soc.* **99:**54–59.
Pantelouris, E. M., 1976, Aspartate aminotransferase variation in the Atlantic eel, *J. Exp. Mar. Biol. Ecol.* **22:**123–130.
Pantelouris, E. M., and Payne, R. H., 1968, Genetic variation in the eel. I. The detection of haemoglobin and esterase polymorphisms, *Genet. Res.* **11:**319–325.
Pantelouris, E. M., Arnason, A., and Tesch, F.-W., 1970, Genetic variation in the eel. II. Transferrins, haemoglobins and esterases in the eastern North Atlantic. Possible interpretations of phenotypic frequency differences, *Genet. Res.* **16:**277–284.
Pantelouris, E. M., Arnason, A., and Tesch, F.-W., 1971, Genetic variation in the eel. III. Comparisons of Rhode Island and Icelandic populations. Implications for the Atlantic eel problem, *Mar. Biol.* **9:**242–249.
Parsons, J., Vickers, K. U., Warden, Y., 1977, Relationship between elver recruitment and changes in the sex ratio of silver eels *Anguilla anguilla* L. migrating from Lough Neagh, Northern Ireland, *J. Fish Biol.* **10:**211–229.
Passakas, T., and Tesch, F.-W., 1980, Karylological and gonadal sex of eels (*Anguilla anguilla*) from the German Bight and the lower River Elbe, *Helgol. Wiss. Meeresunters.* **34:**159–164.
Powers, D. A., and Place, A. R., 1978, Temporal and spatial variation in gene frequencies of *Ldh-B, Mdh-A, Pgi-B,* and *Pgm-A, Biochem. Genet.* **16:**595–608.
Rodino, E., and Comparini, A., 1978, Genetic variability in the European eel, *Anguilla anguilla* L., in: *Marine Organisms: Genetics, Ecology and Evolution* (B. Battaglia and J. A. Beardmore, eds.), Plenum Press, New York, pp. 389–424.
Schmidt, J., 1914, On the classification of the fresh-water eels (*Anguilla*), *Medd. Havunders. Fisk.* **7:**1–19.
Schmidt, J., 1915, Second report on eel investigations, in: *Copenhague, Conseil internat. pour l'exploration de la mer, Rapp. et proces-verben des reunions,* Vol. 23, pp. 1–24.
Schmidt, J., 1916, On the early larval stages of the freshwater eels (*Anguilla*) and some other North Atlantic muraenoids, *Medd. Havunders. Fisk.* **5:**1–20.
Schmidt, J., 1925, The breeding places of the eel, *Smithson. Rep.* **1924:**279–316.
Sick, K., Westergarrd, M., and Frydenberg, O., 1962, Haemoglobin pattern and chromosome number of American, European and Japanese eels (*Anguilla*), *Nature* **193:**1001–1002.
Sick, K., Bahn, E., Frydenberg, O., Nielsin, J. T., and von Wettstein, D., 1967, Haemoglobin polymorphism of the American freshwater eel *Anguilla, Nature* **214:**1141–1142.
Sinha, V. R. P., and Jones, J. W., 1966, On the sex and distribution of the freshwater eel (*Anguilla anguilla*), *J. Zool. Lond.* **150:**371–385.
Sinha, V. R. P., and Jones, J. W., 1967, On the age and growth of the freshwater eel (*Anguilla anguilla*), *J. Zool. Lond.* **153:**99–117.
Slastenenko, I. P., 1936, Revue de la faune ichthyologique de la Mer Noire, *An. Stiint. Univ. Jassy* **22:**280–296.
Smith, D. G., 1968, The occurrence of larvae of the American eel, *Anguilla rostrata,* in the straits of Florida and nearby areas, *Bull. Mar. Sci.* **18:**280–293.
Smith, M. W., and Saunders, J. W., 1955, The American eel in certain freshwaters of the maritime provinces of Canada, *J. Fish. Res. Board Can.* **12:**238–261.
Sola, L., Gentili, G., and Catandella, S., 1980, Eel chromosomes: cytotaxonomical interrelationships and sex chromosomes, *Copeia* **4:**911–913.
Stommel, H., 1965, *The Gulf Stream,* University of California Press, Berkeley.
Taniguchi, N., and Numachi, K., 1978, Genetic variation of 6-phosphogluconate dehydrogenase, isocitrate dehydrogenase, and glutamic-oxaloacetic transaminase in the liver of Japanese eel, *Bull. Jpn. Soc. Sci. Fish.* **44:**1351–1355.

Tesch, F. W., 1973, *Der Aal,* Paul Parey, Hamburg.

Tesch, F. W., 1977, *The Eel. Biology and Management of Anguillid Eels,* Chapman and Hall, London, and Wiley, New York.

Tesch, F. W., 1978, Telemetric observations on the spawning migration of the eel (*Anguilla anguilla*) west of the European continental shelf, *Environ. Biol. Fish.* **3**:203–209.

Timonina, L. G., 1976, A study of blood groups in the eel, *Anguilla anguilla,* from the Courland Lagoon and Vistual Bay, *J. Ichthyol. (USSR)* **16**:348–351.

Tucker, D. W., 1959, A new solution to the Atlantic eel problem, *Nature* **183**:495–501.

Vladykov, V. D., 1964, Quest for the true breeding area of the American eel (*Anguilla rostrata* LeSueur), *J. Fish. Res. Board Can.* **21**:1523–1530.

Vladykov, V. D., 1966, Remarks on the American eel (*Anguilla rostrata* LeSueur). Sizes of elvers entering streams; the relative abundance of adult males and females, and present economic importance of eels in North America, *Verh. Int. Ver. Limnol.* **16**:1007–1017.

Watts, R. D., and Olson, D. B., 1978, Gulf stream ring coalescence with the Gulf Stream off Cape Hatteras, *Science* **202**:971–972.

Wenner, C. A., 1972, Aspects of the biology and systematics of the American eel, *Anguilla rostrata* (LeSueur), M. A. Thesis, College of William and Mary, Virginia.

Williams, G. C., Koehn, R. K., and Mitton, J. B., 1973, Genetic differentiation without isolation in the American eel, *Anguilla rostrata, Evolution* **27**:192–204.

Williams, G. C., Thorsteinsson, V., and Koehn, R. K., 1984, Icelandic eels: Evidence for a single species of *Anguilla* in the North Atlantic, *Copeia* (in press).

Wynne-Edwards, V. C., 1962, *Animal Dispersion in Relation to Social Behaviour,* Oliver and Boyd, Edinburgh.

Yamamoto, K., and Yamauchi, K., 1974, Sexual maturation of Japanese eel and production of eel larvae in the aquarium, *Nature* **251**:220–222.

Yevseyenko, S. A., 1974, New data concerning eel eggs on Georges Bank, *J. Ichthyol.* **14**:294–298.

Zineviei, V., 1967, Considerations on the distribution of the eel (*Anguilla anguilla* L.) in the Romanian sector of the Danube and of its mouths, *Hidrobiologia (Buchar.)* **8**:143–149.

CHAPTER 11

Allozymes of the Cyprinid Fishes

Variation and Application

DONALD G. BUTH

1. Introduction

The cyprinid fishes comprise a major element of the ichthyofauna of Africa, Asia, Europe, and North America. More than 1600 species in over 275 genera make the Cyprinidae the most speciose of fish families (Nelson, 1976). Cyprinid fishes have been the subject of several allozyme studies. However, given the number of species in the family, cyprinids have been underrepresented in such studies to date. Nevertheless, cyprinids have contributed to our understanding of genetic variation in natural populations and a variety of evolutionary processes. Studies of heterozygosity, population differentiation, hybridization including introgression, and rates of evolution have been addressed using cyprinids and are discussed in this chapter.

Electrophoretic data, in allozyme and isozyme form, can be applied in systematic and taxonomic investigations (Avise, 1974; Buth, 1981). As might be expected in any speciose group of organisms, systematic and taxonomic problems are quite numerous among the Cyprinidae. Phylogenetic treatment of allozyme data, especially in congruence studies involving morphological data for hypothesis testing (Mickevich and Johnson, 1976), has yet to see widespread application to these cyprinid problems. Progress has been limited on several levels. A number of basic procedures uniquely problematic to allozyme studies, e.g., documentation and nomenclature, have yet to be standardized in fishes, let alone cyprinids.

DONALD G. BUTH • Department of Biology, University of California at Los Angeles, Los Angeles, California 90024.

Lack of such standardization has limited the communication of research findings, frustrated the comparisons of published investigations, and prevented adequate reproduction of studies. The field of phylogenetic systematics has developed so rapidly in recent years, especially in regard to phylogenetic treatment of allozyme data (Farris, 1981; Mickevich and Mitter, 1981, 1983; Swofford, 1981), that virtually all previous applications to cyprinid systematics should now be reevaluated. Evaluation of these applications and recommendations for standard methods and future research are also presented in this chapter.

2. Methods

2.1. Collection and Documentation

The specimens for morphological studies of cyprinid fishes have, historically, come from two sources: (1) extant museum collections subject to cataloging documentation, and/or (2) new collections of preserved specimens now available to the investigator that are deposited in a museum collection as part of, or immediately subsequent to, the morphological study. The importance of such documentation has long been appreciated by morphological systematists. The need for documentation surpasses the obvious needs of type specimens or series for new taxa; these specimens are indeed a vital part of each study. Their existence allows subsequent investigators to verify all aspects of the original study, i.e., the counts and measurements can be repeated using the original specimens and the identifications can always be confirmed. In a speciose group, such as the Cyprinidae, in which considerable systematic and taxonomic work has yet to be done, documentation via deposition of voucher specimens is of utmost importance.

Studies of allozymes of fishes, especially those with systematic and/or taxonomic components, have routinely failed to provide specimen documentation comparable to that expected in morphological studies. Reasons for these omissions are at least threefold:

1. Fresh, frozen specimens are required for allozyme studies; thus, previously collected museum specimens are of little value to the allozyme investigator (although the collection data associated with the preserved specimens can be invaluable to the recollection efforts!).
2. Most specimens used in allozyme studies are "thoroughly dissected," yielding severely mutilated specimens that are of minimal documentation value and are usually discarded.

3. Many investigators in this field enter via genetics rather than systematics and lack the training in the necessity of specimen documentation associated with the latter discipline.

Allozyme studies are in special need of documentation, as the data, as genotypic arrays or allele frequency distributions, are associated with, at best, named geographic sites of sampling ("populations" of a taxon) rather than with a specific deposition of specimens. Replication of the study utilizing original specimens is not feasible if only for tissue limitations and the temporal limitations of enzyme activity in frozen storage. Without voucher specimens, the identification of taxa is limited to the allozyme investigator alone. Incongruences with subsequent allozyme studies cannot be resolved if these differences are due to misidentifications of specimens that are no longer extant.

Recommendations to alleviate the voucher specimen problem have followed three paths. First, the construction of frozen depositories for dissected specimens of allozyme studies could serve both the needs of specimen reidentification and "character" (= allozyme) reexamination of the original specimens. This approach has several limitations, not the least of which is the expense of essentially duplicating the facilities and curatorial requirements of existing museum collections. Frozen collections are more space-limited and are more vulnerable to destruction (e.g., via power and equipment failures even with appropriate backup equipment). The deposition of tissue samples or extracts without the original specimen is space-efficient but does nothing to address the problem of specimen reidentification. Analyses of allozymes "use up" a portion of a specimen and have obvious repetition limits, limits that do not apply to counts and measures of morphological features. The "shelf-life" of most allozymes has yet to be adequately investigated. While forms of many enzymes may remain active for a number of years under frozen tissue storage conditions, the temporal limits of these characters, based on enzymatic activity, are certainly far less than those of characters of morphological features of preserved specimens. I encourage the implementation of frozen collections, but we should all be aware of the limitations of this approach; it is not a cure-all.

An alternative, which has been practiced by karyologists for years, involves the "careful" dissection of a specimen, followed by its fixation, preservation, and deposition in a museum collection. While this approach may have been suitable for karyological studies in which the dissection damage to specimens was minimal and the number of specimens was small, it is not, in general, suitable for most electrophoretic studies. Dissection damage to specimens is often substantial, involving the removal of eye, brain, liver, kidney, skeletal muscle, heart, etc. This problem is enhanced as more investigators include tissue specificity as a criterion for homology in multilocus systems. As many studies now include geographic sampling (Section 5.1) or enhanced sample sizes, the number of specimens utilized in an allozyme study can be quite substantial. Museum curators are reluctant to fill their shelves with preserved tissue/skeletal remnants of such studies, which would in part fulfill the voucher need but provides little material for any subsequent study.

A third alternative, which I highly recommend, involves additional effort at the planning and specimen acquisition stages of a study. When a sample is to be obtained from a locality, sufficient collection effort is made to obtain a large enough sample to be immediately divided to two subsamples: one sample is fixed on the spot, to be preserved and deposited in a museum collection at a later date, and the other sample is frozen for the allozyme study. The museum subsample should be deposited immediately so species identification can be verified and museum collection numbers will be available when the allozyme study is published. Locality and museum collection numbers should appear in the materials and methods sections of all allozyme papers (e.g., Buth and Mayden, 1981). The lack of 1:1 correspondence between the allozyme-scored sample and the museum sample is balanced by the greater morphological information content of the nondissected museum specimens given that both samples are from the same collection. Errors in identification by the collector are not eliminated by this approach, but may be minimized if the museum sample is as large as or larger than the allozyme sample and is identified by another party. This approach has the limitations of specimen size and species rarity. Large specimens, e.g., those of most shark species, will cause voucher problems whether morphological or allozyme studies are undertaken. Rare species may not be taken in numbers to afford division into substantial subsamples, but even one nondissected voucher specimen is better than none in this case. If the species is so rare that only a single specimen is obtained, its rarity should justify the deposition of the dissected carcass in a museum collection.

As the number of cyprinid allozyme studies grows, so does the need to implement standards for voucher specimens. The responsibility for enforcement of voucher standards lies at least at the level of society journals, e.g., *Malacologia* has established such a requirement for published studies of mollusc biology.

2.2. Enzyme and Locus Nomenclature

A stable and informative system of enzyme and locus nomenclature is essential for the accurate communication of results of allozyme studies.

The causes and consequences of the current lack of such a system in fishes are discussed by Buth (1983). An improved system of locus nomenclature emphasizing information content is advocated by Buth (1983) and builds on the formal, although brief, system of nomenclature proposed by Allendorf and Utter (1978). This improved system includes designations to distinguish; (1) abbreviations for enzymes versus enzymatic loci versus alleles at a particular locus, (2) the mitochondrial–supernatant (cytosolic) relationship of loci in certain multilocus systems, (3) the relationship of loci arising from ancient duplications, and (4) the relationship of loci arising from polyploidization within lineages. Specific recommendations for these and other components of locus and allelic nomenclature are to be presented elsewhere.

Enzyme nomenclature is an issue that needs no additional refinement. All enzyme names should follow the current recommendations of the International Union of Biochemistry (1979). The responsibility for enforcement of this standard should, as should be the case with voucher specimens, lie at least at the level of society journals. Unfortunately, few biological journals have taken the lead in regard to this problem.

Problems dealing with enzyme and locus nomenclature have been minimal among cyprinid studies to date. Such multilocus studies have been relatively few in number, usually utilize the same (few) loci, and come from groups of investigators that enjoy some degree of communication, thus minimizing the development of nomenclatural differences. Recommended names of enzymes commonly used in cyprinid allozyme studies, including locus designations, are listed in Table I.

2.3. Tissue and Buffer Optima

Aquadro and Avise (1982) reemphasized the need for an adequate screening of potential buffers to optimize the chances of resolving allozyme variation. Cyprinid buffer problems may be minimal, as many enzyme systems are known to yield good to excellent resolution using a wide variety of buffers. Further, optimal buffer conditions are often applicable across cyprinid genera, indeed throughout the Cypriniformes, a condition that is problematic among reptilian studies (R. W. Murphy, personal communication).

Table I summarizes recommended tissue sources of enzymes and optimal buffers for cyprinid studies. These recommendations are based on a synthesis of optimal conditions reported for *Campostoma* (Buth and Burr, 1978), *Cyprinus* (Ferris and Whitt, 1977*a*), *Gila* (Crabtree and Buth, 1981), *Hesperoleucus* and *Lavinia* (Avise *et al.*, 1975), *Nocomis* (Ferguson *et al.*, 1981), *Notropis* (Rainboth and Whitt, 1974; Buth, 1979*a*), *Phoxinus*

Table I
Optimal Tissues and Electrophoretic Buffers for Studies of Cyprinid Allozymes

Enzyme	Enzyme commission number	Locus	Tissue source	Electrophoretic buffer(s)[a]
Acid phosphatase	3.1.3.2	*Acp-A*	Brain, liver	A,B,C
Aconitate hydratase (mitochondrial)	4.2.1.3	*M-Acon-A*	Muscle	D
Aconitate hydratase (supernatant)	4.2.1.3	*S-Acon-A*	Liver	D
Adenosine deaminase	3.5.4.4	*Ada-A*	Muscle	E
Adenylate kinase	2.7.4.3	*Ak-A*	Muscle	C
Alcohol dehydrogenase	1.1.1.1	*Adh-A*	Liver	A,F,G,H
Aminopeptidase	3.4.11.1	*Ap-A*	Muscle	C
Aspartate aminotransferase (mitochondrial)	2.6.1.1	*M-Aat-A*	Liver, muscle	C,E,I
Aspartate aminotransferase (supernatant)	2.6.1.1	*S-Aat-A*	Heart, liver, muscle	C,E,I
Creatine kinase	2.7.3.2	*Ck-A*	Muscle	C,G
Creatine kinase	2.7.3.2	*Ck-B*	Brain	G
Fructose-bisphosphate aldolase	4.1.2.13	*Ald-C*	Brain	B,C
Fumarate hydratase	4.2.1.2	*Fum-A*	Brain, muscle	B,E
Glucosephosphate isomerase	5.3.1.9	*Gpi-A*	Brain, liver, muscle	G,H,I
Glucosephosphate isomerase	5.3.1.9	*Gpi-B*	Heart, muscle	G,H,I
Glyceraldehyde-phosphate dehydrogenase	1.2.1.9	*Gapdh-A*	Muscle	C
Glycerol-3-phosphate dehydrogenase	1.1.1.8	*G-3-pdh-A*	Liver, muscle	C,D,H

L-Iditol dehydrogenase	1.1.1.14	*Iddh-A*	Liver	F,J
Isocitrate dehydrogenase (mitochondrial)	1.1.1.42	*M-Icdh-A*	Heart, liver, muscle	A,E,H,K
Isocitrate dehydrogenase (supernatant)	1.1.1.42	*S-Icdh-A*	Liver	A,E
Lactate dehydrogenase	1.1.1.27	*Ldh-A*	Muscle	B,C,D,G,H,I
Lactate dehydrogenase	1.1.1.27	*Ldh-B*	Brain, heart, muscle	B,C,D,G,H,I
Lactate dehydrogenase	1.1.1.27	*Ldh-C*	Liver	B,D,G
Malate dehydrogenase (NAD; mitochondrial)	1.1.1.37	*M-Mdh-A*	Brain, liver, muscle	B,C,D,G,H
Malate dehydrogenase (NAD; supernatant)	1.1.1.37	*S-Mdh-A*	Brain, liver, muscle	B,C,D,G,H
Malate dehydrogenase (NAD; supernatant)	1.1.1.37	*S-Mdh-B*	Muscle	B,C,D,G,H
Malate dehydrogenase (NADP; mitochondrial)	1.1.1.40	*M-Me-A*	Muscle	B,D
Malate dehydrogenase (NADP; supernatant)	1.1.1.40	*S-Me-A*	Liver	A
Phosphoglucomutase	2.7.5.1	*Pgm-A*	Liver, muscle	C,D,H,I
Phosphoglucomutase	2.7.5.1	*Pgm-B*	Brain	B,C
Phosphogluconate dehydrogenase	1.1.1.44	*Pgdh-A*	Brain, liver	B,D,H
Purine nucleoside phosphorylase	2.4.2.1	*Pnp-A*	Brain	B
Pyruvate kinase	2.7.1.40	*Pk-A*	Muscle	C
Superoxide dismutase (supernatant)	1.15.1.1	*S-Sod-A*	Liver	B,G,I
Triosephosphate isomerase	5.3.1.1	*Tpi-B*	Muscle	H
Xanthine dehydrogenase	1.2.1.37	*Xdh-A*	Liver	F

[a]A, Amine–citrate (Clayton and Tretiak, 1972); B, Tris–citrate (Whitt, 1970); C, histidine–citrate (Brewer, 1970); D, Tris–citrate II (Selander *et al.*, 1971); E, phosphate–citrate (Selander *et al.*, 1971); F, borate (Sackler, 1966); G, EDTA–borate–tris (Wilson *et al.*, 1973); H, Tris–citrate–EDTA (Avise *et al.*, 1975); I, ''Poulik'' (Selander *et al.*, 1971); J, Tris–phosphate (Op't Hof *et al.*, 1969); K, phosphate (Wolf *et al.*, 1970).

(Joswiak, 1980; Joswiak *et al.*, 1982) and *Semotilus* (Joswiak, 1980). These studies should be consulted for details regarding *p*H and voltage conditions.

3. Genetic Variation

3.1. Heterozygosity

Initial comparisons of heterozygosity levels among the vertebrates were often based on extremely few species, e.g., 14 species (Selander, 1976) out of 18,000 teleosts (Nelson, 1976) contributed to the initial heterozygosity estimate of 0.078 ± 0.012 (mean proportion of loci heterozygous per individual) for "fish." These few species contributed values that were "highly heterogeneous" relative to other vertebrate groups; however, Selander (1976) suggested that this variability might have been due to the fact that these estimates came from several sources. Cyprinid heterozygosity was later examined in much greater detail by Avise (1977*a*). He examined the gene products of 14–17 loci in eight genomes each of 60 species of eastern North American cyprinids; most were based on single geographic samples. The gene products of 24 loci from considerably larger sample sizes of nine species of western North American cyprinids were also examined. These data reveal the substantial heterogeneity in levels of genic variability within this fish family; indeed within a single speciose cyprinid genus (*Notropis*), heterozygosity estimates ranged from 0.000 (in *N. coccogenis, N. dorsalis, N. spilopterus, Hybopsis lineapunctata,* and *Semotilus atromaculatus*) to 0.154 ± 0.052 in *Notropis texanus*. Avise estimated the mean heterozygosity per cyprinid species at 0.052 ± 0.004. The significance of cyprinid heterozygosity heterogeneity is still in question. Avise (1977*a*) cautioned, "since heterozygosity levels can be influenced by a variety of evolutionary forces, both stochastic and deterministic, the apparent differences among particular species of minnows . . . must be interpreted with extreme caution." A host of additional factors must be considered as variables in the evaluation of cyprinid heterozygosity, including the set of loci studied, the electrophoretic conditions of each study, and the geographic source of the sample. Apparently, sample size is not a critical factor for heterozygosity estimates if the number of loci is sufficiently large (Nei, 1978). Several of these variables, however, may have contributed to the different heterozygosity estimates for *Campostoma anomalum* (Table II).

Despite the questions regarding cyprinid heterozygosity, a number of interesting observations have been made and should be considered. Avise (1977*a*) noticed a taxonomic pattern of distribution of heterozy-

Table II
Heterozygosity Estimates for *Campostoma anomalum*

Number of populations studied	Number of specimens examined: mean (range)	Number of loci scored	Mean heterozygosity per locus: mean (range)	Reference
1	8	16	0.014	Avise (1977*a*)
8	14.7 (4–26)	19	0.063 (0.043–0.111)	Buth and Burr (1978)
12	12.2 (6–19)	17	0.064 (0.020–0.111)	Zimmerman *et al.* (1980)

gosity levels (Table III). Members of the speciose genus *Notropis* appear significantly more heterozygous ($P < 0.01$) than do the species of the relatively depauperate North American genera. Thus, there may be a relationship between the rate of speciation and heterozygosity. On the environmental side, Lin *et al.* (1969) reported increased heterozygosity at the L-iditol (sorbitol) dehydrogenase locus in wild populations of *Carassius auratus* in North America and suggested a condition of heterozygote superiority existing in the wild, but not in the domesticated, state. Zimmerman and Richmond (1981) reported increased heterozygosity at the *S-Mdh-B* locus in *Notropis lutrensis* inhabiting a rapidly fluctuating thermal environment. The unique *S-Mdh-B* allele contributing to higher heterozygosity in these environments may have entered the *N. lutrensis* genome via introgressive hybridization with *N. venustus*.

Table III
Summary of Levels of Heterozygosity per Species in North American Cyprinid Fishes[a]

Group	Number of species assayed	Mean ± SE heterozygosity per species
Notropis	47	0.059 ± 0.006
Other eastern U. S. genera[b]	13	0.036 ± 0.008
Other western U. S. genera[c]	9	0.038 ± 0.008

[a]Data from Avise (1977*a*).
[b]*Campostoma, Dionda, Ericymba, Hybopsis* (three), *Nocomis* (two), *Notomigonus, Phoxinus, Pimephales* (two), and *Semotilus.*
[c]*Gila, Hesperoleucus, Lavinia, Mylopharodon, Orthodon, Pogonichthys, Ptychocheilus, Rhinichthys,* and *Richardsonius.*

3.2. Allozyme Differentiation among Taxa

Genetic differentiation, from the level of initial population differences through the results of the speciation process, has been of great interest to evolutionary biologists and systematists. A number of coefficients of genetic similarity (or, alternatively, genetic difference or distance) have been developed to quantify allozyme differentiation between forms. Two of these genetic similarity coefficients have figured prominently in cyprinid allozyme studies; some, e.g., Menzel (1976), have used Rogers' (1972) coefficient of genetic similarity S, while a large number of investigators have used Nei's (1972) genetic similarity coefficient I to compare populations of taxa. While the taxonomic levels of comparisons vary in this latter set of studies (Table IV), these findings can be summarized to determine the range of genetic similarity values expressed within a given taxonomic category. The genetic similarity coefficients generated from all pairwise comparisons in the studies listed in Table IV are plotted in Fig. 1. As has been shown in comparisons in other groups of organisms, Fig. 1 illustrates the general trend of greater differentiation with higher taxonomic rank. This generalization, combined with mutual exclusion of ranges of genetic similarity coefficients in given taxonomic categories as shown in centrarchids by Avise (1974), would suggest a tremendous taxonomic potential for this divergence measure. However, genetic similarity coefficients for cyprinid populations and taxa show considerable variation. Intrageneric ("species") and intergeneric ("genera") coefficients exhibit a wide range of overlap. At one end of the range, *Hesperleucus symmetricus* and *Lavinia exilicauda* (I = 0.948) have been considered to be congeneric on the basis of their genetic similarity (Avise *et al.*, 1975; Avise and Ayala, 1976). See Section 5.2 for a discussion of the validity of this application.

3.3. Comparisons with Karyotypic Differentiation

Chromosome formulas are known for a relatively large number of both New World and Old World cyprinid fishes. While allozyme differentiation among cyprinids can be quite substantial, karyotypic differentiation has been minimal, with $2n$ values ranging from 44 to 52 in diploid forms (J. R. Gold, personal communication). The vast majority of native North American cyprinids are $2n = 50$ (Gold *et al.*, 1979*a*,*b*, 1981; Joswiak *et al.*, 1980). Old World cyprinids exhibit a wider range of $2n$ values (Nayyar, 1964; Ojima *et al.*, 1972; Chiarelli and Capanna, 1973; Nygren *et al.*, 1975; Cataudella *et al.*, 1977; Taki and Suzuki, 1977; Taki *et al.*, 1977), and it is only among the native Old World cyprinids that tetraploid forms are known (Table V).

Table IV
Studies of Allozyme Differentiation in Cyprinid Fishes[a]

Group	Number of comparisons made				Reference
	Populations	Subspecies	Species	Genera	
Campostoma	14	21[b]	56[b]	—	Buth and Burr (1978)
Campostoma	81	—	72	—	Zimmerman *et al.* (1980)
Nocomis	—	—	1	—	Ferguson *et al.* (1981)
Notropis (Luxilus)	5	4	82	—	Buth (1979*a*)
Phoxinus and *Semotilus*	—	—	21	24	Joswiak (1980)
Nine western U. S. genera[c]	—	—	—	36[d]	Avise and Ayala (1976)

[a]Pairwise comparisons were made between populations, subspecies, species, and genera using Nei's (1972) coefficient of genetic similarity *I*.
[b]Some taxonomic changes may be in order (B. M. Burr, personal communication).
[c]One species each of *Gila, Hesperoleucus, Lavinia, Mylopharodon, Orthodon, Pogonichthys, Ptychocheilus. Rhinichthys,* and *Richardsonius.*
[d]The authors concluded that *Lavinia* and *Hesperoleucus* should be considered congeneric.

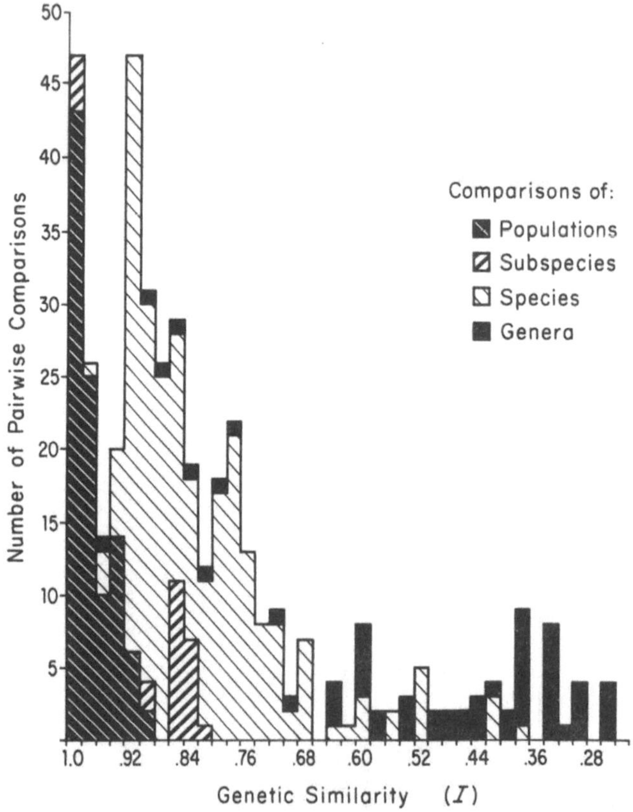

Figure 1. Distribution of Nei's (1972) genetic similarity coefficients *I* calculated from pairwise comparisons in the cyprinid allozyme studies listed in Table IV.

3.3.1. Polyploidization

The duplication of the entire genome has been promoted as a major evolutionary phenomenon (Ohno *et al.*, 1968; Ohno, 1970). The demonstration of tetraploid karyotypes and additional duplication of enzyme loci in certain fishes provided much of the early evidence in support of the polyploid hypothesis (Ohno *et al.*, 1967; Engel *et al.*, 1971; Wolf *et al.*, 1969; Schmidtke and Engel, 1974, 1976; Schmidtke *et al.*, 1976). Despite the nearly even doubling of chromosome numbers of $2n = 98$–104 among the cyprinid tetraploids, White (1978) challenged the polyploid hypothesis as applied to any group of fishes. Duplicate enzyme loci were dismissed by White as support for a polyploid interpretation; however, he did not

Table V
Karyotypic and Electrophoretic Evidence in Support of Tetraploidy in Cyprinid Fishes

Species	Tetraploid karyotype	Duplicate gene expression
Acrossocheilus sumatranus	Suzuki and Taki (1981)	—
Aulopyge hugeli	Berberovic *et al.* (1973)	—
Barbus barbus	Wolf *et al.* (1969)	Klose *et al.* (1969), Engel *et al.* (1971)
Barbus meridionalis	Sofradzija and Berberovic (1973)	Triantaphyllidis *et al.* (1981)
Barbus plebeius	Park (1974)	—
Carassius auratus	Ohno *et al.* (1967), Wolf *et al.* (1969)	Bender and Ohno (1968), Klose *et al.* (1969)
Cyprinus carpio	Ohno *et al.* (1967), Wolf *et al.* (1969)	Bender and Ohno (1968), Klose *et al.* (1969), Ferris and Whitt (1977*a*)
Tor putitora	Khuda-Bukhsh (1980)	—

distinguish between multilocus systems developed via tandem duplication or ancient duplications and those systems, including previous duplicates, that had undergone additional doubling via polyploidy (Buth, 1983). The role of gene silencing (functional diploidization) in polyploid systems was not clear until relatively recently and this form of gene expression undoubtedly contributed to the confusion regarding the relationship between the expression of duplicate enzyme loci and vertebrate polyploidy. Reports of electrophoretic data in support of cyprinid tetraploidy are listed in Table V. A summary of such support among other groups of fishes is provided by Buth (1983).

While the evidence in support of cyprinid tetraploidy is substantial, relatively few tetraploid cyprinids have been identified. As previously mentioned, cyprinid tetraploidy is limited to Old World forms. Of the 100+ Old World cyprinid species that have been karyotyped (J. R. Gold, personal communication), only eight exhibit a tetraploid chromosome set and only four have been electrophoretically examined to demonstrate duplicate gene expression (Table V). The taxonomic diversity of these eight species suggests that tetraploidy was developed independently in several Old World cyprinid lineages. A large number of species closely related to these eight await karyotypic and electrophoretic investigation, so the number of cyprinid tetraploids known may increase with additional studies. However, cyprinids as a group do not express tetraploid expression to the degree that some other cypriniform groups do, e.g., catostomids (Ferris, this volume, Chapter 2).

3.3.2. Functional Diploidization

Duplicate structural gene loci arising via polyploidization will be controlled by separate regulatory systems because these regulatory genes have also been duplicated during the polyploidization event. These independent structural loci are free to vary and to respond to different selective pressures. Thus, as long as one of the duplicate loci continues to produce the required product, the other is free to acquire "forbidden mutations" and possibly acquire new functions (Ohno, 1970). An alternative evolutionary direction, an extreme of regulatory divergence, involves the complete cessation of production of a functional gene product by one of the duplicate loci. Thus, in terms of gene expression, the organism has returned to the diploid, preduplicated, condition. Various mechanisms through which this diploidization might occur have been discussed by Waines (1976), Ferris and Whitt (1977*b*), Takahata and Maruyama (1979), Li (1980), and Takahata (1982). While isolated enzyme systems have been discussed in terms of duplicated-diploidized expression in certain cyprinid species (Klose *et al.*, 1969; Engel *et al.*, 1971; Schmidtke and Engel, 1974), the only comprehensive treatment of multilocus expression comparison is Ferris and Whitt's (1977*a*) study of gene silencing in the tetraploid *Cyprinus carpio*. If comparable studies are conducted on the cyprinid tetraploids, the comparative data base may be quite informative for phylogenetic application, as has been done in catostomids (Ferris and Whitt, 1978; Buth, 1979*b*), or for the study of evolutionary alternatives in the divergence of gene functions.

3.4. Comparisons with Immunological Differentiation

Immunological data, in any form, have rarely been applied to cyprinid studies either as a primary or alternative data base. For example, no cyprinid species have been compared using microcomplement fixation, an omission that is probably not critical given Farris' (1981) exposition of the invalidity of microcomplement fixation and other "distance"-oriented data generating methods. The only direct comparison of cyprinid allozyme and immunological data is provided by English (1980). Trefoil immunodiffusion was used to examine muscle antigens of the four species of the genus *Semotilus* (with two species of *Phoxinus* as outgroups) and these data were compared with the allozyme data for *Semotilus* and *Phoxinus* provided by Joswiak (1980). However, analyses of the two data bases did not yield congruent estimates of relationship within *Semotilus*. English provides several explanations for this disparity, including the observation "that Joswiak's dendrograms were based on phenetic analysis whereas the analysis of immunological data is cladistic." Different analytical methods can indeed contribute to such incongruences (Mickevich and Johnson, 1976). However, analytical differences are only one problem in this comparison. Both Joswiak (1980) and English (1980) converted their particulate allozyme and immunological data, respectively, to "distance" coefficients in attempts to quantify degrees of difference among taxa. (English noted a strong positive correlation between antigenic and allozyme genetic distances, although this relationship was nonlinear.) A subsequent study by Farris (1981) discussed the limitations of using "distances" in any systematic study; both data bases should have remained in particulate form, as characters and character states, for valid analysis. Further, the antigenic dendrograms provided by English are really phenograms, clustered using the unweighted pair-group method with arithmetic averages; these dendrograms can only be considered as cladograms under the assumption of equivalent rates of evolution in each lineage. Thus, while English (1980) and Joswiak (1980) were on the right track by employing a comparative approach and an outgroup perspective, the use of phenetically clustered distance coefficients invalidates this as a phylogenetic comparison.

Piront and Gosslin-Rey (1974) examined the antigenic properties of muscle parvalbumins of several European cyprinids via immunoelectrophoresis. No direct comparison with allozyme or morphological data was intended; rather, this study evaluated these antigens in terms of their potential as characters in systematic studies. A number of interesting observations were made; however, the authors concede that "the use of parvalbumins is complex because of the polymorphism which introduces difficulties in the isolation of pure antigens and of the molecular diversity which strongly limits the extent of immunological cross-reaction among species." While the immunological methods of both Piront and Gosslin-Rey (1974) and English (1980) have the potential to yield particulate data bases that can be phylogenetically treated, immunological data may be quite limited as an alternative data base for comparative studies. The time and effort required to generate a substantial immunological data base far exceed those for allozyme and morphological data. The latter two remain as the most cost-efficient data bases for comparative studies.

3.5. Comparisons with Morphological Differentiation

Have morphological and allozyme characters of fishes evolved concordantly? Mickevich and Johnson (1976) addressed this question using these data bases in their study of the atheriniform genus *Menidia*. These authors noted that it is the analysis, not the data base, that is the important factor in evolutionary studies; cladistic analyses, in theory and by their demonstration, yield congruent phylogenetic estimates regardless of the data bases examined. Thus, the examination of both morphological characters and allozyme characters are especially desirable as two perspectives with which to test phylogenetic hypotheses. Allozyme characters are not to be treated as a necessarily superior data base just because of the technology needed to reveal their expression. While the findings of Mickevich and Johnson (1976) have important implications in systematic treatment of all forms of data, the significance of their study has been clouded by a number of procedural errors as pointed out by Riska (1979) and Colless (1980). Clearly, the congruence study of *Menidia* needs to be repeated, probably with newly generated data bases, given the questions on this level. Studies comparing these data bases in other groups of fishes, especially the speciose Cyprinidae, are needed to evaluate congruence under phylogenetic analysis.

Direct comparisons of allozyme and morphological characters, except in hybridization studies and the study of two species of *Nocomis* by Ferguson *et al.* (1981), are lacking in the cyprinid literature. Allozyme studies usually compare findings with "traditional" morphological studies, i.e., those previous studies in which data were treated in a nonnumerical fashion. Often the "morphological known" was never formally treated; relationships were just read from the extant taxonomy of the group and compared to those generated using a numerical treatment of allozyme data. Phylogenetic treatments of cyprinid morphological data do exist, but these groups have yet to be electrophoretically examined.

While one might expect congruence of morphological and allozyme data in phylogenetic analyses, such congruence will be found only in the *pattern* of differentiation, not necessarily in the *degree* of differentiation. Two cyprinid studies have demonstrated the extremes of variation. Ferguson *et al.* (1981) studied morphological and allozyme differentiation between *Nocomis biguttatus* and *N. micropogon* for the purposes of biochemical identification. Between these species, allozyme differentiation exceeded morphological differentiation; Ferguson *et al.* (1981) found an 0.8665 probability of identifying these species using morphological traits and a 1.000 chance of identification using allozyme characters. On the other hand, Avise *et al.* (1975) found very little allozyme differentiation between *Hesperoleucus symmetricus* and *Lavinia exilicauda*, two species that are morphologically distinctive.

4. Applications

4.1. Hybridization and Introgression

Hybridization among species of freshwater fishes is not uncommon and numerous hybrid combinations have been described for cyprinids. McAllister and Coad (1978) listed all known intergeneric hybrid combinations for North American cyprinids "found in nature in their native range." Many combinations that do not appear in nature can be developed in the laboratory, e.g., *Rhinichthys atratulus* × *R. cataractae* (Clayton and Gee, 1969). Hybrids are expected to express intermediate meristic counts compared to the parental species and this intermediacy may apply to other morphological attributes as well (Hubbs, 1955; Schwartz, 1972; Goodfellow *et al.,* 1982). The identification of hybrids becomes difficult if the parentals express very similar morphologies or if there is a substantial difference between the parentals in terms of dominance/recessiveness of the morphological traits. These problems may be minimized by using allozyme characteristics to assess the interspecific interactions of hybridization and introgression. Allozymes serve not just as an alternative data base in these comparisons, but may be the data base of choice, since the genetic control of various enzyme systems is known and codominant expression is the rule. Hybrids are often revealed via electrophoresis first on the basis of higher than expected levels of variability and second on the shared, additive nature of their allozyme components with the suspected parentals. Allozyme analysis is playing an increasing role, supplementing morphological analysis, in studies of cyprinid hybridization (e.g., Menzel, 1977; Rakocinski, 1980; Goodfellow *et al.,* 1982; Joswiak *et al.,* 1982).

Cyprinid hybridization can extend beyond the development of an F_1 generation to backcrosses and, eventually, to the introgression of alleles into one or both of the parental species. This important evolutionary phenomenon has been reported in the North American cyprinid genera *Campostoma* and *Notropis*. In a comparative study of the allozymes of the species of *Campostoma,* Buth and Burr (1978) noted that two Illinois populations of *C. oligolepis* exhibited higher heterozygosity levels compared to disjunct Missouri populations and that the increased heterozygosity of the Illinois populations was due to alleles that were also present in sympatric populations of *C. anomalum pullum*. As no interbreeding between these forms was thought to be occurring (Burr and Smith, 1976), Buth and Burr (1978) concluded that introgressive hybridization must have occurred in the past. Examining larger population samples and products of an esterase locus that were fixed at different alleles in the two forms, Rakocinski (1980) demonstrated that hybridization between *C. oligolepis* and *C. anomalum pullum* was still occurring in northern Illinois. He concluded that "the occurrence of hybridization seems to be a general rather than localized phenomenon where these species are sympatric in the northern midwest, although ratios of parental population sizes and levels of interbreeding will vary."

A wider range of introgressive relationships has been found in the genus *Notropis*. In an extreme form, introgressive hybridization has the potential to yield new species and Menzel (1976) has suggested such an origin for *Notropis albeolus* from *N. cornutus* and *N. cerasinus*. Additional morphometric and allozyme data have supported Menzel's hypothesis (Menzel, 1977; Buth, 1979*a*). The taxonomic relationship of *Notropis cornutus* and *N. chrysocephalus* has been debated for some time (Gilbert, 1964; Miller, 1968; Menzel, 1976; Buth, 1979*a*). Dowling and Moore (1984) have shown that these species maintain their genetic integrity in spite of localized hybridization. The forms should be recognized as full species in the taxonomic sense, but the lack of complete reproductive isolation suggests that the term "semispecies" best describes this relationship (Buth, 1979*a*; Dowling and Moore, 1984). As Zimmerman and Richmond (1981) have shown in *Notropis lutrensis,* the acquisition of alleles via possible introgression with *Notropis venustus* may have a functional advantage for the former species.

In these cyprinid examples, introgressive hybridization has been revealed during the process of analyzing relationships of particular taxonomic groups. Introgressive hypotheses are advanced upon the resolution of unexpected shared characteristics between sympatric taxa. Introgression may be more common than is currently thought, as most studies concentrate on phyletic relatives rather than sympatric, although distantly related, species.

4.2. Rates of Evolution

The allozyme diversity among cyprinids (Fig. 1) combined with the large number of cyprinid species suggests the considerable potential for the use of this group in studies of evolutionary rates. However, few studies have used cyprinids in this fashion and those that have, have yielded rather limited conclusions.

Avise (1977*a*) examined the relationship between genic heterozygosity and rate of speciation in two groups of North American fishes. He hypothesized that rapidly speciating phylads, e.g., the speciose North American cyprinid fishes, do so due to increased levels of intraspecific genetic variation compared to species-depauperate groups such as the North American perciform family Centrarchidae. However, this variation relationship was not revealed; genic heterozygosity levels are comparable in both groups. While some may debate whether the comparison of orders of fishes is relevant, the question must be raised as to the nature of the hypothesis to be tested. In this study, a *null hypothesis* yields the expectation of no difference in level of heterozygosity between cyprinids and centrarchids (which was observed). Thus, the null hypothesis cannot be rejected and additional studies are necessary to clarify the relationship between heterozygosity and rate of speciation.

Avise (1977*b*), using the cyprinid and centrarchid allozyme data bases, examined the question of gradual evolution versus "rectangular" evolution (= evolution through punctuated equilibria). Avise showed that the rate of protein evolution appears to be decelerated in the rapidly speciating cyprinids. The "results are inconsistent with predictions of rectangular evolution but are not demonstrably incongruent with predictions of phyletic gradualism." Gold's (1980) study of rectangular versus gradual evolution in cyprinids utilizing a chromosomal data base yielded results incompatible with the rectangular mode of divergence, yet inconsistencies were revealed in regard to the gradual mode.

Perhaps the most controversial evolutionary rate comparison is the cyprinid differentiation study of Avise and Ayala (1976) in which the speciose North American cyprinids were compared with the perciform genus *Lepomis* to address the question of whether differentiation is related to time or to the number of cladogenic events in the history of the lineage. Their findings support the time model; no differences in degree of differentiation between cyprinids as a group versus the genus *Lepomis* support the notion that "time since divergence . . . is more important than the number of intermediate cladogenetic events in determining the level of genetic divergence between species." Again, some may debate the relevance of comparing these particular groups, e.g., is an assemblage including a number of monotypic cyprinid genera a speciose group compared to one 11-species genus in another order? These and other issues that are raised in Avise and Ayala's (1976) study are currently being debated (R. L. Mayden, personal communication).

4.3. Biochemical Identification

Both allozyme and isozyme characteristics can be used in taxonomic keys for species identification. Such keys can be developed using external standards (Avise, 1974) or internal standards involving multilocus systems (Buth, 1980). In either case, electrophoretic data have been underutilized as tools for identification of species. Allozyme characters have proven quite useful for the detection of hybrid individuals and for specific identification of the parental forms contributing to the hybrids. However, as data supplementary to morphological characters for the purposes of cyprinid identification, electrophoretic data have only rarely been employed. This lack of application is due to the fact that relatively few cyprinid species have been electrophoretically compared ("characterized") and fewer still have been electrophoretically examined throughout their ranges to add a measure of confidence to the characters used to "characterize" the species. Allozyme studies of cyprinids have not moved in this direction for lack of motivation; there is no real need to develop supplementary characters for identification if external morphological traits will suffice.

Accurate environmental assessment often requires the accounting of all life stages of the biota, not just the adult forms. The identification of larval fishes has become a widespread problem. As a major ichthyofaunal assemblage, the cyprinid fishes would figure prominently in many aquatic assessments. Their larval forms would be especially problematic in this regard, as the adult morphological characteristics used to identify the species are often underdeveloped or insufficiently expressed in the larvae to be used. However, investigators can take advantage of the fact that the allozymes of most structural enzymes used in studies of cyprinid fishes change, at most, quantitatively, not qualitatively, posthatching. Thus, allozyme characters resolved in the adult forms can be applied to biochemically identify the larvae. Allozyme analyses can be much more time and cost effective than evaluations of myomere counts and pigmentation patterns, especially when the latter characters have yet to be studied in most cyprinid species. If the distribution of allozyme characters among taxa is known, it is often the case that relatively few characters need to be used for larval identification in specific geographic contexts; e.g., Ferguson (1981) examined the allozyme characteristics of six cyprinid species inhabiting a particular river system; these six forms were able to be distinguished using just two enzyme systems: lactate dehydrogenase (LDH) and malate dehydrogenase (MDH). Thus, considerable potential exists for allozyme applications in the biochemical identification of cyprinid fishes, especially as applied to larval forms. Future studies involving allozyme comparisons of adult cyprinids will continue to add to the data base that may be later applied to the identification of larvae.

4.4. Systematics

Cyprinid allozyme data have not always been phylogenetically treated (e.g., Menzel, 1976; Joswiak, 1980). Those cyprinid studies that have attempted to use phylogenetic methods are now known to contain key procedural errors. The studies of *Campostoma* and *Notropis* by Buth and Burr (1978) and Buth (1979*a*), respectively, err at the level of encoding the allozyme data. In these studies, the allele was considered as the character and the degree of its expression ("frequency coding" and "presence/absence coding") was considered to determine the character states. A more appropriate (and biologically realistic) coding procedure treats the locus as the character and the allelic composition of the locus as the character state (Mickevich and Mitter, 1981). The allozyme data for *Campostoma* and *Notropis* are currently being recoded and reanalyzed as part of congruence studies using morphological data. Avise and Ayala (1976) used the distance Wagner procedure to phylogenetically treat allozyme data for several western North American cyprinid genera. The application of this procedure is now known to be invalid (Farris, 1981). Particulate allozyme data must remain in particulate form (characters and states) rather than be transformed to distance measures for phylogenetic analysis. Unfortunately, Avise and Ayala (1976) published only the complete distance matrix, not the complete data base, so direct reevaluation of the study is not possible. Recommendations for use of phylogenetic methods are discussed in Section 5.3.

Emphasis should be placed on the systematic relationships of cyprinid populations as well as taxa. Wider geographic sampling (Section 5.1) of *Notropis stramineus* (Koehn *et al.*, 1971) and *Notropis pilsbryi* (Buth and Mayden, 1981) have revealed structuring of these cyprinid populations with instances of intradrainage and interdrainage differentiation, respectively, of products of a few loci. This geographic differentiation may have a selective basis, as suggested by Koehn *et al.* (1971) for *Notropis stramineus* in the Kansas River system, or may have a historical component. Considerable work remains to be done at the population level.

5. Recommendations for Future Research

5.1. Geographic Sampling Strategy: Synthesis of β and γ Analyses

Mayr (1969) recognized three stages in classification:

1. α-Taxonomy, which deals with the description of new species and their preliminary arrangement,
2. β-Taxonomy, which emphasizes the development of a sound classification on the species level and on that of the higher categories, and
3. γ-Taxonomy, which deals with intraspecific variation, speciation, and evolutionary rates and trends.

While Mayr (1969) recognized that these levels overlap and saw a historical trend of α to γ emphasis in given groups, it is at the β and γ levels that allozyme data will play the greatest role in cyprinid systematics. While there are numerous cyprinids that have yet to be described, it is at the β level that most systematic attention should be aimed.

Little is known about the structure of cyprinid populations, and generalizations should not be made based on a few observations in this heterogeneous group. It is premature to propose a typological policy for the application of allozyme data to β-level analyses. On the one hand, examinations of populations of some cyprinid species, e.g., *Hesperoleucus symmetricus* and *Lavinia exilicauda* (Avise *et al.*, 1975), *Ptychocheilus grandis* (Avise and Ayala, 1976), and *Notropis chrysocephalus* (Buth, 1979*a*), have revealed little geographic allozyme differentiation, prompting Avise and Ayala (1976) to conclude that "for purposes of interspecific comparisons, a single population provides an adequate representation of a species." Such representation may be adequate if the comparison between the populations is by way of a genetic similarity coefficient. However, Buth and Crabtree (1982) have noted that significant restrictions to gene flow may be obscured if the pattern of allelic differentiation is quantitative rather than qualitative and if the populations are evaluated by overall similarity. Interdrainage differentiation, as exhibited by *Notropis pilsbryi* (Buth and Mayden, 1981), would not have been revealed in this species under the more restricted sampling strategy of Buth (1979*a*). Indeed, the use of a single population sample of *Notropis cornutus* in the latter study certainly failed to "represent the species" if this population was made up of locally introgressed individuals as suspected by T. Dowling (personal communication).

Clearly there is a tradeoff between accurate representation of species in a β-level analysis and understanding the gene-flow patterns and patterns of population differentiation within species in a γ-level analysis. Avise and Ayala's (1976) generalization for minimal population differentiation may eventually be shown to hold for cyprinids [the generalization may have to be reduced to populations of subspecies following Buth and Burr (1978)], but additional γ-level studies are needed and blanket application of this assumption may discourage investigation at that level. I propose that future β-level cyprinid studies incorporate some degree of geographic sampling rather than assume that no geographic variation exists. This input on the γ level, if financially and/or biologically feasible, can provide the information necessary to pinpoint geographic regions requiring further study or to ultimately draw the conclusion that population differentiation within cyprinid taxa is minimal. Initial geographic sampling need not be extensive; two or three widely separated samples may suffice. A guideline of one sample per major drainage within a species' range may suffice for a more detailed study. Sampling under these guidelines for a study of the subgenus *Luxilus* (genus *Notropis*) is shown in Table VI.

5.2. Reevaluation of "Magnitude" Arguments in Taxonomy

"Magnitude" criteria for taxonomic application have no place in modern taxonomic schemes designed to reflect the phylogeny of the group under study. Statements referring to two forms as "so different they should be placed in separate genera" appear frequently in the earlier allozyme literature and reflect an attitude favoring a phenetically based taxonomy built on degrees of difference rather than that of phylogeny. As more systematists choose a phylogenetic treatment of allozyme data, the "magnitude" arguments may no longer be invoked.

A case study in cyprinid "magnitude" arguments is seen in the taxonomic history of *Notropis pilsbryi* and *N. zonatus,* treated as subspecifically distinct prior to Gilbert's (1964) elevation to full species status. Menzel and Cross (1977) proposed to synonymize these forms using a number of allozyme characteristics. However, additional allozyme data provided by Buth and Mayden (1981) revealed no zone of intergradation and reaffirmed the specific status of these forms. Buth and Mayden (1981) caution against the use of symplesiomorphically distributed alleles, e.g., at *Ldh-A* in *N. pilsbryi* and *N. zonatus,* in similarity "magnitude" arguments and proposed "the geographic pattern of genetic differentiation rather than the absolute magnitude of genetic differentiation as the primary criterion for the evaluation of specific-subspecific status in cyprinid taxonomy."

Avise *et al.* (1975) have recommended congeneric status for *Hesperoleucus symmetricus* and *Lavinia exilicauda* on the basis of the high level of genetic similarity between these forms relative to other cyprinids. Their taxonomic change may be desirable, but should not be based on a "magnitude" criterion. Both of these genera are monotypic and may be the closest relatives of one another if the phylogenetic analysis of Avise and Ayala (1976) holds up under nondistance phylogenetic reanalysis. The fact that these sympatric forms are good species was confirmed by Avise *et al.* (1975) and the specific epithets reflect this distinction. Generic distinction of these forms provides no additional information and, in fact, obscures the relationship of these sister taxa. Their taxonomy would retain more information content if this two-species lineage was treated as a single genus. Thus, the species distinction is maintained and an additional phyletic component is added to the taxonomic scheme. Avise *et al.* (1975) should not have argued for congeneric status for these forms based on

Table VI

Comparison of Sampling Strategies of Menzel (1976) and Buth (1979*a*) and a Geographic Sampling Strategy of One Sample per "Major Drainage" within the Range of Each Taxon for Studies of the Subgenus *Luxilus*, Genus *Notropis*

Species	Number of samples under the "one per major drainage" criterion	Menzel (1976)		Buth (1979*a*)	
		Number of major drainages sampled	(Total sampled)	Number of major drainages sampled	(Total sampled)
N. albeolus	5	2	(3)	1	(2)
N. cerasinus	2	2	(3)	1	(2)
N. chrysocephalus	20+	4	(6)	4	(4)
N. coccogenis	4	1	(1)	1	(1)
N. cornutus	20+	7	(25)	1	(1)
N. pilsbryi	3	—	—	1	(2)
N. zonatus	4	—	—	1	(1)
N. zonistius	2	1	(1)	1	(1)

their similarity, but rather from their close phylogenetic relationship, which could be reflected in the taxonomy under a single generic term.

5.3. Application of Phylogenetic Methods

The full potential of application of allozyme data to problems in cyprinid systematics will not be reached unless such data are afforded phylogenetic treatment. Relationships within a number of speciose North American genera, including *Gila, Hybopsis,* and *Notropis,* await thorough phylogenetic analyses to which allozyme, and perhaps isozyme, data may contribute. Allozyme and isozyme data may play an even greater role in the assessment of intergeneric relationships among both New World and Old World taxa.

Phylogenetic applications should incorporate the following: (1) Electrophoretic data should be treated in a particulate fashion; loci should be considered as characters and the allelic compositions at a given locus as the character states under this encoding scheme (Mickevich and Mitter, 1981). The criteria for recognizing character state differences are, at this point, controversial and may be formalized in quite different ways by various investigators in this field. (2) Phylogenetic clustering should *not* employ the distance Wagner procedure (Farris, 1981), but rather should use any of the currently available "standard" Wagner programs. Newer methods, such as transition series analysis (Mickevich and Mitter, 1981, 1983; Mickevich, 1982), may have much to offer in the way of analytical improvement, but cannot be realistically evaluated without comparative application (which should be encouraged). (3) Character state direction and rooting of Wagner networks (Lundberg, 1972) should be based on outgroup comparisons (Stevens, 1980; Watrous and Wheeler, 1981) as used in *Notropis* (Buth, 1979*a*). (4) Homoplasy in cyprinid allozyme characters should be viewed from the perspective of possible introgression. "Excessive" homoplasy via introgression may represent a problem in phylogenetic analyses that employ parsimony criteria. However, if the introgressive nature of the homoplasy is understood (e.g., Buth and Burr, 1978; Rakocinski, 1980), proper encoding should alleviate problems in phylogenetic analysis. (5) One can pursue phylogenetic analyses in an optimal fashion if two independent data sets are treated and examined for congruence (e.g., Mickevich and Johnson, 1976). The choice of data bases may be investigator-specific, but logically would be those that can be easily generated in terms of finances and effort and that contain a maximal amount of information (usually in terms of numbers of characters). At this time, morphological and allozyme/isozyme characters fulfill these criteria, although karyological data may be preferred in cases of cyprinid polyploidy (Table V).

A taxonomy for cyprinid fishes that is based on phylogenetic relationships may prove to be quite different from that which is currently in use. The chaotic taxonomic history of both New World and Old World cyprinids has obviated any argument of taxonomic retention on the basis of "tradition." Taxonomic changes that reflect the phylogeny of these fishes are improvements that are long overdue.

ACKNOWLEDGMENTS. I would like to thank Robert W. Murphy for his timely assistance in the completion of this chapter, Stephen D. Ferris for his helpful comments, and John R. Gold for providing a wealth of information on cypriniform karyology.

References

Allendorf, F. W., and Utter, F. M., 1978, Population genetics, in: *Fish Physiology,* Vol. 8: *Bioenergetics and Growth* (W. S. Hoar and D. J. Randall, eds.), Academic Press, New York, pp. 407–454.

Aquadro, C. F., and Avise, J. C., 1982, Evolutionary genetics of birds. VI. A reexamination of protein divergence using varied electrophoretic conditions, *Evolution* **36:**1003–1019.

Avise, J. C., 1974, Systematic value of electrophoretic data, *Syst. Zool.* **23:**465–481.

Avise, J. C., 1977*a,* Genic heterozygosity and rate of speciation, *Paleobiology* **3:**422–432.

Avise, J. C., 1977*b,* Is evolution gradual or rectangular? Evidence from living fishes, *Proc. Natl. Acad. Sci. USA* **74:**5083–5087.

Avise, J. C., and Ayala, F. J., 1976, Genetic differentiation in speciose versus depauperate phylads: Evidence from the California minnows, *Evolution* **30:**46–58.

Avise, J. C., Smith, J. J., and Ayala, F. J., 1975, Adaptive differentiation with little genic change between two native California minnows, *Evolution* **29:**411–426.

Bender, K., and Ohno, S., 1968, Duplication of the autosomally inherited 6-phosphogluconate dehydrogenase gene locus in tetraploid species of cyprinid fish, *Biochem. Genet.* **2:**101–107.

Berberovic, L., Hadziselimovic, R., Pavlovic, B., and Sofradzija, A., 1973, Chromosome set of the species *Aulopyge hugeli* Heckel 1841, *Bull. Sci. Acad. RSF Yougosl.* **18:**10–11.

Brewer, G. J., 1970, *An Introduction to Isozyme Techniques,* Academic Press, New York.

Burr, B. M., and Smith, P. W., 1976, Status of the largescale stoneroller, *Campostoma oligolepis, Copeia* **1976:**521–531.

Buth, D. G., 1979*a,* Biochemical systematics of the cyprinid genus *Notropis,* I. The subgenus *Luxilus, Biochem. Syst. Ecol.* **7:**69–79.

Buth, D. G., 1979*b,* Duplicate gene expression in tetraploid fishes of the tribe Moxostomatini (Cypriniformes, Catostomidae), *Comp. Biochem. Physiol.* **63B:**7–12.

Buth, D. G., 1980, Evolutionary genetics and systematic relationships in the catostomid genus *Hypentelium, Copeia* **1980:**280–290.

Buth, D. G., 1981, Cladistic treatment of isozyme (rather than allozyme) data, Presentation at the 2nd Annual Willi Hennig Society Meeting, Ann Arbor, Michigan.

Buth, D. G., 1983, Duplicate isozyme loci in fishes: Origins, distribution, phyletic consequences and locus nomenclature, in: *Isozymes: Current Topics in Biological and Medical Research,* Vol. 10 (M. C. Rattazzi, J. G. Scandalios, and G. S. Whitt, eds.), Alan R. Liss, New York, pp. 381–400.

Buth, D. G., and Burr, B. M., 1978, Isozyme variability in the cyprinid genus *Campostoma, Copeia* **1978:**298–311.

Buth, D. G., and Crabtree, C. B., 1982, Genetic variability and population structure of *Catostomus santaanae* in the Santa Clara drainage, *Copeia* **1982:**439–444.

Buth, D. G., and Mayden, R. L., 1981, Taxonomic status and relationships among populations of *Notropis pilsbryi* and *N. zonatus* (Cypriniformes: Cyprinidae) as shown by the glucosephosphate isomerase, lactate dehydrogenase and phosphoglucomutase enzyme systems, *Copeia* **1981:**583–590.

Cataudella, S., Sola, L., Muratori, R. A., and Capanna, E., 1977, The chromosomes of eleven species of Cyprinidae and one Cobitidae from Italy, with some remarks on the problem of polyploidy in the Cypriniformes, *Genetica* **47:**161–171.

Chiarelli, A. B., and Capanna, E., 1973, Checklist of fish chromosomes, in: *Cytotaxonomy and Vertebrate Evolution* (A. B. Chiarelli and E. Capanna, eds.), Academic Press, New York, pp. 206–232.

Clayton, J. W., and Gee, J. H., 1969, Lactate dehydrogenase isozymes in longnose and blacknose dace (*Rhinichthys cataractae* and *R. atratulus*) and their hybrid, *J. Fish. Res. Board Can.* **26:**3049–3053.

Clayton, J. W., and Tretiak, D. N., 1972, Amine-citrate buffers for *p*H control in starch gel electrophoresis, *J. Fish. Res. Board Can.* **29:**1169–1172.

Colless, D. H., 1980, Congruence between morphometric and allozyme data for *Menidia* species: A reappraisal, *Syst. Zool.* **29:**288–299.

Crabtree, C. B., and Buth, D. G., 1981, Gene duplication and diploidization in tetraploid catostomid fishes *Catostomus fumeiventris* and *C. santaanae, Copeia* **1981:**705–708.

Dowling, T., and Moore, W. S., 1984, The level of reproductive isolation between two cyprinid fishes, *Notropis cornutus* (Mitchill) and *N. chrysocephalus* (Rafinesque), *Copeia* (in press).

Engel, J., Faust, J., and Wolf, U., 1971, Isoenzyme polymorphism of the sorbitol dehydrogenase and the NADP-dependent isocitrate dehydrogenases in the fish family Cyprinidae, *Anim. Blood Groups Biochem. Genet.* **2:**127–133.

English, P. J., 1980, Immunological affinities within a genus of cyprinid fishes (*Semotilus:* Cyprinidae), M. S. dissertation, Wayne State University.

Farris, J. S., 1981, Distance data in phylogenetic analysis, in: *Advances in Cladistics* (V. A. Funk and D. R. Brooks, eds.), The New York Botanical Garden, Bronx, New York, pp. 3–23.

Ferguson, M. M., 1981, Identification and species characterization of some North American minnows by electrophoresis, *Biochem. Syst. Ecol.* **9:**89–91.

Ferguson, M. M., Noakes, D. L., and Danzmann, R. G., 1981, Morphological and biochemical systematics of chubs, *Nocomis biguttatus* and *N. micropogon* (Pisces: Cyprinidae), in southern Ontario, *Can. J. Zool.* **59:**771–775.

Ferris, S. D., and Whitt, G. S., 1977*a,* The evolution of duplicate gene expression in the carp (*Cyprinus carpio*), *Experientia* **33:**1299–1301.

Ferris, S. D., and Whitt, G. S., 1977*b,* Loss of duplicate gene expression after polyploidisation, *Nature* **265:**258–260.

Ferris, S. D., and Whitt, G. S., 1978, Phylogeny of tetraploid catostomid fishes based on the loss of duplicate gene expression, *Syst. Zool.* **27:**189–206.

Gilbert, C. R., 1964, The American cyprinid fishes of the subgenus *Luxilus* (genus *Notropis*), *Bull. Fla. State Mus.* **8**:95–194.

Gold, J. R., 1980, Chromosomal change and rectangular evolution in North American cyprinid fishes, *Genet. Res. Camb.* **35**:157–164.

Gold, J. R., Karel, W. J., and Strand, M. R., 1979*a*, Chromosome formulae of North American Fishes, The Texas Agricultural Experiment Station, Texas A & M University, College Station, Texas.

Gold, J. R., Whitlock, C. W., Karel, W. J., and Barlow, J. A., Jr., 1979*b*, Cytogenetic studies of North American minnows (Cyprinidae): VI. Karyotypes of thirteen species in the genus *Notropis, Cytologica* **44**:457–466.

Gold, J. R., Womac, W. D., Deal, F. H., and Barlow, J. A., Jr., 1981, Cytogenetic studies in North American minnows (Cyprinidae): VII. Karyotypes of thirteen species from the southern United States, *Cytologia* **46**:105–115.

Goodfellow, W. L., Jr., Morgan, R. P., II, Hocutt, C. H., and Stauffer, J. R., Jr., 1982, Electrophoretic analysis of *Campostoma anomalum, Rhinichthys cataractae* and their F_1 offspring, *Biochem. Syst. Ecol.* **10**:95–98.

Hubbs, C. L., 1955, Hybridization between fish species in nature, *Syst. Zool.* **4**:1–20.

International Union of Biochemistry, 1979, *Enzyme Nomenclature, 1978,* Academic Press, New York.

Joswiak, G. R., 1980, Genetic divergence within a genus of cyprinid fish (*Phoxinus:* Cyprinidae), Ph. D. dissertation, Wayne State University.

Joswiak, G. R., Starnes, W. C., and Moore, W. S., 1980, Karyotypes of three species of the genus *Phoxinus* (Pisces: Cyprinidae), *Copeia* **1980**:913–916.

Joswiak, G. R., Stasiak, R. H., and Moore, W. S., 1982, Allozyme analysis of the hybrid *Phoxinus eos* × *Phoxinus neogaeus* (Pisces: Cyprinidae) in Nebraska, *Can. J. Zool.* **60**:968–973.

Khuda-Bukhsh, A. R., 1980, A high number of chromosomes in the hillstream cyprinid, *Tor putitora* (Pisces), *Experientia* **36**:173–174.

Klose, J., Wolf, U., Hitzeroth, H., and Ritter, H., 1969, Polyploidization in the fish family Cyprinidae, order Cypriniformes. II. Duplication of the gene loci coding for lactate dehydrogenase (E.C.: 1.1.1.27) and 6-phosphogluconate dehydrogenase (E.C.: 1.1.1.44) in various species of Cyprinidae, *Humangenetik* **7**:245–250.

Koehn, R. K., Perez, J. E., and Merritt, R. B., 1971, Esterase enzyme function and genetical structure of populations of the freshwater fish, *Notropis stramineus, Am. Nat.* **105**:51–69.

Li, W.-H., 1980, Rate of gene silencing at duplicate loci: A theoretical study and interpretation of data from tetraploid fishes, *Genetics* **95**:237–258.

Lin, C.-C., Schipmann, G., Kittrell, W. A., and Ohno, S., 1969, The predominance of heterozygotes found in wild goldfish of Lake Erie at the gene locus for sorbitol dehydrogenase, *Biochem. Genet.* **3**:603–607.

Lundberg, J. G., 1972, Wagner networks and ancestors, *Syst. Zool.* **21**:398–413.

Mayr, E., 1969, *Principles of Systematic Zoology,* McGraw-Hill, New York.

McAllister, D. E., and Coad, B. W., 1978, A test between relationships based on phenetic and cladistic taxonomic methods, *Can. J. Zool.* **56**:2198–2210.

Menzel, B. W., 1976, Biochemical systematics and evolutionary genetics of the common shiner species group, *Biochem. Syst. Ecol.* **4**:281–293.

Menzel, B. W., 1977, Morphological and electrophoretic identification of a hybrid cyprinid fish, *Notropis cerasinus* × *Notropis c. cornutus,* with implications on the evolution of *Notropis albeolus, Comp. Biochem. Physiol.* **57B**:215–218.

Menzel, B. W., and Cross, F. B., 1977, Systematics of the bleeding shiner species group (Cyprinidae: genus *Notropis,* subgenus *Luxilus*), American Society of Ichthyologists and Herpetologists Meeting, Gainesville, Florida, Abstract 1977.

Mickevich, M. F., 1982, Transformation series analysis, *Syst. Zool.* **31**:169–176.
Mickevich, M. F., and Johnson, M. S., 1976, Congruence between morphological and allozyme data in evolutionary inference and character evolution, *Syst. Zool.* **25**:260–270.
Mickevich, M. F., and Mitter, C., 1981, Treating polymorphic characters in systematics: A phylogenetic treatment of electrophoretic data, in: *Advances in Cladistics* (V. A. Funk and D. R. Brooks, eds.), The New York Botanical Garden, Bronx, New York, pp. 45–58.
Mickevich, M. F., and Mitter, C., 1983, Evolutionary patterns in allozyme data: A systematic approach, in: *Advances in Cladistics II* (N. I. Platnick and V. A. Funk, eds.), Columbia University Press, New York, pp. 169–176.
Miller, R. J., 1968, Speciation in the common shiner: An alternate view, *Copeia* **1968**:640–647.
Nayyar, R. P., 1964, Karyotype studies in seven species of Cyprinidae, *Genetica* **35**:95–104.
Nei, M., 1972, Genetic distance between populations, *Am. Nat.* **106**:283–292.
Nei, M., 1978, Estimation of average heterozygosity and genetic distance from a small number of individuals, *Genetics* **89**:583–590.
Nelson, J. S., 1976, *Fishes of the World,* Wiley, New York.
Nygren, A., Andreasson, J., Jonsson, L., and Jahnke, G., 1975, Cytological studies in Cyprinidae (Pisces), *Hereditas* **81**:165–172.
Ohno, S., 1970, *Evolution by Gene Duplication,* Springer-Verlag, New York.
Ohno, S., Muramoto, J., and Christian, L., 1967, Diploid–tetraploid relationship among Old-World members of the fish family Cyprinidae, *Chromosoma* **23**:1–9.
Ohno, S., Wolf, U., and Atkin, N. B., 1968, Evolution from fish to mammals by gene duplication, *Hereditas* **59**:169–187.
Ojima, Y., Hayashi, M., and Ueno, K., 1972, Cytogenetic studies in lower vertebrates. X. Karyotype and DNA studies in fifteen species of Japanese Cyprinidae, *Jpn. J. Genet.* **47**:431–440.
Op't Hof, J., Wolf, U., and Krone, W., 1969, Studies on isozymes of sorbitol dehydrogenase in some vertebrate species, *Humangenetik* **8**:178–182.
Park, E. H., 1974, A list of the chromosome numbers of fishes, *College Rev. College Liberal Arts Sci. Seoul Nat. Univ.* **20**:346–372.
Piront, A., and Gosselin-Rey, C., 1974, Immunological cross-reactions among Cyprinidae parvalbumins, *Biochem. Syst. Ecol.* **2**:103–107.
Rainboth, W. J., and Whitt, G. S., 1974, Analysis of evolutionary relationships among shiners of the subgenus *Luxilus* (Teleostei, Cypriniformes, *Notropis*) with the lactate dehydrogenase and malate dehydrogenase isozyme systems, *Comp. Biochem. Physiol.* **49B**:241–252.
Rakocinski, C. F., 1980, Hybridization and introgression between *Campostoma oligolepis* and *C. anomalum pullum* (Cypriniformes: Cyprinidae), *Copeia* **1980**:584–594.
Riska, B., 1979, Character variability and evolutionary rate in *Menidia, Evolution* **33**:1001–1004.
Rogers, J. S., 1972, Measures of genetic similarity and genetic distance, University of Texas Publication 7213, pp. 145–153.
Sackler, M. L., 1966, Xanthine oxidase from liver and duodenum of the rat: Histochemical localization and electrophoretic heterogeneity, *J. Histochem. Cytochem.* **14**:326–333.
Schmidtke, J., and Engel, W., 1974, On the problem of regional gene duplication in diploid fish of the orders Ostariophysi and Isospondyli, *Humangenetik* **21**:39–45.
Schmidtke, J., and Engel, W., 1976, Gene action in fish of tetraploid origin, III. Ribosomal DNA amount in cyprinid fish, *Biochem. Genet.* **14**:19–26.
Schmidtke, J., Schulte, B., Kuhl, P., and Engel, W., 1976, Gene action in fish of tetraploid origin. V. Cellular RNA and protein content and enzyme activities in cyprinid, clupeoid, and salmonoid species, *Biochem. Genet.* **14**:975–980.
Schwartz, F. J., 1972, World literature to fish hybrids, with an analysis by family, species,

and hybrid, Gulf Coast Research Laboratory Museum, Ocean Springs, Mississippi, Publication No. 3.

Selander, R. K., 1976, Genic variation in natural populations, in: *Molecular Evolution* (F. J. Ayala, ed.), Sinauer, Sunderland, Massachusetts, pp. 21–45.

Selander, R. K., Smith, M. H., Yang, S. Y., Johnson, W. E., and Gentry, J. B., 1971, IV. Biochemical polymorphism and systematics in the genus *Peromyscus*. I. Variation in the old-field mouse (*Peromyscus polionotus*), in: Studies in Genetics VI, University of Texas Publication 7103, pp. 49–90.

Sofradzija, A., and Berberovic, L., 1973, The chromosome number of *Barbus meridionalis petenyi* Heckel (Cyprinidae, Pisces), *Bull. Sci. Acad. RSF Yougosl.* **18**:77–78.

Stevens, P. F., 1980, Evolutionary polarity of character states, *Annu. Rev. Ecol. Syst.* **11**:333–358.

Suzuki, A., and Taki, Y., 1981, Karyotype of tetraploid origin in a tropical Asian cyprinid, *Acrossocheilus sumatranus*, *Jpn. J. Ichthyol.* **28**:173–176.

Swofford, D. L., 1981, On the utility of the distance Wagner procedure, in: *Advances in Cladistics* (V. A. Funk and D. R. Brooks, eds.), The New York Botanical Garden, Bronx, New York, pp. 25–43.

Takahata, N., 1982, The disappearance of duplicate gene expression, in: *Molecular Evolution, Protein Polymorphism and the Neutral Theory* (M. Kimura, ed.), Japan Scientific Societies Press, Tokyo, pp. 169–190.

Takahata, N., and Maruyama, T., 1979, Polymorphism and loss of duplicate gene expression: A theoretical study with application to tetraploid fish, *Proc. Natl. Acad. Sci. USA* **76**:4521–4525.

Taki, Y., and Suzuki, A., 1977, A comparative chromosome study of *Puntius* (Cyprinidae: Pisces). II, *Proc. Jpn. Acad.* **53**:282–286.

Taki, Y., Urushido, T., Suzuki, A., and Serizawa, C., 1977, A comparative chromosome study of *Puntius* (Cyprinidae: Pisces). I, *Proc. Jpn. Acad.* **53**:231–235.

Triantaphyllidis, C. D., Damianakis, H., Economidis, P. S., and Karakousis, J., 1981, Genetic variation in Greek barbel populations—I. Esterases, LDH, MDH, ME and PGM in *Barbus meridionalis* (Pisces, Cyprinidae), *Comp. Biochem. Physiol.* **70B**:278–293.

Waines, J. G., 1976, A model for the origin of diploidizing mechanisms in polyploid species, *Am. Nat.* **110**:415–430.

Watrous, L. E., and Wheeler, Q. D., 1981, The out-group comparison method of character analysis, *Syst. Zool.* **30**:1–11.

White, M. J. D., 1978, *Modes of Speciation*, Freeman, San Francisco.

Whitt, G. S., 1970, Developmental genetics of the lactate dehydrogenase isozymes of fish, *J. Exp. Zool.* **175**:1–36.

Wilson, F. R., Whitt, G. S., and Prosser, C. L., 1973, Lactate dehydrogenase and malate dehydrogenase isozyme patterns in tissues of temperature acclimated goldfish (*Carassius auratus*), *Comp. Biochem. Physiol.* **46B**:105–116.

Wolf, U., Ritter, H., Atkin, N., and Ohno, S., 1969, Polyploidization in the fish family Cyprinidae, order Cypriniformes, I. DNA content and chromosome set in various species of Cyprinidae, *Humangenetik* **7**:240–244.

Wolf, U., Engel, W., and Faust, J., 1970, Zum mechanismus der diploidisierung in der wirbeltierevolution: Koexistenz von tetrasomen und disomen genloci der isocitrat-dehydrogenasen bei der regenbogenforelle (*Salmo irideus*), *Humangenetik* **9**:150–156.

Zimmerman, E. G., and Richmond, M. C., 1981, Increased heterozygosity at the MDH-B locus in fish inhabiting a rapidly fluctuating thermal environment, *Trans. Am. Fish. Soc.* **110**:410–416.

Zimmerman, E. G., Merritt, R. L., and Wooten, M. C., 1980, Genetic variation and ecology of stoneroller minnows, *Biochem. Syst. Ecol.* **8**:447–453.

CHAPTER 12

Descriptive Genetics of Cichlid Fishes

IRV KORNFIELD

1. Introduction

The evolutionary biology of cichlids is unusual principally because of the extensive adaptive radiations that have occurred in many endemic complexes (Futuyma, 1979). Within the major lakes of East Africa (Fryer and Iles, 1972) and smaller lacustrine systems in both Old and New Worlds (Taylor and Minckley, 1966; Trewavas *et al.*, 1972; Barlow, 1976; Taylor and Miller, 1982), endemic species display dramatic morphological and ecological adaptations. Prodigious numbers of endemic species occur in Lake Malawi (*N* = 500+; McKaye and MacKenzie, 1982; McKaye, personal communication), Lake Tanganyika (*N* = 150+; Bailey and Stewart, 1977), and Lake Victoria (*N* = 300+; Van Oijen *et al.*, 1981). The apparent youth of many species in some of these systems presents at least two fundamental evolutionary questions: first, what mechanisms control divergence in ecology and functional morphology? and second, how do new species arise? While substantial insights have been made in the areas of functional anatomy and ecology by Barel *et al.* (1977), Greenwood (1981), Liem (1980), McKaye (1980), and their coworkers, the basic questions of trophic divergence and speciation have remained controversial (Sage and Selander, 1975; McKaye, 1980; Kornfield *et al.*,1982; McKaye *et al.*, 1983; Trewavas, 1982; Dominey, 1984; Greenwood, 1984).

The Cichlidae is a monophyletic group (Stiassny, 1981) of approximately 1200 species distributed over the entire African continent and much

IRV KORNFIELD • Department of Zoology and Migratory Fish Research Institute, University of Maine, Orono, Maine 04469.

of Central and South America. Additional endemic elements occur in India, Madagascar, and Sri Lanka. A tentative phylogeny has been presented by Chichoki (1976). Intentional introductions, principally of Old World tilapine species, have occurred globally. It is difficult to geologically date the separation of Old and New World faunas, since the ancestral species may have been euryhaline (Darlington, 1957) and thus capable of dispersal after Gondwanian fragmentation. Regardless, considerable evolutionary divergence has occurred both within and between these two major groups. With a few significant exceptions, diversification has primarily occurred in rivers in the New World and in lakes in the Old World. The general biology of many species is reviewed by Fryer and Iles (1972), Wohlfarth and Hulata (1981), and Pullin and Lowe-McConnell (1982).

The taxonomy of cichlids has recently undergone substantial revision. Generic allocations of tilapias and their allies are treated by Trewavas (1982, 1983); haplochromines have received extensive review by Greenwood (1979*b*, 1980, 1981). Additions to the fauna of Lake Tanganyika have been summarized by Bailey and Stewart (1977) and Liem (1981). Future changes in the taxonomy of African cichlids can be anticipated. Major changes have not occurred in New World systems, but a comprehensive treatment of certain elements, particularly the speciose genus *Cichlasoma,* has been warranted for some time. To stimulate thought on the evolutionary biology of cichlids, genetic characterization of the family is presented in the sections that follow. The pertinent literature is reviewed with the incorporation of additional new information. General statements about genomic evolution within the Cichlidae can be made. However, the absence of genetic information about inheritance and phenotypic plasticity, as well as gene flow and population structure, constrain the construction of models of adaptive radiation. Clearly, new genetic studies are needed. A general consideration of speciation and diversification in the family will appear elsewhere.

2. Genome Size

Quantitative estimates of nuclear DNA can be used to index major genomic modifications among related taxa. Given a sensitive assay, polyploidy or substantial deletions or duplications of genetic material can be detected. Within teleost families, haploid DNA content (*C* value) is generally conservative and does not reflect fine-scale genetic changes; homogeneous *C* values may be associated with significant karyotypic heterogeneity and *vice versa* (Park and Kang, 1976). While measurement variations associated with some assay techniques can significantly affect

C values, cytofluorometric analysis of propidium-stained nuclei (Krishan, 1975; Thorgaard *et al.*, 1982; Allen and Stanley, 1983) yields rapid and accurate estimates.

As with many other percoids, published *C* values for cichlids are relatively homogeneous among species, ranging, with one exception, from 1.0 to 1.4 pg (Hinegardner and Rosen, 1972). In the discus, *Symphysodon aequifasciata,* the *C* value is about 1.5 pg, which is 40% greater than typical New World cichlids (Ohno and Atkin, 1966). The high *C* value corresponds to a higher number of chromosomes ($2n = 60$). Thompson (1976) suggested that the karyotype might have been derived by polyploidy or a major duplication event, as opposed to simple fission of ancestral chromosomes.

Related genera of the Old World *Tilapia* lineage exhibit heterogeneous *C* values that parallel heterogeneity in chromosome morphology (Kornfield *et al.*, 1979). However, patterns of *C* values are not correlated with the level of phylogenetic divergence. A small but significant difference in DNA content occurs between species with equivalent karyotypes (*Oreochromis aurus* and *Sarotherodon galilaeus*).

Few DNA values for taxa within the African rift lake complexes have been characterized. The *C* value for six species of the endemic rock-dwelling complex of Lake Malawi (Mbuna), representing the genera *Iodotropheus, Labeotropheus, Melanochromis,* and *Pseudotropheus,* were examined by cytofluorometry. Relative DNA content was homogeneous among species and was similar to estimates for other common New and Old World taxa (Fig. 1). Within this very small sample of Malawi endemics, there is no indication that major genomic modifications have been associated with either speciation or subsequent phyletic divergence. The situation in other species and other lakes remains to be investigated.

3. Electrophoretic Characterization

The characterization of enzymes by electrophoresis has been used to examine species integrity and evolutionary relationships. Despite the inherent interest in the Cichlidae, relatively few electrophoretic studies have been conducted on this group. Various species of tilapias have been examined for a few specific blood proteins (Malecha and Ashton, 1968; Chen, 1969; Chen and Tsuynki, 1970; Avtalion and Wojdani, 1971; Badawi, 1971; Scopes and Hamoir, 1971; Fryer and Iles, 1972; Baron, 1975; Avtalion *et al.*, 1975, 1976; Avtalion and Mires, 1976). This work has demonstrated some species-specific alleles as well as a number of poly-

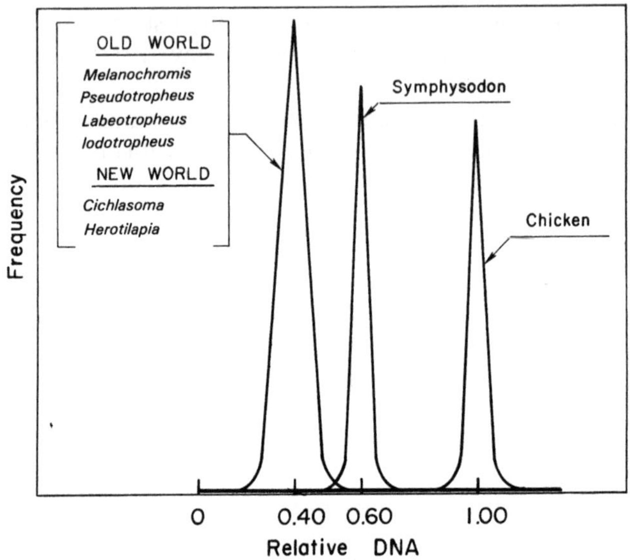

Figure 1. Relative amount of DNA from diploid erythrocyte nuclei of New and Old World cichlids assayed by cytofluorometry. Chicken erythrocytes were used as control (*C* value = 2.6 pg).

morphisms. Because of the global aquaculture of tilapine species (Wohlfarth and Hulata, 1981; Avtalion, 1982), these markers are of considerable value (and are perhaps essential) for stock identification and manipulation. Cruz *et al.* (1982) and McAndrew and Majumdar (1983) have recently summarized an electrophoretic protocol for use with cichlids. Of interest is the demonstration that electrophoretic phenotypes can be scored from surface mucus in cichlids (Herzberg, 1978) or from fin clips (Kornfield, unpublished). Though enzyme activity in those materials is apparently more labile than that in other tissues, ease of collection makes them of considerable value in controlled breeding settings. Aside from a few evolutionary studies noted below, additional electrophoretic characterizations have been limited to preliminary studies of LDH and MDH in five species from Lake Victoria (Basasibwaki, 1975) and LDH in 14 taxa from the New World (Scholl and Holzberg, 1972).

The levels of electrophoretic variation in cichlids are equivalent to those found in many other teleosts (Nevo, 1978); similar levels of polymorphism and heterozygosity are displayed among species from the New and Old World (Table I). Cichlids are often variable for those enzymes that are polymorphic in other fishes, i.e., amniopeptidase, esterase, aconitase, lactate dehydrogenase, phosphoglucoisomerase, and phosphoglucomutase. These enzymes are of particular interest in evolutionary studies, since they are expected to diverge more rapidly than monomorphic proteins (Skibinski and Ward, 1981).

Most characterizations of gene frequencies for specific enzymes are straightforward, but two potential dangers should be noted. First, some esterases possess null alleles, which can significantly distort frequency estimates. Null homozygotes have been observed in two endemic species from Lake Malawi (Kornfield, 1974). Further, expression of some esterases may be developmentally dependent in cichlids (Galman and Carino, 1979). Second, substantial frequency differences may occur between sexes. In *Oreochromis saka,* an abundant Malawi endemic, frequency differences between males and females occur at four enzyme loci, one of which is statistically significant (Kornfield, 1974). Since alleles at these loci are expressed as codominants and heterozygotes are observed in both sexes, the loci are autosomal. The reason for these sex-related differences is speculative, but males and females may be exposed to very different environments when they segregate geographically following spawning (Lowe, 1953). Regardless, caution should be exercised in characterizing gene frequencies.

Similar to comparisons within other vertebrate families, genetic differentiation among cichlids varies enormously; species within some genera are very different, whereas in other genera, species are nearly identical. In the Mbuna from Lake Malawi, haplochromines from Lake Victoria, and *Sarotherodon* of Lake Kinneret, genetic similarities (I_N; Nei, 1972) often exceed 0.90 (Kornfield, 1978; Kornfield *et al.,* 1979; Sage *et al.,* 1984; Kornfield, unpublished; McKaye and Kocher, unpublished). High similarities have also been noted among some New World *Cichlasoma* (Fig. 2). Equivalent similarities have occasionally been reported for other teleosts (Avise *et al.,* 1975; Johnson, 1975), but, in general, more intermediate levels of differentiation (I_N = 0.4–0.8) are typical for cichlids

Table I
Levels of Genetic Variability in Cichlids[a]

	Old World	New World
Number of taxa	8	4
Average number of loci	15	23
Average heterozygosity	0.073	0.054
	(0.044–0.121)	(0.00–0.64)
Average polymorphism (0.05)	0.171	0.141
	(0.077–0.267)	(0.00–0.82)

[a]Data from Kornfield (1979; unpublished) and Sage and Selander (1975). Ranges in parentheses.

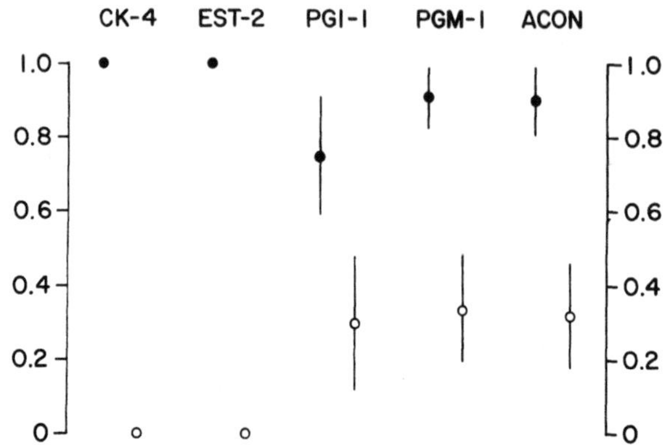

Figure 2. Gene frequencies at five diagnostic enzyme loci in two cichlids, *C. bartoni* (●) and *C. labridens* (○), from Laguna de la Media Luna, San Luis Potosi, Mexico. A systematic treatment of these taxa appears in Taylor and Miller (1983). Species had equal mobilities for an additional 16 monomorphic loci. Genetic similarity between species I_N equals 0.89.

(Fig. 3) and other fishes (Thorpe, 1979). Such intermediate similarities have been noted among endemic Tanganyikan genera (Kornfield, unpublished).

If electrophoretic differences between taxa are related to the duration of reproductive isolation (Wilson *et al.*, 1977), then divergence among some Old World cichlids must have been particularly rapid. An alternative view, that high similarities might reflect deceleration of protein evolution (Avise *et al.*, 1980), has been criticized (Thorpe, 1982). The evidence for rapid differentiation is compelling for some Malawi endemics. In *Pseudotropheus zebra,* sympatric color morphs exhibit complete positive assortative mating (Holzberg, 1978; Schroder, 1980) and are thus biological species. These sibling species are not fixed for alternate alleles, but may be distinguished by differences in gene frequencies at polymorphic enzyme loci (Fig. 4; McKaye *et al.*, 1983). Genetic similarity between the two assortatively mating color group morphs is about 0.97. Color morphs of *Petrotilapia tridentiger,* which apparently do not interbreed in the field (Marsh *et al.*, 1981), also differ in gene frequencies (McKaye *et al.*, 1982). Compared to other teleosts, these endemic Malawi taxa appear to have diverged at atypical rates.

The existence of reproductive isolation may be inferred for sympatric taxa that differ significantly in gene frequencies, but the reverse may not be true; recently isolated cichlids may not have yet diverged. In *Pseudotropheus* sp., a seasonally abundant undescribed cichlid of the Mbuna complex, allele frequencies at three polymorphic loci were identical to

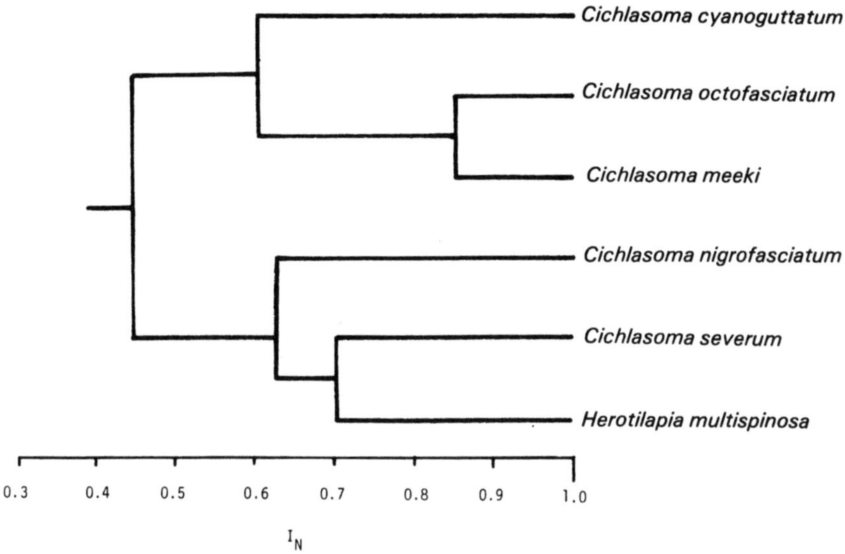

Figure 3. UPGMA phenogram depicting the relative genetic similarities among six New World cichlids. Species were assayed at 15 electrophoretic loci. Cophenetic correlation equals 0.89.

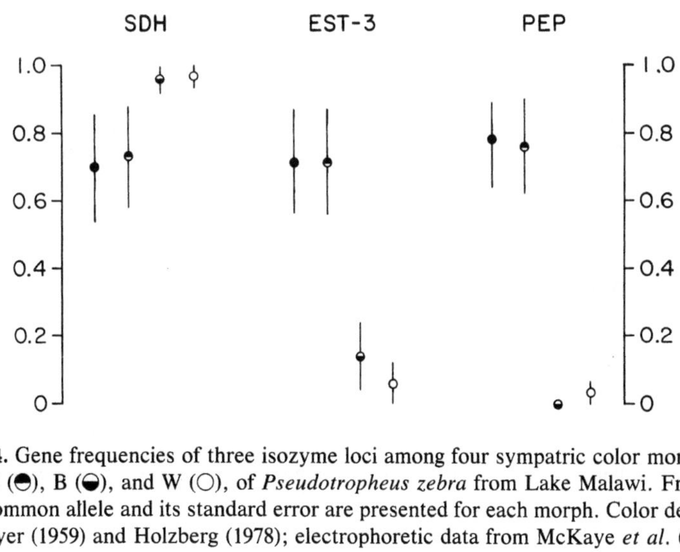

Figure 4. Gene frequencies of three isozyme loci among four sympatric color morphs, BB (●), OB (◓), B (◒), and W (○), of *Pseudotropheus zebra* from Lake Malawi. Frequency of the common allele and its standard error are presented for each morph. Color definitions from Fryer (1959) and Holzberg (1978); electrophoretic data from McKaye *et al.* (1983).

those of a sympatric congener, *P. elegans* (Fig. 5; Kornfield, 1974). These congeners differ significantly in morphology and reproductive coloration, but these differences are not sufficient to allow one to recognize isolation of gene pools. Even greater levels of morphological differentiation occur among sympatric trophic morphs of the polymorphic cichlid, *Cichlasoma minckleyi* in Cuatro Cienegas, Mexico (Sage and Selander, 1975; Kornfield *et al.*, 1982; Kornfield and Taylor, 1983; Liem and Kaufman, 1984).

Future electrophoretic studies will be of value for determining relationships among endemic complexes, detecting the presence of some sibling species, and identifying patterns of spatial variation. However, the technique will probably not be sensitive enough to permit unambiguous confirmation of gene pool isolation among some endemic taxa. Field studies of reproductive behavior, particularly location of spawning sites and determination of breeding seasons, will need to be pursued.

4. Chromosomes

The chromosomal characterization of cichlids is of interest particularly with respect to the idea that alterations may be associated with speciation (Kaina and Rieger, 1979; White, 1978). One mechanism postulated for the development of reproductive isolation involves fixation of a new arrangement in a population followed by strong selection against F_1 individuals with heterozygous karyotypes. Initial postmating isolation

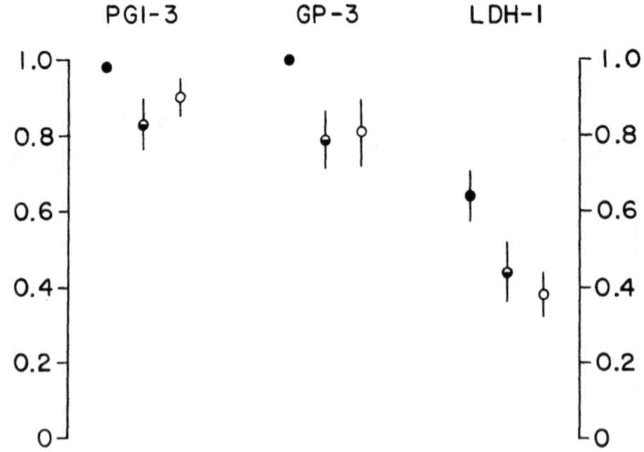

Figure 5. Parallel gene frequency variations among three endemics, *P. livingstoni* (●), *P. elegans* (◒), and *P. elegans* sp. (○), from Lake Malawi. Taxa had equal mobilities for nine additional monomorphic loci. (Data from Kornfield, 1974.)

due to aberrant segregation at meiosis is presumably reinforced by the development of premating isolation mechanisms. Several authors have recently stressed, however, the improbability of such a speciation scenario on theoretical and experimental grounds (Futuyma and Mayer, 1980; Templeton, 1981; Charlesworth *et al.*, 1982). Despite these objections, reproduction isolation via negative heterosis could develop, but only under very restrictive conditions (Walsh, 1982).

The observation that related species have chromosomal differences might imply a functional link between chromosomal modifications (whatever the cause) and development of reproductive isolation (White, 1978). This view has been advanced particularly with respect to differing rates of diversification and chromosomal alteration among some groups of mammals (Wilson *et al.*, 1974, 1975, 1977; Bush *et al.*, 1977; Bengtsson, 1980). However, such associations appear to be absent in other vertebrates (Patton and Baker, 1978; Sites *et al.*, 1981; Shields, 1982), so the relationship is not general. In fishes, karyotypic evolution is generally conservative (Wilson *et al.*, 1975). In several taxa, distant and closely related species often both have identical gross karyotypes (Gold, 1979). In cyprinids and centrarchids there is no association between karyotypic evolution and rate of speciation (Avise and Gold, 1977; Gold, 1980).

About 70 species of cichlids, mostly tilapias and New World genera, have been karyotyped (Table II). Excluding the discus, diploid chromosome number ($2n$) ranges from 32 in *Tilapia macrocephala* to 52 in *Cichlasoma salvini*. The distribution of chromosome numbers in the family is bimodal, presenting a clear difference between New World ($2n = 48$) and Old World ($2n = 44$) species (Fig. 6). Despite the variation, karyotypic evolution in cichlids is relatively conservative, with only minor rearrangements necessary for change in diploid number and morphology during phyletic evolution.

Some variation in chromosome morphology exists among taxa within the few speciose genera that have been studied. In *Cichlasoma*, the typical karyotype ($2n = 48$) consists of four pairs of metacentric/submetacentric chromosomes plus a series of progressively smaller acrocentrics; the arm number (NF) is 54. While the diploid number is fairly stable among species in the genus, variation in arm number is considerable, ranging from 52 to 104. Unfortunately, much of this variation may be artifictitious. Despite the use of standard definitions for chromosomes morphology (Levan *et al.*, 1964), discrepancies are typical for independent studies of the same species. For example, in *C. cyanoguttatum*, Thompson (1979) reported NF = 54, while Zahner (1977) found NF = 94. This difference in characterization probably reflects differential contraction of chromosomes with varied fixation and preparation protocols. Comparisons of chromosome

Table II
Chromosomal Characterizations of the Cichlidae

Species	2*n*	NF	Reference
Acarichthys heckelii	48	54	Thompson (1979)
Aequidens maronii	50	100	Zahner (1977)
Aequidens metae	48	54	Thompson (1979)
Aequidens paraguayensis	44	70	Thompson (1979)
Apistogramma agassizi	46	70	Thompson (1979)
Apistogramma borellii	38	60	Thompson (1979)
Apistogramma ornatipinnis	46	92	Zahner (1977)
Apistogramma ortmanii	46	70	Thompson (1979)
Apistogramma pertense	48	—	Post (1965)
Astatotilapia burtoni	40	—	Thompson (1981)
Astatotilapia flaviijosephi[a]	44	—	Kornfield *et al.* (1979)
Astronotus ocellatus	48	96	Zahner (1977)
	48	54	Thompson (1979)
Cichla temensis	48	48	Thompson (1979)
Cichlasoma beani	48	54	Thompson (1979)
Cichlasoma bimaculatum	48	54	Thompson (1979)
Cichlasoma centrarchus	48	54	Thompson (1979)
Cichlasoma citrinellum	48	—	Nishikawa *et al.* (1973)
	48	96	Zahner (1977)
	48	56	Thomson (1979)
Cichlasoma coryphaenoides	48	54	Thompson (1979)
Cichlasoma cyanoguttatum	48	94	Zahner (1977)
	48	54	Thompson (1979)
Cichlasoma dowi	48	56	Thompson (1979)
Cichlasoma facetum	48	—	Oyhenart-Perera *et al.* (1975)
Cichlasoma festivum	48	96	Zahner (1977)
	48	56	Thompson (1979)
Cichlasoma kraussi	50	56	Thompson (1979)
Cichlasoma labridens	48	54	Thompson (1979)
Cichlasoma managuense	48	96	Zahner (1977)
	48	54	Thompson (1979)
Cichlasoma meeki	48	—	Hinegardner and Rosen (1972)
	48	96	Zahner (1977)
	48	54	Thompson (1979)
Cichlasoma minckleyi[b]	48	54	Thompson (1979)
Cichlasoma nigrofasciatum	48	96	Zahner (1977)
	48	52	Thompson (1979)
Cichlasoma octofasciatum	48	96	Zahner (1977)
	48	54	Thompson (1979)
Cichlasoma salvini	52	104	Zahner (1977)
	52	80	Thompson (1979)
Cichlasoma septemfasciatum	48	54	Thompson (1979)
Cichlasoma severum	48	—	Post (1965)
	48	96	Zahner (1977)
	48	52	Thompson (1979)

Table II (*Continued*)

Species	2*n*	NF	Reference
Cichlasoma trimaculatum	48	54	Thompson (1979)
Crenicara filamentosa	46	58	Thompson (1979)
Crenicichla lepidota	48	54	Thompson (1979)
Crenicichla lucius	48	—	Thompson (1979)
Crenicichla notophthalma	48	54	Thompson (1979)
Crenicichla saxatilis	48	—	Oyhenart-Perera *et al.* (1975)
Crenicichla strigata	48	54	Thompson (1979)
Etroplus maculatus	46	—	Natarajan and Subrahmanyam (1974)
Etroplus suratensis	48	—	Natarajan and Subrahmanyam (1974)
Geophagus brasiliensis	48	—	Michele and Takahashi (1977)
	48	90–92	Zahner (1977)
	48	52	Thompson (1979)
Geophagus jurupari	48	52	Thompson (1979)
Geophagus surinamensis	48	52	Thompson (1979)
Hemichromis bimaculatus	44	—	Post (1965)
	44	88	Zahner (1977)
Herotilapia multispinosa	48	96	Zahner (1977)
	48	54	Thompson (1979)
Lamprologus leleupi	48	—	Post (1965)
Melanochromis auratus	46	56–58	Thompson (1981)
Nannacara anomala	48	—	Post (1965)
	44	62	Thompson (1979)
Neetroplus nematopus	48	56	Thompson (1979)
Oreochromis alcalicus[c]	48	—	Post (1965)
Oreochromis andersonii[d]	44	48	Vervoort (1980)
Oreochromis aurus[d]	44	54	Kornfield *et al.* (1979)
	44	44–50	Thompson (1981)
Oreochromis macrochir[d]	44	—	Jalabert *et al.* (1971)
	44	48	Vervoort (1980)
Oreochromis mossambicus[d]	44	—	Natarajan and Subrahmanyam (1968)
	44	44	Fukuoka and Muramoto (1975)
	44	—	Prasad and Manna (1976)
	44	44–50	Thompson (1981)
Oreochromis niloticus[d]	44	—	Chervinski (1964)
	44	—	Jalabert *et al.* (1971)
	40	—	Badr and El Dib (1976)
	44	62	Arai and Koike (1980)
Papiliochromis ramirezi[e]	48	—	Post (1965)
Pelvicachromis pulcher[f]	48	—	Post (1965)
Pseudocrenilabrus multicolor[g]	44	—	Post (1965)
Pterophylum scalare	48	—	Post (1965)
	48	52	Thompson (1979)
Sarotherodon galilaeus[h]	44	—	Badr and El-Dib (1976)
	44	—	Badr and El-Dib (1977)
	44	54	Kornfield *et al.* (1979)
	44	50	Vervoort (1980)

(*continued*)

Table II (*Continued*)

Species	2*n*	NF	Reference
Symphysodon aequifasciata	60	—	Ohno and Atkin (1966)
	60	118	Thompson (1979)
Tilapia congica	44	54	Vervoort (1980)
Tilapia guineensis	44	52	Vervoort (1980)
Tilapia macrocephala	32	—	Jakowska (1950)
Tilapia mariae	40	44	Thompson (1976)
	40	44–48	Thompson (1981)
	40	44	Vervoort (1980)
Tilapia rendalli	44	52	Michele and Takahashi (1977)
Tilapia sparrmanii	42	46–50	Thompson (1981)
	42	50	Vervoort (1980)
Tilapia zillii	38	—	Badr and El-Dib (1977)
	44	54	Kornfield *et al.* (1979)
Tristramella sacra[i]	44	—	Kornfield *et al.* (1979)
Tristramella simonis[i]	44	—	Kornfield *et al.* (1979)
Uaru amphiacanthoides	46	54	Thompson (1979)

[a]Formerly included in the genus *Haplochromis*.
[b]Known previously as *Cichlasoma* sp.
[c]Formerly known as *Tilapia grahami*.
[d]Formerly included in the genera *Sarotherodon* and *Tilapia*.
[e]Formerly included in the genus *Apistogramma*.
[f]Formerly known as *Pelmatochromis kribensis*.
[g]Formerly included in the genera *Haplochromis* and *Hemihaplochromis*.
[h]Formerly included in the genus *Tilapia*.
[i]Known incorrectly as *Tristamella*.

morphologies, but not diploid numbers, are thus best made on results obtained from single laboratories using highly standardized techniques (see also Arai, 1982; Ihssen *et al.*, 1981).

Differences in NF occur among many of the species of *Cichlasoma* studied by Thompson (1979), but they are restricted in most cases to changes in the morphology of only a few chromosomes. Thus, the karyotype of *C. nigrofaciatum* with four metacentric pairs (NF = 52) can be derived from the ancestral karyotype by a single inversion in one metacentric/submetacentric and unequal translocation between the short arms of the other two msm chromosomes (Thompson, 1976). One karyotypic attribute of *Cichlasoma* is a significant negative correlation between arm ratios of individual chromosomes (*r*) and chromosome size (as percent of total complement length); larger chromosomes tend to be more metacentric. This association may reflect the primitive cichlid karyotype. However, if the ancestral karyotype consisted of 48 uniarmed elements, as is accepted for teleosts in general (Ohno, 1970; Gold, 1979), karyotypic evolution in New World species may primarily involve addition of genetic material to the short arms of individual chromosomes.

In *Cichlasoma,* phyletic divergence is probably unrelated to karyotypic divergence, at least at the gross level. Consider the taxa presented in Fig. 3. The relative duration of temporal isolation among these reproductively isolated species was indexed by electrophoresis and is summarized as a simple UPGMA phenogram. Included is *Herotilapia multispinosa,* a species closely allied to *Cichlasoma* but placed in a monotypic genus to recognize its distinctive oral dentition. The gross karyotype of this species is indistinguishable from those of *Cichlasoma cyanoguttatum, C. meeki,* and *C. octofasciatum,* taxa with which it shows a relatively high level of electrophoretic divergence ($I_N = 0.45$). Similarly, a variety of other *Cichlasoma* species representing different species groups, or perhaps genera, exhibit equivalent gross karyotypes (Thompson, 1979).

The karyotypic differences noted among suprageneric groups in New World cichlids (Thompson, 1979) should be treated with caution. In particular, determination of character state polarity from small alterations in chromosomal morphology (Arai and Yamamoto, 1981) is not advisable.

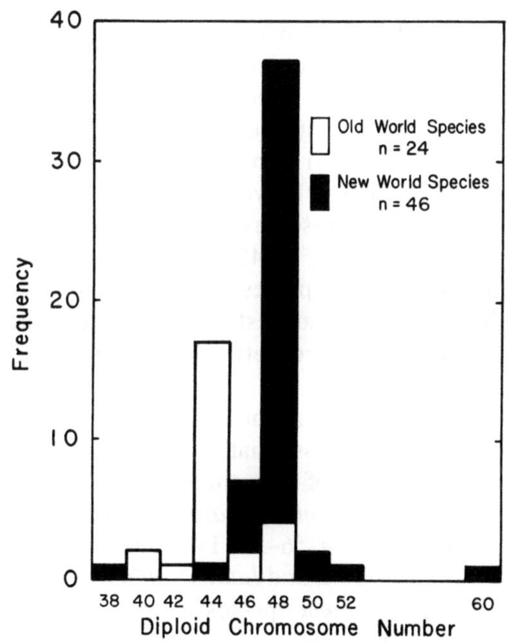

Figure 6. Frequency distribution of published chromosome numbers for 70 species of cichlid fishes.

The karyotype of *Cichla temensis*, a supposedly primitive New World species (Regan, 1906; Liem, 1974), is of interest in this regard. Thompson (1979) noted that the diploid chromosomal complement of this species appeared to have evolved with little detectable change from the presumptive primitive state. However, recent morphological analysis (Stiassny, 1982) suggests that *Cichla* secondarily acquired "primitive" anatomical features and should be more appropriately regarded as a highly specialized taxon. Clearly, phylogenetic inferences based on chromosome morphology alone may be misleading.

In Old World cichlids, the tilapias present a similar picture of relatively conservative karyotypic evolution, particularly in diploid chromosome number (Table II). Species pairs with high electrophoretic similarities may have identical gross karyotypes (Kornfield *et al.*, 1979; Vervoort, 1980). All tilapias examined are characterized by the possession of one very large pair of acrocentric chromosomes. While this pair is probably homologous among species in the lineage, it is unclear whether this marker occurs in other Old World lineages, particularly haplochromines (Thompson, 1981). Despite intrageneric karyotypic conservation, there are subtle differences in chromosome morphology. For example, while their gross karyotypes appear identical, *Tilapia zillii* differs from *Sarotherodon galilaeus* in the distribution and occurrence of C-bands (Kornfield *et al.*, 1979). Chromosomal differentiation in the tilapine lineage is not characteristic of closely related species and thus cannot be associated with speciation or genetic divergence.

The endemic cichlids of Sri Lanka, *Etroplus maculatus* and *E. suratensis*, present quite dissimilar karyotypes, with differences in both the number of metacentric chromosomes as well as simple diploid number (Natarajan and Subrahmanyam, 1974). Significant differences in ecology and breeding biology of these species (Ward and Wyman, 1977) as well as in morphological characters suggest substantial temporal isolation. As expected, the genetic divergence between species is substantial (I_N = 0.50; Kornfield, unpublished).

The chromosome complements of haplochromines from the vast East African lake assemblages are essentially unknown; a systematic survey would be useful. Karyotypes of only three species have been reported. A Malawi endemic, *Melanochromis auratus*, characterized by Thompson (1981), had $2n = 46$ and NF = 56–58. Two other taxa from this Mbuna complex, *Pseudotropheus zebra* and *Labeotropheus fulleborni*, have been recently examined (Fig. 7). These species are very closely related to each other ($I_N = 0.95$) and to *M. auratus* ($I_N = 0.96$) (Kornfield, 1978). While the karyotypes of *P. zebra* and *L. fulleborni* are indistinguishable from each other, they differ from *M. auratus* in possessing two fewer subtel-

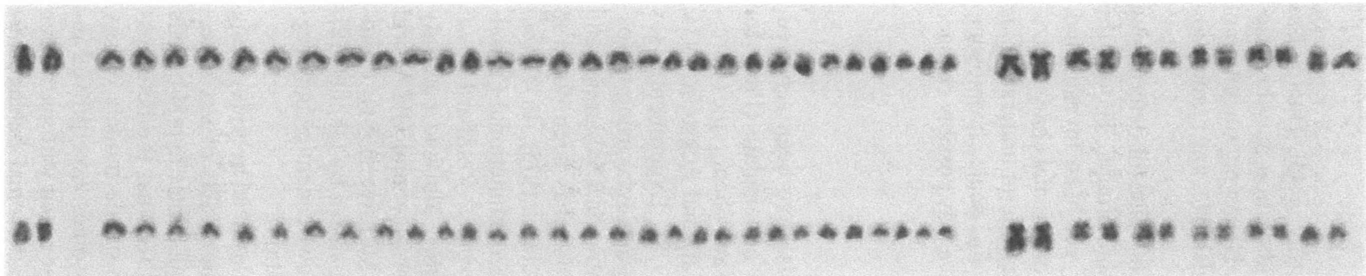

Figure 7. Karyotypes of two endemic cichlids from Lake Malawi. (Top) *Pseudotropheus zebra* (BB morph); (bottom) *Labeotropheus fulleborni*. Giemsa-stained chromosomes from gill epithelium are arranged after Kornfield *et al.* (1979).

ocentric chromosomes in the diploid complement ($2N = 44$). Thus, if this difference in diploid number is real, karyotypic alteration may be associated with diversification of some endemic taxa. However, since two of the species involved are chromosomally indistinguishable, any association will not be absolute.

Future karyotypic studies on cichlids may be of value for determining relationships among major taxonomic or geographic groups. Though the endemic African radiations have effectively not yet been investigated, with the exception of the Tanganyikan fauna, chromosomal characterization will most probably contribute minimally to a general understanding of cichlid evolution. When reliable techniques for fine-scale chromosome banding become available for teleosts, this perspective may possibly change.

5. Sex Markers

Numerous studies concerning sex in cichlids have been reported, with major emphasis on tilapias [see Wohlfarth and Hulata (1981) for recent review]. As for other teleosts, reproduction and phenotypic expression of sex in cichlids is hormonally mediated (Katz and Eckstein, 1974; Fernald, 1976; Terkatin-Shimony and Yaron, 1978). Experimental sex reversal to functional males or females has been achieved using testosterone and estradiol during early gonad differentiation (Nakamura and Takahashi, 1973; Guerrero, 1975). Hermaphroditism has been noted in both New and Old World taxa (Polder, 1971; Peters, 1975). In *Tilapia zillii* it had been suggested that ovatestes may be nonfunctional (Yoshikawa and Oguri, 1978), but in a hybrid *Oreochromis,* Rothbard *et al.* (1982) noted a hermaphrodite with mature gametes. Multiple paternity and cross-fostering of fry have been reported for New and Old World taxa (Hulata *et al.,* 1981; Kornfield *et al.,* 1982; McKaye, 1981). Protogynous sex reversal has been characterized in *Crenicara* by Ohm (1978). It has been suggested that some tilapias and Malawi endemics may regularly change sex (McKaye, personal communication).

Significant departures from equal sex ratios in artificial hybrids among closely related species have stimulated interest in models of sex determination. Because of the production of all-male or all-female hybrid broods in some crosses, it has been inferred (Haldane, 1922) that the heterogametic sex may differ among species. However, simple models of sex determination involving XY and WZ sex chromosomes have proved inadequate to explain certain aberrant sex ratios associated with particular hybrid crosses (e.g., Jalabert *et al.,* 1971). An autosomal balance hypothesis involving two loci (Avtalion and Hammerman, 1978; Hammerman and Avtalion, 1979; Avtalion, 1982) and a single-locus, multiple-allele hypothesis (Moav and Brody, unpublished) have satisfactorily explained most, but not all hybrid sex ratios (Wohlfarth and Hulata, 1981). Since it is possible that some of the presumably pure strains used in some breeding experiments are themselves hybrids (Avtalion, 1982, p. 277), anomalous sex ratios may be encountered despite the validity of particular models. Further, there is a possibility that *p*H of water during reproduction may influence sex ratio. In *Pelvicachromis* and *Apistogramma,* sex ratios are strongly biased toward females in basic water and toward males in acid water (Heiligenberg, 1965; Rubin, 1984.) Both activity and longevity of sperm are affected by *p*H and it is possible that the *p*H of the buccal cavity may be adjusted by special glands in oral incubators (Valenti, 1975). A fully comprehensive hypothesis of sex determination has yet to emerge, but it does seem that closely related species may differ in their complements of sex-determining alleles.

Given this evolutionary malleability of sexuality, it is not surprising that both morphologically distinct sex chromosomes and sex-linked genes have not been discovered. Of karyotyped cichlids, only *Geophagus brasilliensis* has been reported to possess distinct sex chromosomes (Michele and Tabahashi, 1977). However, while heteromorphic pairs were noted, these findings should be treated as tentative, since only one sex was examined.

Of considerable importance is the report by Mehl and Reinboth (1975) on the existence of Barr bodies in liver cells of *Astatotilapia burtoni.* In placental mammals, Barr bodies are associated with interphase nuclei of females and their presence implies the existence of genetically well-differentiated sex chromosomes (Lyon, 1974). Systematic examinations of cells from over 20 species of teleosts, including *Cichlasoma octofasciatum,* have revealed no Barr bodies (Gee and Moore, 1970; Haider, 1973; Anders *et al.,* 1981; Kornfield, unpublished). These surveys included *Salmo gairdneri* and *Xiphophorus maculatus,* two species that possess cytologically distinct sex chromosomes (Foerster and Anders, 1977; Thorgaard, 1977). It is thus highly unlikely that the genetic sex of cichlid fishes can be ascertained from cytological examination of interphase nuclei.

The lack of variation at many allozyme loci in cichlids from natural populations constrains the search for sex-linked markers. Even if high levels of heterozygosity could be induced by mutagenesis or hybridization, the probability of detecting sex linkage may remain low because sex determination may not involve descrete sex chromosomes. Avtalion *et al.* (1976) identified a serum protein (MSP) from *Oreochromis aurus, O. niloticus,* and *Sarotherodon galilaeus* which was present only in males. The absence of MSP in some individuals of *O. aurus* was successfully used by Avtalion and Mires (1976) to identify functional females that presented apparently normal male genital phenotypes. But MSP is probably not a valid genetic marker, since in *O. aurus* it is not observed in immature males (Hardin, 1976), while females display antigenic reaction to antibodies prepared against it (Avtalion, 1982). Like sex-specific serum proteins in other teleosts (e.g., Harva, 1978), it seems probable that MSP is under hormonal control (Wohlfarth and Hulata, 1981).

One possible sex marker in cichlids is the histocompatibility Y antigen, the inductor of the heterogametic gonad. The structural gene for H-Y antigen is located on the Y chromosome of mammals (Ohno, 1978) and both the H-Y antigen and an H-Y gonadal receptor (controlled at an independent locus) have been demonstrated in birds and amphibians (Wachtel *et al.*, 1975; Wachtel, 1981). In cichlids, Pechan *et al.* (1979) demonstrated H-Y antigen in males, but not females of *Haplochromis burtoni* and *Tilapia* spp. However, they express the reservation that H-Y antigen may actually be present in both sexes and might not, therefore, be used to conclusively assign heterogamety to a particular sex (see Müller and Wolf, 1979).

6. Hybridization

Artificial hybridizations have produced fertile F_1 progeny in a number of species [see references in Loiselle (1971) and Wohlfarth and Hulata (1981)]. Viable hybrids have been produced in intergeneric crosses and in crosses involving congeners where genetic distances are appreciable, e.g., *Tilapia* × *Oreochromis* and *Cichlasoma nigrofasciatum* × *C. cyanoguttatum* (see Fig. 3). In many interspecific crosses, however, significantly skewed sex ratios are observed, In the aquarist literature, viable hybrids between various species of Malawi endemics have been reported (Loiselle, 1971; Goldstein, 1973), but a systematic assessment of hybrid viability and fertility has not been made. However, the relative ease of production of artificial intergeneric hybrids suggests that reproductive isolation in nature may be primarily controlled by premating mechanisms.

Gross morphology of hybrids may be intermediate when expressed as a general hybrid index, but a sizable number of morphological characters strongly resemble only one parent (Avault and James, 1971). Unlike morphology, hybrid coloration and color patterns are subjective; nevertheless, hybrids from some crosses obviously resemble one parent while others are intermediate. In anecdotal reports, hybrids have unique colors that were not present in parents (Barel, personal communication). Unfortunately, the genetic control of coloration and morphology has not yet been examined.

There is a clear need for fundamental breeding experiments in cichlids. The variation and inheritence of morphological characteristics have not been studied in detail for any species. Very preliminary data suggest that in some taxa, ecologically important characters may be phenotypically plastic (Greenwood, 1965; Kaufman, unpublished) or polymorphic (Sage and Selander, 1975). The general occurrence of these modes of variation is speculative. The independent evolution of near identical trophic morphologies for taxa within the major endemic complexes might suggest that substantial variation was present in ancestral species, but the heritabilities of the traits of interest are not known. Similarly, the genetics of coloration and other ethological components has not yet been studied.

7. Summary

Significant adaptive radiations of cichlid fishes have occurred within warm water lakes of both the Old and New Worlds. Rapid rates of evolution within endemic faunas suggests that evolutionary processes in cichlids may differ from those of other teleosts. To begin to unravel evolution in these fishes, basic genetic characteristics of the family are examined. However, the lack of information on some fundamental genetic components, particularly the inheritance of trophic and reproductive characters, constrains current understanding.

It appears that evolution in endemic and nonendemic faunas is conservative with respect to large-scale genomic reorganization. With a few exceptions, both the quantity of nuclear DNA and chromosome complement of closely related species are relatively constant. There appears to be little relation between karyotypic diversification and phylogeny within major groups, though chromosomes may be useful for phylogenetic studies among higher level groups within the family. Isozyme studies in endemic and nonendemic faunas have demonstrated levels of heterozygosity similar to those of other teleosts. Electrophoresis appears useful for identifying sibling species, but the absence of structural gene evolution in some endemics limits the ability to define gene pool isolation. Basic breeding studies within and among species are greatly needed to understand the mechanisms controlling diversification in the family.

ACKNOWLEDGMENTS. Bill Glanz, Tim Moerland, and Jon Stanley kindly reviewed the manuscript, and Mike Oliver generously provided advice on taxonomy. Portions of the research were supported by the Faculty Research Fund, University of Maine.

References

Allen, S. K., Jr., and J. G. Stanley, 1983, Ploidy of hybrid grass carp × bighead carp determined by flow cytometry, *Trans. Am. Fish. Soc.* **112:**431–435.

Anders, F., Chatterjee, K., Schwab, M., Scholl, E., and Anders, A., 1981, Tumor gene expression and interphase chromatin appearance in *Xiphophorus, Am. Zool.* **21:**535–548.

Arai, R., 1982, A chromosome study of two cyprinid fishes, *Acrossocheilus labiatus* and *Pseudorasbora pumila pumila,* with notes on eurasian cyprinids and their karyotypes, *Bull. Nat. Sci. Mus. (Tokyo)* **8:**131–152.

Arai, R., and Koike, A., 1980, A karyotype study on two species of freshwater fishes transplanted into Japan, *Bull. Nat. Sci. Mus. (Tokyo)* **6:**275–278.

Arai, R., and Yamamoto, T., 1981, Chromosomes of six species of percoid fishes from Japan, *Bull. Nat. Sci. Mus. (Tokyo)* **7:**87–100.

Avault, J. W., and James, W., 1971, Meristic, morphometric, and physiological comparisons of *Tilapia aurea* Trewavas, *Tilapia mossambica* Peters, and their experimentally produced hybrids, *Proc. La. Acad. Sci.* **33:**46–54.

Avise, J. C., and Gold, J. R., 1977. Chromosomal divergence and speciation in two families of North American fishes, *Evolution* **31:**1–13.

Avise, J. C., Smith, J. J., and Ayala, F. J., 1975, Adaptive differentiation with little genic change between two native California minnows, *Evolution* **29:**411–426.

Avise, J. C., Patton, J. C., and Aquadro, C. F., 1980, Evolutionary genetics of birds: Comparative molecular evolution in New World warblers and rodents, *J. Hered.* **71:**303–310.

Avtalion, R. R., 1982, Genetic markers in *Saratherodon* and their use for sex and species identification, in: *Biology and Culture of Tilapias* (R. H. Lowe-McConnell and R. S. V. Pullin, eds.), ICLARM Conference Proceedings, 7, International Center for Living Aquatic Resources Management, Manila, pp. 269–277.

Avtalion, R. R., and Hammerman, I. S., 1978, Sex determination of *Sarotherodon* (*Tilapia*). I. Introduction to a theory of autosomal influences, *Bamidgeh* **30:**110–115.

Avtalion, R. R., and Mires, D., 1976, Determination of allogeneic and xenogeneic markers in the genus of *Tilapia*. III. Electrophoretic identification of pseudohermaphrodism in *T. aurea* serum, *Aquaculture,* **7:**361–364.

Avtalion, R. R., and Wojdani, A., 1971, Electrophoresis and immunoelectrophoresis of sera from some known F_1 hybrids of *Tilapia, Bamidgeh* **23:**117–124.

Avtalion, R. R., Pruginin, Y., and Rothbard, S., 1975, Determination of allogeneic and xenogeneic markers in the genus of *Tilapia*. I. Identification of sex and hybrids in *Tilapia* by electrophoretic analysis of serum proteins, *Bamidgeh* **27:**8–13.

Avtalion, R. R., Duczyminer, M., Wojdani, A., and Pruginin, Y., 1976, Determination of allogeneic and xenogeneic markers in the genus of *Tilapia*. II. Identification of *T. aurea, T. vulcani* and *T. nilotica* by electrophoretic analysis of their serum proteins, *Aquaculture* **7:**255–265.

Badawi, H. K., 1971, Electrophoretic studies of serum proteins of four *Tilapia* species (Pisces), *Mar. Biol.* **8:**96–98.

Badr, E. A., and El-Dib, S. I., 1976, Effects of water pollution on the cell division cycle and chromosome behavior in *Tilapia* spp., *J. Cell Biol.* **70:**189a.

Badr, E. A., and El-Dib, S. I., 1977, Cytological studies on three species of the cichlid fish, *Egypt. J. Genet. Cytol.* **6**:44–51.

Bailey, R. M., and Stewart, D. J., 1977, Cichlid fishes from Lake Taganyika: Additions to the Zambian fauna including two new species, *Occ. Pap. Mus. Zool. Univ. Mich.* **679**:1–30.

Barel, C. D. N., van Oijen, M. J. O., Witte, F., and Witte-Maas, E. L. M., 1977, An introduction to the taxonomy and morphology of the haplochromine Cichlidae from Lake Victoria. Part A: Text, *Neth. J. Zool.* **27**:333–389.

Barlow, G. W., 1976, The midas cichlid in Nacaragua, in: *Investigations of the Ichthyofauna of Nicaraguan Lakes* (T. B. Thorson, ed.) University of Nebraska-Lincoln, Lincoln, Nebraska, pp. 333–358.

Baron, J.-C., 1975, Note sur le serum de 4 especes de *Tilapia: Tilapia mossambica* Peters, 1852, *T. nilotica* (Linnaeus, 1758), *T. zillii* (Gervais, 1848) et *T. macrochir* Boulenger, 1912, *Hydrobiology* **9**:19–24.

Basasibwaki, P., 1975, Comparative electrophoretic patterns of lactase dehydrogenase and malate dehydrogenase in five Lake Victoria cichlid species, *Afr. J. Trop. Hydrobiol. Fish* **4**:21–26.

Bengtsson, B. O., 1980, Rates of karyotype evolution in placental mammals, *Hereditas* **92**:37–47.

Bush, G. L., Case, S. M., Wilson, A. C., and Patton, J. L., 1977, Rapid speciation and chromosomal evolution in mammals, *Proc. Natl. Acad. Sci. USA* **74**:3942–3946.

Charlesworth, B., Lande, R., and Slatkin, M., 1982, A neo-Darwinian commentary on macroevolution, *Evolution* **36**:474–498.

Chen, F. Y., 1969, Comparative zone electropherograms of hemoglobins and muscle myogens of four *Tilapia* species and their hybrids, *Trop. Fish Cult. Res. Inst. Rep.* **1969**:39–44.

Chen, F. Y., and Tsuyuki, H., 1970, Zone electrophoretic studies on the proteins of *Tilapia mossambica* and *T. hornorum* and their F_1 hybrids, *T. zillii,* and *T. melanopleura, J. Fish. Res. Board Can.* **27**:2167–2177.

Chervinski, J., 1964, Preliminary experiments in cichlid hybrids, *Bamidgeh* **16**:95–105.

Chichoki, F., 1976, Cladistic history of the cichlid fishes and reproductive categories of the American genera, Ph. D. dissertation, University of Michigan, Ann Arbor, Michigan.

Cruz, T. A., Thorpe, J. P., and Pullin, R. S. V., 1982, Enzyme electrophoresis in *Tilapia zillii:* A pattern for determining biochemical genetic markers for use in tilapia stock identification, *Aquaculture* **29**:311–330.

Darlington, P. J., 1957, *Zoogeography: The Geographical Distribution of Animals,* Wiley, New York.

Dominey, W. C., 1984, Species flocks in the cichlid fishes of the Great Lakes of Africa, Hawaiian *Drosophila*, and Lake Baikal amphipods: consequences of sexual selection and life history phenomena. in: *Evolution of Fish Species Flocks* (A. A. Echelle and I. Kornfield, eds.) University of Maine at Orono Press, Orono, Maine (in press).

Fernald, R. D., 1976, The effect of testosterone on the behavior and coloration of adult male cichlid fish *Haplochromis burtoni* Gunther, *Hormone Res.* **7**:172–178.

Foerster, V. W., and Anders, F., 1977, Zytogenetischer Vergleich der Karyotypen verschiedener Rassen und Arten lebendgebarender Zahnkarpfen der Gattung *Xiphophorus, Zool. Anz.* **198**:167–177.

Fryer, G., 1959, The trophic interrelationships and ecology of some littoral communities of Lake Nyasa with especial reference to the fishes, and a discussion of the evolution of a group of rock-frequenting Cichilidae, *Proc. Zool. Soc. Lond.* **132**:153–281.

Fryer, G., and Iles, T. D., 1972, *The Cichlid Fishes of the Great Lakes of Africa,* Tropical Fish Hobbyists Publications, Neptune City, New Jersey.

Fukuoka, H., and Muramoto, J., 1975, Somatic and meiotic chromosomes of *Tilapia mossambica* Peters, *Chromosome Inf. Serv.* **18:**4–6.

Futuyma, D. J., 1979, *Evolutionary Biology,* Sinauer, Sunderland, Massachusetts.

Futuyma, D. J., and Mayer, G. C., 1980, Non-allopatric speciation in animals, *Syst. Zool.* **29:**254–271.

Galman, D. R., and Carino, V. S., 1979, Development expression of esterase isozymes in *Tilapia nilotica, Kalikasan* **8:**13–22.

Gee, P. A., and Moore, K. L., 1970, Chromatin patterns in fishes with emphasis on sexual dimorphism, *Can. J. Zool.* **48:**1283–1286.

Gold, J. R., 1979, Cytogenetics, in: *Fish Physiology,* Vol. 8 (W. S. Hoar, D. J. Randall, and J. R. Brett, eds), Academic Press, New York, pp. 353–405.

Gold, J. R., 1980, Chromosomal change and rectangular evolution in North American cyprinid fishes, *Genet. Res.* **35:**157–164.

Greenwood, P. H., 1965, Environmental effects on the pharyngeal mill of a cichlid, *Astatoreochromis alluandi,* and their taxonomic implications, *Proc. Linn. Soc. Lond.* **176:**1–10.

Greenwood, P. H., 1979*a,* Macroevolution—myth or reality?, *Biol. J. Linn. Soc.* **12:**293–304.

Greenwood, P. H., 1979*b,* Towards a phyletic classification of the genus *Haplochromis* (Pisces, Cichlidae) and related taxa. Part. I, *Bull. Br. Mus. Nat. Hist. (Zool.)* **35:**265–322.

Greenwood, P. H., 1980, Towards a phyletic classification of the genus *Haplochromis* (Pisces, Cichlidae) and related taxa. Part II. The species from Lake Victoria, Nabugabo, Edward, George and Kivu, *Bull. Br. Mus. Nat. Hist. (Zool.)* **39:**1–101.

Greenwood, P. H., 1981, *The Haplochromine Fishes of the East African Lakes,* Cornell University Press, Ithaca, New York.

Greenwood, P. H., 1984, African cichlids and evolutionary theories, in: *Evolution of Fish Species Flocks* (A. A. Echelle, and I. Kornfield, eds.), University of Maine at Orono Press, Orono, Maine (in press).

Guerrero, R. D., 1975, Use of androgens for the production of all-male *Tilapia aurea* (Steindachner), *Trans. Am. Fish. Soc.* **104:**342–348.

Haider, G., 1973, A contribution to the diagnosis of sex by means of heterophyknosis in teleost fishes, *Ind. J. Zool.* **1:**71–76.

Haldane, J. B. S., 1922, Sex ratio and unisexual sterility in hybrid animals, *J. Genet.* **12:**101–109.

Hammerman, I. S., and Avtalion, R. R., 1979, Sex determination in *Sarotherodon* (*Tilapia*). II. The sex ratio as a tool for the determination of genotype—A model of autosomal influence, *Theor. Appl. Genet.* **55:**177–187.

Hardin, S., 1976, Electrophoresis of serum proteins from three species of *Tilapia* and of their F_1 hybrids, M.S. dissertation, Auburn University, Alabama.

Harva, A., 1978, Sexual differences in serum proteins of chum salmon and the purification of female-specific serum protein, *Bull. Jpn. Soc. Sci. Fish.* **44:**689–693.

Heiligenberg, W., 1965, A quantitative analysis of digging movements and their relationship to aggressive behavior in cichlids, *Anim. Behav.* **13:**163–170.

Herzberg, A., 1978, Electrophoretic esterase patterns of the surface mucus for the identification of *Tilapia* species, *Aquaculture* **13:**81–83.

Hinegardner, R., and Rosen, D. E., 1972, Cellular DNA content and evolution of teleost fishes, *Am. Nat.* **106:**621–644.

Holzberg, S., 1978, A field and laboratory study of the behaviour and ecology of *Pseudotropheus zebra* (Boulenger), an endemic cichlid of Lake Malawi (Pisces; Cichlidae), *Z. Zool. Syst. Evolutionsforsch.* **16:**171–187.

Hulata, G., Rothbard, S., and Avtalion, R. R., 1981, Evidence for multiple paternity in *Sarotherodon* broods, *Aquaculture* **25:**281–283.

Ihssen, P. E., Booke, H. E., Casselman, J. M., McGlade, J. M., Payne, N. R., and Utter, R. M., 1981, Stock identification: materials and methods, *Can. J. Fish. Aquat. Sci.* **38:**1838–1855.

Jakowska, S., 1950, Spermatogenesis in the cichlid fish *Tilapia macrocephala* (Beeker), *Trans. Am. Microsc. Soc.* **69:**403–413.

Jalabert, B., Kammacher, P., and Lessent, P., 1971, Determinsme du sexe chez les hybrids entre *Tilapia macrochir* et *Tilapia nilotica*. Étude de la sex-ratio dans les recroisements des hybrids de primiere generation par les especes parentes, *Ann. Biol. Anim. Biochem. Biophys.* **11:**155–165.

Johnson, M. S., 1975, Biochemical systematics of the atherinid genus Menidia, *Copeia* **1975:**662–691.

Kaina, V. B., and Rieger, R., 1979, Chromosomenmutationen, Karyotyp-polymorphismus und Speciation, *Biol. Zentr.* 261. **98:**661–697.

Katz, Y., and Eckstein, B., 1974, Changes in steroid concentration in blood of female *Tilapia aurea* (Teleostei, Cichlidae) during initiation of spawning, *Endocrinology* **95:**963–967.

Kornfield, I. L., 1974, Evolutionary genetics of endemic African cichlids. Ph. D. dissertation, State University of New York, Stony Brook, New York.

Kornfield, I. L., 1978, Evidence for rapid speciation in African cichlid fishes, *Experientia* **34:**335–336.

Kornfield, I. L., Ritte, U., Richler, C., and Wahrman, J., 1979, Biochemical and cytological differentiation among cichlid fishes of the Sea of Galilee, *Evolution* **33:**1–14.

Kornfield, I. L., Smith, D. C., Gagnon, P. S., and Taylor, J. N., 1982, The cichlid fishes of Cuatro Cienegas, Mexico: Direct evidence of conspecificity among distinct trophic morphs, *Evolution* **36:**658–664.

Kornfield, I. L. and Taylor, J. N., 1983, A new species of polymorphic fish, *Cichlasoma minckleyi,* from Cuatro Cienegas, Mexico, *Proc. Biol. Soc. Wash.* **96:**253–269.

Krishan, A., 1975, Rapid flow cytofluorometric analysis of mammalian cell cycle by propidium iodide staining, *J. Cell Biol.* **66:**188–193.

Levan, A., Fredga, K., and Sandberg, A. A., 1964, Nomenclature for centromeric position on chromosomes, *Hereditas* **52:**201–220.

Liem, K. F., 1974, Evolutionary strategies and morphological innovations: Cichlid pharyngeal jaws, *Syst. Zool.* **22:**425–441.

Liem, K. F., 1980, Adaptive significance of intra- and interspecific differences in the feeding repertoires of cichlid fishes, *Am. Zool.* **20:**295–314.

Liem, K. F., 1981, A phyletic study of the Lake Tanganyika cichlid genera *Asprotilapia, Ecotuds, Lestradea, Cunningtonia, Ophthalmochromis,* and *Ophthalmotilapia, Bull. Mus. Comp. Zool. Harv.* **149:**191–214.

Liem, K. F. and Kaufman, L. S., 1984, Intraspecific macroevolution: functional biology of the polymorphic cichlid species *Cichlasoma minckleyi,* in: *Evolution of Fish Species Flocks* (A. A. Echelle, and I. Kornfield, eds.), University of Maine at Orono Press, Orono, Maine (in press).

Loiselle, P. V., 1971, Hybridization in cichlids, *Buntb. Bull.* **27:**9–18.

Lowe (McConnell), R. H., 1953, Notes on the ecology and evolution of Nyasa fishes of the genus *Tilapia,* with a description of *T. saka* Lowe, *Proc. Zool. Soc. Lond.* **122:**1035–1041.

Lyon, M. R., 1974, Mechanisms and evolutionary origins of variable X-chromosome activity in mammals, *Proc. Roy. Soc. Lond.* **187B:**243–268.

Malecha, R. S., and Ashton, G. C., 1968. Inbreeding in fresh water fish populations using transferrins and esterases as markers, in: *XIth European Conference on Animal Blood Groups and Biochemical Polymorphism, Warsaw, 1968,* pp. 523–526.

Marsh, A. C., Ribbink, A. J., and Marsh, B. A., 1981, Sibling species complexes in sympatric populations of *Petrotilapia* Trewavas (Cichlidae, Lake Malawi), *Zool. J. Linn. Soc.* **71**:253–264.

McAndrew, B. J. and Majumdar, K. C., 1983, *Tilapia* stock identification using electrophoretic markers, *Aquaculture* **30**:249–261.

McKaye, K. R., 1980, Seasonality in habitat selection by the bold color morph of *Cichlasoma cintrinellum* and its relevance to sympatric speciation in the family Cichlidae, *Environ. Biol. Fish.* **5**:75–78.

McKaye, K. R., 1981, Natural selection and the evolution of interspecific brood care in fishes, in: *Natural Selection and Social Behavior* (R. Alexander and D. Tinkle, eds.), Chiron, New York, pp. 173–183.

McKaye, K. R., and Mackenzie, C., 1982, *Cyrtocara liemi,* a previously undescribed paedophagous cichlid fish (Teleostei: Cichlidae) from Lake Malawi, Africa, *Proc. Biol. Soc. Wash.* **95**:398–402.

McKaye, K. R., Kocher, T., Reinthal, P., Harrison, R., and Kornfield, I., 1983, Genetic variation among color morphs of a Lake Malawi cichlid fish, *Evolution* (in press).

McKaye, K. R., Kocher, T., Reinthal, P., and Kornfield, I., 1982, Genetic analysis of a sympatric species complex of *Petrotilapia* Trewavas (Cichlidae, Lake Malawi), *J. Linn. Soc.* **76**:91–96.

Mehl, J. A. P., and Reinboth, R., 1975, The possible significance of sex-chromatin for the determination of genetic sex in ambisexual teleost fishes, in: *Intersexuality in the Animal Kingdom* (R. Reinboth, ed.), Springer-Verlag, New York, pp. 243–249.

Michele, J. L., and Takahashi, C. S., 1977, Comparative cytology of *Tilapia rendalli* and *Geophagus brasiliensis* (Cichlidae, Pisces), *Cytologia* **42**:535–537.

Müller, U. and Wolf, U., 1979, Cross-reactivity to mammalian anti- H-Y Antiserum in teleostean fishes, *Differentiation* **14**:185–187.

Nakamura, M. and Takahashi, H., 1973, Gonadal sex differentiation in *Tilapia mossambica,* with special regard to the time of estrogen treatment effective in inducing complete feminization of genetic males, *Bull. Fac. Fish. Hokkaido University* **24**:1–13.

Natarajan, R., and Subrahmanyam, K., 1974, A karyotype study of some teleosts from Portonovo waters, *Proc. Indian Acad. Sci.* **79B**:173–196.

Nei, M., 1972, Genetic distance between populations, *Am. Nat.* **106**:283–292.

Nevo, E., 1978, Genetic variation in natural populations: patterns and theory, *Theor. Popul. Biol.* **13**:121–177.

Nishikawa, S., Kunio, A., and Tsuneo, K., 1973, A preliminary study on the chromosomes of *Cichlasoma citrinella* (Cichlidae: Pisces), *Chromosome Inf. Serv.* **14**:32–33.

Ohm, V. D., 1978, Sexual dimorphismus, Polygamie und Geschlechtswechsel bei *Crenicara punctulata* Günther 1863 (Cichlidae, Teleostei), *Neue. Folge. Silz. Ges. Naturf. Freunde Berlin* **18**:90–104.

Ohno, S., 1970 *Evolution by Gene Duplication,* Springer Verlag, New York, p. 160.

Ohno, S., 1978, The role of H-Y antigen in primary sex determination, *J. Am. Med. Assoc.* **239**:217–220.

Ohno, S., and Atkin, N. B., 1966, Comparative DNA values and chromosome complements of eight species of fishes, *Chromosoma* **18**:455–466.

Oyhenart-Perera, M. F., Luengo, J. A., and Brum-Zorrilla, N., 1975, Estudio citogenetico de *Cichlasoma facetum* (Jenjns) y *Crenicichla sexatilis* (Linn.) (Teleostei, Cichlidae), *Rev. Biol. Urug.* **3**:29–36.

Park, E. H., and Kang, Y. S., 1976, Karyotypic conservation and differences in DNA content in anguilloid fishes, *Science* **193**:64–65.

Patton, J. C., and Baker, R. J., 1978, Chromosomal homology and evolution of phyllostomatoid bats, *Syst. Zool.* **27**:449–462.

Peters, H. M., 1975, Hermaphroditism in cichlid fishes, in: *Intersexuality in the Animal Kingdom* (R. Reinboth, ed.), Springer-Verlag, Berlin, pp. 228–235.
Polder, J. J. W., 1971, On gonads and reproductive behavior in the cichlid fish *Aeguidens portalagrensis* (Hensel), *Neth. J. Zool.* **21**:265–365.
Post, A., 1965, Vergleichende untersuchungen der *Chromosomenzahlen bei Susswasser-Teleosteern, Z. Zool. Syst. Evolutionsforsch.* **3**:47–93.
Prasad, R., and Manna, G. X., 1976, Chromosomes of the fishes, *Tilapia mossambica* and *Notopterus notopterus, Chromosome Inf. Serv.* **21**:11–13.
Pullin, R. S. V., and Lowe-McConnell, R. H., 1982, *The Biology and Culture of Tilapias,* ICLARM Conference Proceedings 7, ICLARM, Manila.
Regan, C. T., 1906, A revision of the fishes of the South-American genera *Cichla, Chaetobranchus,* and *Chaetobranchopsis* with notes on the American Cichlidae, *Ann. Mag. Nat. Hist.* **17**:230–249.
Rubin, D. A., 1984, Effects of pH on sex determination in *Pelvicachromis, Apistogramma* and *Xiphophorus* (Teleostei), *Copeia* (in press).
Rothbard, S., Hulata, G., and Itzkovich, J., 1982, Occurrence of spontaneous hermaphrodism in a *Sarotherodon* hybrid, *Aquaculture* **26**:391–393.
Sage, R. D., and Selander, R. K., 1975, Trophic radiation through polymorphism in cichlid fishes, *Proc. Natl. Acad. Sci. USA* **72**:4669–4673.
Sage, R. D., Loiselle, P. V., Basasibwaki, P., and Wilson, A. C., 1984, Molecular versus morphological change among cichlid fishes (Pisces: Cichlidae) of Lake Victoria, in: *Evolution of Fish Species Flocks* (A. A. Echelle, and I. Kornfield, eds.), University of Maine at Orono Press, Orono, Maine (in press).
Scholl, A., and Holzberg, S., 1972, Zone electrophoretic studies on lactate dehydrogenase isoenzymes in South American cichlids (Teleostei, Percomorphi), *Experientia* **28**:489–491.
Schroder, J. H., 1980, Morphological and behavioral differences between the BB/OB and B/W color morphs of *Pseudotropheus zebra* Boulenger (Pisces: Cichlidae), *Z. Zool. Syst. Evolutionsforsch.* **18**:69–76.
Scopes, R. K., and Hamoir, G., 1971, Methods for detecting glycolytic enzymes after gel electrophoresis, *International Commission for the Exploration of the Sea Rap. Proc. Ver.* **160**:167–169.
Shields, G. F., 1982, Comparative avian cytogenetics: A review, *Condor* **84**:45–58.
Sites, J. W., Bickham, J. W., and Haiduk, M. W., 1981, Conservative chromosomal change in the bat family Mormoupidae, *Can. J. Genet. Cytol.* **23**:459–467.
Skibinski, D. O. F., and Ward, R. D., 1981, Relationship between allozyme heterozygosity and rates of divergence, *Genet. Res. Camb.* **38**:71–92.
Stiassny, M. L. J., 1981, The phyletic status of the family Cichlidae Pisces, Perciformes: A comparative anatomical investigation, *Neth. J. Zool.* **31**:275–314.
Stiassny, M. L. G., 1982, The relationships of the neotropical genus *Cichla* (Perciformes, Cichlidae): A phyletic analysis including some functional considerations, *J. Zool.* **197**:427–453.
Taylor, D. W., and Minckley, W. L., 1966, New world for biologists, *Pac. Discovery* **19**:18–22.
Taylor, J. N., and Miller, R. R., 1983, Cichlid fishes (genus *Cichlasoma*) of the Rio Panuco basin, eastern Mexico, with description of a new species, *Occ. Pap. Mus. Nat. Hist. Univ. Kans* **104**:1–24.
Templeton, A. R., 1981, Mechanism of speciation—A population genetic approach, *Annu. Rev. Ecol. Syst.* **12**:23–48.
Terkatin-Shimony, A., and Yaron, Z., 1978, Estrogens and estrogenic effects in *Tilapia aurea* (Cichlidae, Teleostei), *Ann. Biol. Anim. Biochem. Biophys.* **18**:1007–1012.

Thompson, K. W., 1976, Some aspects of chromosomal evolution of the Cichlidae (Teleostei: Perciformes) with emphasis on neotropical forms, Ph. D. dissertation, University of Texas, Austin, Texas.

Thompson, K. W., 1979, Cytotaxonomy of 41 species of neotropical Cichlidae, *Copeia* **1979**:679–691.

Thompson, K. W., 1981, Karyotypes of six species of African Cichlidae (Pisces: Perceiformes), *Experientia* **37**:351–352.

Thorgaard, G. H., 1977, Heteromorphic sex chromosomes in male rainbow trout, *Science* **196**:900–902.

Thorgaard, G. H., Rabinovitch, P. S., Shen, M. W., Gall, G. A., Propp, J., and Utter, F. M., 1982, Triploid rainbow trout identified by flow cytometry, *Aquaculture* **29**:305–310.

Thorpe, J. P., 1979, Enzyme variation and taxonomy: The estimation of sampling errors in measurements of interspecific genetic similarity, *J. Linn. Soc.* **11**:369–386.

Thorpe, J. P., 1982, The molecular clock hypothesis: Biochemical evolution, genetic differentiation and systematics, *Annu. Rev. Ecol. Syst.* **13**:139–168.

Trewavas, E., 1982, Generic groupings of Tilapinni used in aquaculture, *Aquaculture* **27**:79–82.

Trewavas, E., 1983, A review of the tilapiine fishes of the genera *Sarotherodon, Oreochromis,* and *Danakilia, Bull. Br. Mus. Nat. Hist. (Zool.)* (in press).

Trewavas, E., Green, J., and Corbet, S. A., 1972, Ecological studies on crater lakes in West Cameroon. Fishes of Barumbi, *Mbo. J. Zool.* **167**:41–95.

Valenti, R. J., 1975, Induced polyploidy in *Tilapia aurea* (Steindacrner) by means of temperature shock treatment, *J. Fish Biol.* **7**:519–528.

Van Oijen, M. J. P., Witte, F., and Witte-Maas, E. L. M., 1981, An introduction to ecological and taxonomic investigations on the haplochromine cichlids from the Mwanza Gulf of Lake Victoria, *Neth. J. Zool.* **31**:149–174.

Vervoort, A., 1980, The karyotypes of seven species of *Tilapia* (Teleostei: Cichlidae), *Cytologia* **45**:651–656.

Wachtel, S. S., 1981, Conservation of the H-Y/H-W receptor, *Hum. Genet.* **58**:54–58.

Wachtel, S. S., Koo, G. C., and Boyse, E. A., 1975, Evolutionary conservation of H-Y (male antigen), *Nature* **254**:270–272.

Walsh, J. B., 1982, Rate or accumulation of reproductive isolation by chromosome rearrangements, *Am. Nat.* **120**:510–532.

Ward, J. A., and Wyman, R. L., 1977, Ethology and ecology of cichlid fishes of the genus *Etroplus* in Sri Lanka, *Environ. Biol. Fish.* **2**:137–145.

White, M. J. D., 1978, *Modes of Speciation,* Freeman, San Francisco.

Wilson, A. C., Sarich, V. M., and Maxson, L. R., 1974, The importance of gene rearrangement in evolution: Evidence from studies on rates of chromosomal, protein and anatomical evolution, *Proc. Natl. Acad. Sci. USA* **71**:3028–3030.

Wilson, A. C., Bush, G. L., Case, S. M., and King, M. C., 1975, Social structuring of mammalian populations and rate of chromosomal evolution, *Proc. Natl. Acad. Sci. USA* **72**:5061–5065.

Wilson, A. C., White, T. J., Carlson, S. C., and Cherry, L. M., 1977, Molecular evolution and cytogenetic evolution, in: *Human Cytogenics: ICN-UCLA Symposia on Molecular and Cellular Biology,* Vol. VII (R. S. Sparkes, D. E. Comings, and C. F. Fox, eds), Academic Press, New York, pp. 375–393.

Wohlfarth, G. W., and Hulata, G. I., 1981, Applied genetics of tilapias, *International Center for Living Aquatic Resources Management,* Studies and Reviews No. 6.

Yoshikawa, H., and Oguri, M., 1978, Sex differentiation in a cichlid, *Tilapia zillii, Bull. Jpn. Soc. Sci. Fish.* **44**:313–318.

Zahner, E., 1977, A karyotype analysis of fifteen species of the family Cichlidae, Ph. D. dissertation, St. John's University, New York.

Index

MIX
Papier aus verantwortungsvollen Quellen
Paper from responsible sources
FSC® C105338

FSC
www.fsc.org

If you have any concerns about our products,
you can contact us on
ProductSafety@springernature.com

In case Publisher is established outside the EU,
the EU authorized representative is:
Springer Nature Customer Service Center GmbH
Europaplatz 3, 69115 Heidelberg, Germany

Printed by Libri Plureos GmbH
in Hamburg, Germany